Microsoft® Access® 2013 Programming by Example

with VBA, XML, and ASP

Julitta Korol

Mercury Learning and Information

Dulles, Virginia
Boston, Massachusetts
New Delhi

Publisher: David Pallai
Mercury Learning and Information
22841 Quicksilver Drive
Dulles, VA 20166
info@merclearning.com
www.merclearning.com
1-800-758-3756

This book is printed on acid-free paper.

Julitta Korol. *Microsoft Access 2013 Programming by Example with VBA, XML, and ASP.*
ISBN: 978-1-938549-80-9

Library of Congress Control Number: 2013958001

161718543

Our titles are available for adoption, license, or bulk purchase by institutions, corporations, etc. For additional information, please contact the Customer Service Dept. at 1-800-758-3756 (toll free).

To my husband, Paul, and my parents-in-law, Alfreda and Piotr Korol.
They've been my biggest fans since the first edition of this book.
I am thankful for their love and support.

CONTENTS

PART 4—ERROR HANDLING AND DEBUGGING

PART 9—TAKING YOUR VBA PROGRAMMING SKILLS TO THE WEB

ACKNOWLEDGMENTS

First, I'd like to express my gratitude to everyone at Mercury Learning and Information. A sincere thank-you to my publisher, David Pallai, for offering me the opportunity to update this book to the new 2013 version and tirelessly keeping things on track during this long project.

A whole bunch of thanks go to the editorial team for working so hard to bring this book to print. In particular, I would like to thank the copyeditor, Tracey McCrea, for the thorough review of my writing. To Jennifer Blaney, for her production expertise and keeping track of all the edits and file processing issues. To the compositor, L. Swaminathan at SwaRadha Typesetting, for all the typesetting efforts that gave this book the right look and feel.

Special thanks to my husband, Paul, for his patience during this long project and having to put up with frequent take out dinners.

Finally, I'd like to acknowledge readers like you who cared enough to post reviews of the previous edition of this book online. Your invaluable feedback has helped me raise the quality of this work by including the material that matters to you most. Please continue to inspire me with your ideas and suggestions.

INTRODUCTION

For many years now, Microsoft Access has allowed users to design and develop Windows-based database applications, and Access continues to be the world's most popular database. This book is for people who have already mastered the use of Microsoft® Access® databases and now are ready for the next step — programming. *Microsoft Access 2013 Programming by Example with VBA, XML, and ASP* takes nonprogrammers through detailed steps of creating Access databases from scratch and shows them how to retrieve and manage their data programmatically using various programming languages and techniques. With this book in hand, users can quickly build the toolset required for developing their own database solutions. With this book's approach, programming an Access database from scratch and controlling it via programming code is as easy as designing and maintaining databases with the built-in tools of Access. This book gives a practical overview of many programming languages and techniques necessary in programming, maintaining, and retrieving data from today's Access databases.

PREREQUISITES

You don't need any programming experience to use Microsoft Access 2013 Programming by Example with VBA, XML, and ASP. The only prerequisite is that you already know how to manually design an Access database and perform database tasks by creating and running various types of queries. This book also assumes that you know how to create more complex forms with embedded subforms, combo boxes, and other built-in controls. If you don't have these skills, there are countless books on the market that can teach you step by step how to build simple databases. If you do meet these criteria, this book will take you to the Access programming level by example. You will gain working knowledge immediately by performing concrete tasks and without having to read long descriptions of concepts. True learning by example begins with the first step, followed by the next step, and the next one, and so on. By the time you complete all of the steps in a hands-on exercise or a custom project, you should be able to effectively apply the same technique again and again in your own database projects.

HOW THIS BOOK IS ORGANIZED

This book is divided into nine parts (a total of 32 chapters) that progressively introduce you to programming Access databases.

Part I introduces you to Access 2013 VBA programming. Visual Basic for Applications (VBA) is the programming language for Microsoft Access. In this part of the book, you acquire the fundamentals of VBA that you will use over and over again in building real-life Access database applications.

PART I CONSISTS OF THE FOLLOWING FOUR CHAPTERS:

Chapter 1 — Writing Procedures in Modules
In this chapter you learn about the types of Access procedures you can write and learn how and where they are written.

Chapter 2 — Exploring the Visual Basic Editor (VBE)
In this chapter you learn almost everything you need to know about working with the Visual Basic Editor window, commonly referred to as VBE. Some of the programming tools that are not covered here are discussed and put to use in Chapter 9.

Chapter 3 — Using Variables, Data Types, and Constants
This chapter introduces basic VBA concepts that allow you to store various pieces of information for later use.

Chapter 4 — Passing Arguments to Procedures and Functions
In this chapter you find out how to provide additional information to your procedures and functions before they are run.

While learning the fundamentals of VBA in Part I allows you to write simple functions and procedures that do not require any complex logic, Part II of the book teaches you how decisions and looping statements enable you to control program flow and create more useful VBA procedures and functions.

PART II CONSISTS OF THE FOLLOWING TWO CHAPTERS:

Chapter 5 — Decision Making with VBA
In this chapter you learn how to control your program flow with a number of different decision-making statements.

Chapter 6 — Repeating Actions in VBA
In this chapter you learn how to repeat the same actions in your code by using looping structures.

Although you can use individual variables to store data while your VBA code is executing, many advanced VBA procedures will require that you implement more efficient methods of keeping track of multiple values. In Part III, you'll learn about working with groups of variables (arrays) and storing data in collections.

AGAIN, THERE ARE TWO CHAPTERS IN PART III:

Chapter 7 — Working with Arrays
In this chapter you learn about static and dynamic arrays and how to use them for holding various values.

Chapter 8 — Working with Collections and Class Modules
This chapter teaches you how you can create and use your own objects and collections of objects.

While you may have planned to write a perfect VBA procedure, there is always a chance that incorrectly constructed or typed code, or perhaps logic errors, will cause your program to fail or produce unexpected results. In Part IV of this book, you'll learn about various methods of handling errors, testing, and debugging VBA procedures.

PART IV ALSO CONTAINS TWO CHAPTERS:

Chapter 9 — VBE Tools for Testing and Debugging
In this chapter you begin using built-in debugging tools to test your programming code.

Chapter 10 — Conditional Compilation and Error Trapping
This chapter introduces additional techniques that you will find useful for debugging your Visual Basic applications. In particular, you will learn how conditional compilation can help you to include or ignore certain blocks of code, and how you can add effective error-handling code to your procedures.

The skills obtained in Parts I, II, III, and IV are fairly portable. They can be utilized in programming other Microsoft Office applications that also use VBA as their native programming language such as Excel, Word, PowerPoint, Outlook, and so on.

Part V introduces you to two sets of programming objects known as Data Access Objects (DAO) and ActiveX Data Objects (ADO) that enable Microsoft Access and other client applications to access and manipulate data. In this part of the book, you learn how to use DAO and ADO objects in your VBA code to connect to a data source; create, modify, and secure database objects.

PART V CONSISTS OF THE FOLLOWING EIGHT CHAPTERS:

Chapter 11 — Data Access Technologies in Microsoft Access
In this chapter you get acquainted with two database engines (Jet/ACE) that Access uses, as well as several object libraries that provide objects, properties, and methods for your VBA procedures.

Chapter 12 — Creating and Accessing Database Tables and Fields
This chapter demonstrates how to create, copy, link, and delete database tables programmatically by using objects from the DAO and ADO object libraries. You also learn how to write code to add and delete fields as well as create listings of existing tables in a database and fields in a table.

Chapter 13 — Setting Up Primary Keys, Indexes, and Table Relationships
In this chapter you learn how to write VBA code to add primary keys and indexes to your database tables using objects, properties, and methods from the DAO and ADO object libraries. You also learn how to use objects from the ADOX library to create relationships between your tables.

Chapter 14 — Finding and Reading Records
Here you practice various methods of using programming code to open a set of database records, commonly referred to as a recordset. You learn how to move around in a recordset, and find, filter, and sort the required records, as well as read their contents. This chapter covers both DAO and ADO recordsets.

Chapter 15 — Working with Records
This chapter teaches you essential database operations such as adding, updating, and deleting records. You also learn how to render your database records in three popular formats (Excel, Word, and a text file).

Chapter 16 — Creating and Running Queries with DAO/ADO
In this chapter you learn how to use VBA code instead of the Query Design view to create and run various types of database queries.

Chapter 17 — Using Advanced ADO/DAO Features
This chapter explains several advanced ADO/DAO features such as how to disconnect a recordset from a database, save it in a disk file, clone it, and shape it. You also learn about database transactions.

Chapter 18 — Implementing Database Security
In this chapter you learn about two types of security in Microsoft Access databases: share-level security that applies to both older (MDB) and new (ACCDB) Access databases, and user-level security that can only be used with .mdb files.

You will find the skills obtained in Part V of this book essential for accessing, manipulating, and securing Access databases.

Part VI introduces you to the Data Definition Language (DDL), an important component of the Structured Query Language (SQL). Like ADO and DAO, which were introduced in Part V, DDL is used for defining database objects (tables, views, stored procedures, primary keys, indexes, and constraints) and managing database security. In this part of the book, you learn how to use DDL statements with Jet/ACE databases, ADO, and the Jet OLE DB Provider.

PART VI CONSISTS OF THE FOLLOWING FIVE CHAPTERS:

Chapter 19 — Creating, Modifying, and Deleting Tables and Fields
In this chapter you learn special Data Definition Language commands for creating a new Access database, as well as creating, modifying, and deleting tables. You also learn commands for adding, modifying, and deleting fields and indexes.

Chapter 20 — Enforcing Data Integrity and Relationships between Tables
Here you learn how to define rules regarding the values allowed in table fields to enforce data integrity and relationships between tables.

Chapter 21 — Defining Indexes and Primary Keys
Here you learn DDL commands for creating indexes and primary keys.

Chapter 22 — Database Security
In this chapter you learn how to use DDL commands to manage security in the Microsoft Access database. You learn how to quickly create, modify, and remove a database password, and how to manage user-level accounts.

Chapter 23 — Views and Stored Procedures
This chapter shows you how to work with two powerful database objects known as views and stored procedures. You learn how views are similar to SELECT queries, and how stored procedures can perform various actions similar to Access Action queries and Select queries with parameters.

The skills you learn in Part VI of this book will allow you to create, manipulate, and secure your Access databases using SQL DDL statements. Numerous Access SQL DDL statements and concepts introduced here are important in laying the groundwork for moving into the client/server environment (porting your Microsoft Access database to SQL Server).

Part VII introduces you to responding to events that occur in Access forms and reports. The behavior of Microsoft Access objects such as forms, reports, and controls can be modified by writing programming code known as an event procedure or an event handler. In this part of the book, you learn how you can make your forms, reports, and controls perform useful actions by writing event procedures in class modules. You also learn how to use VBA, macros, and XML to customize the user interface in Access 2013.

PART VII CONSISTS OF THE FOLLOWING SIX CHAPTERS:

Chapter 24 — Enhancing Access Forms
This chapter presents a quick overview of types of forms you can create with Access 2013 and types of formatting you can apply to make your forms more attractive. You learn how you can group form controls using the layouts, implement rich formatting in form controls, professionally format your forms using built-in themes, and enhance forms with images.

Chapter 25 — Using Form Events
In this chapter you learn the types of events that can occur on a Microsoft Access form and write event procedures to handle various form events.

Chapter 26 — Events Recognized by Form Controls
In this chapter you work with a custom application and learn how to write event procedures for various controls that are placed on an Access form.

Chapter 27 — Enhancing Access Reports and Using Report Events
In this chapter you learn about many events that are triggered when an Access report is run. You write your own event procedures to specify what happens when the report is opened, activated/deactivated, or closed.

Chapter 28 — Advanced Event Programming
This chapter teaches advanced concepts in event programming. You learn how to respond to events in standalone class modules to make your code more manageable and portable to other objects. You also learn how to create and raise your own events.

Chapter 29—Programming the User Interface
This chapter provides an overview of the programming elements available in the Ribbon and shows how you can customize the user interface (UI) in your Access database applications. You learn how to create XML Ribbon customization markup and load it in your database. You also learn how Ribbon customizations can be assigned to forms or reports.

The skills acquired in Part VII of this book will allow you to enhance and alter the way users interact with your database application.

Writing VBA code is not the only way to provide rich functionality to your Access database users. Macros have long been used to enhance the user experience without users having to write any VBA code. Access 2013 Macro Designer allows you to include complex logic, business rules, and error handling in your macros. In Part VIII of this book, you are introduced to three types of macros that you can create in Access 2013. In addition, you learn how to convert macros to VBA and get started with built-in templates that extensively use macros.

PART VIII CONTAINS THE FOLLOWING CHAPTER:

Chapter 30 —Macros and Templates
This chapter introduces you to the new features in Access 2013 macros. We take a detailed look at macro security, work with three types of macros (standalone, embedded, and data macros), see examples of using variables in macros, and examine error-handling actions in macros. We also discuss working with the template format in Access 2013.

The skills acquired in Part VIII will allow you to utilize macros in your Access forms and reports, as well as in automating Access Web Applications that are not compatible with VBA.

Part IX introduces you to programming Microsoft Access databases for Internet access. Gone are the times when working with Access required the presence of the Microsoft Access application on a user's desktop. Thanks to the development of Internet technologies, you can publish both static and dynamic Access data to the Web. In this part of the book, you learn how Active Server Pages (ASP) and Extensible Markup Language (XML) are used with Access to develop database solutions for the World Wide Web.

PART IX CONSISTS OF THE FOLLOWING TWO CHAPTERS:

Chapter 31 — Access and Active Server Pages
In this chapter you learn how to use Microsoft's Active Server Pages (ASP) technology to view, insert, delete, and modify records stored in a Microsoft Access database from a Web browser.

Chapter 32 — XML Features in Access 2013
In this chapter you learn how to use the Extensible Markup Language (XML) with Access. You learn how to manually and programmatically export Access data to XML files, as well as import an XML file to Access and display its data in a table. You also learn how to use style sheets and transformations to present Access data to users in a desired format.

The skills acquired in Part IX of this book will make your Access applications Internet and intranet ready. You are now able to connect to, read from, and write to Access databases from within a Web browser using two important Microsoft technologies.

HOW TO WORK WITH THIS BOOK

This book has been designed as a tutorial and should be followed chapter by chapter.

As you read each chapter, perform the tasks that are described. Be an active learner by getting involved in the book's hands-on exercises and custom projects. When you are completely involved, you learn things by doing rather than studying, and you learn faster. Do not move on to new information until you've fully grasped the current topic. Allow your brain to sort things out and put them in proper perspective before you move on. Take frequent breaks between your learning sessions, as some chapters in this book cover lots of material. Do not try to do everything in one sitting. It's always better to divide the material into smaller units than attempt to master all there is to learn at once. However, never stop in the middle of a hands-on exercise; finish it before taking a break. After learning a particular technique or command, try to think of ways to apply it to your own work. As you work with this book, create small sample procedures for yourself based on what you've learned up to a particular point. These procedures will come in handy when you need to review the subject in the future or simply need to steal some ready-made code.

THE COMPANION FILES

The example files for all the hands-on activities in this book are available on the CD-ROM disc included with this book.

Part 1 INTRODUCTION TO ACCESS 2013 VBA PROGRAMMING

Visual Basic® for Applications (VBA) is the programming language for Microsoft® Access®. In this part of the book, you acquire the fundamentals of VBA that you will use over and over again in building real-life Microsoft Access database applications.

Chapter 1 Writing Procedures in Modules

This chapter's objective is to introduce you to writing programming code using the built-in language of Microsoft Access—Visual Basic for Applications (commonly referred to as VBA). By writing your own code you can extend the functionality of existing Microsoft Access database applications or create your own applications from scratch. VBA is a language embedded in all the Microsoft Office applications. Therefore, what you learn in this book about Access programming can be utilized in writing programs that control Excel®, Word®, Outlook®, or Project. Your job as a programmer (at least during the course of this book) will boil down to writing various procedures. A *procedure* is a group of instructions that allows you to accomplish specific tasks when your program runs. When you place instructions (programming code) in a procedure, you can call this procedure whenever you need to perform that particular task. Although many tasks can be automated in Access 2013 by using macro actions, such as opening forms and reports, finding records, and executing queries, you will need VBA skills to perform advanced customizations in your desktop databases.

PROCEDURE TYPES

VBA has the following types of procedures: subroutine procedures, function procedures, event procedures, and property procedures. You create and store procedures in modules. Each procedure in the same module must have a unique name; however, procedures in different modules can have the same name. Let's learn a bit about each procedure type so that you can quickly recognize them when you see them in books, magazine articles, or online.

1. **Subroutine procedures (also called subroutines or subprocedures)**
   Subroutine procedures perform useful tasks but never return values. They begin with the keyword `Sub` and end with the keywords `End Sub`. *Keywords* are words that carry a special meaning in VBA. Let's look at the simple subroutine ShowMessage that displays a message to the user:

```
Sub ShowMessage()
  MsgBox "This is a message box in VBA."
End Sub
```

Notice a pair of empty parentheses after the procedure name. The instruction that the procedure needs to execute is placed on a separate line between the `Sub` and `End Sub` keywords. You may place one or more instructions and even complex control structures within a subroutine procedure. Instructions are also called *statements*. The ShowMessage procedure will always display the same message when executed. `MsgBox` is a built-in VBA function often used for programming user interactions (see Chapter 4, "Passing Arguments to Procedures and Functions," for more information on this function). To execute the preceding subroutine programmatically, you need to write the following line of code:

```
Call ShowMessage()
```

The Call statement followed by the procedure name tells VBA to execute the instructions contained within the named procedure. If you'd like to write a more universal procedure that can display a different message each time the procedure is executed, you will need to write a subroutine that takes arguments. *Arguments* are values that are needed for a procedure to do something. Arguments are placed within the parentheses after the procedure name. Let's look at the following procedure that also displays a message to the user; however, this time we can pass any text string to display:

```
Sub ShowMessage2(strMessage)
  MsgBox strMessage
End Sub
```

This subprocedure requires one text value before it can be run; `strMessage` is the arbitrary argument name. It can represent any text you want. Therefore, if you pass it the text "Today is Monday," that is the text the user will see when the procedure is executed. If you don't pass the value to this procedure, VBA will display an error.

If your subprocedure requires more than one argument, list the arguments within the parentheses and separate them with commas. For example, let's improve the preceding procedure by passing it also a text string containing a user name:

```
Sub ShowMessage3(strMessage, strUserName)
  MsgBox strUserName & ", your message is: " & strMessage
End Sub
```

The ampersand (&) operator is used for concatenating text strings inside the VBA procedure. If we pass to the above subroutine the text "Keep on learning." as the `strMessage` argument and "John" as the `strUserName` argument, the procedure will display the following text in a message box:

```
John, your message is: Keep on learning.
```

To call this subroutine from another procedure, you can write the following line of code:

```
Call ShowMessage3("Keep on learning.", "John")
```

Notice that when calling a procedure with arguments, the arguments are enclosed in parentheses. The arguments are listed in the same order in which they are specified in the procedure definition. See exceptions to this rule and discover more information about passing arguments to subprocedures and functions in Chapter 4.

2. **Function procedures (functions)**

Functions perform specific tasks and can return values. They begin with the keyword `Function` and end with the keywords `End Function`. Let's look at a simple function that adds two numbers:

```
Function addTwoNumbers()
  Dim num1 As Integer
  Dim num2 As Integer

  num1 = 3
  num2 = 2
  addTwoNumbers = num1 + num2
End Function
```

The preceding function procedure always returns the same result, which is the value 5. The `Dim` statements inside this function procedure are used to declare variables that the function will use. A *variable* is a name that is used to refer to an item of data. Because we want the function to perform a calculation, we specify that the variables will hold integer values. Variables and data types used in Access 2013 are covered in detail in Chapter 3, "Using Variables, Data Types, and Constants."

The variable definitions (the lines with the Dim statements) are followed by the variable assignment statements in which we assign specific numbers to the variables `num1` and `num2`. Finally, the calculation is performed by adding together the values held in both variables: `num1 + num2`. To return the result of our calculation, we set the function name to the value or the expression we want to return:

```
addTwoNumbers = num1 + num2
```

Although this function example returns a value, not all functions have to return values. Functions, like subroutines, can perform a number of actions without returning any values.

Similar to procedures, functions can accept arguments. For example, to make our addTwoNumbers function more versatile, we can rewrite it as follows:

```
Function addTwoNumbers2(num1 As Integer, num2 As Integer)
  addTwoNumbers2 = num1 + num2
End Function
```

Now we can pass any two numbers to the preceding function to add them together. For example, we can write the following statement to display the result of the function in a message box:

```
MsgBox("Total=" & addTwoNumbers2(34,80))
```

You will learn more about writing and calling functions in Chapter 4.

3. **Event procedures**
   Event procedures are automatically executed in response to an event initiated by the user or program code, or triggered by the system. Events, event properties, and event procedures are introduced later in this chapter. They are also covered in Part IV, "Error Handling and Debugging."

4. **Property procedures**
   Property procedures are used to get or set the values of custom properties for forms, reports, and class modules. The three types of property procedures (Property Get, Property Let, and Property Set) begin with the `Property`

keyword followed by the property type (Get, Let, or Set), the property name, and a pair of empty parentheses, and end with the `End Property` keywords. Here's an example of a property procedure that retrieves the value of an author's royalty:

```
Property Get Royalty()
  Royalty = (Sales * Percent) - Advance
End Property
```

Property procedures are covered in detail in Chapter 8, "Working with Collections and Class Modules."

MODULE TYPES

As mentioned earlier, procedures are created and stored in modules. Like the previous version, Access 2013 has two types of modules: standard and class.

Standard Modules

Standard modules are used to hold subprocedures and function procedures that can be run from anywhere in the application because they are not associated with any particular form or report.

Writing Procedures in a Standard Module

Because we already have a couple of procedures to try out, let's do a quick hands-on exercise to learn how to open standard modules, write procedures, and execute them.

Please note files for the "Hands-On" project may be found on the companion CD-ROM.

Hands-On 1.1. Working in a Standard Module

1. Create a folder on your hard drive named C:**Access2013_ByExample**.
2. Open Access 2013 and click **Blank Database** (see Figure 1.1).Type**Chap01** in-the File Name box, and click the folder button to set the location for the database to the C:**Access2013_ByExample** folder. Finally, click the **Create** button to create the specified database. Access will create the database in its default .ACCDB format.

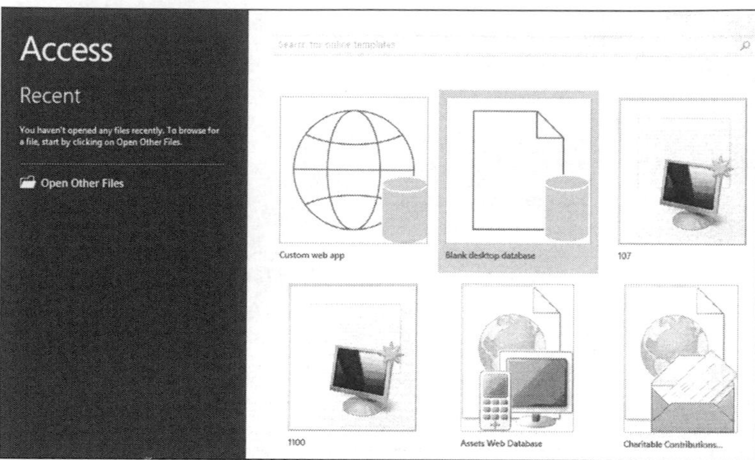

FIGURE 1.1. Creating a blank desktop Access 2013 database.

3. To launch the programming environment in Access 2013, select the **Database Tools** tab and click **Visual Basic** (see Figure 1.2). You can also press **Alt+F11** to get to this screen.

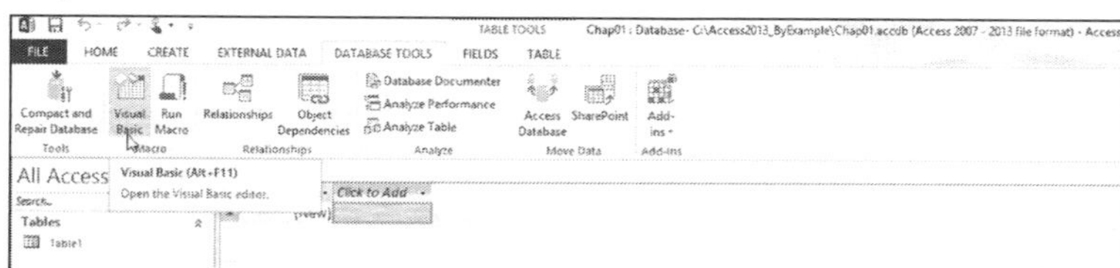

FIGURE 1.2. Switching to a Visual Basic development environment.

4. Insert a standard module by choosing **Module** from the **Insert** menu (see Figure 1.3).

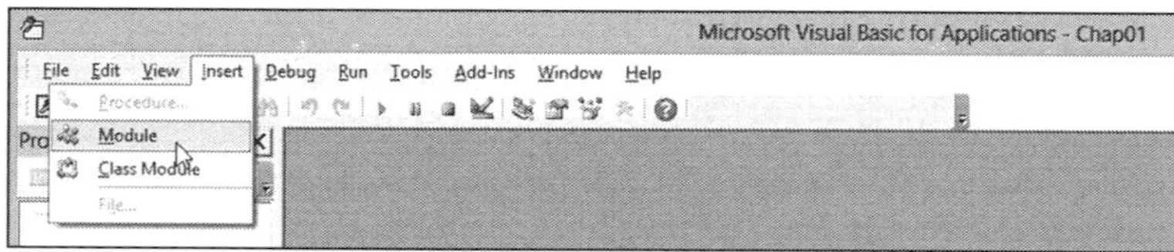

FIGURE 1.3. Inserting a standard module.

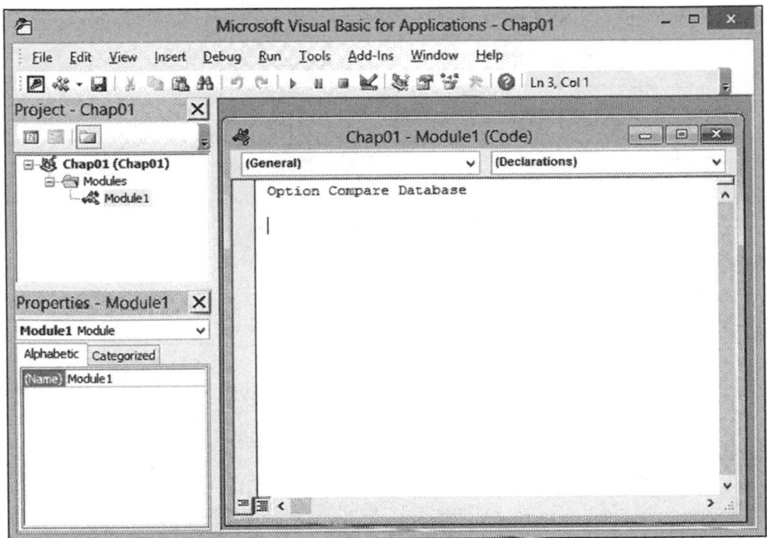

FIGURE 1.4. Standard module.

Each module begins with a declaration section that lists various settings and declarations that apply to every procedure in the module. Figure 1.4 shows the default declaration. `Option Compare` specifies how string comparisons are evaluated in the module—whether the comparison is case sensitive or insensitive. This is a case-insensitive comparison that respects the sort order of the database. This means that "a" is the same as "A." If you delete the `Option Compare Database` statement, the default string comparison setting for the module is `Option Compare Binary` (used for case-sensitive comparisons where "a" is not the same as "A").

Another declaration (not shown here), the `Option Explicit` statement is often used to ensure that all variable used within this module are formally declared. You will learn more about this statement and variables in Chapter 4. Following the declaration section is the procedure section, which holds the module's procedures. You can begin writing your procedures at the cursor position within the Module1 (Code) window.

5. In the Module1 (Code) window, enter the code of subroutines and function procedures as shown in Figure 1.5.

Notice that Access inserts a horizontal line after each `End Sub` or `End Function` keyword to make it easier to identify each procedure. The Procedure drop-down box at the top-right corner of the Module1(Code) window displays the name of the procedure in which the insertion point is currently located.

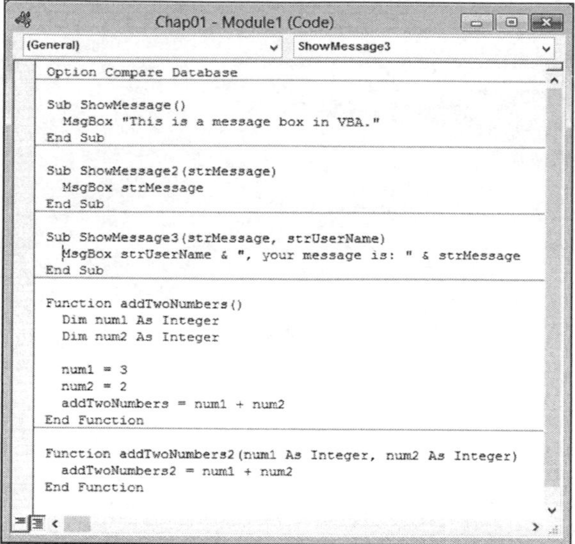

FIGURE 1.5. Standard module with subprocedures and functions.

Executing Your Procedures and Functions

Now that you've filled the standard module with some procedures and functions, let's see how you can run them. There are many ways of running your code. In the next hands-on exercise, you will learn to run your code in four different ways, from the:

- Run menu (Run Sub/UserForm)
- Toolbar button (Run Sub/UserForm)
- Keyboard (F5)
- Immediate window

Hands-On 1.2. Running Procedures and Functions

1. Place the insertion point anywhere within the ShowMessage procedure. The Procedure box in the top-right corner of the Module1 (Code) window should display ShowMessage. Choose **Run Sub/UserForm** from the **Run** menu.
   Access runs the selected procedure and displays the message box with the text "This is a message box in VBA."
2. Click **OK** to close the message box. Try running this procedure again, this time by pressing the **F5** key on the keyboard. Click **OK** to close the message box. If

the Access window seems stuck and you can't activate any menu option, this is often an indication that there is a message box open in the background. Access will not permit you to do any operation until you close the pop-up window.

3. Now, run this procedure for the third time by clicking the Run Sub/UserForm button (▸) on the toolbar. This button has the same tool tip as the Run Sub/UserForm (F5) option on the Run menu.

Procedures that require arguments cannot be executed directly using the methods you just learned. You need to type some input values for these procedures to run. A perfect place to do this is the Immediate window, which is covered in detail in Chapter 2, "Exploring the Visual Basic Editor (VBE)." For now, let's open this window and see how you can use it to run VBA procedures.

4. Select **Immediate Window** from the **View** menu.
   Access opens a small window and places it just below the Module1 (Code) window. You can size and reposition this window as needed. Figure 1.6 shows statements that you will run from the Immediate window in Steps 5–8.

5. Type the following in the Immediate window and press **Enter** to execute.

```
ShowMessage2 "I'm learning VBA."
```

Access executes the procedure and displays the message in a message box. Click **OK** to close the message box. Notice that to execute the ShowMessage2 procedure, you need to type the procedure name, a space, and the text you want to display. The text string must be surrounded by double quotation marks. In a similar way you can execute the ShowMessage3 procedure by providing two required text strings. For example, on a new line in the Immediate window, type the following statement and press **Enter** to execute:

```
ShowMessage3 "Keep on learning.", "John"
```

When you press the Enter key, Access executes the ShowMessage3 procedure and displays the text "John, your message is: Keep on learning." Click **OK** to close this message box.

You can also use the `Call` *statement to run a procedure in the Immediate window. When using this statement you must place the values of arguments within parentheses, as shown here:*

```
Call ShowMessage3("Keep on learning.", "John")
```

Function procedures are executed using different methods. Step 6 demonstrates how to call the add Two Numbers function.

6. On a new line in the Immediate window, type a question mark followed by the name of the function procedure and press **Enter**:

```
?addTwoNumbers
```

Access should display the result of this function (the number 5) on the next line in the Immediate window.

7. Now run the addTwoNumbers2 procedure. Type the following instruction in the Immediate window and press **Enter**:

```
?addTwoNumbers2(56, 24)
```

Access displays the result of adding these two numbers on the next line.

8. If you'd rather see the function result in a message box, type the following instruction in the Immediate window and press **Enter**:

```
MsgBox("Total=" & addTwoNumbers2(34,80))
```

Access displays a message box with the text "Total=114."

See Chapter 2 for more information on running your procedures and functions from the Immediate window.

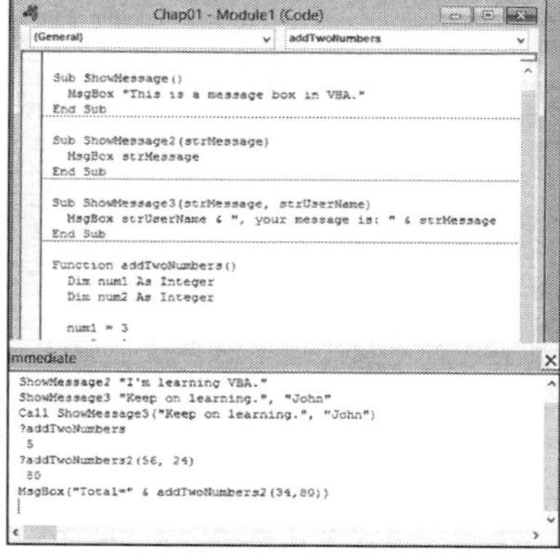

FIGURE 1.6. Running procedures and functions in the Immediate window.

Now that you've familiarized yourself a bit with standard modules, let's move on to another type of module known as the class module.

Class Modules

Class modules come in three varieties: standalone class modules, form modules, and report modules.

- **Standalone class modules**—These modules are used to create your own custom objects with their own properties and methods. You create a standalone class module by choosing **Insert | Class Module** in the Microsoft Visual Basicfor Applications window. Access will create a default class module named Class1 and will list it in the Class modules folder in the Project Explorer window. You will work with standalone class modules in Chapter 8.

- **Form modules** and **report modules**—Each Access form can contain a form module, and each report can contain a report module. These modules are special types of class modules that are saved automatically whenever you save the form or report.

All newly created forms and reports are lightweight by design because they don't have modules associated with them when they're first created. Therefore, they load and display faster than forms and reports with modules. These lightweight forms and reports have their Has Module property set to No (see Figure 1.7). When you open a form or report in Design view and click the View Code button in the Tools section of the Design tab, Access creates a form or report module. The Has Module property of a form or report is automatically set to Yes to indicate that the form or report now has a module associated with it. Note that this happens even if you have not written a single line of VBA code. Access opens a module window and assigns a name to the module that consists of three parts: the name of the object (e.g., form or report), an underscore character, and the name of the form or report. For example, a newly created form that has not been saved is named Form_Form1, a form module in the Customers form is named Form_Customers, and a report module in the Customers report is named Report_Customers (see Figure 1.8).

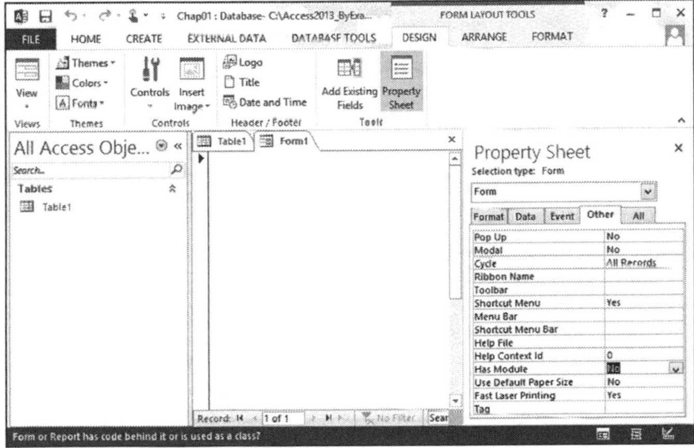

FIGURE 1.7. When you begin designing a new form in the Microsoft Access user interface, the form does not have a module associated with it. Notice that the Has Module property on the form's property sheet is set to No.

As with report modules, form modules store event procedures for events recognized by the form and its controls, as well as general function procedures and subprocedures. You can also write Property Get, Property Let, and Property Set procedures to create custom properties for the form or report. The procedures stored in their class modules areavailable only while you are using that particular form or report.

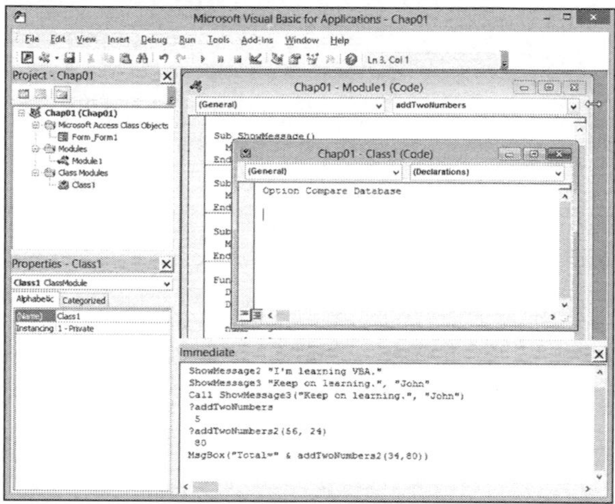

FIGURE 1.8. Database modules are automatically organized in folders. Form and report modules are listed in the Microsoft Access Class Objects folder. Standard modules can be found in the Modules folder. The Class Modules folder organizes standalone class modules.

EVENTS, EVENT PROPERTIES, AND EVENT PROCEDURES

In order to customize your database applications or to deliver products that fit your users' specific needs, you'll be doing quite a bit of event-driven programming. Access 2013 is an *event-driven* application. This means that whatever happens in an Access application is the result of an event that Access has detected. *Events* are things that happen to objects, and can be triggered by the user or by the system, such as clicking a mouse button, pressing a key, selecting an item from a list, or changing a list of items available in a listbox. As a programmer, you will often want to modify the application's built-in response to a particular event. Before the application processes the user's mouseclicks and keypresses in the usual way, you can tell the application how to react to the activity. For example, if a user clicks a Delete button on your form, you can display a custom delete confirmation message to ensure that the user selected the intended record for deletion.

For each event defined for a form, form control, or report, there is a corresponding *event property*. If you open any Microsoft Access form in Design view and choose Properties in the Tools section of the Design tab, and then click the Event tab of the property sheet, you will see a long list of events your form can respond to (see Figure 1.9).

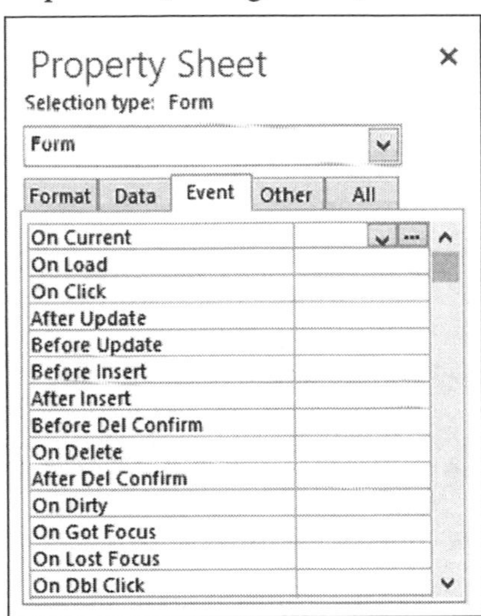

FIGURE 1.9. Event properties for an Access form are listed on the Event tab in the property sheet.

Forms, reports, and the controls that appear on them have various event properties you can use to trigger desired actions. For example, you can open or close a form when a user clicks a command button, or you can enable or disable controls when the form loads.

To specify how a form, report, or control should respond to events, you can write *event procedures*. In your programming code, you may need to describe what should happen if a user clicks on a particular command button or makes a selection from a combo box. For example, when you design a custom form you should anticipate and program events that can occur at runtime (while the form is being used). The most common event is the Click event. Every time a command button is clicked, it triggers an event procedure to respond to the Click event for that button.

When you assign your event procedure to an event property, you set an *event trap*. Event trapping gives you considerable control in handling events because you basically interrupt the default processing that Access would normally carry out in response to the user's keypress or mouseclick. If a user clicks a command button to save a form, whatever code you've written in the Click event of that command button will run. The event programming code is stored as a part of a form, report, or control and is triggered only when user interaction with a form or report generates a specific event; therefore, it cannot be used as a standalone procedure.

Why Use Events?

Events allow you tomake your applications dynamic and interactive. To handle a specific event, you need to select the appropriate event property on the property sheet and then write an event handling procedure. Access will provide its own default response to those events you have not programmed. Events cannot be defined for tables or queries.

Walking Through an Event Procedure

The following hands-on exercise shows you how to get started with writing event procedures. Your task is to change the background color of a text box control on a form when the text box is selected and then return the default background color when you tab or click out of that text box.

Hands-On 1.3. Writing an Event Procedure

1. Close the **Chap01.accdb** database file used in Hands-On 1.1, and save changes to the file when prompted.
2. Copy the **HandsOn_01_3.accdb** database from the **Access2013_HandsOn** folder on the CD-ROM to your **C:\Access2013_ByExample** folder. This file contains copies of the Customers table and the Customers form from the sample Northwind database that comes with an earlier version of Microsoft Access.
3. Open the database **C:\Access2013_ByExample\HandsOn_01_3.accdb**. Upon loading, Access displays a security warning message. In order to use this file, click the **Enable Content** button in the message bar, as shown in Figure 1.10.

The last section of this chapter explains how you can use trusted locations to keep Access 2013 from disabling your VBA code upon opening a database.

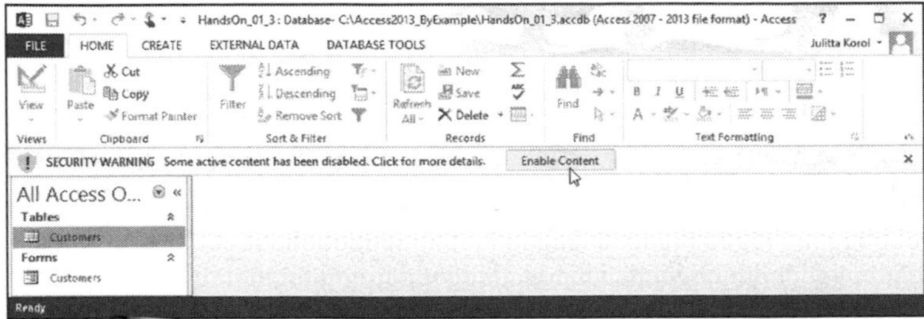

FIGURE 1.10. Active content such as VBA Macros can contain viruses and other security hazards. By default Access displays a Security Warning message when you first load a database file that contains active content. You should enable content only if you trust the contents of the file.

4. Open the Customers form in Design view. To do this, right-click the **Customers** form and choose **Design View** from the shortcut menu.

If the property sheet is not displayed next to the Customers form, click the **Property Sheet** *button in the* **Tools** *group of the* **Form Design Tools** *tab on the Ribbon.*

5. Click the **ContactName** text box control on the Customers form, and then click the **Event** tab in the property sheet. The property sheet will display **ContactName** in the control drop-down box.

The list of event procedures available for the text box control appears, as shown in Figure 1.11.

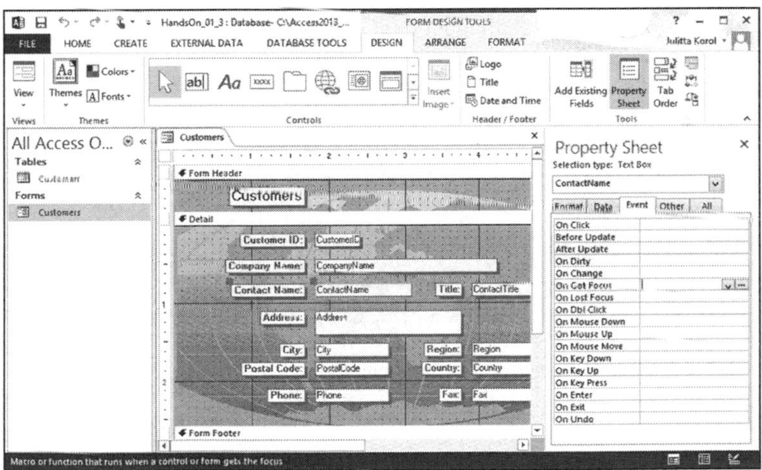

FIGURE 1.11. To create an event procedure for a form control, use the Build button, which is displayed as an ellipsis (...). This button is not available unless an event is selected.

6. Click in the column next to the **On Got Focus** event name, then click the **Build** button (...), as shown in Figure 1.11 in the previous step. This will bring up the Choose Builder dialog box (see Figure 1.12).

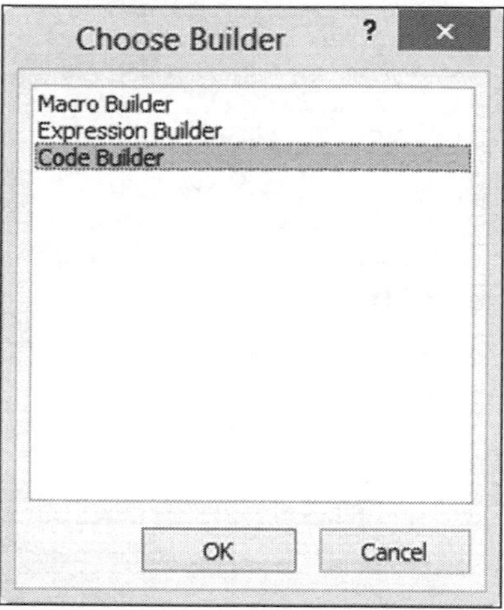

FIGURE 1.12. To write VBA programming code for your event procedure, choose Code Builder in the Choose Builder dialog box.

7. Select **Code Builder** in the Choose Builder dialog box and click **OK**. This will display a VBA code module in the Visual Basic Editor window (see Figure 1.13). This window (often referred to as VBE) is discussed in detail in Chapter 2.

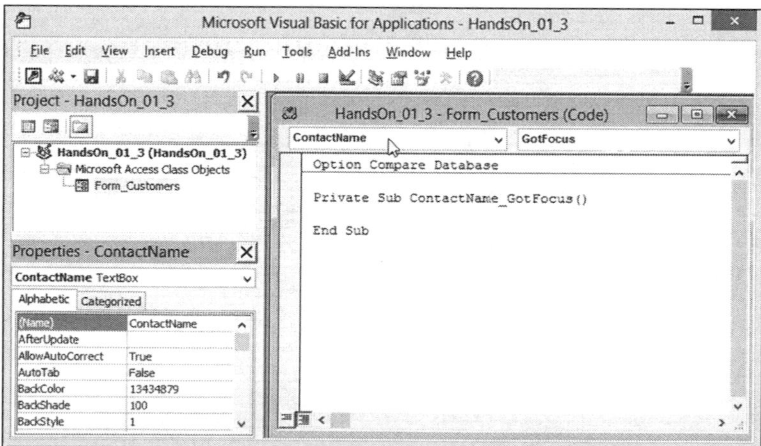

FIGURE 1.13. Code Builder displays the event procedure Code window with a blank event procedure for the selected object. Here you can enter the code for Access to run when the specified GotFocus procedure is triggered.

Take a look at Figure 1.13. Access creates a skeleton of the GotFocus event procedure. The name of the event procedure consists of three parts: the object name (ContactName), an underscore character (_), and the name of the event (GotFocus) occurring to that object. The word `Private` indicates that the event procedure cannot be triggered by an event from another form. The word `Sub` in the first line denotes the beginning of the event procedure. The words `End Sub` in the last line denote the end of the event procedure. The statements to be executed when the event occurs are written between these two lines.

Notice that each procedure name ends with a pair of empty parentheses( ). Words such as `Sub`, `End`, or `Private` have special meaning to Visual Basic and are called *keywords* (reserved words). Visual Basic displays keywords in blue, but you can change the color of your keywords from the Editor Format tab in the Options dialog box (choose Tools | Options in the Visual Basic Editor window). All VBA keywords are automatically capitalized.

At the top of the Code window (see Figure 1.13), there are two drop-down listboxes. The one on the left is called Object. This box displays the currently selected control (ContactName). The box on the right is called Procedure. If you position the mouse over one of these boxes, the tooltip indicates the name of the box. Clicking on the down arrow at the right of the Procedure box displays a list of all possible event procedures associated with the object type selected in the Object box. You can close the drop-down listbox by clicking anywhere in the unused portion of the Code window.

8. To change the background color of a text box control to green, enter the following statement between the existing lines:

```
Me.ContactName.BackColor = 65280
```

Notice that when you type each period, Visual Basic displays a list containing possible item choices. This feature, called List Properties/Methods, is a part of Visual Basic's on-the-fly syntax and programming assistance, and is covered in Chapter 2. When finished, your first event procedure should look as follows:

```
Private Sub ContactName_GotFocus()
  Me.ContactName.BackColor = 65280
End Sub
```

The statement you just entered tells Visual Basic to change the background color of the ContactName text box to green when the cursor is moved into that control. You can also specify the color by using the RGB function, like this:

```
Me.ContactName.BackColor = RGB(0, 255, 0)
```

This statement is equivalent to the statement you used earlier in the ContactName_GotFocus event procedure.

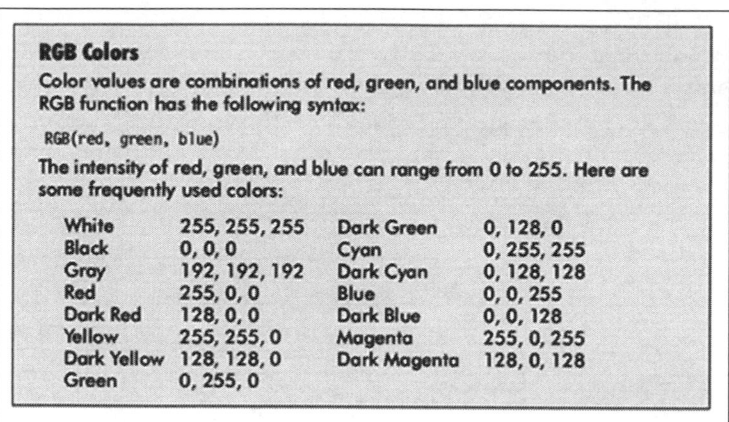

RGB Colors

Color values are combinations of red, green, and blue components. The RGB function has the following syntax:

```
RGB(red, green, blue)
```

The intensity of red, green, and blue can range from 0 to 255. Here are some frequently used colors:

White	255, 255, 255	Dark Green	0, 128, 0
Black	0, 0, 0	Cyan	0, 255, 255
Gray	192, 192, 192	Dark Cyan	0, 128, 128
Red	255, 0, 0	Blue	0, 0, 255
Dark Red	128, 0, 0	Dark Blue	0, 0, 128
Yellow	255, 255, 0	Magenta	255, 0, 255
Dark Yellow	128, 128, 0	Dark Magenta	128, 0, 128
Green	0, 255, 0		

9. In the Visual Basic window, choose **File | Close and Return to Microsoft Access.** Notice that [Event Procedure] now appears next to the On Got Focus event property in the property sheet for the selected ContactName text box control (see Figure 1.14).

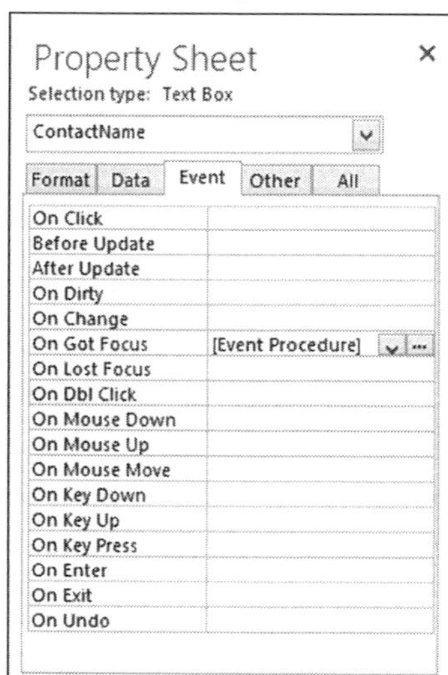

FIGURE 1.14. [Event Procedure] in the property sheet denotes that the text box's On Got Focus event has an event procedure associated with it.

10. To test your GotFocus event procedure, switch from the Design view of the Customers form to Form view by clicking the **View** button on the Ribbon's Design tab.

11. While in the Form view, click in the **Contact Name** text box and notice the change in the background color.
12. Now, click on any other text box control on the Customers form.
    Notice that the Contact Name text box does not return to the original color. So far, you've only told Visual Basic what to do when the specified control receives the focus. If you want the background color to change when the focus moves to another control, there is one more event procedure to write—On Lost Focus.
13. To create the LostFocus procedure, return your form to Design view and click the **Contact Name** control. In the property sheet for this control, select the **Event** tab, then click the **Build** button to the right of the On Lost Focus event property. In the Choose Builder dialog box, select **Code Builder**.
14. To change the background color of a text box control to whatever it was before, enter the following statement inside the ContactName_LostFocus event procedure:

```
Me.ContactName.BackColor = 13434879
```

The completed On Lost Focus procedure is shown in Figure 1.15.

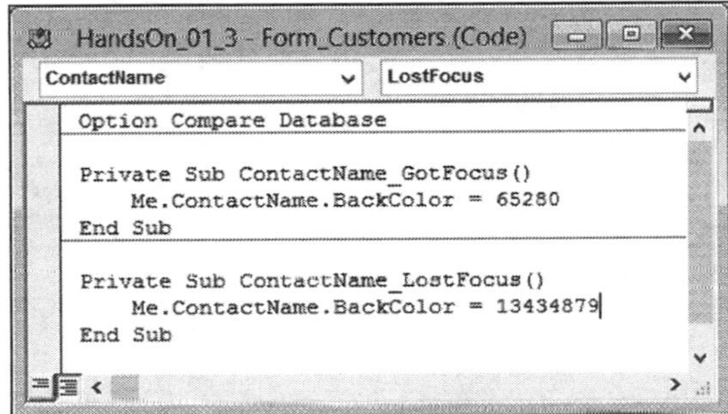

FIGURE 1.15. The GotFocus and LostFocus event procedures will now control the behavior of the ContactName control when the control is in focus and out of focus.

15. In the Visual Basic window, choose **File | Close and Return to Microsoft Access**. Notice that [Event Procedure] now appears next to the On Lost Focus event property in the property sheet for the selected ContactName text box control.
16. Repeat Steps 10–12 to test both of the event procedures you have written.
17. When you are done, close the **HandsOn_01_3** database file and click **OK** when prompted to save the changes.

Because objects recognize a sequence of events, it's important to understand what actions fire the events and the order in which the events occur. In Part IV of this book, you'll learn what events can be used for a particular task to make your application smarter.

Compiling Your Procedures

The VBA code you write in the Visual Basic Editor Code window is automatically compiled by Microsoft Access before you run it. The syntax of your VBA statements is first thoroughly checked for errors, and then your procedures are converted into executable format. If an error is discovered during the compilation process, Access stops compiling and displays an error message. It also highlights the line of code that contains the error. The compiling process can take from seconds to minutes or longer, depending on the number of procedures written and the number of modules used.

To ensure that your procedures have been compiled, you can explicitly compile them after you are done programming. You can do this by choosing Debug | Compile in the Visual Basic Editor window.

Microsoft Access saves all the code in your database in its compiled form. Compiled code runs more quickly when you open it in the future. You should always save your modules after you compile them. In Chapter 9, "VBE Tools for Testing and Debugging," you will learn how to test and troubleshoot your VBA procedures.

PLACING A DATABASE IN A TRUSTED LOCATION

The security features first introduced in Access 2007 made it more complicated for users to work with databases that included VBA code and other executable content such as SQL Pass-Through queries, ActiveX controls, Action queries used to add, modify, or delete data, DDL (Data Definition Language) queries used to create and alter tables, as well as certain unsafe expressions and macro commands. To use the disabled components of the database, a user had to click the Options button on the message bar and then make appropriate selections in the dialog box to either get protection from unknown content or enable this content. Beginning with the previous version of Access (2010), there are fewer annoying prompts. Instead of the Options button, there is the Enable Content button directly in the Security Warning message bar as was shown earlier in Figure 1.10.

To make it easy to work with Access databases in this book, you will not want to bother with enabling content each time you open a database. To trust your databases permanently, you can place them in a *trusted location*—a folder on your local or network drive that you mark as trusted. You can get more information about the Enable Content button and access the Trust Center to set up a trusted folder by choosing File | Info (see Figure 1.16).This screen can also be activated by clicking the text message in the Security Warning message bar—"Some active content has been disabled. Click for more details." (See Figure 1.10 earlier.)

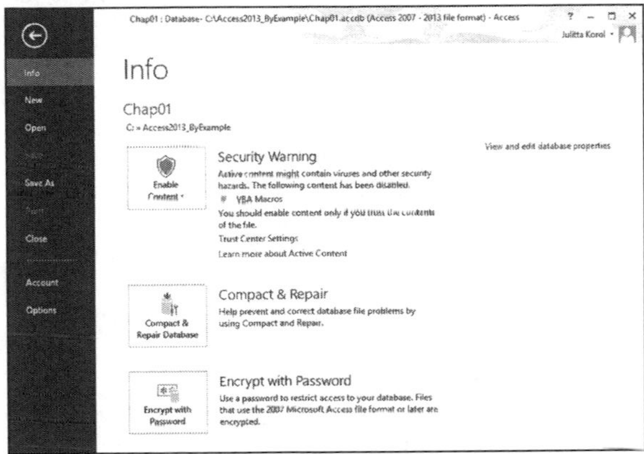

FIGURE 1.16. The Info tab with an explanation of the Security Warning message.

Hands-On 1.4 will step you through the process of setting up a trusted folder for your Access databases by using the Options button.

Hands-On 1.4. Placing an Access Database in a Trusted Location

1. To start the Trust Center, in the Microsoft Access window, activate the **File** tab, and then click the **Options** button.
2. In the left pane of the Access Options dialog box, click **Trust Center**, and then click **Trust Center Settings** in the right pane, as shown in Figure 1.17.

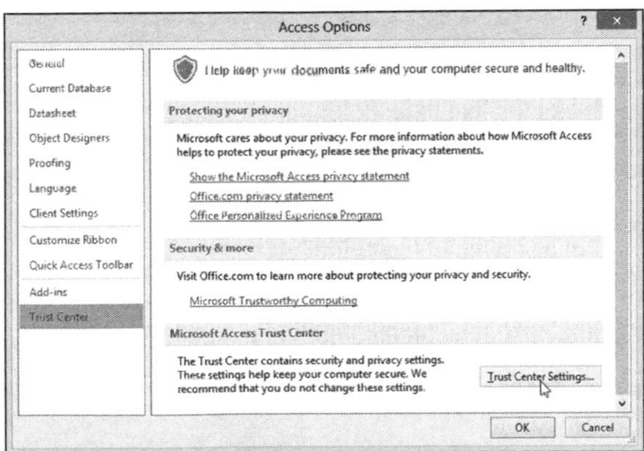

FIGURE 1.17. Working with the Trust Center (Step 1).

3. In the left pane of the Trust Center dialog box, click **Trusted Locations**, as shown in Figure 1.18.

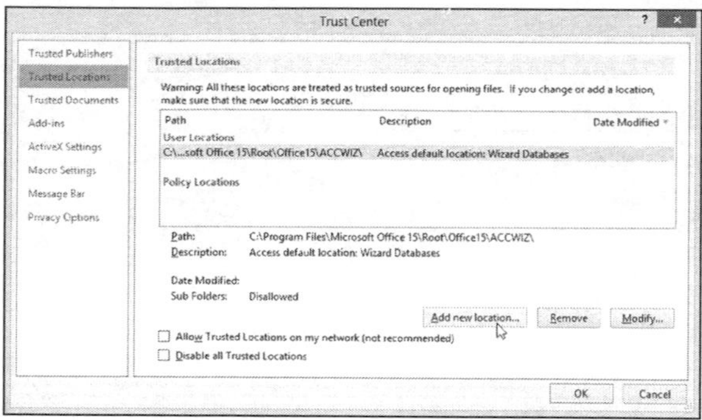

FIGURE 1.18. Working with the Trust Center (Step 2).

4. Click the **Add new location** button, as shown in Figure 1.18.
5. In the Path text box, type the path and folder name of the location on your local drive that you want to set up as a trusted source for opening files. Let's enter **C:\Access2013_ByExample** to designate this folder as a trusted location for this book's database programming exercises, as shown in Figure 1.19.

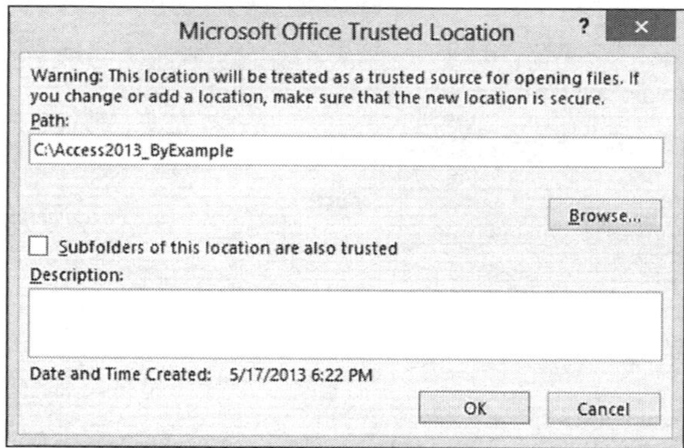

FIGURE 1.19. Working with the Trust Center (Step 3).

6. Click **OK** to close the Microsoft Office Trusted Location dialog box.
7. The Trusted Locations list in the Trust Center dialog box now includes the **C:\Access2013_ByExample** folder as a trusted source (see Figure 1.20). Files put in a trusted location can be opened without being checked by the Trust Center security feature. Click **OK** to close the Trust Center dialog box.

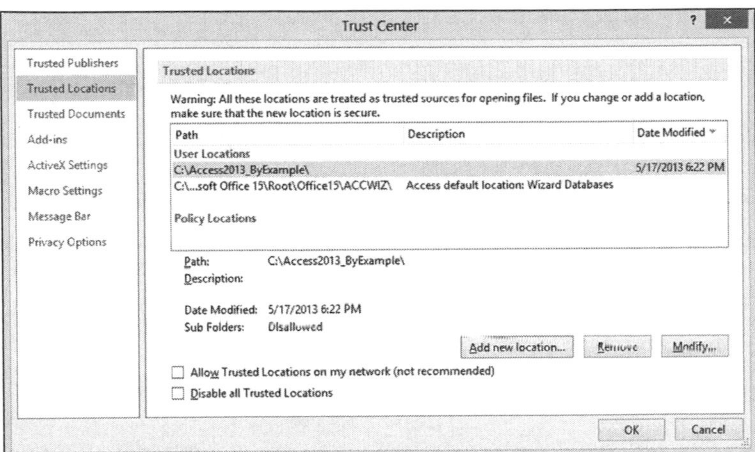

FIGURE 1.20. Working with the Trust Center (Step 4).

8. Click **OK** to close the Access Options dialog box.

CHAPTER SUMMARY

In this chapter, you received a hands-on introduction to writing VBA procedures in Access 2013. You learned about subroutine procedures, function procedures, property procedures, and event procedures. You also learned different ways of executing subroutines and functions. The main hands-on exercise in this chapter walked you through writing two event procedures in the Customers form's class module for a Contact Name text control placed in the form. You finished this chapter by designating a trusted location folder for your Access databases.

This chapter has given you a glimpse of the Microsoft Visual Basic programming environment built into Access. The next chapter will take you deeper into this interface, showing you various windows and shortcuts that you can use to program faster and with fewer errors.

Chapter 2 Exploring the Visual Basic Editor (VBE)

Now that you know how to write procedures and functions in standard modules and event procedures in modules placed behind a form, we'll spend some time in the Visual Basic Editor window to become familiar with the multitude of tools it offers to simplify your programming tasks. With the tools located in the Visual Basic Editor window, you can:

- Write your own VBA procedures
- Create custom forms
- View and modify object properties
- Test and debug VBA procedures and locate errors

You can enter the VBA programming environment in either of the following ways:

- By selecting the Database Tools tab, then Visual Basic in the Macro group
- From the keyboard, by pressing Alt+F11

UNDERSTANDING THE PROJECT EXPLORER WINDOW

The Project Explorer window, located on the left side of the Visual Basic Editor window, provides access to modules behind forms and reports via the Microsoft Access Class Objects folder (see Figure 2.1). The Modules folder lists only standard modules that are not behind a form or report.

In addition to the Microsoft Access Class Objects and Modules folders, the VBA Project Explorer window can contain a Class Modules folder. Class modules are used for creating your own objects, as demonstrated in Chapter 8. Using the Project Explorer window, you can easily move between modules currently loaded into memory.

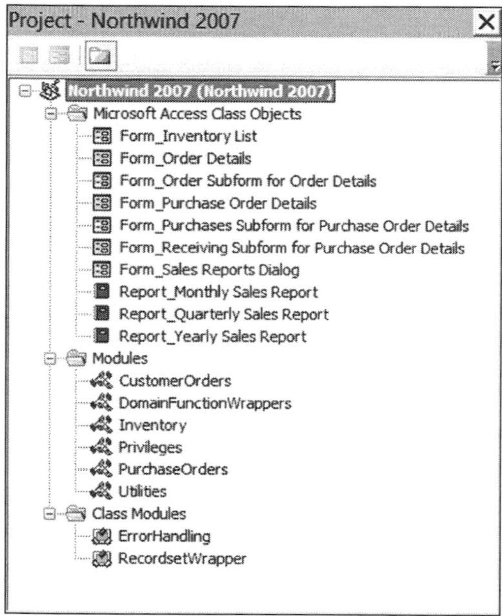

FIGURE 2.1. The Project Explorer window provides easy access to your VBA procedure code.

You can activate the Project Explorer window in one of three ways:

- From the View menu by selecting Project Explorer
- From the keyboard by pressing Ctrl-R
- From the Standard toolbar by clicking the Project Explorer button ( ) as shown in Figure 2.2

If the Project Explorer window is visible but not active, activate it by clicking the Project Explorer titlebar.

Buttons on the Standard toolbar (Figure 2.2) provide a quick way to access many Visual Basic features.

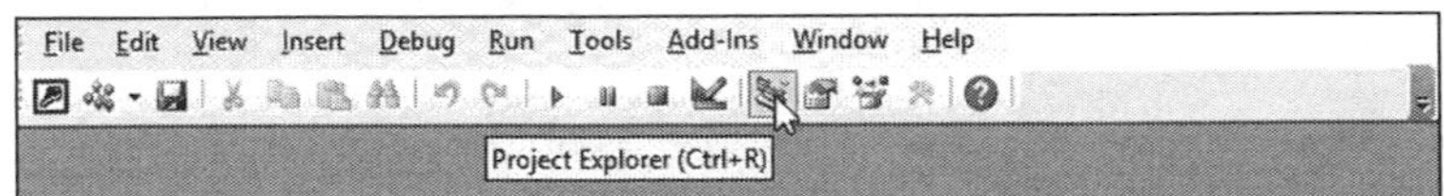

FIGURE 2.2. Use the toolbar buttons to quickly access frequently used features in the VBE window.

The Project Explorer window (see Figure 2.3) contains three buttons:

- **View Code**—Displays the Code window for the selected module.
- **View Object**—Displays the selected form or report in the Microsoft Access Class Objects folder. This button is disabled when an object in the Modules or Class Modules folder is selected.
- **Toggle Folders**—Hides and unhides the display of folders in the Project Explorer window.

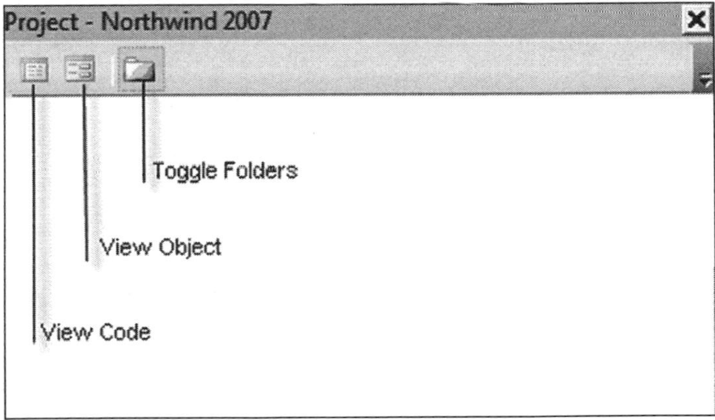

FIGURE 2.3 The VBE Project Explorer window contains three buttons that allow you to view code or objects and toggle folders.

UNDERSTANDING THE PROPERTIES WINDOW

The Properties window allows you to review and set properties for the currently selected Access class or module. The name of the selected object is displayed in the Object box located just below the Properties window titlebar. The Properties window displays the current settings for the selected object. Object properties can be viewed alphabetically or by category by clicking on the appropriate tab.

- **Alphabetic tab**—Lists all properties for the selected object alphabetically. You can change the property setting by selecting the property name, then typing or selecting the new setting.
- **Categorized tab**—Lists all properties for the selected object by category. You can collapse the list so that you see only the category names or you can expand a category to see the properties. The plus (+) icon to the left of the category name indicates that the category list can be expanded. The minus (–) indicates that the category is currently expanded.

  The Properties window can be accessed in the following ways:

- From the View menu by selecting Properties Window
- From the keyboard by pressing F4
- From the Standard toolbar by clicking the Properties Window button ( )located to the right of the Project Explorer button

Figure 2.4 displays the properties of the E-mail Address text box control located in the Form_Order Details form in the Northwind 2007 sample Access database. In order to access properties for a form control, you need to perform the steps outlined in Hands-On 2.1.

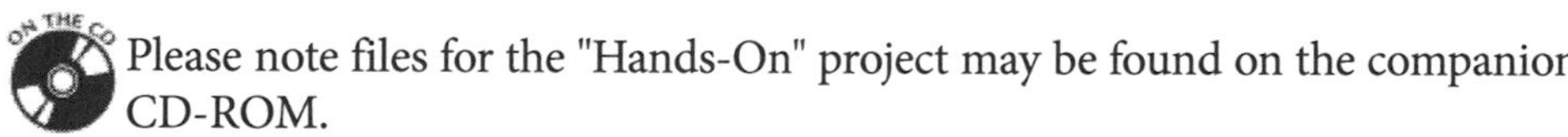

Please note files for the "Hands-On" project may be found on the companion CD-ROM.

Hands-On 2.1. Using the Properties Window to View Control Properties

1. Copy the **Northwind 2007** sample database from the Access2013_HandsOn folder to your **C:\Access2013_ByExample** folder.
2. Open Access 2013 and load the **C:\Access2013_ByExample\Northwind 2007.accdb** file. Log in to the database as **Andrew Cencini**.
3. When Northwind 2007 opens, press **Alt+F11** to activate the Visual Basic Editor window.
4. In the Project Explorer window, click the **Toggle Folders** button ( ) and select the **Microsoft Access Class Objects** folder. Highlight the **Form_Order Details** form (Figure 2.4) and click the **View Object** button ( ). This will open the selected form in Design view.
5. Press **Alt+F11** to return to the Visual Basic Editor. The Properties window will be filled with the properties for the Form_Order Details form. To view the properties of the E-mail Address text box control on this form, as shown in

Figure 2.4, select **E-mail Address** from the drop-down list located below the Properties window titlebar.

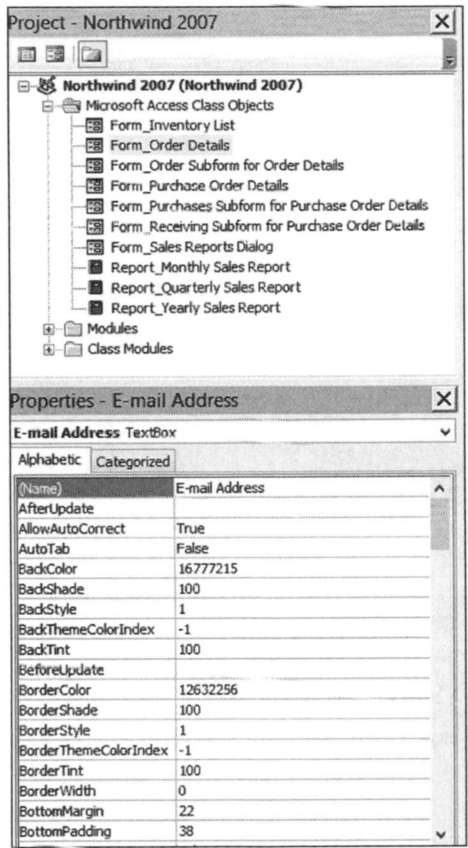

FIGURE 2.4. You can edit object properties in the Properties window, or you can edit them in the property sheet when a form or report is open in Design view.

UNDERSTANDING THE CODE WINDOW

The Code window is used for Visual Basic programming as well as for viewing and modifying the code of existing Visual Basic procedures. Each VBA module can be opened in a separate Code window.

There are several ways to activate the Code window:

- From the Project Explorer window, choose the appropriate module and then click the View Code button ( )

- From the Microsoft Visual Basic menu bar, choose View | Code
- From the keyboard, press F7

At the top of the Code window there are two drop-down listboxes that allow you to move quickly within the Visual Basic code. In the Object box on the left side of the Code window, you can select the object whose code you want to view, as shown in Figure 2.5.

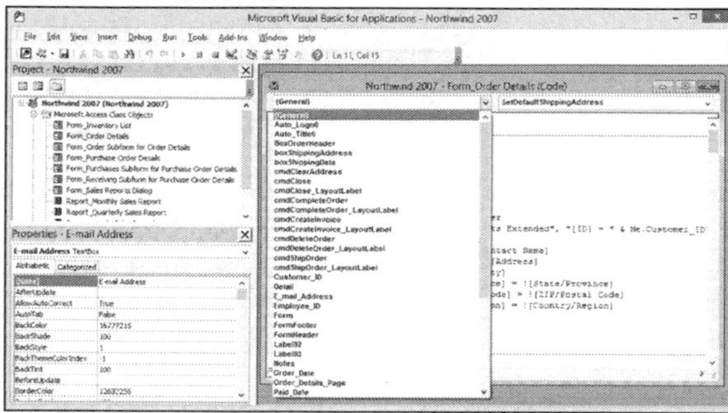

FIGURE 2.5. The Object drop-down box lists objects that are available in the module selected in the Project Explorer window.

The box on the right side of the Code window lets you select a procedure to view. When you click the down arrow at the right of this box, the names of all procedures located in a module are listed alphabetically, as shown in Figure 2.6. When you select a procedure in the Procedure box, the cursor will jump to the first line of that procedure.

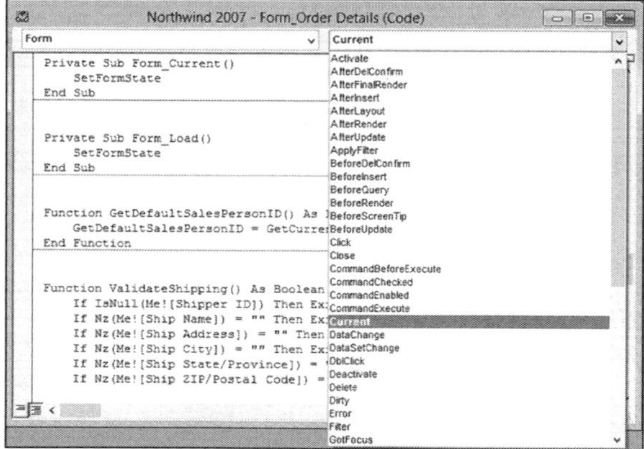

FIGURE 2.6. The Procedure drop-down box lists events to which the object selected in the Object drop-down box can respond. If the selected module contains events written for the highlighted object, the names of these events appear in bold type.

By choosing Window | Split or dragging the split bar down to a selected position in the Code window, you can divide the Code window into two panes, as shown in Figure 2.7.

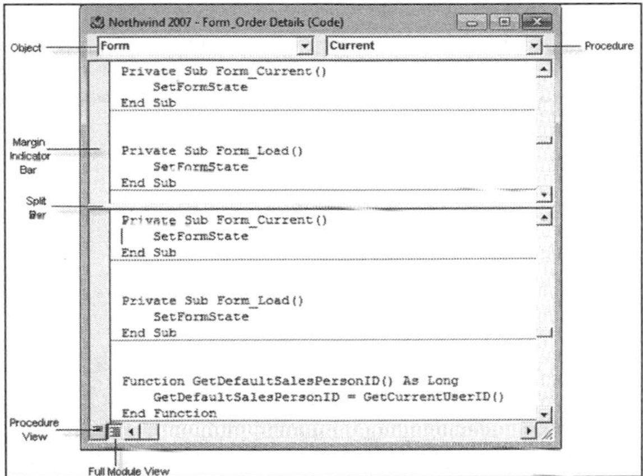

FIGURE 2.7. By splitting the Code window, you can view different sections of a long procedure or a different procedure in each window pane.

Setting up the Code window for the two-pane display is useful for copying, cutting, and pasting sections of code between procedures in the same module. To return to a one-window display, drag the split bar all the way to the top of the Code window or choose Window | Split again.

There are two icons at the bottom of the Code window (see Figure 2.7). The Procedure View icon changes the display to only one procedure at a time in the Code window. To select another procedure, use the Procedure drop-down box. The Full Module View icon changes the display to all the procedures in the selected module. Use the vertical scrollbar in the Code window to scroll through the module's code. The Margin Indicator bar is used by the Visual Basic Editor to display helpful indicators during editing and debugging.

OTHER WINDOWS IN THE VBE

In addition to the Code window, there are several other windows that are frequently used in the Visual Basic environment, such as the Immediate, Locals, Watch, Project Explorer, Properties, and Object Browser windows. The Docking tab in the Options dialog box, shown in Figure 2.8, displays a list of available windows and allows you to choose which windows you want to be dockable. To access this dialog box, select Tools | Options in the Visual Basic Editor window.

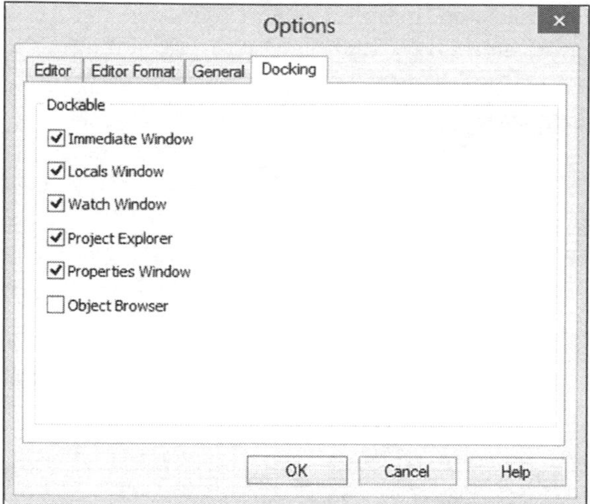

FIGURE 2.8. You can use the Docking tab in the Options dialog box to control which windows are currently displayed in the Visual Basic programming environment.

ASSIGNING A NAME TO THE VBA PROJECT

A VBA Project is a set of Microsoft Access objects, modules, forms, and references.

When you create a Microsoft Access database and later switch to the VBE window, you will see in the Project Explorer window that Access had automatically assigned the database name to the VBA Project. For example, if your database is named Chap01.accdb, the Project Properties window displays Chap01 (Chap01) where the first "Chap01" denotes the VBA Project name and the "Chap01" in the parentheses is the name of the database. You can change the name of the VBA Project in one of the following ways:

- Choose Tools | <database name> Properties, enter a new name in the Project Name box of the Project Properties window (see Figure 2.9) and click OK.
- In the Project Explorer window, right-click the name of the project and select <database name> Properties. Enter a new name in the Project Name box of the Project Properties window (see Figure 2.9) and click OK.

To avoid naming conflicts between projects, make sure that you give your projects unique names.

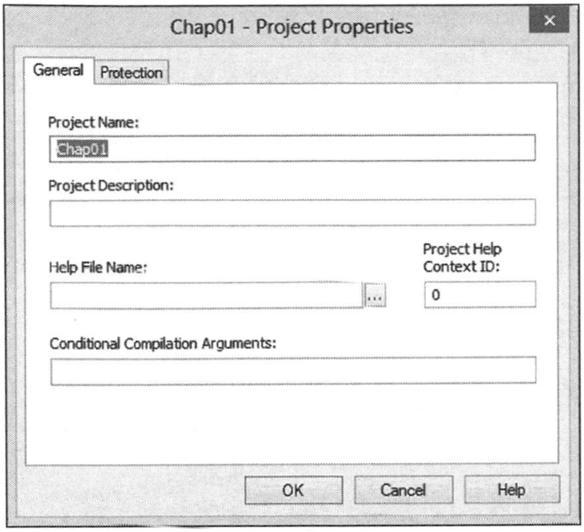

FIGURE 2.9. Use the Project Properties dialog box to rename the VBA Project.

RENAMING THE MODULE

When you insert a new module to your VBA Project, Access generates a default name for the module—Module1, Module2, and so on. You can rename your modules right after you insert them into the VBA project or when your project is being saved for the first time. In the latter case, Access will iterate through all the newly added (not saved) modules and will prompt you with the Save As dialog box to accept or change the module name. You can change the module name at any time via the Properties window. Simply select the module name (for example, Module1) in the Project Explorer window and double-click the Name property in the Properties window. This action will highlight the default module name next to the Name property. Type the new name for the module and press Enter. The module name in the Project Explorer window should now reflect your change.

ON-THE-FLY SYNTAX AND PROGRAMMING ASSISTANCE

Writing procedures in Visual Basic requires that you use hundreds of built-in instructions and functions. Because most people cannot memorize the correct syntax of all the instructions available in VBA, the IntelliSense® technology provides you with syntax and programming assistance on demand while you are entering instructions. While working in the Code window, you can have special tools pop up and guide you through the process of creating correct VBA code. The Edit toolbar in the VBE window, shown in Figure 2.10, contains several buttons that let you enter correctly formatted VBA instructions with speed and ease. If the Edit toolbar isn't currently docked in the Visual Basic Editor window, you can turn it on by choosing View | Toolbars.

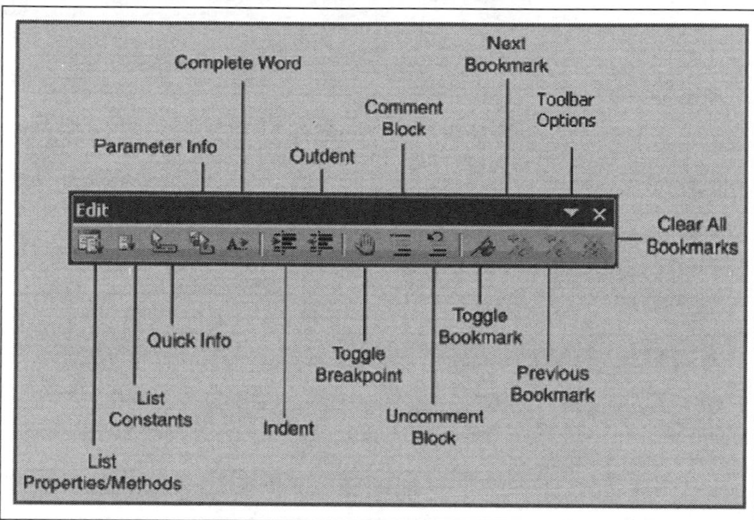

FIGURE 2.10. The Edit toolbar provides timesaving buttons while entering VBA code.

List Properties/Methods

Each object can contain one or more properties and methods. When you enter the name of the object in the Code window followed by a period that separates the name of the object from its property or method, a pop-up menu may appear. This menu lists the properties and methods available for the object that precedes the period. To turn on this automated feature, choose Tools | Options. In the Options dialog box, click the Editor tab, and make sure the Auto List Members checkbox is selected. As you enter VBA instructions, Visual Basic suggests properties and methods that can be used with the particular object, as demonstrated in Figure 2.11.

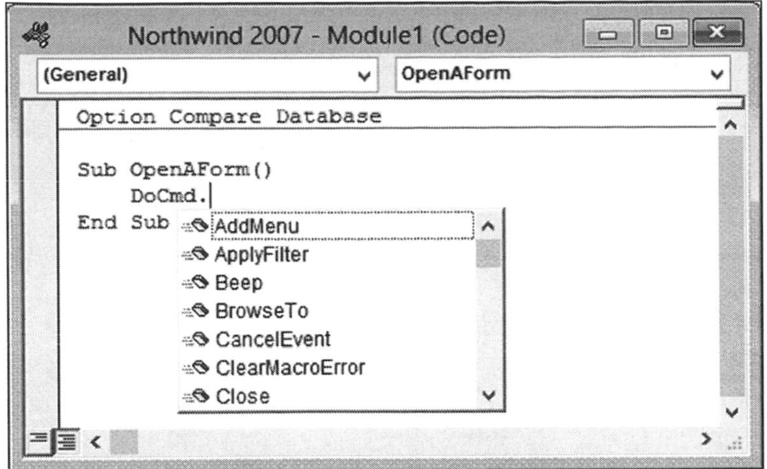

FIGURE 2.11. When Auto List Members is selected, Visual Basic suggests properties and methods that can be used with the object as you are entering the VBA instructions.

To choose an item from the pop-up menu, start typing the name of the property or method you want to use. When the correct item name is highlighted, press Enter to insert the item into your code and start a new line, or press the Tab key to insert the item and continue writing instructions on the same line. You can also double-click the item to insert it in your code. To close the pop-up menu without inserting an item, press Esc. When you press Esc to remove the pop-up menu, Visual Basic will not display the menu for the same object again.

To display the Properties/Methods pop-up menu again, you can:

- Press Ctrl-J
- Use the Backspace key to delete the period, then type the period again
- Right-click in the Code window, and select List Properties/Methods from the shortcut menu
- Choose Edit | List Properties/Methods
- Click the List Properties/Methods button ( ) on the Edit toolbar

Parameter Info

Some VBA functions and methods can take one or more arguments (or parameters). If a Visual Basic function or method requires an argument, you can see the names of required and optional arguments in a tip box that appears just below the cursor as soon as you type the open parenthesis or enter a space. The Parameter Info feature (see Figure 2.12) makes it easy for you to supply correct arguments to a VBA function or method. In addition, it reminds you of two other things that are very important for the function or method to work correctly: the order of the arguments and the required data type of each argument. For example, if you enter in the Code window the instruction `DoCmd.OpenForm` and type a space after the OpenForm method, a tip box appears just below the cursor. Then as soon as you supply the first argument and enter the comma, Visual Basic displays the next argument in bold. Optional arguments are surrounded by square brackets [ ]. To close the Parameter Info window, all you need to do is press Esc.

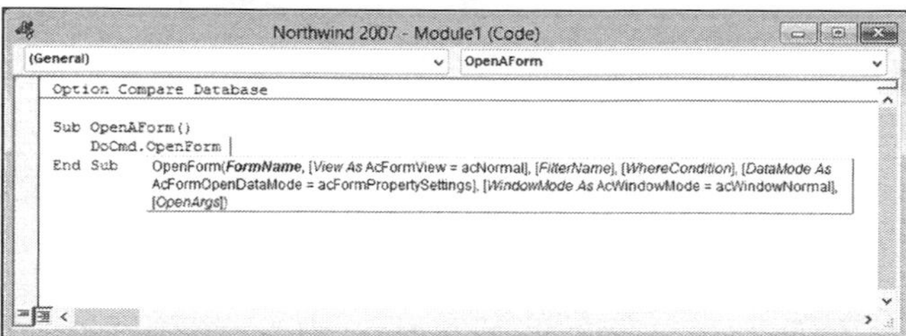

FIGURE 2.12. A tip window displays a list of arguments used by a VBA function or method.

To open the tip box using the keyboard, enter the instruction or function, followed by the open parenthesis, then press Ctrl-Shift-I. You can also click the Parameter Info button ( ) on the Edit toolbar or choose Edit | Parameter Info from the menu bar.

You can also display the Parameter Info box when entering a VBA function. To try this out quickly, choose View | Immediate Window, then type the following in the Immediate window:

```
Mkdir(
```

You should see the MkDir(Path As String) tip box just below the cursor. Now, type `"C:\NewFolder"` followed by the ending parenthesis. When you press Enter, Visual Basic will create a folder named NewFolder in the root directory of your computer. Activate Explorer and check it out!

List Constants

If there is a check mark next to the Auto List Members setting in the Options dialog box (Editor tab), Visual Basic displays a pop-up menu listing the constants that are valid for the property or method. A *constant* is a value that indicates a specific state or result. Access and other members of the Microsoft Office suite have a number of predefined, built-in constants.

Suppose you want to open a form in Design view. In Microsoft Access, a form can be viewed in Design view (acDesign), Datasheet view (acFormDS), PivotChart view (acFormPivotChart), PivotTable view (acFormPivotTable), Form view (acNormal), and Print Preview (acPreview). Each of these options is represented by a built-in constant. Microsoft Access constant names begin with the letters "ac." As soon as you enter a comma and a space following your instruction in the Code window (e.g., `DoCmd.OpenForm "Products",`), a pop-up menu will appear with the names of valid constants for the OpenForm method, as shown in Figure 2.13.

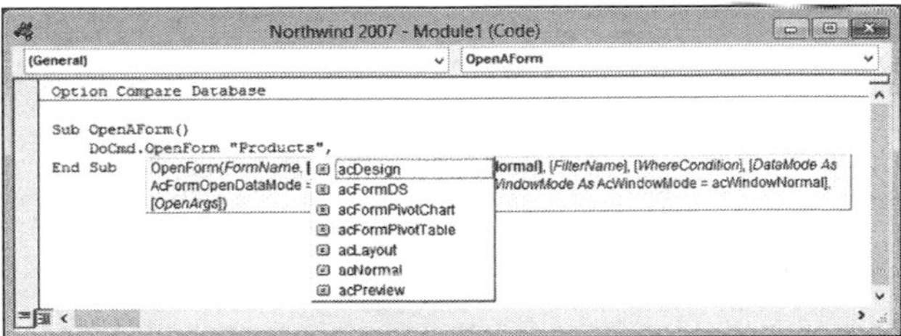

FIGURE 2.13. The List Constants pop-up menu displays a list of constants that are valid for the property or method typed.

The List Constants menu can be activated by pressing Ctrl+Shift+J or by clicking the List Constants button ( ) on the Edit toolbar.

Quick Info

When you select an instruction, function, method, procedure name, or constant in the Code window and then click the Quick Info button ( ) on the Edit toolbar (or press Ctrl+I), Visual Basic will display the syntax of the highlighted item as well as the value of its constant (see Figure 2.14). The Quick Info feature can be turned on or off using the Options dialog box (Tools | Options). To use the feature, click the Editor tab in the Options dialog box, and make sure there is a check mark in the box next to Auto Quick Info.

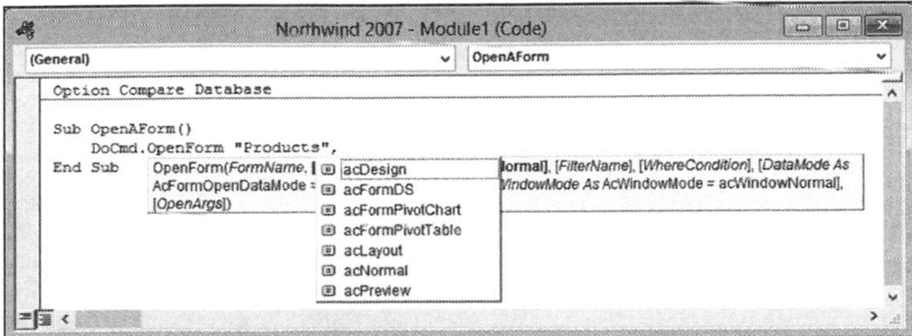

FIGURE 2.14. The Quick Info feature provides a list of function parameters, as well as constant values and VBA statement syntax.

Complete Word

Another way to increase the speed of writing VBA procedures in the Code window is with the Complete Word feature. As you enter the first few letters of a keyword and click the Complete Word button ( ) on the Edit toolbar, Visual Basic will complete the keyword entry for you. For example, if you enter the first three letters of the keyword `DoCmd` (DoC) in the Code window, then click the Complete Word button on the Edit toolbar, Visual Basic will complete the rest of the command. In the place of DoC you will see the entire instruction, DoCmd.

If there are several VBA keywords that begin with the same letters, when you click the Complete Word button on the Edit toolbar Visual Basic will display a pop-up menu listing all of them. To try this, enter only the first three letters of the word Application (App), then press the Complete Word button on the toolbar. You can then select the appropriate word from the pop-up menu.

Indent/Outdent

The Editor tab in the Options dialog box, shown in Figure 2.15, contains many settings you can enable to make automated features available in the Code window.

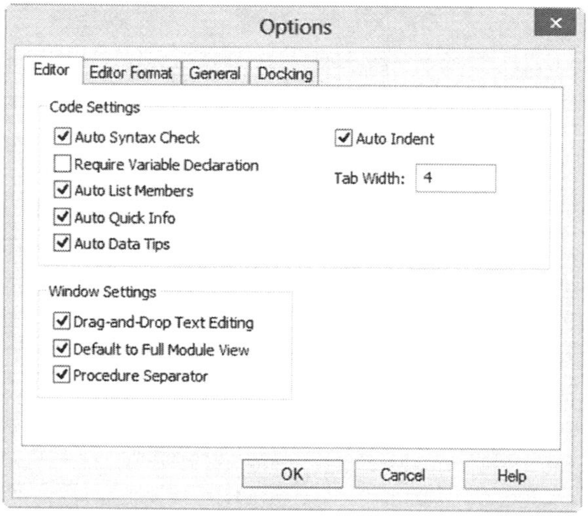

FIGURE 2.15. The Options dialog box lists features you can turn on and off to fit the VBA programming environment to your needs.

When the Auto Indent option is turned on, Visual Basic automatically indents the selected lines of code using the Tab Width value. The default entry for Auto Indent is four characters (see Figure 2.15). You can easily change the tab width by typing a new value in the text box. Why would you want to use indentation in your code? Indentation makes your VBA procedures more readable and easier to understand. Indenting is especially recommended for entering lines of code that make decisions or repeat actions.

Let's see how you can indent and outdent lines of code using the Form_InventoryList form in the Northwind database that you opened in the previous hands-on exercise.

Hands-On 2.2. Using the Indent/Outdent Feature

1. In the Project Explorer window in the Microsoft Access Class Objects folder, double-click on **Form_Inventory List**. The Code window should now show the CmdPurchase_Click event procedure written for this form.
2. In the Code window, select the block of code beginning with the keyword **If** and ending with the keywords **End If**.
3. Click the **Indent** button ( ) on the Edit toolbar or press **Tab** on the keyboard. The selected block of code will move four spaces to the right. You can adjust the number of spaces to indent by choosing **Tools | Options** and entering the appropriate value in the Tab Width box on the Editor tab.
4. Now, click the **Outdent** button ( ) on the Edit toolbar or press **Shift+Tab** to return the selected lines of code to the previous location in the Code window. The Indent and Outdent options are also available from Visual Basic Editor's Edit menu.

Comment Block/Uncomment Block

The apostrophe placed at the beginning of a line of code denotes a comment. Besides the fact that comments make it easier to understand what the procedure does, comments are also very useful in testing and troubleshooting VBA procedures. For example, when you execute a procedure, it may not run as expected. Instead of deleting the lines of code that may be responsible for the problems encountered, you may want to skip the lines for now and return to them later. By placing an apostrophe at the beginning of the line you want to avoid, you can continue checking the other parts of your procedure. While commenting one line of code by typing an apostrophe works fine for most people, when it comes to turning entire blocks of code into comments, you'll find the Comment Block and Uncomment Block buttons on the Edit toolbar very handy and easy to use.

To comment a few lines of code, select the lines and click the Comment Block button ( ). To turn the commented code back into VBA instructions, click the Uncomment Block button ( ). If you click the Comment Block button without selecting a block of text, the apostrophe is added only to the line of code where the cursor is currently located.

USING THE OBJECT BROWSER

If you want to move easily through the myriad of VBA elements and features, examine the capabilities of the Object Browser. This special built-in tool is available in the Visual Basic Editor window.

To access the Object Browser, use any of the following methods:

- Press F2
- Choose View | Object Browser
- Click the Object Browser button ( ) on the toolbar

The Object Browser allows you to browse through the objects available to your VBA procedures, as well as view their properties, methods, and events. With the aid of the Object Browser, you can quickly move between procedures in your database application and search for objects and methods across various type libraries.

The Object Browser window, shown in Figure 2.16, is divided into several sections. The top of the window displays the Project/Library drop-down listbox with the names of all currently available libraries and projects.

A *library* is a special file that contains information about the objects in an application. New libraries can be added via the References dialog box (select Tools | References). The entry for <All Libraries> lists the objects of all libraries installed on your computer. While the Access library contains objects specific to using Microsoft Access, the VBA library provides access to three objects (Debug, Err, and Collection), as well as a number of built-in functions and constants that give you flexibility in programming. You can send output to the Immediate window, get information about runtime errors, work with the Collection object, manage files, deal with text strings, convert data types, set date and time, and perform mathematical operations.

Below the Project/Library drop-down listbox is a search box (Search Text) that allows you to quickly find information in a particular library. This field remembers the last four items you searched for. To find only whole words, right-click anywhere in the Object Browser window, then choose Find Whole Word Only from the shortcut menu. The Search Results section of the Object Browser displays the Library, Class, and Member elements that meet the criteria entered in the Search Text box. When you type the search text and click the Search button, Visual Basic expands the Object Browser window to show the search results. You can hide or show the Search Results section by clicking the button located to the right of the binoculars. In the lower section of the Object Browser window, the Classes listbox displays the available object classes in the selected library. If you select the name of the open database (e.g., Northwind) in the Project/Library listbox, the Classes list will display the objects as listed in the Explorer window.

In Figure 2.16, the Form_Inventory List object class is selected. When you highlight a class, the list on the right side (Members) shows the properties, methods, and events available for that class. By default, members are listed alphabetically. You can, however, organize the Members list by group type (properties, methods, or events) using the Group Members command from the Object Browser shortcut menu (right-click anywhere in the Object Browser window to display this menu).

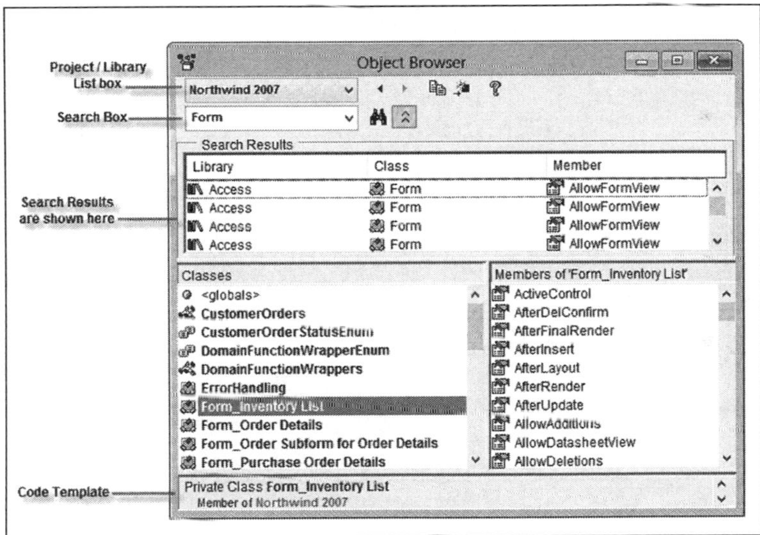

FIGURE 2.16. The Object Browser window allows you to browse through all the objects, properties, and methods available to the current VBA project.

When you select the Northwind 2007 project in the Project/Library listbox, the Members listbox will list all the procedures available in this project. To examine a procedure's code, double-click its name. When you select a VBA library in the Project/Library listbox, you will see the Visual Basic built-in functions and constants. If you need more information on the selected class or member, click the question mark button located at the top of the Object Browser window.

The bottom of the Object Browser window displays a code template area with the definition of the selected member. Clicking the green hyperlink text in the code template lets you jump to the selected member's class or library in the Object Browser window. Text displayed in the code template area can be copied and pasted to a Code window. If the Code window is visible while the Object Browser window is open, you can save time by dragging the highlighted code template and dropping it into the Code window. You can easily adjust the size of the various sections of the Object Browser window by dragging the dividing horizontal and vertical lines.

Let's put the Object Browser to use in VBA programming. Assume that you want to write a VBA procedure to control a checkbox placed on a form and would like to see the list of properties and methods that are available for working with checkboxes.

Hands-On 2.3. Using the Object Browser

1. In the Visual Basic Editor window, press **F2** to display the Object Browser.
2. In the Project/Library listbox (see Figure 2.16), click the drop-down arrow and select the **Access** library.
3. Type **checkbox** in the Search Text box and click the **Search** button (). Make sure you don't enter a space in the search string.

Visual Basic begins to search the Access library and displays the search results. By analyzing the search results in the Object Browser window, you can find the appropriate VBA instructions for writing your VBA procedures. For example, looking at the Members list lets you quickly determine that you can enable or disable a checkbox by setting the Enabled property. To get detailed information on any item found in the Object Browser, select the item and press F1 to activate online help.

USING THE VBA OBJECT LIBRARY

While the Access library contains objects specific to using Microsoft Access, the VBA Object Library provides access to many built-in VBA functions grouped by categories. These functions are general in nature. They allow you to manage files, set the date and time, interact with users, convert data types, deal with text strings, or perform mathematical calculations. In the following exercise, you will see how to use one of the built-in VBA functions to create a new subfolder without leaving Access.

Hands-On 2.4. Using Built-in VBA Functions

1. In the Visual Basic Editor window with the Northwind 2007 database open, choose **Insert | Module** to create a new standard module.
2. In the Code window, enter **Sub NewFolder()** as the name of the procedure and press **Enter**. Visual Basic will enter the ending keywords: **End Sub**.
3. Press **F2** to display the Object Browser.
4. Click the drop-down arrow in the Project/Library listbox and select **VBA**.
5. Enter **file** in the Search Text box and press **Enter**.
6. Scroll down in the Members listbox and highlight the **MkDir** method.
7. Click the **Copy** button in the Object Browser window to copy the selected method name to the Windows clipboard.
8. Close the Object Browser and return to the Code window. Paste the copied instruction inside the **NewFolder** procedure.
9. Now, enter a space, followed by **"C:\Study."** Be sure to enter the name of the entire path and the quotation marks. Your NewFolder procedure should look like the following:

```
Sub NewFolder()
    MkDir "C:\Study"
End Sub
```

10. Choose **Run | Run Sub/UserForm** to run the NewFolder procedure.

After you run the NewFolder procedure, Visual Basic creates a new folder on drive C called Study. To see the folder, activate Windows Explorer. After creating a new folder, you may realize that you don't need it after all. Although you could easily delete the folder while in Windows Explorer, how about getting rid of it programmatically?

The Object Browser contains many other methods that are useful for working with folders and files. The `RmDir` method is just as simple to use as the `MkDir` method. To remove the Study folder from your hard drive, replace the `MkDir` method with the `RmDir` method and rerun the NewFolder procedure. Or create a new procedure called RemoveFolder, as shown here:

```
Sub RemoveFolder()
        RmDir "C:\Study"
End Sub
```

When writing procedures from scratch, it's a good idea to consult the Object Browser for names of the built-in VBA functions.

USING THE IMMEDIATE WINDOW

The Immediate window is a sort of VBA programmer's scratch pad. Here you can test VBA instructions before putting them to work in your VBA procedures. It is a great tool for experimenting with your new language. Use it to try out your statements. If the statement produces the expected result, you can copy the statement from the Immediate window into your procedure (or you can drag it right onto the Code window if the window is visible).

To activate the Immediate window, choose View | Immediate Window in the Visual Basic Editor, or press Ctrl+G while in the Visual Basic Editor window.

The Immediate window can be moved anywhere on the Visual Basic Editor window, or it can be docked so that it always appears in the same area of the screen. The docking setting can be turned on and off from the Docking tab in the Options dialog box (Tools | Options).

To close the Immediate window, click the Close button in the top-right corner of the window.

The following hands-on exercise demonstrates how to use the Immediate window to check instructions and get answers.

Hands-On 2.5. Experiments in the Immediate Window

1. If you are not in the VBE window, press **Alt+F11** to activate it.
2. Press **Ctrl+G** to activate the Immediate window, or choose **View | Immediate Window**.
3. In the Immediate window, type the following instruction and press **Enter**:

```
DoCmd.OpenForm "Inventory List"
```

4. If you entered the preceding VBA statement correctly, Visual Basic opens the Inventory List form, assuming the Northwind database is open.
5. Enter the following instruction in the Immediate window:

```
Debug.Print Forms![Inventory List].RecordSource
```

When you press **Enter**, Visual Basic indicates that `Inventory` is the RecordSource for the Inventory List form. Every time you type an instruction in the Immediate window and press Enter, Visual Basic executes the statement on the line where the insertion point is located. If you want to execute the same instruction again, click anywhere in the line containing the instruction and press Enter. For more practice, rerun the statements shown in Figure 2.17. Start from the instruction displayed in the first line of the Immediate window. Execute the instructions one by one by clicking in the appropriate line and pressing Enter.

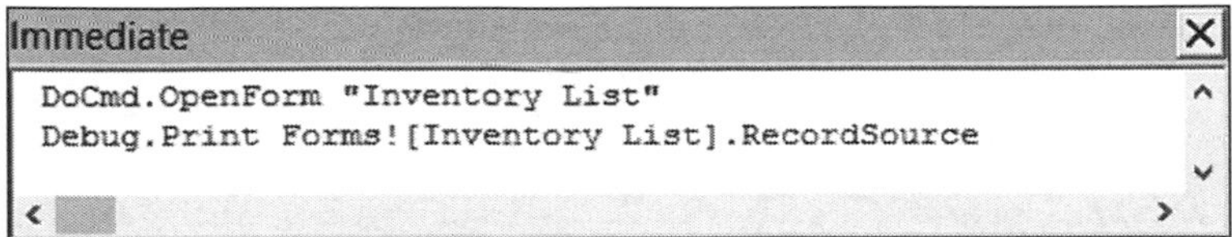

```
DoCmd.OpenForm "Inventory List"
Debug.Print Forms![Inventory List].RecordSource
```

FIGURE 2.17. Use the Immediate window to evaluate and try Visual Basic statements.

So far you have used the Immediate window to perform some actions. The Immediate window also allows you to ask questions. Suppose you want to find out the answers to "How many controls are in the Inventory List form?" or "What's the name of the current application?" When working in the Immediate window, you can easily get answers to these and other questions.

In the preceding exercise, you entered two instructions. Let's return to the Immediate window to ask some questions. Access remembers the instructions entered in the Immediate window even after you close this window. The contents of the Immediate window are automatically deleted when you exit Microsoft Access.

Hands-On 2.6. Asking Questions in the Immediate Window

1. Click in a new line of the Immediate window and enter the following statement to find out the number of controls in the Inventory List form:

   ```
   ?Forms![Inventory List].Controls.Count
   ```

   When you press **Enter**, Visual Basic enters the number of controls on a new line in the Immediate window.
2. Click in a new line of the Immediate window, and enter the following statement:

   ```
   ?Application.Name
   ```

   When you press **Enter**, Visual Basic enters the name of the active application on a new line in the Immediate window.
3. In a new line in the Immediate window, enter the following instruction:

   ```
   ?12/3
   ```

   When you press **Enter**, Visual Basic shows the result of the division on a new line. But what if you want to know the result of 3 + 2 and 12 * 8 right away? Instead of entering these instructions on separate lines, you can enter them on one line as in the following example:

   ```
   ?3+2:?12*8
   ```

   Notice the colon separating the two blocks of instructions. When you press the Enter key, Visual Basic displays the results 5 and 96 on separate lines in the

Immediate window.

Here are a couple of other statements you may want to try out on your own in the Immediate window:

```
?Application.GetOption("Default Database Directory")
?Application.CodeProject.Name
```

Instead of using the question mark, you may precede the statement typed in the Immediate window with the Print command, like this:

```
Print Application.CodeProject.Name
```

To delete the instructions from the Immediate window, highlight all the lines and press **Delete**.

4. In the Visual Basic Editor window, choose **File | Close and Return to Microsoft Access.**
5. Close the **Northwind 2007.accdb** database.

Recall that in Chapter 1 you learned how to run subroutine procedures and functions from the Immediate window. You will find other examples of running procedures and functions from this window in subsequent chapters.

CHAPTER SUMMARY

Programming in Access 2013 requires a working knowledge of objects and collections of objects. In this chapter, you explored features of the Visual Basic Editor window that can assist you in writing VBA code. Here are some important points:

- When in doubt about objects, properties, or methods in an existing VBA procedure, highlight the instruction in question and fire up the online help by pressing F1.
- When you need on-the-fly programming assistance while typing your VBA code, use the shortcut keys or buttons available on the Edit toolbar.
- If you need a quick listing of properties and methods for every available object, or have trouble locating a hard-to-find procedure, go with the Object Browser.
- If you want to experiment with VBA and see the results of VBA commands immediately, use the Immediate window.

In the next chapter, you will learn how you can remember values in your VBA procedures by using various types of variables and constants.

Chapter 3 USING VARIABLES, DATA TYPES, AND CONSTANTS

In Chapter 2, you used the question mark to have Visual Basic return some information in the Immediate window. Unfortunately, when you write Visual Basic procedures outside the Immediate window, you can't use the question mark. So how do you obtain answers to your questions in VBA procedures? To find out what a particular VBA instruction (statement) has returned, you must tell Visual Basic to memorize it. This is done by using variables. This chapter introduces you to many types of variables, data types, and constants that you can and should use in your VBA procedures.

WHAT ARE DATA TYPES?

When you create Visual Basic procedures you have a purpose in mind: You want to manipulate data. Because your procedures will handle different kinds of information, you should understand how Visual Basic stores data.

The *data type* determines how the data is stored in the computer's memory. For example, data can be stored as a number, text, date, object, etc. If you forget to tell Visual Basic the data type, it is assigned the Variant data type. The *Variant* type has the ability to figure out on its own what kind of data is being manipulated and then take on that type. The Visual Basic data types are shown in Table 3.1. In addition to the built-in data types, you can define your own data types; these are known as user-defined data types. Because data types take up different amounts of space in the computer's memory, some of them are more expensive than others. Therefore, to conserve memory and make your procedure run faster, you should select the data type that uses the fewest bytes but at the same time is capable of handling the data that your procedure has to manipulate.

TABLE 3.1. VBA Data Types.

Data Type	Storage Size	Range
Byte	1 byte	A number in the range of 0 to 255.
Boolean	2 bytes	Stores a value of True (0) or False (−1).
Integer	2 bytes	A number in the range of −32,768 to 32,767. The type declaration character for Integer is the percent sign (%).
Long (long integer)	4 bytes	A number in the range of −2,147,483,648 to 2,147,483,647. The type declaration character for Long is the ampersand (&).
LongLong	8 bytes	Stored as a signed 64-bit (8-byte) number ranging in value from −9,223,372,036,854,775,808 to 9,223,372,036,854,775,807. The type declaration character for Long Long is the caret(^). Long Long is a valid declared type only on 64-bit platforms.
LongPtr (Long integer on 32-bit systems;	4 bytes on 32-bit;	Numbers ranging in value from−2,147,483,648 to 2,147,483,647 on 32-bit systems; −9,223,372,036,854,775,808 to

Data Type	Storage Size	Range
LongLong integer on 64-bit systems)	8 bytes on 64-bit	9,223,372,036,854,775,807 on 64-bit systems. UsingLongPtr enables writing code that can run in both 32-bit and 64-bit environments.
Single (single-precision floating-point)	4 bytes	Single-precision floating-point real number ranging in value from –3.402823E38 to–1.401298E–45 for negative values and from 1.401298E–45 to 3.402823E38 for positive values. The type declaration character for Single is the exclamation point (!).
Double (double-precision floating-point)	8 bytes	Double-precision floating-point real number in the range of –1.79769313486231E308 to–4.94065645841247E–324 for negative values and 4.94065645841247E–324 to 1.79769313486231E308 for positive values. The type declaration character for Double is the number sign (#).
Currency (scaled integer)	8 bytes	Monetary values used in fixed-pointcalculations: –922,337,203,685,477.5808 to 922,337,203,685,477.5807. The type declaration character for Currency is theat sign (@).

Decimal	14 bytes	96-bit (12-byte) signed integer scaled by a variable power of 10. The power of 10 scaling factor specifies the number of digits to the right of the decimal point, and ranges from 0 to 28. With no decimal point (scale of 0), the largest value is+/–79,228,162,514,264,337,593,543,950,335.
		With 28 decimal places, the largest value is +/–7.9228162514264337593543950335.
		The smallest nonzero value is +/– 0.0000000000000000000000000001.
		You cannot declare a variable to be of type Decimal. You must use the Variant data type. Use the `CDec`-function to convert a value to a decimal number:
		`Dim numDecimal As Variant`
		`numDecimal = CDec(0.02 * 15.75 * 0.0006)`

(contd..)

Data Type	Storage Size	Range
Date	8 bytes	Date from January 1, 100, to December 31, 9999, and times from 0:00:00 to 23:59:59. Date literals must been closed within number signs (#); for example:#January 1, 2011#
Object	4 bytes	Any Object reference. Use the `Set` statement to declare a variable as an Object.
String (variable-length)	10 bytes + string length	A variable-length string can contain up to approximately 2 billion characters. The type declaration character for String is the dollar sign ($).
String (fixed-length)	Length of string	A fixed-length string can contain 1 to approximately 65,400 characters.
Variant (with numbers)	16 bytes	Any numeric value up to the range of a Double.
Variant (with characters)	22 bytes + string length	Any valid nonnumeric data type in the same range as for a variable-length string.

User-defined (using `Type`)	One or more elements	A data type you define using the `Type` statement. User-defined data types can contain one or more elements of a data type, an array, or a previously defined user-defined type. For example:
		`Type custInfo`
		`    custFullName as String`
		`    custTitle as String`
		`    custBusinessName as String`
		`    custFirstOrderDate as Date`
		`End Type`

WHAT ARE VARIABLES?

A *variable* is a name used to refer to an item of data. Each time you want to remember the result of a VBA instruction, think of a name that will represent it. For example, if you want to keep track of the number of controls on a particular form, you can make up a name such as NumOfControls, TotalControls, or FormsControlCount.

The names of variables can contain characters, numbers, and punctuation marks except for the following:

, # $ % & @ !

The name of a variable cannot begin with a number or contain a space. If you want the name of the variable to include more than one word, use the underscore (_) as a separator. Although a variable name can contain as many as 254 characters, it's best to use short and simple names. Using short names will save you typing time when you need to reuse the variable in your Visual Basic procedure. Visual Basic doesn't care whether you use uppercase or lowercase letters in variable names; however, most programmers use lowercase letters. When the variable name is comprised of more than one word, most programmers capitalize the first letter of each word, as in the following: NumOfControls, First_Name.

Reserved Words Can't Be Used for Variable Names

You can use any label you want for a variable name except for the reserved words that VBA uses. Visual Basic function names and words that have a special meaning in VBA cannot be used as variable names. For example, words such as Name, Len, Empty, Local, Currency, or Exit will generate an error message if used as a variable name.

Give your variables names that can help you remember their roles. Some programmers use a prefix to identify the variable's type. A variable name preceded with "str," such as strName, can be quickly recognized within the procedure code as the variable holding the text string.

CREATING VARIABLES

You can create a variable by declaring it with a special command or by just using it in a statement. When you declare your variable, you make Visual Basic aware of the variable's name and data type. This is called *explicit variable declaration.*

Advantages of Explicit Variable Declaration

Explicit variable declaration:

- Speeds up the execution of your procedure. Since Visual Basic knows the data type, it reserves only as much memory as is absolutely necessary to store the data.
- Makes your code easier to read and understand because all the variables are listed at the very beginning of the procedure.
- Helps prevent errors caused by misspelling a variable name. Visual Basic automatically corrects the variable name based on the spelling used in the variable declaration.

If you don't let Visual Basic know about the variable prior to using it, you are implicitly telling VBA that you want to create this variable. *Implicit variables* are automatically assigned the Variant data type (see Table 3.1 earlier in the chapter). Although implicit variable declaration is convenient (it allows you to create variables on the fly and assign values to them without knowing in advance the data type of the values being assigned), it can cause several problems.

Disadvantages of Implicit Variable Declaration

- If you misspell a variable name in your procedure, Visual Basic may display a runtime error or create a new variable. You are guaranteed to waste some time troubleshooting problems that could easily have been avoided had you declared your variable at the beginning of the procedure.
- Since Visual Basic does not know what type of data your variable will store, it assigns it a Variant data type. This causes your procedure to run slower because Visual Basic has to check the data type every time it deals with your variable. And because Variant variables can store any type of data, Visual Basic has to reserve more memory to store your data.

DECLARING VARIABLES

You declare a variable with the `Dim` keyword. Dim stands for "dimension." The `Dim` keyword is followed by the variable's name and type.

Suppose you want the procedure to display the age of an employee. Before you can calculate the age, you must feed the procedure the employee's date of birth. To do this, you declare a variable called `dateOfBirth`, as follows:

```
Dim dateOfBirth As Date
```

Notice that the `Dim` keyword is followed by the name of the variable (`dateOfBirth`). If you don't like this name, you are free to replace it with another word, as long as the word you are planning to use is not one of the VBA keywords. You specify the data type the variable will hold by including the `As` keyword followed by one of the data types from Table 3.1. The Date data type tells Visual Basic that the variable `dateOfBirth` will store a date.

To store the employee's age, you declare the variable as follows:

```
Dim intAge As Integer
```

The `intAge` variable will store the number of years between today's date and the employee's date of birth. Becauseage is displayed as a whole number, the `intAge` variable has been assigned the Integer data type. You may also want your procedure to keep track of the employee's name, so you declare another variable to hold the employee's first and last name:

```
Dim strFullName As String
```

Because the word Name is on the VBA list of reserved words, using it in your VBA procedure would guarantee an error. To hold the employee's full name, we used the variable `strFullName` and declared it as the String data type because the data it will hold is text. Declaring variables is regarded as good programming practice because it makes programs easier to read and helps prevent certain types of errors.

Informal (Implicit) Variables

Variables that are not explicitly declared with `Dim` statements are said to be implicitly declared. These variables are automatically assigned a data type called Variant. They can hold numbers, strings, and other types of information. You can create an informal variable by assigning some value to a variable name anywhere in your VBA procedure. For example, you implicitly declare a variable in the following way: `intDaysLeft = 100`.

Now that you know how to declare your variables, let's write a procedure that uses them.

Please note files for the "Hands-On" project may be found on the companion CD-ROM.

Hands-On 3.1. Using Variables

1. Start Microsoft Access 2013 and create a new database named **Chap03.accdb** in your **C:\Access2013_ByExample** folder.
2. Once your new database is opened, press **Alt+F11** to switch to the Visual Basic Editor window.
3. Choose **Insert | Module** to add a new standard module, and notice Module1 under the Modules folder in the Project Explorer window.
4. In the Module1 (Code) window, enter the following **AgeCalc** procedure:

```
Sub AgeCalc()
  ' variable declaration
  Dim strFullName As String
  Dim dateOfBirth As Date
  Dim intAge As Integer

  ' assign values to variables
  strFullName = "John Smith"
  dateOfBirth = #1/3/1967#

  ' calculate age
  IntAge = Year(Now()) - Year(dateOfBirth)

  ' print results to the Immediate window
  Debug.Print strFullName & " is " & intAge & " years old."
End Sub
```

Notice that in the AgeCalc procedure the variables are declared on separate lines at the beginning of the procedure. You can also declare several variables on the same line, separating each variable name with a comma, as shown here (be sure to enter this on one line):

```
Dim strFullName As String, dateOfBirth As Date, intAge As Integer
```

When you list all your variables on one line, the `Dim` keyword appears only once at the beginning of the variable declaration line.

5. If the Immediate window is not open, press **Ctrl+G** or choose **View | Immediate Window**. Because the example procedure writes the results to the Immediate window, you should ensure that this window is open prior to executing Step 6.
6. To run the AgeCalc procedure, click any line between the `Sub` and `End Sub` keywords and press **F5**.

What Is the Variable Type?

You can find out the type of a variable used in your procedure by right-clicking the variable name and selecting Quick Info from the shortcut menu.

When Visual Basic executes the variable declaration statements, it creates the variables with the specified names and reserves memory space to store their values. Then specific values are assigned to these variables. To assign a value to a variable, you begin with a variable name followed by an equal sign. The value entered to the right of the equal sign is the data you want to store in the variable. The data you enter here must be of the type stated in the variable declaration. Text data should be surrounded by quotation marks and dates by # characters.

Using the data supplied by the `dateOfBirth` variable, Visual Basic calculates the age of an employee and stores the result of the calculation in the variable called `intAge`. Then, the full name of the employee and the age are printed to the Immediate window using the instruction `Debug.Print`.

Concatenation

You can combine two or more strings to form a new string. The joining operation is called *concatenation*. You saw an example of concatenated strings in the AgeCalc procedure in Hands-On 3.1. Concatenation is represented by an ampersand character (&). For instance, `"His name is " &strFirstName` will produce a string like: His name is John or His name is Michael. The name of the person is determined by the contents of the `strFirstName` variable. Notice that there is an extra space between "is" and the ending quotation mark: `"His name is "`. Concatenation of strings can also be represented by a plus sign (+); however, many programmers prefer to restrict the plus sign to numerical operations to eliminate ambiguity.

SPECIFYING THE DATA TYPE OF A VARIABLE

If you don't specify the variable's data type in the `Dim` statement, you end up with the *untyped* variable. Untyped variables in VBA are always assigned the Variant data type. Variant data types can hold all the other data types (except for user-defined data types). This feature makes Variant a very flexible and popular data type. Despite this flexibility, it is highly recommended that you create typed variables. When you declare a variable of a certain data type, your VBA procedure runs faster because Visual Basic does not have to stop to analyze the variable to determine its type.

Visual Basic can work with many types of numeric variables. Integer variables can only hold whole numbers from –32,768 to 32,767. Other types of numeric variables are Long, Single, Double, and Currency. The Long variables can hold whole numbers in the range –2,147,483,648 to 2,147,483,647. As opposed to Integer and Long variables, Single and Double variables can hold decimals.

String variables are used to refer to text. When you declare a variable of the String data type, you can tell Visual Basic how long the string should be. For instance, `Dim strExtension As String * 3` declares the fixed-length String variable named `strExtension` that is three characters long. If you don't assign a specific length, the String variable will be *dynamic*. This means that Visual Basic will make enough space in computer memory to handle whatever text length is assigned to it.

After a variable is declared, it can only store the type of information that you stated in the declaration statement.

Assigning string values to numeric variables or numeric values to string variables results in the error message "Type Mismatch" or causes Visual Basic to modify the value. For example, if your variable was declared to hold whole numbers and your data uses decimals, Visual Basic will disregard the decimals and use only the whole part of the number.

Let's use the MyNumber procedure in Hands-On 3.2 as an example of how Visual Basic modifies the data according to the assigned data types.

Hands-On 3.2. Understanding the Data Type of a Variable

This hands-on exercise uses the **C:\Access2013_ByExample\Chap03.accdb** database that you created in Hands-On 3.1.

1. In the Visual Basic Editor window, choose **Insert | Module** to add a new module.

2. Enter the following procedure code for **MyNumber** in the new module's Code window:

```
Sub MyNumber()
  Dim intNum As Integer

  intNum = 23.11
  MsgBox intNum
End Sub
```

3. To run the procedure, click any line between the `Sub` and `End Sub` keywords and press **F5** or choose **Run | Run Sub/UserForm**.

   When you run this procedure, Visual Basic displays the contents of the variable `intNum` as 23, and not 23.11, because the `intNum` variable was declared as an Integer data type.

USING TYPE DECLARATION CHARACTERS

If you don't declare a variable with a `Dim` statement, you can still designate a type for it by using a special character at the end of the variable name. For example, to declare the `FirstName` variable as String, you append the dollar sign to the variable name:

```
Dim FirstName$
```

This is the same as `Dim FirstName As String`. Other type declaration characters are shown in Table 3.2. Notice that the type declaration characters can only be used with six data types. To use the type declaration character, append the character to the end of the variable name.

TABLE 3.2. Type declaration characters

Data Type	Character
Integer	%
Long	&
Single	!
Double	#
Currency	@
String	$

Declaring Typed Variables

The variable type can be indicated by the `As` keyword or by attaching a type symbol. If you don't add the type symbol or the `As` command, VBA will default the variable to the Variant data type.

Hands-On 3.3. Using Type Declaration Characters in Variable Names

This hands-on exercise uses the Chap03.accdb database that you created in Hands-On 3.1.

1. In the Visual Basic window, choose **Insert | Module** to add a new module.
2. Enter the **AgeCalc2** procedure code in the new module's Code window.

```
Sub AgeCalc2()
  ' variable declaration
  Dim FullName$
  Dim DateOfBirth As Date
  Dim age%

  ' assign values to variables
  FullName$ = "John Smith"
  DateOfBirth = #1/3/1967#

  ' calculate age
  age% = Year(Now()) - Year(DateOfBirth)
  ' print results to the Immediate window
  Debug.Print FullName$ & " is " & age% & " years old."
End Sub
```

3. To run the procedure, click any line between the `Sub` and `End Sub` keywords and press **F5** or choose **Run | Run Sub/UserForm**.

ASSIGNING VALUES TO VARIABLES

Now that you know how to correctly name and declare variables, it's time to learn how to initialize them.

Hands-On 3.4. Assigning Values to Variables

This hands-on exercise uses the **C:\Access2013_ByExample\Chap03.accdb** database that you created in Hands-On 3.1.

1. In the Visual Basic window, choose **Insert | Module** to add a new module.
2. Enter the code of the **CalcCost** procedure in the new module's Code window.

```
Sub CalcCost()
  slsPrice = 35
  slsTax = 0.085
  cost = slsPrice + (slsPrice * slsTax)
  strMsg = "The calculator total is " & "$" & cost & "."
  MsgBox strMsg
End Sub
```

3. To run the procedure, click any line between the `Sub` and `End Sub` keywords and press **F5** or choose **Run | Run Sub/UserForm**.

The CalcCost procedure uses four variables: `slsPrice`, `slsTax`, `cost`, and `strMsg`. Because none of these variables have been explicitly declared with the `Dim` keyword and a specific data type, they all have the same data type—Variant. The variables `slsPrice` and `slsTax` were created by assigning some values to the variable names at the beginning of the procedure. The `cost` variable was assigned the value resulting from the calculation `slsPrice` + (`slsPrice` * `slsTax`). The `cost` calculation uses the values supplied by the `slsPrice` and `slsTax` variables. The `strMsg` variable puts together a text message to the user. This message is then displayed with the `MsgBox` function.

When you assign values to variables, you follow the name of the variable with the equals sign. After the equals sign you enter the value of the variable. This can be text surrounded by quotation marks, a number, or an expression. While the values assigned to the variables `slsPrice`, `slsTax`, and `cost` are easily understood, the value stored in the `strMsg` variable is a little more involved.

Let's examine the content of the `strMsg` variable:

```
strMsg = "The calculator total is " & "$" & cost & "."
```

- The string `"The calculator total is "` begins and ends with quotation marks. Notice the extra space before the ending quotation mark.
- The & symbol allows one string to be appended to another string or to the contents of a variable and must be used every time you want to append a new piece of information to the previous string.
- The $ character is used to denote the type of currency. Because it is a character, it is surrounded by quotation marks.
- The & symbol attaches another string.

- The `cost` variable is a placeholder. The actual cost of the calculator will be displayed here when the procedure runs.
- The & symbol attaches yet another string.
- The period (.) is a character and must be surrounded by quotation marks. When you require a period at the end of the sentence, you must attach it separately when it follows the name of a variable.

Variable Initialization

Visual Basic automatically initializes a new variable to its default value when it is created. Numerical variables are set to zero (0), Boolean variables are initialized to False, string variables are set to the empty string (""), and Date variables are set to December 30, 1899.

Notice that the cost displayed in the message box has three decimal places. To display the cost of a calculator with two decimal places, you need to use a function. VBA has special functions that allow you to change the format of data. To change the format of the cost variable you should use the Format function. This function has the following syntax:

```
Format(expression, format)
```

where `expression` is a value or variable you want to format and `format` is the type of format you want to apply.

Hands-On 3.4. (Continued)

4. Change the calculation of the `cost` variable in the **CalcCost** procedure as follows:

```
cost = Format(slsPrice + (slsPrice * slsTax), "0.00")
```

5. To run the modified procedure, click any line between the `Sub` and `End Sub` keywords and press **F5** or choose **Run | Run Sub/UserForm**.

After having tried the CalcCost procedure, you may wonder why you should bother declaring variables if Visual Basic can handle undeclared variables so well. The CalcCost procedure is very short, so you don't need to worry about how many bytes of memory will be consumed each time Visual Basic uses the Variant variable. In short procedures, however, it is not the memory that matters but the mistakes you are bound to make when typing variable names. What will happen if the second time you use the `cost` variable you omit the "o" and refer to it as `cst`?

```
strMsg = "The calculator total is " & "$" & cst & "."
```

And what will you end up with if, instead of `slsTax`, you use the word `tax` in the formula?

```
cost = Format(slsPrice + (slsPrice * tax), "0.00")
```

When you run the procedure with the preceding errors introduced, Visual Basic will not show the cost of the calculator because it does not find the assignment statement for the `cst` variable. And because Visual Basic does not know the sales tax, it displays the price of the calculator as the total cost. Visual Basic does not guess—it simply does what you tell it to do. This brings us to the next section, which explains how to make sure that errors of this sort don't occur.

Before you continue with this chapter, be sure to replace the names of the variables `cst` *and* `tax` *with* `cost` *and* `slsTax`.

FORCING DECLARATION OF VARIABLES

Visual Basic has an `Option Explicit` statement that you can use to automatically remind yourselfto formally declare all your variables. This statement must be entered at the top of each of your modules. The `Option Explicit` statement will cause Visual Basic to generate an error message when you try to run a procedure that contains undeclared variables.

Hands-On 3.5. Forcing Declaration of Variables

1. Return to the Code window where you entered the **CalcCost** procedure (see Hands-On 3.4).
2. At the top of the module window (below the Option Compare Database statement), enter

   ```
   Option Explicit
   ```

   and press **Enter**. Visual Basic will display the statement in blue.
3. Position the insertion point anywhere within the CalcCost procedure and press **F5** to run it. Visual Basic displays the error message: "Compile error: Variable not defined."

4. Click **OK** to exit the message box. Visual Basic selects the name of the variable, `slsPrice`, and highlights in yellow the name of the procedure, `Sub CalcCost()`. The titlebar displays "Microsoft Visual Basic for Applications—Chap03 [break]—[Module4 (Code)]." The Visual Basic Break mode allows you to correct the problem before you continue. Now you have to formally declare the `slsPrice` variable.
5. Enter the declaration statement

```
Dim slsPrice As Currency
```

   on a new line just below `Sub CalcCost()` and press **F5** to continue. When you declare the `slsPrice` variable and rerun your procedure, Visual Basic will generate the same compile error as soon as it encounters another variable name that was not declared. To fix the remaining problems with the variable declaration in this procedure, choose **Run | Reset** to exit the Break mode.
6. Enter the following declarations at the beginning of the CalcCost procedure:

```
' declaration of variables
Dim slsPrice As Currency
Dim slsTax As Single
Dim cost As Currency
Dim strMsg As String
```

7. To run the procedure, click any line between the `Sub` and `End Sub` keywords and press **F5** or choose **Run | Run Sub/UserForm**. Your revised CalcCost procedure looks like this:

```
' revised CalcCost procedure with variable declarations
Sub CalcCost()
  ' declaration of variables
  Dim slsPrice As Currency
  Dim slsTax As Single
  Dim cost As Currency
  Dim strMsg As String
  slsPrice = 35
  slsTax = 0.085

  cost = Format(slsPrice + (slsPrice * slsTax), "0.00")
  strMsg = "The calculator total is " & "$" & cost & "."

  MsgBox strMsg
End Sub
```

The `Option Explicit` statement you entered at the top of the module Code window (see Step 2) forced you to declare variables. Because you must include the `Option Explicit` statement in each module where you want to require variable declaration, you can have Visual Basic enter this statement for you each time you insert a new module.

To automatically include `Option Explicit` in every new module you create, follow these steps:

8. Choose **Tools | Options**.
9. Ensure that the **Require Variable Declaration** checkbox is selected in the Options dialog box (Editor tab).
10. Choose **OK** to close the Options dialog box.

From now on, every new module will be added with the `Option Explicit` statement. If you want to require variables to be explicitly declared in a module you created prior to enabling Require Variable Declaration in the Options dialog box, you must enter the `Option Explicit` statement manually by editing the module yourself.

More about `Option Explicit`

`Option Explicit` forces formal (explicit) declaration of all variables in a particular module. One big advantage of using `Option Explicit` is that misspellings of variable names will be detected at compile time (when Visual Basic attempts to translate the source code to executable code). The `Option Explicit` statement must appear in a module before any procedures.

UNDERSTANDING THE SCOPE OF VARIABLES

Variables can have different ranges of influence in a VBA procedure. *Scope* defines the availability of a particular variable to the same procedure or other procedures.

Variables can have the following three levels of scope in Visual Basic for Applications:

- Procedure-level scope
- Module-level scope
- Project-level scope

Procedure-Level (Local) Variables

From this chapter you already know how to declare a variable using the `Dim` statement. The position of the `Dim` statement in the module determines the scope of a variable. Variables declared with the `Dim` statement within a VBA procedure have a *procedure-level* scope. Procedure-level variables can also be declared by using the `Static` statement (see "Understanding and Using Static Variables" later in this chapter).

Procedure-level variables are frequently referred to as *local* variables, which can only be used in the procedure where they were declared. Undeclared variables always have a procedure-level scope.

A variable's name must be unique within its scope. This means that you cannot declare two variables with the same name in the same procedure. However, you can use the same variable name in different procedures. In other words, the CalcCost procedure can have the `slsTax` variable, and the ExpenseRep procedure in the same module can have its own variable called `slsTax`. Both variables are independent of each other.

Local Variables: with `Dim` or `Static`?

When you declare a local variable with the `Dim` statement, the value of the variable is preserved only while the procedure in which it is declared is running. As soon as the procedure ends, the variable dies. The next time you execute the procedure, the variable is reinitialized.

When you declare a local variable with the `Static` statement, the value of the variable is preserved after the procedure in which the variable was declared has finished running. Static variables are reset when you quit the Microsoft Access application or when a runtime error occurs while the procedure is running.

Module-Level Variables

Often you want the variable to be available to other VBA procedures in the module after the procedure in which the variable was declared has finished running. This situation requires that you change the variable's scope to *module-level.*

Module-level variables are declared at the top of the module (above the first procedure definition) by using the `Dim` or `Private` statement. These variables are available to all of the procedures in the module in which they were declared, but are not available to procedures in other modules.

For instance, to make the `slsTax` variable available to any other procedure in the module, you could declare it by using the `Dim` or `Private` statement:

```
Option Explicit
Dim slsTax As Single' module-level variable declared with
  'Dim statement

Sub CalcCost()
    ...Instructions of the procedure...
End Sub
```

Notice that the `slsTax` variable is declared at the top of the module, just below the `Option Explicit` statement and before the first procedure definition. You could also declare the `slsTax` variable like this:

```
Option Explicit
Private slsTax As Single ' module-level variable declared with
                         ' Private statement
Sub CalcCost()
...Instructions of the procedure...
End Sub
```

There is no difference between module-level variables declared with `Dim` or `Private` statements.

Before you can see how module-level variables actually work, you need another procedure that also uses the `slsTax` variable.

Hands-On 3.6. Understanding Module-Level Variables

This hands-on exercise requires the prior completion of Hands-On 3.4 and 3.5.

1. In the Code window, in the same module where you entered the CalcCost procedure, cut the declaration line **Dim slsTax As Single** and paste it at the top of the module sheet, below the `Option Explicit` statement.
2. Enter the following code of the **ExpenseRep** procedure in the same module where the CalcCost procedure is located (see Figure 3.1):

```
Sub ExpenseRep()
  Dim slsPrice As Currency
  Dim cost As Currency

  slsPrice = 55.99
```

```
    cost = slsPrice + (slsPrice * slsTax)

    MsgBox slsTax
    MsgBox cost
End Sub
```

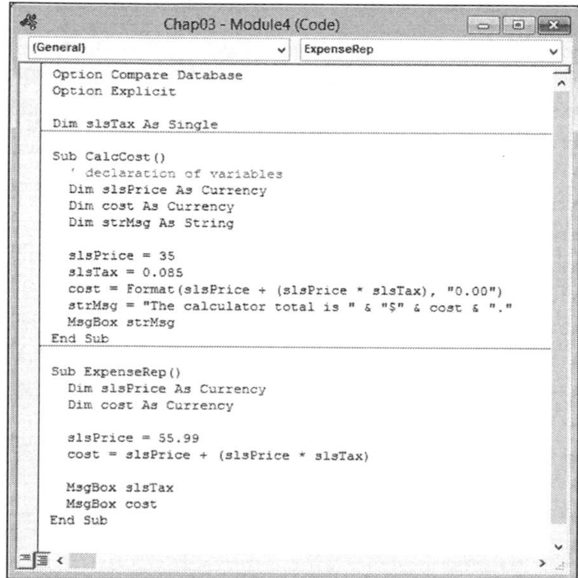

```
Option Compare Database
Option Explicit

Dim slsTax As Single

Sub CalcCost()
  ' declaration of variables
  Dim slsPrice As Currency
  Dim cost As Currency
  Dim strMsg As String

  slsPrice = 35
  slsTax = 0.085
  cost = Format(slsPrice + (slsPrice * slsTax), "0.00")
  strMsg = "The calculator total is " & "$" & cost & "."
  MsgBox strMsg
End Sub

Sub ExpenseRep()
  Dim slsPrice As Currency
  Dim cost As Currency

  slsPrice = 55.99
  cost = slsPrice + (slsPrice * slsTax)

  MsgBox slsTax
  MsgBox cost
End Sub
```

FIGURE 3.1. Using module-level variables.

The ExpenseRep procedure declares two Currency type variables: `slsPrice` and `cost`. The `slsPrice` variable is then assigned a value of 55.99. The `slsPrice` variable is independent of the `slsPrice` variable declared within the CalcCost procedure.

The ExpenseRep procedure calculates the cost of a purchase. The cost includes the sales tax. Because the sales tax is the same as the one used in the CalcCost procedure, the `slsTax` variable has been declared at the module level. After Visual Basic executes the CalcCost procedure, the contents of the `slsTax` variable equals 0.085. If `slsTax` were a local variable, the contents of this variable would be empty upon the termination of the CalcCost procedure. The ExpenseRep procedure ends by displaying the value of the `slsTax` and `cost` variables in two separate message boxes.

After running the CalcCost procedure, Visual Basic erases the contents of all the variables except for the `slsTax` variable, which was declared at a module level. As soon as you attempt to calculate the cost by running the ExpenseRep procedure, Visual Basic retrieves the value of the slsTax variable and uses it in the calculation.

3. Click anywhere inside the revised CalcCost procedure and press **F5** to run it.
4. As soon as the CalcCost procedure finishes executing, run the ExpenseRep procedure.

Project-Level Variables

In the previous sections, you learned that declaring a variable with the Dim or Private keyword at the top of the module makes it available to other procedures in that module. Module-level variables that are declared with the Public keyword (instead of Dim or Private) have project-level scope. This means that they can be used in any Visual Basic for Applications module. When you want to work with a variable in all the procedures in all the open VBA projects, you must declare it with the Public keyword. For instance:

```
Option Explicit
Public gslsTax As Single
Sub CalcCost()
    ...Instructions of the procedure...
End Sub
```

Notice that the gslsTax variable declared at the top of the module with the Public keyword will now be available to any VBA modules that your code references.

A variable declared in the declaration section of a module using the Public keyword is called a *global variable*. This variable can be seen by all procedures in the database's modules. It is customary to use the prefix "g" to indicate this type of variable.

When using global variables, it's important to keep in mind the following:

- The value of the global variable can be changed anywhere in your program. An unexpected change in the value of a variable is a common cause of problems. Be careful not to write a block of code that modifies a global variable. If you need to change the value of a variable within your application, make sure you are using a local variable.
- Values of all global variables declared with the Public keyword are cleared when Access encounters an error. Since the release of the Access 2007 database format (ACCDB), you can use the TempVars collection for your global variable needs (see "Understanding and Using Temporary Variables" later in this chapter).

- Don't put your global variable declaration in a form class module. Variables in the code module behind the form are never global even if you declare them as such. You must use a standard code module (Insert | Module) to declare variables to be available in all modules and forms. Variables declared in a standard module can be used in the code for any form.
- Use constants as much as possible whenever your application requires global variables. Constants are much more reliable because their values are static. Constants are covered later in this chapter.

Public Variables and the `Option Private Module` Statement

Variables declared using the `Public` keyword are available to all procedures in all modules across all applications. To restrict a public module-level variable to the current database, include the `Option Private Module` statement in the declaration section of the standard or class module in which the variable is declared.

UNDERSTANDING THE LIFETIME OF VARIABLES

In addition to scope, variables have a *lifetime*. The *lifetime* of a variable determines how long a variable retains its value. Module-level and project-level variables preserve their values as long as the project is open. Visual Basic, however, can reinitialize these variables if required by the program's logic. Local variables declared with the `Dim` statement lose their values when a procedure has finished. Local variables have a lifetime as long as a procedure is running, and they are reinitialized every time the program is run. Visual Basic allows you to extend the lifetime of a local variable by changing the way it is declared.

UNDERSTANDING AND USING TEMPORARY VARIABLES

In the previous section, you learned that you can declare a global variable with the `Public` keyword and use it throughout your entire application. You also learned that these variables can be quite problematic, especially when you or another programmer accidentally changes the value of the variable or your application encounters an error and the values of the variables you have initially set for your application to use are completely wiped out. To avoid such problems, many programmers resort to using a separate global variables form to hold their global variables. And if they need certain values to be available the next time the application starts, they create a separate database table to store these values. A *global variables form* is simply a blank Access form where you can place both bound and unbound controls. Bound controls are used to pull the data from the table where global variables have been stored. You can use unbound controls on a form to store values of global variables that are not stored in a separate table. Simply set the ControlSource property of the unbound control by typing a value in it or use a VBA procedure to set the value of the ControlSource. The form set up as a global variables form must be open while the application is running for the values of the bound and unbound controls to be available to other forms, reports, and queries in the database. A global variables form can be hidden if the values of the global variables are pulled from a database table or set using VBA procedures or macro actions.

If your database is in the ACCDB format, instead of using a database table or a global variables form, you can use the TempVars collection to store the Variant values you want to reuse. TempVars stands for *temporary variables*. Temporary variables are global. You can refer to them in VBA modules, event procedures, queries, expressions, add-ins, and in any referenced databases. Access .ACCDB databases allow you to define up to 255 temporary variables at one time. These variables remain in memory until you close the database (unless you remove them when you are finished working with them). Unlike public variables, temporary variable values are not cleared when an error occurs.

Creating a Temporary Variable with a TempVars Collection Object

Let's look at some examples of using the TempVars collection first introduced in Access 2007. Assume your application requires three variables named `gtvUserName`, `gtvUserFolder`, and `gtvEndDate`.

To try this out, open the Immediate window and type the following statements. The variable is created as soon as you press Enter after each statement.

```
TempVars("gtvUserName").Value = "John Smith"

TempVars("gtvUserFolder").Value = Environ("HOMEPATH")

TempVars("gtvEndDate").Value = Format(now(),"mm/dd/yyyy")
```

Notice that to create a temporary variable all you have to do is specify its value. If the variable does not already exist, Access adds it to the TempVars collection. If the variable exists, Access modifies its value.

You can explicitly add a global variable to the TempVars collection by using the `Add` method, like this:

```
TempVars.Add "gtvCompleted", "true"
```

Retrieving Names and Values of TempVar Objects

Each TempVar object in the TempVars collection has Name and Value properties that you can use to access the variable and read its value from any procedure. By default, the items in the collection are numbered from zero (0), with the first item being zero, the second item being one, the third two, and so on. Therefore, to find the value of the second variable in the TempVars you have entered (`gtvUserFolder`), type the following statement in the Immediate window:

```
?TempVars(1).Value
```

When you press Enter, you will see the location of the user's private folder on the computer. In this case, it is your private folder. The folder information was returned by passing the "HOMEPATH" parameter to the built-in `Environ` function. Functions and parameter passing are covered in Chapter 4.

You can also retrieve the value of the variable from the TempVars collection by using its name, like this:

```
?TempVars("gtvUserFolder").Value
```

You can iterate through the TempVars collection to see the names and values of all global variables that you have placed in it. To do this from the Immediate window, you need to use the colon operator (:) to separate lines of code. Type the following statement all on one line to try this out:

```
For Each gtv in TempVars : Debug.Print gtv.Name & ":"
& gtv.Value : Next
```

When you press Enter, the `Debug.Print` statement will write to the Immediate window a name and value for each variable that is currently stored in the TempVars collection:

```
gtvUserName:John Smith
gtvUserFolder:\Documents and Settings\John
gtvEndDate:09/12/2011
gtvCompleted:true
```

The popular VBA programming construct, `For Each...Next` statement, is covered in detail in Chapter 6. The "gtv" is an object variable used as an iterator. An *iterator* allows you to traverse through all the elements of a collection. You can use any variable name as an iterator as long as it is not a VBA keyword. Object variables are discussed later in this chapter. For more information on working with collections, see Chapter 8.

Using Temporary Global Variables in Expressions

You can use temporary global variables anywhere expressions can be used. For example, you can set the value of the unbound text box control on a form to display the value of your global variable by activating the property sheet and typing the following in the ControlSource property of the text box:

```
=[TempVars]![gtvCompleted]
```

You can also use a temporary variable to pass selection criteria to queries:

```
SELECT * FROM Orders WHERE Order_Date = TempVars!gtvEndDate
```

Removing a Temporary Variable from a TempVars Collection Object

When you are done using a variable, you can remove it from the TempVars collection with the `Remove` method, like this:

```
TempVars.Remove "gtvUserFolder"
```

To check the number of the TempVar objects in the TempVars collection, use the Count property in the Immediate window:

```
?TempVars.Count
```

Finally, to quickly remove all global variables (TempVar objects) from the TempVars collection, simply use the `RemoveAll` method, like this:

```
TempVars.RemoveAll
```

The TempVars Collection Is Exposed to Macros

The following three macros in Access 2007/2010/2013 allow macro users to set and remove TempVar objects:

- SetTempVar—Sets a TempVar to a given value. You must specify the name of the temporary variable and the expression that will be used to set the value of this variable. Expressions must be entered without an equal sign (=).

- RemoveTempVar—Removes the TempVar from the TempVars collection. You must specify the name of the temporary variable you want to remove.
- RemoveAllTempVars—Clears the TempVars collection.

The values of TempVar objects can be used in the arguments and in the condition columns of macros. For more information on macro programming in Access 2013, see Chapter30, "Macros and Templates."

UNDERSTANDING AND USING STATIC VARIABLES

A variable declared with the `Static` keyword is a special type of local variable. *Static variables* are declared at the procedure level. Unlike the local variables declared with the `Dim` keyword, static variables remain in existence and retain their values when the procedure in which they were declared ends.

The CostOfPurchase procedure (see Hands-On 3.7) demonstrates the use of the static variable `allPurchase`. The purpose of this variable is to keep track of the running total.

Hands-On 3.7. Using Static Variables

This hands-on exercise uses the **C:\Access2013_ByExample\Chap03.accdb** database that you created in Hands-On 3.1.

1. In the Visual Basic window, choose **Insert | Module** to add a new module.
2. Enter the following **CostOfPurchase** procedure code in the new module's Code window:

```
Sub CostOfPurchase()
  ' declare variables
  Static allPurchase
  Dim newPurchase As String
  Dim purchCost As Single

  newPurchase = InputBox("Enter the cost of a purchase:")
  purchCost = CSng(newPurchase)
  allPurchase = allPurchase + purchCost

  ' display results
  MsgBox "The cost of a new purchase is: " & newPurchase
```

```
  MsgBox "The running cost is: " & allPurchase
End Sub
```

This procedure begins with declaring a static variable named `allPurchase` and two local variables named `newPurchase` and `purchCost`. The `InputBox` function is used to get a user's input while the procedure is running. As soon as the user inputs the value and clicks OK, Visual Basic assigns the value to the `newPurchase` variable. Because the result of the `InputBox` function is always a string, the `newPurchase` variable was declared as the String data type. You cannot use strings in mathematical calculations, so the next instruction uses a *type conversion* function (`CSng`) to translate the text value into a numeric value, which is stored as a Single data type in the variable `purchCost`. The `CSng` function requires only one argument: the value you want to translate. Refer to Chapter 4 for more information about converting data types.

The next instruction, `allPurchase = allPurchase + purchCost`, adds the new value supplied by the `InputBox` function to the current purchase value. When you run this procedure for the first time, the value of the `allPurchase` variable is the same as the value of the `purchCost` variable. During the second run, the value of the static variable is increased by the new value entered in the dialog box. You can run the CostOfPurchase procedure as many times as you want. The `allPurch` variable will keep the running total for as long as the project is open.

3. To run the procedure, position the insertion point anywhere within the CostOfPurchase procedure and press **F5**.
4. When the dialog box appears, enter a number. For example, type **100** and press **Enter**. Visual Basic displays the message "The cost of a new purchase is: 100."
5. Click **OK** in the message box. Visual Basic displays the second message "The running cost is: 100."
6. Rerun the same procedure.
7. When the input box appears, enter another number. For example, type **50** and press **Enter**. Visual Basic displays the message "The cost of a new purchase is: 50."
8. Click **OK** in the message box. Visual Basic displays the second message "The running cost is: 150."
9. Run the procedure a couple of times to see how Visual Basic keeps track of the running total.

Type Conversion Functions

To learn more about the `CSng` function, position the insertion point anywhere within the word `CSng` and press F1.

UNDERSTANDING AND USING OBJECT VARIABLES

The variables you've learned about so far are used to store data, which is the main reason for using "normal" variables in your procedures. There are also special variables that refer to the Visual Basic objects. These variables are called *object variables.* Object variables don't store data; they store the location of the data. You can use them to reference databases, forms, and controls as well as objects created in other applications. Object variables are declared in a similar way as the variables you've already seen. The only difference is that after the `As` keyword, you enter the type of object your variable will point to. For instance:

```
Dim myControl As Control
```

This statement declares the object variable called `myControl` of type Control.

```
Dim frm As Form
```

This statement declares the object variable called `frm` of type Form.

You can use object variables to refer to objects of a generic type such as Application, Control, Form, or Report, or you can point your object variable to specific object types such as TextBox, ToggleButton, CheckBox, CommandButton, ListBox, OptionButton, Subform or Subreport, Label, BoundObjectFrame or UnboundObjectFrame, and so on. When you declare an object variable, you also have to assign it a specific value before you can use it in your procedure. You assign a value to the object variable by using the `Set` keyword followed by the equals sign and the value that the variable refers to. For example:

```
Set myControl = Me!CompanyName
```

The preceding statement assigns a value to the object variable called `myControl`. This object variable will now point to the CompanyName control on the active form. If you omit the word `Set`, Visual Basic will display the error message "Runtime error 91: Object variable or With block variable not set."

Again, it's time to see a practical example. The HideControl procedure in Hands-On 3.8 demonstrates the use of the object variables `frm` and `myControl`.

Hands-On 3.8. Working with Object Variables

1. Close the currently open Access database **Chap03.accdb**. When prompted to save changes in the modules, click **OK**. Save the modules with the suggested default names Module1, Module2, and so on.
2. Copy the **HandsOn_03_8.accdb** database from the **C:\Access2013_HandsOn** folder to your **C:\Access2013_ByExample** folder. This database contains a Customer table and a simple Customer form imported from the Northwind.mdb sample database that shipped with an earlier version of Microsoft Access.
3. Open Access and load the **C:\Access2013_ByExample\HandsOn_03_8.accdb** database file.
4. Open the **Customers** form in Form view.
5. Press **Alt+F11** to switch to the Visual Basic Editor window.
6. Choose **Insert | Module** to add a new module.
7. Enter the following **HideControl** procedure code in the new module's Code window:

```
Sub HideControl()
  ' this procedure is run against the open Customers form
  Dim frm As Form
  Dim myControl As Control

  Set frm = Forms!Customers
  Set myControl = frm.CompanyName
  myControl.Visible = False
End Sub
```

8. To run the procedure, click any line between the `Sub` and `End Sub` keywords and press **F5** or choose **Run | Run Sub/UserForm**.

The procedure begins with the declaration of two object variables called `frm` and `myControl`. The object variable `frm` is set to reference the Customers form. For the procedure to work, the referenced form must be open. Next, the `myControl` object variable is set to point to the CompanyName control located on the Customers form.

Instead of using the object's entire address, you can use the shortcut—the name of the object variable. For example, the statement

```
Set myControl = frm.CompanyName
```

is the same as

```
Set myControl = Forms!Customers.CompanyName
```

The purpose of this procedure is to hide the control referenced by the object variable `myControl`. After running the HideControl procedure, switch to the Microsoft Access window containing the open Customers form. The CompanyName control should not be visible on the form.

To make the CompanyName text box visible again, modify the last line of this procedure by setting the Visible property of `myControl` *to True and rerun the procedure.*

Advantages of Using Object Variables

The advantages of object variables are:

- They can be used instead of the actual object.
- They are shorter and easier to remember than the actual values they point to.
- You can change their meaning while your procedure is running.

Disposing of Object Variables

When the object variable is no longer needed, you should assign Nothing to it. This frees up memory and system resources:

```
Set frm = Nothing
Set myControl = Nothing
```

FINDING A VARIABLE DEFINITION

When you find an instruction that assigns a value to a variable in a VBA procedure, you can quickly locate the definition of the variable by selecting the variable name and pressing Shift+F2. Alternately, you can choose View | Definition. Visual Basic will jump to the variable declaration line. To return your mouse pointer to its previous position, press Ctrl+Shift+F2 or choose View | Last Position. Let's try it out.

Hands-On 3.9. Finding a Variable Definition

This hands-on exercise requires prior completion of Hands-On 3.8.

1. Locate the code of the procedure **HideControl** you created in Hands-On 3.8.

2. Locate the statement **myControl.Visible** = .
3. Right-click the **myControl** variable name and choose **Definition** from the shortcut menu.
4. Press **Ctrl+Shift+F2 to** return to the previous location in the procedure code (myControl.Visible =).

DETERMINING THE DATA TYPE OF A VARIABLE

Visual Basic has a built-in `VarType` function that returns an integer indicating the variable's type. Let's see how you can use this function in the Immediate window.

Hands-On 3.10. Asking Questions about the Variable Type

1. Open the Immediate window (**View | Immediate Window**) and type the following statements that assign values to variables:

```
age = 28
birthdate = #1/1/1981#
firstName = "John"
```

2. Now, ask Visual Basic what type of data each variable holds:

```
?varType(age)
```

When you press **Enter**, Visual Basic returns 2. The number 2 represents the Integer data type, as shown in Table 3.3.

```
?varType(birthdate)
```

Now Visual Basic returns 7 for Date. If you make a mistake in the variable name (let's say you type `birthday` instead of `birthdate`), Visual Basic returns zero (0).

```
?varType(firstName)
```

Visual Basic tells you that the value stored in the `firstName` variable is a String (8).

TABLE 3.3. Values returned by the `VarType` function

Constant	Value	Description
vbEmpty	0	Empty (uninitialized)
vbNull	1	Null (no valid data)
vbInteger	2	Integer
vbLong	3	Long integer
vbSingle	4	Single-precision floating-point number
vbDouble	5	Double-precision floating-point number
vbCurrency	6	Currency value
vbDate	7	Date value
vbString	8	String
vbObject	9	Object
vbError	10	Error value
vbBoolean	11	Boolean value
vbVariant	12	Variant (used only with arrays of variants)
vbDataObject	13	Data access object
vbDecimal	14	Decimal value
vbByte	17	Byte value
vbLongLong	20	Long Long integer (on 64-bit platform only)
vbUserDefinedType	36	Variants that contain user-defined types
vbArray	8192	Array

USING CONSTANTS IN VBA PROCEDURES

The value of a variable can change while your procedure is executing. If your procedure needs to refer to unchanged values over and over again, you should use constants. A *constant* is like a named variable that always refers to the same value. Visual Basic requires that you declare constants before you use them.

You declare constants by using the `Const` statement, as in the following examples:

```
Const dialogName = "Enter Data" As String
Const slsTax = 8.5
Const Discount = 0.5
Const ColorIdx = 3
```

A constant, like a variable, has a scope. To make a constant available within a single procedure, you declare it at the procedure level, just below the name of the procedure. For instance:

```
Sub WedAnniv()
  Const Age As Integer = 25
...instructions...
End Sub
```

If you want to use a constant in all the procedures of a module, use the `Private` keyword in front of the `Const` statement. For instance:

```
Private Const dsk = "B: " As String
```

The `Private` constant has to be declared at the top of the module, just before the first `Sub` statement.

If you want to make a constant available to all modules in your application, use the `Public` keyword in front of the `Const` statement. For instance:

```
Public Const NumOfChar As Integer= 255
```

The `Public` constant has to be declared at the top of the module, just before the first `Sub` statement.

When declaring a constant, you can use any one of the following data types: Boolean, Byte, Integer, Long, Currency, Single, Double, Date, String, or Variant.

Like variables, constants can be declared on one line if separated by commas. For instance:

```
Const Age As Integer = 25, PayCheck As Currency = 350
```

Using constants makes your VBA procedures more readable and easier to maintain. For example, if you need to refer to a certain value several times in your procedure, use a constant instead of using a value. This way, if the value changes (for example, the sales tax rate goes up), you can simply change the value in the declaration of the `Const` statement instead of tracking down every occurrence of the value.

INTRINSIC CONSTANTS

Both Microsoft Access and Visual Basic for Applications have a long list of predefined (intrinsic) constants that do not need to be declared. These built-in constants can be looked up using the Object Browser window, which was discussed in detail in Chapter 2.

Let's open the Object Browser to take a look at the list of constants in Access.

Hands-On 3.11. Exploring Access's Constants

1. In the Visual Basic Editor window, choose **View | Object Browser**.
2. In the Project/Library list box, click the drop-down arrow and select the **Access** library.
3. Enter **constants** as the search text in the Search Text box and either press **Enter** or click the **Search** button. Visual Basic shows the results of the search in the Search Results area. The right side of the Object Browser window displays a list of all built-in constants available in the Microsoft Access Object Library (see Figure 3.2). Notice that the names of all the constants begin with the prefix "ac."

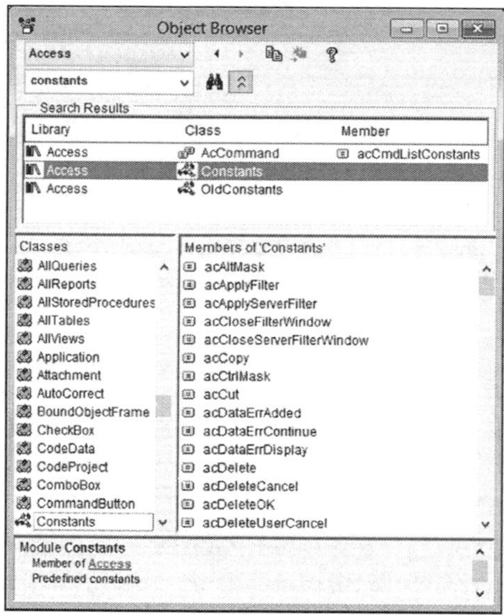

FIGURE 3.2. Use the Object Browser to look up any intrinsic constant.

4. To look up VBA constants, choose **VBA** in the Project/Library list box. Notice that the names of the VBA built-in constants begin with the prefix "vb."

Hands-On 3.12 illustrates how to use the intrinsic constants `acFilterByForm` and `acFilterAdvanced` to disable execution of filtering on a form.

Hands-On 3.12. Using Intrinsic Constants in a VBA Procedure

This hands-on exercise uses the HandsOn_03_8.accdb database file used in Hands-On 3.8.

1. Open the **Customers** form in Design view.
2. If the property sheet is not visible, activate it by pressing **Alt+Enter**.
3. In the property sheet, click the **Event** tab. Make sure that **Form** is selected in the drop-down box on the top of the property sheet.
4. Click to the right of the **On Filter** property and select the **Build** button (...).
5. In the Choose Builder dialog box, select **Code Builder** and click **OK**.
6. In the Code window, enter the following **Form_Filter** event procedure code.

```
Private Sub Form_Filter(Cancel As Integer, FilterType As Integer)
  If FilterType = acFilterByForm Or _
   FilterType = acFilterAdvanced Then
    MsgBox "You need authorization to filter records."
    Cancel = True
  End If
End Sub
```

7. Press **Alt+F11** to switch back to Design view in the Customers form.
8. Right-click the Customers form tab and choose **Form View**. You can also use the Views section of the Design tab to activate the Form view.
9. Choose **Home | Sort & Filter | Advanced Filter Options | Filter By Form.** Access displays the message "You need authorization to filter records." The same message appears when you choose Advanced Filter/Sort from the Advanced Filter Options.

GROUPING CONSTANTS WITH THE ENUM TYPE

If your application needs to use a set of related constants, consider declaring them with an enumerated data type. VBA has an enumerated data type (Enum) that you can use to define your constant values. The following code entered in a standard module creates an Enum that defines various status types:

```
Public Enum StatusType
  Rejected = 0
  MoreInfo = 1
  Approved = 2
  Completed = 3
End Enum
```

Notice that the `Enum` statement is always followed by the name of the enumeration. Within the `Enum` statement you simply list the names of constants you need to use. You can assign values to the constants using the equals sign (=). If you don't assign values, the first constant in the enumeration will be given the value zero (0), the second constant the value 1, and so on. The values assigned within the `Enum` statement are constant and can't be changed at runtime.

To end the enumeration, use the `End Enum` statement.

The StatusType enumeration will be visible throughout the entire VBA project. Enum types are Public by default. To limit the visibility of the Enum type to the module in which it was declared, use the `Private` keyword.

Once you have defined your custom Enum object, you can refer to the specific constant by typing the name of the enumeration followed by a dot operator and the name of the constant. As soon as you enter the Enum name followed by a period, Visual Basic will display a list of available constants, as shown in Figure 3.3.

The GetStatusType procedure shows how to use the constants listed in the StatusType enumeration. You can enter this procedure in the same standard module where the Enum StatusType was defined.

```
Sub GetStatusType()
  Dim result As Integer
  result = InputBox("Please enter a number from 0 to 3", _
    "Status Type", 0)

  If result = StatusType.MoreInfo Then
    MsgBox ("Please provide more details on a separate form.")
  ElseIf result = StatusType.Rejected Then
    MsgBox ("Please apply again next year.")
  ElseIf result >= StatusType.Approved Then
    MsgBox ("Thank you for applying.")
  End If
End Sub
```

The precedingVBA procedure uses two popular built-in functions (`InputBox` and `MsgBox`) that allow interaction with the procedure's end users. These functions are covered in detail in the next chapter. The `InputBox` function prompts the user to type a number for the status type. Zero (0) is specified as the default value. When the user enters a value and clicks OK in the input box, the supplied value will be stored in a local variable named `result`. The remaining code of the GetStatusType procedure uses the `If...Then...ElseIf` conditional statement (see Chapter 5, "Decision Making with VBA") to determine which message text should be displayed to the user. The `MsgBox` function used in this example specifies just a simple text message. Before a specific message is displayed, the value in the `result` variable is compared to the value of the enumerated constant.

By using enumerated data types instead of hard-coding values, you make your programming code easier to understand and reduce the chance of accidental errors.

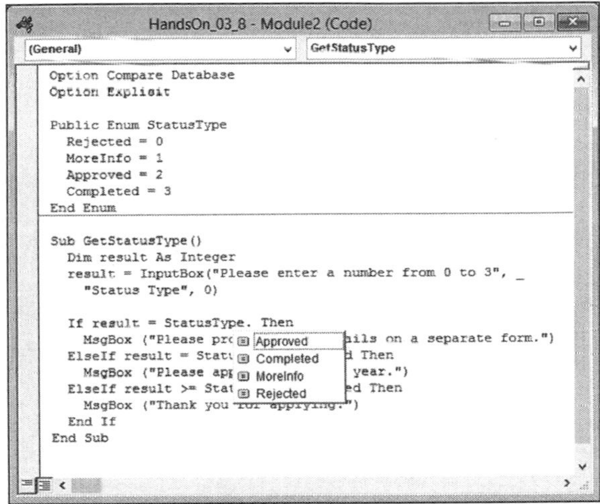

FIGURE 3.3. VBA provides a list of constants to choose from after you type your custom Enum name followed by a dot operator.

CHAPTER SUMMARY

This chapter has introduced you to several VBA concepts such as data types, variables, and constants. You learned how to declare various types of variables and define their types. You also saw the difference between a variable and a constant.

In the next chapter, you will expand your knowledge of Visual Basic for Applications by writing procedures and functions with arguments. In addition, you will learn about functions that allow your VBA procedures to interact with the user.

This chapter concludes the final part of this book, which focused on working with an Access database over the Internet by writing Active Server Pages and XML files.

Chapter 4

Passing Arguments to Procedures and Functions

As you already know from Chapter 1, VBA subroutines and function procedures often require arguments to perform certain tasks. In this chapter, you learn various methods of passing arguments to procedures and functions.

WRITING A FUNCTION PROCEDURE

Function procedures can perform calculations based on data received through arguments. When you declare a function procedure, you list the names of arguments inside a set of parentheses, as shown in Hands-On 4.1.

Please note files for the "Hands-On" project may be found on the companion CD-ROM.

Hands-On 4.1. Writing a Function Procedure with Arguments

1. Start Microsoft Access 2013 and create a new database named **Chap04.accdb** in your C:**Access2013_ByExample** folder.
2. Once your new database is opened, press **Alt+F11** to switch to the Visual Basic Editor **window.**
3. Choose **Insert | Module** to add a new standard module and notice that Module1 appears under the Modules folder in the Project Explorer window.
4. In the Module1 (Code) window, enter the code of the **JoinText** function procedure as shown here.

```
Function JoinText(k, o)
  JoinText = k + " " + o
End Function
```

Note that there is a space character in quotation marks concatenated between the two arguments of the JoinText function's result: `JoinText = k + " " + o`.

A better way of adding a space is by using one of the following built-in functions:

```
JoinText = k + Space(1) + o
```

or:

```
JoinText = k + Chr(32) + o
```

The `Space` function returns a string of spaces as indicated by the number in the parentheses. The `Chr` function returns a string containing the character associated with the specified character code.

Other control characters you may need to use when writing your VBA procedures include:

Tab	`Chr(9)`
Linefeed	`Chr(10)`
Carriage Return	`Chr(13)`

You can execute a function procedure from the Immediate window, or you can write a subroutine to call the function. See Hands-On 4.2 and 4.3 for instructions on how to run the JoinText function procedure using these two methods.

Hands-On 4.2. Executing a Function Procedure from the Immediate Window

This hands-on exercise requires prior completion of Hands-On 4.1.

1. Choose **View | Immediate Window** or press **Ctrl+G**, and enter the following statement:

   ```
   ?JoinText("function", " procedure")
   ```

   Notice that as soon as you type the opening parenthesis, Visual Basic displays the arguments that the function expects. Type the value of the first argument, enter the comma, and supply the value of the second argument. Finish by entering the closing parenthesis.
2. Press **Enter** to execute this statement from the Immediate window. When you press Enter, the string "function procedure" appears in the Immediate window.

You can also execute a function procedure programmatically, as demonstrated in Hands-On 4.3.

Hands-On 4.3. Executing a Function Procedure from a Subroutine

This hands-on exercise requires prior completion of Hands-On 4.1.

1. In the same module where you entered the JoinText function procedure, enter the following **EnterText** subroutine:

   ```
   Sub EnterText()
     Dim strFirst As String, strLast As String, strFull As String
     strFirst = InputBox("Enter your first name:")
     strLast = InputBox("Enter your last name:")
     strFull = JoinText(strFirst, strLast)

     MsgBox strFull
   End Sub
   ```

2. Place the cursor anywhere inside the code of the EnterText procedure and press **F5** to run it.
   As Visual Basic executes the statements of the EnterText procedure, it uses the `InputBox` function to collect the data from the user, then stores the data (the val-

ues of the first and last names) in the variables `strFirst` and `strLast`. Then these values are passed to the JoinText function. Visual Basic substitutes the variables' contents for the arguments of the JoinText function and assigns the result to the name of the function (JoinText). When Visual Basic returns to the EnterText procedure, it stores the function's value in the `strFull` variable. The `MsgBox` function then displays the contents of the `strFull` variable in a message box. The result is the full name of the user (first and last name separated by a space).

More about Arguments

Argument names are like variables. Each argument name refers to whatever value you provide at the time the function is called. You write a subroutine to call a function procedure. When a subroutine calls a function procedure, the required arguments are passed to the procedure as variables. Once the function does something, the result is assigned to the function name. Notice that the function procedure's name is used as if it were a variable.

SPECIFYING THE DATA TYPE FOR A FUNCTION'S RESULT

Like variables, functions can have types. The data type of your function's result can be a String, Integer, Long, etc. To specify the data type for your function's result, add the `As` keyword and the name of the desired data type to the end of the function declaration line. For example:

```
Function MultiplyIt(num1, num2) As Integer
```

If you don't specify the data type, Visual Basic assigns the default type (Variant) to your function's result. When you specify the data type for your function's result, you get the same advantages as when you specify the data type for your variables—your procedure uses memory more efficiently, and therefore runs faster.

Let's take a look at an example of a function that returns an integer, even thoughthe arguments passed to it are declared as Single in a calling subroutine.

Hands-On 4.4. Calling a Function from a Procedure

1. In the Visual Basic Editor window, choose **Insert | Module** to add a new module.

2. Enter the following **HowMuch** subroutine in the Code window:

```
Sub HowMuch()
  Dim num1 As Single
  Dim num2 As Single
  Dim result As Single

  num1 = 45.33
  num2 = 19.24
  result = MultiplyIt(num1, num2)

  MsgBox result
End Sub
```

3. Enter the following **MultiplyIt** function procedure in the Code window below the HowMuch subroutine:

```
Function MultiplyIt(num1, num2) As Integer
  MultiplyIt = num1 * num2
End Function
```

4. Click anywhere within the HowMuch procedure and press **F5** to run it. Because the values stored in the variables `num1` and `num2` are not whole numbers, you may want to assign the Integer type to the result of the function to ensure that the result of the multiplication is a whole number. If you don't assign the data type to the MultiplyIt function's result, the HowMuch procedure will display the result in the data type specified in the declaration line of the `result` variable. Instead of 872, the result of the multiplication will be 872.1492. To make the MultiplyIt function more useful, instead of hard-coding the values to be used in the multiplication, you can pass different values each time you run the procedure by using the `InputBox` function.
5. Take a few minutes to modify the HowMuch procedure on your own, following the example of the EnterText subroutine that was created in Hands-On 4.3.
6. To pass a specific value from a function to a subroutine, assign the value to the function name. For example, the NumOfDays function shown here passes the value of 7 to the subroutine DaysInAWeek.

```
Function NumOfDays()
  NumOfDays = 7
End Function

Sub DaysInAWeek()
  MsgBox "There are " & NumOfDays & " days in a week."
End Sub
```

Subroutines or Functions: Which Should You Use?

Create a subroutine when you:

- Want to perform some actions
- Want to get input from the user
- Want to display a message on the screen

Create a function when you:

- Want to perform a simple calculation more than once
- Must perform complex computations
- Must call the same block of instructions more than once
- Want to check whether a certain expression is true or false

PASSING ARGUMENTS BY REFERENCE AND BY VALUE

In some procedures, when you pass arguments as variables, Visual Basic can suddenly change the value of the variables. To ensure that the called function procedure does not alter the value of the passed arguments, you should precede the name of the argument in the function's declaration line with the `ByVal` keyword. Let's practice this in the following example.

Hands-On 4.5. Passing Arguments to Subroutines and Functions

1. In the Visual Basic Editor window, choose **Insert | Module** to add a new module.
2. In the Code window, type the following**ThreeNumbers** subroutine and the **MyAverage** function procedure:

```
Sub ThreeNumbers()
  Dim num1 As Integer, num2 As Integer, num3 As Integer
  num1 = 10
  num2 = 20
  num3 = 30

  MsgBox MyAverage(num1, num2, num3)
  MsgBox num1
  MsgBox num2
  MsgBox num3
End Sub
```

```
Function MyAverage(ByVal num1, ByVal num2, ByVal num3)
  num1 = num1 + 1
  MyAverage = (num1 + num2 + num3) / 3
End Function
```

3. Click anywhere within the ThreeNumbers procedure and press **F5** to run it. The ThreeNumbers procedure assigns values to three variables, and then calls the MyAverage function to calculate and return the average of the numbers stored in these variables. The function's arguments are the names of the variables: `num1`, `num2`, and `num3`. Notice that all variable names are preceded with the `ByVal` keyword. Also, notice that prior to the calculation of the average, the MyAverage function changes the value of the `num1` variable. Inside the function procedure, the `num1` variable equals 11 (10 + 1). Therefore, when the function passes the calculated average to the ThreeNumbers procedure, the `MsgBox` function displays the result as 20.3333333333333 and not 20, as expected. The next three functions show the contents of each of the variables. The values stored in these variables are the same as the original values assigned to them: 10, 20, and 30.

   What will happen if you omit the `ByVal` keyword in front of the `num1`1 argument in the MyAverage function's declaration line? The function's result will still be the same, but the content of the `num1` variable displayed by the `MsgBox-num1` is now 11. The MyAverage function has not only returned an unexpected result (20.3333333333333 instead of 20), but also has modified the original data stored in the `num1` variable. To prevent Visual Basic from permanently changing the values supplied to the function, use the `ByVal` keyword.

Know Your Keywords: ByRef and ByVal

Because any of the variables passed to a function procedure (or a subroutine) can be changed by the receiving procedure, it is important to know how to protect the original value of a variable. Visual Basic has two keywords that give or deny the permission to change the contents of a variable: `ByRef` and `ByVal`.

By default, Visual Basic passes information to a function procedure (or a subroutine) by reference (`ByRef` keyword), referring to the original data specified in the function's argument at the time the function is called. So, if the function alters the value of the argument, the original value is changed. You will get this result if you omit the `ByVal` keyword in front of the `num1` argument in the MyAverage function's declaration line. If you want the function procedure to change the original value, you don't need to explicitly insert the `ByRef` keyword because passed variables default to `ByRef`.

When you use the `ByVal` keyword in front of an argument name, Visual Basic passes the argument by value, which means that Visual Basic makes a copy of the original data. This copy is then passed to a function. If the function changes the value of an argument passed by value, the original data does not change—only the copy changes. That's why when the MyAverage function changed the value of the `num1` argument, the original value of the `num1` variable remained the same.

USING OPTIONAL ARGUMENTS

At times, you may want to supply an additional value to a function. Let's say you have a function that calculates the price of a meal per person. Sometimes, however, you'd like the function to perform the same calculation for a group of two or more people. To indicate that a procedure argument isn't always required, precede the name of the argument with the `Optional` keyword. Arguments that are optional come at the end of the argument list, following the names of all the required arguments. Optional arguments must always be the Variant data type. This means that you can't specify the optional argument's type by using the `As` keyword.

In the preceding section, you created a function to calculate the average of three numbers. Suppose that sometimes you would like to use this function to calculate the average of two numbers. You could define the third argument of the MyAverage function as optional. To preserve the original MyAverage function, let's create the Avg function to calculate the average for two or three numbers.

Hands-On 4.6. Using Optional Arguments

1. In the Visual Basic Editor window, choose **Insert | Module** to add a new module.
2. Type the following **Avg** function procedure in the Code window:

```
Function Avg(num1, num2, Optional num3)
  Dim totalNums As Integer

  totalNums = 3
  If IsMissing(num3) Then
  num3 = 0
  totalNums = totalNums - 1
  End If
  Avg = (num1 + num2 + num3) / totalNums
End Function
```

3. Call this function from the Immediate window by entering the following instruction and pressing **Enter**:

```
?Avg(2, 3)
```

As soon as you press Enter, Visual Basic displays the result: 2.5.

4. Now, type the following instruction and press **Enter**:

```
?Avg(2, 3, 5)
```

This time the result is: 3.3333333333333.

As you've seen, the Avg function is used to calculate the average of two or three numbers. You decide what values and how many values (two or three) you want to average. When you start typing the values for the function's arguments in the Immediate window, Visual Basic displays the name of the optional argument enclosed in square brackets.

Let's take a few minutes to analyze the Avg function. This function can take up to three arguments. Arguments `num1` and `num2` are required. Argument `num3` is optional. Notice that the name of the optional argument is preceded by the `Optional` keyword. The optional argument is listed at the end of the argument list. Because the types of the `num1`, `num2`, and `num3` arguments are not declared, Visual Basic treats all three arguments as Variants.

Inside the function procedure, the `totalNums` variable is declared as an Integer and then assigned a beginning value of 3. Because the function has to be capable of calculating an average of two or three numbers, the handy built-in function `IsMissing` checks for the number of supplied arguments. If the third (optional) argument is not supplied, the `IsMissing` function puts the value of zero (0) in its place, and deducts the value of 1 from the value stored in the `totalNums` variable. Hence, if the optional argument is missing, `totalNums` is 2. The next statement calculates the average based on the supplied data, and the result is assigned to the name of the function.

USING THE ISMISSING FUNCTION

The `IsMissing` function called from within Hands-On 4.6 allows you to determine whether or not the optional argument was supplied. This function returns the logical value of True if the third argument is not supplied, and returns False when the third argument is given. The `IsMissing` function is used here with the decision-making statement `If...Then` (discussed in Chapter 5). If the `num3` argument is missing (`IsMissing`), then Visual Basic supplies a zero (0) for the value of the third argument (`num3 = 0`), and reduces the value stored in the argument `totalNums` by 1 (`totalNums = totalNums - 1`).

Testing a Function Procedure

To test whether a custom function does what it was designed to do, write a simple subroutine that will call the function and display its result. In addition, the subroutine should show the original values of arguments. This way, you'll be able to quickly determine when the argument value was altered. If the function procedure uses optional arguments, you'll also need to check those situations in which the optional arguments may be missing.

BUILT-IN FUNCTIONS

VBA comes with numerous built-in functions that can be looked up in the Visual Basic online help. To access an alphabetical listing of all VBA functions, choose Help | Microsoft Visual Basic for Applications Help in the Visual Basic Editor window. In the Table of Contents, choose Visual Basic for Applications Language Reference | Visual Basic Language Reference | Functions. Each function is described in detail, and is often illustrated with a code fragment or a complete function procedure that shows how to use it in a specific context. After completing this chapter, be sure to launch the VBA help, and browse through the built-in functions to familiarize yourself with their names and usage. You can also search for the function name in your favorite browser to get more information.

If you are working with Access 2013 via the Office 365 subscription service, you will need an active Internet connection to access the Visual Basic for Applications language reference for Office 2013. You will find the list of all VBA functions under this link:

http://msdn.microsoft.com/en-us/library/office/jj692811.aspx

The following link will bring up the Visual Basic for Applications language reference for Office 2013:

http://msdn.microsoft.com/en-us/library/office/gg264383.aspx

One of the features of a good program is its interaction with the user. When you work with Microsoft Access, you interact with the application by using various dialog boxes such as message boxes and input boxes. When you write your own procedures, you can use the `MsgBox` function to inform users about an unexpected error or the result of a specific calculation. So far you have seen a simple implementation of this function. In the next section, you will find out how to control the appearance of your message. Then you will learn how to get information from the user with the `InputBox` function.

USING THE MSGBOX FUNCTION

The `MsgBox` function you have used thus far was limited to displaying a message to the user in a simple, one-button dialog box. You closed the message box by clicking the OK button or pressing the Enter key. You can create a simple message box by following the `MsgBox` function name with the text of the message enclosed in quotation marks. In other words, to display the message "The procedure is complete." you use the following statement:

```
MsgBox "The procedure is complete."
```

You can try this instruction by entering it in the Immediate window. When you type this instruction and press Enter, Visual Basic displays the message box shown in Figure 4.1.

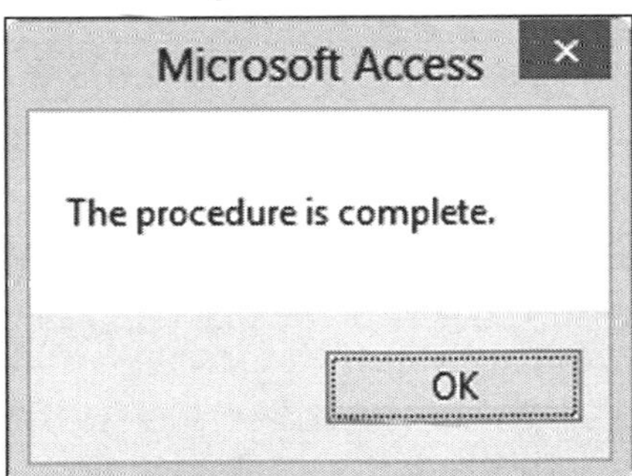

FIGURE 4.1. To display a message to the user, place the text as the argument of the MsgBox function.

The `MsgBox` function allows you to use other arguments that make it possible to determine the number of buttons that should be available in the message box or to change the title of the message box from the default. You can also assign your own help topic. The syntax of the `MsgBox` function is shown here.

```
MsgBox (prompt [, buttons] [, title], [, helpfile, context])
```

Notice that while the `MsgBox` function has five arguments, only the first one, `prompt`, is required. The arguments listed in square brackets are optional.

When you enter a long text string for the `prompt` argument, Visual Basic decides how to break the text so it fits the message box. Let's do some exercises in the Immediate window to learn various text formatting techniques.

Hands-On 4.7. Formatting the Message Box

1. In the Visual Basic Editor window, activate the Immediate window and enter the following instruction. Be sure to enter the entire text string on one line, and then press **Enter**.

```
MsgBox "All done. Now open ""Test.doc"" and place an empty CD
or DVD in your computer'sCD/DVD drive. The following procedure
will copy this file to the disc."
```

As soon as you press **Enter**, Visual Basic shows the resulting dialog box (see Figure 4.2). If you get a compile error, click **OK**. Then make sure that the name of the file is surrounded by double quotation marks (`""Test.doc""`).

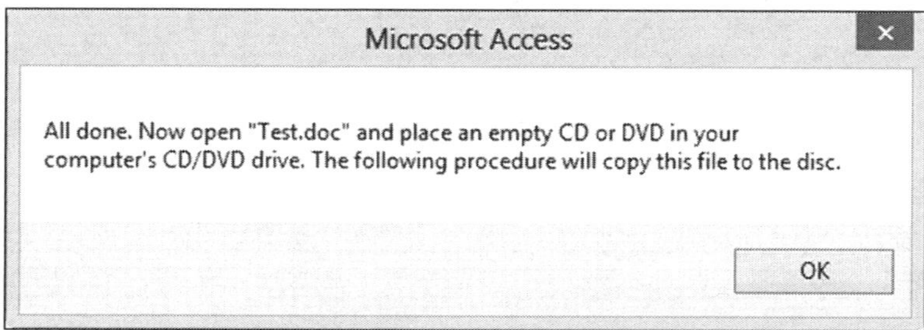

FIGURE 4.2. This long message will look more appealing to the user when you take the text formatting into your own hands.

When the text of your message is particularly long, you can break it into several lines using the VBA `Chr` function. The `Chr` function's argument is a number from 0 to 255, which returns a character represented by this number. For example, `Chr(13)` returns a carriage return character (this is the same as pressing the Enter key), and `Chr(10)` returns a linefeed character (this is useful for adding spacing between the text lines).

2. Modify the instruction entered in the previous step in the following way and make sure it stays on the same line in the Immediate window:

```
MsgBox "All done." & Chr(13) & "Now open ""Test.doc"" and place
an empty" & Chr(13) & "CD or DVD in your computer's CD/DVD
drive." & Chr(13) & "The following procedure will copy this
file to the disc."
```

Your result should look like Figure 4.3.

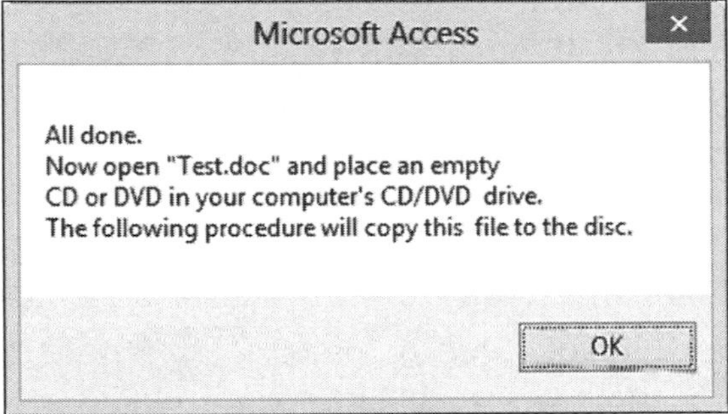

FIGURE 4.3. You can break a long text string into several lines by using the Chr(13) function.

You must surround each text fragment with quotation marks. Quoted text embedded in a text string requires an additional set of quotation marks, as in `""Test.doc""`. The `Chr(13)` function indicates a place where you'd like to start a new line. The concatenate character (&) is used to combine the strings. When you enter exceptionally long text messages on one line, it's easy to make a mistake. An underscore (_) is a special line continuation character in VBA that allows you to break a long VBA statement into several lines. Unfortunately, the line continuation character cannot be used in the Immediate window. A better place to try out various formatting of your long strings for the MsgBox function is within a VBA procedure.

3. Add a new module by choosing **Insert | Module**.
4. In the Code window, enter the following**MyMessage** subroutine. Be sure to precede each line continuation character (_) with a space.

```
Sub MyMessage()
    MsgBox "All done." & Chr(13) _
    & "Now open ""Test.doc"" and place an empty" & Chr(13) _
    & "CD or DVD in your computer's CD/DVD drive." & Chr(13) _
    & "The following procedure will copy this file to the disc."
End Sub
```

5. Position the insertion point within the code of the MyMessage procedure and press **F5** to run it.

When you run the MyMessage procedure, Visual Basic displays the same message as the one illustrated earlier in Figure 4.3.

As you can see, the text entered on several lines is more readable, and the code is easier to maintain. To improve the readability of your message, you may want to add more spacing between the text lines by including blank lines. To do this, use two `Chr(13)` functions, as shown in the following step.

6. Enter the following **MyMessage2** procedure:

```
Sub MyMessage2()
    MsgBox "All done." & Chr(13) & Chr(13) _
    & "Now open ""Test.doc"" and place an empty" & Chr(13) _
    & "CD or DVD in your computer's CD/DVD drive." & Chr(13) _
    & Chr(13) & "The following procedure will copy this " & _
    "file to the disc."
End Sub
```

7. Position the insertion point within the code of the MyMessage2 procedure and press **F5** to run it. The result should look like Figure 4.4.

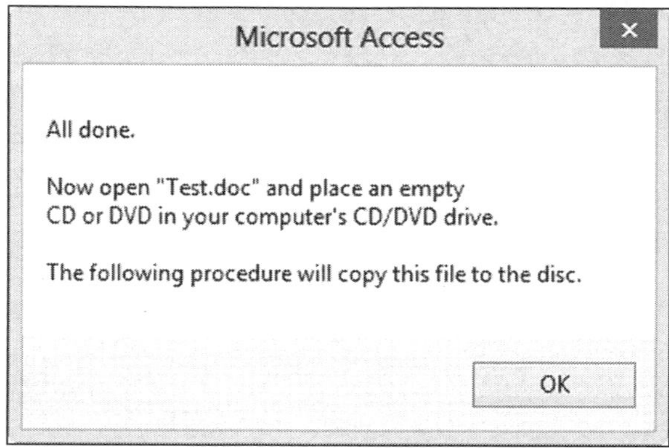

FIGURE 4.4. You can increase the readability of your message by increasing spacing between selected text lines.

Now that you have mastered the text formatting techniques, let's take a closer look at the next argument of the `MsgBox` function. Although the `buttons` argument is optional, it is frequently used. The `buttons` argument specifies how many and what types of buttons you want to appear in the message box. This argument can be a constant or a number (see Table 4.1). If you omit this argument, the resulting message box contains only the OK button, as you've seen in the preceding examples.

TABLE 4.1. The `MsgBox buttons` argument settings

Constant	Value	Description
Button settings		
vbOKOnly	0	Displays only an OK button. This is the default.
vbOKCancel	1	OK and Cancel buttons

Constant	Value	Description
`vbAbortRetryIgnore`	2	Abort, Retry, and Ignore buttons
`vbYesNoCancel`	3	Yes, No, and Cancel buttons
`vbYesNo`	4	Yes and No buttons
`vbRetryCancel`	5	Retry and Cancel buttons
Icon settings		
`vbCritical`	16	Displays the Critical Message icon
`vbQuestion`	32	Displays the Question Message icon
`vbExclamation`	48	Displays the Warning Message icon
`vbInformation`	64	Displays the Information Message icon
Default button settings		
`vbDefaultButton1`	0	The first button is default.
`vbDefaultButton2`	256	The second button is default.
`vbDefaultButton3`	512	The third button is default.
`vbDefaultButton4`	768	The fourth button is default.
Message box modality		
`vbApplicationModal`	0	The user must respond to the message before continuing to work in the current application.
`vbSystemModal`	4096	On Win16 systems, this constant is used to prevent the user from interacting with any other window until he or she dismisses the message box. On Win32 systems, this constant works like the `vbApplicationModal` constant with the following exception: The message box always remains on top of any other programs you may have running.
Other MsgBox display settings		
`vbMsgBoxHelpButton`	16384	Adds the Help button to the message box
`vbMsgBoxSetForeground`	65536	Specifies the message box window as the foreground window
`vbMsgBoxRight`	524288	Text is right-aligned.
`vbMsgBoxRtlReading`	1048576	Text appears as right-to-left reading on Hebrew and Arabic systems.

When should you use the `buttons` argument? Suppose you want the user of your procedure to respond to a question with Yes or No. Your message box will-then require two buttons. If a message box includes more than one button, one of them is considered a default button. When the user presses Enter, the default button is selected automatically.

Because you can display various types of messages (critical, warning, information), you can visually indicate the importance of the message by including the graphical representation (icon). In addition to the type of message, the `buttons` argument can include a setting to determine whether the message box must be closed before the user switches to another application. It's quite possible that the user may want to switch to another program or perform another task before he responds to the question posed in your message box. If the message box is application modal (`vbApplicationModal`), then the user must close the message box before continuing to use your application.

For example, consider the following message box:

```
MsgBox "How are you?", vbOKOnly + vbApplicationModal,"Please Close Me"
```

If you type the preceding statement in the Immediate window and press Enter, a message box will pop up and you won't be able to work with your currently open Microsoft Access application until you respond to the message box.

On the other hand, if you want to keep the message box visible while the user works with other open applications, you must include the `vbSystemModal` setting in the `buttons` argument, like this:

```
MsgBox "How are you?", vbOKOnly + vbSystemModal, "System Modal"
```

Use the `vbSystemModal` constant when you want to ensure that your message box is always visible (not hidden behind other windows).

The `buttons` argument settings are divided into five groups: button settings, icon settings, default button settings, message box modality, and other `MsgBox` display settings (see Table 4.1). Only one setting from each group can be included in the `buttons` argument. To create a `buttons` argument, you can add up the values for each setting you want to include. For example, to display a message box with two buttons (Yes and No), the question mark icon, and the No button as the default button, look up the corresponding values in Table 4.1, and add them up. You should arrive at 292 (4 + 32 + 256).

To see the message box using the calculated message box argument, enter the following statement in the Immediate window:

```
MsgBox "Do you want to proceed?", 292
```

The resulting message box is shown in Figure 4.5.

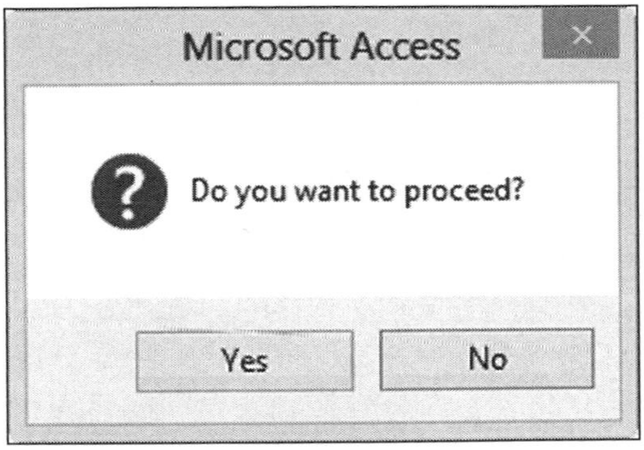

FIGURE 4.5. You can specify the number of buttons to include, their text, and an icon in the message box by using the optional buttons argument.

When you derive the `buttons` argument by adding up the constant values, your procedure becomes less readable. There's no reference table where you can check the hidden meaning of 292. To improve the readability of your `MsgBox` function, it's better to use the constants instead of their values. For example, enter the following revised statement in the Immediate window:

```
MsgBox "Do you want to proceed?", vbYesNo + vbQuestion + vbDefaultBut-
ton2
```

The preceding statement produces the result shown in Figure 4.5. The following example shows how to use the `buttons` argument inside a Visual Basic procedure.

Hands-On 4.8. Using the MsgBox Function with Arguments

1. In the Visual Basic Editor window, choose **Insert | Module** to add a new module.
2. In the Code window, enter the **MsgYesNo** subroutine shown here:

```
Sub MsgYesNo()
  Dim question As String
  Dim myButtons As Integer

  question = "Do you want to open a new report?"
  myButtons = vbYesNo + vbQuestion + vbDefaultButton2
  MsgBox question, myButtons
End Sub
```

3. Run the MsgYesNo procedure by pressing **F5**.

In this subroutine, the `question` variable stores the text of your message. The settings for the `buttons` argument are placed in the `myButtons` variable. Instead of using the names of constants, you can use their values, as in the following:

```
myButtons = 4 + 32 + 256
```

The `question` and `myButtons` variables are used as arguments for the `MsgBox` function. When you run the procedure, you see a result similar to the one shown in Figure 4.5. Note that the No button is selected, indicating that it's the default button for this dialog box. If you press Enter, Visual Basic removes the message box from the screen. Nothing happens because your procedure does not have any instructions following the `MsgBox` function. To change the default button, use the `vbDefaultButton1` setting instead.

The third argument of the `MsgBox` function is `title`. While this is also an optional argument, it's very handy because it allows you to create procedures that don't provide visual clues to the fact that you programmed them with Microsoft Access. Using this argument, you can set the titlebar of your message box to any text you want.

Suppose you want the MsgYesNo procedure to display the text "New report" in its title. The following MsgYesNo2 procedure demonstrates the use of the `title` argument.

```
Sub MsgYesNo2()
  Dim question As String
  Dim myButtons As Integer
  Dim myTitle As String

  question = "Do you want to open a new report?"
  myButtons = vbYesNo + vbQuestion + vbDefaultButton2
  myTitle = "New report"
  MsgBox question, myButtons, myTitle
End Sub
```

The text for the `title` argument is stored in the `myTitle` variable. If you don't specify the value for the `title` argument, Visual Basic displays the default text "Microsoft Access." Notice that the arguments are listed in the order determined by the `MsgBox` function.

If you would like to list the arguments in any order, you must precede the value of each argument with its name, as shown here:

```
MsgBox title:=myTitle, prompt:=question, buttons:=myButtons
```

The last two `MsgBox` arguments, `helpfile` and `context`, are used by more advanced programmers who are experienced with using help files in the Windows environment. The `helpfile` argument indicates the name of a special help file that contains additional information you may want to display to your VBA application user. When you specify this argument, the Help button will be added to your message box. When you use the `helpfile` argument, you must also use the `context` argument. This argument indicates which help subject in the specified help file you want to display. Suppose HelpX.hlp is the help file you created and 55 is the context topic you want to use. To include this information in your `MsgBox` function, you would use the following instruction:

```
MsgBox title:=myTitle, _
  prompt:=question, _
  buttons:=myButtons, _
  helpfile:= "HelpX.hlp", _
  context:=55
```

The precedingis a single VBA statement broken down into several lines using the line continuation character.

Returning Values from the MsgBox Function

When you display a simple message box dialog with one button, clicking the OK button or pressing the Enter key removes the message box from the screen. However, when the message box has more than one button, your procedure should detect which button was pressed. To do this, you must save the result of the message box in a variable. Table 4.2 lists values that the `MsgBox` function returns.

TABLE 4.2. Values returned by the `MsgBox` function

Button Selected	Constant	Value
OK	`vbOK`	1
Cancel	`vbCancel`	2
Abort	`vbAbort`	3
Retry	`vbRetry`	4
Ignore	`vbIgnore`	5
Yes	`vbYes`	6
No	`vbNo`	7

The MsgYesNo3 procedure in Hands-On 4.9 is a revised version of MsgYesNo2. It demonstrates how to store the user's response in a variable.

Hands-On 4.9. Returning Values from the MsgBox Function

1. In the Visual Basic Editor window, choose **Insert | Module** to add a new module.
2. In the Code window, enter the following code of the **MsgYesNo3** procedure:

```
Sub MsgYesNo3()
  Dim question As String
  Dim myButtons As Integer
  Dim myTitle As String
  Dim myChoice As Integer

  question = "Do you want to open a new report?"
  myButtons = vbYesNo + vbQuestion + vbDefaultButton2
  myTitle = "New report"
  myChoice = MsgBox(question, myButtons, myTitle)
  MsgBox myChoice
End Sub
```

3. Position the insertion point within the MsgYesNo3 procedure and press **F5** to run it.

   In this procedure, you assigned the result of the `MsgBox` function to the variable `myChoice`. Notice that the arguments of the `MsgBox` function are now listed in parentheses:

```
myChoice = MsgBox(question, myButtons, myTitle)
```

   When you run the MsgYesNo3 procedure, a two-button message box is displayed. By clicking on the Yes button, the statement `MsgBox myChoice` displays the number 6. When you click the No button, the number 7 is displayed.

MsgBox Function—With or without Parentheses?

Use parentheses around the `MsgBox` function argument list when you want to use the result returned by the function. By listing the function's arguments without parentheses, you tell Visual Basic that you want to ignore the function's result. Most likely, you will want to use the function's result when the message box contains more than one button.

USING THE INPUTBOX FUNCTION

The `InputBox` function displays a dialog box with a message that prompts the user to enter data. This dialog box has two buttons: OK and Cancel. When you click OK, the `InputBox` function returns the information entered in the text box. When you select Cancel, the function returns the empty string (""). The syntax of the `InputBox` function is as follows:

```
InputBox(prompt [, title] [, default] [, xpos] [, ypos]
  [, helpfile, context])
```

The first argument, `prompt`, is the text message you want to display in the dialog box. Long text strings can be entered on several lines by using the `Chr(13)` or `Chr(10)` functions. (See examples of using the `MsgBox` function earlier in this chapter.) All the remaining `InputBox` arguments are optional.

The second argument, `title`, allows you to change the default title of the dialog box. The default value is "Microsoft Access."

The third argument of the `InputBox` function, `default`, allows the display of a default value in the text box. If you omit this argument, the empty text box is displayed.

The following two arguments, `xpos` and `ypos`, let you specify the exact position where the dialog box should appear on the screen. If you omit these arguments, the input box appears in the middle of the current window. The `xpos` argument determines the horizontal position of the dialog box from the left edge of the screen. When omitted, the dialog box is centered horizontally. The `ypos` argument determines the vertical position from the top of the screen. If you omit this argument, the dialog box is positioned vertically approximately one-third of the way down the screen. Both `xpos` and `ypos` are measured in special units called *twips*. One twip is the equivalent of approximately 0.0007 inches.

The last two arguments, `helpfile` and `context`, are used in the same way as the corresponding arguments of the `MsgBox` function discussed earlier in this chapter.

Now that you know the meaning of the `InputBox` arguments, let's see some examples of using this function.

Hands-On 4.10. Using the InputBox Function

1. In the Visual Basic Editor window, choose **Insert | Module** to add a new module.

2. In the Code window, type the following **Informant** subroutine:

```
Sub Informant()
  InputBox prompt:="Enter your place of birth:" & Chr(13) _
  & " (e.g., Boston, Great Falls, etc.) "
End Sub
```

3. Position the insertion point within the Informant procedure and press **F5** to run it.
   This procedure displays a dialog box with two buttons. The input prompt is displayed on two lines (see Figure 4.6). Similar to using the `MsgBox` function, if you plan on using the data entered by the user in the dialog box, you should store the result of the `InputBox` function in a variable.

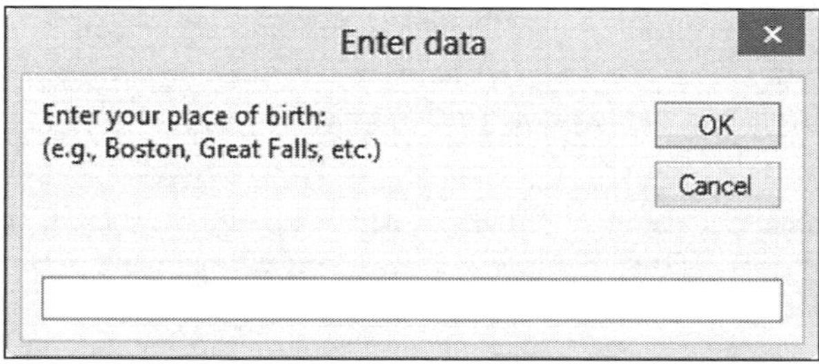

FIGURE 4.6. A dialog box generated by the Informant procedure.

4. Now, in the same module, enter the following code of the **Informant2** procedure:

```
Sub Informant2()
  Dim myPrompt As String
  Dim town As String

  Const myTitle = "Enter data"
  myPrompt = "Enter your place of birth:" & Chr(13) _
  & "(e.g., Boston, Great Falls, etc.)"
  town = InputBox(myPrompt, myTitle)

MsgBox "You were born in " & town & ".", , "Your response"
End Sub
```

5. Position the insertion point within the Informant2 procedure and press **F5** to run it.
   Notice that the Informant2 procedure assigns the result of the `InputBox` function to the `town` variable.

This time, the arguments of the `InputBox` function are listed in parentheses. Parentheses are required if you want to use the result of the `InputBox` function later in your procedure. The Informant2 subroutine uses a constant to specify the text to appear in the titlebar of the dialog box. Because the constant value remains the same throughout the execution of your procedure, you can declare the input box title as a constant. However, if you'd rather use a variable, you still can.

When you run a procedure using the `InputBox` function, the dialog box generated by this function always appears in the same area of the screen. To change the location of the dialog box, you must supply the `xpos` and `ypos` arguments, which were explained earlier.

6. To display the dialog box in the top lefthand corner of the screen, modify the `InputBox` function in the Informant2 procedure as follows:

```
town = InputBox(myPrompt, myTitle, , 1, 200)
```

Notice that the argument `myTitle` is followed by two commas. The second comma marks the position of the omitted `default` argument. The next two arguments determine the horizontal and vertical position of the dialog box. If you omit the second comma after the `myTitle` argument, Visual Basic will use the number 1 as the value of the `default` argument. If you precede the values of arguments by their names (for example, `prompt:=myPrompt, title:=myTitle, xpos:=1, ypos:=200`), you won't have to remember to insert a comma in the place of each omitted argument.

What will happen if, instead of the name of a town, you enter a number? Because users often supply incorrect data in the input box, your procedure must verify that the data the user entered can be used in further data manipulations. The `InputBox` function itself does not provide a facility for data validation. To validate user input, you must use other VBA instructions, which are discussed in Chapter 5.

CONVERTING DATA TYPES

The result of the `InputBox` function is always a string. So if a user enters a number, its *string* value must be converted to a *numeric* value before your procedure can use the number in mathematical computations. Visual Basic is able to automatically convert many values from one data type to another.

Hands-On 4.11. Converting Data Types

1. In the Visual Basic Editor window, choose **Insert | Module** to add a new module.
2. In the Code window, enter the following**AddTwoNums** procedure:

```
Sub AddTwoNums()
  Dim myPrompt As String
  Dim value1 As String
  Dim mySum As Single

  Const myTitle = "Enter data"

  myPrompt = "Enter a number:"
  value1 = InputBox(myPrompt, myTitle, 0)
  mySum = value1 + 2

  MsgBox mySum & " (" & value1 & " + 2)"
End Sub
```

3. Place the cursor anywhere inside the code of the AddTwoNums procedure and press **F5** to run it.

This procedure displays the dialog box shown in Figure 4.7. Notice that this dialog box has two special features that are obtained by using the `InputBox` function's optional arguments: `title` and `default`. Instead of the default title "Microsoft Access," the dialog box displays a text string as defined by the contents of the `myTitle` constant. The zero (0) entered as the default value in the edit box suggests that the user enter a number instead of text. Once the user provides the data and clicks OK, the input is assigned to the variable `value1`.

```
value1 = InputBox(myPrompt, myTitle, 0)
```

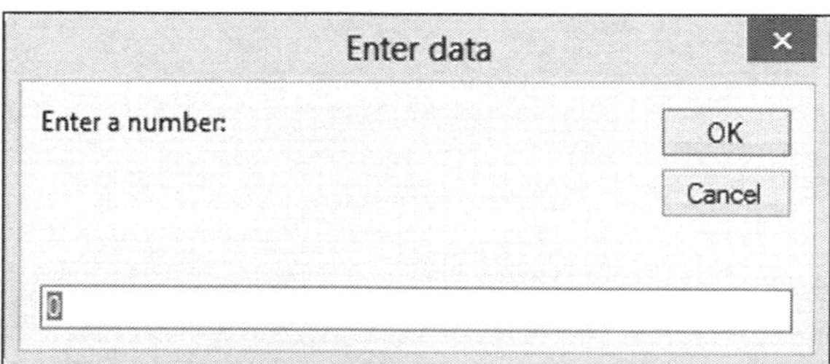

FIGURE 4.7. To suggest that the user enter a specific type of data, you may want to provide a default value in the edit box.

The data type of the variable `value1` is String. You can check the data type easily if you follow the preceding instruction with this statement:

```
MsgBox varType(value1)
```

When Visual Basic runs this line, it will display a message box with the number 8. Recall that this number represents the String data type. The next line,

```
mySum = value1 + 2
```

adds 2 to the user's input and assigns the result of the calculation to the variable `mySum`. Because the `value1` variable's data type is String, Visual Basic goes to work behind the scenes to perform the data type conversion. Visual Basic has the brains to understand the need for conversion. Without it, the two incompatible data types (text and number) would generate a Type Mismatch error.

The procedure ends with the `MsgBox` function displaying the result of the calculation and showing the user how the total was derived.

Define a Constant

To ensure that all the titlebars in a particular VBA procedure display the same text, assign the title text to a constant. By doing so, you will save yourself the time of typing the title text in more than one place.

Avoid a Type Mismatch Error

If you attempt to run the AddTwoNums procedure in older versions of Microsoft Access (prior to version 2000), you will get a Type Mismatch error when Visual Basic tries to execute the following line of code:

```
mySum= value1 + 2
```

To avoid the Type Mismatch error, use the built-in `CSng` function to convert the string stored in the `value1` variable to a Single type number. You would write the following statement:

```
mySum= CSng(value1) + 2
```

USING MASTER PROCEDURES AND SUBPROCEDURES

When your VBA procedure gets larger, it may be difficult to maintain its many lines of code. To make your program easier to write, understand, and change, you should use a structured approach. When you create a structured program, you break a large problem into small problems that can be solved one at a time. In VBA, you do this by creating a master procedure and one or more subordinate procedures. Because both master procedures and subordinate procedures are subroutines, you declare them with the `Sub` keyword. The master procedure can call the required subroutines and pass arguments to them. It may also call functions.

The following hands-on exercise demonstrates the AboutUser procedure. The procedure requests the user's full name, then extracts the first and last name from the `fullName` string. The last statement displays the user's last name followed by a comma and the first name. As you read further, this procedure will be broken down into several tasks to demonstrate the concept of using master procedures, subprocedures, and functions.

Hands-On 4.12. Breaking Up Large Procedures

1. In the Visual Basic Editor window, choose **Insert | Module** to add a new module.
2. In the Code window, enter the following code of the **AboutUser** procedure:

```
Sub AboutUser()
  Dim fullName As String
  Dim firstName As String
  Dim lastName As String
  Dim space As Integer

  ' get input from user
  fullName = InputBox("Enter first and last name:")

  ' get first and last name strings
  space = InStr(fullName, " ")
  firstName = Left(fullName, space - 1)
  lastName = Right(fullName, Len(fullName) - space)

  ' display last name, first name
  MsgBox lastName & ", " & firstName
End Sub
```

3. Position the insertion point within the code of the AboutUser procedure and press **F5** to run it.

   The AboutUser procedure can be divided into smaller tasks. The first task is obtaining the user's full name. The next task requires that you divide the user-supplied data into two strings: last name and first name. These tasks can be delegated to separate functions (for example: GetLast and GetFirst). The last task is displaying a message showing the reordered full name string.

   Now that you know what tasks you should focus on, let's see how you can accomplish each one.
4. In the Visual Basic Editor window, choose **File | Save Chap04**.
5. In the Save dialog box, click **Yes** to save changes to all the listed modules. Click **OK** to confirm the default module name in each Save As prompt.
6. In the Visual Basic Editor window, choose **Insert | Module** to add a new module.
7. Choose **File | Save Chap04** and type the new module name, **MasterProcedures**, in the Save As dialog box. Click **OK** when done.
8. Enter the following **AboutUserMaster** procedure in the MasterProcedures module window:

```
Sub AboutUserMaster()
  Dim first As String, last As String, full As String

  Call GetUserName(full)

  first = GetFirst(full)
  last = GetLast(full)

  Call DisplayLastFirst(first, last)
End Sub
```

   The master procedure shown here controls the general flow of your program by calling appropriate subprocedures and functions. The master procedure begins with variable declarations. The first statement, `Call GetUserName(full)`, calls the GetUserName subroutine (see Step 9), and passes an argument to it—the contents of the `full` variable. Because the variable `full` is not assigned any value prior to the execution of the `Call` statement, it has the value of an empty string (""). Notice that the name of the subprocedure is preceded by the `Call` statement. Although you are not required to use the `Call` keyword when calling a procedure, you must use it when the call to the procedure requires arguments. The argument list must be enclosed in parentheses.

9. In the MasterProcedures module, enter the following **GetUserName** subroutine:

```
Sub GetUserName(fullName As String)
  fullName = InputBox("Enter first and last name:")
End Sub
```

The GetUserName procedure demonstrates two very important Visual Basic programming concepts: how to pass arguments to a subprocedure and how to pass values back from a subprocedure to a calling procedure.

In the master procedure in Step 8, you called the GetUserName procedure and passed one argument to it—the variable `full`. This variable is received by a `fullName` parameter declared in the GetUserName subprocedure's `Sub` statement. Because the variable `full` contained an empty string, when Visual Basic called the GetUserName subprocedure, the `fullName` parameter received the same value—an empty string (""). When Visual Basic displays the dialog box and gets the user's full name, the name is assigned to the `fullName` parameter. A value assigned to a parameter is passed back to the matching argument after the subprocedure is executed. Therefore, when Visual Basic returns to the master procedure, the `full` variable will contain the user's full name.

Arguments passed to a subprocedure are received by *parameters*. Notice that the parameter name (`fullName`) is followed by the declaration of the data type (`As String`). Although the parameter's data type must agree with the data type of the matching argument, different names may be used for an argument and its corresponding parameter.

10. In the MasterProcedures module, enter the following **GetFirst** function procedure:

```
Function GetFirst(fullName As String)
  Dim space As Integer

  space = InStr(fullName, " ")
  GetFirst = Left(fullName, space - 1)
End Function
```

The second statement in the master procedure (see step 8), `first = GetFirst(full)`, passes the value of the full variable to the GetFirst function. This value is received by the function's parameter—`fullName`. To extract the first name from the user-provided `fullName` string, you must find the location of the space separating the first name and last name. Therefore, the function begins with a declaration of a local variable—`space`. The next statement uses the VBA built-in function `InStr` to return the position of a space character (" ") in the `fullName` string. The obtained number is then assigned to the variable `space`. Finally, the `Left` function is used to extract the specified number of characters (`space -1`) from the left side of the `fullName` string. The length of the first name is one character less than the value stored in the variable `space`. The result of the function (user's first name) is then assigned to the function's name. When Visual Basic returns to the master procedure, it places the result in the variable `first`.

11. In the MasterProcedures module, enter the following **GetLast** function procedure:

```
Function GetLast(fullName As String)
  Dim space As Integer

  space = InStr(fullName, " ")
  GetLast = Right(fullName, Len(fullName) - space)
End Function
```

The third statement in the master procedure (see Step 8), `last = GetLast(full)`, passes the value of the `full` variable to the GetLast function. This function's purpose is to extract the user's last name from the user-supplied `fullName` string. The GetLast function uses the built-in `Len` function to calculate the total number of characters in the `fullName` string. The `Right` function extracts the specified number of characters (`Len(fullName) - space`) from the right side of the `fullName` string. The obtained string is then assigned to the function name and is stored in the variable `last` upon returning to the master procedure.

12. In the MasterProcedures module, enter the following **DisplayLastFirst** subroutine:

```
Sub DisplayLastFirst(firstName As String, lastName As String)
  MsgBox lastName & ", " & firstName
End Sub
```

The fourth statement in the master procedure (see step 8), `Call DisplayLastFirst(first, last)`, calls the DisplayLastFirst subroutine and passes two arguments to it: `first` and `last`. To receive these arguments, the DisplayLastFirst subprocedure is declared with two matching parameters: `firstName` and `lastName`. Recall that different names can be used for arguments and their corresponding parameters. The DisplayLastFirst subprocedure then displays the message box showing the user's last name followed by a comma and the first name.

13. Position the insertion point within the code of the AboutUserMaster procedure and press **F5** to run it.
14. Choose **File | Save Chap04** to save changes to the module.
15. Choose **File | CloseandReturn to Microsoft Access**.
16. Close the **Chap04.accdb** database and exit Microsoft Access.

Arguments versus Parameters

- An argument is a variable, constant, or expression that is passed to a subprocedure.
- A parameter is a variable that receives a value passed to a subprocedure.

Advantages of Using Subprocedures

- It's easier to maintain several subprocedures than one large procedure.
- A task performed by a subprocedure can be used by other procedures.
- Each subprocedure can be tested individually before being placed in the main program.
- Different people can work on individual subprocedures that constitute a larger procedure.

CHAPTER SUMMARY

In this chapter, you learned the difference between subroutine procedures that perform actions and function procedures that return values. You saw examples of function procedures called from another Visual Basic procedure. You learned how to pass arguments to functions and how to determine the data type of a function's result. You increased your repertoire of VBA keywords with the `ByVal`, `ByRef`, and `Optional` keywords. Finally, you learned how subprocedures can pass values back to the calling procedures with the help of parameters.

After working through this chapter, you should be able to create some custom functions of your own that are suited to your specific needs. You should also be able to interact easily with your users by employing the `MsgBox` and `InputBox` functions.

Part 2 Controlling Program Execution

In Part I of this book, you learned the fundamentals of VBA. You wrote simple functions and procedures that did not require any complex logic. In this part of the book, you'll find out how decisions and looping statements enable you to control program flow and create more useful VBA procedures and functions.

Chapter 5 Decision Making with VBA

Visual Basic for Applications offers special statements called conditional statements, or "control structures," which allow you to include decision points in your procedures. In a conditional expression, a relational operator (see Table 5.1), a logical operator (see Table 5.2), or a combination of both, evaluates the expression to determine whether it is true or false. If the answer is true, the procedure executes a specified block of instructions. If the answer is false, the procedure either executes a different block of instructions or simply doesn't do anything. In this chapter, you will learn how to use these VBA conditional statements to alter the flow of your program.

RELATIONAL AND LOGICAL OPERATORS

You can make decisions in your VBA procedures by using conditional expressions inside the special control structures. A *conditional expression* is an expression that uses a relational operator (see Table 5.1), a logical operator (see Table 5.2), or a combination of both. When Visual Basic encounters a conditional expression in your program, it evaluates the expression to determine whether it is true or false.

TABLE 5.1. Relational operators in VBA

Operator	Description
=	Equal to
<>	Not equal to
>	Greater than
<	Less than
>=	Greater than or equal to
<=	Less than or equal to

TABLE 5.2. Logical operators in VBA

Operator	Description
AND	All conditions must be true before an action can be taken.
OR	At least one of the conditions must be true before an action can be taken.
NOT	If a condition is true, NOT makes it false. If a condition is false, NOT makes it true.

Boolean Expressions

Conditional expressions and logical operators are also known as *Boolean*. George Boole was a nineteenth-century British mathematician who made significant contributions to the evolution of computer programming. Boolean expressions can be evaluated as true or false.

For example:

One meter equals 10 inches.	False
Two is less than three.	True

IF...THEN STATEMENT

The simplest way to get some decision making into your VBA procedure is by using the `If...Then` statement. Suppose you want to choose an action depending on a condition. You can use the following structure:

```
If condition Then statement
```

For example, a quiz procedure might ask the user to guess the number of weeks in a year. If the user's response is other than 52, the procedure should display the message "Try Again."

Please note files for the "Hands-On" project may be found on the companion CD-ROM.

Hands-On 5.1. Using the If...Then Statement

1. Start Microsoft Access 2013 and create a new database named **Chap05.accdb** in your **C:\Access2013_ByExample** folder.
2. Once your new database is opened, press **Alt+F11** to switch to the Visual Basic Editor window.
3. Choose **Insert | Module** to add a new standard module.
   In the Module1 Code window, enter the following **SimpleIfThen** procedure:

```
Sub SimpleIfThen()
  Dim weeks As String

  weeks = InputBox("How many weeks are in a year:", "Quiz")
  If weeks<>52 Then MsgBox "Try Again"
End Sub
```

   The SimpleIfThen procedure stores the user's answer in the `weeks` variable. The variable's value is then compared with the number 52. If the result of the comparison is true (that is, if the value stored in the variable `weeks` is not equal to 52), Visual Basic will display the message "Try Again."
4. Run the SimpleIfThen procedure and enter a number other than 52.
5. Rerun the SimpleIfThen procedure and enter the number **52**. When you enter the correct number of weeks, Visual Basic does nothing. The procedure ends. It would be nice to also display a message when the user guesses right.
6. Enter the following instruction on a separate line before the `End Sub` keywords:

```
If weeks = 52 Then MsgBox "Congratulations!"
```

7. Run the SimpleIfThen procedure again and enter the number **52**. When you enter the correct answer, Visual Basic does not execute the "Try Again" statement. When the procedure is executed, the statement to the right of the `Then` keyword is ignored if the result from evaluating the supplied condition is false. As you recall, a VBA procedure can call another procedure. Let's see if it can also call itself.
8. Modify the first `If` statement in the SimpleIfThen procedure as follows:

```
If weeks <> 52 Then MsgBox "Try Again" : SimpleIfThen
```

We added a colon and the name of the SimpleIfThen procedure to the end of the existing `If...Then` statement. If you enter the incorrect answer, you'll see a message. After clicking the OK button in the message box, you'll get another chance to supply the correct answer. You'll be able to keep on guessing for a long time. In fact, you won't be able to exit the procedure gracefully until you've supplied the correct answer. After clicking Cancel, you'll have to deal with the unfriendly "Type Mismatch" error message. For now (until you learn other ways of handling errors in VBA), let's revise your SimpleIfThen procedure as follows:

```
Sub SimpleIfThen()
  Dim weeks As String

  On Error GoTo VeryEnd

  weeks = InputBox("How many weeks are in a year:", "Quiz")
  If weeks <> 52 Then MsgBox "Try Again" : SimpleIfThen
  If weeks = 52 Then MsgBox "Congratulations!"

  VeryEnd:
End Sub
```

If Visual Basic encounters an error, it will jump to the `VeryEnd` label placed at the end of the procedure. The statements placed between `On Error GoTo VeryEnd` and the `VeryEnd` labels are ignored. Later in this chapter, you will find other examples of trapping errors in your VBA procedures.
9. Run your revised SimpleIfThen procedure a few times by supplying incorrect answers. The error trap that you added to your procedure will allow you to quit guessing without having to deal with the ugly error message.

MULTI-LINE IF...THEN STATEMENT

Sometimes you may want to perform several actions when the condition is true. Although you could add other statements on the same line by separating them with colons, your code will look clearer if you use the multi-line version of the `If...Then` statement, as shown here:

```
If condition Then
  statement1
  statement2
  statementN
End If
```

For example, let's modify the SimpleIfThen procedure to include additional statements.

Hands-On 5.2. Using the Multi-Line If...Then Statement

1. Insert a new module and enter the following **SimpleIfThen2** procedure:

```
Sub SimpleIfThen2()
  Dim weeks As String
  Dim response As String

  On Error GoTo VeryEnd
  weeks = InputBox("How many weeks are in a year?", "Quiz")
  If weeks <> 52 Then
  response = MsgBox("This is incorrect. Would you like " _
    & " to try again?", vbYesNo + vbInformation _
    + vbDefaultButton1, _
    "Continue Quiz?")
  If response = vbYes Then
    Call SimpleIfThen2
  End If
  End If
  VeryEnd:
End Sub
```

2. Run the SimpleIfThen2 procedure and enter any number other than 52. In this example, the statements between the first `Then` and the first `End If` keywords don't get executed if the variable `weeks` is equal to 52. Notice that the multiline `If...Then` statement must end with the keywords `End If`. How does

Visual Basic make a decision? Simply put, it evaluates the condition it finds between the `If...Then` keywords.

Two Formats of the If...Then Statement

The `If...Then` statement has two formats: a single-line format and a multiline format. The short format is good for statements that fit on one line, like:

```
If secretCode <> "01W01" Then MsgBox "Access denied"
```

or

```
If secretCode = "01W01" Then alpha = True : beta = False
```

In these examples, `secretCode`, `alpha`, and `beta` are the names of variables. In the first example, Visual Basic displays the message "Access denied" if the value of the `secretCode` variable is not equal to 01W01. In the second example, Visual Basic will set the value of the variable `alpha` to True and the value of the variable `beta` to False when the `secretCode` value is equal to 01W01. Notice that the second statement to be executed is separated from the first one by a colon. The multi-line `If...Then` statement is more clear when there are more statements to be executed when the condition is true, or when the statement to be executed is extremely long.

DECISIONS BASED ON MORE THAN ONE CONDITION

The SimpleIfThen procedure you worked with earlier evaluated only a single condition in the `If...Then` statement. This statement, however, can take more than one condition. To specify multiple conditions in an `If...Then` statement, you use the logical operators AND and OR (see Table 5.2 at the beginning of the chapter). Here is the syntax of the `If...Then` statement with the AND operator:

```
If condition1 AND condition2 Then statement
```

In this syntax, both `condition1` and `condition2` must be true for Visual Basic to execute the statement to the right of the `Then` keyword. For example:

```
If sales = 10000 AND salary < 45000 Then SlsCom = sales * 0.07
```

In this example, `condition1` is sales = 10000, and `condition2` is salary < 45000.

When AND is used in the conditional expression, both conditions must be true before Visual Basic can calculate the sales commission (`SlsCom`). If any of these conditions is false or both are false, Visual Basic ignores the statement after `Then`. When it's good enough to meet only one of the conditions, you should use the OR operator. Here is the syntax:

```
If condition1 OR condition2 Then statement
```

The OR operator is more flexible. Only one of the conditions must be true before Visual Basic can execute the statement following the `Then` keyword. Let's look at this example:

```
If dept = "S" OR dept = "M" Then bonus = 500
```

In this example, if at least one condition is true, Visual Basic assigns 500 to the `bonus` variable. If both conditions are false, Visual Basic ignores the rest of the line.

Now, let's look at a complete procedure example. Suppose you can get a 10% discount if you purchase 50 units of a product priced at $7.00. The IfThenAnd procedure demonstrates the use of the AND operator.

Hands-On 5.3. Using the If...Then...AND Conditional Statement

1. Insert a new module and enter the following **IfThenAnd** procedure in the module's Code window:

```
Sub IfThenAnd()
  Dim price As Single
  Dim units As Integer
  Dim rebate As Single

  Const strMsg1 = "To get a rebate, buy an additional "
  Const strMsg2 = "Price must equal $7.00"

  units = 234
  price = 7

  If price = 7 And units >= 50 Then
         rebate = (price * units) * 0.1
         MsgBox "The rebate is: $" & rebate
  End If

  If price = 7 And units < 50 Then
         MsgBox strMsg1 & "50 - units."
  End If
```

```
    If price <> 7 And units >= 50 Then
           MsgBox strMsg2
    End If
    If price <> 7 And units < 50 Then
           MsgBox "You didn't meet the criteria."
    End If
End Sub
```

2. Run the IfThenAnd procedure.

   The IfThenAnd procedure has four `If...Then` statements that are used to evaluate the contents of two variables: `price` and `units`. The AND operator between the keywords `If...Then` allows more than one condition to be tested. With the AND operator, all conditions must be true for Visual Basic to run the statements between the `Then...End If` keywords.

Indenting If Block Instructions

To make the `If` blocks easier to read and understand, use indentation. Compare the following:

If *condition* Then
action
End If
	If *condition* Then
	  action
	End If

Looking at the block statement on the right side, you can easily see where the block begins and where it ends.

IF...THEN...ELSE STATEMENT

Now you know how to display a message or take an action when one or more conditions are true or false. What should you do, however, if your procedure needs to take one action when the condition is true and another action when the condition is false? By adding the `Else` clause to the simple `If...Then` statement, you can direct your procedure to the appropriate statement depending on the result of the test.

The `If...Then...Else` statement has two formats: single-line and multiline. The single-line format is as follows:

```
If condition Then statement1 Else statement2
```

The statement following the `Then` keyword is executed if the condition is true, and the statement following the `Else` clause is executed if the condition is false. For example:

```
If sales > 5000 Then Bonus = sales * 0.05 Else MsgBox "No Bonus"
```

If the value stored in the variable `sales` is greater than 5000, Visual Basic will calculate the bonus using the following formula: `sales * 0.05`. However, if the variable `sales` is not greater than 5000, Visual Basic will display the message "No Bonus."

The `If...Then...Else` statement should be used to decide which of two actions to perform. When you need to execute more statements when the condition is true or false, it's better to use the multiline format of the `If...Then...Else` statement:

```
If condition Then
   statements to be executed if condition is True
Else
 statements to be executed if condition is False
End If
```

Notice that the multi-line (block) `If...Then...Else` statement ends with the `End If` keywords. Use the indentation as shown to make this block structure easier to read.

```
If Me.Dirty Then
  Me!btnUndo.Enabled = True
Else
  Me!btnUndo.Enabled = False
End If
```

In this example, if the condition (`Me.Dirty`) is true, Visual Basic will execute the statements between `Then` and `Else`, and will ignore the statement between `Else` and `End If`. If the condition is false, Visual Basic will omit the statements between `Then` and `Else`, and will execute the statement between `Else` and `End If`. The purpose of this procedure fragment is to enable the Undo button when the data on the form has changed and keep the Undo button disabled if the data has not changed. Let's look at a procedure example.

Hands-On 5.4. Using the If...Then...Else Conditional Statement

1. Insert a new module and enter the following **WhatTypeOf Day** procedure in the module's Code window:

```
Sub WhatTypeOfDay()
  Dim response As String
```

```
    Dim question As String
    Dim strMsg1 As String, strMsg2 As String
    Dim myDate As Date

    question = "Enter any date in the format mm/dd/yyyy:" _
     & Chr(13) & " (e.g., 07/06/2011)"
    strMsg1 = "weekday"
    strMsg2 = "weekend"
    response = InputBox(question)
    myDate = Weekday(CDate(response))

    If myDate >= 2 And myDate <= 6 Then
      MsgBox strMsg1
    Else
      MsgBox strMsg2
    End If
End Sub
```

2. Run the WhatTypeOfDay procedure.
   This procedure asks the user to enter any date. The user-supplied string is then converted to the Date data type with the built-in `CDate` function. Finally, the `Weekday` function converts the date into an integer that indicates the day of the week (see Table 5.3). The integer is stored in the variable `myDate`. The conditional test is performed to check whether the value of the variable `myDate` is greater than or equal to 2 (`>=2`) and less than or equal to 6 (`<=6`). If the result of the test is true, the user is told that the supplied date is a weekday; otherwise, the program announces that it's a weekend.
3. Run the procedure a few more times, each time supplying a different date. Check the Visual Basic answers against your desktop or wall calendar.

TABLE 5.3. The `Weekday` function values

Constant	Value
vbSunday	1
vbMonday	2
vbTuesday	3
vbWednesday	4
vbThursday	5
vbFriday	6
vbSaturday	7

IF...THEN...ELSEIF STATEMENT

Quite often you will need to check the results of several different conditions. To join a set of `If` conditions together, you can use the `ElseIf` clause. Using the `If...Then...ElseIf` statement, you can evaluate more conditions than is possible with the `If...Then...Else` statement that was the subject of the preceding section. Here is the syntax of the `If...Then...ElseIf` statement:

```
If condition1 Then
  statements to be executed if condition1 is True
ElseIf condition2 Then
  statements to be executed if condition2 is True
ElseIf condition3 Then
  statements to be executed if condition3 is True
ElseIf conditionN Then
  statements to be executed if conditionN is True
Else
  statements to be executed if all conditions are False
End If
```

The `Else` clause is optional; you can omit it if there are no actions to be executed when all conditions are false.

ElseIf Clause

Your procedure can include any number of `ElseIf` statements and conditions. The `ElseIf` clause always comes before the `Else` clause. The statements in the `ElseIf` clause are executed only if the condition in this clause is true.

Let's look at the following procedure fragment:

```
If myNumber = 0 Then
  MsgBox "You entered zero."
ElseIf myNumber > 0 Then
  MsgBox "You entered a positive number."
ElseIf myNumber < 0 Then
  MsgBox "You entered a negative number."
End If
```

This example checks the value of the number entered by the user and stored in the variable `myNumber`. Depending on the number entered, an appropriate message (zero, positive, negative) is displayed. Notice that the `Else` clause is not used. If the result of the first condition (`myNumber = 0`) is false, Visual Basic jumps to the next `ElseIf` statement and evaluates its condition (`myNumber > 0`). If the value is not greater than zero, Visual Basic skips to the next `ElseIf` and the condition `myNumber < 0` is evaluated.

NESTED IF...THEN STATEMENTS

You can make more complex decisions in your VBA procedures by placing an `If...Then` or `If...Then...Else` statement inside another `If...Then` or `If...Then...Else` statement. Structures in which an `If` statement is contained inside another `If` block are referred to as *nested* `If` statements. To understand how nested `If...Then` statements work, it's time for another hands-on exercise.

Hands-On 5.5. Using Nested If...Then Statements

1. In the database **Chap05.accdb**, create a blank form by choosing **Blank form** in the Forms section of the Create tab (Microsoft Access 2013 window). When Access opens the new form in Layout view, switch to Design view.
2. Use the text box control in the Controls section of the Design tab to add two text boxes to the form (see Figure 5.1).

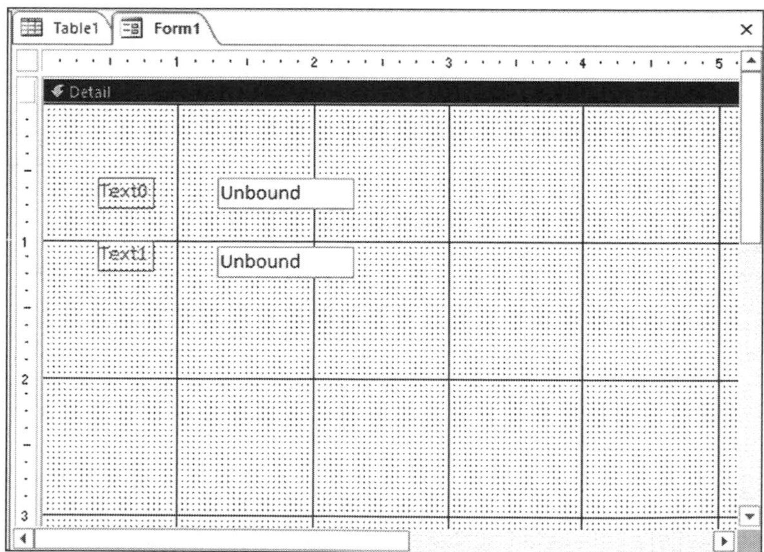

FIGURE 5.1. Placing text box controls on an Access form for Hands-On 5.5.

3. Click the **Property Sheet** button in the **Tools** section of the Design tab.

4. In the property sheet, change the Caption property for the label in front of the first text box to **User** and the Caption property for the label in front of the second text box to **Password**.
5. Click the Unbound text box to the right of the User label. In the property sheet on the Other tab, set the Name property of this control to **txtUser**. Click the Unbound text box to the right of the Password label. In the property sheet on the Other tab, set the Name property of this text box to **txtPwd** (see Figure 5.2).
6. In the property sheet on the Data tab, type **Password** next to the Input Mask property of the txtPwd text box control.
7. Click the **Button** (Form Control) in the **Controls** section of the Design tab, and add a button to the form. When the Command Button Wizard dialog box appears, click **Cancel**. With the Command button selected, set the Caption and Name properties of this button by typing the following values in the property sheet next to the shown property name (see Figure 5.3):
   Name property: **cmdOK**
   Caption property: **OK**
8. Right-click the **OK** button and choose **Build Event** from the shortcut menu. In the Choose Builder dialog box, select **Code Builder** and click **OK**.
9. Enter the following code for the **cmdOK_Click** event procedure. To make the procedure easier to understand, the conditional statements are shown with different formatting (bold and underlined).

```
Private Sub cmdOK_Click()
  If txtPwd = "FOX" Then
    MsgBox "You're not authorized to run this report."
  ElseIf txtPwd = "DOG" Then
    If txtUser = "John" Then
      MsgBox "You're logged on with restricted privileges."
    ElseIf txtUser = "Mark" Then
      MsgBox "Contact the Admin now."
    ElseIf txtUser = "Anne" Then
      MsgBox "Go home."
    Else
      MsgBox "Incorrect user name."
    End If
  Else
    MsgBox "Incorrect password or user name"
  End If
  Me.txtUser.SetFocus
End Sub
```

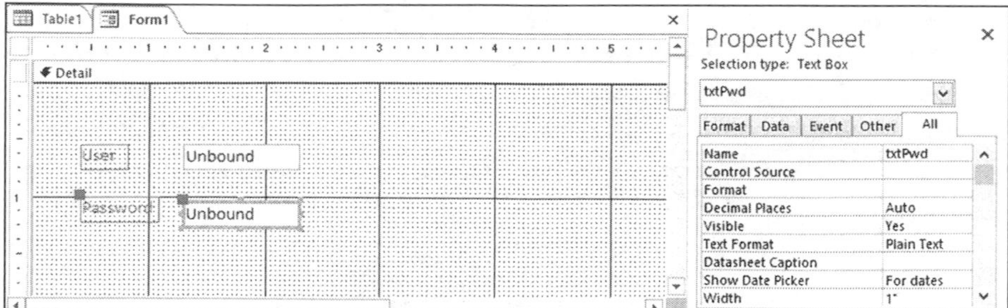

FIGURE 5.2. Setting the Name property of the text box control for Hands-On 5.5.

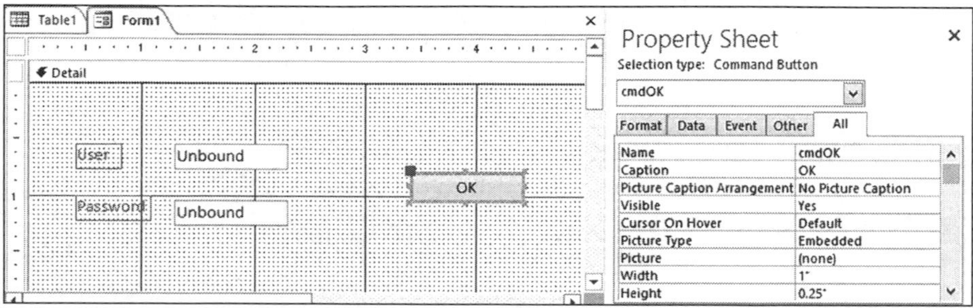

FIGURE 5.3. Setting the Command button properties for Hands-On 5.5.

10. Choose **File | Close and Return to Microsoft Access**. Save your form as **frm-TestNesting**. When prompted to save standard modules you created in earlier exercises, save these objects with default names.
11. Switch to **Form** view. Enter any data in the User and Password text boxes, then click **OK**.

The procedure first checks if the `txtPwd` text box on the form holds the text string "FOX." If this is true, the message is displayed, and Visual Basic skips over the `ElseIf` and `Else` clauses until it finds the matching `End If` (see the bolded conditional statement).

If the `txtPwd` text box holds the string "DOG," we use a nested `If...Then...Else` statement (underlined) to check if the content of the `txtUser` text box is set to John, Mark, or Anne, and then display the appropriate message. If the user name is not one of the specified names, then the condition is false and we jump to the underlined `Else` to display a message stating that the user entered an incorrect user name.

The first `If` block (in bold) is called the *outer* `If` statement. This outer statement contains one *inner* `If` statement (underlined).

Nesting Statements

Nesting means placing one type of control structure inside another control structure. You will see more nesting examples with the looping structures discussed in Chapter 6, "Repeating Actions in VBA."

SELECT CASE STATEMENT

To avoid complex nested `If` statements that are difficult to follow, you can use the `Select Case` statement instead. The syntax of this statement is as follows:

```
Select Case testExpression
  Case expressionList1
    statements to be executed
    if expressionList1 matches testExpression
  Case expressionList2
    statements to be executed
    if expressionList2 matches testExpression
  Case expressionListN
    statements to be executed
    if expressionListN matches testExpression
  Case Else
    statements to be executed
    if no values match testExpression
End Select
```

You can place any number of cases to test between the keywords `Select Case` and `End Select`. The `Case Else` clause is optional. Use it when you expect that there may be conditional expressions that return False. In the `Select Case` statement, Visual Basic compares each `expressionList` with the value of `testExpression`.

Here's the logic behind the `Select Case` statement. When Visual Basic encounters the `Select Case` clause, it makes note of the value of `testExpression`. Then it proceeds to test the expression following the first `Case` clause. If the value of this expression (`expressionList1`) matches the value stored in `testExpression`, Visual Basic executes the statements until another `Case` clause is encountered, and then jumps to the `End Select` statement. If, however, the expression tested in the first `Case` clause does not match `testExpression`, Visual Basic checks the value of each `Case` clause until it finds a match. If none of the `Case` clauses contain the expression that matches the value stored in `testExpression`, Visual Basic jumps to the `Case Else` clause and executes the statements until it encounters the `End Select` keywords. Notice that the `Case Else` clause is optional. If your procedure does not use `Case Else`, and none of the `Case` clauses contain a value matching the value of `testExpression`, Visual Basic jumps to the statements following `End Select` and continues executing your procedure.

Let's look at an example of a procedure that uses the `Select Case` statement. As you already know, the `MsgBox` function allows you to display a message with one or more buttons. You also know that the result of the `MsgBox` function can be assigned to a variable. Using the `Select Case` statement, you can decide which action to take based on the button the user pressed in the message box.

Hands-On 5.6. Using the Select Case Statement

1. Press **Alt+F11** to switch from the Microsoft Access 2013 application window to the Visual Basic Editor window.
2. Insert a new module and enter the following **TestButtons** procedure in the module's Code window:

```
Sub TestButtons()
  Dim question As String
  Dim bts As Integer
  Dim myTitle As String
  Dim myButton As Integer

  question = "Do you want to preview the report now?"
  bts = vbYesNoCancel + vbQuestion + vbDefaultButton1
  myTitle = "Report"
  myButton = MsgBox(prompt:=question, buttons:=bts, _
   Title:=myTitle)

  Select Case myButton
    Case 6
      DoCmd.OpenReport "Sales by Year", acPreview
    Case 7
      MsgBox "You can review the report later."
    Case Else
      MsgBox "You pressed Cancel."
  End Select
End Sub
```

3. Run the TestButtons procedure three times, each time selecting a different button. (Because there is no Sales by Year report in the current database, an error message will pop up when you select Yes. Click **End** to exit the error message.)

The first part of the TestButtons procedure displays a message with three buttons: Yes, No, and Cancel. The value of the button selected by the user is assigned to the variable `myButton`.

If the user clicks Yes, the variable `myButton` is assigned the `vbYes` constant or its corresponding value 6. If the user selects No, the variable `myButton` is assigned the constant `vbNo` or its corresponding value 7. Lastly, if Cancel is pressed, the content of the variable `myButton` equals `vbCancel`, or 2.

The `Select Case` statement checks the values supplied after the `Case` clause against the value stored in the variable `myButton`. When there is a match, the appropriate `Case` statement is executed.

The TestButtons procedure will work the same if you use constants instead of button values:

```
Select Case myButton
  Case vbYes
    DoCmd.OpenReport "Sales by Year", acPreview
  Case vbNo
    MsgBox "You can review the report later."
  Case Else
    MsgBox "You pressed Cancel."
End Select
```

You can omit the `Else` clause. Simply revise the `Select Case` statement as follows:

```
Select Case myButton
  Case vbYes
    DoCmd.OpenReport "Sales by Year", acPreview
  Case vbNo
    MsgBox "You can review the report later."
  Case vbCancel
    MsgBox "You pressed Cancel."
End Select
```

Capture Errors with Case Else

Although using `Case Else` in the `Select Case` statement isn't required, it's always a good idea to include one just in case the variable you are testing has an unexpected value. The `Case Else` clause is a good place to put an error message.

Using Is with the Case Clause

Sometimes a decision is made based on whether the test expression uses the greater than, less than, equal to, or some other relational operator (see Table 5.1). The `Is` keyword lets you use a conditional expression in a `Case` clause. The syntax for the `Select Case` clause using the `Is` keyword is as follows:

```
Select Case testExpression
  Case Is condition1
    statements if condition1 is true
  Case Is condition2
    statements if condition2 is true
  Case Is conditionN
    statements if conditionN is true
End Select
```

Let's look at an example:

```
Select Case myNumber
  Case Is <= 10
    MsgBox "The number is less than or equal to 10."
  Case 11
    MsgBox "You entered 11."
  Case Is >= 100
    MsgBox "The number is greater than or equal to 100."
  Case Else
    MsgBox "The number is between 12 and 99."
End Select
```

Assuming that the variable `myNumber` holds 120, the third `Case` clause is true, and the only statement executed is the one between `Case Is >= 100` and the `Case Else` clause.

Specifying a Range of Values in a Case Clause

In the preceding example, you saw a simple `Select Case` statement that uses one expression in each `Case` clause. Many times, however, you may want to specify a range of values in a `Case` clause. You do this by using the `To` keyword between the values of expressions, as in the following example:

```
Select Case unitsSold
  Case 1 To 100
    Discount = 0.05
  Case Is <= 500
    Discount = 0.1
  Case 501 To 1000
    Discount = 0.15
  Case Is >1000
    Discount = 0.2
End Select
```

Let's analyze this `Select Case` block with the assumption that the variable `unitsSold` currently has a value of 99. Visual Basic compares the value of the variable `unitsSold` with the conditional expression in the `Case` clauses. The first and third `Case` clauses illustrate how to use a range of values in a conditional expression by using the `To` keyword.

Because `unitsSold` equals 99, the condition in the first `Case` clause is true; thus, Visual Basic assigns the value 0.05 to the variable `Discount`. Well, how about the second `Case` clause, which is also true? Although it's obvious that 99 is less than or equal to 500, Visual Basic does not execute the associated statement `Discount = 0.1`. The reason for this is that once Visual Basic locates a `Case` clause with a true condition, it doesn't bother to look at the remaining `Case` clauses. It jumps over them and continues to execute the procedure with the instructions that may follow the `End Select` statement.

For more practice with the Select Case statement, let's use it in a function procedure. As you recall from Chapter 4, function procedures allow you to return a result to a subroutine. Suppose a subroutine has to display a discount based on the number of units sold. You can get the number of units from the user and then run a function to figure out which discount applies.

Hands-On 5.7. Using the Select Case Statement in a Function

1. Insert a new module and enter the following **DisplayDiscount** procedure in the Code window:

```
Sub DisplayDiscount()
  Dim unitsSold As Integer
  Dim myDiscount As Single
```

```
    unitsSold = InputBox("Units Sold:")
    myDiscount = GetDiscount(unitsSold)
    MsgBox myDiscount
End Sub
```

2. In the same module, enter the following **GetDiscount** function procedure:

```
Function GetDiscount(unitsSold As Integer)
  Select Case unitsSold
    Case 1 To 200
      GetDiscount = 0.05
    Case 201 To 500
      GetDiscount = 0.1
    Case 501 To 1000
      GetDiscount = 0.15
    Case Is > 1000
      GetDiscount = 0.2
  End Select
End Function
```

3. Place the insertion point anywhere within the code of the DisplayDiscount procedure and press **F5** to run it.

   The DisplayDiscount procedure passes the value stored in the variable `unitsSold` to the GetDiscount function. When Visual Basic encounters the `Select Case` statement, it checks whether the value of the first `Case` clause expression matches the value stored in the `unitsSold` parameter. If there is a match, Visual Basic assigns a 5% discount (0.05) to the function name, and then jumps to the `End Select` keywords. Because there are no more statements to execute inside the function procedure, Visual Basic returns to the calling procedure, DisplayDiscount. Here it assigns the function's result to the variable `myDiscount`. The last statement displays the value of the retrieved discount in a message box.
4. Choose **File | Save Chap05** and click OK when prompted to save the changes to the modules you created during the Hands-On exercises.
5. Choose **File | Close and Return to Microsoft Access**.
6. Close the **Chap05.accdb** database and exit Microsoft Access.

Specifying Multiple Expressions in a Case Clause

You may specify multiple conditions within a single `Case` clause by separating each condition with a comma:

```
Select Case myMonth
  Case "January", "February", "March"
    Debug.Print myMonth & ": 1st Qtr."
```

```
  Case "April", "May", "June"
    Debug.Print myMonth & ": 2nd Qtr."
  Case "July", "August", "September"
    Debug.Print myMonth & ": 3rd Qtr."
  Case "October", "November", "December"
    Debug.Print myMonth & ": 4th Qtr."
End Select
```

Multiple Conditions within a Case Clause
The commas used to separate conditions within a `Case` *clause have the same meaning as the OR operator used in the* `If` *statement. The* `Case` *clause is true if at least one of the conditions is true.*

CHAPTER SUMMARY

Conditional statements, introduced in this chapter, let you control the flow of your VBA procedure. By testing the truth of a condition, you can decide which statements should be run and which should be skipped over. In other words, instead of running your procedure from top to bottom, line by line, you can execute only certain lines. Here are a few guidelines to help you determine which conditional statement you should use:

- If you want to supply only one condition, the simple `If…Then` statement is the best choice.
- If you need to decide which of two actions to perform, use the `If…Then…Else` statement.
- If your procedure requires two or more conditions, use the `If…Then…ElseIf` or `Select Case` statements.
- If your procedure has many conditions, use the `Select Case` statement. This statement is more flexible and easier to comprehend than the `If…Then…ElseIf` statement.

Sometimes decisions have to be repeated. The next chapter teaches you how your procedures can perform the same actions over and over again.

Chapter 6 Repeating Actions in VBA

Now that you've learned how conditional statements can give your VBA procedures decision-making capabilities, it's time to get more involved. Not all decisions are easy. Sometimes you will need to perform a number of statements several times to arrive at a certain condition. On other occasions, however, after you've reached the decision, you may need to run the specified statements as long as a condition is true or until a condition becomes true. In programming, performing repetitive tasks is called *looping*. VBA has various looping structures that allow you to repeat a sequence of statements a number of times. In this chapter, you learn how to loop through your code.

What Is a Loop?

A *loop* is a programming structure that causes a section of program code to execute repeatedly. VBA provides several structures to implement loops in your procedures: `Do...While`, `Do...Until`, `For...Next`, and `For Each...Next`.

USING THE DO...WHILE STATEMENT

Visual Basic has two types of `Do` loop statements that repeat a sequence of statements either as long as or until a certain condition is true: `Do...While` and `Do...Until`.

The `Do...While` statement lets you repeat an action as long as a condition is true. This statement has the following syntax:

```
Do While condition
  statement1
  statement2
  statementN
Loop
```

When Visual Basic encounters this loop, it first checks the truth value of the condition. If the condition is false, the statements inside the loop are not executed, and Visual Basic will continue to execute the program with the first statement after the `Loop` keyword or will exit the program if there are no more statements to execute. If the condition is true, the statements inside the loop are run one by one until the `Loop` statement is encountered. The `Loop` statement tells Visual Basic to repeat the entire process again as long as the testing of the condition in the `Do...While` statement is true.

Let's see how you can put the `Do...While` loop to good use in Microsoft Access. You will find out how to continuously display an input box until the user enters the correct password. The following hands-on exercise demonstrates this.

Please note files for the "Hands-On" project may be found on the companion CD-ROM.

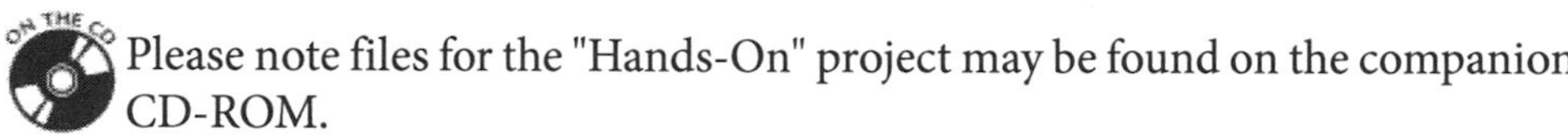

Hands-On 6.1. Using the Do...While Statement

1. Start Microsoft Access 2013 and create a new database named **Chap06.accdb** in your **C:\Access2013_ByExample** folder.
2. Once your new database is opened, press **Alt+F11** to switch to the Visual Basic Editor window.

3. Choose **Insert | Module** to add a new standard module.
4. In the Module1 Code window, enter the following **AskForPassword** procedure:

```
Sub AskForPassword()
  Dim pWord As String

  pWord = ""
  Do While pWord <> "DADA"
    pWord = InputBox("What is the report password?")
  Loop
  MsgBox "You entered the correct report password."
End Sub
```

5. Run the AskForPassword procedure.

   In this procedure, the statement inside the `Do...While` loop is executed as long as the variable `pWord` is not equal to the string "DADA." If the user enters the correct password ("DADA"), Visual Basic leaves the loop and executes the `MsgBox` statement after the `Loop` keyword.

   To allow the user to exit the procedure gracefully and cancel out of the input box if he does not know the correct password, add the following statement on an empty line before the `Loop` keyword:

```
If pWord = "" Then Exit Do
```

   The `Exit Do` statement tells Visual Basic to exit the `Do` loop if the variable `pWord` does not hold any value (please see the section titled "Exiting Loops Early" later in this chapter). Therefore, when the input box appears, the user can leave the text field empty and click OK or Cancel to stop the procedure. Without the `Exit Do` statement, the procedure will keep on asking the user to enter the password until the correct value is supplied.

   To forgo displaying the informational message when the user has not provided the correct password, you may want to use the conditional statement `If...Then` that you learned in the previous chapter. Below is the revised AskForPassword procedure:

```
Sub AskForPassword() ' revised procedure
  Dim pWord As String

  pWord = ""
  Do While pWord <> "DADA"
    pWord = InputBox("What is the report password?")
    If pWord = "" Then
      MsgBox "You did not enter a password."
```

```
      Exit Do
    End If
  Loop
  If pWord <> "" Then
    MsgBox "You entered the correct report password."
  End If
End Sub
```

ANOTHER APPROACH TO THE DO...WHILE STATEMENT

The `Do...While` statement has another syntax that lets you test the condition at the bottom of the loop:

```
Do
    statement1
    statement2
    statementN
Loop While condition
```

When you test the condition at the bottom of the loop, the statements inside the loop are executed at least once. Let's try this in the next hands-on exercise.

Hands-On 6.2. Using the Do...While Statement with a Condition at the Bottom of the Loop

1. In the Visual Basic Editor window, insert a new module and enter the following **SignIn** procedure:

```
Sub SignIn()
  Dim secretCode As String

  Do
    secretCode = InputBox("Enter your secret code:")
    If secretCode = "sp1045" Then Exit Do
  Loop While secretCode <> "sp1045"
End Sub
```

2. Run the SignIn procedure.
   Notice that by the time the condition is evaluated, Visual Basic has already executed the statements one time. In addition to placing the condition at the end of the loop, the SignIn procedure shows again how to exit the loop when a condition is reached. When the `Exit Do` statement is encountered, the loop ends immediately.

To exit the loop in the SignIn procedure without entering the password, you may revise it as follows:

```
Sub SignIn() 'revised procedure
  Dim secretCode As String

  Do
    secretCode = InputBox("Enter your secret code:")
    If secretCode = "sp1045" Or secretCode = "" Then
      Exit Do
    End If
  Loop While secretCode <> "sp1045"
End Sub
```

Avoid Infinite Loops

If you don't design your loop correctly, you can get an *infinite loop*—a loop that never ends. You will not be able to stop the procedure by using the Esc key. The following procedure causes the loop to execute endlessly because the programmer forgot to include the test condition:

```
Sub SayHello()
  Do
    MsgBox "Hello."
  Loop
End Sub
```

To stop the execution of the infinite loop, you must press Ctrl+Break. When Visual Basic displays the message box "Code execution has been interrupted," click End to end the procedure.

USING THE DO...UNTIL STATEMENT

Another handy loop is `Do...Until`, which allows you to repeat one or more statements until a condition becomes true. In other words, `Do...Until` repeats a block of code as long as something is false. Here is the syntax:

```
Do Until condition
  statement1
  statement2
  statementN
Loop
```

Using the preceding syntax, you can now rewrite the AskForPassword procedure (written in Hands-On 6.1) as shown in the following hands-on exercise.

Hands-On 6.3. Using the Do...Until Statement

1. In the Visual Basic Editor window, insert a new module and type the **AskForPassword2** procedure:

```
Sub AskForPassword2()
  Dim pWord As String

  pWord = ""
  Do Until pWord = "DADA"
    pWord = InputBox("What is the report password?")
  Loop
End Sub
```

2. Run the AskForPassword2 procedure.
The first line of this procedure says: Perform the following statements until the variable `pWord` holds the value "DADA." As a result, until the correct password is supplied, Visual Basic executes the `InputBox` statement inside the loop. This process continues as long as the condition `pWord = "DADA"` evaluates to false.
You could modify this procedure to allow the user to cancel the input box without supplying the password, as follows:

```
Sub AskForPassword2() 'revised procedure
  Dim pWord As String

  pWord = ""
  Do Until pWord = "DADA"
    pWord = InputBox("What is the report password?")
    If pWord = "" Then Exit Do
  Loop
End Sub
```

Variables and Loops

All variables that appear in a loop should be assigned default values before the loop is entered.

ANOTHER APPROACH TO THE DO...UNTIL STATEMENT

Similar to the `Do...While` statement, the `Do...Until` statement has a second syntax that lets you test the condition at the bottom of the loop:

```
Do
  statement1
  statement2
  statementN
Loop Until condition
```

If you want the statements to execute at least once, no matter what the value of the condition, place the condition on the line with the `Loop` statement. Let's try out the following example that prints 27 numbers to the Immediate window.

Hands-On 6.4. Using the Do...Until Statement with a Condition at the Bottom of the Loop

1. In the Visual Basic Editor window, insert a new module and type the **PrintNumbers** procedure shown here:

```
Sub PrintNumbers()
  Dim num As Integer

  num = 0
  Do
    num = num + 1
    Debug.Print num
  Loop Until num = 27
End Sub
```

2. Make sure the Immediate window is open in the Visual Basic Editor window (choose **View | Immediate Window** or press **Ctrl+G**).
3. Run the PrintNumbers procedure.
   The variable `num` is initialized at the beginning of the procedure to zero (0). When Visual Basic enters the loop, the content of the variable `num` is increased by one, and the value is written to the Immediate window with the `Debug.Print` statement. Next, the condition tells Visual Basic that it should execute the statements inside the loop until the variable `num` equals 27.
4. Return to the Microsoft Access application window by choosing **File | Close and Return to Microsoft Access**. When prompted, save the changes to all the modules.

Counters

A *counter* is a numeric variable that keeps track of the number of items that have been processed. The preceding PrintNumbers procedure declares the variable `num` to keep track of numbers that were printed. A counter variable should be initialized (assigned a value) at the beginning of the program. This ensures that you always know the exact value of the counter before you begin using it. A counter can be incremented or decremented by a specified value.

FOR...NEXT STATEMENT

The `For...Next` statement is used when you know how many times you want to repeat a group of statements. The syntax of a `For...Next` statement looks like this:

```
For counter = start To end [Step increment]
  statement1
  statement2
  statementN
Next [counter]
```

The code in the brackets is optional. `Counter` is a numeric variable that stores the number of iterations. `Start` is the number at which you want to begin counting. `End` indicates how many times the loop should be executed. For example, if you want to repeat the statements inside the loop five times, use the following `For` statement:

```
For counter = 1 To 5
  statements
Next
```

When Visual Basic encounters the `Next` statement, it will go back to the beginning of the loop and execute the statements inside the loop again, as long as the counter hasn't reached the end value. As soon as the value of `counter` is greater than the number entered after the `To` keyword, Visual Basic exits the loop. Because the variable `counter` automatically changes after each execution of the loop, sooner or later the value stored in the counter exceeds the value specified in `end`.

By default, every time Visual Basic executes the statements inside the loop, the value of the variable `counter` is increased by one. You can change this default setting by using the `Step` clause. For example, to increase the variable `counter` by three, use the following statement:

```
For counter = 1 To 5 Step 3
  statements
Next counter
```

When Visual Basic encounters this statement, it executes the statements inside the loop twice. The first time the loop runs, the counter equals 1. The second time the loop runs, the counter equals 4 (1+3). The loop does not run a third time, because now the counter equals 7 (4+3), causing Visual Basic to exit the loop.

Note that the `Step` increment is optional. Optional statements are always shown in square brackets (see the syntax at the beginning of this section). The `Step` increment isn't specified unless it's a value other than 1. You can place a negative number after `Step` in order to decrement this value from the counter each time it encounters the `Next` statement. The name of the variable (`counter`) after the `Next` statement is also optional; however, it's good programming practice to make your `Next` statements explicit by including the `counter` variable's name.

How can you use the `For...Next` loop in Microsoft Access? Suppose you want to retrieve the names of the text boxes located on an active form. The procedure in the next hands-on exercise demonstrates how to determine whether a control is a text box and how to display its name if a text box is found.

Hands-On 6.5. Using the For...Next Statement

1. Import the **Customers** table from the **Northwind 2007.accdb** database located in the Access2013_HandsOn folder. To do this, click **Access** in the Import & Link section of the External Data tab. In the File name text box of the Get External Data dialog box, enter **C:\Access2013_HandsOn\Northwind 2007.accdb** and click **OK**. In the Import Objects dialog box, select the **Customers** table and click **OK**. Click **Close** to exit the Get External Data dialog box.
2. Now, create a simple **Customers** form based on the Customers table. To do this, select the Customers table in the navigation pane by clicking on its name. Next, click the **Form** button in the Forms section of the Create tab. Access creates a form as shown in Figure 6.1.

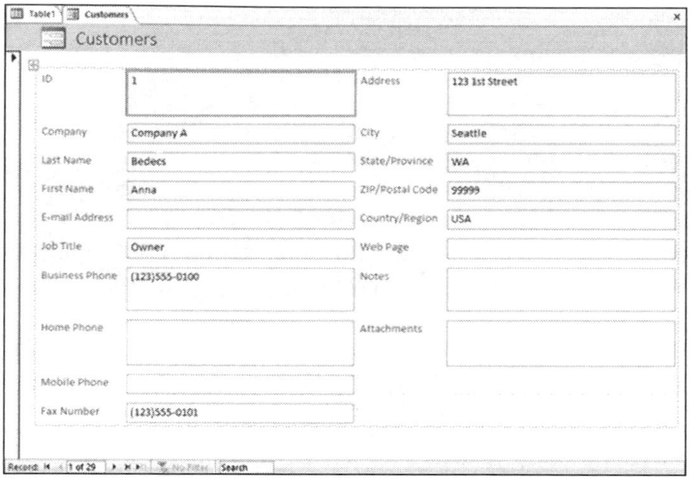

FIGURE 6.1. Automatic data entry form created by Microsoft Access 2013.

Press **Alt+F11** to switch to the Visual Basic Editor window and insert a new module.

3. In the module's Code window, enter the following **GetTextBoxNames** procedure:

```
Sub GetTextBoxNames()
  Dim myForm As Form
  Dim myControl As Control
  Dim c As Integer

  Set myForm = Screen.ActiveForm
  Set myControl = Screen.ActiveControl

  For c = 0 To myForm.Count - 1
    If TypeOf myForm(c) Is TextBox Then
      MsgBox myForm(c).Name
    End If
  Next c
End Sub
```

The conditional statement (`If...Then`) nested inside the `For...Next` loop tells Visual Basic to display the name of the active control only if it is a text box.

4. Run the GetTextBoxNames procedure.

Paired Statements

`For` and `Next` must be paired. If one is missing, Visual Basic generates the following error message: "For without Next."

FOR EACH...NEXT STATEMENT

When your procedure needs to loop through all of the objects of a collection or all of the elements in an array (arrays are the subject of the next chapter), the `For Each...Next` statement should be used. This loop does not require a counter variable. Visual Basic can figure out on its own how many times the loop should execute. The `For Each...Next` statement looks like this:

```
For Each element In Group
  statement1
  statement2
  statementN
Next [element]
```

`Element` is a variable to which all the elements of an array or collection will be assigned. This variable must be of the Variant data type for an array and of the Object data type for a collection. `Group` is the name of a collection or an array. Let's now see how to use the `For Each...Next` statement to print the names of the controls in the Customers form to the Immediate window.

Hands-On 6.6. Using the For Each...Next Statement

This hands-on exercise requires the completion of Steps 1 and 2 of Hands-On 6.5.

1. Ensure that the Customers form you created in Hands-On 6.5 is still open in Form view.
2. Switch to the Visual Basic Editor window and insert a new module.
3. In the Code window, enter the **GetControls** procedure shown here:

```
Sub GetControls()
  Dim myControl As Control
  Dim myForm As Form

  DoCmd.OpenForm "Customers"
  Set myForm = Screen.ActiveForm

  For Each myControl In myForm
    Debug.Print myControl.Name
```

```
  Next
End Sub
```

4. Run the GetControls procedure.
5. The results of the procedure you just executed will be displayed in the Immediate window. If the window is not visible, press **Ctrl+G** in the Visual Basic Editor window to open the Immediate window or choose **View | Immediate Window**.

EXITING LOOPS EARLY

Sometimes you might not want to wait until the loop ends on its own. It's possible that a user will enter the wrong data, a procedure will encounter an error, or perhaps the task will complete and there's no need to do additional looping. You can leave the loop early without reaching the condition that normally terminates it. Visual Basic has two types of `Exit` statements:

- The `Exit For` statement is used to end either a `For...Next` or a `For Each...Next` loop early.
- The `Exit Do` statement immediately exits any of the VBA `Do` loops.

The following hands-on exercise demonstrates how to use the `Exit For` statement to leave the `For Each...Next` loop early.

Hands-On 6.7. Early Exit from a Loop

1. In the Visual Basic Editor window, choose **Insert | Module**.
2. In the module's Code window, enter the following **GetControls2** procedure:

```
Sub GetControls2()
  Dim myControl As Control
  Dim myForm As Form

  DoCmd.OpenForm "Customers"
  Set myForm = Screen.ActiveForm

  For Each myControl In myForm
    Debug.Print myControl.Name
    If myControl.Name = "Address" Then
      Exit For
    End If
  Next
End Sub
```

3. Run the GetControls2 procedure.
   The GetControls2 procedure examines the names of the controls in the open Customers form. If Visual Basic encounters the control named "Address," it exits the loop.
4. Return to the Microsoft Access application window by choosing **File | Close and Return to Microsoft Access.**

Exiting Procedures

If you want to exit a subroutine earlier than normal, use the `Exit Sub` statement. If the procedure is a function, use the `Exit Function` statement instead.

NESTED LOOPS

So far in this chapter you have tried out various loops. Each procedure demonstrated the use of an individual looping structure. In programming practice, however, one loop is often placed inside another. Visual Basic allows you to "nest" various types of loops (`For` and `Do` loops) within the same procedure. When writing nested loops, you must make sure that each inner loop is completely contained inside the outer loop. Also, each loop must have a unique counter variable. When you use nesting loops, you can often execute specific tasks more effectively.

The GetFormsAndControls procedure shown in the following hands-on exercise illustrates how one `For Each...Next` loop is nested within another `For Each...Next` loop.

Hands-On 6.8. Using Nested Loops

1. Import the **Employees** table from the **Northwind 2007.accdb** database located in the Access2013_HandsOn folder. To do this, click **Access** in the Import section of the External Data tab. In the File name text box of the Get External Data dialog box, enter **C:\Access2013_HandsOn\Northwind 2007.accdb** and click **OK**. In the Import Objects dialog box, select the **Employees** table and click **OK**. Click **Close** to exit the Get External Data dialog box.
2. Now, create a simple **Employees** form based on the Employees table. To do this, select the Employees table in the navigation pane by clicking on its name.

Next, click the **Form** button in the Forms section of the Create tab. Access creates a simple Employees data entry form.

3. Leave the Employees form in Form view and press **Alt+F11** to switch to the Visual Basic Editor window.
4. Choose **Insert | Module** to add a new module. In the module's Code window, enter the **GetFormsAndControls** procedure shown here:

```
Sub GetFormsAndControls()
  Dim accObj As AccessObject
  Dim myControl As Control

  For Each accObj In CurrentProject.AllForms
    Debug.Print accObj.Name & " Form"
    If Not accObj.IsLoaded Then
      DoCmd.OpenForm accObj.Name
    End If
    For Each myControl In Forms(accObj.Name).Controls
      Debug.Print Chr(9) & myControl.Name
    Next
    DoCmd.Close , , acSaveYes
  Next
End Sub
```

5. Run the GetFormsAndControls procedure.

The GetFormsAndControls procedure uses two `For Each...Next` loops to print the name of each currently open form and its controls to the Immediate window. To enumerate through the form's controls, the form must be open. Notice the use of the Access built-in function `IsLoaded`. The procedure will only open the form if it is not yet loaded. The control names are indented in the Immediate window using the `Chr(9)` function. This is like pressing the Tab key once. To get the same result, you can replace `Chr(9)` with a VBA constant: `vbTab`.

After reading the names of the controls, the form is closed and the next form is processed in the same manner. The procedure ends when no more forms are found in the AllForms collection of CurrentProject.

6. Choose **File | Save Chap06** to save changes to the modules.
7. Choose **File | Close and Return to Microsoft Access**.
8. Close the **Chap06.accdb** database and click **Yes** when prompted to save changes. You do not need to save Table1 that Access automatically created for you when you chose to create a blank desktop database.
9. Exit Microsoft Access.

CHAPTER SUMMARY

In this chapter, you learned how to repeat certain groups of statements in VBA procedures by using loops. While working with several types of loops, you saw how each loop performs repetitions in a slightly different way. As you gain experience, you'll find it easier to choose the appropriate flow control structure for your task.

In the following chapters of this book, there are many additional examples of using loops. In the next chapter, for instance, you will see how, by using arrays and nested loops, you can create your own VBA procedure to pick lottery numbers. The next chapter also shows you how to code procedures that require a large number of variables.

Part 3

Keeping Track of Multiple Values

Although you can use individual variables to store data while your VBA code is executing, many advanced VBA procedures will require that you implement more efficient methods of keeping track of multiple values.

In this part of the book, you'll learn about working with groups of variables (arrays) and storing data in collections.

Chapter 7 Working with Arrays

In previous chapters, you worked with many VBA procedures that used variables to hold specific information about an object, property, or value. For each single value you wanted your procedure to manipulate, you declared a variable. But what if you have a series of values? If you had to write a VBA procedure to deal with larger amounts of data, you would have to create enough variables to handle all of the data. Can you imagine the nightmare of storing currency exchange rates for all the countries in the world in your program? To create a table to hold the necessary data, you'd need at least three variables for each country: country name, currency name, and exchange rate. Fortunately, Visual Basic has a way to get around this problem. By clustering the related variables together, your VBA procedures can manage a large amount of data with ease. In this chapter, you'll learn how to manipulate lists and tables of data with arrays.

WHAT IS AN ARRAY?

In Visual Basic, an *array* is a special type of variable that represents a group of similar values that are of the same data type (String, Integer, Currency, Date, etc.). The two most common types of arrays are one-dimensional arrays (lists) and two-dimensional arrays (tables).

A one-dimensional array is sometimes referred to as a *list*. A shopping list, a list of the days of the week, and an employee list are examples of one-dimensional arrays or, simply, numbered lists. Each element in the list has an index value that allows you to access that element. For example, in the following illustration we have a one-dimensional array of six elements indexed from 0 to 5:

(0)	(1)	(2)	(3)	(4)	(5)

You can access the third element of this array by specifying index (2). By default, the first element of an array is indexed zero (0). You can change this behavior by using the `Option Base 1` statement or by explicitly coding the lower bound of your array as explained later in this chapter.

All elements of the array should be of the same data type. In other words, if you declare an array to hold textual data you cannot store in it both strings and integers. If you want to store values of *different* data types in the same array, you must declare the array as Variant as discussed later. Following are two examples of one-dimensional arrays: an array named `cities` that is populated with text (String data type—$) and an array named `lotto` that contains six lottery numbers stored as integers (Integer data type—%).

A one-dimensional array: cities$

cities(0)	Baltimore
cities(1)	Atlanta
cities(2)	Boston
cities(3)	Washington
cities(4)	New York
cities(5)	Trenton

A one-dimensional array: lotto%

lotto(0)	25
lotto(1)	4
lotto(2)	31
lotto(3)	22
lotto(4)	11
lotto(5)	5

As you can see, the contents assigned to each array element match the array type. Storing values of different data types in the same array requires that you declare the array as Variant. You will learn how to declare arrays in the next section.

A two-dimensional array may be thought of as a table or matrix. The position of each element in a table is determined by its row and column numbers.

For example, an array that holds the yearly sales data for each product your company sells has two dimensions: the product name and the year. The following is a diagram of an empty two-dimensional array.

(0,0)	(0,1)	(0,2)	(0,3)	(0,4)	(0,5)
(1,0)	(1,1)	(1,2)	(1,3)	(1,4)	(1,5)
(2,0)	(2,1)	(2,2)	(2,3)	(2,4)	(2,5)
(3,0)	(3,1)	(3,2)	(3,3)	(3,4)	(3,5)
(4,0)	(4,1)	(4,2)	(4,3)	(4,4)	(4,5)
(5,0)	(5,1)	(5,2)	(5,3)	(5,4)	(5,5)

You can access the first element in the second row of this two-dimensional array by specifying indices (1, 0). Following are two examples of two-dimensional arrays: an array named `yearlyProductSales` that stores yearly product sales using the Currency data type (@) and an array named `exchange` (of Variant data type) that stores the name of the country, its currency, and the U.S. dollar exchange rate.

A two-dimensional array: yearlyProductSales@

Walking Cane (0,0)	$25,023 (0,1)
Pill Crusher (1,0)	$64,085 (1,1)
Electric Wheelchair (2,0)	$345,016 (2,1)
Folding Walker (3,0)	$85,244 (3,1)

A two-dimensional array: exchange (not actual rates)

Japan (0,0)	Japanese Yen (0,1)	108.83 (0,2)
Australia (1,0)	Australian Dollar (1,1)	1.28601 (1,2)
Canada (2,0)	Canadian Dollar (2,1)	1.235 (2,2)
Norway (3,0)	Norwegian Krone (3,1)	6.4471 (3,2)
Europe (4,0)	Euro (4,1)	0.816993 (4,2)

In these examples, the `yearlyProductSales` array can hold a maximum of 8 elements (4 rows * 2 columns = 8) and the `exchange` array will allow a maximum of 15 elements (5 rows * 3 columns = 15).

Although VBA arrays can have up to 60 dimensions, most people find it difficult to picture dimensions beyond 3D. A three-dimensional array is an array of two-dimensional arrays (tables) where each table has the same number of rows and columns. A three-dimensional array is identified by three indices: table, row, and column. The first element of a three-dimensional array is indexed (0, 0, 0).

DECLARING ARRAYS

Because an array is a variable, you must declare it in a similar way that you declare other variables (by using the keywords `Dim`, `Private`, or `Public`). For fixed-length arrays, the array bounds are listed in parentheses following the variable name. The *bounds* of an array are its lowest and highest indices. If a variable-length, or dynamic, array is being declared, the variable name is followed by an empty pair of parentheses.

The last part of the array declaration is the definition of the data type that the array will hold. An array can hold any of the following data types: Integer, Long, Single, Double, Variant, Currency, String, Boolean, Byte, or Date. Let's look at some examples:

Array Declaration (one-dimensional)	Description
`Dim cities(5) as String`	Declares a 6-element array, indexed 0 to 5
`Dim lotto(1 To 6) as String`	Declares a 6-element array, indexed 1 to 6
`Dim supplies(2 To 11)`	Declares a 10-element array, indexed 2 to 11
`Dim myIntegers(-3 To 6)`	Declares a 10-element array, indexed –3 to 6
`Dim dynArray() as Integer`	Declares a variable-length array whose bounds will be determined at runtime (see examples later in this chapter)

Array Declaration (two-dimensional)	Description
`Dim exchange(4,2) as Variant`	Declares a two-dimensional array (five rows by three columns)

Array Declaration (two-dimensional)	Description
`Dim yearlyProductSales(3, 1) as Currency`	Declares a two-dimensional array (four rows by two columns)
`Dim my2Darray(1 To 3, 1 To 7) as Single`	Declares a two-dimensional array (three rows indexed 1 to 3 by seven columns indexed 1 to 7)

When you declare an array, Visual Basic automatically reserves enough memory space for it. The amount of memory allocated depends on the array's size and data type. For a one-dimensional array with six elements, Visual Basic sets aside 12 bytes—2 bytes for each element of the array (recall that the size of the Integer data type is 2 bytes, hence 2 * 6 = 12). The larger the array, the more memory space is required to store the data. Because arrays can eat up a lot of memory and impact your computer's performance, it's recommended that you declare arrays with only as many elements as you think you'll use.

What Is an Array Variable?

An *array* is a group of variables that have a common name. While a typical variable can hold only one value, an *array variable* can store a large number of individual values. You refer to a specific value in the array by using the array name and an index number.

Subscripted Variables

The numbers inside the parentheses of the array variables are called *subscripts*, and each individual variable is called a subscripted variable or element. For example, `cities(5)` is the sixth subscripted variable (element) of the array `cities()`.

ARRAY UPPER AND LOWER BOUNDS

By default VBA assigns zero (0) to the first element of the array. Therefore, number 1 represents the second element of the array, number 2 represents the third, and so on. With numeric indexing starting at 0, the one-dimensional array `cities(5)` contains six elements numbered from 0 to 5. If you'd rather start counting your array's elements at 1, you can explicitly specify a lower bound of the array by using an `Option Base 1` statement. This instruction must be placed in the declaration section at the top of a VBA module before any `Sub` statements. If you don't specify `Option Base 1` in a procedure that uses arrays, VBA assumes that the statement `Option Base 0` is to be used and begins indexing your array's elements at 0. If you'd rather not use the `Option Base 1` statement and still have the array indexing start at a number other than 0, you must specify the bounds of an array when declaring the array variable. As mentioned in the previous section, the *bounds* of an array are its lowest and highest indices. Let's take a look at the following example:

```
Dim cities(3 To 6) As Integer
```

This statement declares a one-dimensional array with four elements. The numbers enclosed in parentheses after the array name specify the lower (3) and upper (6) bounds of the array. The index of the first element of this array is 3, the second 4, the third 5, and the fourth 6. Notice the keyword `To` between the lower and upper indices.

INITIALIZING AN ARRAY

After you declare an array, you must assign values to its elements. This is often referred to as "initializing an array," "filling an array," or "populating an array." The three methods you can use to load data into an array are discussed in this section.

Filling an Array Using Individual Assignment Statements

Assume you want to store the names of your six favorite cities in a one-dimensional array named `cities`. After declaring the array with the `Dim` statement:

```
Dim cities(5) as String
```

or

```
Dim cities$(5)
```

you can assign values to the array variable like this:

```
cities(0) = "Baltimore"
cities(1) = "Atlanta"
```

```
cities(2) = "Boston"
cities(3) = "San Diego"
cities(4) = "New York"
cities(5) = "Denver"
```

Filling an Array Using the Array Function

VBA's built-in `Array` function returns an array of Variants. Because Variant is the default data type, the `As Variant` clause is optional in the array variable declaration:

```
Dim cities() as Variant
```

or

```
Dim cities()
```

Notice that you don't specify the number of elements between the parentheses.

Next, use the `Array` function as shown here to assign values to your `cities` array:

```
cities = Array("Baltimore", "Atlanta", "Boston", _
"San Diego", "New York", "Denver")
```

When using the `Array` function to populate a six-element array like `cities`, the lower bound of the array is 0 or 1 and the upper bound is 5 or 6, depending on the setting of `Option Base` (see the previous section titled "Array Upper and Lower Bounds").

Filling an Array Using the For...Next Loop

The easiest way to learn how to use loops to populate an array is by writing a procedure that fills an array with a specific number of integer values. Let's look at the following example procedure:

```
Sub LoadArrayWithIntegers()
  Dim myIntArray(1 To 10) As Integer
  Dim i As Integer

  ' Initialize random number generator
  Randomize

  ' Fill the array with 10 random numbers between 1 and 100
  For i = 1 To 10
    myIntArray(i) = Int((100 * Rnd) + 1)
  Next

  ' Print array values to the Immediate window
  For i = 1 To 10
```

```
    Debug.Print myIntArray(i)
  Next
End Sub
```

This procedure uses a `For...Next` loop to fill `myIntArray` with 10 random numbers between 1 and 100. The second loop is used to print out the values from the array. Notice that the procedure uses the `Rnd` function to generate a random number. This function returns a value less than 1 but greater than or equal to 0. You can try it out in the Immediate window by entering:

```
x=rnd
?x
```

Before calling the `Rnd` function, the `LoadArrayWithIntegers` procedure uses the `Randomize` statement to initialize the random number generator. To become more familiar with the `Randomize` statement and `Rnd` function, be sure to follow up with the Access online help. For an additional example of using loops, `Randomize`, and `Rnd`, see Hands-On 7.4.

USING ARRAYS IN VBA PROCEDURES

Having learned the basics of array variables, let's write a couple of VBA procedures to make arrays a part of your new skill set. The procedure in Hands-On 7.1 uses a one-dimensional array to programmatically display a list of six North American cities.

Please note files for the "Hands-On" project may be found on the companion CD-ROM.

Hands-On 7.1. Using a One-Dimensional Array

1. Start Microsoft Access 2013 and create a new Microsoft Access 2013 database named **Chap07.accdb** in your **C:\Access2013_ByExample** folder.
2. Once your new database is opened, press **Alt+F11** to switch to the Visual Basic Editor window.
3. Choose **Insert | Module** to add a new standard module.
4. In the Module1 Code window, enter the following **FavoriteCities** procedure. Be sure to enter the `Option Base 1` statement at the top of the module.

```
Option Base 1
Sub FavoriteCities()
  ' declare the array
  Dim cities(6) As String
```

```
    ' assign the values to array elements
    cities(1) = "Baltimore"
    cities(2) = "Atlanta"
    cities(3) = "Boston"
    cities(4) = "San Diego"
    cities(5) = "New York"
    cities(6) = "Denver"

    ' display the list of cities
    MsgBox cities(1) & Chr(13) & cities(2) & Chr(13) _
     & cities(3) & Chr(13) & cities(4) & Chr(13) _
     & cities(5) & Chr(13) & cities(6)
End Sub
```

5. Choose **Run | Run Sub/UserForm** to execute the FavoriteCities procedure. Before the FavoriteCities procedure begins, the default indexing for an array is changed. Notice the `Option Base 1` statement at the top of the module window before the `Sub` statement. This statement tells Visual Basic to assign the number 1 instead of the default 0 to the first element of the array. The array `cities()` is declared with six elements of the String data type. Each element of the array is then assigned a value. The last statement in this procedure uses the `MsgBox` function to display the list of cities in a message box. When you run this procedure, each city name will appear on a separate line (see Figure 7.1). You can change the order of the displayed data by switching the index values.

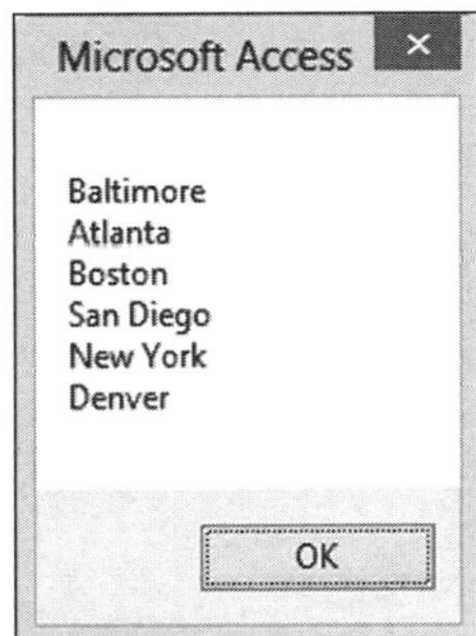

FIGURE 7.1. You can display the elements of a one-dimensional array with the MsgBox function.

6. Click **OK** to close the message box.
7. On your own, modify the FavoriteCities procedure so that it displays the names of the cities in reverse order (from 6 to 1).

The Range of the Array

The spread of the elements specified by the Dim statement is called the *range* of the array. For example: `Dim mktgCodes(5 To 15)`.

ARRAYS AND LOOPING STATEMENTS

Several of the looping statements you learned about in Chapter 6 (`For...Next` and `For Each...Next`) will come in handy now that you're ready to perform such tasks as populating an array and displaying the elements of an array. It's time to combine the skills you've learned so far.

How can you rewrite the FavoriteCities procedure so each city name is shown in a separate message box? To answer this question, notice how in the FavoriteCities2 procedure in Hands-On 7.2 we are replacing the last statement of the original procedure with the `For Each...Next` loop.

Hands-On 7.2. Using the For Each...Next Statement to List the Array Elements

1. In the Visual Basic Editor window, insert a new module.
2. Enter the **FavoriteCities2** procedure in the Code window. Be sure to enter the `Option Base 1` statement at the top of the module.

```
Option Base 1
Sub FavoriteCities2()
  ' declare the array
  Dim cities(6) As String
  Dim city As Variant

  ' assign the values to array elements
  cities(1) = "Baltimore"
  cities(2) = "Atlanta"
  cities(3) = "Boston"
  cities(4) = "San Diego"
  cities(5) = "New York"
  cities(6) = "Denver"
  ' display the list of cities in separate messages
  For Each city In cities
    MsgBox city
  Next
End Sub
```

3. Choose **Run | Run Sub/UserForm** to execute the FavoriteCities2 procedure. Notice that the `For Each...Next` loop uses the variable `city` of the Variant data type. As you recall from the previous chapter, the `For Each...Next` loop allows you to loop through all of the objects in a collection or all of the elements of an array and perform the same action on each object or element. When you run the FavoriteCities2 procedure, the loop will execute as many times as there are elements in the array.

In Chapter 4, you practiced passing arguments as variables to subroutines and functions. The CityOperator procedure in Hands-On 7.3 demonstrates how you can pass elements of an array to another procedure.

Hands-On 7.3. Passing Elements of an Array to Another Procedure

1. In the Visual Basic Editor window, insert a new module.
2. Enter the following two procedures (**CityOperator** and **Hello**) in the module's Code window. Be sure to enter the `Option Base 1` statement at the top of the module.

```
Option Base 1
Sub CityOperator()
  ' declare the array
  Dim cities(6) As String
  ' assign the values to array elements
  cities(1) = "Baltimore"
  cities(2) = "Atlanta"
  cities(3) = "Boston"
  cities(4) = "San Diego"
  cities(5) = "New York"
  cities(6) = "Denver"

  ' call another procedure and pass
  ' the array as argument
  Hello cities()
End Sub

Sub Hello(cities() As String)
  Dim counter As Integer

  For counter = 1 To 6
    MsgBox "Hello, " & cities(counter) & "!"
  Next
End Sub
```

Notice that the last statement in the CityOperator procedure calls the Hello procedure and passes to it the array `cities()` that holds the names of our favorite cities. Also notice that the declaration of the Hello procedure includes an array type argument—`cities()`—passed to this procedure as String. In order to iterate through the elements of an array, you need to know how many elements are included in the passed array. You can easily retrieve this information via two array functions—`LBound` and `UBound`. These functions are discussed later in this chapter. In this procedure example, `LBound(cities())` will return 1 as the first element of the array, and `UBound(cities())` will return 6 as the last element of the `cities()` array. Therefore, the statement `For counter = LBound(cities()) To UBound(cities())` will boil down to `For counter = 1 To 6`.

3. Execute the CityOperator procedure (choose **Run | Run Sub/UserForm**).

Passing array elements from a subroutine to a subroutine or function procedure allows you to reuse the same array in many procedures without unnecessary duplication of the program code.

Here's how you can put to work your newly acquired knowledge about arrays and loops in real life. If you're an avid lotto player who is getting tired of picking your own lucky numbers, have Visual Basic do the picking. The Lotto procedure in Hands-On 7.4 populates an array with six numbers from 1 to 54. You can adjust this procedure to pick numbers from any range.

Hands-On 7.4. Using Arrays and Loops in Real Life

1. In the Visual Basic Editor window, insert a new module.
2. Enter the following **Lotto** procedure in the module's Code window:

```
Sub Lotto()
  Const spins = 6
  Const minNum = 1
  Const maxNum = 54
  Dim t As Integer ' looping variable in outer loop
  Dim i As Integer ' looping variable in inner loop
  Dim myNumbers As String ' string to hold all picks
  Dim lucky(spins) As String ' array to hold generated picks

  myNumbers = ""
  For t = 1 To spins
    Randomize
    lucky(t) = Int((maxNum - minNum + 1) * Rnd) + minNum

    ' check if this number was picked before
```

```
      For i = 1 To (t - 1)
        If lucky(t) = lucky(i) Then
          lucky(t) = Int((maxNum - minNum + 1) * Rnd) + minNum
          i = 0
        End If
      Next i
        MsgBox "Lucky number is " & lucky(t), , "Lucky number " & t
        myNumbers = myNumbers & " -" & lucky(t)
    Next t
    MsgBox "Lucky numbers are " & myNumbers, , "6 Lucky Numbers"
  End Sub
```

The `Randomize` statement initializes the random number generator. The instruction `Int((maxNum - minNum + 1) * Rnd + minNum)` uses the `Rnd` function to generate a random value from the specified `minNum` to `maxNum`. The `Int` function converts the resulting random number to an integer. Instead of assigning constant values for `minNum` and `maxNum`, you can use the `InputBox` function to get these values from the user.

The inner `For...Next` loop ensures that each picked number is unique—it may not be any one of the previously picked numbers. If you omit the inner loop and run this procedure multiple times, you'll likely see some occurrences of duplicate numbers.

3. Execute the Lotto procedure (choose **Run | Run Sub/UserForm**) to get the computer-generated lottery numbers.

Initial Value of an Array Element

Until a value is assigned to an element of an array, the element retains its default value. Numeric variables have a default value of zero (0), and string variables have a default value of empty string ("").

Passing Arrays between Procedures

When an array is declared in a procedure, it is local to this procedure and unknown to other procedures. However, you can pass the local array to another procedure by using the array's name followed by an empty set of parentheses as an argument in the calling statement. For example, the statement `Hello cities()` calls the procedure named Hello and passes to it the array `cities`.

USING A TWO-DIMENSIONAL ARRAY

Now that you know how to programmatically produce a list (a one-dimensional array), it's time to take a closer look at how you can work with tables of data. The following procedure creates a two-dimensional array that will hold country name, currency name, and exchange rate for three countries.

Hands-On 7.5. Using a Two-Dimensional Array

1. In the Visual Basic Editor window, insert a new module.
2. Enter the **Exchange** procedure in the module's Code window:

```
Sub Exchange()
  Dim t As String
  Dim r As String
  Dim Ex(3, 3) As Variant

  t = Chr(9) & Chr(9) ' 2 Tabs
  r = Chr(13) ' Enter

  Ex(1, 1) = "Japan"
  Ex(1, 2) = "Yen"
  Ex(1, 3) = 81.0379
  Ex(2, 1) = "Europe"
  Ex(2, 2) = "Euro"
  Ex(2, 3) = 0.698422
  Ex(3, 1) = "Canada"
  Ex(3, 2) = "Dollar"
  Ex(3, 3) = 0.966127

  MsgBox "Country " & t & "Currency" & t & _
   "1 USD" & r & r _
   & Ex(1, 1) & t & Ex(1, 2) & t & Ex(1, 3) & r _
   & Ex(2, 1) & t & Ex(2, 2) & t & Ex(2, 3) & r _
   & Ex(3, 1) & t & Ex(3, 2) & t & Ex(3, 3), , _
   "Exchange Rates"
End Sub
```

3. Execute the Exchange procedure (choose **Run | Run Sub/UserForm**).
When you run the Exchange procedure, you will see a message box with the information presented in three columns, as shown in Figure 7.2.

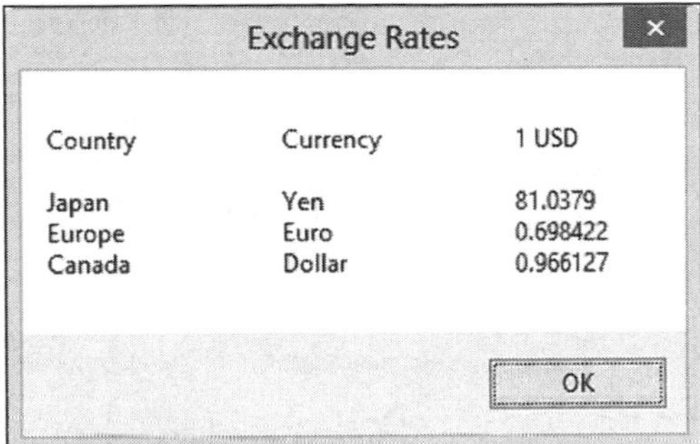

FIGURE 7.2. The text displayed in the message box can be custom formatted. (Note that these are fictitious exchange rates for demonstration only.)

4. Click **OK** to close the message box.

STATIC AND DYNAMIC ARRAYS

The arrays introduced thus far are static. A *static array* is an array of a specific size. You use a static array when you know in advance how big the array should be. The size of the static array is specified in the array's declaration statement. For example, the statement `Dim Fruits(10) As String` declares a static array called `Fruits` that is made up of 10 elements.

But what if you're not sure how many elements your array will contain? If your procedure depends on user input, the number of user-supplied elements might vary every time the procedure is executed. How can you ensure that the array you declare is not wasting memory?

You may recall that after you declare an array, VBA sets aside enough memory to accommodate the array. If you declare an array to hold more elements than what you need, you'll end up wasting valuable computer resources. The solution to this problem is making your arrays dynamic. A *dynamic array* is an array whose size can change. You use a dynamic array when the array size will be determined each time the procedure is run.

Fixed-Dimension Arrays

A static array contains a fixed number of elements. The number of elements in a static array will not change once it has been declared.

A dynamic array is declared by placing empty parentheses after the array name. For example:

```
Dim Fruits() As String
```

Before you use a dynamic array in your procedure, you must use the `ReDim` statement to dynamically set the lower and upper bounds of the array.

The `ReDim` statement redimensions arrays as the procedure code executes. The `ReDim` statement informs Visual Basic about the new size of the array. This statement can be used several times in the same procedure. Now let's write a procedure that demonstrates the use of a dynamic array.

Hands-On 7.6. Using a Dynamic Array

1. Insert a new module and enter the following **DynArray** procedure in the module's Code window:

```
Sub DynArray()
  Dim counter As Integer
  Dim myArray() As Integer ' declare a dynamic array
  ReDim myArray(5) ' specify the initial size of the array
  Dim myValues As String

  ' populate myArray with values
  For counter = 1 To 5
    myArray(counter) = counter + 1
    myValues = myValues & myArray(counter) & Chr(13)
  Next

  ' change the size of myArray to hold 10 elements
  ReDim Preserve myArray(10)

  ' add new values to myArray
  For counter = 6 To 10
    myArray(counter) = counter * counter
    myValues = myValues & myArray(counter) & Chr(13)
  Next counter

  MsgBox myValues
  For counter = 1 To 10
    Debug.Print myArray(counter)
  Next counter
End Sub
```

In the DynArray procedure, the statement `Dim myArray() As Integer` declares a dynamic array called `myArray`. Although this statement declares the array, it does not allocate any memory to the array. The first `ReDim` statement specifies the initial size of `myArray` and reserves for it 10 bytes of memory to hold its five elements. As you know, every Integer value requires 2 bytes of memory. The `For...Next` loop populates `myArray` with data and writes the array's elements to the variable `myValues`. The value of the variable `counter` equals 1 at the beginning of the loop.

The first statement in the loop (`myArray(counter) = counter +1`) assigns the value 2 to the first element of `myArray`. The second statement (`myValues = myValues & myArray(counter) & Chr(13)`) enters the current value of `myArray`'s element followed by a carriage return (`Chr(13)`) into the variable `myValues`. The statements inside the loop are executed five times. Visual Basic places each new value in the variable `myValues` and proceeds to the next statement: `ReDim Preserve myArray(10)`.

Normally, when you change the size of the array, you lose all the values that were in that array. When used alone, the `ReDim` statement reinitializes the array. However, you can append new elements to an existing array by following the `ReDim` statement with the `Preserve` keyword. In other words, the `Preserve` keyword guarantees that the redimensioned array will not lose its existing data.

The second `For...Next` loop assigns values to the 6th through 10th elements of `myArray`. This time the values of the array's elements are obtained by multiplication: `counter * counter`.

2. Execute the **DynArray** procedure (choose **Run | Run Sub/UserForm**).

Dimensioning Arrays

You can't assign a value to an array element until you have declared the array with the `Dim` or `ReDim` statement. (An exception to this is if you use the `Array` function discussed in the next section.)

ARRAY FUNCTIONS

You can manipulate arrays with five built-in VBA functions: `Array`, `IsArray`, `Erase`, `LBound`, and `UBound`. The following sections demonstrate the use of each of these functions in VBA procedures.

The Array Function

The `Array` function allows you to create an array during code execution without having to first dimension it. This function always returns an array of Variants. You can quickly place a series of values in a list by using the `Array` function.

The CarInfo procedure in the following hands-on exercise creates a fixed-size, one-dimensional, three-element array called `auto`.

Hands-On 7.7. Using the Array Function

1. Insert a new module and enter the following **CarInfo** procedure in the module's Code window:

```
Option Base 1
Sub CarInfo()
  Dim auto As Variant

  auto = Array("Ford", "Black", "2013")
  MsgBox auto(2) & " " & auto(1) & ", " & auto(3)

  auto(2) = "4-door"
  MsgBox auto(2) & " " & auto(1) & ", " & auto(3)
End Sub
```

2. Run the **CarInfo** procedure and examine the results.
When you run this procedure, you get two message boxes. The first one displays the following text: "Black Ford, 2013." After changing the value of the second array element, the second message box will say: "4-door Ford, 2013."

Be sure to enter `Option Base 1` *at the top of the module before running the CarInfo procedure. If this statement is missing in your module, Visual Basic will display runtime error 9—"Subscript out of range."*

The IsArray Function

The `IsArray` function lets you test whether a variable is an array. The `IsArray` function returns True if the variable is an array or False if it is not an array. Let's do another hands-on exercise.

Hands-On 7.8. Using the IsArray Function

1. Insert a new module and enter the code of the **IsThisArray** procedure in the module's Code window:

```
Sub IsThisArray()
  ' declare a dynamic array
  Dim tblNames() As String
  Dim totalTables As Integer
  Dim counter As Integer
  Dim db As Database

  Set db = CurrentDb

  ' count the tables in the open database
  totalTables = db.TableDefs.Count

  ' specify the size of the array
  ReDim tblNames(1 To totalTables)

  ' enter and show the names of tables
  For counter = 1 To totalTables - 1
    tblNames(counter) = db.TableDefs(counter).Name
    Debug.Print tblNames(counter)
  Next counter

  ' check if this is indeed an array
   If IsArray(tblNames) Then
   MsgBox "The tblNames is an array."
  End If
End Sub
```

2. Run the IsThisArray procedure to examine its results.
   When you run this procedure, the list of tables in the current database is written to the Immediate window. A message box displays whether the `tblNames` array is indeed an array.

The Erase Function

When you want to remove the data from an array, you should use the `Erase` function. This function deletes all the data held by static or dynamic arrays. In addition, the `Erase` function reallocates all of the memory assigned to a dynamic array. If a procedure has to use the dynamic array again, you must use the `ReDim` statement to specify the size of the array. The next hands-on exercise demonstrates how to erase the data from the array `cities`.

Hands-On 7.9. Removing Data from an Array

1. Insert a new module and enter the code of the **FunCities** procedure in the module's Code window:

```
' start indexing array elements at 1
Option Base 1

Sub FunCities()
  ' declare the array
  Dim cities(1 To 5) As String

  ' assign the values to array elements
  cities(1) = "Las Vegas"
  cities(2) = "Orlando"
  cities(3) = "Atlantic City"
  cities(4) = "New York"
  cities(5) = "San Francisco"

  ' display the list of cities
  MsgBox cities(1) & Chr(13) & cities(2) & Chr(13) _
   & cities(3) & Chr(13) & cities(4) & Chr(13) _
   & cities(5)

  Erase cities

  ' show all that was erased
  MsgBox cities(1) & Chr(13) & cities(2) & Chr(13) _
   & cities(3) & Chr(13) & cities(4) & Chr(13) _
   & cities(5)
End Sub
```

2. Run the FunCities procedure to examine its results.
3. Click **OK** to close the message box.
   Visual Basic should now display an empty message box because all values were deleted from the array by the `Erase` function.
4. Click **OK** to close the empty message box.

The LBound and UBound Functions

The `LBound` and `UBound` functions return whole numbers that indicate the lower bound and upper bound indices of an array.

Hands-On 7.10. Finding the Lower and Upper Bounds of an Array

1. Insert a new module and enter the code of the **FunCities2** procedure in the module's Code window:

```
Sub FunCities2()
  ' declare the array
  Dim cities(1 To 5) As String

  ' assign the values to array elements
  cities(1) = "Las Vegas"
  cities(2) = "Orlando"
  cities(3) = "Atlantic City"
  cities(4) = "New York"
  cities(5) = "San Francisco"

  ' display the list of cities
  MsgBox cities(1) & Chr(13) & cities(2) & Chr(13) _
   & cities(3) & Chr(13) & cities(4) & Chr(13) _
   & cities(5)

  ' display the array bounds
  MsgBox "The lower bound: " & LBound(cities) & Chr(13) _
   & "The upper bound: " & UBound(cities)
End Sub
```

2. Run the FunCities2 procedure.
3. Click **OK** to close the message box that displays the favorite cities.
4. Click **OK** to close the message box that displays the lower and upper bound indices.

   To determine the upper and lower indices in a two-dimensional array, you may want to add the following statements at the end of the Exchange procedure that was prepared in Hands-On 7.5 (add these lines just before the `End Sub` keywords):

```
MsgBox "The lower bound (first dimension) is " & LBound(Ex, 1) & "."
MsgBox "The upper bound (first dimension) is " & UBound(Ex, 1) & "."
MsgBox "The lower bound (second dimension) is " & LBound(Ex, 2) & "."
MsgBox "The upper bound (second dimension) is " & UBound(Ex, 2) & "."
```

When determining the lower and upper bound indices of a two-dimensional array, you must specify the dimension number: 1 for the first dimension and 2 for the second dimension.

ERRORS IN ARRAYS

When working with arrays, it's easy to make a mistake. If you try to assign more values than there are elements in the declared array, Visual Basic will display the error message "Subscript out of range" (see Figure 7.3).

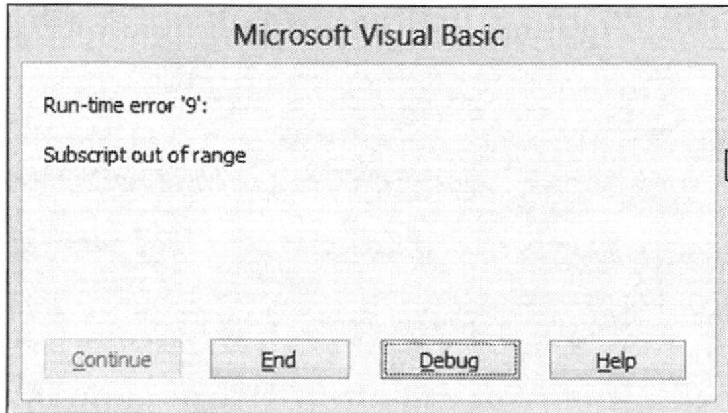

FIGURE 7.3. This error was caused by an attempt to access a nonexistent array element.

Suppose you declared a one-dimensional array that consists of three elements, and you are trying to assign a value to the fourth element. When you run the procedure, Visual Basic can't find the fourth element, so it displays the error message shown in Figure 7.3. If you click the Debug button, Visual Basic will highlight the line of code that caused the error (see Figure 7.4).

```
(General)                                   Zoo1
Option Compare Database
Option Explicit

Sub Zoo1()
   ' this procedure triggers an error
   ' "Subscript out of range"
   Dim zoo(3) As String
   Dim i As Integer
   Dim response As String

   i = 0
   Do
      i = i + 1
      response = InputBox("Enter a name of animal:")
      zoo(i) = response
   Loop i = 4 l response = ""
End Sub

Sub Zoo2()
   ' this procedure avoids the error
   ' "Subscript out of range"
   Dim zoo(3) As String
   Dim i As Integer
   Dim response As String
```

FIGURE 7.4. The statement that triggered the error shown in Figure 7.3. is highlighted.

The error *Subscript out of range* is often triggered in procedures using loops. The procedure Zoo1 shown in Hands-On 7.11 serves as an example of such a situation.

Hands-On 7.11. Understanding Errors in Arrays

1. Insert a new module and enter the following **Zoo1** and **Zoo2** procedures in the module's Code window:

```
Sub Zoo1()
  ' this procedure triggers an error
  ' "Subscript out of range"
  Dim zoo(3) As String
  Dim i As Integer
  Dim response As String

  i = 0
  Do
    i = i + 1
    response = InputBox("Enter a name of animal:")
    zoo(i) = response
  Loop Until response = ""
End Sub

Sub Zoo2()
  ' this procedure avoids the error
  ' "Subscript out of range"
  Dim zoo(3) As String
  Dim i As Integer
  Dim response As String

  i = 1
  Do While i >= LBound(zoo) And i <= UBound(zoo)
    response = InputBox("Enter a name of animal:")
    If response = "" Then Exit Sub
      zoo(i) = response
      Debug.Print zoo(i)
      i = i + 1
  Loop
End Sub
```

2. Run the Zoo1 procedure and enter your favorite animal names when prompted. Do not cancel the procedure until you see the error.
   While executing this procedure, when the variable `i` equals 4, Visual Basic will not be able to find the fourth element in a three-element array, so the error message will appear.
3. Click the **Debug** button in the error message.
   Visual Basic will highlight the code that caused the error.
4. Position the cursor over the variable `i` in the highlighted line of code to view the variable's value.
   Visual Basic displays: `i=4`
   Notice that at the top of the Zoo1 procedure `zoo` has been declared as an array containing only three elements:
   ```
   Dim zoo(3) As String
   ```
   Because Visual Basic could not find the fourth element, it displayed the "Subscript out of range" error.
   The Zoo2 procedure demonstrates how, by using the `LBound` and `UBound` functions introduced in the preceding section, you can avoid errors caused by an attempt to access a nonexistent array element.
5. Choose **Run | Reset** to terminate the debugging session and exit the procedure. You will learn more about debugging procedures in Chapter 9.

Another frequent error you may encounter while working with arrays is a *Type Mismatch* error. To avoid this error, keep in mind that each element of an array must be of the same data type. Therefore, if you attempt to assign to an element of an array a value that conflicts with the data type of the array, you will get a Type Mismatch error during the code execution. If you need to hold values of different data types in an array, declare the array as Variant.

PARAMETER ARRAYS

In Chapter 4, you learned that values can be passed between subroutines or functions as either required or optional arguments. If the passed argument is not absolutely required for the procedure to execute, the argument's name is preceded by the keyword `Optional`. Sometimes, however, you don't know in advance how many arguments you want to pass. A classic example is addition. One time you may want to add 2 numbers together, another time you may want to add 3, 10, or 15 numbers.

Using the keyword `ParamArray`, you can pass an array consisting of any number of elements to your subroutines and functions. The following hands-on exercise uses the AddMultipleArgs function to add as many numbers as you may require. This function begins with the declaration of an array `myNumbers`. Notice the use of the `ParamArray` keyword.

The array must be declared as type Variant, and it must be the last argument in the procedure definition.

Hands-On 7.12. Working with Parameter Arrays

1. Insert a new module and enter the following **AddMultipleArgs** function procedure in the module's Code window:

```
Function AddMultipleArgs(ParamArray myNumbers() As Variant)
  Dim mySum As Single
  Dim myValue As Variant

  For Each myValue In myNumbers
   mySum = mySum + myValue
  Next
  AddMultipleArgs = mySum
End Function
```

2. Choose **View | Immediate Window** and type the following instruction, then press **Enter** to execute it:

```
?AddMultipleArgs(1, 23.24, 3, 24, 8, 34)
```

When you press Enter, Visual Basic returns the total of all the numbers in the parentheses: 93.24. You can supply an unlimited number of arguments. To add more values, enter additional values in the parentheses after the function name in the Immediate window, then press Enter. Notice that each function argument must be separated by a comma.

PASSING ARRAYS TO FUNCTION PROCEDURES

You can pass an array to a function procedure and return an array from a function. For example, let's assume you have a list of countries. You want to convert the country names stored in your array to uppercase and keep the original array intact. You can delegate the conversion process to a function procedure. When the array is passed using the `ByVal` keyword, the function will work with the copy of the original array. Any modifications performed within the function will only affect the copy. Therefore, the array in the calling procedure will not be modified.

Hands-On 7.13. Passing an Array to a Function Procedure

1. Insert a new module and enter the following procedure and function in the module's Code window:

```
Sub ManipulateArray()
  Dim countries(1 To 6) As Variant
  Dim countriesUCase As Variant
  Dim i As Integer

  ' assign the values to array elements
  countries(1) = "Bulgaria"
  countries(2) = "Argentina"
  countries(3) = "Brazil"
  countries(4) = "Sweden"
  countries(5) = "New Zealand"
  countries(6) = "Denmark"

  countriesUCase = ArrayToUCase(countries)
  For i = 1 To 6
    Debug.Print countriesUCase(i)
    Debug.Print countries(i) & " (Original Entry)"
  Next i
End Sub

Public Function ArrayToUCase(ByVal myValues _
 As Variant) As String()
  Dim i As Integer
  Dim Temp() As String
  If IsArray(myValues) Then
    ReDim Temp(LBound(myValues) To UBound(myValues))
    For i = LBound(myValues) To UBound(myValues)
      Temp(i) = CStr(UCase(myValues(i)))
    Next i
    ArrayToUCase = Temp
  End If
 End Function
```

2. Run the ManipulateArray procedure and check its results in the Immediate window.

SORTING AN ARRAY

We all find it easier to work with sorted data. Some operations on arrays, like finding maximum and minimum values, require that the array is sorted. Once it is sorted, you can find the maximum value by assigning the upper bound index to the sorted array, as in the following:

```
y = myIntArray(UBound(myIntArray))
```

The minimum value can be obtained by reading the first value of the sorted array:

```
x = myIntArray(1)
```

So how can you sort an array? Hands-On 7.14 demonstrates how to delegate the sorting task to a classic bubble sort routine. A *bubble sort* is a comparison sort. To create a sorted set, you step through the list to be sorted, compare each pair of adjacent items, and swap them if they are in the wrong order. As a result of this sorting algorithm, the smaller values "bubble" to the top of the list. In the next procedure, we will sort the list of countries alphabetically in ascending order.

Hands-On 7.14. Sorting an Array

This Hands-On requires prior completion of Hands-On 7.13.

1. In the same module where you entered the ArrayToUCase function procedure, enter the following BubbleSort function procedure:

```
Sub BubbleSort(myArray As Variant)
  Dim i As Integer
  Dim j As Integer
  Dim uBnd As Integer
  Dim Temp As Variant
  uBnd = UBound(myArray)
    For i = LBound(myArray) To uBnd - 1
      For j = i + 1 To uBnd
        If UCase(myArray(i)) > UCase(myArray(j)) Then
          Temp = myArray(j)
          myArray(j) = myArray(i)
          myArray(i) = Temp
        End If
      Next j
    Next i
End Sub
```

2. Add the following statements to the ManipulateArray procedure, placing them just above the `For...Next` statement block (see Figure 7.5):

```
' call function to sort the array
  BubbleSort countriesUCase
```

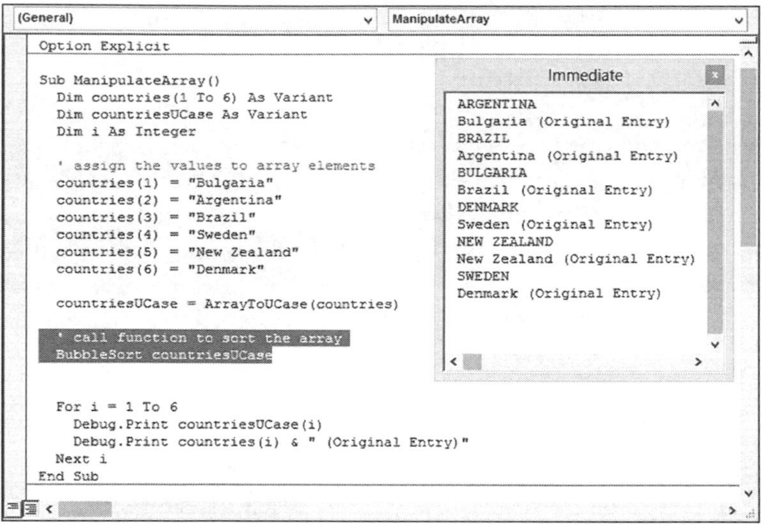
```
Option Explicit

Sub ManipulateArray()
  Dim countries(1 To 6) As Variant
  Dim countriesUCase As Variant
  Dim i As Integer

  ' assign the values to array elements
  countries(1) = "Bulgaria"
  countries(2) = "Argentina"
  countries(3) = "Brazil"
  countries(4) = "Sweden"
  countries(5) = "New Zealand"
  countries(6) = "Denmark"

  countriesUCase = ArrayToUCase(countries)

  ' call function to sort the array
  BubbleSort countriesUCase

  For i = 1 To 6
    Debug.Print countriesUCase(i)
    Debug.Print countries(i) & " (Original Entry)"
  Next i
End Sub
```


FIGURE 7.5. Calling the BubbleSort function procedure from the ManipulateArray procedure.

3. Run the ManipulateArray procedure and check its results in the Immediate window. Notice that the countries that appear in uppercase letters are shown in alphabetic order.
4. Choose **File | Save Chap07** and save changes to the modules when prompted.
5. Choose **File | Close and Return to Microsoft Access.**
6. Close the **Chap07.accdb** database and exit Microsoft Access.

CHAPTER SUMMARY

In this chapter, you learned how, by creating an array, you can write procedures that require a large number of variables. You worked with examples of procedures that demonstrated how to declare and use a one-dimensional array (list) and a two-dimensional array (table). You learned the difference between static and dynamic arrays. This chapter introduced you to five built-in VBA functions that are frequently used with arrays (`Array`, `IsArray`, `Erase`, `LBound`, and `UBound`), as well as the `ParamArray` keyword. You also learned how to pass one array and return another array from a function procedure. Finally, you saw how to sort an array. You now know all the VBA control structures that can make your code more intelligent: conditional statements, loops, and arrays.

In the next chapter, you will learn how to use collections instead of arrays to manipulate large amounts of data.

Chapter 8 WORKING WITH COLLECTIONS AND CLASS MODULES

Microsoft Access 2013 offers a large number of built-in objects that you can access from your VBA procedures to automate many aspects of your databases. You are not limited to using these built-in objects, however. VBA allows you to create your own objects and collections of objects, complete with their own methods and properties. While writing your own VBA procedures, you may come across a situation where there's no built-in collection to handle the task at hand. The solution is to create a custom collection object. You already know from the previous chapter how to work with multiple items of data by using dynamic or static arrays. Because collections have built-in properties and methods that allow you to add, remove, and count their elements, they make working with multiple data items much easier. In this chapter, you learn how to work with collections, including how to declare a custom Collection object. Using class modules to create user-defined objects will also be discussed. Before diving into theory and this chapter's hands-on examples, let's review the following terms:

Collection—An object that contains a set of related objects.
Class—A definition of an object that includes its name, properties, methods, and events. The class acts as a sort of object template from which an instance of an object is created at runtime.

Class module—A module that contains the definition of a class, including its property and method definitions.
Event—An action recognized by an object, such as a mouseclick or a keypress, for which you can define a response. Events can be triggered by a user action, a VBA statement, or the system.
Event procedure—A procedure that is automatically executed in response to an event triggered by the user, program code, or the system.
Form module—A module that contains the VBA code for all event procedures triggered by events occurring in a user form or its controls. A form module is a type of class module.
Instance—A specific object that belongs to a class is referred to as an *instance of the class.* When you create an instance, you create a new object that has the properties and methods defined by the class.
Module—A structure containing subroutine and function procedures that are available to other VBA procedures and are not related to any object in particular.

WORKING WITH COLLECTIONS

Collections are objects that contain other similar objects. For example, a Microsoft Access database has a collection of Tables, and each table has a collection of Fields and Indexes. In Microsoft Excel, all open workbooks belong to the Workbooks collection, and all the sheets in a particular workbook are members of the Worksheets collection. In Microsoft Word, all open documents belong to the Documents collection, and each paragraph in a document is a member of the Paragraphs collection.

No matter what collection you want to work with, you can do the following:

- Insert new items into the collection by using the `Add` method.

  The following example uses the Immediate window to create a collection named myTestCollection and adds three items to the collection. To try out these examples, type the statements in the Immediate window, then press Enter after each line:

  ```
  set myTestCollection = New Collection
  myTestCollection.Add "first member"
  myTestCollection.Add "second member"
  myTestCollection.Add "third member"
  ```

- Determine the number of items in the collection by using the Count property.

For example, when you type this statement in the Immediate window, then press Enter:

```
?myTestCollection.Count
```

it returns the total number of items stored in the `myTestCollection` object variable.

- Refer to a specific object in a collection by using an index value.

  For example, to find out the names of the collection members, you can type the following statement in the Immediate window, then press Enter:

  ```
  ?myTestCollection.Item(1)
  ```

  Because the `Item` method is a default method of the collection, you may omit it from the statement, as shown here:

  ```
  ?myTestCollection(1)
  ```

- Remove an object from a collection by using the `Remove` method.

  For example, to remove the first object from the `myTestCollection` object variable, enter the following statement, then press Enter:

  ```
  myTestCollection.Remove 1
  ```

- Cycle through every object in the collection by using the `For Each...Next` loop.

  For example, to remove all objects from the `myTestCollection` object variable, type the following looping structure in the Immediate window, then press Enter:

  ```
  For Each m in myTestCollection : myTestCollection.Remove 1 : Next
  ```

Note that a colon is used to separate one statement from the next. You can write two or more statements on a single line by separating them with a colon (:). This is very convenient when testing statements in the Immediate window. Because collections are reindexed, the preceding statement will remove the first member of the collection on each iteration. When you press Enter, `myTestCollection` should have zero objects. However, to be sure, type the following statement in the Immediate window, then press Enter:

```
?myTestCollection.Count
```

Now that you have learnt the basics of working with built-in collections, let's move on to declaring and using custom collections.

Declaring a Custom Collection

To create a user-defined collection, you should begin by declaring an object variable of the Collection type. This variable is declared with the `New` keyword in the `Dim` statement:

```
Dim collection Fruits As New Collection
```

Adding Objects to a Custom Collection

After you've declared the Collection object, you can insert new items into the collection by using the `Add` method. The objects with which you populate your collection do not have to be of the same data type. The `Add` method looks as follows:

```
object.Add item[, key, before, after]
```

For example, the following statement adds a new item to the previously declared Fruits collection:

```
Fruits.Add "apples"
```

You are only required to specify `object` and `item`. `object` is the collection name, such as Fruits. This is the same name that was used in the declaration of the Collection object. The Item such as "apples" is the object you want to add to the collection (Fruits).

Although the other arguments are optional, they are quite useful. It's important to understand that the items in a collection are automatically assigned numbers starting with 1. However, they can also be assigned a unique key value. Instead of accessing a specific item with an index (1, 2, 3, and so on) at the time an object is added to a collection, you can assign a key for that object. For instance, to identify an individual in a collection of students or employees, you could use Social Security numbers as a key. If you want to specify the position of the object in the collection, you should use either the `before` or `after` argument (but not both). The `before` argument is the object before which the new object is added. The `after` argument is the object after which the new object is added.

The NewEmployees procedure in the following hands-on exercise declares the custom Collection object called `colEmployees`.

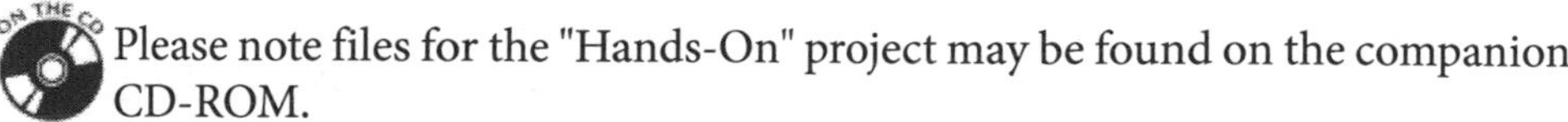
Please note files for the "Hands-On" project may be found on the companion CD-ROM.

Hands-On 8.1. Creating a Custom Collection

1. Start Microsoft Access 2013 and create a new database named **Chap08.accdb** in your **C:\Access2013_ByExample** folder.
2. Once your new database is opened, press **Alt+F11** to switch to the Visual Basic Editor window.
3. Choose **Insert | Module** to add a new standard module.
4. In the Module1 Code window, enter the following **NewEmployees** procedure. Be sure to enter the `Option Base 1` statement before this procedure.

```
Option Base 1      ' ensure that there is only one
                   ' Option Base 1 statement
                   ' at the top of the module

Sub NewEmployees()
  ' declare the employees collection
  Dim colEmployees As New Collection
  ' declare a variable to hold each element of a collection
  Dim emp As Variant

  ' Add 3 new employees to the collection
  With colEmployees
    .Add Item:="John Collins", Key:="128634456"
    .Add Item:="Mary Poppins", Key:="223998765"
    .Add Item:="Karen Loza", Key:="120228876", Before:=2
  End With

  ' list the members of the collection
  For Each emp In colEmployees
    Debug.Print emp
  Next
  MsgBox "There are " & colEmployees.Count & " employees."
End Sub
```

Note that the control variable used in the `For Each...Next` loop must be declared as Variant or Object. When you run this procedure, you will notice that the order of employee names stored in the `colEmployees` collection (as displayed in the Immediate window) may be different from the order in which these employees were entered in the program code. This is the result of using the optional `before` argument with Karen Loza's entry. This argument's value tells Visual Basic to place Karen before the second item in the collection.

5. Choose **Run | Run Sub/UserForm** to execute the NewEmployees procedure.

Removing Objects from a Custom Collection

Removing an item from a custom collection is as easy as adding an item. To remove an item, use the `Remove` method in the following format:

```
object.Remove index
```

`object` is the name of the custom collection that contains the object you want to remove. `index` is an expression specifying the position of the object in the collection.

To demonstrate the process of removing an item from a collection, let's work with the following hands-on exercise that modifies the NewEmployees procedure that you prepared in Hands-On 8.1.

Hands-On 8.2. Removing Objects from a Collection

This hands-on exercise requires the prior completion of Hands-On 8.1.

1. Add the following lines to the **NewEmployees** procedure just before the `End Sub` keywords:

```
' remove the third item from the collection
colEmployees.Remove 3
MsgBox colEmployees.Count & " employees remain."
```

2. Rerun the NewEmployees procedure.

Reindexing Collections

Collections are reindexed automatically when an item is removed. Therefore, to remove all items from a custom collection you can use 1 for the `Index` argument, as in the following example:

```
Do While myCollection.Count > 0
  myCollection.Remove Index:=1
Loop
```

CREATING CUSTOM OBJECTS

There are two module commands available in the Visual Basic Editor's Insert menu: Module and Class Module. So far you've used a standard module to create subprocedures and function procedures. You'll use the class module for the first time in this chapter to create a custom object and define its properties and methods.

Creating a new VBA object involves inserting a class module into your project and adding code to that module. However, before you do so you need a basic understanding of what a class is.

If you refer back to the list of terms at the beginning of this chapter, you will find out that the *class* is a sort of object template. A frequently used analogy is comparing an object class to a cookie cutter. Just like a cookie cutter defines what a particular cookie will look like, the definition of the class determines how a particular object should look and how it should behave. Before you can actually use an object class, you must first create a new *instance* of that class. Object instances are the cookies. Each object instance has the characteristics (properties and methods) defined by its class. Just as you can cut out many cookies using the same cookie cutter, you can create multiple instances of a class. You can change the properties of each instance of a class independently of any other instance of the same class.

A *class module* lets you define your own custom classes, complete with custom properties and methods. A *property* is an attribute of an object that defines one of its characteristics, such as shape, position, color, title, etc. A *method* is an action that the object can perform. You can create the properties for your custom objects by writing property procedures in a class module. The object methods are also created in a class module by writing subprocedures or function procedures.

After building your object in the class module, you can use it in the same way you use other built-in objects. You can also export the object class outside the VBA project to other VBA-capable applications.

Creating a Class

The following sections of this chapter walk you through the process of creating and working with a custom object called `CEmployee`. This object will represent an employee. It will have properties such as ID, FirstName, LastName, and Salary. It will also have a method to modify the current salary.

Custom Project 8.1. (Step 1) Creating a Class Module

1. In the Visual Basic Editor window, choose **Insert | Class Module.**
2. In the Project Explorer window, highlight the **Class1** module and use the Properties window to rename the class module **CEmployee** (see Figure 8.1).

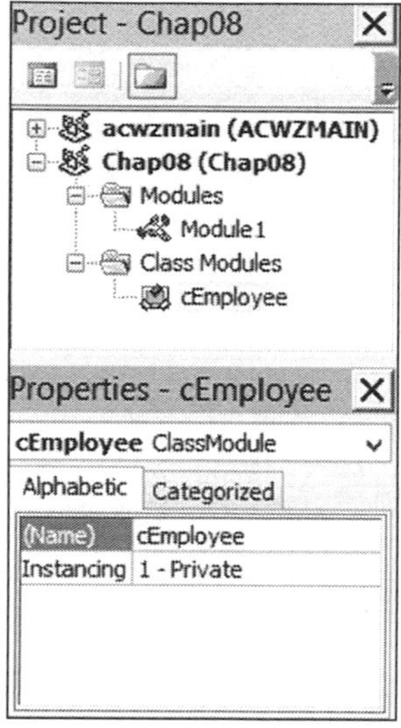

FIGURE 8.1. Use the Name property in the Properties window to rename the Class module.

Naming a Class Module

Every time you create a new class module, give it a meaningful name. Set the name of the class module to the name you want to use in your VBA procedures using the class. The name you choose for your class should be easily understood and should identify the "thing" the object class represents. As a rule, the object class name is prefaced with an uppercase "C."

Variable Declarations

After adding and renaming the class module, the next step is to declare the variables that will hold the data you want to store in the custom CEmployee object. Each item of data you want to store in an object should be assigned a variable. Class variables are called *data members* and are declared with the `Private` keyword. Using the `Private` keyword in a class module hides the data members and prevents other parts of the application from referencing them. Only the procedures within the class module in which the private variables were defined can modify the value of these variables.

Because the name of a variable also serves as a property name, use meaningful names for your object's data members. It's traditional to preface the class variable names with "m_" to indicate that they are data members of a class.

Custom Project 8.1. (Step 2) Declaring Class Members

1. Type the following declaration lines at the top of the **CEmployee** class module's code window:

```
Option Explicit
' declarations
Private m_LastName As String
Private m_FirstName As String
Private m_Salary As Currency
Private m_ID As String
```

Notice that the name of each data member variable begins with the prefix "m_."

Defining the Properties for the Class

Declaring the variables with the `Private` keyword ensures that they cannot be directly accessed from outside the object. This means that the VBA procedures outside the class module will not be able to set or read data stored in those variables. To enable other parts of your VBA application to set or retrieve the employee data, you must add special property procedures to the CEmployee class module. There are three types of property procedures:

- **Property Let**—This type of procedure allows other parts of the application to set the value of a property.
- **Property Get**—This type of procedure allows other parts of the application to get or read the value of a property.
- **Property Set**—This type of procedure is used instead of Property Let when setting the reference to an object.

Property procedures are executed when an object property needs to be set or retrieved. The Property Get procedure can have the same name as the Property Let procedure. You should create property procedures for each property of the object that can be accessed by another part of your VBA application.

The easiest of the three types of property statements to understand is the Property Get procedure. Let's examine the syntax of the property procedures by taking a close look at the Property Get LastName procedure.

Property procedures contain the following parts:

- A procedure declaration line
- An assignment statement
- The `End Property` keywords

A procedure declaration line specifies the name of the property and the data type:

```
Property Get LastName() As String
```

LastName is the name of the property and `As String` determines the data type of the property's return value.

An assignment statement is similar to the one used in a function procedure:

```
LastName = m_LastName
```

LastName is the name of the property and *m_LastName* is the data member variable that holds the value of the property you want to retrieve or set. The `m_LastName` variable should be defined with the `Private` keyword at the top of the class module. Here's the complete Property Get procedure:

```
Property Get LastName() As String
  LastName = m_LastName
End Property
```

The Property Get procedure can return a result from a calculation, like this:

```
Property Get Royalty()
  Royalty = (Sales * Percent) - Advance
End Property
```

The `End Property` keywords specify the end of the property procedure.

Immediate Exit from Property Procedures

Just as the `Exit Sub` and `Exit Function` keywords allow you to exit early from a subroutine or a function procedure, the `Exit Property` keywords give you a way to immediately exit from a property procedure. Program execution will continue with the statements following the statement that called the Property Get, Property Let, or Property Set procedure.

Creating the Property Get Procedures

The CEmployee class object has four properties that need to be exposed to VBA procedures that we will write later in a standard module named EmpOperations. When working with the CEmployee object, you would certainly like to get information about the employee ID, first and last name, and current salary.

Custom Project 8.1. (Step 3) Writing Property Get Procedures

1. Type the following Property Get procedures in the **CEmployee** class module, just below the declaration section that you entered in Step 2 of this custom project:

```
Property Get ID() As String
  ID = m_ID
End Property

Property Get LastName() As String
  LastName = m_LastName
End Property

Property Get FirstName() As String
  FirstName = m_FirstName
End Property

Property Get Salary() As Currency
  Salary = m_Salary
End Property
```

Notice that each employee information type requires a separate Property Get procedure. Each of the preceding Property Get procedures returns the current value of the property. Notice also how a Property Get procedure is similar to a function procedure. Similar to function procedures, the Property Get procedures contain an assignment statement. As you recall from Chapter 4, to return a value from a function procedure, you must assign it to the function's name.

Creating the Property Let Procedures

In addition to retrieving values stored in data members (private variables) with Property Get procedures, you must prepare corresponding Property Let procedures to allow other procedures to change the values of these variables as needed. The only time you don't define a Property Let procedure is when the value stored in a private variable is meant to be *read-only*.

Suppose you don't want the user to change the employee ID. To make the ID read-only, you simply don't write a Property Let procedure for it. Hence, the CEmployee class will only have three properties (LastName, FirstName, and Salary). Each of these properties will require a separate Property Let procedure. The employee ID will be assigned automatically with a return value from a function procedure.

Let's continue with our project and write the required Property Let procedures for our custom CEmployee object.

Custom Project 8.1. (Step 4) Writing Property Let Procedures

1. Type the following Property Let procedures in the **CEmployee** class module below the Property Get procedures:

```
Property Let LastName(L As String)
  m_LastName = L
End Property

Property Let FirstName(F As String)
  m_FirstName = F
End Property

Property Let Salary(ByVal dollar As Currency)
  m_Salary = dollar
End Property
```

The Property Let procedures require at least one parameter that specifies the value you want to assign to the property. This parameter can be passed by *value* (note the `ByVal` keyword in the preceding Property Let Salary procedure) or by *reference* (`ByRef` is the default). If you need a refresher on the meaning of these keywords, see the section titled "Passing Arguments by Reference and by Value" in Chapter 4.

The data type of the parameter passed to the Property Let procedure must be exactly the same data type as the value returned from the Property Get or Set procedure with the same name. Notice that the Property Let procedures have the same names as the Property Get procedures prepared in the preceding section. By skipping the Property Let procedure for the ID property, you created a read-only ID property that can be retrieved but not set.

Defining the Scope of Property Procedures

You can place the `Public`, `Private`, or `Static` keyword before the name of a property procedure to define its scope. To indicate that the Property Get procedure is accessible to procedures in all modules, use the following statement format:

```
Public Property Get FirstName() As String
```

To make the Property Get procedure accessible only to other procedures in the module where it is declared, use the following statement format:

```
Private Property Get FirstName() As String
```

To preserve the Property Get procedure's local variables between procedure calls, use the following statement format:

```
Static Property Get FirstName() As String
```

If not explicitly specified using either `Public` or `Private`, property procedures are public by default. Also, if the `Static` keyword is not used, the values of local variables are not preserved between procedure calls.

Creating the Class Methods

Apart from properties, objects usually have one or more methods. A *method* is an action that the object can perform. Methods allow you to manipulate the data stored in a class object. Methods are created with subroutines or function procedures. To make a method available outside the class module, use the `Public` keyword in front of the sub or function definition. The CEmployee object that you create in this chapter has one method that allows you to calculate the new salary. Assume that the employee salary can be increased or decreased by a specific percentage or amount.

Let's continue with our project by writing a class method that calculates the employee salary.

Custom Project 8.1. (Step 5) Writing Class Methods

1. Type the following **CalcNewSalary** function procedure in the **CEmployee** class module:

```
Public Function CalcNewSalary(choice As Integer, _
  curSalary As Currency, amount As Long) As Currency
```

```
    Select Case choice
      Case 1 ' by percent
        CalcNewSalary = curSalary + ((curSalary * amount) / 100)
      Case 2 ' by amount
        CalcNewSalary = curSalary + amount
    End Select
  End Function
```

The CalcNewSalary function defined with the `Public` keyword in a class module serves as a method for the CEmployee class. To calculate a new salary, a VBA procedure from outside the class module must pass three arguments: `choice`, `CurSalary`, and `amount`. The `choice` argument specifies the type of the calculation. Suppose you want to increase the employee salary by 5% or by $5.00. The first option will increase the salary by the specified percent, and the second option will add the specified amount to the current salary. The `curSalary` argument is the current salary figure for an employee, and `amount` determines the value by which the salary should be changed.

About Class Methods

- Only those methods that will be accessed from outside of the class should be declared as `Public`. All others should be declared as `Private`.
- Methods perform some operation on the data contained within the class.
- If a method needs to return a value, write a function procedure. Otherwise, create a subprocedure.

Creating an Instance of a Class

After typing all the necessary Property Get, Property Let, sub, or function procedures for your VBA application in the class module, you are ready to create a new instance of a class, which is called an *object*.

Before an object can be created, an object variable must be declared in a standard module to store the reference to the object. If the name of the class module is CEmployee, then a new instance of this class can be created with the following statement:

```
Dim emp As New CEmployee
```

The `emp` variable will represent a reference to an object of the CEmployee class. When you declare the object variable with the `New` keyword, VBA creates the object and allocates memory for it. However, the object isn't instanced until you refer to it in your procedure code by assigning a value to its property or by running one of its methods.

You can also create an instance of the object by declaring an object variable with the data type defined to be the class of the object, as in the following:

```
Dim emp As CEmployee
Set emp = New CEmployee
```

If you don't use the `New` keyword with the `Dim` statement, VBA does not allocate memory for your custom object until your procedure actually needs it.

Custom Project 8.1. (Step 6) Creating an Instance of a Class

1. Activate the Visual Basic Editor window and choose **Insert | Module** to add a standard module to your application.
2. Use the Name property in the Properties window to change the name of the new module to **EmpOperations.**
3. Type the following declarations at the top of the EmpOperations module:

```
Dim emp As New CEmployee
Dim CEmployee As New Collection
```

The first declaration statement (`Dim`) declares the variable `emp` as a new instance of the CEmployee class. The second statement declares a custom collection. The CEmployee collection will be used to store all employee data.

Event Procedures in the Class Module

An *event* is basically an action recognized by an object. Custom classes recognize only two events: `Initialize` and `Terminate`. These events are triggered when an instance of the class is created and destroyed, respectively. The `Initialize` event is generated when an object is created from a class (see the preceding section on creating an instance of a class).

In the CEmployee class example, the `Initialize` event will also fire the first time that you use the `emp` variable in code. Because the statements included inside the `Initialize` event are the first ones to be executed for the object before any properties are set or any methods are executed, the `Initialize` event is a good place to perform initialization of the objects created from the class. As you recall, the ID is read-only in the CEmployee class. You can use the `Initialize` event to assign a unique five-digit number to the `m_ID` variable.

The Class_Initialize procedure uses the following syntax:

```
Private Sub Class_Initialize()
  [code to perform tasks as the object is created goes here]
End Sub
```

The `Terminate` event occurs when all references to an object have been released. This is a good place to perform any necessary cleanup tasks. The Class_Terminate procedure uses the following syntax:

```
Private Sub Class_Terminate()
  [cleanup code goes here]
End Sub
```

To release an object variable from an object, use the following syntax:

```
Set objectVariable = Nothing
```

When you set the object variable to `Nothing`, the `Terminate` event is generated. Any code in this event is executed then.

CREATING THE USER INTERFACE

Implementing our custom CEmployee object requires that you design a form to enter and manipulate employee data.

Custom Project 8.1. (Step 7) Designing a User Form

1. Choose **File | Close and Return to Microsoft Access**.
2. Click the **Blank form** in the Forms section of the Create tab. Access will display a blank form in the Form view.
3. Switch to the form's Design view by choosing **Design View** from the Views section.
4. Save the form as **frmEmployeeSalaries**.
5. Use the tools in the Controls section of the Design tab to place controls on the form as shown in Figure 8.2.

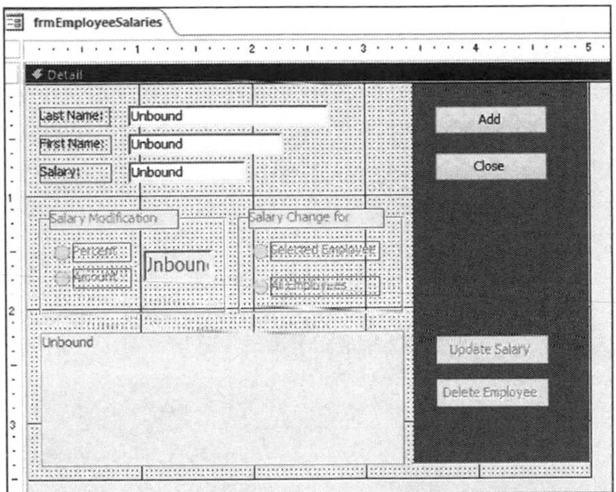

FIGURE 8.2. This form demonstrates the use of the CEmployee custom object.

6. Activate the property sheet and set the following properties for the form controls. To set the specified property, first click the control on the form to select it. Then, in the property sheet type the information shown in the Setting column next to the property indicated in the Property column.

Object	Property	Setting
Label1	Caption	Last Name
Text box next to the Last Name label	Name	txtLastName
Label2	Caption	First Name
Text box next to the First Name label	Name	txtFirstName
Label3	Caption	Salary
Text box next to the Salary label	Name	txtSalary
Option group 1	Name Caption	frSalaryMod Salary Modification
Text box in the option group titled "Salary Modification"	Name	txtRaise
Option button 1	Name Caption	optPercent Percent
Option button 2	Name Caption	optAmount Amount

(contd.)

Object	Property	Setting
Option group 2	Name Caption	frSalaryFor Salary Change for
Option button 3	Name Caption	optSelected Selected Employee
Option button 4	Name Caption	optAll All Employees
Listbox	Name Row Source Type Column Count Column Widths	lboxPeople Value List 4 0.5";0.9";0.7";0.5"
Command Button 1	Name Caption	cmdAdd Add
Command Button 2	Name Caption	cmdClose Close
Command Button 3	Name Caption	cmdUpdate Update Salary
Command Button 4	Name Caption	cmdDelete Delete Employee

Now that the form is ready, you need to write a few event procedures to handle various events, such as clicking a command button or loading the form.

Custom Project 8.1. (Step 8) Writing Event Procedures

1. Activate the Code window behind the form by choosing the **View Code** button in the **Tools** section of the **Design** tab.
2. Enter the following variable declarations at the top of the form's Code window:

```
' variable declarations
Dim choice As Integer
Dim amount As Long
```

Please ensure that the `Option Explicit` *statement appears at the top of the module, above the variable declaration statements.*

3. Type the following **UserForm_Initialize** procedure to enable or disable controls on the form:

```
Private Sub UserForm_Initialize()
    txtLastName.SetFocus
    cmdUpdate.Enabled = False
```

```
    cmdDelete.Enabled = False
    lboxPeople.Enabled = False
    frSalaryFor.Enabled = False
    frSalaryFor.Value = 0
    frSalaryMod.Enabled = False
    frSalaryMod.Value = 0
    txtRaise.Enabled = False
    txtRaise.Value = ""
End Sub
```

4. Type the following **Form_Load** event procedure:

```
Private Sub Form_Load()
  Call UserForm_Initialize
End Sub
```

When the form loads, the UserForm_Initialize procedure will run.

5. Enter the following **cmdAdd_Click** procedure to add the employee to the collection:

```
Private Sub cmdAdd_Click()
  Dim strLast As String
  Dim strFirst As String
  Dim curSalary As Currency

  ' Validate data entry
  If IsNull(txtLastName.Value) Or txtLastName.Value = "" _
   Or IsNull(txtFirstName.Value) Or txtFirstName.Value = "" _
   Or IsNull(txtSalary.Value) Or txtSalary.Value = "" Then
    MsgBox "Enter Last Name, First Name and Salary."
    txtLastName.SetFocus
    Exit Sub
  End If
  If Not IsNumeric(txtSalary) Then
    MsgBox "You must enter a value for the Salary."
    txtSalary.SetFocus
    Exit Sub
  End If
  If txtSalary < 0 Then
    MsgBox "Salary cannot be a negative number."
    Exit Sub
  End If
```

```
    ' assign text box values to variables
    strLast = txtLastName
    strFirst = txtFirstName
    curSalary = txtSalary

    ' enable buttons and other controls
    cmdUpdate.Enabled = True
    cmdDelete.Enabled = True
    lboxPeople.Enabled = True
    frSalaryFor.Enabled = True
    frSalaryMod.Enabled = True
    txtRaise.Enabled = True
    txtRaise.Value = ""
    lboxPeople.Visible = True

    ' enter data into the CEmployees collection
    EmpOperations.AddEmployee strLast, strFirst, curSalary

    ' update listbox
    lboxPeople.RowSource = GetValues

    ' delete data from text boxes
    txtLastName = ""
    txtFirstName = ""
    txtSalary = ""
    txtLastName.SetFocus
End Sub
```

The cmdAdd_Click procedure starts off by validating the user's input in the Last Name, First Name, and Salary text boxes. If the user entered correct data, the text box values are assigned to the variables `strLast`, `strFirst`, and `curSalary`. Next, a number of statements enable buttons and other controls on the form so that the user can work with the employee data. The following statement calls the AddEmployee procedure in the EmpOperations standard module and passes the required parameters to it:

```
EmpOperations.AddEmployee strLast, strFirst, curSalary
```

Once the employee is entered into the collection, the employee data is added to the listbox (see Figure 8.3) with the following statement:

```
lboxPeople.RowSource = GetValues
```

GetValues is the name of a function procedure in the EmpOperations module (see Step 12 further on). This function cycles through the CEmployee collection to create a string of values for the listbox row source.

The cmdAdd_Click procedure ends by clearing the text boxes, then setting the focus to the Last Name text box so the user can enter new employee data.

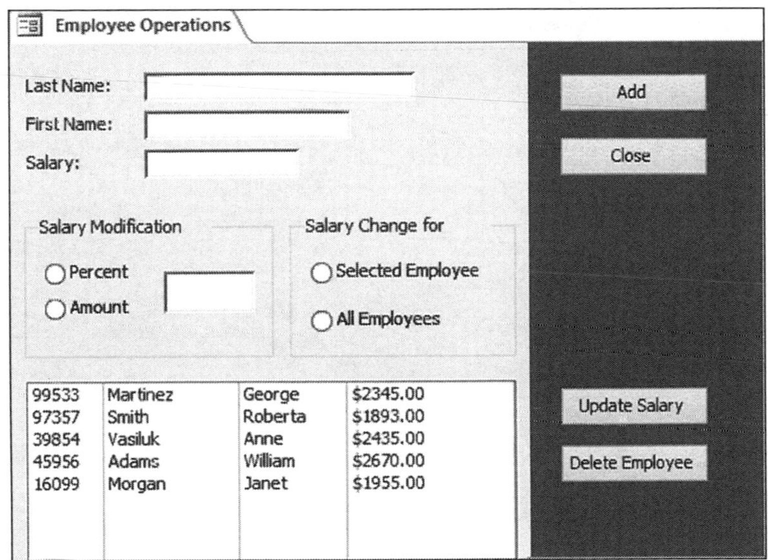

FIGURE 8.3. The listbox control displays employee data as entered in the custom collection CEmployee.

6. Enter the following **cmdClose_Click** procedure to close the form:

```
Private Sub cmdClose_Click()
  DoCmd.Close
End Sub
```

7. Write the following **Click** procedure for the cmdUpdate button:

```
Private Sub cmdUpdate_Click()
    Dim numOfPeople As Integer
    Dim colItem As Integer

    'validate user selections
    If frSalaryFor.Value = 0 Or frSalaryMod.Value = 0 Then
        MsgBox "Please choose appropriate option button in " & _
          vbCr & "the 'Salary Modification' and " & _
          "'Change the Salary for' areas.", vbOKOnly, _
          "Insufficient selection"
        Exit Sub
```

```
    ElseIf Not IsNumeric(txtRaise) Or txtRaise = "" Then
        MsgBox "You must enter a number."
        txtRaise.SetFocus
        Exit Sub
    ElseIf frSalaryMod.Value = 1 And _
        lboxPeople.ListIndex = -1 Then
        MsgBox "Click the employee name.", , _
          "Missing selection in the List box"
        Exit Sub
    End If

    If frSalaryMod.Value = 1 And lboxPeople.ListIndex = -1 Then
       MsgBox "Enter data or select an option."
       Exit Sub
    End If
    'get down to calculations
    amount = txtRaise
    colItem = lboxPeople.ListIndex + 1
    If frSalaryFor.Value = 1 And frSalaryMod.Value = 1 Then
      'by percent, one employee
      choice = 1
      numOfPeople = 1
    ElseIf frSalaryFor.Value = 2 And frSalaryMod.Value = 1 Then
      'by amount, one employee
       choice = 2
       numOfPeople = 1
    ElseIf frSalaryFor.Value = 1 And frSalaryMod.Value = 2 Then
       'by percent, all employees
       choice = 1
       numOfPeople = 2
    ElseIf frSalaryFor.Value = 2 And frSalaryMod.Value = 2 Then
       'by amount, all employees
       choice = 2
       numOfPeople = 2
    End If
    UpdateSalary choice, amount, numOfPeople, colItem
    lboxPeople.RowSource = GetValues
End Sub
```

When the Update Salary button is clicked, the procedure checks to see whether the user selected the appropriate option buttons and entered the adjusted figure in the text box. The update can be done for the selected employee or for all the employees listed in the listbox control and collection. You can increase the salary by the specified percentage or amount (see Figure 8.4). Depending on which options are specified, values are assigned to the variables `choice`, `amount`, `numOfpeople`, and `colItem`. These variables serve as parameters for the Update-Salary procedure located in the EmpOperations module (see Step 13 further on). The last statement in the cmdUpdate_Click procedure sets the row source property of the listbox control to the result obtained from the GetValues function, which is located in the EmpOperations standard module.

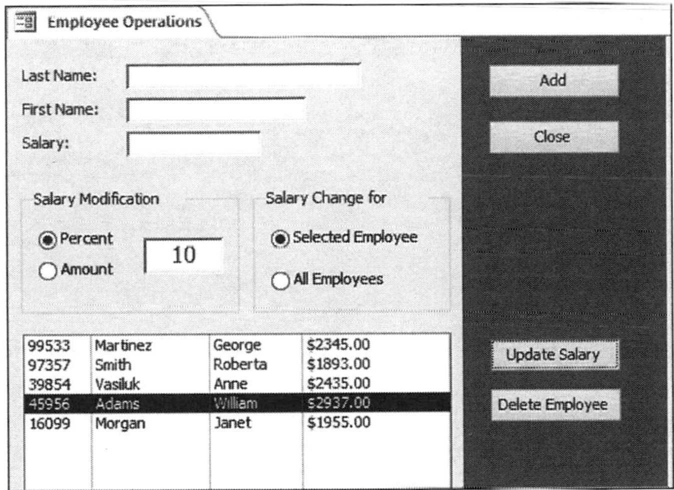

FIGURE 8.4. The employee salary can be increased or decreased by the specified percentage or amount.

8. Enter the following **cmdDelete_Click** procedure:

```
Private Sub cmdDelete_Click()
  ' make sure an employee row is highlighted
  ' in the listbox control
  If lboxPeople.ListIndex > -1 Then
    DeleteEmployee lboxPeople.ListIndex + 1
    If lboxPeople.ListCount = 1 Then
      lboxPeople.RowSource = GetValues
      UserForm_Initialize
    Else
      lboxPeople.RowSource = GetValues
    End If
  Else
    MsgBox "Click the item you want to remove."
  End If
End Sub
```

The cmdDelete_Click procedure lets you remove an employee from the custom collection CEmployee. If you click an item in the listbox and then click the Delete Employee button, the DeleteEmployee procedure is called. This procedure requires an argument that specifies the index number of the item selected in the listbox. After the employee is removed from the collection, the row source of the listbox control is reset to display the remaining employees. When the last employee is removed from the collection, the UserForm_Initialize procedure is called to tackle the task of disabling controls that cannot be used until at least one employee is entered into the CEmployee collection.

9. To activate the **EmpOperations** module that you created earlier, double-click its name in the Project Explorer window. The top of the module should contain the following declaration lines, the first two automatically added by Access:

```
Option Compare Database
Option Explicit
Dim emp As New CEmployee
Dim CEmployee As New Collection
```

10. In the **EmpOperations** standard module, enter the following **AddEmployee** procedure:

```
Sub AddEmployee(empLast As String, empFirst As String, _
 empSalary As Currency)
  With emp
    .ID = SetEmpId
    .LastName = empLast
    .FirstName = empFirst
    .Salary = CCur(empSalary)
    If .Salary = 0 Then Exit Sub
    CEmployee.Add emp
  End With
End Sub
```

The AddEmployee procedure is called from the cmdAdd_Click procedure attached to the form's Add button. This procedure takes three arguments. When Visual Basic for Applications reaches the `With emp` construct, a new instance of the CEmployee class is created. The LastName, FirstName, and Salary properties are set with the values passed from the cmdAdd_Click procedure. The ID property is set with the number generated by the result of the SetEmpId function (see the following step). Each time VBA sees the reference to the instanced emp object, it will call upon the appropriate Property Let procedure located in the class module. (The next section of this chapter demonstrates how to walk through this procedure step by step to see exactly when the Property procedures are executed.) The last statement inside the With emp construct adds the user-defined object emp to the custom collection called CEmployee.

11. In the **EmpOperations** standard module, enter the following **SetEmpID** function procedure:

```
Function SetEmpID() As String
  Dim ref As String
  Randomize
  ref = Int((99999 - 10000) * Rnd + 10000)
  SetEmpId = ref
End Function
```

This function will assign a unique five-digit number to each new employee. To generate a random integer between two given integers where ending_number = 99999 and beginning_number = 10000, the following formula is used:

```
= Int((ending_number - beginning_number) * Rnd + beginning_number)
```

The SetEmpId function procedure also uses the Randomize statement to reinitialize the random number generator. For more information on using the Rnd and Integer functions, as well as the Randomize statement, refer to the online help.

12. Enter the following **GetValues** function procedure. This function, which is called from the cmdAdd_Click, cmdUpdate_Click, and cmdDelete_Click procedures, provides the values for the listbox control to synchronize it with the current values in the CEmployee collection.

```
Function GetValues()
  Dim myList As String

  myList = ""
  For Each emp In CEmployee
    myList = myList & emp.ID & ";" & _
    emp.LastName & ";" & _
    emp.FirstName & "; $" & _
    Format(emp.Salary, "0.00") & ";"
  Next emp
  GetValues = myList
End Function
```

13. Enter the following **UpdateSalary** procedure:

```
Sub UpdateSalary(choice As Integer, myValue As Long, _
 peopleCount As Integer, colItem As Integer)
  Set emp = New CEmployee

  If choice = 1 And peopleCount = 1 Then
    CEmployee.Item(colItem).Salary = _
     emp.CalcNewSalary(1, CEmployee.Item( _
     colItem).Salary, myValue)
  ElseIf choice = 1 And peopleCount = 2 Then
    For Each emp In CEmployee
      emp.Salary = emp.Salary + ((emp.Salary * myValue) _
       / 100)
    Next emp
  ElseIf choice = 2 And peopleCount = 1 Then
    CEmployee.Item(colItem).Salary = _
     CEmployee.Item(colItem).Salary + myValue
  ElseIf choice = 2 And peopleCount = 2 Then
    For Each emp In CEmployee
      emp.Salary = emp.Salary + myValue
    Next emp
  Else
    MsgBox "Enter data or select an option."
  End If
End Sub
```

The UpdateSalary procedure is called from the cmdUpdate_Click procedure, which is assigned to the Update Salary button on the form. The click procedure passes four parameters that the UpdateSalary procedure uses for the salary calculations. When a salary for the selected employee needs to be updated by a percentage or amount, the CalcNewSalary method residing in the class module is called. For modification of salary figures for all the employees, we iterate over the CEmployee collection to obtain the value of the Salary property of each `emp` object, then perform the required calculation by using a formula. By entering a negative number in the form's txtRaise text box, you can decrease the salary by the specified percentage or amount.

14. Enter the **DeleteEmployee** procedure:

```
Sub DeleteEmployee(colItem As Integer)
  Dim getcount As Integer
  CEmployee.Remove colItem
End Sub
```

The DeleteEmployee procedure uses the `Remove` method to delete the selected employee from the CEmployee custom collection. Recall that the `Remove` method requires one argument, which is the position of the item in the collection. The value of this argument is obtained from the cmdDelete_Click procedure. The class module procedures were called from the standard module named EmpOperations. This was done to avoid creating a new instance of a user-defined class every time we needed to call it.

RUNNING THE CUSTOM APPLICATION

Now that you have finished writing the necessary VBA code, let's load frmEmployeeSalaries to enter and modify employee information.

Custom Project 8.1. (Step 9) Running the Custom Project

1. Choose **File | Save Chap08** to save all the objects in the VBA project.
2. Switch to the Microsoft Office Access window and activate **frmEmployeeSalaries** in the Form view.
3. Enter the employee last and first name and salary, and click the **Add** button.
   The employee information now appears in the listbox. Notice that an employee ID is automatically entered in the first column. All the disabled form controls are now enabled.
4. Enter data for another employee, then click the **Add** button.
5. Enter information for at least three more people.
6. Increase the salary of the third employee in the listbox by 10%. To do this, click the employee name in the listbox, click the **Percent** option button, and type **10** in the text box in the Salary Modification section of the form. In the Change the Salary for section of the form, click the **Selected Employee** option button. Finally, click the **Update Salary** button to perform the update operation.
7. Now increase the salary of all the employees by **$5**.
8. Remove the fourth employee from the listbox. To do this, select the employee in the listbox and click the **Delete Employee** button.
9. Close frmEmployeeSalaries by clicking the **Close** button.

WATCHING THE EXECUTION OF YOUR VBA PROCEDURES

To help you understand what's going on when your code runs and how the custom object works, let's walk through the cmdAdd_Click procedure. Treat this exercise as a brief introduction to the debugging techniques that are covered in detail in the next chapter.

Custom Project 8.1. (Step 10) Custom Project Code Walkthrough

1. Open **frmEmployeeSalaries** in Design view and click **View Code** in the Tools section of the Design tab.
2. Select **cmdAdd** from the combo box at the top left of the Code window.
3. Set a breakpoint by clicking in the left margin next to the following line of code, as shown in Figure 8.5:

```
If IsNull(txtLastName.Value) Or txtLastName.Value = "" _
 Or IsNull(txtFirstName.Value) Or txtFirstName.Value = "" _
 Or IsNull(txtSalary.Value) Or txtSalary.Value = "" Then
```

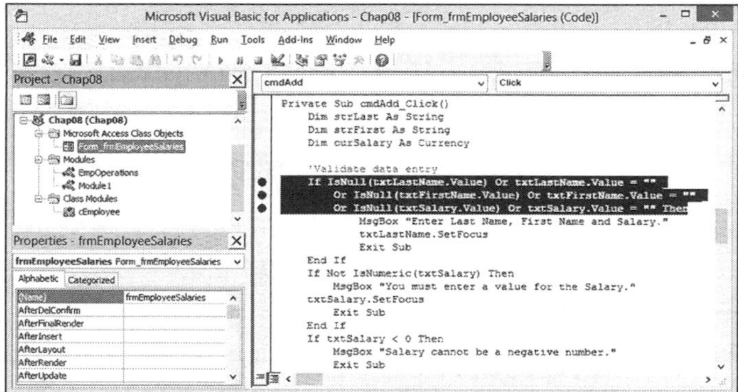

FIGURE 8.5. A red circle in the margin indicates a breakpoint. The statement with a breakpoint is displayed as white text on a red background.

4. Press **Alt+F11** to return to the form **frmEmployeeSalaries**, then switch to the Form view.
5. Enter data in the Last Name, First Name, and Salary text boxes, then click the form's **Add** button. Visual Basic should now switch to the Code window because it came across the breakpoint in the first line of the cmdAdd_Click procedure (see Figure 8.6).
6. Step through the code one statement at a time by pressing **F8**. Visual Basic runs the current statement, then automatically advances to the next statement and suspends execution. The current statement is indicated by a yellow arrow in the margin and a yellow background. Keep pressing **F8** to execute the pro-

cedure step by step. After Visual Basic switches to the EmpOperations module to run the AddEmployee procedure and encounters the `With emp` statement, it will run the function to set the employee ID and will go out to execute the Property Let procedures in the CEmployee class module (see Figure 8.7).

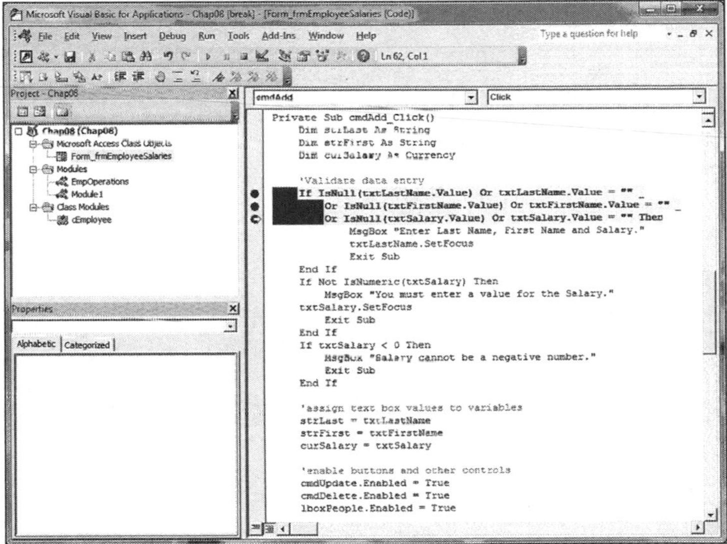

FIGURE 8.6. When Visual Basic encounters a breakpoint while running a procedure, it switches to the Code window and displays a yellow arrow in the margin to the left of the statement at which the procedure is suspended.

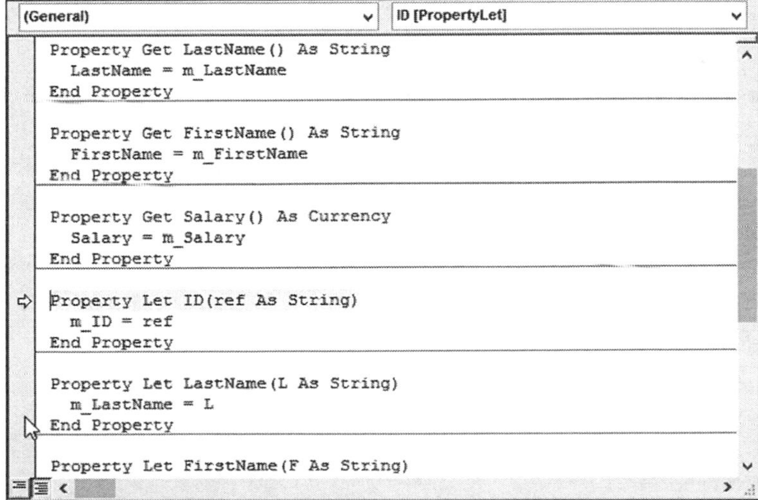

FIGURE 8.7. Setting the properties of your custom object is accomplished through the Property Let procedures.

7. Using the **F8** key, continue executing the cmdAdd_Click procedure code to the end. When VBA encounters the end of the procedure (`End Sub`), the yellow highlighter will be turned off. At this time, press **F5** to finish execution of the remaining code. Next, switch back to the active form by pressing **Alt+F11**.

To activate the form, you may need to first click the Table1 tab and then reselect the Employee Operations tab (see Figure 8.3)

8. Enter data for a new employee, then click the **Add** button. When Visual Basic displays the Code window, choose **Debug | Clear All Breakpoints**. Now press **F5** to run the remaining code without stepping through it.
9. In the Visual Basic Editor window, choose **File | Save Chap08**, then save changes to the modules when prompted.
10. Choose **File | Close and Return to Microsoft Access**.
11. Close the **Chap08.accdb** database and exit Microsoft Access.

VBA Debugging Tools

Visual Basic provides a number of debugging tools to help you analyze how your application operates, as well as to locate the source of errors in your procedures. See the next chapter for details on working with these tools.

CHAPTER SUMMARY

In this chapter, you learned how to create and use your own objects and collections in VBA procedures. You used a class module to create a user-defined (custom) object. You saw how to define your custom object's properties using the Property Get and Property Let procedures. You also learned how to write a method for your custom object and saw how to make the class module available to the user with a custom form. Finally, you learned how to analyze your VBA application by stepping through its code.

As your procedures become more complex, you will need to start using special tools for tracing errors, which are covered in the next chapter.

Part 4 Error Handling and Debugging

While you may have planned to write a perfect VBA procedure, there is always a chance that incorrectly constructed or typed code, or perhaps logic errors, will cause your program to fail or produce unexpected results. In this part of the book, you'll learn about various methods of handling errors, and testing and debugging VBA procedures.

Chapter 9

VBE Tools for Testing and Debugging

In the course of writing or editing VBA procedures, no matter how careful you are, you're likely to make some mistakes. For example, you may misspell a word, misplace a comma or quotation mark, or forget a period or ending parenthesis. These kinds of mistakes are known as *syntax errors*. Fortunately, Visual Basic for Applications is quite helpful in spotting these kinds of errors. To have VBA automatically check for correct syntax after you enter a line of code, choose Tools | Options in the VBE window. Make sure the Auto Syntax Check setting is selected on the Editor tab, as shown in Figure 9.1.

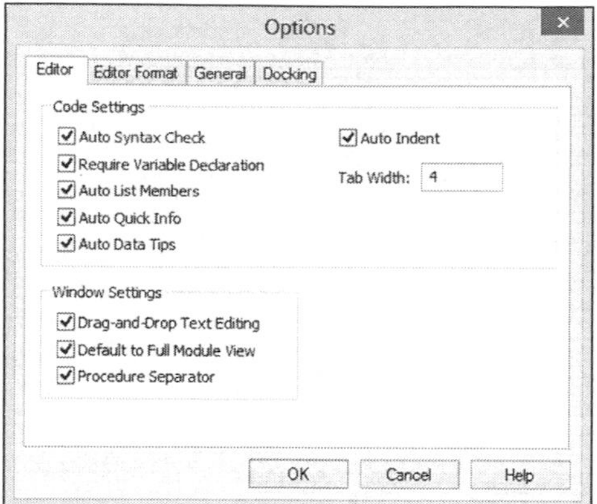

FIGURE 9.1. The Auto Syntax Check setting on the Editor tab of the Options dialog box helps you find typos in your VBA procedures.

When VBA finds a syntax error, it displays an error message box and changes the color of the incorrect line of code to red, or another color as indicated on the Editor Format tab in the Options dialog box.

If the explanation of the error in the error message isn't clear, you can click the Help button for more help. If Visual Basic for Applications cannot point you in the right direction, you must return to your procedure and carefully examine the offending instruction for missed letters, quotation marks, periods, colons, equals signs, and beginning and ending parentheses. Finding syntax errors can be aggravating and time consuming. Certain syntax errors can be caught only during the execution of the procedure. While attempting to run your procedure, VBA can find errors that were caused by using invalid arguments or omitting instructions that are used in pairs, such as `If...End` statements and looping structures.

You've probably heard that computer programs are "full of bugs." In programming, errors are called bugs, and *debugging* is a process of eliminating errors from your programs. Visual Basic for Applications provides a myriad of tools for tracking down and eliminating bugs. The first step in debugging a procedure is to correct all syntax errors. In addition to syntax errors, there are two other types of errors: runtime and logic. *Runtime errors*, which occur while the procedure is running, are often caused by unexpected situations the programmer did not think of while writing the code. For example, the program may be trying to access a drive or a file that does not exist on the user's computer. Or it may be trying to copy a file to a CD-ROM disk without first determining whether the user had inserted a CD.

The third type of error, a logic error, often does not generate a specific error message. Even though the procedure has no flaws in its syntax and runs without errors, it produces incorrect results. *Logic errors* happen when your procedure simply does not do what you want it to do. Logic errors are usually very difficult to locate. Those that happen intermittently are sometimes so well concealed that you can spend long hours—even days—trying to locate the source of the error.

TESTING VBA PROCEDURES

When testing your VBA procedure, use the following guidelines:

- If you want to analyze your procedure, step through your code one line at a time by pressing F8 or by choosing **Debug | Step Into.**
- If you suspect that an error may occur in a specific place in your procedure, use a breakpoint.
- If you want to monitor the value of a particular variable or expression used by your procedure, add a watch expression.
- If you are tired of scrolling through a long procedure to get to sections of code that interest you, set up a bookmark to quickly jump to the desired location (see "Navigating with Bookmarks" in Chapter 10).

STOPPING A PROCEDURE

VBA offers four methods of stopping your procedure and entering into a so-called *break mode*:

- Pressing Ctrl+Break
- Setting one or more breakpoints
- Inserting the `Stop` statement
- Adding a watch expression

A break occurs when execution of your VBA procedure is temporarily suspended. Visual Basic remembers the values of all variables and the statement from which the execution of the procedure should resume when you decide to continue.

You can resume a suspended procedure in one of the following ways:

- Click the Run Sub/UserForm button on the toolbar
- Choose Run | Run Sub/UserForm from the menu bar

- Click the Continue button in the error message box (see Figure 9.2)

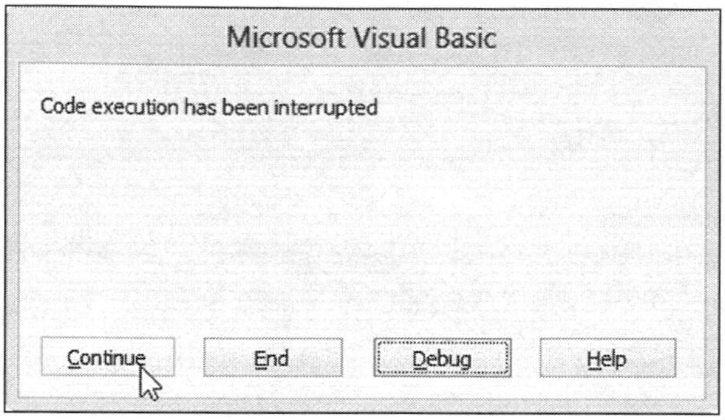

FIGURE 9.2. This message appears when you press Ctrl+Break while your VBA procedure is running.

The error message box shown in Figure 9.2 informs you that the procedure was halted. The description of each button is provided in Table 9.1.

TABLE 9.1. Error message box buttons

Button Name	Description
Continue	Click this button to resume code execution. This button will be grayed out if an error was encountered.
End	Click this button if you do not want to troubleshoot the procedure at this time. VBA will stop code execution.
Debug	Click this button to enter break mode. The Code window will appear, and VBA will highlight the line at which the procedure execution was suspended. You can examine, debug, or step through the code.
Help	Click this button to view the online help that explains the cause of this error message.

Using Breakpoints

If you know more or less where there may be a problem in your procedure code, you should suspend code execution at that location (on a given line). Set a breakpoint by pressing F9 when the cursor is on the desired line of code. When VBA gets to that line while running your procedure, it will display the Code window immediately. At this point you can step through the procedure code line by line by pressing F8 or choosing Debug | Step Into.

To see how this works, let's look at the following scenario. Assume that during the execution of the ListEndDates function procedure (see Custom Project 9.1) the following line of code could get you into trouble:

```
ListEndDates = Format(((Now() + intOffset) - 35) + 7 * row, _
 "MM/DD/YYYY")
```

Please note files for the "Hands-On" project may be found on the companion CD-ROM.

Custom Project 9.1. Debugging a Function Procedure

1. Start Microsoft Access 2013 and create a new database named **Chap09.accdb** in your **C:\Access2013_ByExample** folder.
2. Create the form shown in Figure 9.3.

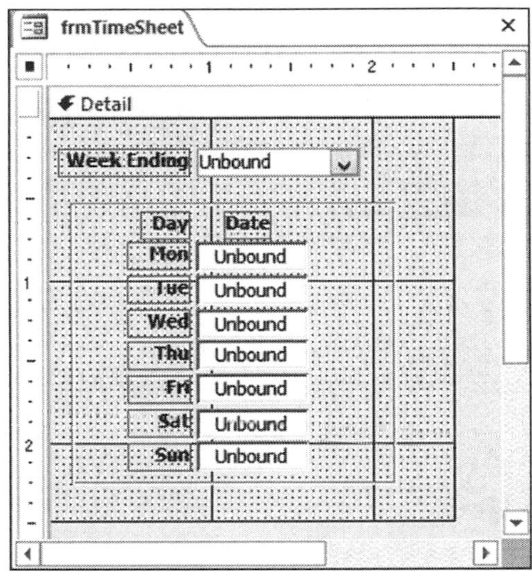

FIGURE 9.3. The combo box control shown on this form will be filled with the result of the ListEndDates function.

3. Use the property sheet to set the following control properties:

Control Name	Property Name	Property Setting
combo box	Name Row Source Type Column Count	cboEndDate ListEndDates 1
text box controls	Name	txt1 txt2 txt3 txt4 txt5 txt6 txt7

4. Save the form as **frmTimeSheet**.
5. In the property sheet, select **Form** from the drop-down listbox. Click the **Event** tab. Choose **[Event Procedure]** from the drop-down list next to the **On Load** property, and then click the **Build** button (…). Complete the following **Form_Load** procedure when the Code window appears:

```
Private Sub Form_Load()
  With Me.cboEndDate
    .SetFocus
    .ListIndex = 5 ' Select current end date
  End With
End Sub
```

6. Select the combo box control (cboEndDate) on the form. In the property sheet, click the **Event** tab. Choose **[Event Procedure]** from the drop-down list next to the **On Change** property, and then click the **Build** button (…). Enter the following code:

```
Private Sub cboEndDate_Change()
  Dim endDate As Date
  endDate = Me.cboEndDate.Value
  With Me
    .txt1 = Format(endDate - 6, "mm/dd")
    .txt2 = Format(endDate - 5, "mm/dd")
    .txt3 = Format(endDate - 4, "mm/dd")
    .txt4 = Format(endDate - 3, "mm/dd")
    .txt5 = Format(endDate - 2, "mm/dd")
    .txt6 = Format(endDate - 1, "mm/dd")
    .txt7 = Format(endDate - 0, "mm/dd")
  End With
End Sub
```

7. In the Visual Basic Editor window, choose **Insert | Module** to add a new standard module.
8. In the Properties window, change the Name property of Module1 to **TimeSheetProc**.
9. Enter the **ListEndDates** function procedure in the TimeSheetProc module:

```
Function ListEndDates(fld As Control, id As Variant, _
 row As Variant, col As Variant, _
 code As Variant) As Variant

  Dim intOffset As Integer
```

```
    Select Case code
      Case acLBInitialize
        ListEndDates = True
      Case acLBOpen
        ListEndDates = Timer
      Case acLBGetRowCount
        ListEndDates = 11
      Case acLBGetColumnCount
        ListEndDates = 1
      Case acLBGetColumnWidth
        ListEndDates = -1
      Case acLBGetValue
        ' days till end date
        intOffset = Abs((8 - Weekday(Now)) Mod 7)
        ' start 5 weeks prior to current week end date
        ' (7 days * 5 weeks = 35 days before next end date)
        ' and show 11 dates

        ListEndDates = Format(((Now() + intOffset) - 35) _
         + 7 * row, "MM/DD/YYYY")
    End Select
  End Function
```

10. In the ListEndDates function procedure, click anywhere on the line containing the following statement:

```
ListEndDates = Format(((Now() + intOffset) - 35) _
 + 7 * row, "MM/DD/YYYY")
```

11. Press **F9** (or choose **Debug | Toggle Breakpoint**) to set a breakpoint on the line where the cursor is located.

When you set the breakpoint, Visual Basic displays a red dot in the margin. At the same time, the line that has the breakpoint will change to white text on a red background (see Figure 9.4). The color of the breakpoint can be changed on the Editor Format tab in the Options dialog box (choose Tools | Options).

Another way of setting a breakpoint is to click in the margin indicator to the left of the line on which you want to stop the procedure.

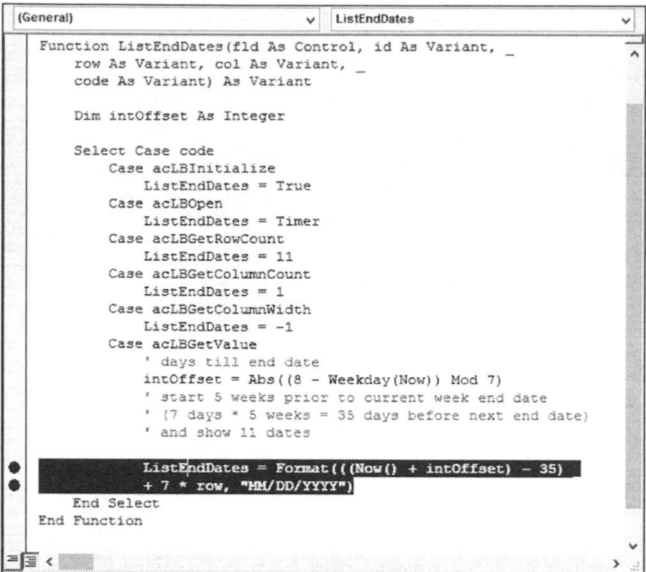
```
Function ListEndDates(fld As Control, id As Variant, _
    row As Variant, col As Variant, _
    code As Variant) As Variant

    Dim intOffset As Integer

    Select Case code
        Case acLBInitialize
            ListEndDates = True
        Case acLBOpen
            ListEndDates = Timer
        Case acLBGetRowCount
            ListEndDates = 11
        Case acLBGetColumnCount
            ListEndDates = 1
        Case acLBGetColumnWidth
            ListEndDates = -1
        Case acLBGetValue
            ' days till end date
            intOffset = Abs((8 - Weekday(Now)) Mod 7)
            ' start 5 weeks prior to current week end date
            ' (7 days * 5 weeks = 35 days before next end date)
            ' and show 11 dates

            ListEndDates = Format(((Now() + intOffset) - 35)
            + 7 * row, "MM/DD/YYYY")
    End Select
End Function
```

FIGURE 9.4. The line of code where the breakpoint is set is displayed in the color specified on the Editor Format tab in the Options dialog box.

12. Press Alt+F11 to switch to the Microsoft Access application window and open the form **frmTimeSheet** in the Form view.

When the form is opened, Visual Basic for Applications will call the ListEndDates function to fill the combo box, executing all the statements until it encounters the breakpoint you set in Steps 10–11. Once the breakpoint is reached, the code is suspended and the screen displays the Code window in break mode (notice the word "break" surrounded by square brackets in the Code window's titlebar), as shown in Figure 9.5. VBA displays a yellow arrow in the margin to the left of the statement at which the procedure was suspended. At the same time, the statement appears inside a box with a yellow background. The arrow and the box indicate the current statement, or the statement that is about to be executed. If the current statement also contains a breakpoint, the margin displays both indicators overlapping one another (the circle and the arrow).

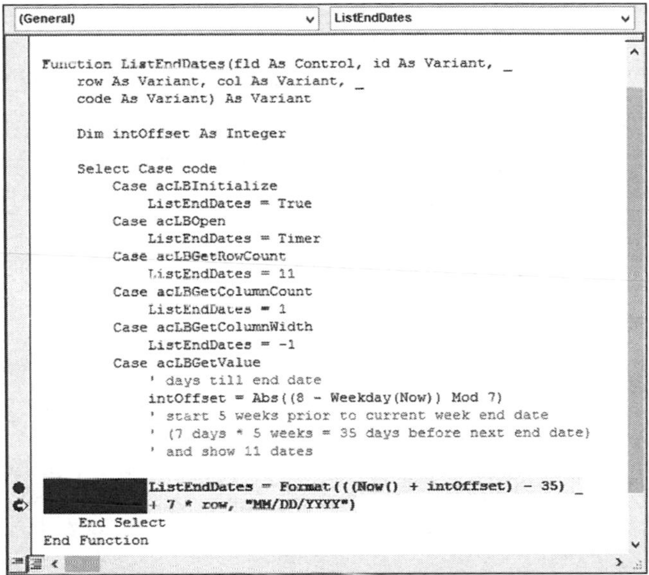

FIGURE 9.5. Code window in break mode. A yellow arrow appears in the margin to the left of the statement at which the procedure was suspended. Because the current statement also contains a breakpoint (indicated by a red circle), the margin displays both indicators overlapping one another (the circle and the arrow).

13. Finish running the ListEndDates function procedure by pressing **F5** to continue without stopping, or press **F8** to execute the procedure line by line.

When you step through your procedure code line by line by pressing F8, you can use the Immediate window to further test your procedure (see the section titled "Using the Immediate Window in Break Mode"). To learn more about stepping through a procedure, refer to the section titled "Stepping Through VBA Procedures" later in this chapter.

You can set any number of breakpoints in a procedure. This way you can suspend and continue the execution of your procedure as you please. Press F5 to quickly move between the breakpoints. You can analyze the code of your procedure and check the values of variables while code execution is suspended. You can also perform various tests by typing statements in the Immediate window. Consider setting a breakpoint if you suspect that your procedure never executes a certain block of code.

Removing Breakpoints

When you finish running the procedure in which you had set breakpoints, VBA does not automatically remove them. To remove the breakpoint, choose Debug

| Clear All Breakpoints or press Ctrl+Shift+F9. All the breakpoints are removed. If you had set several breakpoints in a given procedure and would like to remove only some of them, click on the line containing the breakpoint you want to remove and press F9 (or choose Debug | Clear Breakpoint). You should clear the breakpoints when they are no longer needed. The breakpoints are automatically removed when you exit Microsoft Access.

Note:

Remove the breakpoint you set in Custom Project 9.1.

Using the Immediate Window in Break Mode

When the procedure execution is suspended, the Code window appears in break mode. This is a good time to activate the Immediate window and type VBA instructions to find out, for instance, the name of the open form or the value of a certain control. You can also use the Immediate window to change the contents of variables in order to correct values that may be causing errors. By now, you should be an expert when it comes to working in the Immediate window. Figure 9.6 shows the suspended ListEndDates function procedure and the Immediate window with the questions that were asked of Visual Basic for Applications while in break mode.

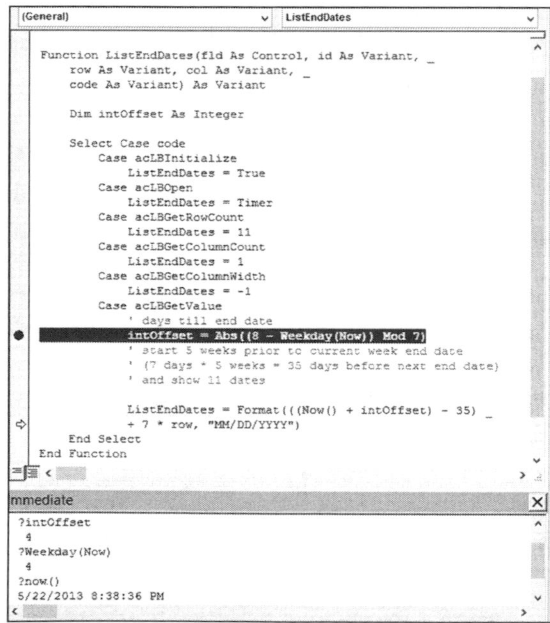
(General)
ListEndDates
Function ListEndDates(fld As Control, id As Variant, _
row As Variant, col As Variant, _
code As Variant) As Variant
Dim intOffset As Integer
Select Case code
Case acLBInitialize
ListEndDates = True
Case acLBOpen
ListEndDates = Timer
Case acLBGetRowCount
ListEndDates = 11
Case acLBGetColumnCount
ListEndDates = 1
Case acLBGetColumnWidth
ListEndDates = -1
Case acLBGetValue
' days till end date
intOffset = Abs((8 - Weekday(Now)) Mod 7)
' start 5 weeks prior to current week end date
' (7 days * 5 weeks = 35 days before next end date)
' and show 11 dates
ListEndDates = Format(((Now() + intOffset) - 35) _
+ 7 * row, "MM/DD/YYYY")
End Select
End Function
mmediate
?intOffset
4
?Weekday(Now)
4
?now()
5/22/2013 8:38:36 PM

FIGURE 9.6. When code execution is suspended, you can check current values of variables and expressions by entering appropriate statements in the Immediate window.

In break mode, you can also hold the mouse pointer over any variable in a running procedure to see the variable's value. For example, in the ListEnd-Dates function procedure shown in Figure 9.7, the breakpoint has been set on the statement just before the `End Select` keywords. When Visual Basic for Applications encounters this statement, the Code window appears in break mode. Because the statement that stores the value of the variable `intOffset` has already been executed, you can quickly find out the value of this variable by resting the mouse pointer over its name. The name of the variable and its current value appear in a floating frame. To show the values of several variables used in a procedure, you should use the Locals window, which is discussed later in this chapter.

```
        Case acLBGetValue
            ' days till end date
            intOffset = Abs((8 - Weekday(Now)) Mod 7)
            ' start 5 weeks prior to current week end date
            ' (7 days * 5 weeks = 35 days before next end date)
            ' and show 11 dates

            ListEndDates = Format(((Now() + intOffset) - 35) _
            + 7 * row, "MM/DD/YYYY")            intOffset = 4
    End Select
End Function
```

FIGURE 9.7. In break mode, you can find out the value of a variable by resting the mouse pointer on that variable.

Working in a Code Window in Break Mode

While in break mode, you can change code, add new statements, execute the procedure one line at a time, skip lines, set the next statement, use the Immediate window, and more. When the procedure is in break mode, all of the options on the Debug menu are available. You can enter break mode by pressing Ctrl+Break or F8 or by setting a breakpoint. In break mode, if you change a certain line of code, VBA will prompt you to reset the project by displaying the message "This action will reset your project, proceed anyway?" Click OK to stop the program's execution and proceed editing your code, or click Cancel to delete the new changes and continue running the code from the point where it was suspended. For example, change the variable declaration. As you press F5 to resume code execution, you'll be prompted to reset your project.

Using the Stop Statement

Sometimes you won't be able to test your procedure right away. If you set up your breakpoints and then close the database file, the breakpoints will be removed; next time, when you are ready to test your procedure, you'll have to begin by setting up your breakpoints again. If you need to postpone the task of testing your procedure until later, you can take a different approach by inserting a `Stop` statement into your code wherever you want to halt a procedure.

Figure 9.8 shows the `Stop` statement before the `With...End With` construct. VBA will suspend the execution of the cboEndDate_Change event procedure when it encounters the `Stop` statement, and the screen will display the Code window in break mode. Although the `Stop` statement has exactly the same effect as setting a breakpoint, it does have one disadvantage: All `Stop` statements stay in the procedure until you remove them. When you no longer need to stop your procedure, you must locate and remove all the `Stop` statements.

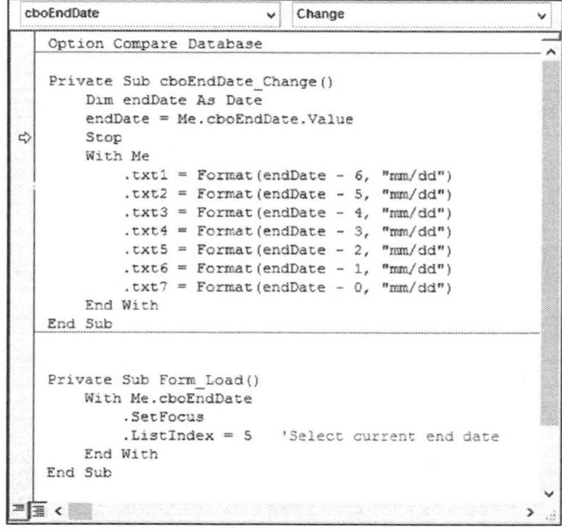

```
cboEndDate                    Change

Option Compare Database

Private Sub cboEndDate_Change()
    Dim endDate As Date
    endDate = Me.cboEndDate.Value
    Stop
    With Me
        .txt1 = Format(endDate - 6, "mm/dd")
        .txt2 = Format(endDate - 5, "mm/dd")
        .txt3 = Format(endDate - 4, "mm/dd")
        .txt4 = Format(endDate - 3, "mm/dd")
        .txt5 = Format(endDate - 2, "mm/dd")
        .txt6 = Format(endDate - 1, "mm/dd")
        .txt7 = Format(endDate - 0, "mm/dd")
    End With
End Sub

Private Sub Form_Load()
    With Me.cboEndDate
        .SetFocus
        .ListIndex = 5    'Select current end date
    End With
End Sub
```

FIGURE 9.8. You can insert a Stop statement anywhere in your VBA procedure code. The procedure will halt when it gets to the Stop statement, and the Code window will appear with the code line highlighted.

Using the Assert Statement

A very powerful and easy to apply debugging technique is utilizing `Debug.Assert` statements. Assertions allow you to write code that checks itself while running. By including assertions in your programming code you can verify that a particular condition or assumption is true. Assertions give you immediate feedback when an error occurs. They are great for detecting logic errors early in the development phase instead of hearing about them later from your end users. Just because your procedure ran on your system without generating an error does not mean that there are no bugs in that procedure. Don't assume anything—always test for validity of expressions and variables in your code. The `Debug.Assert` statement takes any expression that evaluates to True or False and activates the break mode when that expression evaluates to False. The syntax for `Debug.Assert` is as follows:

```
Debug.Assert condition
```

where *condition* is a VBA code or expression that returns True or False. If *condition* evaluates to False or 0 (zero), VBA will enter break mode. For example, when running the following looping structure, the code will stop executing when the variable `i` equals 50:

```
Sub TestDebugAssert()
  Dim i As Integer
  For i = 1 To 100
      Debug.Assert i <> 50
  Next
End Sub
```

Keep in mind that `Debug.Assert` does nothing if the condition is False or zero (0). The execution simply stops on that line of code and the VBE screen opens with the line containing the false statement highlighted so that you can start debugging your code. You may need to write an error handler to handle the identified error. Error-handling procedures are covered in Chapter 10. While you can stop the code execution by using the `Stop` statement (see the previous section), `Debug.Assert` differs from the `Stop` statement in its conditional aspect; it will stop your code only under specific conditions. Conditional breakpoints can also be set by using the Watches window (see the next section). After you have debugged and tested your code, comment out or remove the `Debug.Assert` statements from your final code. The easiest way to do this is to use Edit | Replace in the VBE editor screen. To comment out the statements, in the Find What box, enter `Debug.Assert`. In the Replace With box, enter an apostrophe followed by `Debug.Assert`.

To remove the `Debug.Assert` *statements from your code, enter* `Debug.Assert` *in the Find What box. Leave the Replace With box empty, but be sure to mark the Use Pattern Matching checkbox.*

Using the Add Watch Window

Many errors in procedures are caused by variables that assume unexpected values. If a procedure uses a variable whose value changes in various locations, you may want to stop the procedure and check the current value of that variable. VBA offers a special Watches window that allows you to keep an eye on variables or expressions while your procedure is running. To add a watch expression to your procedure, select the variable whose value you want to monitor in the Code window, then choose Debug | Add Watch. The screen will display the Add Watch dialog box, as shown in Figure 9.9.

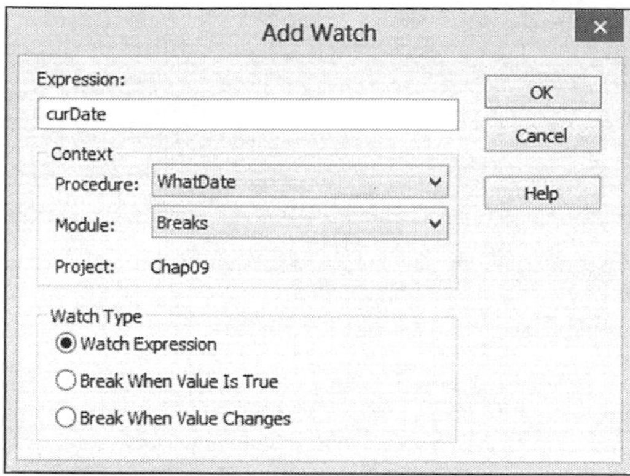

FIGURE 9.9. The Add Watch dialog box allows you to define conditions you want to monitor while a VBA procedure is running.

The Add Watch dialog box contains three sections, which are described in Table 9.2.

TABLE 9.2. Add Watch dialog box sections

Section	Description
Expression	Displays the name of a variable you have highlighted in your procedure. If you opened the Add Watch dialog box without selecting a variable name, type the name of the variable you want to monitor in the Expression text box.
Context	In this section, indicate the name of the procedure that contains the variable and the name of the module where this procedure is located.
Watch Type	Specifies how to monitor the variable. If you choose: • The Watch Expression option button, you can read the value of the variable in the Add Watch window while in break mode. • Break When Value Is True, Visual Basic will automatically stop the procedure when the variable evaluates to True (nonzero). • Break When Value Changes, Visual Basic will automatically stop the procedure each time the value of the variable or expression changes.

You can add a watch expression before running a procedure or after suspending the execution of your procedure.

The difference between a breakpoint and a watch expression is that the breakpoint always stops a procedure in a specified location, but the watch stops the procedure only when the specified condition (Break When Value Is True or Break When Value Changes) is met. Watches are extremely useful when you are not sure where the variable is being changed. Instead of stepping through many lines of code to find the location where the variable assumes the specified value, you can put a watch breakpoint on the variable and run your procedure as normal. Let's see how this works.

Hands-On 9.1. Watching the Values of VBA Expressions

1. In the Visual Basic Editor window, choose **Insert | Module** to insert a new standard module.
2. Use the Properties window to change the name of the module to **Breaks**.
3. In the Breaks Code window, type the following **WhatDate** procedure:

```
Sub WhatDate()
  Dim curDate As Date
  Dim newDate As Date
  Dim x As Integer

  curDate = Date
  For x = 1 To 365
    newDate = Date + x
  Next x
End Sub
```

The WhatDate procedure uses the `For...Next` loop to calculate the date that is `x` days in the future. You won't see any result when you run this procedure unless you insert the following instruction in the procedure code just before the `End Sub` keywords:

```
MsgBox "In " & x & " days, it will be " & NewDate
```

However, you don't want to display the individual dates, day after day. Suppose that you want to stop the program when the value of the variable `x` reaches 211. In other words, you want to know what date will be 211 days from now. To get the answer, you could insert the following statement into your procedure before the `Next x` statement:

```
If x = 211 Then MsgBox "In " & x & " days it will be " & _
 NewDate
```

But this time, you want to get the answer without introducing any new statements into your procedure. If you add watch expressions to the procedure, Visual Basic for Applications will stop the `For...Next` loop when the specified condition is met, and you'll be able to check the values of the desired variables.

4. Choose **Debug | Add Watch.**
5. In the Expression text box, enter the following expression: **x = 211.**
6. In the Context section, choose **WhatDate** from the Procedure combo box and **Breaks** from the Module combo box.
7. In the Watch Type section, select the **Break When Value Is True** option button.
8. Click **OK** to close the Add Watch dialog box. You have now added your first watch expression.
9. In the Code window, position the insertion point anywhere within the name of the **curDate** variable.
10. Choose **Debug | Add Watch** and click **OK** to set up the default watch type with the **Watch Expression** option.
11. In the Code window, position the insertion point anywhere within the name of the **newDate** variable.
12. Choose **Debug | Add Watch** and click **OK** to set up the default watch type with the **Watch Expression** option.

After performing these steps, the WhatDate procedure contains the following three watches:

`x = 211`	Break When Value Is True
`curDate`	Watch Expression
`newDate`	Watch Expression

13. Position the cursor anywhere inside the code of the WhatDate procedure and press **F5.**

Visual Basic stops the procedure when x = 211 (see Figure 9.10). Notice that the value of the variable `x` in the Watches window is the same as the value you specified in the Add Watch dialog box.

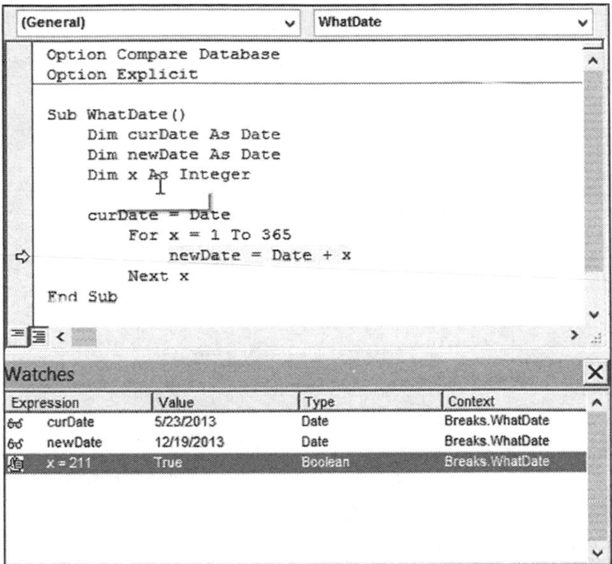

FIGURE 9.10. Using the Watches window.

In addition, the Watches window shows the value of the variables `curDate` and `newDate`. The procedure is in break mode. You can press F5 to continue, or you can ask another question: What date will be in 277 days? The next step shows how to do this.

14. Choose **Debug | Edit Watch** and enter the following expression: **x = 277**.
    You can also display the Edit Watch dialog box by double-clicking the expression in the Watches window.
15. Click **OK** to close the Edit Watch dialog box. Notice that the Watches window now displays a new value of the expression. `x` is now false.
16. Press **F5**. The procedure stops again when the value of x = 277. The value of `curDate` is the same; however, the `newDate` variable now contains a new value—a date that is 277 days from now. You can change the value of the expression again or finish the procedure.
17. Press **F5** to finish the procedure without stopping.

When your procedure is running and a watch expression has a value, the Watches window displays the value of the Watch expression. If you open the Watches window after the procedure has finished, you will see the error "<out of context>" instead of the variable values. In other words, when the watch expression is out of context, it does not have a value.

Removing Watch Expressions

To remove a watch expression, click on the expression you want to remove from the Watches window and press Delete. Remove all the watch expressions you defined in the preceding exercise.

Using Quick Watch

To check the value of an expression not defined in the Watches window, you can use Quick Watch (see Figure 9.11).

To access the Quick Watch dialog box while in break mode, position the insertion point anywhere inside a variable name or an expression you want to watch and choose Debug | Quick Watch, or press Shift+F9.

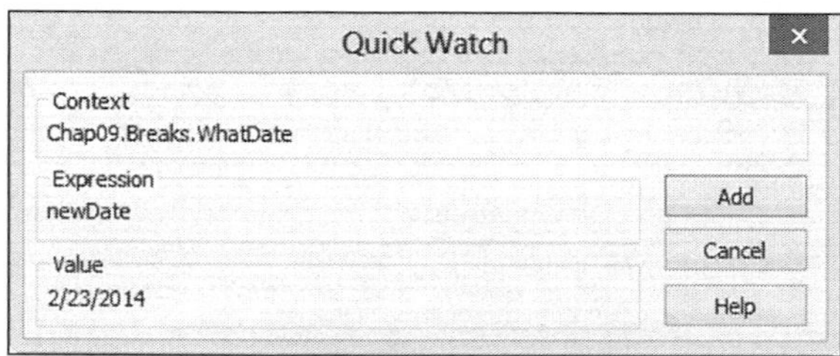

FIGURE 9.11. The Quick Watch dialog box shows the value of the selected expression in a VBA procedure.

The Quick Watch dialog box contains an Add button that allows you to add the expression to the Watches window. Let's see how to take advantage of Quick Watch.

Hands-On 9.2. Using the Quick Watch Dialog Box

Remove all the watch expressions you defined in Hands-On 9.1. See the preceding section on how to remove a watch expression from the Watches window.

1. In the **WhatDate** procedure, position the insertion point on the name of the variable **x**.
2. Choose **Debug | Add Watch**.
3. Enter the expression **x = 50**.
4. Choose the **Break When Value Is True** option button, and click **OK**.
5. Run the WhatDate procedure.

Visual Basic will suspend procedure execution when x = 50. Notice that the Watches window does not contain either the `newDate` or the `curDate` variables. To check the values of these variables, you can position the mouse pointer over the appropriate variable name in the Code window, or you can invoke the Quick Watch dialog box.

6. In the Code window, position the mouse inside the **newDate** variable and press **Shift+F9**, or choose **Debug | Quick Watch**.
   The Quick Watch dialog box shows the name of the expression and its current value.
7. Click **Cancel** to return to the Code window.
8. In the Code window, position the mouse inside the **curDate** variable and press **Shift+F9**, or choose **Debug | Quick Watch**.
9. The Quick Watch dialog box now shows the value of the variable `curDate`.
10. Click **Cancel** to return to the Code window.
11. Press **F5** to continue running the procedure.

USING THE LOCALS WINDOW

If you need to keep an eye on all the declared variables and their current values during the execution of a VBA procedure, choose View | Locals Window before you run your procedure. While in break mode, VBA will display a list of variables and their corresponding values in the Locals window (see Figure 9.12).

The Locals window contains three columns: Expression, Value, and Type.

The Expression column displays the names of variables that are declared in the current procedure. The first row displays the name of the module preceded by the plus sign. When you click the plus sign, you can check if any variables have been declared at the module level. Here the class module will show the system variable `Me`. In the Locals window, global variables and variables used by other projects aren't displayed.

The second column, Value, shows the current variable values. In this column, you can change the value of a variable by clicking on it and typing the new value. After changing the value, press Enter to register the change. You can also press Tab, Shift+Tab, or the up or down arrows, or click anywhere within the Locals window after you've changed the variable value.

Type, the third column, displays the type of each declared variable.

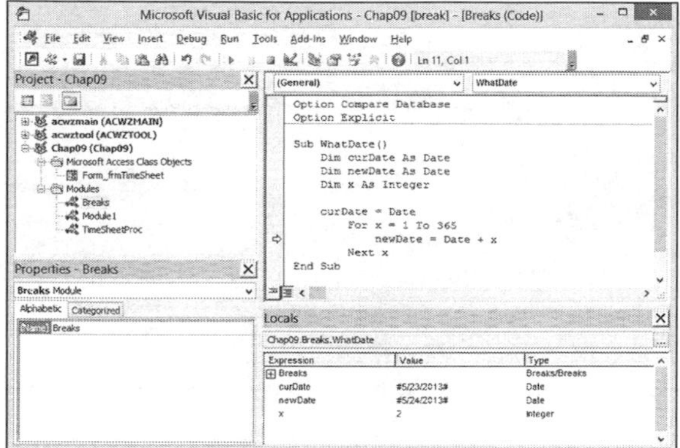

FIGURE 9.12. The Locals window displays the current values of all the declared variables in the current VBA procedure.

To observe the variable values in the Locals window, let's proceed to the following hands-on exercise.

Hands-On 9.3. Using the Locals Window

1. Choose **View | Locals Window**.
2. Click anywhere inside the **WhatDate** procedure and press **F8**.
   Pressing F8 places the procedure in break mode. The Locals window displays the name of the current module, the local variables, and their beginning values.
3. Press **F8** a few more times while keeping an eye on the Locals window.
4. Press **F5** to continue running the procedure.

UNDERSTANDING THE CALL STACK DIALOG BOX

The Locals window (see Figure 9.13) contains a button with an ellipsis (...). This button opens the Call Stack dialog box (see Figure 9.14), which displays a list of all active procedure calls. An *active procedure call* is a procedure that is started but not completed. You can also activate the Call Stack dialog box by choosing View | Call Stack. This option is only available in break mode.

Locals

Chap09.Module1.MyProcedure ...

Expression	Value	Type
⊞ Module1		Module1/Module1
myName	"Label1"	String

FIGURE 9.13. While in break mode, you can display the Call Stack dialog box by clicking the ellipsis (...) button in the Locals window.

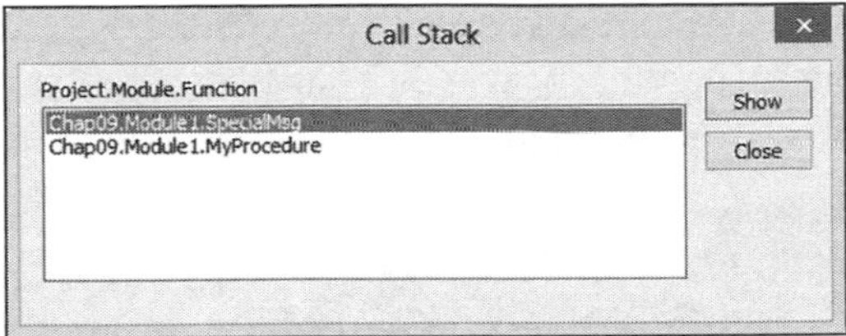

FIGURE 9.14. The Call Stack dialog box displays a list of procedures that are started but not completed.

The Call Stack dialog box is especially helpful for tracing nested procedures. Recall that a nested procedure is a procedure that is being called from within another procedure (see Hands-On 9.5). If a procedure calls another, the name of the called procedure is automatically added to the Calls list in the Call Stack dialog box. When VBA has finished executing the statements of the called procedure, the procedure name is automatically removed from the Call Stack dialog box. You can use the Show button in the Call Stack dialog box to display the statement that calls the next procedure listed in the Call Stack dialog box.

STEPPING THROUGH VBA PROCEDURES

Stepping through the code means running one statement at a time. This allows you to check every line in every procedure that is encountered. To start stepping through the procedure from the beginning, place the cursor anywhere inside the code of your procedure and choose Debug | Step Into, or press F8. The Debug menu contains several options that allow you to execute a procedure in step mode (see Figure 9.15).

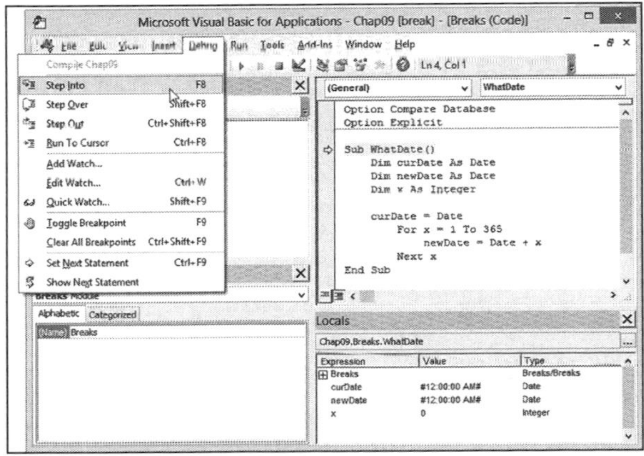

FIGURE 9.15. The Debug menu offers many commands for stepping through VBA procedures. Certain commands on this menu are only available in break mode.

When you run a procedure one statement at a time, VBA executes each statement until it encounters the `End Sub` keywords. If you don't want to step through every statement, you can press F5 at any time to run the remaining code of the procedure without stepping through it.

Hands-On 9.4. Stepping Through a Procedure

1. Place the cursor anywhere inside the procedure you want to trace.
2. Press **F8** or choose **Debug | Step Into**.
   Visual Basic for Applications executes the current statement, then automatically advances to the next statement and suspends execution. While in break mode, you can activate the Immediate window, the Watches window, or the Locals window to see the effect of a particular statement on the values of variables and expressions. And, if the procedure you are stepping through calls other procedures, you can activate the Call Stack dialog box to see which procedures are currently active.
3. Press **F8** again to execute the selected statement. After executing this statement, VBA will select the next statement, and again the procedure execution will be halted.
4. Continue stepping through the procedure by pressing **F8**, or press **F5** to continue running the code without stopping.
5. You can also choose **Run | Reset** to stop the procedure at the current statement without executing the remaining statements.

   When you step over procedures (**Shift+F8**), VBA executes each procedure as if it were a single statement. This option is particularly useful if a procedure contains calls to other procedures you don't want to step into because they have already been tested and debugged, or because you want to concentrate only on the new code that has not been debugged yet.

Stepping Over a Procedure

Suppose that the current statement in MyProcedure calls the SpecialMsg procedure. If you choose Debug | Step Over (Shift+F8) instead of Debug | Step Into (F8), VBA will quickly execute all the statements inside the SpecialMsg procedure and select the next statement in the calling procedure, MyProcedure. While the SpecialMsg procedure is being executed, VBA continues to display the current procedure in the Code window.

Hands-On 9.5. Stepping Over a Procedure

This hands-on exercise refers to the Access form named frmTimeSheet that you created in Custom Project 9.1 at the beginning of this chapter.

1. In the Visual Basic Editor window, choose **Insert | Module** to add a new standard module.
2. In the module's Code window, enter the **MyProcedure** and **SpecialMsg** procedures as shown here:

```
Sub MyProcedure()
  Dim myName As String

  myName = Forms!frmTimeSheet.Controls(1).Name

  ' choose Step Over to avoid stepping through the
  ' lines of code in the called procedure - SpecialMsg
  SpecialMsg myName
End Sub

Sub SpecialMsg(n As String)
  If n = "Label1" Then
    MsgBox "You must change the name."
  End If
End Sub
```

3. Add a breakpoint within MyProcedure at the following statement:

```
SpecialMsg myName
```

4. Place the insertion point anywhere within the code of **MyProcedure** and press **F5** to run it.
   Visual Basic halts execution when it reaches the breakpoint.
5. Press **Shift+F8** or choose **Debug | Step Over**.
   Visual Basic runs the SpecialMsg procedure, and then execution advances to the statement immediately after the call to the SpecialMsg procedure.
6. Press **F5** to finish running the procedure without stepping through its code.
   Now suppose you want to execute MyProcedure to the line that calls the SpecialMsg procedure.
7. Click anywhere inside the statement **SpecialMsg myName**.
8. Choose **Debug | Run to Cursor**.
   Visual Basic will stop the procedure when it reaches the specified line.

9. Press **Shift+F8** to step over the SpecialMsg procedure.
10. Press **F5** to execute the rest of the procedure without single stepping.

    Stepping over a procedure is particularly useful when you don't want to analyze individual statements inside the called procedure (SpecialMsg).

Stepping Out of a Procedure

Another command on the Debug menu, Step Out (Ctrl+Shift+F8), is used when you step into a procedure and then decide that you don't want to step all the way through it. When you choose this option, Visual Basic will execute the remaining statements in this procedure in one step and proceed to activate the next statement in the calling procedure.

In the process of stepping through a procedure, you can switch between the Step Into, Step Over, and Step Out options. The option you select depends on which code fragment you wish to analyze at a given moment.

Running a Procedure to Cursor

The Debug menu Run To Cursor command (Ctrl+F8) lets you run your procedure until the line you have selected is encountered. This command is really useful if you want to stop the execution before a large loop or you intend to step over a called procedure.

SETTING THE NEXT STATEMENT

At times, you may want to rerun previous lines of code in the procedure or skip over a section of code that is causing trouble. In each of these situations, you can use the Set Next Statement option on the Debug menu. When you halt execution of a procedure, you can resume the procedure from any statement you want. VBA will skip execution of the statements between the selected statement and the statement where execution was suspended.

Skipping Lines of Code

Although skipping lines of code can be very useful in the process of debugging your VBA procedures, it should be done with care. When you use the Next Statement option, you tell Visual Basic for Applications that this is the line you want to execute next. All lines in between are ignored. This means that certain things you may have expected to occur don't happen, which can lead to unexpected errors.

SHOWING THE NEXT STATEMENT

If you are not sure where procedure execution will resume, you can choose Debug | Show Next Statement, and VBA will place the cursor on the line that will run next. This is particularly useful when you have been looking at other procedures and are not sure where execution will resume. The Show Next Statement option is available only in break mode.

STOPPING AND RESETTING VBA PROCEDURES

At any time while stepping through the code of a procedure in the Code window, you can press F5 to execute the remaining instructions without stepping through them, or choose Run | Reset to finish the procedure without executing the remaining statements. When you reset your procedure, all the variables lose their current values. Numeric variables assume the initial value of zero `(0)`, variable-length strings are initialized to a zero-length string (`""`), and fixed-length strings are filled with the character represented by the ASCII character code 0, or `Chr(0)`. Variant variables are initialized to `Empty`, and the value of Object variables is set to `Nothing`.

CHAPTER SUMMARY

In this chapter, you learned how to test your VBA procedures to make sure they perform as planned. You debugged your code by stepping through it using breakpoints and watches. You learned how to work with the Immediate window in break mode; you found out how the Locals window can help you monitor the values of variables; and you learned how the Call Stack dialog box can be helpful in keeping track of where you are in a complex program. By using the built-in debugging tools, you can quickly pinpoint the problem spots in your procedures. Try to spend more time getting acquainted with the Debug menu options and debugging tools discussed in this chapter. Mastering the art of debugging can save you hours of trial and error. In the next chapter, you will learn additional debugging techniques for your VBA procedures.

Chapter 10 Conditional Compilation and Error Trapping

If you've worked through Chapter 9, you should now know how to debug your code using the built-in VBE debugging tools. With these tools, you can easily set a breakpoint to stop a program while it is running and use step options to run your code one statement at a time. You can review the values of variables by using Watch expressions and the Locals and Immediate windows. With more complex procedures, you can use the Call Stack dialog box to view all active procedure calls and trace the execution of nested procedures. This chapter introduces additional techniques that you will find useful for debugging your Visual Basic applications. In particular, you will learn how conditional compilation can help you include or ignore certain blocks of code, and how you can add effective error-handling code to your procedures.

UNDERSTANDING AND USING CONDITIONAL COMPILATION

When you run a procedure for the first time, your VBA statements are converted into the machine code understood by the computer. This process is called *compiling*. You can also perform the compilation of your entire VBA project before you run the procedure by choosing Debug | Compile (*name of the current VBA project*). You can tell VBA to include or ignore certain blocks of code when compiling and running by using so-called *conditional compilation*.

To enable conditional compilation, use special expressions called directives. Use the `#Const` directive to declare a Boolean (True or False) constant. Next, check this constant inside the `#If...Then...#Else` directive. The portion of code that you want to compile conditionally must be surrounded by these directives. Notice that the `If` and `Else` keywords are preceded by a number sign (#). If a portion of code is to be run, the value of the conditional constant must be set to True (–1). Otherwise, the value of this constant should be set to False (0). Declare the conditional constant in the declaration section of the module like this:

```
#Const User = True
```

This declares the conditional constant named `User`.

Conditional compilation can be used to compile an application that will be run on different platforms (Windows or Macintosh, Win16-, Win32-, Win64-bit). It is also useful in localizing an application for different languages or excluding certain debugging statements before the VBA application is sent off for distribution. The program code excluded during the conditional compilation is omitted from the final file; thus it has no effect on the size or performance of the program.

In the procedure that follows, data is displayed in Spanish when the conditional constant named `verSpanish` is True. The WhatDay procedure calls the DayOfWeek function, which returns the name of the day based on the supplied date. To compile the program in English, all you have to do is change the conditional constant to False and Visual Basic will jump to the block of instructions located after the `#Else` directive.

Please note files for the "Hands-On" project may be found on the companion CD-ROM.

Hands-On 10.1. Using Conditional Compilation

1. Start Microsoft Access 2013 and create a new database named **Chap10.accdb** in your **C:\Access2013_ByExample** folder.
2. Switch to the VBE window and insert a new module. In the Properties window, rename the module **Conditional.**
3. Enter the following **WhatDay** and **DayOfWeek** function procedures:

```
' declare a conditional compiler constant
#Const verSpanish = True

Sub WhatDay()
  Dim dayNr As Integer
  #If verSpanish = True Then
    dayNr = Weekday(InputBox("Entre la fecha, " _
     & "por ejemplo 10/01/2013"))
    MsgBox "Sera " & DayOfWeek(dayNr) & "."
  #Else
    WeekdayName
  #End If
End Sub

Function DayOfWeek(dayNr As Integer) As String
  DayOfWeek = Choose(dayNr, "Domingo", "Lunes", "Martes", _
   "Miercoles", "Jueves", "Viernes", "Sabado")
End Function

Function WeekdayName() As String
  Select Case Weekday(InputBox("Enter date, e.g., 10/01/2013"))
  Case 1
    WeekdayName = "Sunday"
  Case 2
    WeekdayName = "Monday"
  Case 3
    WeekdayName = "Tuesday"
  Case 4
    WeekdayName = "Wednesday"
  Case 5
    WeekdayName = "Thursday"
  Case 6
    WeekdayName = "Friday"
```

```
    Case 7
      WeekdayName = "Saturday"
    End Select
    MsgBox "It will be " & WeekdayName & "."
End Function
```

4. Run the WhatDay procedure.
   Because the conditional compilation constant (`verSpanish`) is set to True at the top of the module, Visual Basic runs the Spanish version of the WhatDay procedure. It asks for the user's input in Spanish and displays the result in Spanish. To run the English version of the code, change the `verSpanish` constant to False, and rerun the procedure.
   Instead of declaring the conditional compiler constants at the top of a module, you can choose Tools | (Debugging) Properties to open the Project Properties window, as shown in Figure 10.1. When you use the Project Properties window, use the following syntax in the Conditional Compilation Arguments text box to enable the English version of the WhatDay procedure:

```
verSpanish = 0
```

   To enable the Spanish version of the WhatDay procedure, use the following conditional compilation argument in the Project Properties window:

```
verSpanish = - 1
```

   If there are more conditional compilation constants, each of the constants must be separated by a colon.

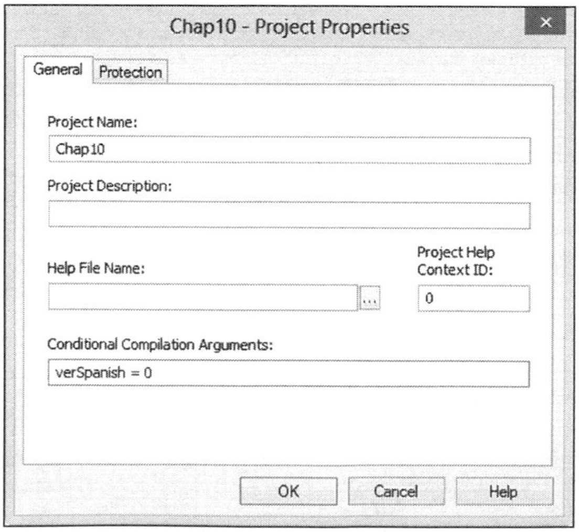

FIGURE 10.1. The conditional compilation constant can be declared either at the top of the module or in the Project Properties window, but never in both places.

5. Comment out the `#Const verSpanish` directive at the top of the module, and enter the conditional compilation constant in the Project Properties window as shown in Figure 10.1. Then run the WhatDay procedure to see how the `Else` branch of your program is now executed for English-speaking users.

NAVIGATING WITH BOOKMARKS

In the process of analyzing or reviewing your VBA procedures, you will often find yourself jumping to certain areas of code. Using the built-in bookmark feature, you can easily mark the spots you want to navigate between.

To set up a bookmark:

1. Click anywhere in the statement you want to define as a bookmark.
2. Choose **Edit | Bookmarks | Toggle Bookmark** (or click the **Toggle Bookmark** button on the Edit toolbar).

Visual Basic will place a blue, rounded rectangle in the left margin beside the statement, as shown in Figure 10.2.

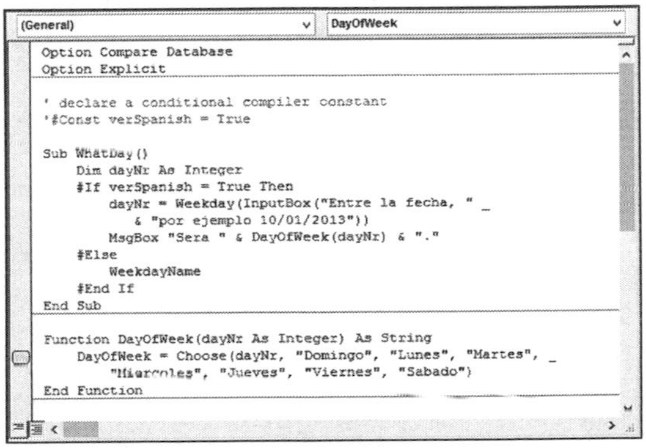

```
(General)                                DayOfWeek
Option Compare Database
Option Explicit

' declare a conditional compiler constant
'#Const verSpanish = True

Sub WhatDay()
    Dim dayNr As Integer
    #If verSpanish = True Then
        dayNr = Weekday(InputBox("Entre la fecha, " _
            & "por ejemplo 10/01/2013"))
        MsgBox "Sera " & DayOfWeek(dayNr) & "."
    #Else
        WeekdayName
    #End If
End Sub

Function DayOfWeek(dayNr As Integer) As String
    DayOfWeek = Choose(dayNr, "Domingo", "Lunes", "Martes", _
        "Miercoles", "Jueves", "Viernes", "Sabado")
End Function
```

FIGURE 10.2. Using bookmarks, you can quickly jump between often-used sections of your procedures.

Once you've set up two or more bookmarks, you can jump between the marked locations of your code by choosing Edit | Bookmarks | Next Bookmark or simply clicking the Next Bookmark button on the Edit toolbar. You may also right-click anywhere in the Code window and select Next Bookmark from the shortcut menu. To go to the previous bookmark, select Previous Bookmark. You can remove bookmarks at any time by choosing Edit | Bookmarks | Clear All or by clicking the Clear All Bookmarks button on the Edit toolbar. To remove a single bookmark, click anywhere in the bookmarked statement and choose Edit | Bookmarks | Toggle Bookmark, or click the Toggle Bookmark button on the Edit toolbar.

TRAPPING ERRORS

No one writes bug-free programs the first time. For this reason, when you create VBA procedures you have to determine how your program will respond to errors. Many unexpected errors happen at runtime. For example, your procedure may try to give a new file the same name as an open file.

Runtime errors are often discovered not by a programmer but by the user who attempts to do something that the programmer has not anticipated. If an error occurs when the procedure is running, Visual Basic displays an error message and the procedure is stopped. The error message that VBA displays to the user is often quite cryptic.

You can keep users from seeing many runtime errors by including error-handling code in your VBA procedures. This way, when Visual Basic encounters an error, instead of displaying a default error message, it will show a much friendlier, more comprehensive error message, perhaps advising the user how to correct the error.

How do you implement error handling in your VBA procedure? The first step is to place the `On Error` statement in your procedure. This statement tells VBA what to do if an error happens while your program is running. In other words, VBA uses the `On Error` statement to activate an error-handling procedure that will trap runtime errors. Depending on the type of procedure, you can exit the error trap by using one of the following statements: `Exit Sub`, `Exit Function`, `Exit Property`, `End Sub`, `End Function`, or `End Property`.

You should write an error-handling routine for each procedure. Table 10.1 shows how the `On Error` statement can be used.

TABLE 10.1. On Error statement options

On Error Statement	Description
`On Error GoTo Label`	Specifies a label to jump to when an error occurs. This label marks the beginning of the error-handling routine. An *error handler* is a routine for trapping and responding to errors in your application. The label must appear in the same procedure as the `On Error GoTo` statement.

On Error Statement	Description
`On Error Resume Next`	When a runtime error occurs, Visual Basic ignores the line that caused the error and continues the procedure with the next line. An error message is not displayed.
`On Error GoTo 0`	Turns off error trapping in a procedure. When VBA runs this statement, errors are detected but not trapped within the procedure.

Is This an Error or a Mistake?

In programming, mistakes and errors are not the same thing. A mistake—such as a misspelled or missing statement, a misplaced quotation mark or comma, or an assignment of a value of one type to a variable of a different (and incompatible) type—can be removed from your program through proper testing and debugging. But even though your code may be free of mistakes, errors can still occur. An *error* is a result of an event or operation that doesn't work as expected. For example, if your VBA procedure accesses a particular file on disk and someone deleted this file or moved it to another location, you'll get an error no matter what. An error prevents the procedure from carrying out a specific task.

USING THE ERR OBJECT

Your error-handling code can utilize various properties and methods of the Err object. For example, to check which error occurred, check the value of `Err.Number`. The Number property of the Err object will tell you the value of the last error that occurred, and the Description property will return a description of the error. You can also find the name of the application that caused the error by using the Source property of the Err object (this is very helpful when your procedure launches other applications). After handling the error, use the `Err.Clear` statement to reset the error number. This will set `Err.Number` back to zero.

To test your error-handling code you can use the `Raise` method of the Err object. For example, to raise the "Disk not ready" error, use the following statement:

```
Err.Raise 71
```

The following OpenToRead procedure demonstrates the use of the `On Error` statement and the Err object.

Hands-On 10.2. Error Trapping Techniques

1. In the Visual Basic Editor window, insert a new module and rename it **ErrorTraps.**
2. In the Code window, enter the following **OpenToRead** procedure:

```
Sub OpenToRead()
  Dim strFile As String
  Dim strChar As String
  Dim strText As String
  Dim FileExists As Boolean

  FileExists = True

  On Error GoTo ErrorHandler

  strFile = InputBox("Enter the name of file to open:")
  Open strFile For Input As #1

  If FileExists Then
    Do While Not EOF(1) ' loop until the end of file
      strChar = Input(1, #1) ' get one character
      strText = strText + strChar
    Loop
    Debug.Print strText
    ' Close the file
    Close #1
  End If
  Exit Sub

ErrorHandler:
  FileExists = False
  Select Case Err.Number
    Case 71
      MsgBox "The CD/DVD drive is empty."
    Case 53
      MsgBox "This file can't be found on the specified drive."
    Case 76
      MsgBox "File Path was not found."
```

```
        Case Else
          MsgBox "Error " & Err.Number & " :" & Err.Description
          Exit Sub
      End Select
      Resume Next
    End Sub
```

Before continuing with this Hands-On, let's examine the code of the OpenToRead procedure. The purpose of the OpenToRead procedure is to read the contents of the user-supplied text file character by character. When the user enters a filename, various errors can occur. For example, the filename may be wrong, the user may attempt to open a file from a CD-ROM or DVD disc without actually placing the disc in the drive, or he may try to open a file that is already open. To trap these errors, the error-handling routine at the end of the OpenToRead procedure uses the Number property of the Err object. The Err object contains information about runtime errors. If an error occurs while the procedure is running, the statement `Err.Number` will return the error number.

If errors 71, 53, or 76 occur, Visual Basic will display the user-friendly messages given inside the `Select Case` block and then proceed to the `Resume Next` statement, which will send it to the line of code following the one that had caused the error. If another (unexpected) error occurs, Visual Basic will return its error code (`Err.Number`) and error description (`Err.Description`).

At the beginning of the procedure, the variable `FileExists` is set to True. If the program doesn't encounter an error, all the instructions inside the `If FileExists Then` block will be executed. However, if VBA encounters an error, the value of the `FileExists` variable will be set to False (see the first statement in the error-handling routine just below the `ErrorHandler` label).

If you comment the `Close #1` instruction, Visual Basic will encounter the error on the next attempt to open the same file. Notice the `Exit Sub` statement before the `ErrorHandler` block. Put the `Exit Sub` statement just above the error-handling routine. You don't want Visual Basic to carry out the error handling if there are no errors.

How does this procedure accomplish the read operation? The `Input` function allows you to return any character from a sequential file. *Sequential access files are files* where data is retrieved in the same order as it is stored, such as files stored in the CSV format (comma-delimited text), TXT format (text separated by tabs), or PRN format (text separated by spaces). Configuration files, error logs, HTML files, and all sorts of plain text files are all sequential files. These files are stored on disk as a sequence of characters. The beginning of a new text line is indicated by two special characters: the carriage return and the linefeed. When you work with sequential files, start at the beginning of the file and move forward character by character, line by line, until you encounter the end of the file. Sequential access files can be easily opened and manipulated by just about any text editor.

If you use the VBA function named `LOF` (length of file) as the first argument of the `Input` function, you can quickly read the contents of the sequential file without having to loop through the entire file.

For example, instead of the following `Do...While` loop statement block:

```
Do While Not EOF(1) ' loop until the end of file
  strChar = Input(1, #1) ' get one character
  strText = strText + strChar
Loop
```

you can simply write the following statement to get the contents of the file at once:

```
strText = Input(LOF(1), #1)
```

The `LOF` function returns the number of bytes in a file. Each byte corresponds to one character in a text file.

To read data from a file, you must first open the file with the `Open` statement using the following syntax:

```
Open pathname For mode[Access access][lock] As [#]filenumber _
    [Len=reclength]
```

The `Open` statement has three required arguments: `pathname`, `mode`, and `filenumber`. `Pathname` is the name of the file you want to open. The filename may include the name of a drive and folder.

`Mode` is a keyword that determines how the file was opened. Sequential files can be opened in one of the following modes: `Input`, `Output`, or `Append`. Use `Input` to read the file, `Output` to write to a file and overwrite any existing file, and `Append` to write to a file by adding to any existing information.

`Filenumber` is a number from 1 to 511. This number is used to refer to the file in subsequent operations. You can obtain a unique file number using the VBA built-in `FreeFile` function.

The optional `Access` clause can be used to specify permissions for the file (Read, Write, or Read Write). The optional `lock` argument determines which file operations are allowed for other processes. For example, if a file is open in a network environment, `lock` determines how other people can access it. The following `lock` keywords can be used: `Shared`, `Lock Read`, `Lock Write`, or `Lock Read Write`. The last element of the `Open` statement, `reclength`, specifies the buffer size (total number of characters) for sequential files.

Therefore, to open a sequential file in order to read its data, the example procedure uses the following instruction:

```
Open strFile For Input As #1
```

And, to close the sequential file, the following statement is used:

```
Close #1
```

3. Click anywhere within the **OpenToRead** procedure and press **F5** to run it. When prompted for the file to open, type **C:\Access2013_HandsOn\Vacation.txt** in the input dialog box and click **OK**. The procedure reads the contents of the **Vacation.txt** file into the Immediate window.
4. Run the **OpenToRead** procedure again. When prompted for the file to open, type **P:\Access2013_HandsOn\Vacation.txt** in the input dialog box and click **OK**. This time Visual Basic cannot find the specified file, so it displays the message "File Path was not found."
5. Run the **OpenToRead** procedure again. This time, when prompted for the filename, enter the name of any file that references your CD/DVD drive (when the drive slot is empty). This should trigger error 71 and result in the message "The CD/DVD drive is empty."
6. Comment the **Close #1** statement and run **OpenToRead**. When prompted for the file, enter **C:\Access2013_HandsOn\Vacation.txt** as the filename. Run the same procedure again supplying the same filename. The second run will cause the statements within the `Case Else` block to run. You should get an error 55 – "File already open" message because the text file will still be open in memory. To remove the file from memory, type **Close #1** in the Immediate window and press **Enter**. Next, uncomment the **Close # 1** statement in the OpenToRead procedure to return it to the original state.

PROCEDURE TESTING

You are responsible for the code you produce. Before you give your procedure to others to test, you should test it yourself. After all, you understand best how it is supposed to work. Some programmers think testing their own code is some sort of degrading activity, especially when they work in an organization that has a team devoted to testing. *Don't make this mistake.* The testing process at the programmer level is as important as the code development itself. After you've tested the procedure yourself, you should give it to the users to test. Users will provide you with answers to questions such as: Does the procedure produce the expected results? Is it easy and fun to use? Does it follow the standard conventions? Also, it is a good idea to give the entire application to someone who knows the least about using this particular application, and ask them to play around with it and try to break it.

You can test the ways your program responds to runtime errors by causing them on purpose:

- Generate any built-in error by entering the following syntax:

  ```
  Error error_number
  ```

  For example, to display the error that occurs on an attempt to divide by zero (0), type the following statement in the Immediate window:

  ```
  Error 11
  ```

  When you press Enter, Visual Basic will display the error message saying, "Run-time error 11. Division by zero."
- To check the meaning of the generated error, use the following syntax:

  ```
  Error(error_number)
  ```

  For example, to find out what error number 7 means, type the following in the Immediate window:

  ```
  ?Error(7)
  ```

  When you press Enter, Visual Basic returns the error description:

  ```
  "Out of memory"
  ```

  To generate the same error at runtime in the form of a message box like the one in Figure 10.3, enter in the Immediate window or in your procedure code:

  ```
  Err.Raise 7
  ```

When you finish debugging your VBA procedures, make sure you remove all statements that raise errors.

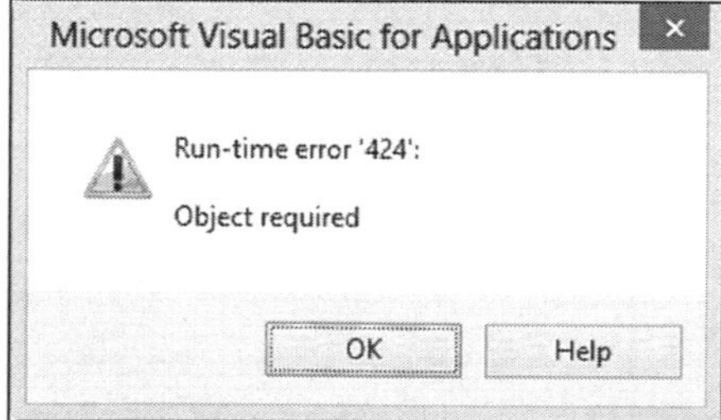

FIGURE 10.3. To test your error-handling code, use the Raise method of the Err object. This will generate a runtime error during the execution of your procedure.

SETTING ERROR TRAPPING OPTIONS IN YOUR VISUAL BASIC PROJECT

You can specify the error-handling settings for your current Visual Basic project by choosing Tools | Options and selecting the General tab (shown in Figure 10.4). The Error Trapping area located on the General tab determines how errors are handled in the Visual Basic environment. The following options are available:

- Break on All Errors

  This setting will cause Visual Basic to enter the break mode on any error, no matter whether an error handler is active or whether the code is in a class module (class modules were covered in Chapter 8).

- Break in Class Module

  This setting will trap any unhandled error in a class module. Visual Basic will activate the break mode when an error occurs and will highlight the line of code in the class module that produced this error.

- Break on Unhandled Errors

  This setting will trap errors for which you have not written an error handler. The error will cause Visual Basic to activate the break mode. If the error occurs in a class module, the error will cause Visual Basic to enter break mode on the line of code that called the offending procedure of the class.

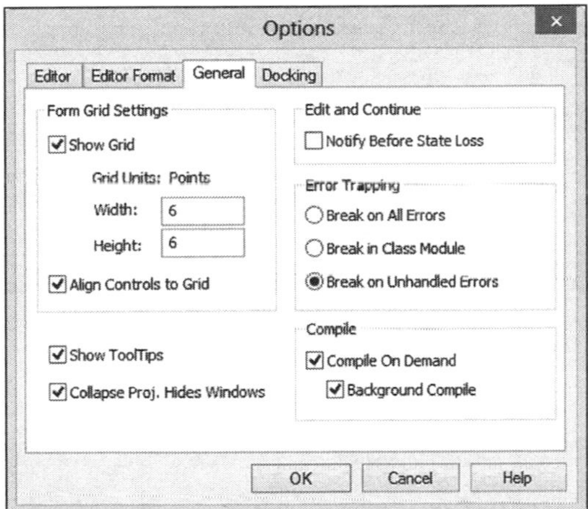

FIGURE 10.4. Setting the error trapping options in the Options dialog box will affect all instances of Visual Basic started after you change the setting.

CHAPTER SUMMARY

In this chapter, you've learned how conditional compilation can enable you to run various parts of code based on the specified condition. You also learned how to mark your code with bookmarks so you can easily navigate between sections of your procedure. Additionally, this chapter showed you how to trap errors by including an error-handling routine inside your VBA procedure and how to use the VBA Err object. This chapter concludes Part IV of this book.

In the next chapter, you will learn how to use data access models to access and manipulate Microsoft Access databases with VBA.

Part 5

ACCESS VBA PROGRAMMING WITH DAO AND ADO

There are two sets of programming objects known as Data Access Objects (DAO) and ActiveX® Data Objects (ADO) that enable Microsoft Access and other client applications to access and manipulate data. In this part of the book, you learn how to use DAO and ADO objects in your VBA procedures to connect to a data source; create, modify, and secure database objects; and read, add, update, and delete data.

Chapter 11

Data Access Technologies in Microsoft Access

Microsoft Access has been effectively used by people all over the world for organizing and accessing data. While each new software release brings numerous changes in the design of the user interface and offers simpler ways of performing common database tasks, database access methods have evolved at a little slower pace.

This chapter begins with the introduction of the older (Jet) and the newer (ACE) database engines and proceeds to an overview of Access versions and file formats supported by Microsoft Access 2013. This is followed by a review of data access methods that programmers and database developers can use to read, write, and manipulate data in Access databases (in .mdb and the .accdb file formats). In addition, this chapter demonstrates various ways of opening both native Microsoft Jet databases and external data sources. You will also learn how to establish a connection to the currently open database, connect to an SQL Server, create a new database, set database properties, and handle database errors.

UNDERSTANDING DATABASE ENGINES: JET/ACE

Since version 1.0 (1992), an integrated part of Microsoft Access has been its database engine, commonly referred to as *Microsoft Jet* (Joint Engine Technology (JET)) or *Jet database engine*. Microsoft Jet is a multiuser relational database engine that provides support for the standard DBMS (Database Management System) functionality such as data definition, data manipulation, querying, security, and maintenance, as well as remote data access.

Jet stores data in the Microsoft Access database file format (.mdb) according to the Indexed Sequential Access Method (ISAM). Queries are performed by the Jet query engine. A replication engine is used to create copies (replicas) of database structures on multiple systems with periodic synchronization. Jet provides password-protected security and different levels of access via the user and group accounts. The user information is kept in a separate system database (MDW). Security is also built into the database tables in the form of object permissions.

The Microsoft Jet database engine enables you to access data that resides in Microsoft Jet databases (.mdb files), external data sources (dBASE files, Microsoft Excel spreadsheets, SharePoint lists, Microsoft Outlook folders, text files, XML files, or HTML documents), and Open Database Connectivity (ODBC) data sources (SQL Server®, Oracle®, or Sybase®). To access external data via ODBC, you need a specific ODBC driver installed on the computer containing the data source.

The main component of the Microsoft Jet database engine is a dynamic link library file (.dll) installed in the \Windows\System32 folder (see Table 11.1). On the Windows platform, DLLs are libraries of common code that can be used by more than one application. The Jet DLL provides a simple interface to the data. If the data source is an .mdb file, then Jet reads and writes directly to the file. If the data source is external, Jet calls on the appropriate ODBC driver to perform the request.

Different versions of Access use different versions of Jet (see Table 11.1).

Beginning with Office Access 2007, Microsoft made many enhancements to the database engine, making it private for Microsoft Office suite applications. This private version of the database engine, called the Access Connectivity Engine (ACE), uses the file extension .accdb and offers many useful features to Access users and developers alike (see Table 11.2).

TABLE 11.1. Database engine versions in Access 2013 and earlier

MS Access Version	Database Engine Used	Dynamic Link Library (DLL) File
Access 2013/2010/2007	ACE 12	ace.dll
Access 2003/2002/2000	Jet 4	msjet40.dll
Access 97	Jet 3.5	msjet35.dll

UNDERSTANDING ACCESS VERSIONS AND FILE FORMATS

In Microsoft Access 2013/2010/2007, the default file format is .accdb; however, you can still directly open and use Jet databases (.mdb files) created in Access 2003/2002/2000. Jet databases created with Access 97 or earlier must be either enabled or converted for use in Access 2013/2010/2007. When an older database is enabled, it is made compatible with Access 2013/2010/2007 so that you can make changes to the data. However, any design changes must be made in the version of Access that was used when the database was first created. When you opt to convert an Access 97 or earlier database to the .accdb file format, you must first convert it to Access 2003/2002/2000. Table 11.2 lists various file formats that are supported by Access 2013/2010/2007.

TABLE 11.2. File formats supported in Access 2013/2010/2007

File Format	Description	Additional Notes
`.accdb`	File format first introduced in Access 2007 (default). This file format is not readable by Access versions prior to 2007. DO NOT use this file format if you need to support: • Replication • User-level security	This file format is also the default format in Access 2013. ***Note:*** Access 2013 does not support replicated databases. Use Access 2010/2007 to create a replica of an MDB database formatted in Access 2003/2002/2000 file format. (See the "Database Replication" document on the CD-ROM disk.)
`.accde`	File extension for Access 2013/2010/2007 .accdb files that are in execute only mode. These files have all VBA source code removed.	Users can only execute VBA code; they cannot view or modify it. In addition, users do not have permissions to make design changes to forms or reports.

(Contd.)

File Format	Description	Additional Notes
	This file extension replaces the .mde file extension used in earlier versions of Access.	If you need to save the Access 2013 database in .accde format, open the database and choose File | Save As | Save Database As. Select Make ACCDE and click the Save As button.
`.accdt`	This is an Access Database Template file. Access 2013/2010/2007 all come with professionally designed database templates.	Templates provide you with predefined tables/table relationships, forms, reports, queries, and macros. To save the Access 2013 database as a template, open the database and choose File | Save As | Save Database As. Select Template (.accdt) and click the Save As button.
`.accdr`	This file extension denotes an Access 2013/2010/2007 database functioning in runtime mode.	To create a "locked-down" version of your Access 2013database, simply change the file extension from .accdb to .accdr. To restore the full database functionality, do the reverse: change the file extension from .accdr to .accdb.
`.mdb (Access 97, Access 2000, Access 2002, Access 2003)`	Access database file format used in versions prior to 2007. ***Note:*** In Access 2013/2010/2007 you can create files in either the Access 2000 format or the Access 2002-2003 database format. These files will have the extension .mdb.	Use the .mdb file format if the database will be used in earlier versions of Access to: • Support replication • Support user-level security
`.mde (Access 97, Access 2000, Access 2002, Access 2003)`	An .mde file is a compiled version of an .mdb database without any VBA code. This change prevents a database user from reading or changing your VBA code. Users cannot edit the design of forms, reports, or modules.	An .accde file is the Access 2013/2010/2007 version of the .mde file in earlier versions of Access.

File Format	Description	Additional Notes
`.adp`	This is a file extension for a Microsoft Access Data Project file that lets you connect to an SQL Server database or the Microsoft Data Engine (MSDE) on your PC and create client/server applications. A project file does not contain any data or data definition objects such as tables, views, stored procedures, or user-defined functions. All database objects are stored in the SQL Server database. An .adp file stores only database frontend forms, reports, and other application objects (macros, modules).	Access 2013 no longer supports the .adp file format. If you need to open and edit an existing ADP database that was created in an earlier version of Access or create a new ADP database, use Access 2010/2007.
`.ade`	This is a file format for a Microsoft Access project (.adp) file with all modules compiled and all editable source code removed. Similar to .mde files, projects stored in the .ade file format prevent users from making design changes to the frontend and gaining access to your VBA source code.	Access 2013 does not support the .ade file format. To create an .ade file from your Access Data Project (ADP), use Access 2010/2007.
`.mdw (Access 97, Access 2000, Access 2002, Access 2003)`	This file format is used by a Workgroup Information Fle. The .mdw files store information for secured MDB databases.	There are no changes to the .mdw file format in Access 2013. The .mdw files created in earlier versions of Access (2000 through 2003) can be used by Access 2013. When an MDB database is opened in Access 2013, you can choose File | Info | Users & Permissions | User-Level Security Wizard to create a new Workgroup Information File (.mdw).

(Contd.)

File Format	Description	Additional Notes
`.ldb`	This is a locking file extension for the MDB database. This file prevents users from writing data to pages that have been locked by other users and lets you determine which computer/user has a file or record locked. The .ldb file keeps track of usernames/computer names of the people who are currently logged into the MDB.	A locking file is created automatically when the database is opened and is deleted automatically when the last user closes a shared database. Note: You can view the information stored in this file by opening it with Windows Notepad.
`.laccdb`	This is the file extension for a locking file used by Access 2013/2010/2007 (.accdb file format).	As with the .ldb file, the .laccdb file is created automatically when the database is opened and is deleted automatically when the last user closes a shared database. Note: Because different locking files are created for MDB and ACCDB databases in Access 2013/2010/2007, .mdb and .accdb files can be open in Access 2013/2010/2007 without causing conflicts in the locking file.

UNDERSTANDING LIBRARY REFERENCES

A Microsoft Access database consists of various types of objects stored in different object libraries. *Libraries* are components that provide specific functionality. They are listed in the References dialog box, shown in Figure 11.1, which can be opened from the Visual Basic Editor window by selecting Tools | References. If you create an Access 2013 database in the default .accdb file format, you will see the following default references in the References dialog box:

- Visual Basic For Applications
- Microsoft Access 15.0 Object Library
- OLE Automation
- Microsoft Office 15.0 Access database engine Object Library

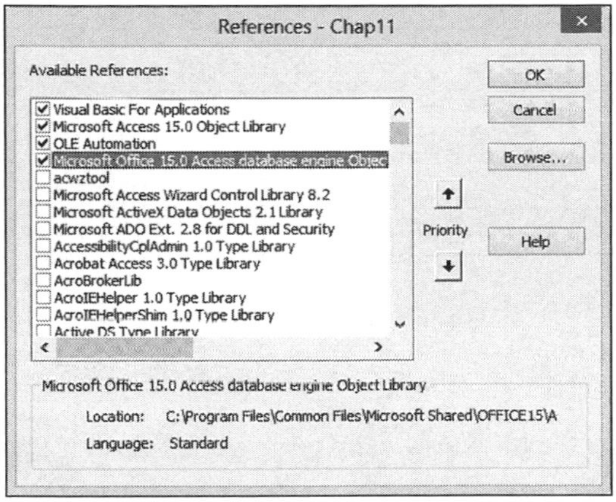

FIGURE 11.1. The default object libraries for Access 2013.

The Visual Basic for Applications and Access libraries that appear at the top of the References dialog box are built in. Access will not allow you to remove them from the database. The references that are checked are listed by priority. References that are not checked are listed alphabetically except for a few exceptions that are seen in Figure 11.1. When your VBA procedure references an object, Visual Basic searches each referenced object library in the order in which the libraries are displayed in the References dialog box. If the referenced libraries have objects with the same name, Visual Basic uses the object definition provided by the library listed higher in the Available References list. You can change the priority of an object library by selecting its name and clicking the up or down arrow button in the References dialog box. To help Visual Basic resolve library references, specify in your code the name of the library you intend to use. For example, to specify that the DAO Recordset should be used, declare it like this:

```
Dim rst As DAO.Recordset
```

To use the ADO Recordset, use the following declaration:

```
Dim rst As ADODB.Recordset
```

You can reference additional libraries in your Access database if your VBA application requires features that are not provided by the default libraries. For example, if your VBA procedures need to access files and folders on the computer, you may want to check the box next to Microsoft Scripting Runtime.

Do not add references to libraries you don't plan to use as they consume memory and may make your Access VBA project more time-consuming to compile and harder to debug.

Missing Library

If the library is marked as Missing in the References dialog box, click the Browse button, and locate the correct library file. You can disable a missing reference by clearing the checkbox to the left of the reference labeled "Missing."

Library Does Not Show in the References Dialog Box

If the library you want to reference is not shown in the Available References list box, you may need to unregister and reregister it with Windows.

To unregister a library, close Microsoft Access. In Windows 7, choose Run from the Start menu, and enter `regsvr32-u` followed by a space and the full path to the library file surrounded by quotation marks. For example:

```
regsvr32-u "C:\Program Files\Common Files\System\Ado\msjro.dll"
```

To register a library, choose Run from the Start menu, and enter `regsvr32` followed by a space and the full path to the library file surrounded by quotation marks. For example:

```
regsvr32 "C:\Program Files\Common Files\System\Ado\msjro.dll"
```

The next time you open Access the library name should be listed in the References dialog box.

Note: In Windows 8, press Windows Key + R to quickly access the Run dialog box.

If you move a library file from where it was originally installed, be sure to reregister it with the operating system or things may not work as expected.

Because referencing a wrong library for the version of Access used can cause data corruption, it is important to know which library files were designed for a particular version of Access. The next section introduces you to library files that you will find useful in creating and manipulating MDB and ACCDB databases using VBA code.

OVERVIEW OF OBJECT LIBRARIES IN MICROSOFT ACCESS

The object library contains information about its objects, properties, and methods. To work with the VBA programming examples included in this book, you will need to access objects from the libraries listed in the following subsections.

The Visual Basic for Applications Object Library (VBA)

Objects contained in this library allow you to access your computer's file system, work with date and time functions, perform mathematical and financial computations, interact with users, convert data, and read text files. The VBA library is stored in the vbe7.dll file located in the C:\Program Files\Common Files\Microsoft Shared\VBA\VBA7.1 folder.

The Microsoft Access 15.0 Object Library

This library provides objects that are used to display data and work with the Microsoft Access 2013 application. In Access 2013, the Access library is stored in the msacc.olb file and can be found in the C:\Program Files\Microsoft Office 15\Root\Office15 folder. The following list provides the names and locations of Access library files in earlier versions of Access:

Access 2010	C:\Program Files\Microsoft Office\Office14\msacc.olb
Access 2007	C:\Program Files\Microsoft Office\Office12\msacc.olb
Access 2003	C:\Program Files\Microsoft Office\Office11\msacc.olb
Access 2002	C:\Program Files\Microsoft Office\Office10\msacc.olb
Access 2000	C:\Program Files\Microsoft Office\Office\msacc9.olb
Access 97	C:\Program Files\Microsoft Office\Office\msacc8.olb

The Microsoft Office 15.0 Access Database Engine Object Library

This library is the enhanced version of the DAO Object Library. It was built specifically for working with the ACE database engine. In Access 2013, the library is stored in the acedao.dll file in the C:\Program Files\Common Files\Microsoft Shared\Office15 folder. This library is used when you open an Access database in the default Access 2013 format (.accdb).

The Microsoft DAO 3.6 Object Library

This library is stored in the dao360.dll file located in the C:\Program Files\Common Files\Microsoft Shared\DAO folder and is used by Access MDB databases created in Access 2000 through 2013. Access 97 uses dao350.dll, which is located in the same folder.

DAO provides programmatic access to Jet Access databases. It consists of a hierarchy of objects that supply methods and properties for designing and manipulating databases. The DBEngine object positioned at the top of the DAO object hierarchy is often referred to as the Jet engine and is used to reference the database engine as a whole. All the other objects and collections in the DAO object hierarchy fall under DBEngine. The DBEngine contains the following two collections of objects:

- The Errors collection, which stores a list of errors that have occurred in the DBEngine. These errors are represented by the Error objects and should not be confused with the Err object, which stores runtime errors generated in Visual Basic.
- The Workspaces collection (the default collection of the DBEngine object), which contains the Workspace objects and is used for database security in multiuser applications. The Workspace object is used in conjunction with User and Group objects.

Each open database is represented by the Database object. The Database object is used to reference a Microsoft Access database file (.mdb) or another external database represented by an ODBC data source. The Databases collection contains all currently open databases. The Containers, QueryDefs, Relations, and TableDefs collections contain objects that are used to reference various components of the Database object. For example, the TableDef object represents a table or a linked table in a Microsoft Jet workspace. The QueryDef object represents a query in DAO. If values are supplied to a query, they are represented in DAO by a Parameter object. The Parameters collection contains all of the Parameter objects defined for a QueryDef object. The Relation object represents a relationship between fields in tables and queries. The Container object is used to access collections of saved objects that represent databases, tables, queries, and relationships.

The Recordsets collection contains all open Recordset objects. Each Recordset object represents a set of records within a database. You will use Recordset objects for retrieving, adding, editing, and deleting records from a database.

The Field object represents a field in a table, query, index, relation, or recordset. The Fields collection is the default collection of a TableDef, QueryDef, Index, Relation, or Recordset object.

Some DAO objects have a Properties collection. The Properties collection contains a separate object for each property of the DAO object that is referenced. You can use an object's Properties collection to enumerate its properties or to return their settings. You can also define your own custom properties on DAO objects.

The Microsoft ActiveX Data Objects 6.1 Library (ADO)

This library is stored in msado15.dll and can be found in the C:\Program Files\Common Files\System\ADO folder. ActiveX Data Objects (ADO) that are provided by this library are used for accessing and manipulating data from a variety of sources through an OLE DB provider.

If you scroll down the list of the Available References (Figure 11.1), you may get confused to see several different versions of the Microsoft ActiveX Data Objects Library. Which version should you use depends on whether the users of your Access applications are on Windows 7 or 8, Vista or XP. For Windows 7 and above, use version 6.1 of this library. For Windows Vista, stick to version 6.0 which came with Vista, and for Windows XP SP3 or Windows Server 2003 SP1, select version 2.8 or lower.

ADO works with the technology known as *OLE DB*. This technology is object-based, but it is not limited to relational databases. OLE DB can access both relational and nonrelational data sources such as directory services, mail stores, multimedia, and text files, as well as mainframe data (VSAM and MVS). You do not need any specific drivers installed on your computer to access external data with OLE DB because OLE DB does not use drivers; it uses data providers to communicate with data stores. *Data providers* are programs that enable access to data. OLE DB has many providers, such as Microsoft OLE DB Provider for SQL Server and Microsoft Jet 4.0 OLE DB Provider. There are also providers for Oracle, NT 5 Active Directory®, and ODBC.

Similar to DAO, ADO objects make it possible to establish a connection with a data source in order to read, insert, modify, and delete data. ADO offers to programmers many advanced features that are not available in DAO. For example, the ADO Connection object's State property lets you determine whether the connection is closed (adStateClosed), open and ready (adStateOpen), still trying to connect (adStateConnecting), processing a command (adStateExecuting), or fetching data (adStateFetching). The ADO Recordsets can be hierarchical, fabricated, disconnected, or persisted on disk.

ADO consists of three object models, each providing a different area of functionality (see Table 11.3). Because of this, only the objects necessary for a specific task need to be loaded at any given time.

TABLE 11.3. Components of ADO

Object Model	What It's Used For
ADODB (ActiveX Data Objects)	Data manipulation Access and manipulate data through an OLE DB provider. With ADO objects you can connect to a data source and read, add, update, or delete data. Library Name: Microsoft ActiveX Data Objects 6.1 Library Library File: msado15.dll Library Folder: C:\Program Files\Common Files\System\ADO
ADOX (ADO Extensions for DDL and Security)	Data definition and security With ADOX objects you can define data such as tables, views, indexes, or relationships, as well as create and modify user and group accounts, and grant and revoke permissions on objects. Library Name: Microsoft ADO Ext. 6.0 for DDL and Security (ADOX) Library File: msadox.dll Library Folder: C:\Program Files\Common Files\System\ADO
JRO (Jet and Replication Objects)	Replication (used with .mdb databases only) With JRO objects you can compact a Jet database, and create, modify, and synchronize replicas. JRO can be used only with Microsoft Jet databases. Library Name: Microsoft Jet and Replication Objects 2.6 Library (JRO) Library File: msjro.dll Library Folder: C:\Program Files\Common Files\System\ADO

Later in this book you will learn how ADO can be used from a scripting language such as Microsoft Visual Basic Scripting Edition (VBScript).

Important Note: Access 2000 was the first version to support ADO. In an attempt to promote universal data access, Microsoft made ADO the default library in Access 2000 and 2002. DAO was to be phased out and Access programmers were advised to move their application code from DAO to ADO. Since then, having found out that DAO still performed faster in most cases, was easier to use, and offered features that were specifically designed with Jet/ODBC databases in mind, Microsoft has returned to DAO as the main data access layer. In Access 2007, DAO was enhanced to use the new data types and other improvements available in the .accdb format. This enhanced version of DAO was offered as the Microsoft Office 12.0 Access database engine Object Library. In Access 2013, it is offered as the Microsoft Office 15.0 Access database engine Object Library.

ADO Classic versus ADO.NET

The classic ADO used in VBA in Microsoft Access and other Microsoft Office applications is a completely different object model from ADO.NET used with the Microsoft .NET framework. ADO.NET is not built on ActiveX technology and its objects cannot be used directly in a VBA project.

CREATING A REFERENCE TO THE ADO LIBRARY

Prior to declaring variables as ADO objects in your VBA procedures, make sure that the reference to the library you are intending to use is set in the References dialog box. Hands-On 11.1 demonstrates how to create a reference to the Microsoft ActiveX Data Objects 6.1 Object Library.

Please note files for the "Hands-On" project may be found on the companion CD-ROM.

Hands-On 11.1. Setting Up a Reference to the ADO Object Library

1. Start Microsoft Access 2013 and create a new database named **Chap11.accdb** in your **C:\Access2013_ByExample** folder.
2. Press **Alt+F11** to switch to the Visual Basic Editor window, and then choose **Tools | References**.

3. Scroll down the list of available references until you locate the **Microsoft ActiveX Data Objects 6.1 Library**. Click the checkbox to the left of the name to select it.
4. Click **OK** to close the References dialog box.

All libraries that are checked in the References dialog box can be browsed using the Object Browser. This is a good way to become familiar with the names of objects that are available in a specific library and their various properties and methods (see Figure 11.2).

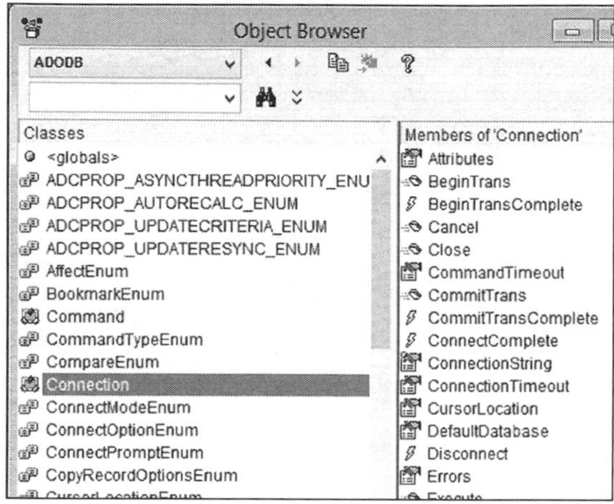

FIGURE 11.2. Use the Object Browser to find the objects available in a specific library.

UNDERSTANDING CONNECTION STRINGS

Needless to say, to retrieve or write data to a database, you will need to open it. There are many ways to connect to a database or an external data source from Microsoft Access 2013. The first thing to know about establishing database connections from your VBA procedures is how to prepare and use connection strings.

A *connection string* is a string variable that tells your VBA application how to establish a connection to a data source. There are two types of connection strings:

- ODBC connection strings (used by ODBC drivers)
- OLE DB connection strings (used by the OLE DB provider)

The syntax of ODBC and OLE DB connection strings is very similar. The connection string consists of a series of keyword and value pairs separated by semicolons:

```
Keyword1=value; Keyword2=value
```

Please note that the connection string does not contain spaces before or after the equal sign (=). The parameters in the connection string may vary depending on the ODBC driver or OLE DB provider used and the data store that you are connecting to (e.g., Microsoft Access, SQL Server, and so forth).

Let's examine the connection string you would need to connect to an older Microsoft Access database in the .mdb file format. For the ODBC connection, the following connection string will allow you to connect to an Access database called Northwind.mdb:

```
"Driver={Microsoft Access Driver (*.mdb)};" & _
"DBQ=C:\Access2013_ByExample\Northwind.mdb;"
```

In the preceding connection string, `Driver` specifies what type of database you're using. `DBQ` is the physical path to the database. If the Northwind.mdb file is protected with a password, you must provide additional information in the connection string:

```
"Driver={Microsoft Access Driver (*.mdb)};" & _
"DBQ=C:\Access2013_ByExample\Northwind.mdb;" & _
"UID=admin;PWD=secret;"
```

`UID` specifies the username. `PWD` specifies the user password.

To create an OLE DB connection to the same Northwind.mdb database that uses standard security, you will need to write the connection string as follows:

```
"Provider=Microsoft.Jet.OLEDB.4.0;" & _
"Data Source=C:\Access2013_ByExample\Northwind.mdb;" & _
"User Id=Admin;Password=;"
```

`Provider` identifies the OLE DB provider for your database; in this case, we want to use the Jet OLE DB Provider. `Data Source` specifies the full path and filename of the .mdb database file.

To create an OLE DB connection to the SQL database called Northwind, use the following connection string:

```
"Provider=SQLOLEDB; Data Source=(local);" & _
"Integrated Security=SSPI;Initial Catalog=Northwind"
```

In this connection string, `SQLOLEDB` is the name of the OLE DB provider for SQL Server databases. The `Data Source` parameter specifies the name or address of the SQL Server. To connect with an SQL Server running on the same computer, use the keyword `(local)` for the `Data Source`. For a trusted connection (Microsoft Windows NT integrated security), set the `Integrated Security` parameter to `SSPI`. Use the `Initial Catalog` parameter to specify which database you want to connect to.

If the `Provider` keyword is not included in the connection string, the OLE DB provider for ODBC (`MSDASQL`) is the default value. This provides backward compatibility with ODBC connection strings.

USING ODBC CONNECTION STRINGS

When you choose to connect to a data source via the ODBC, you must specify the connection information. You do this by creating a DSN (Data Source Name) or DSN-less connection. *DSN* connections store the connection information in the Windows Registry or in a .dsn file. In a *DSN-less* connection, all connection information is specified in the connection string. The following subsections explain each ODBC connection type in detail.

Creating and Using ODBC DSN Connections

Windows uses an ODBC Data Source Administrator (see Figure 11.3) to manage ODBC drivers and data sources available on the computer. In Windows 7, you can access this tool by opening Control Panel | System and Security | Administrative Tools | Data Sources (ODBC). In Windows XP, go straight to Administrative Tools in the Control Panel and choose Data Sources (ODBC).

The DSN contains information about database configuration, location, and user security. There are three types of DSNs:

- **User DSN**—A User DSN is stored locally in the Windows Registry and limits database connectivity to the user who creates it. In other words, if you create a User DSN under your user account, no other user will be able to see it or use it. Hands-On 11.2 demonstrates how to create this type of DSN so that you can run the example code on your computer.
- **File DSN**—A File DSN is a special type of file that stores all the connection settings. File DSNs are saved by default in the Program Files\Common Files\Odbc\Data Sources folder. Because the connection parameters

and values are stored in a file, they can be easily shared with other users. If other users require the same connection, simply send them the DSN file and you won't need to configure a DSN for each system.

- **System DSN**—A System DSN is stored locally in the Windows Registry and allows any logged-on user, process, and service to see it and use it. System DSNs are often used in establishing connections to external data sources from Active Server Pages (ASP).

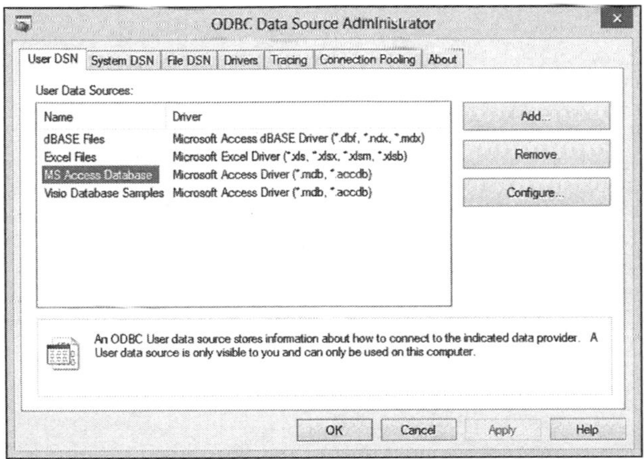

FIGURE 11.3. The ODBC Data Source Administrator allows you to set up appropriate connections with the required data provider via the User, System, or File DSN.

Hands-On 11.2 will get you started with the ODBC Data Source Administrator by walking you through the creation of a User DSN named MyDbaseFile to access data in a legacy dBASE database file (Customer.dbf). You will then use this data source name to programmatically open a dBASE file with ADO using the ODBC DSN connection.

Hands-On 11.2. Creating and Using the ODBC DSN Connection to Read Data from a dBASE File

The procedure code in this Hands-On relies on the reference to the ActiveX Data Objects Library that was set in Hands-On 11.1.

1. Copy the **Customer.dbf** file from the **C:\Acccess2013_HandsOn** folder to your **C:\Access2013_ByExample** folder.
2. Open the Control Panel, activate **Administrative Tools**, and double-click **Data Sources (ODBC)**.

   The ODBC Data Source Administrator dialog box appears, as shown earlier in Figure 11.3.

3. With the **User DSN** tab selected, click the **Add** button.
4. Select **Microsoft dBASE driver (*.dbf)** and click **Finish.**
5. Enter **MyDbaseFile** as the Data Source Name and choose **dBASE 5.0** for the database version, as shown in Figure 11.4. Make sure you clear the **Use Current Directory** checkbox, then click the **Select Directory** button.

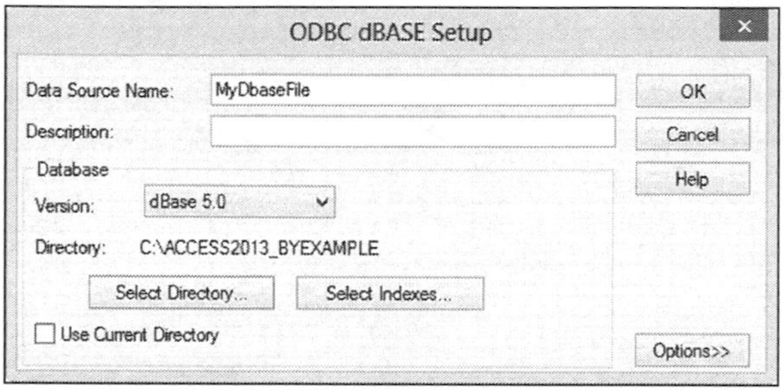

FIGURE 11.4. Creating a Data Source Name (DSN) to access a dBASE file.

6. In the Select Directory dialog box, select the **C:\Access2013_ByExample** folder where the Customer.dbf file is located, and click the **OK** button.
7. Click **OK** to exit the ODBC dBASE Setup dialog box.
   The MyDbaseFile data source now appears in the list of User Data Sources in the ODBC Data Source Administrator dialog box.
8. Click **OK** to close the ODBC Data Source Administrator dialog box.
9. Activate the Visual Basic Editor window and choose **Insert | Module.**
10. In the module's Code window, enter the following **Open_AndRead_dBaseFile** procedure:

```
Sub Open_AndRead_dBaseFile()
  Dim conn As ADODB.Connection
  Dim rst As ADODB.Recordset

  Set conn = New ADODB.Connection
  conn.Open "Provider=MSDASQL;DSN=MyDbaseFile;"

  Debug.Print conn.ConnectionString

  Set rst = New ADODB.Recordset
  rst.Open "Customer.dbf", conn

  Do Until rst.EOF
    Debug.Print rst.Fields(1).Value
    rst.MoveNext
  Loop
```

```
    rst.Close
    Set rst = Nothing
    conn.Close
    Set conn = Nothing
End Sub
```

11. Choose **Run | Run Sub/UserForm** to execute the procedure.
12. Open the Immediate window to view the data returned by the procedure.

If Visual Basic displays the runtime error "Data source name not found and no default driver specified," make sure there are no extra spaces in the connection string:

```
conn.Open "Provider=MSDASQL;DSN=MyDbaseFile;"
```

This is a very common error and it's hard to trace because spaces are difficult to spot.

The Open_AndRead_dBaseFile procedure uses the ADO Connection object to establish a connection with the data source. Prior to using ADO objects in your VBA procedures, make sure that the References dialog box contains the reference to the ActiveX Data Objects Library (see Hands-On 11.1). The procedure begins by declaring an object variable of Connection type, like this:

```
Dim conn As ADODB.Connection
```

The Connection object variable can be declared at procedure level or at module level. By declaring the variable at the top of the module, you can reuse it in multiple procedures in your module.

To handle data retrieval, an object variable of Recordset type is also declared:

```
Dim rst As ADODB.Recordset
```

Before you can use the declared ADO Connection object, you must initialize the object variable by using the `Set` keyword:

```
Set conn = New ADODB.Connection
```

At this point you can proceed to opening the data source by using the ADO Connection object's `Open` method. The required database connection information is passed to the `Open` method in the connection string, like this:

```
conn.Open "Provider=MSDASQL;DSN=MyDbaseFile;"
```

MSDASQL is the Microsoft OLE DB provider for all ODBC data sources. The names of common data providers used with ADO are listed in Table 11.4.

The Provider property of the ADO Connection object is used in the connection string as the provider name. DSN is the name of the data source that you specified for your connection settings in the ODBC Data Source Administrator dialog box. Since MSDASQL is the default provider for ODBC, it's okay to leave it off, like this:

```
conn.Open "DSN=MyDbaseFile;"
```

TABLE 11.4. Common data providers used with ADO

Provider Name	Provider Property	Description
Microsoft ACE	Microsoft.ACE.OLEDB.14.0	Used by Access 2013/2010 databases in .accdb file format. By default, this provider opens databases in Read/Write mode.
Microsoft Jet	Microsoft.Jet.OLEDB.4.0	Used for Jet 4.0 databases (in .mdb file format). By default, this provider opens databases in Read/Write mode.
Microsoft SQL Server	SQLOLEDB	Used to access SQL Server databases.
Oracle	MSDAORA	Used to access Oracle databases.
ODBC	MSDASQL	Used to access ODBC data sources without a specific OLE DB provider. This is the default provider for ADO.
Active Directory Service	ADSDSOObject	Used to access Windows NT 4.0 directory services, Novell® directory services, and LDAP-compliant directory services.
Index Server	MSIDXS	Read-only access to Web data.

Once the connection to the dBASE database file is open, the procedure initializes the `rst` object variable using the `Set` keyword in order to gain access to its data:

```
Set rst = New ADODB.Recordset
```

ADO Recordsets are covered in detail in Chapter 14, "Finding and Reading Records." The ADO Recordset object's `Open` method is used to open the Customer.dbf file, like this:

```
rst.Open "Customer.dbf", conn
```

When you open the recordset, you need to specify at the minimum the data you want to retrieve (Customer.dbf) and how to connect to that data (`conn`). Once the recordset is open, you can start reading its data. The `Do Until` loop will iterate through the recordset until the `EOF` (End of File) is reached. Each time through the loop, VBA will write to the Immediate window the value of the first field. When the procedure ends you should see in the Immediate window the names of all customers from the Customer.dbf file.

When you are done reading the records, the procedure uses the `Close` method to close the recordset and destroy the `rst` object variable by setting it to `Nothing`:

```
Set rst = Nothing
```

This statement completely releases the resources used by the Recordset object. The same should be done with the Connection object variable (`conn`) when it is no longer needed:

```
conn.Close
Set conn = Nothing
```

Creating and Using DSN-less ODBC Connections

It is possible that your VBA application that relies on database access via ODBC DSN (Data Source Name) may suddenly fail because the DSN was modified or deleted. Therefore, it may be a better idea to use a so-called DSN-less connection. Instead of setting up a DSN as you did in Hands-On 11.2, specify your ODBC driver name and all driver-specific information in your connection string. Different types of databases can require that you specify different parameters. Because the ODBC DSN setup is not required, this type of connection is called "DSN-less."

You can rewrite the procedure in Hands-On 11.2 to use a DSN-less ODBC connection. See the Access2013_HandsOn\HandsOn11.2_Supplement.txt on the CD-ROM disk.

TABLE 11.5. ODBC connection strings for common data sources

Data Source Driver	ODBC Connection String (used in DSN-less connections)
Microsoft Access 2013/2010/2007 (accessing .mdb or .accdb files from Access 2013/2007)	"Driver={Microsoft Access Driver (*.mdb, *.accdb)}; DBQ=path to mdb/accdb file;UID=admin;PWD=;"
Microsoft Access (accessing .mdb files from Access 2003/2002/2000/97)	*Using standard security:* "Driver={Microsoft Access Driver (*.mdb)}; DBQ=C:\filepath\myDb.mdb;UID=admin;PWD=;" *Using user-level security (workgroup information file)*: "Driver={Microsoft Access Driver (*.mdb)}; DBQ=C:\filepath\myDb.mdb;SystemDB=C:\filepath\myDb.mdw;UID=myUserName;PWD=myPassword;"
Microsoft Excel 2013/2010/2007 (accessing .xls, .xlsx, .xlsm, and .xlsb files from Excel 2013/2010/2007)	"Driver={Microsoft Excel Driver (*.xls, *.xlsx, *.xlsm, *.xlsb)};DBQ=path to xls/xlsx/xlsm/xlsb file; "
Microsoft Excel (accessing .xls files from Excel 2003/2002/2000/97)	"Driver={Microsoft Excel Driver (*.xls)}; DBQ=C:\filepath\Spreadsheet.xls;"
dBASE	"Driver={Microsoft dBASE Driver (*.dbf)}; DBQ=C:\filepath;"
Text	"Driver={Microsoft Text Driver (*.txt, *.csv)}; DefaultDir=C:\filepath\myText.txt;"
Microsoft SQL Server	*Using Trusted Connection security:* "Driver={SQL Server}; Server=myServerName; Database=myDatabaseName; UID=;PWD=;" *Using standard security:* "Driver={SQL Server}; Server=myServerName;

Data Source Driver	ODBC Connection String (used in DSN-less connections)
	Trusted_Connection=no; Database=myDatabaseName; UID=myUserName;PWD=myPassword;"
Oracle	"Driver={Microsoft ODBC for Oracle}; Server=OracleServer.World; UID=myUserName;PWD=myPassword;"

USING OLE DB CONNECTION STRINGS

In numerous VBA procedures in this chapter, we'll use an OLE DB provider to communicate with a data source. See Table 11.4. earlier in this chapter for the names of common OLE DB providers used with ADO. Table 11.6. shows OLE DB connection strings for common data sources.

TABLE 11.6. OLE DB connection strings for common data sources

Data Source	OLE DB Connection String
Microsoft Access 2013/2010/2007	"Provider=Microsoft.ACE.OLEDB.12.0; Data Source=C:\Access2013_ByExample\Northwind 2007.accdb"
Microsoft Access (prior to 2007)	"Provider=Microsoft.Jet.OLEDB.4.0; Data Source=C:\Access2013_ByExample\Northwind.mdb;"
Microsoft Excel 2013/2010	"Provider=Microsoft.ACE.OLEDB.12.0; Data Source=C:\Access2013_ByExample\Report2013.xlsx; Extended Properties="" Excel 12.0;HDR=Yes"";"
Microsoft Excel 2007	"Provider=Microsoft.ACE.OLEDB.12.0; Data Source=C:\Access2013_ByExample\Report2013.xlsx; Extended Properties=""Excel 12.0;HDR=Yes"";"

(Contd.)

Data Source	OLE DB Connection String
Microsoft Excel (prior to 2007)	"Provider=Microsoft.Jet.OLEDB.4.0; Data Source=C:\Access2013_ByExample\Report.xls; Extended Properties=""Excel 8.0;HDR=Yes"";"
Microsoft SQL Server	"Provider=SQLOLEDB;Data Source=myServerName;Network Library=DBMSSOCN;Initial Catalog=Pubs;"
Oracle	"Provider=MSDAORA;Data Source=myTable;"

CONNECTION STRING VIA A DATA LINK FILE

If you are using the Windows operating system and are looking for an easy way to create and test a connection string that uses an ODBC driver or OLE DB provider, you may want to use the Data Link Properties dialog box, which is shown in Figure 11.5.

A universal data link file (.udl) is a text file containing the connection information. Hands-On 11.3 demonstrates how to create the .udl file to connect to a Microsoft Access 2013 database. You can use the same technique to create a valid connection string to other external data sources as long as the ADO provider is installed on your computer.

Hands-On 11.3. Creating and Using a Universal Data Link File

1. In Windows Explorer, select the **C:\Access2013_ByExample** folder. Make sure that the Show file name extension is checked in the folder options. Choose **File | New** and select **Text document**.
2. A new file named New Text Document.txt appears in the **Access2013_ByExample** folder. Rename this file **ConnectToAccdb.udl.**
   When changing the filename, be sure to type the new extension (.udl) as indicated.
3. Windows will display a warning message that changing the file extension can cause the file to become unusable. Ignore this message and click **OK**.
   Windows creates an empty universal data link file. Notice that the file size is 0 Kb.
4. Double-click the **ConnectToAccdb.udl** file.
   Windows opens the Data Link Properties dialog box, which contains the following four tabs:

Data Link Tab	Description
Provider	Lists the names of the ADO providers installed on your computer. The provider name you select must be appropriate for the data source you want to use. For example, if you select Microsoft Jet 4.0 OLE DB provider, you must select an Access database in .mdb format.
Connection	Allows you to define a data source name for the selected provider type. The entries shown here are specific to the provider type selected via the Provider tab. The Connection tab is active by default when you activate the Data Link Properties dialog box.
Advanced	Allows you to view and set other initialization properties for your data connection.
All	Allows you to review and edit all OLE DB initialization properties available for the selected OLE DB provider.

5. Click the **Provider** tab and select **Microsoft Office 12.0 Access Database Engine OLE DB Provider**, as shown in Figure 11.6.
6. Click the **Next** button or activate the **Connection** tab.
   The entries shown on the Connection tab are related to the type of provider you selected in Step 5.
7. In the Data Source box, type the location and filename of the Access 2013 database you want to connect to: **C:\Access2013_ByExample\Northwind 2007.accdb** (see Figure 11.7).
8. Click the **Test Connection** button to test whether you can connect to the specified database using the chosen data provider.
9. Click **OK** to the message box "Test connection succeeded."
   If you misspelled a filename or Windows cannot locate the file in the specified folder, you will get an error.

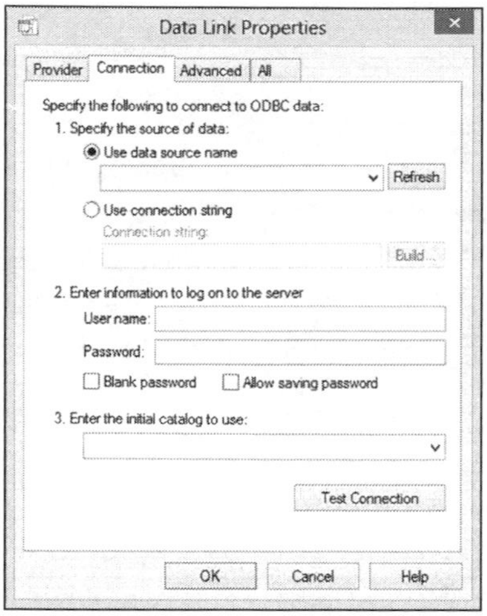

FIGURE 11.5. The Data Link Properties dialog box appears after you click on the .udlfile.

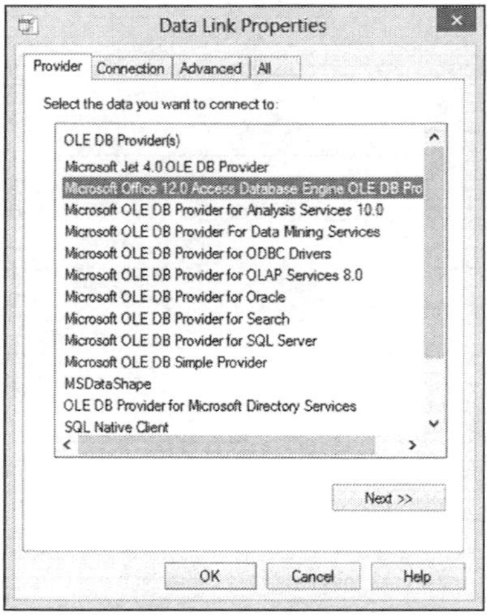

FIGURE 11.6. The Provider tab in the Data Link Properties dialog box lists the names of the ADO providers installed on your computer.

FIGURE 11.7. Using the Data Link Properties dialog box to define a data source name for the selected provider type. Be sure to enter .accdb as the extension for the Northwind 2007 database (the Data Source text box is too short to capture the entire path in this image).

At this point your connection string is ready to use.

10. Click **OK** to close the Data Link Properties dialog box.

When writing a VBA procedure to connect to the Northwind 2007.accdb database, you can simply pass the .udl filename to the Connection object's `Open` method:

```
Dim conn As ADODB.Connection
Set conn As New ADODB.Connection
conn.Open "File Name=C:\Access2013_ByExample\ConnectToAccdb.udl;"
```

When you use .udl files to store connection information, it is very easy to switch your procedure's data source without having to make changes to your code. Simply double-click the .udl file and make desired modifications in the Data Link Properties dialog box.

If you'd rather use the connection string in your VBA procedure, then go ahead and copy the string from the .udl file. You can open this file in Notepad in one of the following ways (see Figure 11.8):

- Right-click the .udl filename and choose Open With, then select Notepad. If Notepad is not available in the shortcut menu, select Choose Program, then choose Notepad, and click OK.
- Make a copy of the .udl file. Change the .udl extension of the created copy to .txt. Double-click the file to open it in Notepad.

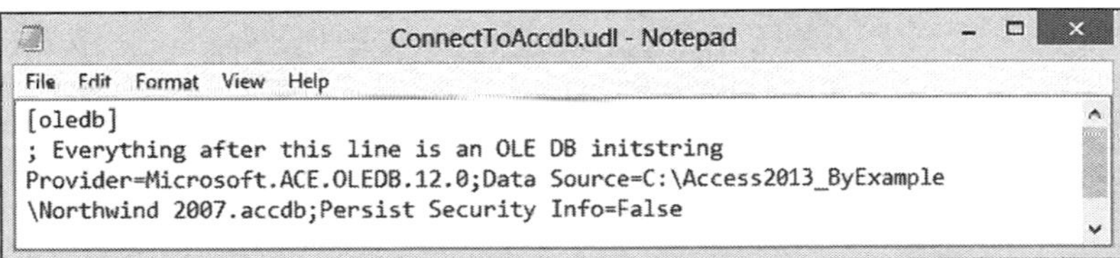

FIGURE 11.8. You can obtain the connection string from the universal data link (.udl) file by opening the file in Windows Notepad.

OPENING MICROSOFT ACCESS DATABASES

In this section, you will learn how to use DAO and ADO to open Microsoft Access ACCDB and MDB databases in read/write mode and in read-only mode. You will also learn how to open Access databases that have been protected with database passwords or user-level security.

Opening a Microsoft Jet Database in Read/Write Mode with DAO

The easiest way to open an existing Microsoft Access database from a VBA procedure is by using the Microsoft Access database engine's `OpenDatabase` method.

This method requires that you provide at least one parameter—the name of the existing database. When you open the database with the `OpenDatabase` method, always remember to close it. The `Close` method removes the database from the Databases collection.

Hands-On 11.4 demonstrates how to open an Access database in .accdb or .mdb format using the DAO's `OpenDatabase` method. This example will list containers and documents in the open database. Each Database object has a Containers collection that consists of built-in Container objects. The Containers collection is used for storing Microsoft Access's own objects. The Jet engine creates the following Container objects: Databases, Tables, and Relations. Other Container objects are created by Microsoft Access (Forms, Reports, Macros, and Modules).

Table 11.7 lists the Container objects and the type of information they contain.

TABLE 11.7. Container objects

Container Name	Type of Information Stored
Databases	Saved databases
Tables	Saved tables and queries
Relations	Saved relationships
Forms	Saved forms
Modules	Saved modules
Reports	Saved reports
Scripts	Saved scripts

Each Container object contains a Documents collection. Each document in this collection represents an object that can be found in an Access database. For example, the Forms container stores a list of all saved forms in a database, and each form is represented by a Document object. You cannot create new Container and Document objects; you can only retrieve the information about them.

Hands-On 11.4. Opening a Database with DAO in Read/Write Mode

1. In the **Chap11.accdb** database that you created in Hands-On 11.1 (in the Visual Basic Editor window), choose **Insert | Module** to add a new module to the current VBA project.

2. In the module's Code window, type the following **openDB_DAO** procedure:

```
Sub openDB_DAO()
  Dim db As DAO.Database
  Dim dbName As String
  Dim c As Container
  Dim doc As Document

  dbName = InputBox("Enter a name of an existing database:", _
   "Database Name")

  If dbName = "" Then Exit Sub
  If Dir(dbName) = "" Then
    MsgBox dbName & " was not found."
    Exit Sub
  End If

  Set db = OpenDatabase(dbName)
  With db
    ' list the names of the Container objects
    For Each c In .Containers
      Debug.Print c.Name & " container:" & _
       c.Documents.Count
      ' list the document names
      ' in the specified Container
      If c.Documents.Count > 0 Then
        For Each doc In c.Documents
          Debug.Print vbTab & doc.Name
        Next doc
      End If
    Next c
    .Close
  End With
End Sub
```

This procedure uses the `OpenDatabase` method of the DBEngine object to open the specified database in the default workspace. The database is opened as shared with read/write access. By supplying additional arguments to the `OpenDatabase` method you could open the database exclusively (a database opened exclusively can be accessed by a single user at a time) or as read-only.

The openDB_DAO procedure uses a `For Each...Next` loop to retrieve the names of all the Container objects in the opened database. If the specified container is not empty, the inner `For Each...Next` loop will print the name of each Document object in the Immediate window.

3. Position the insertion point anywhere within the code of openDB_DAO and press **F5** or choose **Run | Run Sub/UserForm** to execute the procedure. When you run this procedure you are prompted to enter the name of the Access database.
4. Enter **C:\Access2013_HandsOn\Northwind 2007.accdb** or **C:\Access2013_HandsOn\Northwind.mdb** and press **OK**. Check the procedure output in the Immediate window.

Opening a Microsoft Jet Database in Read/Write Mode with ADO

You can use ADO to open a Microsoft Access database for shared access (read/write). To connect to an older Microsoft Access database in the .mdb format, use the Microsoft.Jet.OLEDB.4.0 provider. To connect to an Access database in the .accdb format, use the Microsoft.ACE.OLEDB.12.0 provider. The names of common data providers used with ADO are listed in Table 11.4 earlier in this chapter.

To specify the data source name, use the Connection object's ConnectionString property. As you recall from earlier discussion, connection strings describe how to access data. Here's a code fragment that specifies the minimum required connection information:

```
With conn
  .Provider = "Microsoft.Jet.OLEDB.4.0;"
  .ConnectionString = "Data Source=" & CurrentProject.Path & _
   "\Northwind.mdb"
End With
```

In the preceding example, the data source includes the full path to the database file you are going to open. Change the Provider string to "Microsoft.ACE.OLEDB.12.0" if you are planning to open an Access 2013\2010\2007 database:

```
With conn
  .Provider = "Microsoft.ACE.OLEDB.12.0;"
  .ConnectionString = "Data  Source="  &  CurrentProject.Path  &  _
                     "\Northwind 2007.accdb"
End With
```

Once you've specified the minimum connection information, you may proceed to open the database.

Use the Connection object's `Open` method to open the connection to a data source:

```
conn.Open
```

ADO syntax is quite flexible. A connection to a database can also be opened like this:

```
conn.Open "Provider = Microsoft.Jet.OLEDB.4.0;" & "Data Source=" & _
                        CurrentProject.Path & "\Northwind.mdb"
```

As you can see in the preceding code fragment, the Provider name and the data source (in this example, path to the database) are supplied as arguments when you call a Connection object's `Open` method.

Or you could open the database connection like this:

```
With conn
  .Provider = "Microsoft.ACE.OLEDB.12.0;"
  .Mode = adModeReadWrite
  .ConnectionString = "Data Source=" & CurrentProject.Path & _
   "\Northwind 2007.accdb"
  .Open
End With
```

By default, the Connection object's `Open` method opens a database for shared access. You can use the Connection object's Mode property to explicitly specify the type of access to a database. The Mode property must be set prior to opening the connection because it is read-only once the connection is open. Connections can be opened read-only, write-only, or read/write. You can also specify whether other applications should be prevented from opening a connection. The value for the Mode property can be one of the constants/values specified in Table 11.8.

TABLE 11.8. Intrinsic constants of the Connection object's Mode property

Constant Name	Value	Type of Permission
`adModeUnknown`	0	Permissions have not been set yet or cannot be determined. This is the default setting.
`adModeRead`	1	Read-only permissions.
`adModeWrite`	2	Write-only permissions.
`adModeReadWrite`	3	Read/write permissions.

(Contd.)

Constant Name	Value	Type of Permission
adModeShareDenyRead	4	Prevents others from opening the connection with read permissions.
adModeShareDenyWrite	8	Prevents others from opening the connection with write permissions.
adModeShareExclusive	12	Prevents others from opening the connection.
adModeShareDenyNone	16	Prevents others from opening the connection with any permissions.

Hands-On 11.5 demonstrates how to use ADO to open an Access database for shared access (read/write).

Hands-On 11.5. Opening a Database with ADO in Read/Write Mode

1. In the Visual Basic Editor window, choose **Insert | Module** to add a new module to the currently open Chap11.accdb database.
2. In the module's Code window, type the following **openDB_ADO** procedure:

```
Sub openDB_ADO()
  Dim conn As ADODB.Connection
  Dim strDb As String

  On Error GoTo ErrorHandler

  strDb = CurrentProject.Path & "\Northwind 2007.accdb"
  Set conn = New ADODB.Connection

  With conn
    .Provider = "Microsoft.ACE.OLEDB.12.0;"
    .Mode = adModeReadWrite
    .ConnectionString = "Data Source=" & strDb
    .Open
  End With

  If conn.State = adStateOpen Then
    MsgBox "Connection was opened."
  End If

  conn.Close
  Set conn = Nothing
```

```
  MsgBox "Connection was closed."
  Exit Sub
ErrorHandler:
  MsgBox Err.Number & ": " & Err.Description
End Sub
```

3. Position the insertion point anywhere within the code of the openDB_ADO procedure and press **F5** or choose **Run | Run Sub/UserForm** to execute the procedure.

The ADO Connection object's State property returns a value that describes whether the connection is open, closed, connecting, executing, or retrieving data (see Table 11.9).

```
If conn.State = adStateOpen Then
  MsgBox "Database connection was established."
End If
```

TABLE 11.9. Intrinsic constants of the Connection object's State property

Constant	Value	Description
adStateClosed	0	Connection is closed.
adStateOpen	1	Connection is open.
adStateConnecting	2	Connection is connecting.
adStateExecuting	4	Connection is executing a command.
adStateFetching	8	Connection is retrieving data.

If an error occurs during the procedure execution (for example, when a database with the specified name or path cannot be found), the statement `On Error GoTo ErrorHandler` will pass the program control to the error-handling code located at the `ErrorHandler` label at the bottom of the procedure. Errors that occur in ADO are reported to the VBA Err object. You can find out the details about the error that occurred by using various properties of the Err object (Name, Description, Source, HelpFile, or HelpContext). The code in the error handler will execute only if an error occurs. If the procedure executes without an error, the `Exit Sub` statement will cause the procedure to finish without running the error code. You will find more information on database errors near the end of this chapter.

Opening a Microsoft Access Database in Read-Only Mode with DAO

You can open a Microsoft Access database in read-only mode by providing settings for optional arguments in the `OpenDatabase` method.

Additional Code on CD-ROM

File Name: Access2013_HandsOn\ openDB_DAOReadOnly.txt
Description: Open a database for shared, read-only access using DAO

Opening a Microsoft Jet Database in Read-Only Mode with ADO

If you'd like to open a database for read-only access, simply set the ADO Connection object's Mode property to the `adModeRead` constant (see Table 11.8 earlier).

Opening a Microsoft Jet Database Secured with a Password

Using passwords to secure the database or objects in the database is known as *share-level security*. When you set a password on the database, users will be required to enter a password in order to gain access to the data and database objects. Keep in mind that passwords are case-sensitive. You must use Data Access Objects (DAO) or ActiveX Data Objects (ADO) to programmatically open a password-protected Microsoft Access database. When using DAO to change the password of an existing Microsoft Access database in a VBA procedure, follow these steps:

1. Open the database in exclusive mode by setting the second argument of the `OpenDatabase` method to `True`.
2. To set a database password, use the NewPassword property of the Database object. This property requires that you first specify the old password and then the new one. Passwords can be up to 20 characters long and can include any characters except the ASCII character 0 (Null). To specify that the database does not have a password, use a zero-length string ("") in the first parameter of the NewPassword property. To clear the password, use the zero-length string for the second parameter of the New Password property.

To open a password-protected database using DAO, you must specify the database password in the `Connect` parameter of the `OpenDatabase` method as shown in Hands-On 11.6.

Hands-On 11.6. Setting a Database Password and Opening a Password-Protected Database with DAO

1. In the Visual Basic Editor window, choose **Insert | Module** to add a new module to the currently open Chap11.accdb database.
2. In the module's Code window, type the **setPass_AndOpenDB_withDAO** procedure shown here:

```
Sub setPass_AndOpenDB_withDAO()
  Dim db As DAO.Database
  Dim strDb As String

  ' strDb = "C:\Access2013_ByExample\Northwind 2007.accdb"
  strDb = "C:\Access2013_ByExample\Northwind.mdb"

  ' open the database in exclusive mode
  ' to set database password
  Set db = DBEngine.OpenDatabase(strDb, True)
  db.NewPassword "", "secret"
  MsgBox "Access Database version: " & Int(db.Version)
  db.Close

  ' open password-protected database
  Set db = DBEngine.OpenDatabase(Name:=strDb, _
   Options:=False, _
   ReadOnly:=False, _
   Connect:=";PWD=secret")

  MsgBox "Successfully opened a password-protected database."
  db.Close
  MsgBox "Password-protected database was closed."

  ' remove password protection from the database
  Set db = DBEngine.OpenDatabase(Name:=strDb, _
   Options:=True, _
   ReadOnly:=False, _
   Connect:=";PWD=secret")
   db.NewPassword "secret", ""

  MsgBox "Password protection was removed."
  db.Close
End Sub
```

3. Position the insertion point anywhere within the code of the setPass_AndOpenDB_withDAO procedure and press **F5** or choose **Run | Run Sub/UserForm** to execute the procedure.

When you run this procedure, Access displays the version number of the Microsoft Jet or Microsoft Access database engine using the Version property of DBEngine. The version number consists of the version number, a period, and the release number. The procedure uses the VBA `Int` function to display only the integer portion of the number. Microsoft Access 2013/2010/2007 files use the Microsoft Access database engine 12.0. Databases created in versions 2000, 2002, and 2003 use Microsoft Jet 4.0. Microsoft Access 97 uses Microsoft Jet 3.5.

If a VBA procedure uses a method or a property that requires two or more parameters, you can make the procedure more readable by specifying the names of the parameters like this:

```
Set db = DBEngine.OpenDatabase(Name:=strDb, _
 Options:=False, _
 ReadOnly:=False, _
 Connect:=";PWD=secret")
```

Use the Microsoft Visual Basic help to find the names of methods and properties and the names of the required and optional parameters.

Hands-On 11.7 demonstrates how to use ADO to set a database password for a Microsoft Access database in the .mdb format and then open it using the new password. This technique will not work for the Access 2013/2010/2007 databases in the .accdb format. To set a database password on an .mdb database file, use the JRO JetEngine object's `CompactDatabase` method and specify the `Password` parameter. The JRO JetEngine object is a member of the Microsoft Jet and Replication Objects (JRO) Library.

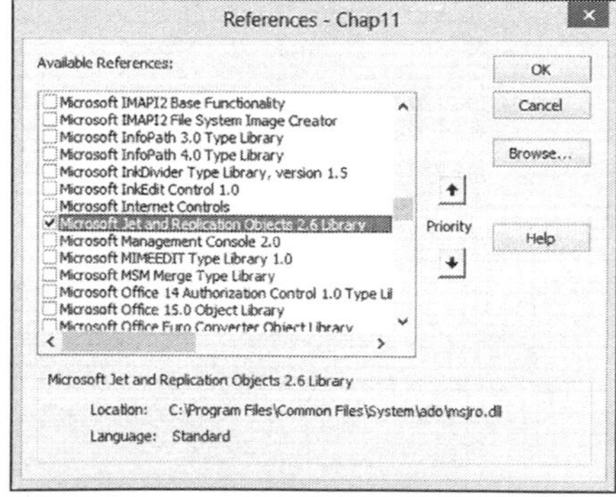

FIGURE 11.9. Before writing procedures that set or change the database password using ADO, you must set a reference to the Microsoft Jet and Replication Objects Library. To do this, in the Visual Basic Editor window, choose Tools | References and select the required library in the list of Available References.

Hands-On 11.7. Setting a Database Password and Opening a Password-Protected Database with ADO

1. Copy the **Northwind.mdb** database from the **C:\Access2013_HandsOn** folder to the **C\Access2013_ByExample** folder.
2. In the Visual Basic Editor window, choose **Tools | References** and select the **Microsoft Jet and Replication Objects 2.6 Library** as shown in Figure 11.9, then click **OK.**
3. In the Visual Basic Editor window, choose **Insert | Module** to add a new module to the currently open Chap11.accdb database.
4. In the module's Code window, type the following **setPass_AndOpenDB_withADO** procedure:

```
Sub setPass_AndOpenDB_withADO()
  Dim jetEng As JRO.JetEngine
  Dim conn As ADODB.Connection
  Dim strCompactFrom As String
  Dim strCompactTo As String
  Dim strPath As String

  strPath = CurrentProject.Path & "\"

  strCompactFrom = "Northwind.mdb"
  strCompactTo = "Northwind_P.mdb"

  On Error GoTo ErrorHandler

  Set jetEng = New JRO.JetEngine

  ' Compact the database specifying
  ' the new database password
  jetEng.CompactDatabase "Data Source=" & _
   strPath & strCompactFrom & ";", _
   "Data Source=" & strPath & strCompactTo & ";" & _
   "Jet OLEDB:Database Password=welcome"

  MsgBox "The database file " & strPath & strCompactTo & _
   " has been protected with password."
  Set jetEng = Nothing

  ' now open the password-protected MDB database
  Set conn = New ADODB.Connection
```

```
    With conn
      .Provider = "Microsoft.Jet.OLEDB.4.0;"
      .ConnectionString = "Data Source=" & _
       strPath & strCompactTo & ";" & _
       "Jet OLEDB:Database Password=welcome;"
      .Open
    End With

    If conn.State = adStateOpen Then
      MsgBox "Password-protected database was opened."
    End If

    conn.Close
    MsgBox "Password-protected database was closed."

    Set conn = Nothing
    Exit Sub
  ErrorHandler:
    If Err.Number = -2147217897 Then
      Kill strPath & strCompactTo
    ElseIf Err.Number = -2147467259 Then
      MsgBox "Make sure to close the " & strCompactFrom & _
       " database file prior to compacting it."
      Exit Sub
    Else
      MsgBox Err.Number & ": " & Err.Description
      Exit Sub
    End If
    Resume
  End Sub
```

5. Position the insertion point anywhere within the code of the setPass_AndOpenDB_withADO procedure and press **F8** or choose **Debug | Step Into** to execute the procedure one line at a time. Keep pressing **F8** until the procedure ends.

The procedure demonstrated here uses the JetEngine object's `CompactDatabase` method to compact a Microsoft Jet database (MDB) and password-protect it. By compacting the database, you can greatly improve its performance and reduce its file size. The `CompactDatabase` method requires that you provide the name of the .mdb file you want to compact and the name for the resulting compacted file. There are a number of connection properties that you can use with the `CompactDatabase` method. This procedure illustrates how to use the Jet OLEDB:Database Password property to set the password for the compacted database. The database password is set using the following code:

```
jetEng.CompactDatabase "Data Source=" & _
 strPath & strCompactFrom & ";", _
 "Data Source=" & strPath & strCompactTo & ";" & _
 "Jet OLEDB:Database Password=welcome"
```

You must close the database before attempting to compact it or Visual Basic will generate error -2147467259. The `ErrorHandler` code is used to trap errors that may occur during procedure execution. For example, if the database file cannot be found in the specified path, Visual Basic will display the error number and error description and will immediately exit the procedure. If you run this procedure more than once, Visual Basic will encounter error –2147217897: database already exists. To allow the procedure to run again, use the VBA `Kill` statement. This statement tells VBA to delete the file. The `Resume` statement will pass the procedure execution back to the line of code that caused the error and Visual Basic will proceed to execute this line and the remaining lines of code that follow.

Notice that to open a Microsoft Jet database (an .mdb file) secured with a password, you must specify the Jet OLEDB:Database Password property as part of the Connection object's ConnectionString property, like this:

```
With conn
  .Provider = "Microsoft.Jet.OLEDB.4.0;"
  .ConnectionString = "Data Source=" & _
   strPath & strCompactTo & ";" & _
   "Jet OLEDB:Database Password=welcome;"
  .Open
End With
```

Opening a Microsoft Jet Database with User-Level Security

Microsoft Access 2013/2010/2007 do not provide user-level security for databases that are created in the new file format (.accdb and .accde). The following discussion and Hands-On 11.8 apply only to Access databases in the .mdb file format.

User-level security secures the code and objects in your MDB database so that users can't accidentally modify or change them. With this type of security you can provide the most restrictive access over the database and objects it contains. When you implement user-level security, the Microsoft Jet Engine uses a workgroup information file named System.mdw to determine who can open a database and to secure its objects.

The workgroup information file holds group and user information, including passwords. The information contained in this file determines not only who can open the database but also the permissions users and groups have on the objects in the database. The workgroup information file contains built-in groups (Admins and Users) and a generic user account (Admin) with unlimited privileges on the database and the objects it contains. When an .mdb file is open in Access 2013/2010/2007, the Access user interface provides commands that allow you to manually implement user-level security (see Figure 11.10).

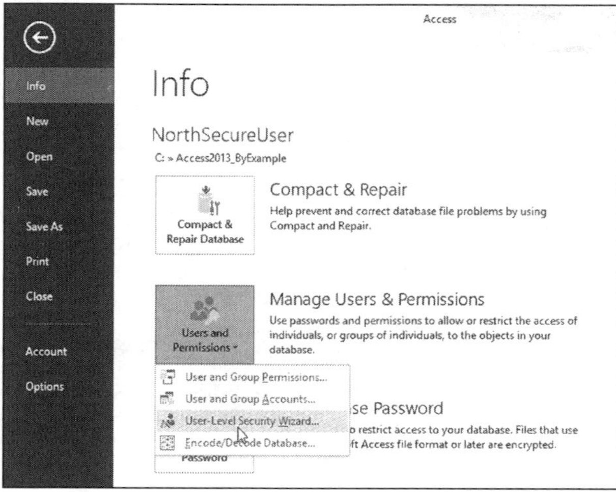

FIGURE 11.10. Setting user-level security in Access 2013 for earlier versions of Access in the .mdb file format.

To open an MDB database that is secured at the user level, you must supply the following:

- Full path to the workgroup information file (system database)
- User ID
- Password

Hands-On 11.8. Opening a Database Secured at the User Level

1. Use Windows Explorer to create a copy of the **C:\Access2013_ByExample\Northwind.mdb** database file and name it **NorthSecureUser.mdb**.
2. Open the **C:\Access2013_ByExample\NorthSecureUser.mdb** database. On the Info page, click the down arrow in the Users and Permissions button and select

User-Level Security Wizard to begin creating a new workgroup information file.

3. Follow the steps of the Security Wizard. Do not change anything until you get to the screen asking for username and password. Set up a user account named **Developer** with a password of **WebMaster**, and click the **Add This User to The List** button. Click the **Next** button, and assign Developer to the **Admins** group. To do this, begin by selecting **Developer** from the Group and User Name drop-down, then click the checkbox next to the Admins group. When done, press the **Finish** button. Access will display the One-Step Security Wizard Report. Print it out for your reference, then close it. Follow the Access prompts to create a snapshot of the data and close the NorthSecureUser database.
4. Open the **Chap11.accdb** database, switch to the Visual Basic Editor window, and choose **Insert | Module.**
5. In the module's Code window, type the following **Open_WithUserSecurity** procedure:

```
Sub Open_WithUserSecurity()
  Dim conn As ADODB.Connection
  Dim strDb As String
  Dim strSysDb As String

  On Error GoTo ErrorHandler
  strDb = CurrentProject.Path & "\NorthSecureUser.mdb"
  strSysDb = CurrentProject.Path & "\Security.mdw"
  Set conn = New ADODB.Connection
  With conn
    .Provider = "Microsoft.Jet.OLEDB.4.0;"
    .ConnectionString = "Data Source=" & strDb & ";" & _
     "Jet OLEDB:System Database=" & strSysDb
    .Open, "Developer", "WebMaster"
  End With
  MsgBox "Secured database was opened."
  conn.Close
  Set conn = Nothing
  MsgBox "Database was closed."
  Exit Sub

ErrorHandler:
  MsgBox Err.Number & ": " & Err.Description
End Sub
```

6. Position the insertion point anywhere within the procedure code and press **F5** or choose **Run | Run Sub/UserForm** to execute the procedure.

The Security Wizard places a shortcut to the NorthSecureUser.mdb file on your desktop to make it easy for you to start the secured database using the new workgroup information file (Security.mdw). The path to the file is as follows:

"C:\Program Files\Microsoft Office 15\root\office15\MSACCESS.EXE" "C:\Access2013_ByExample\NorthSecureUser.mdb" /WRKGRP "C:\Access2013_ByExample\Security.mdw
For more information on ,mdw files and implementing database security with ADOX and JRO, see Chapter 18, "Implementing Database Security."

CONNECTING TO THE CURRENT ACCESS DATABASE

Microsoft Access provides a quick way to access the current DAO database by using the `CurrentDb` method. This method returns an object variable of type Database that represents the database currently open in the Microsoft Access window. In ADO, however, use the `CurrentProject.Connection` statement to access the currently open database. The CurrentProject object refers to the project for the current Microsoft Access database. These statements work only in VBA procedures created in Microsoft Access. If you'd like to reuse your VBA procedures in other Microsoft Office Visual Basic applications, you will be better off creating a connection via an appropriate OLE DB provider.

The procedure in Hands-On 11.9 uses the `CurrentProject.Connection` statement to return a reference to the current database. Once the connection to the current database is established, the example procedure loops through the Properties collection of the Connection object to retrieve its property names and settings. The results are written both to the Immediate window and to a text file named C:\Access2013_ByExample\Propfile.txt.

Hands-On 11.9. Establishing a Connection to the Current Access Database

1. In the Visual Basic Editor window, choose **Insert | Module**.
2. In the module's Code window, type the **Connect_ToCurrentDB** procedure shown here:

```
Sub Connect_ToCurrentDB()
  Dim conn As ADODB.Connection
  Dim fs As Object
  Dim txtfile As Object
  Dim i As Integer
  Dim strFileName As String

  strFileName = "C:\Access2013_ByExample\Propfile.txt"

  Set conn = CurrentProject.Connection
  Set fs = CreateObject("Scripting.FileSystemObject")
  Set txtfile = fs.CreateTextFile(strFileName, True)

  For i = 0 To conn.Properties.Count - 1
    Debug.Print conn.Properties(i).Name & "=" & _
     conn.Properties(i).Value
    txtfile.WriteLine (conn.Properties(i).Name & _
     "=" & conn.Properties(i).Value)
  Next i
  MsgBox "Please check results in the " & _
   "Immediate window." & vbCrLf _
   & "The results have also been written to the " _
   & Chr(13) & strFileName & " file."

  txtfile.Close

  Set fs = Nothing
  conn.Close
  Set conn = Nothing
End Sub
```

The Connect_ToCurrentDB procedure uses the `CurrentProject.Connection` statement to get a reference to the currently open database. To create a text file from a VBA procedure, the `CreateObject` function is used to access the Scripting.FileSystemObject. This function returns the FileSystemObject (`fs`). The `CreateTextFile` method of the FileSystemObject creates the TextStream object that represents a text file (txtfile). The `WriteLine` method writes each property and the corresponding setting to the newly created text file (C:\Propfile.txt). Finally, the `Close` method closes the text file.

3. Choose **Run | Run Sub/UserForm** to execute the procedure.

OPENING OTHER DATABASES, SPREADSHEETS, AND TEXT FILES FROM ACCESS 2013

Microsoft Access Jet/ACE database engine can be used to access other databases, spreadsheets, and text files. The following subsections of this chapter demonstrate how to connect to SQL Server, Excel spreadsheets, and text files.

Connecting to an SQL Server Database

The ADO provides a number of ways of connecting to an SQL Server database. To access data residing on Microsoft SQL Server, use SQLOLEDB, which is the native Microsoft OLE DB provider for SQL.

You can also connect to an SQL database using the MSDASQL provider. This provider allows you to access any existing ODBC data sources. You can open a connection to the SQL Server by using an ODBC DSN (Data Source Name) or an ODBC DSN-less connection. Both of these connection types were discussed earlier in this chapter. The following code snippet opens and then closes a connection with the SQL Server database based on a DSN named Pubs.

```
With conn
  .Open "Provider=MSDASQL; DSN=Pubs"
  .Close
End With
```

Recall that you can skip setting the Provider property because MSDASQL is the default provider for ODBC. All you really need to establish a connection in this case is a DSN.

Additional Code on CD-ROM

Filename: Access2013_HandsOn/Hands-On 11.10.txt
Description: Connecting to an SQL Server Database Using SQLOLEDB Provider

Opening a Microsoft Excel Workbook

You can open a Microsoft Excel workbook from Access by writing procedures that use DAO or ADO objects.

To open a Microsoft Excel 2013/2010/2007 workbooks with the .xlsx file format using DAO, use the `OpenDatabase` method like this:

```
Dim db As DAO.Database
```

```
Set db = OpenDatabase("C:\Access2013_ByExample\" & _
        "Report2013.xlsx", _
        False, True, "Excel 12.0; HDR=YES;")
```

In the first parameter, specify the path and filename to your workbook. The second parameter of the `OpenDatabase` method (`False`) indicates that the file is to be opened in shared mode (this is the default). The third parameter is set to `True`, which means the workbook file opens in read-only mode. The fourth parameter is the connection information. It specifies the version of the Excel sheet. For Excel 2013/2010/2007, set it to Excel 12.0; for Excel 2000/2002/2003, set it to Excel 8.0: and for Excel version 97, set it to Excel 5.0. `HDR=YES;` indicates that the first row contains column names. To indicate that the workbook does not contain column names, set this to `NO`.

To open Microsoft Excel 2013/2010/2007 workbook files with the .xlsx file format using ADO, use the Microsoft ACE OLEDB 12.0 provider and use the Extended Properties of the ADO Connection object to pass the connection string like this:

```
Dim conn As ADODB.Connection
Set conn = New ADODB.Connection
conn.Open "Provider=Microsoft.ACE.OLEDB.12.0;" & _
 "Data Source=C:\Access2013_ByExample\" & _
 "Report2013.xlsx;" & _
 "Extended Properties=""Excel 12.0; HDR=YES"";"
```

To open workbook files created in Excel 2000/2002/2003, use the Microsoft Jet OLE DB 4.0 provider and Excel 8.0 in the Extended Properties:

```
Dim conn As ADODB.Connection
conn.Open "Provider=Microsoft.Jet.OLEDB.4.0;" & _
  "Data Source=C:\Access2013_ByExample\Report.xls;" & _
  "Extended Properties=""Excel 8.0; HDR=YES"";"
```

You can also use ODBC to open an Excel workbook file. For example, the following code snippet establishes an ODBC DSN-less connection:

```
Dim conn As ADODB.Connection
Set conn = New ADODB.Connection
With conn
  .ConnectionString = "Driver={Microsoft Excel Driver " & _
   "(*.xls, *.xlsx, *.xlsm, *.xlsb)};" & _
   "DBQ=C:\Access2013_ByExample\Report2013.xlsx;"
  .Open
End With
```

Hands-On 11.11 demonstrates how to use DAO to open a Microsoft Excel workbook.

Hands-On 11.11. Opening an Excel Workbook with DAO

1. Copy the **Report2013.xlsx** and **Report.xls** workbook files from **C:\Access2013_HandsOn** to your **C:\Access2013_ByExample** folder.
2. In the Visual Basic Editor window, choose **Insert | Module.**
3. In the module's Code window, type the following **Open_Excel_DAO** procedure:

```
Sub Open_Excel_DAO(strFileName)
  Dim db As DAO.Database
  Dim rst As DAO.Recordset
  Dim strHeader As String
  Dim strValues As String
  Dim fld As Variant

  strHeader = ""
  strValues = ""

  If Right(strFileName, 1) = "x" Then
    Set db = OpenDatabase(CurrentProject.Path & _
     "\Report2013.xlsx", False, True, _
     "Excel 12.0; HDR=YES;")
  Else
    Set db = OpenDatabase(CurrentProject.Path & _
    "\Report.xls", False, True, _
    "Excel 8.0; HDR=YES;")
  End If

  Set rst = db.OpenRecordset("Sheet1$")

  ' get column names
  For Each fld In rst.Fields
    strHeader = strHeader & fld.Name & vbTab
  Next

  Debug.Print strHeader

  ' get cell values
  Do Until rst.EOF
```

```
      For Each fld In rst.Fields
        strValues = strValues & fld.Value & _
         vbTab & vbTab
      Next
      Debug.Print strValues
      strValues = ""
      rst.MoveNext
    Loop

    rst.Close
    Set rst = Nothing
    db.Close
    Set db = Nothing
  End Sub
```

4. In the Visual Basic Editor window, press **Ctrl+G** to open the Immediate window or choose **View | Immediate Window**.
5. To run the Open_Excel_DAO procedure, type **Open_Excel_DAO "Report.xls"** in the Immediate window and press **Enter**.
6. Run the procedure again, supplying **Report2013.xlsx** as the parameter.

To run the Open_Excel_DAO procedure, you must provide the name of the workbook file to open. If the last character in the file extension is "x" (this is determined with the VBA `Right` function), then the procedure uses the connection string designed for opening Excel 2013/2010/2007 files. After making a connection to the Excel file, the procedure goes on to retrieve information stored in the desired worksheet. Using the DAO's `OpenRecordset` method, we can access the data on the Sheet1 worksheet. Notice a dollar sign ($) appended to the sheet name. You must use the dollar sign syntax, Sheet1$, to refer to a sheet. The procedure uses the `For Each...Next` loop to obtain the names of all worksheet columns. The heading string is then written to the Immediate window. Next, the `Do Until...Loop block` loops through the records until the end of file (`EOF`) is reached. Cell values from each worksheet row are written to the `strValues` variable and then to the Immediate window. Once the data retrieval is completed, the Recordset is closed and its variable is destroyed. The same is done with the Connection object.

Hands-On 11.12 demonstrates how to open an Excel workbook with ADO and modify its data.

Hands-On 11.12. Opening an Excel Workbook with ADO

1. In the same module where you entered the procedure in Hands-On 11.11, type the following **Open_Excel_ADO** procedure:

```
Sub Open_Excel_ADO(strFileName As String)
  Dim conn As ADODB.Connection
  Dim rst As ADODB.Recordset
  Dim strFindWhat As String

  Set conn = New ADODB.Connection

  If Right(strFileName, 1) = "x" Then
    With conn
      .Provider = "Microsoft.ACE.OLEDB.12.0;"
      .ConnectionString = "Data Source=" & _
       CurrentProject.Path & "\" & strFileName & _
       ";Extended Properties=""Excel 12.0;HDR=Yes;IMEX=0"";"""
      .Open
    End With
  Else
    conn.Open "Provider=Microsoft.Jet.OLEDB.4.0;" & _
     "Data Source=" & CurrentProject.Path & _
     "\" & strFileName & _
     ";Extended Properties=""Excel 8.0;HDR=Yes;IMEX=0"";"""
  End If

  Set rst = New ADODB.Recordset
  rst.Open "SELECT * FROM [Sheet1$]", conn, _
    adOpenStatic, adLockOptimistic

  strFindWhat = "[Excel Version] = 'Excel 2000'"
  rst.Find strFindWhat
  rst(1).Value = "500"
  rst.Update
  rst.Close
  Set rst = Nothing
  MsgBox "Excel workbook was opened and updated."

  conn.Close
  Set conn = Nothing
End Sub
```

2. In the Visual Basic Editor window, press **Ctrl+G** to open the Immediate window or choose **View | Immediate Window**.
3. To run the Open_Excel_ADO procedure, type **Open_Excel_ADO "Report2013.xlsx"** in the Immediate window and press **Enter**.

Notice how the Open_Excel_ADO procedures passed the connection string to the ADO Connection object's `Open` method. Depending on the version of Microsoft Excel used, the provider name is set to Microsoft Jet OLEDB 4.0 or Microsoft ACE OLEDB 12.0, and Extended Properties is set to use either Excel 8.0 or Excel 12.0. Notice, the IMEX option, which stands for Import Export mode, is set to zero (IMEX=0). This setting will allow the data in the worksheet to be updatable. When IMEX=1, the file becomes read-only and you'll get an error on attempt to update the recordset. Once the connection to the workbook file is open, an ADO Recordset is opened. We instruct the procedure to select all data from Sheet1 worksheet using the following SQL statement:

```
"SELECT * FROM [Sheet1$]"
```

Notice that in the `SELECT` statement, the sheet name must be enclosed in square brackets and have a dollar sign ($) appended to it. The Recordset is opened using the open connection (`conn`). The procedure uses the ADO constants `adOpenStatic` (`Cursor Type` parameter) and `adLockOptimistic` (`Lock Type` parameter) to ensure that the Recordset is updatable. See Chapter 14, "Finding and Reading Records," for using various parameters when opening a Recordset. Before you can modify data in a worksheet, you must find it. The search criteria string is defined in the `strFindWhat` variable. To find the data, the procedure uses the `Find` method of the Recordset object. Once the searched data is located, we simply assign a new value to the Recordset field using the Value property:

```
rst(1).Value = "500"
```

The ADO Recordset fields are counted beginning with zero (0). Therefore, the preceding statement sets the value in the second column in the worksheet. To save the changes to the file, call the `Update` method, like this:

```
rst.Update
```

The remaining code in this procedure performs the standard cleanup: closing the objects and releasing the memory used by the object variables (`rst`, `conn`).

Opening a Text File Using ADO

There are several ways to open text files programmatically. This section demonstrates how to gain access to a text file by using the Microsoft Text Driver. Notice that this is a DSN-less connection (as explained earlier in this chapter). Hands-On 11.13 demonstrates how to open a Recordset based on a comma-separated file format and write the file contents to the Immediate window.

Hands-On 11.13. Opening a Text File with ADO

1. Copy the **C:\Access2013_HandsOn\Employees.txt** file to your **C:\Access2013_ByExample** folder, or prepare the text file from scratch by typing the following in Notepad and saving the file as **C:\Access2013_ByExample\Employees.txt**:

   "Last Name", "First Name", "Birthdate", "Years Worked"
   "Krawiec","Bogdan",#1963-01-02#,3
   "Gorecka","Jadwiga",#1948-05-12#,1
   "Olszewski","Stefan",#1957-04-07#,0

2. In the Visual Basic Editor window, choose **Insert | Module.**
3. In the module's Code window, type the following **Open_TextFile** procedure:

```
Sub Open_TextFile()
  Dim conn As ADODB.Connection
  Dim rst As ADODB.Recordset
  Dim fld As ADODB.Field

  Set conn = New ADODB.Connection
  Debug.Print conn.ConnectionString
  conn.Open "DRIVER={Microsoft Text Driver (*.txt; *.csv)};" & _
   "DBQ=" & CurrentProject.Path & "\"

  Set rst = New ADODB.Recordset
  rst.Open "SELECT * FROM [Employees.txt]", conn, adOpen   Static, _
    adLockReadOnly, adCmdText
  Do Until rst.EOF
    For Each fld In rst.Fields
      Debug.Print fld.Name & "=" & fld.Value
    Next fld
    rst.MoveNext
  Loop
  rst.Close
  Set rst = Nothing

  conn.Close
```

```
    Set conn = Nothing
    MsgBox "Open the Immediate window to view the data."
  End Sub
```

4. Make sure that the C:\Access2013_ByExample\Employees.txt file is closed and choose **Run | Run Sub/UserForm** to execute the procedure.
5. Open the Immediate window to view the procedure results.

If you worked through the previous exercises in this chapter, you should have no problem following the code of the Open_TextFile procedure. Because you are only reading the records, you can open the Recordset using the `adOpenStatic` and `adLockReadOnly` ADO constants. Notice that the ADO constant `adCmdText` is used as the last parameter of the Recordset's `Open` method (see Chapter 16, "Creating and Running Queries with DAO/ADO," for SQL examples):

```
rst.Open "SELECT * FROM [Employees.txt]", conn, adOpenStatic, _
adLockReadOnly, adCmdText
```

The last parameter in the preceding statement can be any valid option. You can indicate the type of source you are using with the `adCmdText` constant (for an SQL statement), `adCmdTable` (to retrieve all the rows in a table), or `adCmdStored-Proc` (to get records via a stored procedure). If you do not specify the type of source, `adCmdUnknown` is used as the default.

CREATING A NEW ACCESS DATABASE

You can create a new Microsoft Access database programmatically by using DAO or ADO. This section explains how to use both methods.

Creating a Database with DAO

When you start Microsoft Access, the program automatically creates a default workspace named DBEngine.Workspaces(0). The Workspace object has several useful methods, and the most frequently used are `CreateDatabase` (for creating a new database) and `OpenDatabase` (for opening an existing database). The `CreateDatabase` method requires that you specify the name and path of your database as well as the built-in constant indicating a collating order for creating the database. Use the built-in constant `dbLangGeneral` for English, German, French, Portuguese, Italian, and Modern Spanish.

The procedure in Hands-On 11.14 creates a new Access 2013 database and displays the number of system tables that Access automatically creates for its own use.

Hands-On 11.14. Creating a Database Using DAO

1. In the Visual Basic Editor window, choose **Insert | Module**.
2. In the module's Code window, type the following **CreateNewDB_DAO** procedure:

```
Sub CreateNewDB_DAO()
  Dim db As DAO.Database
  Dim dbName As String

  dbName = "C:\Access2013_ByExample\TestDAO.accdb"

  On Error GoTo ErrorHandler

  Set db = CreateDatabase(dbName, dbLangGeneral)

  MsgBox "The database contains " & _
  db.TableDefs.Count & " tables."
  db.Close
  Set db = Nothing
  Exit Sub
ErrorHandler:
  MsgBox Err.Description
End Sub
```

3. Choose **Run | Run Sub/UserForm** to execute the procedure.

To create a Database object in code, first declare an object variable of type Database. Once the Database object variable is defined, set the variable to the object returned by the `CreateDatabase` method:

```
Set db = CreateDatabase(dbName, dbLangGeneral)
```

The `CreateDatabase` method creates a new Database object and appends it to the Databases collection. The new database contains several system tables that Access creates for its own use. If the database already exists, an error occurs. You can check for the existence of the database by using an `If` statement in combination with the VBA `Dir` function (see Hands-On 11.16) and then use the VBA `Kill` statement to delete the database (see Hands-On 11.15).

Creating a Database with ADO

To create a new Access database using ADO, you must use the ADOX Catalog object's `Create` method. The ADOX library is discussed in Chapter 18. The `Cre-`

`ate` method creates and opens a new ADO connection to the data source. An error will occur if the provider does not support creating new catalogs.

The procedure in Hands-On 11.15 creates a new blank database named TestADO.mdb in your C:\Access2013_ByExample folder. The error trap ensures that the procedure works correctly even if the specified file already exists. The VBA `Kill` statement is used to delete the file from your hard disk when the error is encountered.

Hands-On 11.15. Creating a Database Using ADO

1. In the Visual Basic Editor window, choose **Tools | References**. In the References dialog box, select the **Microsoft ADO Ext. 6.0 for DDL and Security Object Library** and click **OK**.
2. In the same module where you entered the procedure in Hands-On 11.14, type the **CreateNewDB_ADO** procedure shown here:

```
Sub CreateNewDB_ADO()
  ' you must make sure that a reference to
  ' Microsoft ADO Ext. 6.0 for DDL and Security
  ' Object Library is set in the References dialog box

  Dim cat As ADOX.Catalog
  Dim strDb As String

  Set cat = New ADOX.Catalog
  strDb = "C:\Access2013_ByExample\TestADO.mdb"

  On Error GoTo ErrorHandler
  cat.Create "Provider=Microsoft.Jet.OLEDB.4.0;" & _
   "Data Source=" & strDb
  MsgBox "The database was created (" & strDb & ")."
  Set cat = Nothing
  Exit Sub

ErrorHandler:
  If Err.Number = -2147217897 Then
    Kill strDb
    Resume 0
  Else
    MsgBox Err.Number & ": " & Err.Description
  End If
End Sub
```

3. Choose **Run | Run Sub/UserForm** to execute the procedure.
   This procedure uses the error handler to detect whether a database of the specified name already exists. When error –2147217897 occurs, the procedure deletes the database file using the VBA `Kill` statement and returns to the statement that caused the error.

   While creating a database, you may specify that the database should be encrypted by setting the Jet OLEDB:Encrypt Database property to `True`. You can also include the database version information with the Jet OLEDB:Engine Type property. Simply include these properties in the connection string, as shown in the following example:

```
cat.Create "Provider=Microsoft.Jet.OLEDB.4.0;" & _
 "Data Source=" & strDb & _
 "Jet OLEDB:Encrypt Database=True;" & _
 "Jet OLEDB:Engine Type=1;"
```

To create a Microsoft Access database (ACCDB) in Access 2013/2010/2007, change the name of the provider to Microsoft.ACE.OLEDB.12.0.

COPYING A DATABASE

At times you may want to duplicate your database programmatically. This can be easily done in DAO with the DBEngine object's `CompactDatabase` method. ADO does not have a special method for copying files. However, you can set up a reference to the File Scripting object (the Microsoft Scripting Runtime Library) to gain access to your computer filesystem, or use the `CreateObject` function to access this library without setting up a reference.

Copying a Database with DAO

Before using the `CompactDatabase` method, make sure the source database is closed and there is enough disk space to create a duplicate copy. Creating a copy of your database in code requires that you define two string variables: one to hold the name of the source database and the other to specify the name for the duplicate version.

Hands-On 11.16 shows how to use the `CompactDatabase` method to copy a database.

Hands-On 11.16. Copying a Database with DAO

This hands-on exercise makes a copy of the TestDAO.accdb database created in Hands-On 11.14.

1. In the Visual Basic Editor window, choose **Insert | Module.**
2. In the module's Code window, type the following **CopyDB_DAO** procedure:

```
Sub CopyDB_DAO()
  Dim dbName As String
  Dim dbNewName As String

  dbName = InputBox("Enter the name of the database you " & _
   "want to copy: " & Chr(13) & _
   "(example: C:\Access2013_ByExample\TestDAO.accdb)", _
   "Create a copy of")

  If dbName = "" Then Exit Sub

  If Dir(dbName) = "" Then
    MsgBox dbName & " was not found. " & Chr(13) _
     & "Check the database name or path."
    Exit Sub
  End If

  dbNewName = InputBox("Enter the name of the duplicate " & _
   "database:" & Chr(13) _
   & "(example: C:\Access2013_ByExample\Copy_TestDAO.accdb)", _
   "Save As")

  If dbNewName = "" Then Exit Sub

  If Dir(dbNewName) <> "" Then
    Kill dbNewName
  End If

  DBEngine.CompactDatabase dbName, dbNewName
End Sub
```

3. Choose **Run | Run Sub/UserForm** to execute the procedure. You will be prompted to specify the name of the database you want to copy and the name for the copy.

This procedure uses the VBA Dir function to check for the existence of the database with the specified name:

```
If Dir(dbNewName) <> "" Then
  Kill dbNewName
End If
```

Because the database cannot be deleted programmatically using DAO, the VBA Kill statement is used to perform the deletion. The last statement in the CopyDB_DAO procedure uses the CompactDatabase method of the DBEngine object to create a copy of a database using the user-supplied arguments: a source database name (dbName) and a destination database name (dbNewName).

Copying a Database with FileSystemObject

You can use the CopyFile method of the FileSystemObject from the Microsoft Scripting Runtime Library to copy any file. This method allows you to copy one or more files and requires that you specify the source and destination. The source is the name of the file you want to copy or the file specification. For example, to copy all your MDB databases located in a specific directory, you can include wildcard characters to specify the source like this: C:\Access2013_ByExample*.mdb. The destination is the string specifying where the file or files are to be copied. You cannot use wildcard characters in the destination string. The third argument of the CopyFile method is optional. It indicates whether existing files in the destination are to be overwritten. If True, files are overwritten; if False, they are not. The default is True.

Hands-On 11.17 demonstrates how to copy a file from one directory to another using this method.

Hands-On 11.17. Copying a File Using FileSystem Object

This hands-on exercise makes a copy of the TestADO.mdb database created in Hands-On 11.15.

1. In the Visual Basic Editor window, choose **Insert | Module**.
2. In the module's Code window, type the following **Copy_AnyFile** procedure:

```
Sub Copy_AnyFile()
  Dim fso As Object
  Dim strFolder As String
  Dim strFolderNew As String
  Dim strDb As String
```

```
    On Error GoTo ErrorHandler
    strFolder = "C:\Access2013_ByExample\"
    strFolderNew = strFolder & "TestFolder"
    strDb = strFolder & "TestADO.mdb"
    Set fso = CreateObject("Scripting.FileSystemObject")
    fso.CreateFolder strFolderNew
    fso.CopyFile strDb, strFolderNew & "\TestADO.mdb"

    Set fso = Nothing
    Exit Sub
  ErrorHandler:
    MsgBox Err.Number & ":" & Err.Description
  End Sub
```

3. Choose **Run | Run Sub/UserForm** to execute the procedure.

This procedure uses the `CreateObject` method to return a reference to a FileSystemObject from the Microsoft Scripting Runtime Library. The `CreateFolder` method of the FileSystemObject is used to create a new folder named TestFolder in your Access2013_ByExample folder. The `CopyFile` method of the FileSystemObject is then used to copy the specified database to the newly created folder.

DATABASE ERRORS

So far in this book you've seen several procedures that incorporated error handling. You already know that an *error handler* is a block of code that is executed when a runtime error occurs. The procedure execution is transferred to error-handling code via the `On Error GoTo <Label>` statement. Recall that there are three types of `On Error` statements:

- **On Error GoTo <Label>**—This statement tells VBA to jump to the specified label when an error occurs. A label is any unreserved word followed by a colon and is placed on a separate line in the same procedure as the `On Error` statement. The code between the line that caused the error and the line with the label is simply ignored. The execution of the procedure continues from the line following the label. The error-handling code is placed at the very bottom of the procedure. To ensure that the error handler is not executed if there are no errors, place an `Exit Sub` or `Exit Function` statement on a separate line just before the label.

- **On Error Resume Next**—This statement tells VBA to resume the procedure execution at the line following the statement that caused the error. Place this statement in your code anywhere you think the error might occur. The runtime error will be trapped and stored in the VBA Err object. You should check the error number of the Err object immediately after that statement to determine how to handle the error.
- **On Error GoTo 0**—This statement disables the error handler in the current procedure. When an error occurs, VBA will display its standard runtime error message box in which you can click the End button to terminate the procedure or press Debug to enter the break mode for troubleshooting.

In Chapter 10, you learned that VBA has a built-in Err object that has several properties useful for determining the type of error that occurred. You can use the Err object's Number property to determine the error number. The Description property contains the text description of the error. You can also find out the source of an error by using the Source property.

When using ADO to access data, you can get information about the errors from both the VBA Err object and the ADO Error object. When an error occurs in an application that uses the ADO Object Model, an Error object is appended to the ADO Errors collection of the Connection object and you are advised about the error via a message box.

While the VBA Err object holds information only about the most recent error, the ADO Errors collection can contain several entries regarding the last ADO error. You can count the errors caused by an invalid operation by using the Count property of the Errors collection. By checking the contents of the Errors collection you can learn more information about the nature of the error. The Errors collection is available only from the Connection object. Errors that occur in ADO itself are reported to the VBA Err object. Errors that are provider-specific are appended to the Errors collection of the ADO Connection object. These errors are reported by the specific OLE DB provider when ADO objects are being used to access data.

The DBError2 procedure in Hands-On 11.18 attempts to open a nonexistent database to demonstrate the capabilities of the VBA Err object and the ADO Errors collection.

Hands-On 11.18. Using the VBA Err Object and ADO Errors Collection

1. In the Visual Basic Editor window, choose **Insert | Module**.
2. In the module's Code window, type the following **DBError2** procedure:

```
Sub DBError2()
  Dim conn As New ADODB.Connection
  Dim errADO As ADODB.Error

  On Error GoTo CheckErrors
  conn.Open "Provider=Microsoft.ACE.OLEDB.12.0;" _
   & "Data Source=C:\my.accdb"
  Debug.Print CurrentProject.Path

CheckErrors:
  Debug.Print "VBA error number: " _
   & Err.Number & vbCrLf _
   & " (" & Err.Description & ")"
  Debug.Print "Listed below is information " _
   & "regarding this error " & vbCrLf _
   & "contained in the ADO Errors collection."
  For Each errADO In conn.Errors
    Debug.Print vbTab & _
     "Error Number: " & errADO.Number
    Debug.Print vbTab & _
     "Error Description: " & errADO.Description
    Debug.Print vbTab & _
     "Jet Error Number: " & errADO.SQLState
    Debug.Print vbTab & _
     "Native Error Number: " & errADO.NativeError
    Debug.Print vbTab & _
     "Source: " & errADO.Source
    Debug.Print vbTab & _
     "Help Context: " & errADO.HelpContext
    Debug.Print vbTab & _
     "Help File: " & errADO.HelpFile
  Next
  MsgBox "Errors were written to the Immediate window."
End Sub
```

3. Choose **Run | Run Sub/UserForm** to execute the procedure.

In this procedure, an error is encountered when VBA attempts to open a database file that does not exist in the specified directory. The `On Error GoTo CheckErrors` statement tells VBA to jump to the line labeled `CheckErrors`. The line that prints the current project path is never executed. The `CheckErrors` handler reads the content of the VBA Err object and prints the error number and its description to the Immediate window. After that, we retrieve more information about the encountered errors by looping through the ADO Errors collection.

To trace errors in your VBA procedures, don't forget to use the Step commands in the Visual Basic Debug menu (see Chapter 9 for more information).

COMPACTING A DATABASE

With frequent use over a period of time, the performance of your database may deteriorate. When objects are deleted from a database but the space isn't reclaimed, fragmentation may occur. To improve database performance and reduce the database file size, you can compact or repair Microsoft Access databases. To compact a database, use one of the following methods:

- **CompactDatabase** (Microsoft Jet and Replication Objects (JRO) Library)—To use the JRO library, choose Tools | References in the Visual Basic application window and select Microsoft Jet and Replication Objects 2.6 Library. We will use this method to compact a database in Hands-On 11.19.
- **CompactDatabase** (DBEngine object)—This method requires that you specify the full path and filename of the database you want to compact, and the full path and filename of the compacted database, as follows:

```
DBEngine.CompactDatabase "C:\Access2013_ByExample\Northwind
2007.accdb","C:\Access2013_ByExample\CompNorthwind.accdb"
```

- **CompactRepair** (Application object)—The Application object refers to the active Microsoft Access application. This method requires that you specify the full path and filename of the database you want to compact, and the full path and filename of the compacted database. You may also specify an optional argument to indicate whether a log file should be created. `True` means that if corruption is detected in the source file, the log file will be created in the destination directory. If you omit the third argument or set it to `False`, no log file is created. Here's an example that uses this method:

```
Application.CompactRepair "C:\Access2013_ByExample\Northwind.mdb",
"C:\Access2013_ByExample\TestFolder\NorthwindRepaired.mdb", false
```

The preceding statement entered on one line in the Immediate window will create a compacted and repaired version of the Northwind.mdb database in the specified folder. Recall the destination folder was created in Hands-On 11.17. Upon running this statement, Access displays a Security Warning message. Click the Open button to proceed with the creation of the database.

When compacting or repairing a database, keep in mind the following:

- You cannot compact or repair a database that is currently open.
- You cannot compact a database to the same filename. You must specify a new name for the compacted database. After the compact/repair process is complete, simply delete the original database file and rename the compacted database using the original name.
- You cannot compact a database if it is secured and you don't have appropriate permissions.

Hands-On 11.19 demonstrates how to compact the Northwind database using JRO.

Hands-On 11.19. Compacting a Database Using JRO

1. In the Visual Basic Editor window, choose **Insert | Module.**
2. In the module's Code window, type the following **CompactDb** procedure:

```
' use Tools|References to set up a reference
' to the Microsoft Jet and Replication Objects Library

Sub CompactDb()
  Dim jetEng As JRO.JetEngine
  Dim strCompactFrom As String
  Dim strCompactTo As String
  Dim strPath As String

  strPath = CurrentProject.Path & "\"
  strCompactFrom = "Northwind.mdb"
  strCompactTo = "NorthwindComp.mdb"

  ' Make sure there isn't already a file with the
  ' name of the compacted database.
  On Error GoTo HandleErr
```

```
    ' Compact the database
    Set jetEng = New JRO.JetEngine
    jetEng.CompactDatabase "Data Source=" & _
     strPath & strCompactFrom & ";", _
     "Data Source=" & _
     strPath & strCompactTo & ";"

    ' Delete the original database
    Kill strPath & strCompactFrom

    ' Rename the file back to the original name
    Name strPath & strCompactTo As strPath & strCompactFrom

  ExitHere:
    Set jetEng = Nothing
    MsgBox "Compacting completed."
    Exit Sub
  HandleErr:
    MsgBox Err.Number & ": " & Err.Description
    Resume ExitHere
  End Sub
```

3. Choose **Run | Run Sub/UserForm** to execute the procedure.

CHAPTER SUMMARY

In this chapter, you were introduced to the two database engines that Microsoft Access 2013 uses (Jet and ACE) as well as several object libraries that provide objects, properties, and methods for your VBA procedures. You mastered the art of programmatically connecting to native Microsoft Access databases and external databases and files using various connection strings and connection methods (ODBC/OLE DB, DSN, and DSN-less connections, and .udl files). You also learned how to write VBA code to create, open, copy, compact, and delete a Microsoft Access database. Finally, you learned about statements (`On Error GoTo...`) and objects (Err and Error) that are helpful in trapping and troubleshooting database errors.

In the next chapter, you will learn the DAO and ADO techniques for creating and linking tables, and adding and modifying fields. In other words, you will learn which DAO/ADO objects can give you access to the structure of the database.

Chapter 12 Creating and Accessing Database Tables and Fields

Now that you know how to create a Microsoft Access database programmatically and connect to it using multiple methods, it's time to fill it with some useful objects. The first object you will create is a table. Typical operations you may want to perform on database tables and fields include the following:

- Setting field properties
- Making a copy of a table
- Deleting a table
- Listing table properties
- Adding new fields to an existing table
- Changing field properties
- Deleting a field from a table
- Linking a table to a database
- Listing tables in a database
- Changing the AutoNumber
- Listing data types

In this chapter, you will write VBA procedures that use both DAO and ADO objects to perform these database tasks.

CREATING A MICROSOFT ACCESS TABLE AND SETTING FIELD PROPERTIES (DAO METHOD)

Each saved table in an Access database is an object called a TableDef object. The TableDef object has a number of properties that characterize it, such as Name, RecordCount, DateCreated, and DateUpdated. The TableDef object also has methods that act on the object. For example, the `CreateField` method creates a new field for the TableDef object and the `OpenRecordset` method creates an object called Recordset that is used to manipulate the data in the table.

The procedure in Hands-On 12.1 illustrates how to create a table in the current database using DAO.

Please note files for the "Hands-On" project may be found on the companion CD-ROM.

Hands-On 12.1. Creating a Table (DAO)

1. Start Microsoft Access 2013 and create a new database named **Chap12.accdb** in your **C:\Access2013_ByExample** folder.
2. Press **Alt+F11** to switch to the Visual Basic Editor window and choose **Insert | Module.**
3. In the module's Code window, type the following **CreateTableDAO** procedure:

```
Sub CreateTableDAO()
    Dim db As DAO.Database
    Dim tblNew As DAO.TableDef
    Dim fld As DAO.Field
    Dim prp As DAO.Property

    On Error GoTo ErrorHandler
    Set db = CurrentDB
    Set tblNew = db.CreateTableDef("Agents")

    Set fld = tblNew.CreateField("AgentID", dbText, 6)
    fld.ValidationRule = "Like 'A*'"
```

```
        fld.ValidationText = "Agent ID must begin with the " & _
        "letter 'A' and cannot contain more than 6 characters."
        tblNew.Fields.Append fld

        Set fld = tblNew.CreateField("Country", dbText)
        fld.DefaultValue = "USA"
        tblNew.Fields.Append fld

        Set fld = tblNew.CreateField("DateOfBirth", dbDate)
        fld.Required = True
        tblNew.Fields.Append fld
        db.TableDefs.Append tblNew

        ' Create Caption property and set its value
        ' add it to the collection of field properties
        Set prp = tblNew.Fields("DateOfBirth"). _
            CreateProperty("Caption")
        prp.Type = dbText
        prp.Value = "Date of Birth"
        fld.Properties.Append prp
        MsgBox fld.Properties("Caption").Value

        Set prp = tblNew.CreateProperty("Description")
        prp.Type = dbText
        prp.Value = "Sample table created with DAO code"
        tblNew.Properties.Append prp

    ExitHere:
        Set fld = Nothing
        Set tblNew = Nothing
        Set db = Nothing
        Exit Sub
    ErrorHandler:
        MsgBox Err.Number & ": " & Err.Description
        Resume ExitHere
    End Sub
```

4. Choose **Run | Run Sub/UserForm** to execute the CreateTableDAO procedure.
5. Choose **File | Save** and click **OK** to save the Module1 when prompted. This will ensure that Access refreshes the application window and makes the newly created table visible in the navigation bar.

The CreateTableDAO procedure uses the `CurrentDb` method to define an object variable (`db`) to point to the database that is currently open in the Microsoft Access window. This method allows you to access the current database from Visual Basic without having to know the database name. Next, a table is created using the `CreateTableDef` method of a DAO Database object. This method requires that you specify a string or string variable to hold the name of the new TableDef object. For instance, the following line sets the object variable `tblNew` to point to a table named Agents:

```
Set tblNew = db.CreateTableDef("Agents")
```

Because a table must have at least one field, the next step in the table creation process is to use the `CreateField` method of the TableDef object to create fields. For instance, in the following statement:

```
Set fld = tblNew.CreateField("AgentID", dbText, 6)
```

- `tblNew` is a table definition variable.
- `"AgentID"` is a string specifying the name for the new field object.
- `dbText` is an integer constant that determines the data type of the new Field object (see Table 12.1).
- 6 is an integer indicating the maximum size in bytes for a text field. Text fields can hold from 1 to 255 bytes. This argument is ignored for other types of fields.

TABLE 12.1. Constants for the Type property in DAO Object Library (DataTypeEnum enumeration)

Data Type Name	Value	Description
dbAttachment	101	Attachment data
dbBigInt	16	Big integer data
dbBinary	9	Binary data
dbBoolean	1	Boolean (True/False) data
dbByte	2	Byte (8-bit) data
dbChar	18	Text data (fixed width)
dbComplexByte	102	Multivalue byte data
dbComplexDecimal	108	Multivalue decimal data
dbComplexDouble	106	Multivalue double-precision floating-point data
dbComplexGUID	107	Multivalue GUID data

Data Type Name	Value	Description
`dbComplexInteger`	103	Multivalue integer data
`dbComplexLong`	104	Multivalue long integer data
`dbComplexSingle`	105	Multivalue single-precision floating-point data
`dbComplexText`	109	Multivalue text data (variable width)
`dbCurrency`	5	Currency data
`dbDate`	8	Date value data
`dbDecimal`	20	Decimal data (ODBCDirect only)
`dbDouble`	7	Double-precision floating-point data
`dbFloat`	21	Floating-point data (ODBCDirect only)
`dbGUID`	15	GUID data
`dbInteger`	3	Integer data
`dbLong`	4	Long integer data
`dbLongBinary`	11	Binary data (bitmap)
`dbMemo`	12	Memo data (extended text)
`dbNumeric`	19	Numeric data (ODBCDirect only)
`dbSingle`	6	Single-precision floating-point data
`dbText`	10	Text data (variable width)
`dbTime`	22	Data in time format (ODBCDirect only)
`dbTimeStamp`	23	Data in time and date format (ODBCDirect only)
`dbVarBinary`	17	Variable binary data (ODBCDirect only)

ODBCDirect workspaces are not supported since the release of Access 2007. Use ADO if you want to access external data sources without using the Microsoft Access database engine.

Constants for complex data types and the dbAttachment data type do not apply to versions prior to Access 2007.

When creating fields for your table, you may want to set certain field properties such as Validation Rule, Validation Text, Default Value, and Required. The Validation Rule property is a text string that describes the rule for validation. In the CreateTableDAO procedure, we require that each entry in the AgentID field begin with the letter "A."

The Validation Text property is a string that is displayed to the user when the validation fails; that is, when the user attempts to enter data that does not comply with the specific validation rule.

The Default Value property sets or returns the default value of a Field object. In this example procedure, we make the data entry easier for the user by specifying "USA" as the default value in the Country field. Each new record will automatically have an entry of USA in the Country field. Because certain fields should not be left blank, you can ensure that the user enters data in a particular field by setting the Required property of that field to True.

In addition to built-in properties of an object, there are two other types of properties:

- Application-defined properties
- User-defined properties

The application-defined property is created only if you assign a value to that property. A classic example of such a property is the Description property of the TableDef object. To set the Description property of a table in the Access user interface, simply right-click on the table name and choose Table Properties, then type the text you want in the Description field. Access will create a Description property for the table and will append it automatically to the Properties collection for that TableDef object. If you do not type a description in the Description field, Access will not create a Description property. Therefore, if you use the Description property in your code in this case, Access will display an error. For this reason, it is a good idea to check beforehand whether a referenced property exists. Users may create their own properties to hold additional information about an object.

The CreateTableDAO procedure demonstrates how to use the `CreateProperty` method of the TableDef object to create application-defined or user-defined properties. To create a property you will need to supply the name for the property, the property type, and the property value. For example, here's how to use the `CreateProperty` method to create a Caption property for the DateOfBirth field in the newly created table Agents:

```
Set prp = tblNew.Fields("DateOfBirth").CreateProperty("Caption")
```

Next, the data type of the Property object is defined:

```
prp.Type = dbText
```

See Table 12.1 earlier in the chapter for the names of the Type property constants in VBA.

Finally, a value is assigned to the new property:

```
prp.Value = "Date of Birth"
```

Instead of writing three separate lines of code, you can create a new property of an object with the following line:

```
Set prp = tblNew.Fields("DateOfBirth").CreateProperty("Caption", dbText,
"Date of Birth")
```

A user-defined property must be appended to the Properties collection of the corresponding object. In this example procedure, the Caption property is appended to the Properties collection of the Field object, and the Description property is appended to the Properties collection of the TableDef object:

```
fld.Properties.Append prp
tblNew.Properties.Append prp
```

After creating a field and setting its built-in, application-defined, or user-defined properties, the `Append` method is used to add the field to the Fields collection, as in the following example:

```
tblNew.Fields.Append fld
```

Once all the fields have been created and appended to the Fields collection, remember to append the new table to the TableDefs collection, as in the following example:

```
db.TableDefs.Append tblNew
```

You can delete user-defined properties from the Properties collection, but you can't delete built-in properties. If you set a property in the user interface, you don't need to create and append the property in code because the property is automatically included in the Properties collection.

After running the procedure code, a new table named Agents appears in the Microsoft Access window.

To check the value of the Description property for the Agents table that was set as a result of running the example procedure, right-click the Agents table in the database window, and choose Table Properties from the shortcut menu.

To check the properties that were set and defined in this procedure, activate the Agents table in Design view, click the field name for which you set or created a custom property in the code, and examine the corresponding field properties. Figure 12.1 shows the current settings of the Validation Rule and Validation Text properties for the AgentID field.

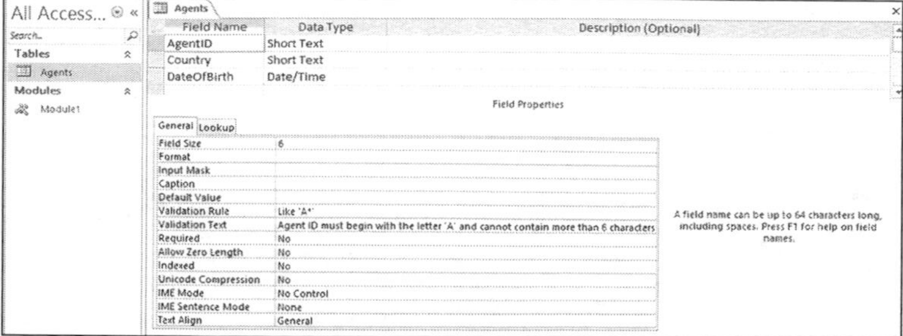

FIGURE 12.1. You can create a database table like this one using VBA code. You can also set appropriate field properties programmatically.

CREATING A MICROSOFT ACCESS TABLE AND SETTING FIELD PROPERTIES (ADO METHOD)

You can also get going with your database design by using Access objects contained in the ADOX library. The full name of this library is ActiveX Data Object Extensions for DDL and Security. To use ADOX in your VBA procedures, choose Tools | References from your Visual Basic Editor window and select Microsoft ADO Ext. 6.0 for DDL and Security. The ADOX Object Model is an extension of the ADODB library.

The most important ADOX object is called Catalog. It represents an entire database and contains database tables, columns, indexes, groups, users, procedures, and views. You will use the ADOX Catalog object in your VBA procedures to create a table.

The following steps outline the process of creating a new Microsoft Access table:

1. Declare the variables representing the Connection, Catalog, and Table objects:

```
Dim conn As ADODB.Connection
Dim cat As ADOX.Catalog
Dim tbl As ADOX.Table
```

2. Open the connection to your database:

```
Set conn = New ADODB.Connection
conn.Open "Provider=Microsoft.Jet.OLEDB.4.0;" & _
 "Data Source=C:\Access2013_ByExample\Chap11b.mdb"
```

3. Supply the open connection to the ActiveConnection property of the ADOX Catalog object:

```
Set cat = New ADOX.Catalog
Set cat.ActiveConnection = conn
```

4. Create a new Table object:

```
Set tbl = New ADOX.Table
```

5. Provide the name for your table:

```
tbl.Name = "tblAssets"
```

   The Table object is a member of the Tables collection, which in turn is a member of the Catalog object. Each Table object has a Name property and a Type property. The Type property specifies whether a Table object is a standard Microsoft Access table, a linked table, a system table, or a view. To see an example of using the Type property, refer to the section titled "Listing Database Tables" later in this chapter.

6. Append the Table object to the Catalog object's Tables collection:

```
cat.Tables.Append tbl
```

   At this point your table is empty.

7. Add new fields (columns) to your new table:

```
With tbl.Columns
  .Append "SiteID", adVarWChar, 10
  .Append "Category", adSmallInt
  .Append "InstallDate", adDate
End With
```

The preceding code fragment creates three fields named SiteID, Category, and InstallDate. You can create new fields in a table by passing the Column object's Name, Type, and DefinedSize properties as arguments of the Columns collection's `Append` method. Notice that ADOX uses different data types than those used in the Access user interface (see Table 12.2 for a comparison of the data types).

The Table object contains the Columns collection that contains Column objects. To add a new field to a table, you could create a Column object and write the code like this:

```
Dim col As ADOX.Column
set col = New ADOX.Column
With col
  .Name = "SiteID"
  .DefinedSize = 10
End With
tbl.Columns.Append col
```

The last statement in the preceding example appends the new Column object (field) to the Columns collection of a table. The Name property specifies the name of the column. The DefinedSize property designates the maximum size of an entry in the column. To create another field, you would have to create a new Column object and set its properties. Creating fields in this manner takes longer and is less efficient than using the method demonstrated earlier.

The complete procedure is shown here:

```
Sub CreateTableADO()
Dim conn As ADODB.Connection
Dim cat As ADOX.Catalog
Dim tbl As ADOX.Table

' make sure to set up a reference to
' the Microsoft ActiveX Data Objects 6.1 Library
' and ADO Ext. 6.0 for DDL and Security

' copy Chap11b.mdb from C:\Access2013_HandsOn folder
' to your C:\Access2013_ByExample folder

Set conn = New ADODB.Connection
conn.Open "Provider=Microsoft.Jet.OLEDB.4.0;" & _
  "Data Source=C:\Access2013_ByExample\Chap11b.mdb"

Set cat = New ADOX.Catalog
Set cat.ActiveConnection = conn

Set tbl = New ADOX.Table
tbl.Name = "tblAssets"
```

```
cat.Tables.Append tbl

With tbl.Columns
  .Append "SiteID", adVarWChar, 10
  .Append "Category", adSmallInt
  .Append "InstallDate", adDate
End With
Set cat = Nothing
conn.Close
Set conn = Nothing
End Sub
```

TABLE 12.2. ADO data types versus Microsoft Access data types

ADO Data Type	Corresponding Data Type in Access
adBoolean	Yes/No
adUnsignedTinyInt	Number (FieldSize = Byte)
adSmallInt	Number (FieldSize = Integer)
adSingle	Number (FieldSize = Single)
adDouble	Number (FieldSize = Double)
adDecimal	Number (FieldSize = Decimal)
adInteger	Number (FieldSize = LongInteger) AutoNumber
adCurrency	Currency
adVarWChar	Text
adDate	Date/Time
adLongVarBinary	OLE object
adLongVarWChar	Memo
adLongVarWChar	Hyperlink

ADO does not support the Attachment data type, multiselect lookup fields, and the Append Only and Rich Text memo fields that were first introduced in Access 2007. To programmatically access these features in Access 2013/2010/2007, you must rely on DAO.

COPYING A TABLE

The procedure in Hands-On 12.2 uses the SQL SELECT...INTO statement to select all records from the Customers table in the Northwind database and place them into a new table called CustomersCopy. The `SELECT...INTO` statement is equivalent to a MakeTable query in the Microsoft Access user interface. This statement creates a new table and inserts data from other tables. To copy a table, the SQL statement is passed as the first argument of the `Execute` method of the ADO Connection object. Note that the copied table will not have the indexes that may exist in the original table.

Hands-On 12.2. Making a Copy of a Table (ADO)

1. In the Visual Basic Editor window, choose **Insert | Module**.
2. In the module's Code window, type the following **Copy_Table** procedure:

```
' make sure to set up a reference to
' the Microsoft ActiveX Data Objects 6.1 Library

Sub Copy_Table()
  Dim conn As ADODB.Connection
  Dim strTable As String
  Dim strSQL As String

  On Error GoTo ErrorHandler

  strTable = "Customers"
  strSQL = "SELECT " & strTable & ".* INTO "
  strSQL = strSQL & strTable & "Copy "
  strSQL = strSQL & "FROM " & strTable

  Debug.Print strSQL
  Set conn = New ADODB.Connection
  conn.Open "Provider=Microsoft.Jet.OLEDB.4.0;" & _
    "Data Source=" & CurrentProject.Path & _
    "\Northwind.mdb"

  conn.Execute strSQL
  conn.Close
  Set conn = Nothing
  MsgBox "The " & strTable & " table was copied."
  Exit Sub

ErrorHandler:
  If Err.Number = -2147217900 Then
    conn.Execute "DROP Table " & strTable
```

```
    Resume
  Else
    MsgBox Err.Number & ": " & Err.Description
  End If
End Sub
```

3. Choose **Run | Run Sub/UserForm** to execute the procedure.
   When you run this procedure, Access creates a copy of the Customers table named CustomersCopy in the Northwind.mdb database.

DELETING A DATABASE TABLE

You can use ADO to delete a table programmatically by opening the ADOX Catalog object, accessing its Tables collection, and calling the `Delete` method. The procedure in Hands-On 12.3 requires a parameter that specifies the name of the table you want to delete.

Hands-On 12.3. Deleting a Table from a Database (ADO)

1. In the Visual Basic Editor window, choose **Insert | Module**.
2. In the module's Code window, type the **Delete_Table** procedure shown here:

```
Sub Delete_Table(strTblName As String)
  Dim conn As ADODB.Connection
  Dim cat As ADOX.Catalog

  On Error GoTo ErrorHandler

  Set conn = New ADODB.Connection
  conn.Open "Provider=Microsoft.Jet.OLEDB.4.0;" & _
    "Data Source=" & CurrentProject.Path & _
    "\Northwind.mdb"

  Set cat = New ADOX.Catalog

  cat.ActiveConnection = conn
  cat.Tables.Delete strTblName
  Set cat = Nothing
  conn.Close
  Set conn = Nothing

  Exit Sub
```

```
ErrorHandler:
  MsgBox "Table '" & strTblName & _
    "' cannot be deleted " & vbCrLf & _
    "because it does not exist."
  Resume Next
End Sub
```

3. To run this procedure, type the following statement in the Immediate window and press **Enter**:

   ```
   Delete_Table "CustomersCopy"
   ```

   The CustomersCopy table was created by running the Copy_Table procedure in Hands-On 12.2. When you press Enter, Visual Basic will delete the specified table from the Northwind.mdb database. If the table does not exist, an appropriate message is displayed.

ADDING NEW FIELDS TO AN EXISTING TABLE

At times you may want to programmatically add a new field to an existing table. The procedure in Hands-On 12.4 adds a new text field called MyNewField to a table located in the Northwind database.

Hands-On 12.4. Adding a New Field to a Table (ADO)

The procedure demonstrated in this hands-on exercise uses the CustomersCopy table in the Northwind database.

1. In the Visual Basic Editor window, choose **Insert | Module**.
2. In the module's Code window, type the following **Add_NewFields** procedure:

```
Sub Add_NewFields()
  Dim conn As ADODB.Connection
  Dim cat As New ADOX.Catalog
  Dim myTbl As New ADOX.Table

  Set conn = New ADODB.Connection
  conn.Open "Provider=Microsoft.Jet.OLEDB.4.0;" & _
   "Data Source=" & CurrentProject.Path & _
   "\Northwind.mdb"
  Set cat = New ADOX.Catalog
  cat.ActiveConnection = conn
  cat.Tables("CustomersCopy").Columns.Append _
```

```
    "MyNewField", adVarWChar, 15

   Set cat = Nothing
   conn.Close
   Set conn = Nothing
 End Sub
```

3. Run the **Copy_Table** procedure in Hands-On 12.2 to ensure that the CustomersCopy table exists in the Northwind database.
4. Choose **Run | Run Sub/UserForm** to run the Add_NewFields procedure.

In DAO, use the `CreateField` and `Append` methods to add new fields to the existing table.

Hands-On 12.5. Adding a New Field to a Table (DAO)

1. In the Visual Basic Editor window, choose **Insert | Module**.
2. In the module's Code window, type the following **Add_NewFieldsDAO** procedure:

```
Sub Add_NewFieldsDAO()
  Dim db As DAO.Database
  Dim tdf As DAO.TableDef
  Dim tblName As String

  tblName = "CustomersCopy"

  On Error GoTo ErrorHandler
  Set db = OpenDatabase("C:\Access2013_ByExample\Northwind.mdb")
  Set tdf = db.TableDefs(tblName)

  MsgBox "Number of fields in the table: " & _
   db.TableDefs(tblName).Fields.Count

  With tdf
    .Fields.Append .CreateField("NoOfMeetings", dbInteger)
    .Fields.Append .CreateField("Result", dbMemo)
  End With

  MsgBox "Number of fields in the table: " & _
   db.TableDefs(tblName).Fields.Count
  db.Close
  Exit Sub
```

```
ErrorHandler:
  MsgBox Err.Number & ": " & Err.Description
End Sub
```

3. Choose **Run | Run Sub/UserForm** to run the Add_NewFieldsDAO procedure.

   The Add_NewFieldsDAO procedure uses the following `With...End With` construct to quickly add two new fields to an existing table:

```
With tdf
  .Fields.Append .CreateField("NoOfMeetings", dbInteger)
  .Fields.Append .CreateField("Result", dbMemo)
End With
```

   Each new field is appended to the Fields collection of the specified DAO TableDef object. In this example, we create a new field on the fly while calling the `Append` method. Be sure to include a space between the `Append` method and the dot operator in front of the `CreateField` method. To add two new fields to an existing table without using the `With...End With` construct, you would use the following statements:

```
tdf.Fields.Append tdf.CreateField("NoOfMeetings", dbInteger)
tdf.Fields.Append tdf.CreateField("Result", dbMemo)
```

   However, using the `With...End With` construct makes the code both clearer and faster to execute.

CREATING CALCULATED FIELDS

Access has the ability to store calculated values in tables via a so-called *calculated field*. A classic example of the calculated field is a person's full name. A person's first and last names are stored in separate fields in an Access table. In versions of Access prior to 2010, the full name was generally obtained via a query by writing an expression that concatenated the first and last name:

```
Select [FirstName] & " " & [LastName] AS FullName
```

In Access 2013/2010, you can define the expression for the calculation in the calculated field and Access will store the calculated values in the table. With this feature, there is no need to calculate the person's full name in multiple locations in your Access application. When the underlying values change (for example, a female employee got married and the last name field used in the expression was updated), the expression will automatically update the value that is stored in the calculated field.

Calculated columns can be added to Access tables manually or with VBA. To create a calculated column using the manual method, open the table in Design view and enter the field name. In the Data Type column, select Calculated. At this point, Access will display the Expression Builder dialog box where you can enter the expression (see Figure 12.2).

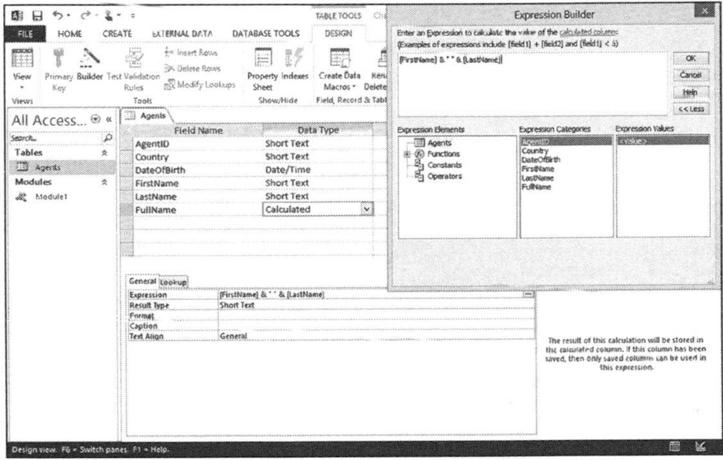

FIGURE 12.2. You can add a calculated field to a table using the table Design view and the Expression Builder.

Certain calculations should never be stored in a calculated field in a table. For example, expressions based on the results of the date and time functions such as `Date()` *and* `Now()` *will return different values each time they are called, and therefore should be left in queries. Also, expressions that use domain aggregate functions (such as* `DCount()`, `DSum()`, `DAvg()`, *and so on) are not good candidates for use in calculated fields because checking changes in underlying values requires going beyond one record which, depending on the number of records that have to be accessed, can hinder database performance.*

To create a calculated field in DAO, you will need to set the Expression property of the DAO.Field2 object to the expression you'd like to use for the calculated field, as shown in Hands-On 12.6. A Field2 object represents a column of data in an Access table. It contains all of the same properties and methods as the Field object with the addition of several properties and methods that support field types added in Access 2007 (multivalue lookup fields and attachment fields) and Access 2010 (calculated fields).

Hands-On 12.6. Creating a Calculated Field with DAO

1. In the VBE screen, choose **Insert | Module** and enter the following procedure in the module's Code window:

```
Sub CreateCalcField()
  Dim db As DAO.Database
  Dim tdf As DAO.TableDef
  Dim fld As DAO.Field2

  On Error GoTo ErrorHandler

  Set db = CurrentDb
  Set tdf = db.TableDefs("Agents")

  ' add two text fields
  tdf.Fields.Append tdf.CreateField("FirstName", dbText, 25)
  tdf.Fields.Append tdf.CreateField("LastName", dbText, 25)

  ' add a calculated field
  Set fld = tdf.CreateField("FullName", dbText, 50)
  fld.Expression = "[FirstName] & "" "" & [LastName]"
  tdf.Fields.Append fld

ExitHere:
  Set fld = Nothing
  Set tdf = Nothing
  Set db = Nothing
  Exit Sub
ErrorHandler:
  If Err.Number = 3211 Then
    ' table is open; need to close it to continue
    DoCmd.Close acTable, "Agents", acSaveYes
    Resume
  Else
    MsgBox Err.Number & ": " & Err.Description
    Resume ExitHere
  End If
End Sub
```

1. Run the **CreateCalcField** procedure.

   The CreateCalcField procedure adds three new fields (FirstName, LastName, and FullName) to the existing Agents table in the current database. To append fields to the table, Access needs exclusive access to the table definition. The included ErrorHandler executes the statement that closes the table if it is found open:

   ```
   DoCmd.Close acTable, "Agents", acSaveYes
   ```

   Notice that before you can create a calculated field you need to ensure that the fields the calculation is based upon are also present in the table. After adding the required fields to the table, the calculated field is added and its expression for the calculation is defined as follows:

   ```
   fld.Expression = "[FirstName] & "" "" & [LastName]"
   ```

   Figure 12.2 earlier in this section shows the Agents table in Design view displaying the properties of the calculated field. To manually change the calculation expression, click the ellipsis button to the right of the Expression property to bring up the Expression Builder.

CREATING MULTIVALUE LOOKUP FIELDS WITH DAO

Thanks to the introduction of the complex multivalue data type in the .accdb file format, table columns can store more than one value. This makes it easy for an Access user to create a lookup field without having to know much about setting table relationships. Access will automatically store the values entered in multivalue fields in hidden system tables and create proper table relationships if necessary. The source data for a multivalue field can be one of the following: value list, field list, or table/query. To have Access guide you in the creation of a multivalue field, choose Lookup Wizard in the Data Type column of the table's Design view.

Multivalue lookup fields are often referred to as complex fields because they use data types that begin with dbComplex (see Table 12.3).

TABLE 12.3. Data types used by multivalue lookup fields

Data Type	Value	Description
dbComplexByte	102	Multivalue byte data
dbComplexDecimal	108	Multivalue decimal data

(Contd.)

Data Type	Value	Description
dbComplexDouble	106	Multivalue double-precision floating-point data
dbComplexGUID	107	Multivalue GUID data
dbComplexInteger	103	Multivalue integer data
dbComplexLong	104	Multivalue long integer data
dbComplexSingle	105	Multivalue single-precision floating-point data
dbComplexText	109	Multivalue text data (variable width)

The following hands-on exercise demonstrates how to use VBA to add a multivalue field named Literature to the Northwind 2007.accdb database's Customers table.

Hands-On 12.7. Creating a Multivalue Lookup Field with DAO

1. In the VBE screen, choose **Insert | Module** and enter the following **CreateMultiValueFld** procedure in the module's Code window:

```
Sub CreateMultiValueFld()
  Dim db As DAO.Database
  Dim tdf As DAO.TableDef
  Dim fld As DAO.Field
  Dim strDbName As String
  Dim strTblName As String
  Dim strLitItems As String

  On Error GoTo ErrorHandler

  strDbName = "C:\Access2013_ByExample\Northwind 2007.accdb"
  strLitItems = "Product Brochure;Product Flyer A;"
  strLitItems = strLitItems & "Product Flyer B"
  strTblName = "Customers"

  Set db = OpenDatabase(strDbName)
  Set tdf = db.TableDefs(strTblName)
  Set fld = tdf.CreateField("Literature", dbComplexText)
  tdf.Fields.Append fld

  With fld
    .Properties.Append .CreateProperty( _
     "DisplayControl", dbText, acComboBox)
    .Properties.Append .CreateProperty( _
```

```
        "RowSourceType", dbText, "Value List")
      .Properties.Append .CreateProperty( _
        "RowSource", dbText, strLitItems)
      .Properties.Append .CreateProperty( _
        "BoundColumn", dbInteger, 1)
      .Properties.Append .CreateProperty( _
        "ColumnCount", dbInteger, 1)
      .Properties.Append .CreateProperty( _
        "ColumnWidths", dbText, "1")
      .Properties.Append .CreateProperty( _
        "ListWidth", dbText, "1.5")
      .Properties.Append .CreateProperty( _
        "AllowMultipleValues", dbBoolean, True)
      .Properties.Append .CreateProperty( _
        "AllowValueListEdits", dbBoolean, True)
    End With

  ExitHere:
    db.Close
    Set fld = Nothing
    Set tdf = Nothing
    Set db = Nothing
    Exit Sub
  ErrorHandler:
    MsgBox Err.Number & ": " & Err.Description
    Resume ExitHere
  End Sub
```

2. Run the **CreateMultiValueFld** procedure.
3. Open the **C:\Access2013_ByExample\Northwind 2007.accdb** database and check the newly created Literature field in the Customers table (see Figures 12.3 and 12.4).

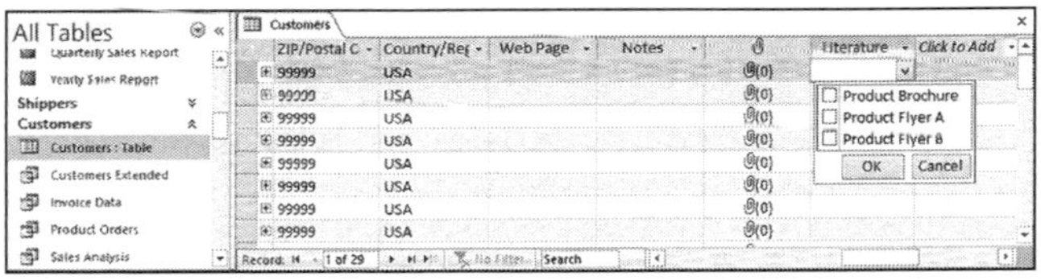

FIGURE 12.3. The multivalue lookup field (Literature) created by the VBA procedure in Hands-On 12.7 displays a combo box.

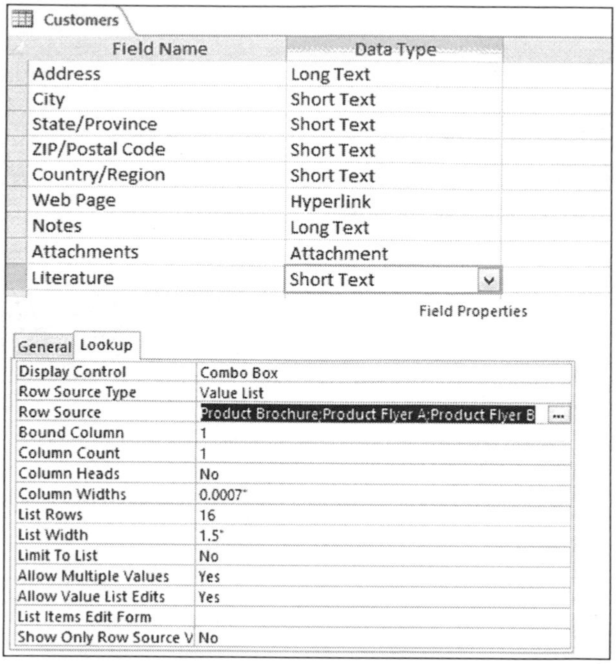

Customers

Field Name	Data Type
Address	Long Text
City	Short Text
State/Province	Short Text
ZIP/Postal Code	Short Text
Country/Region	Short Text
Web Page	Hyperlink
Notes	Long Text
Attachments	Attachment
Literature	Short Text

Field Properties

General Lookup

Display Control	Combo Box
Row Source Type	Value List
Row Source	Product Brochure;Product Flyer A;Product Flyer B
Bound Column	1
Column Count	1
Column Heads	No
Column Widths	0.0007"
List Rows	16
List Width	1.5"
Limit To List	No
Allow Multiple Values	Yes
Allow Value List Edits	Yes
List Items Edit Form	
Show Only Row Source V	No

FIGURE 12.4. The Field Properties Lookup tab contains numerous properties that tell Access how to display values in the Literature field.

CREATING ATTACHMENT FIELDS WITH DAO

The Attachment data type makes it possible to store various types of external files directly in the database. This data type is only available in Access databases created in the .accdb file format in Access 2013/2010/2007. Earlier versions of Access used the OLE Object data type for embedding external files within MDB databases, and this format continues to be available in Access 2010 for backward compatibility. The Attachment data type eliminates the bloating issues that plagued Access MDB databases whenever the OLE Object data type was used. To keep .accdb files as small as possible, Access compresses the uncompressed files in the attachments before storing them in a database.

To avoid the bloating issue in Access 2003 and earlier, do not embed external files directly in the Access MDB database. Simply store the external filename in a table using the Text data type. Store your files in the same folder as the database and write VBA code to retrieve the files as needed. The sample Northwind.mdb file included with Access 2003 and earlier has VBA code in the Employee form's Current event to display an employee picture as the user moves through the form.

The Attachment data type allows you to add multiple attachments to a single record. However, keep in mind that the maximum size of an attached data file cannot exceed 256 MB (megabytes). You can store as many external files as you want as long as you stay within 2 GB (gigabytes) of data, which is the maximum size of an Access database. You cannot restrict how many attachments are allowed in a database field. Also, some attachment file types are not supported. (You can see the list of blocked file extensions in the Access online help.)

You can work with attachments manually via the Attachments dialog box (see Figure 12.5) or programmatically using the Attachment object. Hands-On 12.8 demonstrates how to create an Attachment field. You will find more details about working with attachments in Chapter 15, "Working with Records."

Hands-On 12.8. Adding an Attachment Field to an Existing Table

1. In the VBE screen, choose **Insert | Module** and enter the following **CreateAttachmentFld** procedure in the module's Code window:

```
Sub CreateAttachmentFld()
  Dim db As DAO.Database
  Dim tdf As DAO.TableDef
  Dim fld As DAO.Field2
  On Error GoTo ErrorHandler

  Set db = CurrentDb
  Set tdf = db.TableDefs("Agents")

  ' add an attachment field
  Set fld = tdf.CreateField("AttachLiterature", dbAttachment)
  tdf.Fields.Append fld

ExitHere:
  Set fld = Nothing
  Set tdf = Nothing
```

```
    Set db = Nothing
    Exit Sub
  ErrorHandler:
    If Err.Number = 3211 Then
      ' table is open; need to close it to continue
      DoCmd.Close acTable, "Agents", acSaveYes
       Resume
    Else
      MsgBox Err.Number & ": " & Err.Description
      Resume ExitHere
    End If
  End Sub
```

2. Run the **CreateAttachmentFld** procedure.
   After running this procedure, the Agents table in the current database contains an extra field as shown in Figure 12.5. To add attachments, double-click the @(0) in the record to bring up the Attachments dialog box. To add and manipulate attachments programmatically, refer to Chapter 15.

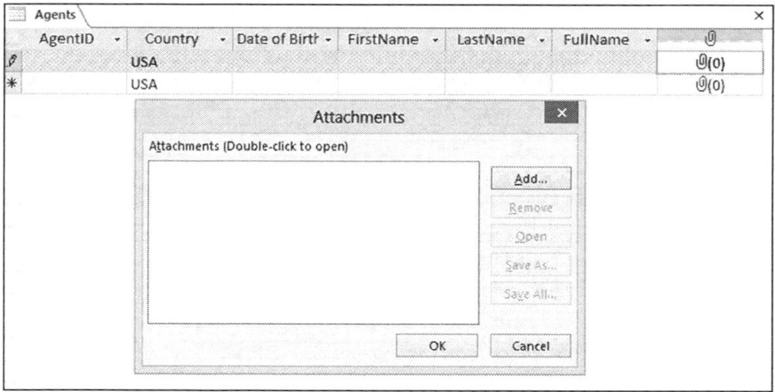

FIGURE 12.5. The attachment field added with the VBA procedure in Hands-On 12.8 currently does not contain any attachments.

CREATING APPEND ONLY MEMO FIELDS WITH DAO

Another type of complex multivalue field available in the .accdb file format is the Append Only memo field (see Figure 12.6). When the Append Only property is set to Yes, you can append data to the field, but you are not allowed to change the data that has been previously entered into this field. This feature is useful for keeping track of the changes made to the field. Let's say you want to preserve the history of problems submitted by users. Every time you edit the data in the Append Only memo field, the date and time stamp and your changes are automatically saved to the version history of the field (see Figure 12.7). You can view the history of an Append Only memo field by right-clicking a value in the field, and selecting Show column history from the shortcut menu. Custom Project 12.1 demonstrates how to create a table with an Append Only memo field and how to retrieve the history of data changes from this field.

Agents

Field Name	Data Type
AgentID	Short Text
Country	Short Text
DateOfBirth	Date/Time
FirstName	Short Text
LastName	Short Text
FullName	Calculated
AttachLiterature	Attachment
FieldNotes	Long Text

Field Properties

General | Lookup

Format	
Caption	
Default Value	
Validation Rule	
Validation Text	
Required	No
Allow Zero Length	No
Indexed	No
Unicode Compression	No
IME Mode	No Control
IME Sentence Mode	None
Text Format	Plain Text
Text Align	General
Append Only	Yes

FIGURE 12.6. To collect history on a memo field, you must set the field's Append Only property to Yes.

In Access 2013 user interface, there is no "memo" data type in the Data Type list. The Long Text data type replaces the memo data type found in prior versions of Access. From now on, the Long Text data type should be used for longer text fields (see FieldNotes field in Figure 12.6) and Short Text data type (which is the replacement for the Text data type found in previous versions of Access) should be used for storing up to 255 characters.

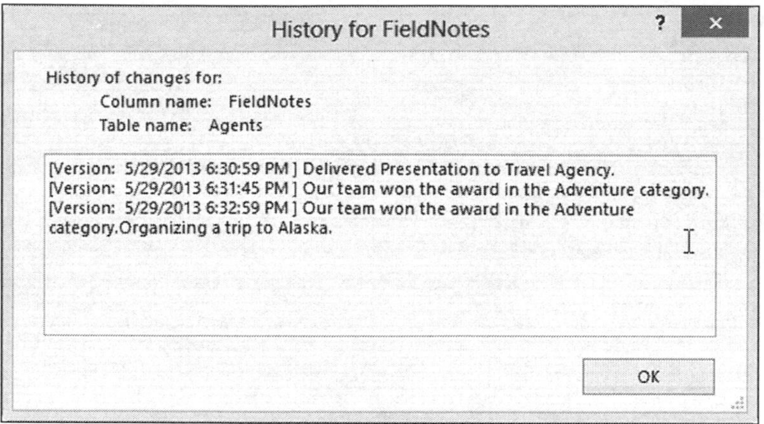

FIGURE 12.7. To access the memo field's history, right-click the field and select Show column history from the shortcut menu.

Custom Project 12.1. Working with Append Only Memo Fields

1. In the VBE screen, choose **Insert | Module** and enter the following **CreateAppendOnlyFld** procedure in the module's Code window:

```
Sub CreateAppendOnlyFld()
  Dim db As DAO.Database
  Dim tdf As DAO.TableDef
  Dim fld As DAO.Field2
  On Error GoTo ErrorHandler

  Set db = CurrentDb
  Set tdf = db.TableDefs("Agents")

   ' create a memo field
  Set fld = tdf.CreateField("FieldNotes", dbMemo)
  tdf.Fields.Append fld

  ' set the memo field to track version history
  fld.AppendOnly = True

ExitHere:
  Set fld = Nothing
  Set tdf = Nothing
  Set db = Nothing
  Exit Sub
```

```
ErrorHandler:
  If Err.Number = 3211 Then
    ' table is open; need to close it to continue
    DoCmd.Close acTable, "Agents", acSaveYes
     Resume
  Else
    MsgBox Err.Number & ": " & Err.Description
    Resume ExitHere
  End If
End Sub
```

2. Run the **CreateAppendOnlyFld** procedure.
   Notice that after creating the FieldNotes memo field, you need to set the AppendOnly property of this field to True to ensure that Access keeps the history of changes for this field.
   Let's enter some data in one or two records.
3. Switch to the Microsoft Access window, and double-click the Agents table in the navigation pane.
4. Type in the data as shown in Figure 12.8. Recall that you don't need to enter data in the FullName field because this is a calculated field and Access will perform the required calculation based on the defined expression.

AgentID	Country	Date of Birth	FirstName	LastName	FullName
A100	USA	7/12/1969	Barbara	McDonald	Barbara McDonald
A1001	USA	7/22/1981	Ronald	Sepia	Ronald Sepia
*	USA				

FIGURE 12.8. Entering sample data in a Datasheet view.

5. In the FieldNotes field for Barbara McDonald, type the following text:

   Delivered Presentation to Travel Agency.

6. In the FieldNotes field for Ronald Sepia, type the following text:

   Mr. Brook invited us to a dinner party tomorrow.

7. Move back to Barbara McDonald's record and in the FieldNotes enter the following text, overwriting the previously written text:

   Our team won the award in the Adventure category.

8. Press **Enter** to record the changes to the field.

9. Move back to the FieldNotes field of Barbara McDonald's record and type the following text:

   Organizing a trip to Alaska.

10. Continue to enter more data in the first two records of the Agents table so that you can build up some history in the FieldNotes Append Only memo field.
11. After you are done with the data entry, right-click on the FieldNotes field in each record and choose **Show Column History** to check out the version history. The record history for the first record is shown in Figure 12.7 earlier.
12. Close the Agents table.

    You can retrieve the history of values that have been stored in a memo field by using the `ColumnHistory` method of the Application object. For example, the following statement entered on one line in the Immediate window will print all the notes for an agent whose AgentID equals A100 (see Figure 12.9).

```
Debug.Print Application.ColumnHistory("Agents", "FieldNotes",
  "AgentID='A100'")
```

```
Immediate
Debug.Print Application.ColumnHistory("Agents", "FieldNotes", "AgentID='A100'")
[Version:  5/29/2013 6:30:59 PM ] Delivered Presentation to Travel Agency.
[Version:  5/29/2013 6:31:45 PM ] Our team won the award in the Adventure category.
[Version:  5/29/2013 6:32:59 PM ] Our team won the award in the Adventure category.Organizing
```

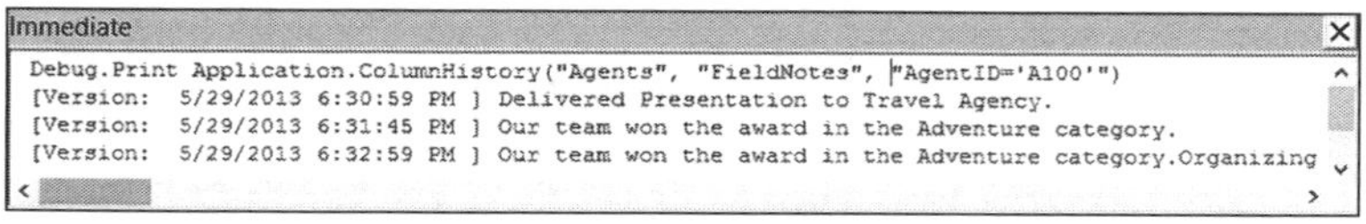

FIGURE 12.9. Retrieving the history of values stored in a memo field.

    Notice that the `ColumnHistory` method requires three parameters: the name of the table that contains the Append Only memo field, the name of the memo field, and a string used to locate a record in the table. Let's now write a complete procedure that will retrieve the history data into three separate items: MemoDate, MemoTime, and MemoText.

13. In the same module where you entered the previous procedure in this project, enter the following **RetrieveMemoHistory** procedure:

```
Sub RetrieveMemoHistory()
  Dim arrayString() As String
  Dim MemoText As String
  Dim i As Integer
  Dim strSearch As String
  Dim startPos As Integer
  Dim EndDatePos As Integer
  Dim EndTimePos As Integer
  Dim MemoDate As Date
  Dim MemoTime As Date
```

```
  arrayString = Split(Application.ColumnHistory( _
    "Agents", "FieldNotes", _
    "AgentID='A100'"), "[Version:  ")

  If UBound(arrayString) = -1 Then
    MsgBox "There is no history data for this field."
    Exit Sub
  End If
  For i = 1 To UBound(arrayString)
    startPos = 1
    strSearch = arrayString(i)
    EndDatePos = InStr(startPos, strSearch, " ")
    MemoDate = CDate(Left(strSearch, EndDatePos - 1))
    startPos = EndDatePos + 1
    EndTimePos = InStr(startPos, arrayString(i), "]") - 3
    MemoTime = CDate(Mid(strSearch, startPos, _
        EndTimePos - startPos))
    startPos = EndTimePos + 3
    strSearch = Trim(Replace(strSearch, vbCrLf, ""))
    MemoText = Right(strSearch, Len(strSearch) - startPos)
    Debug.Print MemoDate, MemoTime, MemoText
  Next
End Sub
```

As mentioned earlier, the Application object's `ColumnHistory` method is used in VBA to retrieve memo column history data. Because Access returns this data in a single string and we want to divide it into separate items, we use the `Split` function. This function is ideal for breaking a long string into an array of substrings based on a specified delimiter. The `Split` function returns a zero-based, one-dimensional array where each substring is an element. To hold the result of this function, the RetrieveMemoHistory procedure defines an array variable of the String data type named `arrayString`.

The first argument of the `Split` function specifies the string expression you want to split. The string that will be returned by the `ColumnHistory` method of the Application object is as follows:

```
arrayString = Split(Application.ColumnHistory( _
 "Agents", "FieldNotes", _
 "AgentID='A100'"), "[Version:  ")
```

The second argument of the `Split` function specifies a string that is used to identify substring limits. You can split a string on a single character, a space, or a group of characters. Because each history item is separated by a line feed and a carriage return (vbCrLf), you might think that it's a good idea to break the Access-generated history string into separate lines by using the vbCrLf delimiter. Well, it isn't, simply because memo fields allow carriage returns. A better delimiter is something that does not conflict with anything the user may enter into the memo field. You should be able to use the "[Version: " string that Access adds to each line of the history string without having to worry about unexpected results. Notice that there are two spaces after the colon that we also want to include in the delimiting string. Now that we have eliminated from the history string extraneous text ([Version:), we need to iterate through the array elements using the `For...Next` loop. However, there is no point doing this if the `arrayString` variable does not contain any elements. To check this out, we can use the UBound function, which will return –1 when the array is empty. While enumerating the history data, the procedure uses several variables to determine the character position where the date and time strings end (`EndDatePos`, `EndTimePos`). We also use the `startPos` variable to specify at which position in the search string the search should begin. Before extracting the date and time strings, we find the end character positions for these strings using the `InStr` function:

```
EndDatePos = InStr(startPos, strSearch, " ")
```

The `InStr` function returns the position of the first occurrence of one string within another. The first parameter is optional. It indicates the character position where the search should start. Obviously, we want to start at the first position so that we can examine the entire string. The second parameter is the string to search in. We are storing it in the `strSearch` variable. The third parameter of the `InStr` function is the string you want to find. In this case, we want to find a single space after the date. Notice that the single space separates the date from the time (see Figure 12.9 earlier). The `InStr` function also has an optional fourth argument that specifies the type of string comparison. When omitted, Access performs a binary comparison where each character matches only itself. This is the default. The `InStr` function will return a zero (0) when the string you are looking for is not found in the string you searched.

We also use other text functions (`Right`, `Left`, `Mid`) to extract a specified number of characters from the string. The `Right` function is used to extract characters from the right side of the string; the `Left` function does the same but from the left side of the string; and the `Mid` function extracts characters from the middle of the string. Notice that the text operations also require the use of the built-in `Len` function that returns the total number of characters in the specified string.

We defined the `MemoDate` and `MemoTime` variables as Date; thus, after extracting the date and time strings from the searched string, we use the `CDate` function to convert them into the Date format.

14. Run the **RetrieveMemoHistory** procedure. The procedure prints to the Immediate window the history string broken into three columns as shown in Figure 12.10.

```
Immediate
5/29/2013     6:30:59 AM     Delivered Presentation to Travel Agency.
5/29/2013     6:31:45 AM     Our team won the award in the Adventure category.
5/29/2013     6:32:59 AM     Our team won the award in the Adventure category.Organizing a tri
```

FIGURE 12.10. The history data from the FieldNotes column is output to the Immediate window via a VBA procedure.

CREATING RICH TEXT MEMO FIELDS WITH DAO

The .accdb file format boasts the Rich Text feature in memo fields. This allows you to format your memo fields in a datasheet with the bold, italics, underline, and other formatting options that are available via the Ribbon. To enable the Rich Text feature, open a table in Design view; in the Field Properties area for the selected memo field, set the Text Format property to Rich Text (see Figure 12.11). When you use the Rich Text feature in a memo field, Access stores the data in HTML format. Figure 12.12 shows an example of rich text formatting for a field in the Northwind database.

Employees

Field Name	Data Type	Description (Optional)
State/Province	Short Text	
ZIP/Postal Code	Short Text	
Country/Region	Short Text	
Web Page	Hyperlink	
Notes	Long Text	
Attachments	Attachment	

Field Properties

General | Lookup

Format	
Caption	
Default Value	
Validation Rule	
Validation Text	
Required	No
Allow Zero Length	No
Indexed	No
Unicode Compression	Yes
IME Mode	No Control
IME Sentence Mode	Phrase Predict
Text Format	Rich Text
Text Align	General
Append Only	No

Choose Rich text to store text as HTML and allow rich formatting. Choose Plain text to store only text. Avoid using Rich text if data might be used in Microsoft Access 2003 or earlier.

FIGURE 12.11. Enabling Rich Text for a memo field.

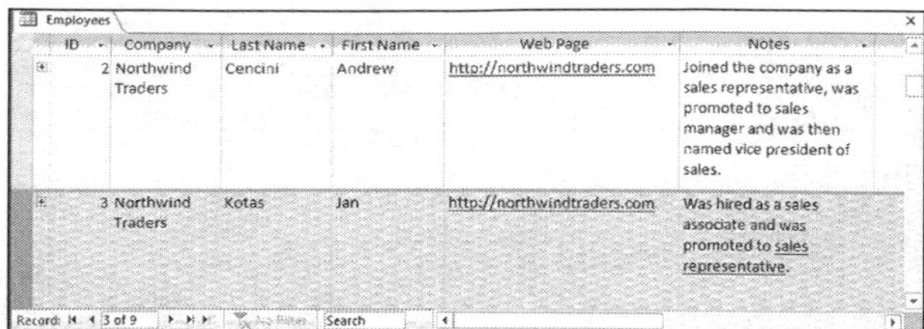

FIGURE 12.12. The Notes field for Jan Kotas in the Employees table of the Northwind 2007.accdb database is shown here with the rich text formatting.

Hands-On 12.9. Creating a Rich Text Memo Field

1. In the VBE screen, choose **Insert | Module** and enter the following two procedures in the module's Code window:

```
Sub CreateRichMemoFld()
    Dim db As DAO.Database
    Dim tdf As DAO.TableDef
    Dim fld As DAO.Field2
    Dim strTbl As String
    Dim strFld As String

    On Error GoTo ErrorHandler

    strTbl = "Agents"
    strFld = "PersonalNotes"

    Set db = CurrentDb
    Set tdf = db.TableDefs(strTbl)

    ' add an attachment field
    Set fld = tdf.CreateField(strFld, dbMemo)
    tdf.Fields.Append fld

    ConvertToRichText strTbl, strFld
ExitHere:
    Set fld = Nothing
    Set tdf = Nothing
    Set db = Nothing
    Exit Sub
ErrorHandler:
    If Err.Number = 3211 Then
```

```
            ' table is open; need to close it to continue
              DoCmd.Close acTable, strTbl, acSaveYes
            Resume
        Else
              MsgBox Err.Number & ": " & Err.Description
              Resume ExitHere
        End If
End Sub
Sub ConvertToRichText(strTbl As String, strFld As String)
        With CurrentDb
            With .TableDefs(strTbl)
                With .Fields(strFld)
                    On Error Resume Next
                     .Properties("TextFormat") = 1
                     If Err.Number = 3270 Then _
                         .Properties.Append .CreateProperty( _
                         "TextFormat", dbByte, 1)
                End With
            End With
        End With
End Sub
```

Run the **CreateRichMemoFld** procedure.

The CreateRichMemoFld procedure begins by creating a memo field called PersonalNotes in the current database's Agents table. Once the field is appended to the TableDefs collection of the Agents table, we need to set its TextFormat property to RichText. We do this by calling the ConvertToRichText procedure. If the TextFormat property already exists, this procedure will set the TextFormat property of the FieldNotes field to 1, which denotes the Rich Text setting. The default value of the TextFormat property is 0 (Plain Text). If the property is not found, error 3270 will occur, and at this point we want Access to execute the statement that will create the new property called TextFormat and then append it to the Properties collection:

```
If Err.Number = 3270 Then
  .Properties.Append .CreateProperty( _
   "TextFormat", dbByte, 1)
```

The last argument in the `CreateProperty` method specifies the type of setting for the Rich Text memo field. As mentioned earlier, 1 represents Rich Text, and 0 represents Plain Text.

Open the Agents table in Design view and verify the changes made by the CreateRichMemoFld procedure in this hands-on exercise.

Close the **Agents** table.

To modify the setting of the TextFormat property of the memo field in another Access database, use the `OpenDatabase()` *method of the DBEngine object to open the database. See the revised Convert-ToRichText procedure on the CD-ROM.*

REMOVING A FIELD FROM A TABLE

You may remove any field from an existing table, whether or not this field contains data. However, you can't delete a field after you have created an index that references that field. You must first delete the index.

The procedure in Hands-On 12.10 illustrates how to access the ADOX Columns collection of a Table object and use the Columns collection's `Delete` method to remove a field from a table. This procedure will fail if the field you want to delete is part of an index.

Hands-On 12.10. Removing a Field from a Table (ADO)

This hands-on exercise requires that you created the Add_NewFields procedure in Hands-On 12.4.

1. In the Visual Basic Editor window, choose **Insert | Module**.
2. In the module's Code window, type the **Delete_Field** procedure shown here:

```
Sub Delete_Field()
  Dim conn As ADODB.Connection
  Dim cat As New ADOX.Catalog
  Set conn = New ADODB.Connection
  conn.Open "Provider=Microsoft.Jet.OLEDB.4.0;" & _
   "Data Source=" & CurrentProject.Path & _
   "\Northwind.mdb"
  Set cat = New ADOX.Catalog
  cat.ActiveConnection = conn
  cat.Tables("CustomersCopy").Columns.Delete "MyNewField"
  Set cat = Nothing
  conn.Close
  Set conn = Nothing
End Sub
```

3. Choose **Run | Run Sub/UserForm** to execute the procedure.

In DAO, use the Fields collection's `Delete` method to remove a field from an existing table.

Hands-On 12.11. Removing a Field from a Table (DAO)

The following procedure removes from the CustomersCopy table two fields that were added by the procedure in Hands-On 12.5.

1. In the Visual Basic Editor window, choose **Insert | Module**.
2. In the module's Code window, type the following **DeleteFields_DAO** procedure:

```
Sub DeleteFields_DAO()
  Dim db As DAO.Database
  Dim tdf As DAO.TableDef
  Dim strDBName As String
  Dim strTblName As String
  On Error GoTo ErrorHandler
  strDBName = "C:\Access2013_ByExample\Northwind.mdb"
  strTblName = "CustomersCopy"
  Set db = OpenDatabase(strDBName)
  Set tdf = db.TableDefs(strTblName)
  MsgBox "Number of fields in the table: " & _
   db.TableDefs(strTblName).Fields.Count
  With tdf
    .Fields.Delete "NoOfMeetings"
    .Fields.Delete "Result"
  End With
  MsgBox "Number of fields in the table: " & _
   db.TableDefs(strTblName).Fields.Count
  db.Close
  Exit Sub
ErrorHandler:
  MsgBox Err.Number & ": " & Err.Description
End Sub
```

3. Choose **Run | Run Sub/UserForm** to execute the procedure.

RETRIEVING TABLE PROPERTIES

You can set or retrieve table properties using the Properties collection of an ADOX Table object. The Properties collection exposes standard ADO properties as well as properties specific to the data provider. You can iterate through all of the properties of an object using the `For Each...Next` programming structure.

The procedure in the following hands-on exercise accesses the CustomersCopy table and lists its properties and their values in the Immediate window (see Figure 12.13).

Hands-On 12.12. Listing Table Properties

1. In the Visual Basic Editor window, choose **Insert | Module**.
2. In the module's Code window, type the following **List_TableProperties** procedure:

```
Sub List_TableProperties()
  Dim conn As ADODB.Connection
  Dim cat As ADOX.Catalog
  Dim tbl As ADOX.Table
  Dim pr As ADOX.Property

  Set conn = New ADODB.Connection
  conn.Open "Provider=Microsoft.Jet.OLEDB.4.0;" & _
   "Data Source=" & CurrentProject.Path & _
   "\Northwind.mdb"

  Set cat = New ADOX.Catalog
  cat.ActiveConnection = conn

  Set tbl = cat.Tables("CustomersCopy")

  ' retrieve table properties
  For Each pr In tbl.Properties
    Debug.Print tbl.Name & ": " & _
     pr.Name & "= "; pr.Value
  Next
  Set cat = Nothing
  conn.Close
  Set conn = Nothing
End Sub
```

3. Choose **Run | Run Sub/UserForm** to execute the procedure.

```
Immediate
CustomersCopy: Temporary Table= False
CustomersCopy: Jet OLEDB:Table Validation Text=
CustomersCopy: Jet OLEDB:Table Validation Rule=
CustomersCopy: Jet OLEDB:Cache Link Name/Password= False
CustomersCopy: Jet OLEDB:Remote Table Name=
CustomersCopy: Jet OLEDB:Link Provider String=
CustomersCopy: Jet OLEDB:Link Datasource=
CustomersCopy: Jet OLEDB:Exclusive Link= Falszze
CustomersCopy: Jet OLEDB:Create Link= False
CustomersCopy: Jet OLEDB:Table Hidden In Access= False
```

FIGURE 12.13. You can list the names of table properties and their values programmatically as shown in Hands-On 12.12.

Additional Code on CD-ROM

File Name: Access2013_HandsOn\ListTableProperties_DAO.txt
Description: Use the Properties collection of the DAO TableDef object to list properties of the Agents table in the Chap12.accdb database.

RETRIEVING FIELD PROPERTIES

The procedure in Hands-On 12.13 retrieves the field properties of the field named AgentID located in the Agents table in the current database and prints them to the Immediate window, as shown in Figure 12.14.

Hands-On 12.13. Listing Field Properties

1. In the Visual Basic Editor window, choose **Insert | Module**.
2. In the module's Code window, type the **List_FieldProperties** procedure shown here:

```
Sub List_FieldProperties()
  Dim cat As ADOX.Catalog
  Dim col As ADOX.Column
  Dim pr As ADOX.Property

  Set cat = New ADOX.Catalog
  Set cat.ActiveConnection = CurrentProject.Connection
  Set col = New ADOX.Column
  Set col = cat.Tables("Agents").Columns("AgentID")
```

```
    Debug.Print "Properties of the AgentID field " & _
     "(" & col.Properties.Count & ")"
      ' retrieve Field properties
    For Each pr In col.Properties
      Debug.Print pr.Name & "="; pr.Value
    Next

    Set cat = Nothing
  End Sub
```

3. Choose **Run | Run Sub/UserForm** to execute the procedure.

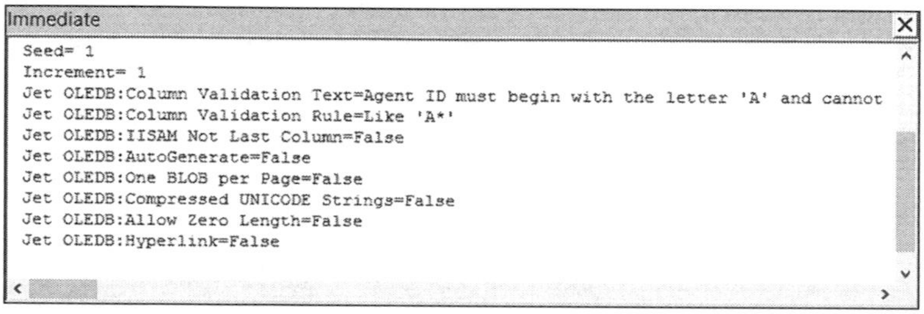

```
Immediate
Seed= 1
Increment= 1
Jet OLEDB:Column Validation Text=Agent ID must begin with the letter 'A' and cannot
Jet OLEDB:Column Validation Rule=Like 'A*'
Jet OLEDB:IISAM Not Last Column=False
Jet OLEDB:AutoGenerate=False
Jet OLEDB:One BLOB per Page=False
Jet OLEDB:Compressed UNICODE Strings=False
Jet OLEDB:Allow Zero Length=False
Jet OLEDB:Hyperlink=False
```

FIGURE 12.14. Running the procedure in Hands-On 12.13 generates a list of field properties and their values in the Immediate window.

LINKING A MICROSOFT ACCESS TABLE

You can create links to tables in Access databases as well as other data formats such as Excel, dBASE, Paradox, Exchange/Outlook, Lotus, and text and HTML files.

To create a linked Access table, you must set the following table properties:

```
Jet OLEDB:LinkDatasource
Jet OLEDB:Remote Table Name
Jet OLEDB:CreateLink
```

The procedure in Hands-On 12.14 demonstrates how to establish a link to the Customers table located in the Northwind.mdb database.

Hands-On 12.14. Linking a Microsoft Jet Table

1. In the Visual Basic Editor window, choose **Insert | Module**.
2. In the module's Code window, type the following **Link_JetTable** procedure:

```
Sub Link_JetTable()
  Dim cat As ADOX.Catalog
  Dim lnkTbl As ADOX.Table
```

```
    Dim strDb As String
    Dim strTable As String

    On Error GoTo ErrorHandler

    strDb = CurrentProject.Path & "\Northwind.mdb"
    strTable = "Customers"
    Set cat = New ADOX.Catalog
    cat.ActiveConnection = CurrentProject.Connection

    Set lnkTbl = New ADOX.Table
    With lnkTbl
      ' Name the new Table and set its ParentCatalog property to the
      ' open Catalog to allow access to the Properties collection.
      .Name = strTable
      Set .ParentCatalog = cat

     ' Set the properties to create the link
      .Properties("Jet OLEDB:Create Link") = True
      .Properties("Jet OLEDB:Link Datasource") = strDb
      .Properties("Jet OLEDB:Remote Table Name") = strTable
    End With

    ' Append the table to the Tables collection
    cat.Tables.Append lnkTbl

    Set cat = Nothing
    MsgBox "The current database contains a linked " & _
     "table named " & strTable
    Exit Sub

ErrorHandler:
    MsgBox Err.Number & ": " & Err.Description
End Sub
```

3. Choose **Run | Run Sub/UserForm** to execute the procedure.
   To access the linked Customers table after running this procedure, be sure to refresh the Access application window by pressing **Ctrl+F5**.

LINKING A DBASE TABLE

You can also link tables that reside in other programs or have different file formats such as Microsoft Excel, dBASE, or Paradox. In DAO, to link a table to an Access database, use the `CreateTableDef` method to create a new table:

```
Set myTable = db.CreateTableDef("TableDBASE")
```

Next, specify the Connect property of the TableDef object. For example, the following statement specifies the connect string:

```
myTable.Connect = "dBase 5.0;Database=C:\Access2013_ByExample"
```

Next, specify the SourceTableName property of the TableDef object to indicate the actual name of the table in the source database:

```
myTable.SourceTableName = "Customer.dbf"
```

Finally, use the `Append` method to append the TableDef object to the TableDefs collection:

```
db.TableDefs.Append myTable
```

File Name: Access2013_HandsOn\LinkDBaseTable_DAO.txt
Description: Linking a dBASE table to the current database

LINKING A MICROSOFT EXCEL WORKSHEET

You can link an Excel worksheet to a Microsoft Access database by using the `TransferSpreadsheet` method of the DoCmd object, as shown in Hands-On 12.15. Note, however, that neither the DoCmd object nor its `TransferSpreadsheet` method are members of the ADO Object Model. The DoCmd object is built into the Microsoft Access library.

Hands-On 12.15. Linking an Excel Worksheet

This hands-on exercise uses the Regions.xls workbook file provided on the CD-ROM disk. You can revise the procedure code to use any workbook file that you have available; however, you must match the name of the spreadsheet constant with the Excel version. Table 12.4 shows the constant names and values if you need a different format.

TABLE 12.4. Spreadsheet constants

Constant	Value	Description
acSpreadsheetTypeExcel3	0	Microsoft Excel 3.0 format
acSpreadsheetTypeExcel4	6	Microsoft Excel 4.0 format
acSpreadsheetTypeExcel5	5	Microsoft Excel 5.0 format
acSpreadsheetTypeExcel7	5	Microsoft Excel 95 format
acSpreadsheetTypeExcel8	8	Microsoft Excel 97 format
acSpreadsheetTypeExcel9	8	Microsoft Excel 2000-2003 format
acSpreadsheetTypeExcel12	9	Microsoft Excel 2010/2007 format (.xls)
acSpreadsheetTypeExcel12Xml	10	Microsoft Excel 2013/2010/2007 format (.xml)

1. Copy the **Regions.xls** workbook from the **C:\Access2013_HandsOn** folder to your **C:\Access2013_ByExample** folder.
2. In the Visual Basic Editor window, choose **Insert | Module**.
3. In the module's Code window, type the following **Link_ExcelSheet** procedure:

```
Sub Link_ExcelSheet()
    Dim rst As ADODB.Recordset

    DoCmd.TransferSpreadsheet acLink, _
      acSpreadsheetTypeExcel12, _
      "mySheet", _
      CurrentProject.Path & "\Regions.xls", _
      -1, "Regions!A1:B15"

    Set rst = New ADODB.Recordset
    With rst
        .ActiveConnection = CurrentProject.Connection
        .CursorType = adOpenKeyset
        .LockType = adLockOptimistic
        .Open "mySheet", , , , adCmdTable
    End With

    Do Until rst.EOF
        Debug.Print rst.Fields(0).Value, rst.Fields(1).Value
        rst.MoveNext
```

```
        Loop
        rst.Close
        Set rst = Nothing
    End Sub
```

4. Choose **Run | Run Sub/UserForm** to execute the procedure.

The Link_ExcelSheet procedure begins by creating a linked table named mySheet from the specified range of cells (A1:B15) in the Regions worksheet in the Regions.xls file. The first argument in the `DoCmd` statement indicates that the first row of the spreadsheet contains column headings. Next, the procedure uses the ADO Recordset object to retrieve the data from the mySheet table into the Immediate window. Notice that prior to opening the Recordset object, several properties of the Recordset object must be set:

- The ActiveConnection property sets the reference to the current database.
- The CursorType property specifies how the Recordset object should interact with the data source.

The `adOpenKeyset` setting tells Visual Basic that instead of retrieving all the records from the data source, only the keys are to be retrieved. The data for these keys is retrieved only as you scroll through the recordset. This guarantees better performance than retrieving big chunks of data at once.

- The LockType property determines how to lock the data while it is being manipulated.

The `adLockOptimistic` setting locks the record only when you attempt to save it.

- Opening the Recordset object also requires that you specify the data source. The data source in this procedure is the linked table named mySheet. The parameter passed depends on the source type used.

  The `adCmdTable` setting indicates that all rows from the source table should be included.

You could also open the Recordset object by passing all the required parameters at once, as follows:

```
rst.Open "mySheet", _
  CurrentProject.Connection, adOpenKeyset, _
  adLockOptimistic, adCmdTable
```

LISTING DATABASE TABLES

The procedure in Hands-On 12.16 generates a list of tables in the Northwind database. It uses the ADOX Catalog object to gain access to the database, then iterates through the Tables collection to retrieve the names of Access tables, system tables, and views. The ADOX Tables collection stores various types of Table objects, as shown in Table 12.5.

TABLE 12.5. Types of tables in the ADOX Tables collection

Name	Description
ACCESS TABLE	An Access system table
LINK	A linked table from a non-ODBC data source
PASS-THROUGH	A linked table from an ODBC data source
SYSTEM TABLE	A Microsoft Jet system table
TABLE	A Microsoft Access table
VIEW	A table from a row-returning, nonparameterized query

Hands-On 12.16. Creating a List of Database Tables

1. In the Visual Basic Editor window, choose **Insert | Module**.
2. In the module's Code window, type the following **ListTbls** procedure:

```
Sub ListTbls()
  Dim cat As ADOX.Catalog
  Dim tbl As ADOX.Table

  Set cat = New ADOX.Catalog
  cat.ActiveConnection = "Provider=Microsoft.Jet.OLEDB.4.0;" & _
   "Data Source=" & CurrentProject.Path & _
   "\Northwind.mdb"

  For Each tbl In cat.Tables
    If tbl.Type <> "VIEW" And _
     tbl.Type <> "SYSTEM TABLE" And _
     tbl.Type <> "ACCESS TABLE" Then Debug.Print tbl.Name
  Next tbl
  Set cat = Nothing
  MsgBox "View the list of tables in the Immediate window."
End Sub
```

3. Choose **Run | Run Sub/UserForm** to execute the procedure.

You can also use the `OpenSchema` method of the ADO Connection object to list tables in your database (see the following section).

LISTING TABLES AND FIELDS

Earlier in this chapter you learned how to enumerate tables in the Northwind database by accessing the Tables collection of the ADOX Catalog object. The procedures in Hands-On 12.17 and Hands-On 12.18 demonstrate how to use the `OpenSchema` method of the ADO Connection object to obtain more information about a database table and its fields.

Hands-On 12.17. Using the OpenSchema Method to List Database Tables

1. In the Visual Basic Editor window, choose **Insert | Module**.
2. In the module's Code window, type the following **ListTbls2** procedure:

```
Sub ListTbls2()
  ' This procedure lists database tables using
  ' the OpenSchema method
  Dim rst As ADODB.Recordset
  Set rst = CurrentProject.Connection.OpenSchema(adSchemaTables)

  Do Until rst.EOF
    Debug.Print rst.Fields("TABLE_TYPE") & " ->" _
     & rst.Fields("TABLE_NAME")
    rst.MoveNext
  Loop
End Sub
```

3. Choose **Run | Run Sub/UserForm** to execute the procedure.

Obtaining the names of fields requires that you use `adSchemaColumns` as the parameter for the `OpenSchema` method. The ListTblsAndFields procedure in Hands-On 12.18 retrieves the names of fields in each table of the Northwind database.

Hands-On 12.18. Listing Tables and Their Fields Using the OpenSchema Method

1. In the Visual Basic Editor window, choose **Insert | Module**.
2. In the module's Code window, type the following **ListTblsAndFields** procedure:

```
Sub ListTblsAndFields()
  Dim conn As ADODB.Connection
  Dim rst As ADODB.Recordset
  Dim curTable As String
  Dim newTable As String
  Dim counter As Integer

  Set conn = New ADODB.Connection
  conn.Open "Provider=Microsoft.Jet.OLEDB.4.0;" _
   & "Data Source=" & CurrentProject.Path & _
   "\Northwind.mdb"

  Set rst = conn.OpenSchema(adSchemaColumns)
  curTable = ""
  newTable = ""
  counter = 1
  Do Until rst.EOF
    curTable = rst!table_Name
    If (curTable <> newTable) Then
      newTable = rst!table_Name
      Debug.Print "Table: " & rst!table_Name
      counter = 1
    End If
  Debug.Print "Field" & counter & ": " & rst!Column_Name
  counter = counter + 1
  rst.MoveNext
  Loop
  rst.Close
  conn.Close
  Set rst = Nothing
  Set conn = Nothing
End Sub
```

3. Choose **Run | Run Sub/UserForm** to execute the procedure.

LISTING DATA TYPES

The ListDataTypes procedure in Hands-On 12.19 uses the `adSchemaProvider-Types` parameter of the ADO Connection object's `OpenSchema` method to list the data types supported by the Microsoft Jet OLE DB 4.0 provider.

Hands-On 12.19. Listing Data Types

1. In the Visual Basic Editor window, choose **Insert | Module**.
2. In the module's Code window, type the **ListDataTypes** procedure shown below.

```
Sub ListDataTypes()
  Dim conn As ADODB.Connection
  Dim rst As ADODB.Recordset

  Set conn=New ADODB.Connection
  conn.Open "Provider=Microsoft.Jet.OLEDB.4.0;" _
   & "Data Source=" & CurrentProject.Path & _
   "\Northwind.mdb"
  Set rst=conn.OpenSchema(adSchemaProviderTypes)
  Do Until rst.EOF
    Debug.Print rst!Type_Name & vbTab _
     & "Size: " & rst!Column_Size
    rst.MoveNext
  Loop

  rst.Close
  conn.Close
  Set rst = Nothing
  Set conn = Nothing
End Sub
```

3. Choose **Run | Run Sub/UserForm** to execute the procedure.

CHANGING THE AUTONUMBER

When you create a table in a Microsoft Access database, you can assign an AutoNumber data type to a primary key field manually using the Access user interface. The AutoNumber is a unique sequential number (incremented by 1) or a random number assigned by Microsoft Access whenever a new record is added to a table. You can set the start and step value of auto-increment fields programmatically by using Jet 4.0 SQL statements (see Chapter 19, "Creating, Modifying, and Deleting Tables and Fields," for more information).

The procedure in Hands-On 12.20 opens the ADO Recordset object based on the Shippers table in the Northwind database, retrieves the last used AutoNumber value, and determines the current step (increment) value in effect.

Hands-On 12.20. Changing the Value of an AutoNumber

1. In the Visual Basic Editor window, choose **Insert | Module.**
2. In the module's Code window, type the **ChangeAutoNumber** procedure shown here:

```
Sub ChangeAutoNumber()
  Dim conn As ADODB.Connection
  Dim rst As ADODB.Recordset
  Dim strSQL As String
  Dim beginNum As Integer
  Dim stepNum As Integer

  Set conn = New ADODB.Connection
  conn.Open "Provider = Microsoft.Jet.OLEDB.4.0;" & _
   "Data Source=" & CurrentProject.Path & _
   "\Northwind.mdb"

  Set rst = New ADODB.Recordset
  With rst
    .CursorType = adOpenKeyset
    .LockType = adLockReadOnly
    .Open "Shippers", conn
    .MoveLast
  End With
  beginNum = rst(0)
  rst.MovePrevious
  stepNum = beginNum - rst(0)

  MsgBox "Last Auto Number Value = " & beginNum & vbCr & _
   "Current Step Value = " & stepNum, vbInformation, _
   "AutoNumber"
```

```
    rst.Close
    conn.Close
    Set conn = Nothing
End Sub
```

3. Choose **Run | Run Sub/UserForm** to execute the procedure.

CHAPTER SUMMARY

This chapter has shown you how to programmatically create Microsoft Access tables by using ADO and DAO objects. You learned how to add fields to your tables and define field data types and field properties. You found out how to list both tables and fields, and investigate their properties. In addition to creating new tables from scratch, you discovered how to work with linked tables. You also learned how to copy and delete tables.

The next chapter will demonstrate how to create indexes and set up table relationships using VBA procedures.

Chapter 13

Setting up Primary Keys, Indexes, and Table Relationships

After defining the fields for your tables, take the time to set up primary keys, indexes, and relationships between tables. This chapter focuses on the DAO and ADOX objects that are designed to work with these features.

CREATING A PRIMARY KEY INDEX

Indexes determine the order in which records are accessed from database tables and whether or not duplicate records are accepted. While indexes can speed up access to specific records in large tables, too many indexes can also slow down updates to the database. Each table in your database should include a field (or set of fields) that uniquely identifies each individual record in a table. Such a field or set of fields is called a *primary key*. A primary key is an index with its Unique and Primary properties set to `True`. There can be only one primary key per table.

CREATING INDEXES USING ADO

In ADO, indexes are created using the Key object from the ADOX library. The Type property of the Key object allows you to determine whether the key is primary, foreign, or unique. For example, to create a primary key, set the Key object's Type property to `adKeyPrimary`.

The procedure in Hands-On 13.1 demonstrates how to add a primary key to the tblFilters table.

Please note files for the "Hands-On" project may be found on the companion CD-ROM.

Hands-On 13.1. Creating a Primary Key (ADO)

1. Start Microsoft Office Access 2010 and create a new database named **Chap13.accdb** in your **C:\Access2013_ByExample** folder.
2. In the Access window, press **Alt+F11** to switch to the Visual Basic Editor window.
3. In the Visual Basic Editor window, choose **Insert | Module**.
4. Choose **Tools | References** and add a reference to the **Microsoft ActiveX Data Objects 6.1** and **Microsoft ADO Ext. 6.0 for DDL and Security** libraries.
5. In the module's Code window, type the following **Create_PrimaryKey** procedure:

```
' make sure to set up a reference to
' the Microsoft ActiveX Data Objects 6.1
' and Microsoft ADO Ext. 6.0 for DDL and Security
```

```
Sub Create_PrimaryKey()
Dim cat As ADOX.Catalog
  Dim tbl As ADOX.Table
  Dim pKey As ADOX.Key

  On Error GoTo ErrorHandler

  Set cat = New ADOX.Catalog
  cat.ActiveConnection = CurrentProject.Connection

  Set tbl = New ADOX.Table
  tbl.Name = "tblFilters"

  cat.Tables.Append tbl

  With tbl.Columns
    .Append "ID", adVarWChar, 10
    .Append "Description", adVarWChar, 255
    .Append "Type", adInteger
  End With

  SetKey:
  Set pKey = New ADOX.Key
  With pKey
    .Name = "PrimaryKey"
    .Type = adKeyPrimary
  End With

  pKey.Columns.Append "ID"
  tbl.Keys.Append pKey

  Set cat = Nothing
  Exit Sub

  ErrorHandler:
    If Err.Number = -2147217856 Then
      MsgBox "The " & tbl.Name & " is open.", _
       vbCritical, "Please close the table"
    ElseIf Err.Number = -2147217857 Then
      MsgBox Err.Description
      Set tbl = cat.Tables(tbl.Name)
      Resume SetKey
```

```
        ElseIf Err.Number = -2147217767 Then
          tbl.Keys.Delete pKey.Name
          Resume
        Else
          MsgBox Err.Number & ": " & Err.Description
        End If
    End Sub
```

6. Choose **Run | Run Sub/UserForm** to execute the procedure.
   The Create_PrimaryKey procedure begins by creating a table named tblFilters in the currently open database and proceeds to set the primary key index on the ID field. If the tblFilters table already exists, the error handler code displays the error message and sets an object variable (`tbl`) to point to this table. The `Resume SetKey` statement refers the procedure execution to the label `SetKey`. The code that follows that label defines the primary key using the Name and Type properties of the Key object. Next, the procedure appends the ID column to the Columns collection of the Key object, and the Key object itself is appended to the Keys collection of the table. Because errors could occur if a table is open or it already contains the primary key, the error handler is included to ensure that the procedure runs as expected.
7. Run this procedure again using step mode (press **F8**).

CREATING INDEXES USING DAO

In DAO, indexes are created using the `CreateIndex` method for a TableDef object. The following statement creates an index named PrimaryKey:

```
Set idx = tdf.CreateIndex("PrimaryKey")
```

To ensure that the correct type of index is created, you need to set index properties. For example, the Primary property of an index indicates that the index fields constitute the primary key for the table:

```
idx.Primary = True
```

Use the Unique property to specify whether or not the values in an index must be unique:

```
idx.Unique = True
```

The Required property indicates whether the index can accept Null values. When you set this property to `True`, nulls will not be accepted:

```
idx.Required = True
```

Use the IgnoreNulls property to determine whether a record with a Null value in the index fields should be included in the index:

```
idx.IgnoreNulls = False
```

To actually index a table, you must use the `CreateField` method on the Index object to create a Field object for each field you want to include in the index:

```
Set fld = idx.CreateField("AgentID", dbText)
```

Note that in the Microsoft Access 2013 User Interface the AgentID field will display its data type as Short Text when you open the table in the Design view. The Short Text replaces the Text data type used in previous versions of Access. In programming, use the dbText to indicate that the field should hold character data. This has not changed from the prior versions.

Once the Field object is created, you need to append it to the Fields collection:

```
idx.Fields.Append fld
```

The last step in index creation is appending the Index object to the Indexes collection:

```
tdf.Indexes.Append idx
```

The procedure in Hands-On 13.2 uses DAO to create a primary key index in the Agents table.

Hands-On 13.2. Creating a Primary Key (DAO)

The procedure in this hands-on exercise uses the Agents table in the Chap12.accdb database that you created in Chapter 12.

1. In the Visual Basic Editor window, choose **Insert | Module**.
   In the module's Code window, type the following **Create_PrimaryKeyDAO** procedure:

```
Sub Create_PrimaryKeyDAO()
  Dim db As DAO.Database
  Dim tdf As DAO.TableDef
  Dim fld As DAO.Field
  Dim idx As DAO.Index
  Dim strDB As String
```

```
    strDB = "C:\Access2013_ByExample\Chap12.accdb"
    Set db = OpenDatabase(strDB)
    Set tdf = db.TableDefs("Agents")

    ' create a Primary Key
    Set idx = tdf.CreateIndex("PrimaryKey")
    idx.Primary = True
    idx.Required = True
    idx.IgnoreNulls = False
    Set fld = idx.CreateField("AgentID", dbText)
    idx.Fields.Append fld

    ' add the index to the Indexes collection in the Agents table
    tdf.Indexes.Append idx
    db.Close
    Set db = Nothing
End Sub
```

2. Choose **Run | Run Sub/UserForm** to execute the procedure.

   To verify that the index was created, open the Agents table in Chap12.accdb database. Activate the Design view and click the Indexes button on the Ribbon. The result of running the Create_PrimaryKeyDAO procedure is shown in Figure 13.1.

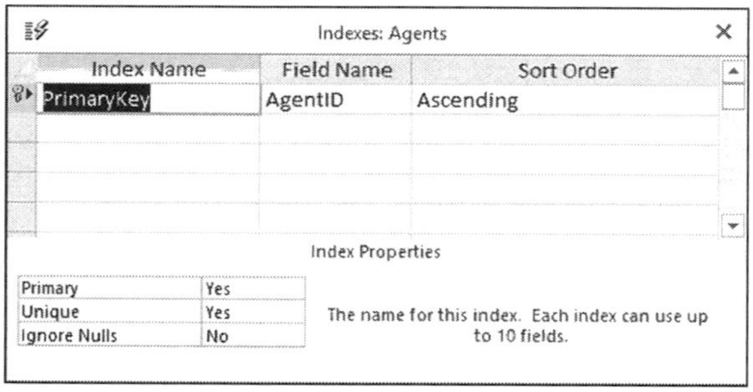

FIGURE 13.1. The Indexes window after running the procedure in Hands-On 13.2.

CREATING A SINGLE-FIELD INDEX USING ADO

In ADO, you can add an index to a table by using the ADOX Index object. Before creating an index, make sure the table is not open and that it does not already contain an index with the same name.

To define an index, perform the following:

- Append one or more columns to the index by using the `Append` method.
- Set the Name property of the Index object and define other index properties, if necessary.
- Use the `Append` method to add the Index object to the table's Indexes collection.

You can use the Unique property of the Index object to specify whether the index keys must be unique. The default value of the Unique property is `False`. Another property, IndexNulls, lets you specify whether Null values are allowed in the index. This property can be set to one of the constants shown in Table 13.1.

TABLE 13.1. Intrinsic constants for the IndexNulls property of the ADOX Index object (see the AllowNullsEnum in the ADOX Library)

Constant Name	Description
`adIndexNullsAllow`	You can create an index if there is a Null value in the index field (an error will not occur).
`adIndexNullsDisallow` (This is the default value)	You cannot create an index if there is a Null value in the index field for the column (an error will occur).
`adIndexNullsIgnore`	You can create an index if there is a Null value in the index field (an error will not occur). The Ignore Nulls property in the Indexes window in the user interface will be set to Yes.
`adIndexNullsIgnoreAny` (This value is not supported by the Microsoft Jet Provider)	You can create an index if there is a Null value in the index field. The Ignore Nulls property in the Indexes window in the user interface will be set to No.

The Add_SingleFieldIndex procedure in Hands-On 13.3 demonstrates how to add a single-field index called idxDescription to the table tblFilters.

Hands-On 13.3. Adding a Single-Field Index to an Existing Table (ADO)

This procedure uses the tblFilters table created in Hands-On 13.1 (Chap13.accdb database).

1. In the Visual Basic Editor window, choose **Insert | Module**.

2. In the module's Code window, type the following **Add_SingleFieldIndex** procedure:

```
Sub Add_SingleFieldIndex()
  Dim cat As New ADOX.Catalog
  Dim myTbl As New ADOX.Table
  Dim myIdx As New ADOX.Index
  Dim strTblName As String

  On Error GoTo ErrorHandler

  strTblName = "tblFilters"
  cat.ActiveConnection = CurrentProject.Connection
  Set myTbl = cat.Tables(strTblName)

  With myIdx
    .Name = "idxDescription"
    .Unique = False
    .IndexNulls = adIndexNullsIgnore
    .Columns.Append "Description"
    .Columns(0).SortOrder = adSortAscending
  End With
  myTbl.Indexes.Append myIdx

  Set cat = Nothing
  Exit Sub
ErrorHandler:
  If Err.Number = -2147217856 Then
    MsgBox strTblName & " will be closed.", _
     vbCritical, "Warning: Table is Open"
    DoCmd.Close acTable, strTblName, acSaveYes
    Resume
  ElseIf Err.Number = -2147217868 Then
    myTbl.Indexes.Delete myIdx.Name
    Resume
  Else
    MsgBox Err.Number & ": " & Err.Description
  End If
End Sub
```

3. Choose **Run | Run Sub/UserForm** to execute the procedure.

After the index properties are set, the Description column is appended to the index, and the index sort order is set to the default (`adSortAscending`). To set the index field's sort order to descending, use the `adSortDescending` constant. Next, the index is appended to the Indexes collection of the Table object.

ADDING A MULTIPLE-FIELD INDEX TO A TABLE USING DAO

The procedure in Hands-On 13.3 demonstrated adding a single-field index to an existing table by using the ADOX Index object. The procedure in the next hands-on exercise shows how to use DAO to add a multiple-field index to the Employees table in the Northwind database.

Hands-On 13.4. Adding a Multiple-Field Index to an Existing Table (DAO)

1. In the Visual Basic Editor window, choose **Insert | Module.**
   In the module's Code window, type the **Add_MultiFieldIndex** procedure shown here:

```
Sub Add_MultiFieldIndex()
  Dim db As DAO.Database
  Dim tdf As DAO.TableDef
  Dim fld As DAO.Field
  Dim idx As DAO.Index
  Dim strDB As String
  Dim strTblName As String

  strDB = "C:\Access2013_ByExample\Northwind.mdb"
  strTblName = "Employees"
  Set db = OpenDatabase(strDB)
  Set tdf = db.TableDefs(strTblName)

  Set idx = tdf.CreateIndex("Location")
  Set fld = idx.CreateField("City", dbText)
  idx.Fields.Append fld

  Set fld = idx.CreateField("Region", dbText)
  idx.Fields.Append fld
  tdf.Indexes.Append idx

  db.Close
```

```
    Set db = Nothing
    Debug.Print "New index (Location) was created."
End Sub
```

2. Choose **Run | Run Sub/UserForm** to execute the procedure.

   The Add_MultiFieldIndex procedure creates a two-field index in the Employees table. To create an index, use the `CreateIndex` method on a TableDef object. Next, use the `CreateField` method on the Index object to create the first field to be included in the index, and then append this field to the Fields collection. Repeat the same steps for the second field you want to include in the index. It is important to remember that the order in which the fields are appended has an effect on the index order. If you open the Indexes window in the Employees table of the Northwind database after running this procedure, the Location index will consist of two fields, as shown in Figure 13.2.

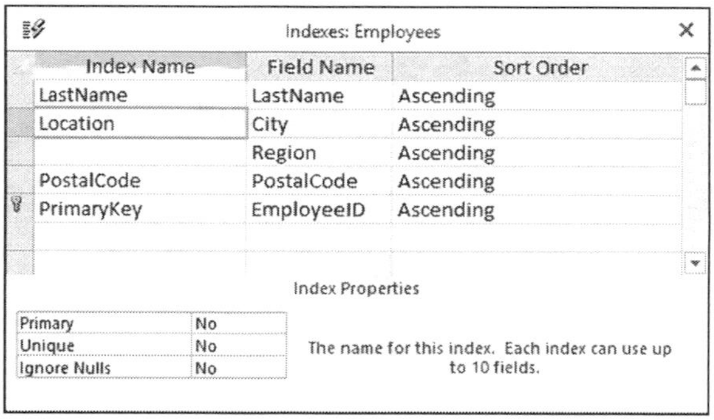

Indexes: Employees

Index Name	Field Name	Sort Order
LastName	LastName	Ascending
Location	City	Ascending
	Region	Ascending
PostalCode	PostalCode	Ascending
PrimaryKey	EmployeeID	Ascending

Index Properties

Primary	No
Unique	No
Ignore Nulls	No

The name for this index. Each index can use up to 10 fields.

FIGURE 13.2. The Location index was created by running the procedure in Hands-On 13.4.

LISTING INDEXES IN A TABLE

The ADO Indexes collection contains all Index objects of a table. You can retrieve all the index names from the Indexes collection. The procedure in the next hands-on exercise demonstrates how to list the names of indexes available in the Northwind database's Employees table in the Immediate window.

Hands-On 13.5. Listing Indexes in a Table (ADO)

1. In the Visual Basic Editor window, choose **Insert | Module**.
2. In the module's Code window, type the following **List_Indexes** procedure:

```
Sub List_Indexes()
    Dim conn As New ADODB.Connection
```

```
    Dim cat As New ADOX.Catalog
    Dim tbl As New ADOX.Table
    Dim idx As New ADOX.Index

    With conn
      .Provider = "Microsoft.Jet.OLEDB.4.0"
      .Open "Data Source=" & CurrentProject.Path & _
       "\Northwind.mdb"
    End With
    cat.ActiveConnection = conn
    Set tbl = cat.Tables("Employees")

    For Each idx In tbl.Indexes
      Debug.Print idx.Name
    Next idx

    conn.Close
    Set conn = Nothing
    MsgBox "Indexes are listed in the Immediate window."
  End Sub
```

3. Choose **Run | Run Sub/UserForm** to execute the procedure.

In DAO, you can use the `For Each...Next` loop to retrieve the names of indexes from the Indexes collection of the TableDef object, as illustrated in the following procedure:

```
Sub List_IndexesDAO()
  Dim db As DAO.Database
  Dim tdf As DAO.TableDef
  Dim idx As DAO.Index

  Set db = CurrentDb
  Set tdf = db.TableDefs("tblFilters")

  For Each idx In tdf.Indexes
    Debug.Print idx.Name
  Next

  ' show Immediate window
  SendKeys "^g"
  Set db = Nothing
End Sub
```

DELETING TABLE INDEXES

Although you can delete unwanted or obsolete indexes from the Indexes window in the Microsoft Access 2013 user interface, it is much faster to remove them programmatically. The procedure in Hands-On 13.6 illustrates how to delete all but the primary key index from the Employees table located in the Northwind database.

Hands-On 13.6. Deleting Indexes from a Table (ADO)

The procedure in this hands-on exercise will delete all but the primary key index from the Employees table in the Northwind database. It is recommended that you prepare a backup copy of the original Northwind.mdb database prior to running this code.

1. In the Visual Basic Editor window, choose **Insert | Module.**
2. In the module's Code window, type the following **Delete_Indexes** procedure:

```
Sub Delete_Indexes()
  ' This procedure deletes all but the primary key index
  ' from the Employees table in the Northwind database.
  ' Prior to running this procedure make a backup copy of
  ' the original Northwind.mdb database.

  Dim conn As New ADODB.Connection
  Dim cat As New ADOX.Catalog
  Dim tbl As New ADOX.Table
  Dim idx As New ADOX.Index
  Dim count As Integer

  With conn
    .Provider = "Microsoft.Jet.OLEDB.4.0"
    .Open "Data Source=" & CurrentProject.Path & _
     "\Northwind.mdb"
  End With

  cat.ActiveConnection = conn
  Setup:
  Set tbl = cat.Tables("Employees")

  Debug.Print tbl.Indexes.count
  For Each idx In tbl.Indexes
```

```
      If idx.PrimaryKey <> True Then
        tbl.Indexes.Delete (idx.Name)
        GoTo Setup
      End If
    Next idx

    conn.Close
    Set conn = Nothing
    MsgBox "All Indexes but Primary Key were deleted."
End Sub
```

3. Choose **Run | Run Sub/UserForm** to execute the procedure.

   Notice that each time you delete an index from the table's Indexes collection you must set the reference to the table because current settings are lost when an index is deleted. Hence, the `GoTo Setup` statement sends Visual Basic to the `Setup` label to get the new reference to the Table object.

CREATING TABLE RELATIONSHIPS USING ADO

This book assumes that you are familiar with the manual method of creating various types of relationships between Microsoft Access tables. If you need a refresher, you can either check the online help or peruse other materials on this topic. This section demonstrates how you can relate two tables via VBA code. We will establish the most common relationship, known as a *parent-child relationship*. In database terms, this relationship is also called a *one-to-many relationship*. We will create a Publishers table as a parent table and a Titles table as a child table. Then we will link them by a parent-child relationship. Recall that in this type of relationship, a record in the parent table can have multiple child records in the other table. In other words, when the term *one-to-many* is used, the parent is the *one* (single record) and *many* represents the children (multiple child records) in the other table.

In ADO, to establish a one-to-many relationship between tables, you'll need to perform the following steps:

1. Use the ADOX Key object to create a foreign key and set the Type property of the Key object to `adKeyForeign`. A *foreign key* consists of one or more fields in a foreign table that uniquely identify all rows in a primary table.
2. Use the RelatedTable property to specify the name of the related table.

3. Use the `Append` method to add appropriate columns in the foreign table to the foreign key. A foreign table is usually located on the "many" side of a one-to-many relationship and provides a foreign key to another table in a database.
4. Set the RelatedColumn property to the name of the corresponding column in the primary table.
5. Use the `Append` method to add the foreign key to the Keys collection of the table containing the primary key.

The procedure in Hands-On 13.7 illustrates how to create a one-to-many relationship between two tables: Titles and Publishers.

Hands-On 13.7. Creating a One-to-Many Relationship

1. In the current database (Chap13.accdb), create the Titles and Publishers tables and add the fields as shown in the following table:

Table Name	Field Name	Data Type	Size
Titles	TitleID	Number	
Titles	PubID	Number	
Titles	Title	Short Text	100
Titles	Price	Currency	
Publishers	PubID	Number	
Publishers	PubName	Short Text	40
Publishers	City	Short Text	25
Publishers	Country	Short Text	25

Instead of building the Titles and Publishers tables manually, try to write a VBA procedure based on what you learned about creating table definitions in Chapter 12.

2. Make **TitleID** the primary key for the Titles table and **PubID** the primary key for the Publishers table.
3. In the Visual Basic Editor window, choose **Insert | Module**.
4. In the module's Code window, type the **CreateTblRelation** procedure shown here:

```
Sub CreateTblRelation()
  Dim cat As New ADOX.Catalog
  Dim fKey As New ADOX.Key
```

```
    On Error GoTo ErrorHandler

    cat.ActiveConnection = CurrentProject.Connection

    With fKey
      .Name = "fkPubID"
      .Type = adKeyForeign
      .RelatedTable = "Publishers"
      .Columns.Append "PubID"
      .Columns("PubID").RelatedColumn = "PubID"
    End With
    cat.Tables("Titles").Keys.Append fKey
    MsgBox "Relationship was created."

    Set cat = Nothing
    Exit Sub

  ErrorHandler:
    cat.Tables("Titles").Keys.Delete "fkPubID"
    Resume
  End Sub
```

5. Choose **Run | Run Sub/UserForm** to execute the procedure.

   If you receive an error while running this procedure, make sure that both tables are closed.

   You can view the relationship between the Publishers and Titles tables that was created by the CreateTblRelation procedure in the Relationships window. To activate this window, switch to the Access application window and choose Database Tools | Relationships. You should see the Publishers and Titles tables in the Relationships window linked with a one-to-many relationship (see Figure 13.3).

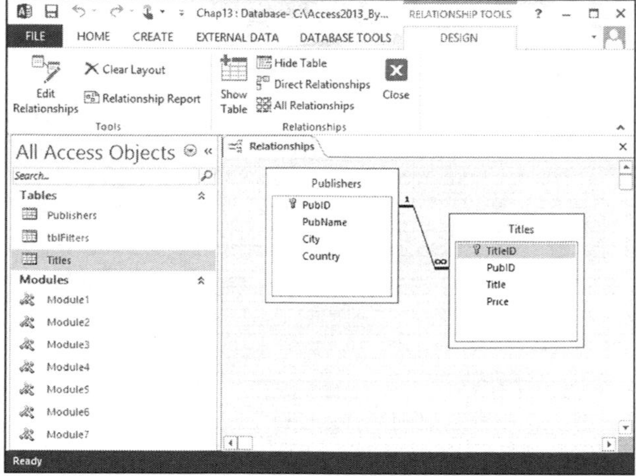

FIGURE 13.3. The one-to-many relationship between the Publishers and Titles tables was created programmatically by accessing objects in the ADOX library (see the code in the CreateTblRelation procedure in Hands-On 13.7).

CHAPTER SUMMARY

In this short chapter, you acquired programming skills that enable you to create keys (primary keys and indexes) in Microsoft Access tables. You also learned how to establish a one-to-many relationship between tables.

In the next chapter, you will learn how to find and read database records.

Chapter 14

Finding and Reading Records

In order to work with data, you need to learn how to use the Recordset object. The *Recordset object* represents a set of records in a table or a set of records returned by executing a stored query or an SQL statement. Each column of a Recordset represents a field, and each row represents a record. The Recordset is a temporary object and is not saved in the database. All Recordset objects cease to exist after the procedure ends. All open Recordset objects are contained in the Recordsets collection. Creating and using the Recordset objects depends on the type of object library (DAO/ADO) that you've selected for your programming task. In this chapter, you will learn various methods of opening the Recordset object. You will also find out how to navigate in the Recordset, and how to find, filter, read, and count the records. Both DAO and ADO Recordsets will be covered.

INTRODUCTION TO DAO RECORDSETS

In DAO there are five types of Recordset objects: Table-type, Dynaset-type, Snapshot-type, Forward-only-type, and Dynamic-type. Each of these recordsets offers a different functionality (see Table 14.1). You create a Recordset object using the `OpenRecordset` method. The type of the Recordset is specified by the `type` argument of the `OpenRecordset` method. If the Recordset's type is not specified, DAO will attempt to create a Table-type Recordset. If this type isn't available, attempts are made to create a Dynaset, Snapshot, or Forward-only-type Recordset object.

TABLE 14.1. Types of DAO Recordsets

Recordset Type	Description
Table-type	Used to access records in a table stored in an Access database. You can retrieve, add, update, and delete records in a single table.
Dynaset-type	Used to retrieve, add, update, and delete records from one or more tables in a database as well as any table that is linked to the Access database.
Snapshot-type	Used to access records from a local table stored in an Access database as well as any linked table or a query. Snapshot recordsets contain a copy of the records in RAM (random access memory) and provide no direct access to the underlying data. They are used for reading data only—you can't use them to add, update, or delete records.
Forward-only-type	This is a special type of a Snapshot recordset that only allows you to scroll forward through the records. It provides the fastest access when you want to make a single pass through the data.
Dynamic-type	This recordset is generated by a query based on one or more tables. It allows you to add, change, or delete records from a row-returning query. In addition, it includes the records that other users may have added, modified, or deleted.

In order to find and read database records, you must understand how to navigate through the recordset. When you open a Recordset object, the first record is the current record. All recordsets have a current record.

- To move to subsequent records, use the `MoveNext` method.
- To move to the previous record, use the `MovePrevious` method.

- The MoveFirst and MoveLast methods move the cursor to the first and last records, respectively.
- If you call the MoveNext method when the cursor is already pointing to the last record, the cursor will move off the last record to the area known as end of file (EOF), and the EOF property will be set to True.
- If you call the MoveNext method when the EOF property is True, an error is generated because you cannot move past the end of the file. Similarly, by calling the MovePrevious method when the cursor is pointing to the first record, you will move the cursor to the area known as beginning of file (BOF). This will set the BOF property to True. When the BOF property is True and you call the MovePrevious method, an error will be generated.

When navigating through a recordset you may want to mark a specific record in order to return to it at a later time. You can use the Bookmark property to obtain a unique identification for a specific record.

The Recordset object has numerous properties and methods. We will discuss only those properties and methods that are required for performing a specific task, as demonstrated in the example procedures.

Opening Various Types of Recordsets

Use the OpenRecordset method to create or open a Recordset. For example, to open a Table-type recordset on a table named tblClients, use the following statement:

```
Set rst = CurrentDb.OpenRecordset("tblClients", dbOpenTable)
```

Notice that the second argument in the OpenRecordset method specifies the type of recordset. The RecordsetTypeEnum constants (shown in Table 14.2) can be used here.

TABLE 14.2. Constants used to specify the type of a DAO Recordset object

Type Constant	Value	Description
dbOpenTable	1	Opens a Table-type recordset
dbOpenDynaset	2	Opens a Dynaset-type recordset
dbOpenSnapshot	4	Opens a Snapshot-type recordset
dbOpenForwardOnly	8	Opens a Forward-only-type recordset
dbOpenDynamic	16	Opens a Dynamic-type recordset

In the preceding example, if you don't specify a recordset type, a Table-type recordset will be created based on tblClients. A Table-type recordset represents the records in a single table in a database.

The `OpenRecordset` method opens a new Recordset object for reading, adding, updating, or deleting records from a database. The `OpenRecordset` method can also be performed on a query. Note that a query can only be opened as a Dynaset or Snapshot Recordset object. For example, to open a Recordset based on a query, use the following statements:

```
Dim db As DAO.Database
Dim rst As DAO.Recordset
Set db = CurrentDb()
Set rst = db.OpenRecordset("qryMyQuery", dbOpenSnapshot)
```

The procedure in Hands-On 14.1 demonstrates how to open various types of DAO Recordsets on the Customers table in the Northwind 2007.accdb database and return the total number of records.

Please note files for the "Hands-On" project may be found on the companion CD-ROM.

Hands-On 14.1. Opening Table-, Dynaset-, and Snapshot-Type Recordsets (DAO)

1. Start Microsoft Access 2010 and create a new database named **Chap14.accdb** in your **C:\Access2013_ByExample** folder.
2. In the Access window, press **Alt+F11** to switch to the Visual Basic Editor window.
3. In the Visual Basic Editor window, choose **Insert | Module.**
4. In the module's Code window, type the following **ThreeRecordsetsDAO** procedure:

```
Sub ThreeRecordsetsDAO()
    Dim db As DAO.Database
    Dim tblRst As DAO.Recordset
    Dim dynaRst As DAO.Recordset
    Dim snapRst As DAO.Recordset
    Dim strDb As String

    strDb = "C:\Access2013_ByExample\Northwind 2007.accdb"
    Set db = OpenDatabase(strDb)
    Set tblRst = db.OpenRecordset("Customers", dbOpenTable)
    Debug.Print "Records in a table: " & tblRst.RecordCount

    Set dynaRst = db.OpenRecordset("Customers", dbOpenDynaset)
```

```
    Debug.Print "Records in a Dynaset: " & dynaRst.RecordCount
    dynaRst.MoveLast
    Debug.Print "Records in a Dynaset: " & dynaRst.RecordCount

    Set snapRst = db.OpenRecordset("Customers", dbOpenSnapshot)
    Debug.Print "Records in a Snapshot: " & snapRst.RecordCount
    snapRst.MoveLast
    Debug.Print "Records in a Snapshot: " & snapRst.RecordCount

    tblRst.Close
    dynaRst.Close
    snapRst.Close
    db.Close
    Set db = Nothing

    SendKeys "^g"
End Sub
```

5. Choose **Run | Run Sub/UserForm** to execute the procedure.

Notice that to get the correct count of records in Dynaset and Snapshot recordsets, you need to invoke the `MoveLast` method to access all the records. Counting records in covered in more detail in the next section.

The last statement in this procedure (`SendKeys "^g"`) activates the Immediate window so that you can see the results for yourself.

Opening a Snapshot and Counting Records

When you want to search tables or queries, you will get the fastest results by opening a Snapshot-type recordset. A snapshot is simply a non-updatable set of records that contain fields from one or more tables or queries. Snapshot-type Recordset objects can be used only for retrieving data. Use the `OpenRecordset` method to create or open a recordset. For example, to open a Snapshot-type recordset on a table named Customers, use the following statement:

```
Set rst = CurrentDb.OpenRecordset("Customers", dbOpenSnapshot)
```

At times, you may need to know where you are in a recordset. There are two properties that can be used to determine your position in the recordset:

- The AbsolutePosition property allows you to position the current record pointer at a specific record based on its ordinal position in a Dynaset- or Snapshot-type Recordset object. This property lets you determine the current record number. Zero (0) refers to the first record in the Recordset object. If there is no current record, the AbsolutePosition property returns –1. However, because the position of a record changes when preceding records are deleted, you should rely more on bookmarks to position the current record. The AbsolutePosition property can be used only with Dynasets and Snapshots. Because the AbsolutePosition property value is zero-based, 1 is added to the AbsolutePosition value to display current record information:

```
MsgBox "Current record: " & rst.AbsolutePosition + 1
```

- The PercentPosition property shows the current position relative to the number of records that have been accessed. Both AbsolutePosition and PercentPosition are not accurate until you move to the last record.

The procedure in Hands-On 14.2 attempts to get the total number of records in a Snapshot-type recordset by using the RecordCount property.

Hands-On 14.2. Opening a Snapshot-Type Recordset and Retrieving the Number of Records (DAO)

1. In the Visual Basic Editor window, choose **Insert | Module**.
2. In the module's Code window, type the **OpenSnapshot** procedure shown here:

```
Sub OpenSnapshot()
  Dim db As DAO.Database
  Dim rst As DAO.Recordset

  Set db = OpenDatabase("C:\Access2013_ByExample\Northwind 2007.accdb")
  Set rst = db.OpenRecordset("Customers", dbOpenSnapshot)

  MsgBox "Current record: " & rst.AbsolutePosition + 1
  MsgBox "Number of records: " & rst.RecordCount
  rst.MoveLast
  MsgBox "Current record: " & rst.AbsolutePosition + 1
  MsgBox "Number of records: " & rst.RecordCount
  rst.Close
  Set rst = Nothing
  db.Close
  Set db = Nothing
End Sub
```

3. Choose **Run | Run Sub/UserForm** to execute the procedure.

The RecordCount property of the Recordset object returns the number of records that have been accessed.

Zero (0) is returned if there are no records in the recordset, and 1 is returned if there are records in the recordset. If you open a Table-type recordset and check the RecordCount property, it will return the total number of records in a table. However, if you open a Dynaset- or Snapshot-type recordset, the RecordCount property will return 1, indicating that the recordset contains records. To find out the total number of records in a Dynaset or Snapshot, call the `MoveLast` method prior to retrieving the RecordCount property value. The record count becomes accurate after you've visited all the records in the recordset.

Retrieving the Contents of a Specific Field in a Table

To retrieve the contents of any field, start by creating a recordset based on the desired table or query, then loop through the recordset, printing the field's contents for each record to the Immediate window.

The procedure in Hands-On 14.3 generates a listing of all clients in the Customers table. Customer names are retrieved starting from the last record (see the `MoveLast` method). The BOF property of the Recordset object determines when the beginning of your recordset is reached.

Hands-On 14.3. Retrieving Field Values (DAO)

1. In the Visual Basic Editor window, choose **Insert | Module**.
2. In the module's Code window, type the **ReadFromEnd** procedure shown here:

```
Sub ReadFromEnd()
  Dim db As DAO.Database
  Dim rst As DAO.Recordset
  Dim strDb As String

  strDb = "C:\Access2013_ByExample\Northwind 2007.accdb"
  Set db = OpenDatabase(strDb)
  Set rst = db.OpenRecordset("Customers", dbOpenTable)
  rst.MoveLast
  Do Until rst.BOF
    Debug.Print rst!Company
    rst.MovePrevious
  Loop
  SendKeys "^g"
  rst.Close
  Set rst = Nothing
```

```
    db.Close
    Set db = Nothing
End Sub
```

3. Choose **Run | Run Sub/UserForm** to execute the procedure.

Moving between Records in a Table

All recordsets have a current position and a current record. The current record is usually the record at the current position. However, the current position can be before the first record and after the last record. You can use one of the `Move` methods in Table 14.3 to change the current position.

TABLE 14.3. Move methods used with DAO Recordsets

Method Name	Description
`MoveFirst`	Moves to the first record
`MoveLast`	Moves to the last record
`MoveNext`	Moves to the next record
`MovePrevious`	Moves to the previous record
`Move` *n*	Moves forward or backward *n* positions

The procedure in Hands-On 14.4 demonstrates how to move between records in the Employees table using the Table-type or Dynaset-type recordset.

Hands-On 14.4. Moving between Records in a Table (DAO)

1. In the Visual Basic Editor window, choose **Insert | Module**.
2. In the module's Code window, type the following **NavigateRecords** procedure:

```
Sub NavigateRecords()
    Dim db As DAO.Database
    Dim tblRst As DAO.Recordset
    Dim dynaRst As DAO.Recordset
    Dim strDb As String

    strDb = "C:\Access2013_ByExample\Northwind 2007.accdb"

    Set db = OpenDatabase(strDb)
    Set tblRst = db.OpenRecordset("Employees")
    tblRst.MoveFirst

    Do While Not tblRst.EOF
```

```
    Debug.Print "Employee: " & tblRst![Last Name]
    tblRst.MoveNext
  Loop

  Set dynaRst = db.OpenRecordset("Employees", dbOpenDynaset)
  dynaRst.MoveFirst

  Do While Not dynaRst.EOF
    Debug.Print "Hello " & dynaRst![Last Name]
    dynaRst.MoveNext
  Loop

  tblRst.Close
  dynaRst.Close

  Set tblRst = Nothing
  Set dynaRst = Nothing
  db.Close
  Set db = Nothing
  SendKeys "^g"
End Sub
```

3. Choose **Run | Run Sub/UserForm** to execute the procedure.

Finding Records in a Table-Type Recordset

While the `Move` methods are convenient for looping through records in a Recordset object, you should use `Seek` or `Find` methods to look for specific records. When you know exactly which record you want to find in a Table-type recordset and the field you are searching is indexed, the quickest way to find that record is to use the `Seek` method. One thing to remember with the `Seek` method is that the table must contain an index. The Index property must be set before the `Seek` method can be used. If you try to use the `Seek` method on a Table-type recordset without first setting the current index, a runtime error will occur. The `Seek` method searches through the recordset and locates the first matching record. Once the record is found, it is made the current record and the NoMatch property is set to `False`. If the record is not found, the NoMatch property is set to `True` and the current record is undefined. Table 14.4 lists comparison operators that you can use with the `Seek` method.

TABLE 14.4. Comparison operators used with the Seek method

Operator	Description
"="	Finds the first record whose indexed field is equal to the specified value
">="	Finds the first record whose indexed field is greater than or equal to the specified value
">"	Finds the first record whose indexed field is greater than the specified value
"<="	Finds the first record whose indexed field is less than or equal to the specified value
"<"	Finds the first record whose indexed field is less than the specified value

The comparison operator used with the `Seek` method must be enclosed in quotes. If there are several records that match your criteria, the `Seek` method returns the first record it finds. The `Seek` method cannot be used to search for records in a linked table. You must use the `Find` methods (see the next section) for locating specific records in linked tables, as well as Dynaset- and Snapshot-type recordsets. The procedure in Hands-On 14.5 searches for an employee whose last name begins with the letter "K."

Hands-On 14.5. Finding Records in a Table-Type Recordset (DAO)

1. In the Visual Basic Editor window, choose **Insert | Module**.
2. In the module's Code window, type the following **FindRecordsInTable** procedure:

```
Sub FindRecordsInTable()
  Dim db As DAO.Database
  Dim tblRst As DAO.Recordset
  Dim strDb As String

  strDb = "C:\Access2013_ByExample\Northwind 2007.accdb"

  Set db = OpenDatabase(strDb)

  Set tblRst = db.OpenRecordset("Employees", dbOpenTable)
  ' find the first employee in the table whose name
  ' begins with the letter "K"

  tblRst.Index = "Last Name"
```

```
    tblRst.Seek ">=", "K"

    If Not tblRst.NoMatch Then
      MsgBox "Found the following employee: " & tblRst![Last Name]
    Else
      MsgBox "There is no employee with such a name."
    End If

    tblRst.Close
    Set tblRst = Nothing
    db.Close
    Set db = Nothing
End Sub
```

3. Choose **Run | Run Sub/UserForm** to execute the procedure.

Finding Records in Dynasets or Snapshots

Use the `Find` methods to search for a record in Dynaset-type and Snapshot-type recordsets. Table 14.5 lists the available `Find` methods.

TABLE 14.5. Find methods in a DAO Recordset

Method Name	Description
`FindFirst`	Finds the first matching record in the recordset
`FindNext`	Finds the next matching record, starting at the current record
`FindPrevious`	Finds the previous matching record, starting at the current record
`FindLast`	Finds the last matching record in the recordset

If a record is not found for the given criteria, the NoMatch property of the Recordset object is set to `True`. Before searching for records, set a bookmark at the current record. If the search fails, you will be able to use the bookmark to return to the current record; otherwise, you will get the error "No current record." Each record in a Recordset object has a unique bookmark that you can use to locate that record. To get the current record's bookmark, move the cursor to that record and assign the value of the Bookmark property of the Recordset object to a Variant variable:

```
Dim mySpot As Variant
mySpot = dynaRst.Bookmark
```

In Hands-On 14.6, the bookmark is set on the first record of a Dynaset-type recordset. The procedure then searches for employees whose name ends with the string "er." The asterisk (*) in the search string is a wildcard character representing any number of letters (*er).

To return to the bookmarked record, set the Bookmark property to the value held by the Variant variable:

```
dynaRst.Bookmark = mySpot
```

While recordsets based on local Microsoft Access tables support bookmarks, non-Access databases may not support them. To determine whether a Recordset object supports bookmarks, you can check the Bookmarkable property. Bookmarks are supported if this property is `True`.

```
If dynaRst.Bookmarkable Then
 mySpot = dynaRst.Bookmark
End If
```

If the Recordset object does not support bookmarks, an error occurs. You can set as many bookmarks as you wish. Bookmarks can be created for a record other than the current record by moving to the desired record and assigning the value of the Bookmark property to a String variable that identifies that record.

Hands-On 14.6. Finding a Record in a Dynaset-Type Recordset (DAO)

1. In the Visual Basic Editor window, choose **Insert | Module**.
2. In the module's Code window, type the following **FindRecInDynaset** procedure:

```
Sub FindRecInDynaset()
  Dim db As DAO.Database
  Dim dynaRst As DAO.Recordset
  Dim mySpot As Variant
  Dim strDb As String

  strDb = "C:\Access2013_ByExample\Northwind 2007.accdb"

  Set db = OpenDatabase(strDb)
  Set dynaRst = db.OpenRecordset("Employees", dbOpenDynaset)

  MsgBox "Current employee: " & dynaRst![Last Name]
  mySpot = dynaRst.Bookmark
```

```
    ' find clients whose name ends with the string "er"
    dynaRst.FindFirst "[Last Name] Like '*er'"

    Do While Not dynaRst.NoMatch
      Debug.Print dynaRst![Last Name]
      dynaRst.FindNext "[Last Name] Like '*er'"
    Loop

    dynaRst.Bookmark = mySpot
    MsgBox "Back to record: " & dynaRst![Last Name]
    dynaRst.Close

    Set dynaRst = Nothing
    db.Close
    Set db = Nothing
    SendKeys "^g"
  End Sub
```

3. Choose **Run | Run Sub/UserForm** to execute the procedure.
   The names of all employees that match the search criteria are printed to the Immediate window.

Finding the nth Record in a Snapshot

The procedure in Hands-On 14.7 demonstrates how to locate the *n*th record in a Snapshot-type recordset.

Hands-On 14.7. Finding the *n*th Record in a Snapshot-Type Recordset (DAO)

1. In the Visual Basic Editor window, choose **Insert | Module**.
2. In the module's Code window, type the **FindNthRecord** procedure shown here:

```
Sub FindNthRecord()
  Dim db As DAO.Database
  Dim rst As DAO.Recordset
  Dim fld As DAO.Field
  Dim totalRec As Integer
  Dim nth As String
  Dim strDb As String
```

```
    strDb = "C:\Access2013_ByExample\Northwind 2007.accdb"

    Set db = OpenDatabase(strDb)
    Set rst = db.OpenRecordset("Employees", dbOpenSnapshot)
    rst.MoveLast

    totalRec = rst.RecordCount
    rst.MoveFirst
    nth = InputBox("Enter the number of positions to move forward:")

    On Error Resume Next
    If totalRec > nth Then
      rst.Move nth
      For Each fld In rst.Fields
        Debug.Print fld.Name & ": " & fld.Value
      Next fld
    Else
      MsgBox "You must enter a value that is less than " _
       & totalRec & "."
    End If

    rst.Close
    Set rst = Nothing
    db.Close
    Set db = Nothing
End Sub
```

3. Choose **Run | Run Sub/UserForm** to execute the procedure.

Notice that immediately after opening the recordset, the `MoveLast` method is used to ensure that all records have been visited. The total number of records is then retrieved with the RecordCount property of the Recordset object and stored in the `totalRec` variable. Next, the `MoveFirst` method is used to return to the first record and the `InputBox` method is used to prompt the user for the number of positions to move forward in the recordset. If the user-supplied value is less than the total number of records, the cursor moves to the specified record and the `For...Each` loop is used to print this record's field names and values to the Immediate window. An attempt to move beyond the end of the recordset will cause an error. Therefore, the procedure displays a message if the user-supplied position to move to is greater than the total number of records.

INTRODUCTION TO ADO RECORDSETS

The Recordset object is one of the three most-used ADO objects (the other two are Connection and Command). What you can do with a recordset depends entirely on the built-in capabilities of its OLE DB provider. You can open an ADO Recordset by using the Recordset object's `Open` method. The information needed to open a recordset can be provided by first setting properties and then calling the `Open` method, or by using the `Open` method's parameters like this:

```
rst.Open [Source], [ActiveConnection], [CursorType], [LockType],
[CursorLocation], [Options]
```

Notice that all the parameters are optional (they appear in square brackets). If you decide that you don't want to pass parameters, then use a different syntax to open a recordset. For example, examine the following code block:

```
With rst
  .Source = strSQL
  .ActiveConnection = strConnect
  .CursorType = adOpenStatic
  .LockType = adLockOptimistic
  .CursorLocation = adUseClient
  .Open Options := adCmdText
End with
```

The preceding code segment opens a recordset by first setting properties of the Recordset object, then calling its `Open` method. Notice that the names of the required Recordset properties are equivalent to the parameter names listed earlier. The values assigned to each property are discussed later. You will become familiar with both methods of opening a recordset as you work with the example procedures that follow.

Let's return to the syntax of the Recordset's `Open` method, which specifies the parameters. Needless to say, you need to know what each parameter is and how it is used. The `Source` parameter determines where you want your records to come from. The data source can be an SQL string, a table, a query, a stored procedure or view, a saved file, or a reference to a Command object. Later in this chapter you will learn how to open a recordset based on a table, a query, and an SQL statement.

The `ActiveConnection` parameter can be an SQL string that specifies the connection string or a reference to a Connection object. This parameter tells where to find the database as well as what security credentials to use.

Before we discuss the next three parameters, you need to know that the ADO Recordsets are controlled by a cursor. The *cursor* determines whether the recordset is scrollable (backward and forward or forward only), whether it is read-only or updatable, and whether changes made to the data are visible to other users.

The ADO cursors have three functions specified by the following parameters:

- `CursorType`
- `LockType`
- `CursorLocation`

Before you choose the cursor, you need to think of how your application will use the data. Some cursors yield better performance than others. It's important to determine where the cursor will reside and whether changes made while the cursor is open need to be visible immediately. The following subsection should assist you in choosing the correct cursor.

Cursor Types

The `CursorType` parameter specifies how the recordset interacts with the data source and what is allowed or not allowed when it comes to data changes or movement within the recordset. This parameter can take one of four constants: `adOpenForwardOnly` (0), `adOpenKeyset` (1), `adOpenDynamic` (2), and `adOpenStatic` (3).

You can find out what types of cursors are available by using the Object Browser. Before proceeding, make sure that the Chap14.accdb database file you created at the beginning of this chapter has a reference to the ActiveX Data Objects library. Set this reference by switching to the Visual Basic Editor window and choosing Tools | References. Find and select Microsoft ActiveX Data Objects 6.1 Library in the References dialog box and click OK. Next, activate the Object Browser window by pressing F2 or choosing View | Object Browser. Select ADODB from the Project/Library drop-down listbox and type CursorType in the Search text box, as shown in Figure 14.1.

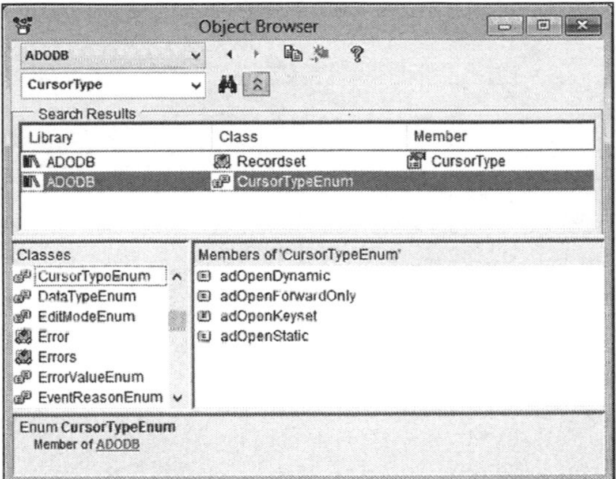

FIGURE 14.1. The Object Browser lists four predefined constants you can use to specify the cursor type to be retrieved.

- When the cursor type is dynamic (`adOpenDynamic`), users are allowed to view changes other users made to the database. The dynamic cursor is not supported by the Jet 4.0 engine in Microsoft Access. To use this cursor, you must use other OLE DB providers, such as MSDASQL or SQLOLEDB. Using the dynamic cursor you can move back and forth in the recordset.
- When the cursor type is forward-only (`adOpenForwardOnly`), additions, changes, or deletions made by other users are not visible. This is both the default and the fastest cursor because it only allows you to scroll forward in the recordset.
- When the cursor type is keyset driven (`adOpenKeyset`), you can scroll back and forth in the recordset; however, you cannot view records added or deleted by another user. Use the Recordset's `Requery` method to overcome this limitation.
- When the cursor type is static (`adOpenStatic`), all the data is retrieved as it was at a point in time. This cursor is desirable when you need to find data or generate a report. You can scroll back and forth within a recordset, but additions, changes, or deletions by other users are not visible. Use this cursor to retrieve an accurate record count.
- You must set the `CursorType` before opening the recordset with the `Open` method. Otherwise, Access will create a Forward-only recordset. You may use a constant name or its value in your VBA procedures.

Lock Types

After you choose a cursor type, it is important to specify how the ADO should lock the row when you make a change. The `LockType` specifies whether the recordset is updatable. The default setting for `LockType` is read-only. The `LockType` predefined constants are listed in the Object Browser, as shown in Figure 14.2.

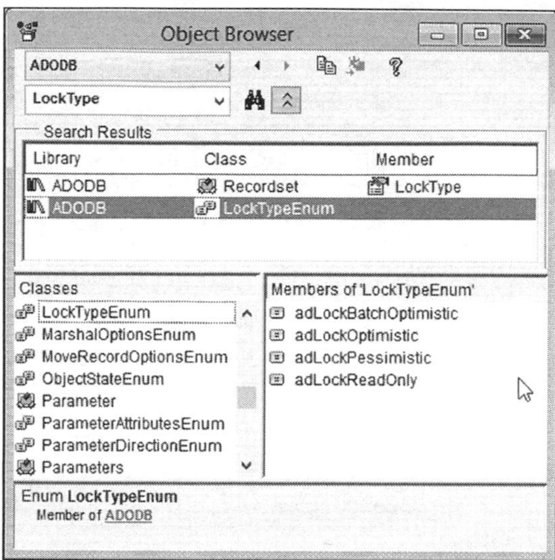

FIGURE 14.2. The Object Browser lists four predefined constants that you can use to specify what type of locking ADO should use when you make a change to the data.

When the `LockType` property is batch optimistic (`adLockBatchOptimistic`), batch updates made to the data are stored locally until the `UpdateBatch` method is called, during which all pending updates are committed all at once. Until the `UpdateBatch` method is called, no locks are placed on edited data. Batch optimistic locking eliminates network roundtrips that normally occur with optimistic locking (`adLockOptimistic`) when users make changes to one record and move to another. With batch optimistic locking, a user can make all the changes to all the records and then submit them as a single operation.

- When the LockType property is optimistic (`adLockOptimistic` = 3), no locks are placed on the data until you attempt to save a row. Records are locked only when you call the `Update` method, and the lock is released as soon as the Save operation is completed. Two users are allowed to update a record at the same time. Optimistic locking allows you to work with one row at a time. If you need to make multiple updates, it's better to save them all at once by using batch optimistic locking.

- When the LockType property is pessimistic (`adLockPessimistic` = 2), all the records are locked as soon as you begin editing a record. The record remains locked until the edit is committed or canceled. This type of lock guarantees that two users will not make changes to the same record. If you use pessimistic locking, ensure that your code does not require any input from the users. You certainly don't want a scenario where a user opens a record and makes a change, then leaves for lunch without saving the record. In that case, the record is locked until the user comes back and saves or discards the edit. In this situation, it is better to use optimistic locking.
- When the LockType property is read-only (`adLockReadOnly` = 1), you will not be able to alter any data. This is the default setting.

Cursor Location

The `CursorLocation` parameter determines whether ADO or the SQL Server database engine manages the cursor. Cursors use temporary resources to hold the data. These resources can be memory, a disk paging file, temporary disk files, or even temporary storage in the database.

- When a cursor is created and managed by ADO, the recordset is said to be using a *client-side cursor* (`adUseClient`). With the client-side cursor, all the data is retrieved from the server in one operation and is placed on the client computer. Because all the requested data is available locally, the connection to the database can be closed and reopened only when another set of data is needed. Since the entire result set has been downloaded to the client computer, browsing through the rows is very fast.
- When a cursor is managed by a database engine, the recordset is said to be using a *server-side cursor* (`adUseServer`). With the server side cursor, all the data is stored on the server and only the requested data is sent over the network to the user's computer. This type of cursor can provide better performance than the client-side cursor when excessive network traffic is an issue. However, it's important to point out that a server-side cursor consumes server resources for every active client and, because it provides only single-row access to the data, it can be quite slow.

It is recommended that you use the server-side cursor when working with local Access databases, and the client-side cursor when working with remote Access databases or SQL Server databases.

The `CursorLocation` predefined constants are listed in the Object Browser, as shown in Figure 14.3.

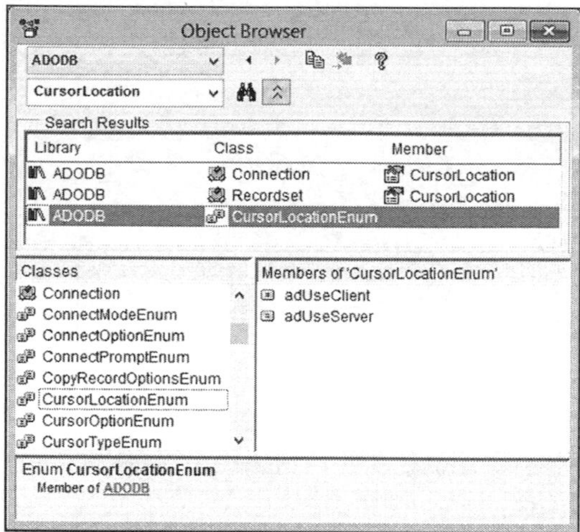

FIGURE 14.3. The CursorLocation parameter of the Recordset's Open method can be set by using the adUseClient or adUseServer constant.

The Options Parameter

The `Options` parameter specifies the data source type being used. Similar to the parameters related to cursors, the `Options` parameter can take one of many values, as shown in Figure 14.4.

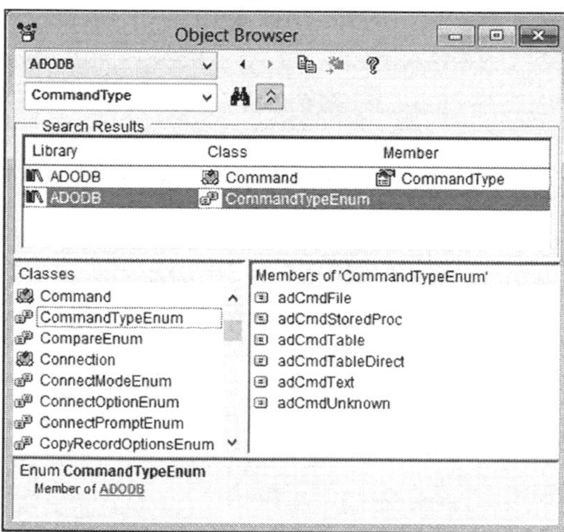

FIGURE 14.4. The Options parameter of the Recordset's Open method is supplied by the constant values listed under the CommandType property of the Command object.

- When the `Options` parameter is set to `adCmdFile` (256), it tells the ADO that the source of the recordset is a path or filename. ADO can open recordsets based on files in different formats.
- When the `Options` parameter is set to `adCmdStoredProc` (4), it tells the ADO that the source of the recordset is a stored procedure or parameterized query.
- When the `Options` parameter is set to `adCmdTable` (2), it tells the ADO that the source of the recordset is a table or view. The `adCmdTable` constant will cause the provider to generate an SQL query to return all rows from a table or view by prepending `SELECT * FROM` in front of the specified table or view name.
- When the `Options` parameter is set to `adCmdTableDirect` (512), it tells the ADO that the `Source` argument should be evaluated as a table name. How does this constant differ from `adCmdTable`? The `adCmdTableDirect` constant is used by OLE DB providers that support opening tables directly by name, using an interface called IOpenRowset instead of an ADO Command object. Since the IOpenRowset method does not need to build and execute a Command object, its use results in increased performance and functionality.
- When the `Options` parameter is set to `adCmdText` (1), it tells the ADO that you are using an SQL statement to open the recordset.
- When the `Options` parameter is set to `adCmdUnknown` (8), it tells the ADO that the command type in the `Source` argument is unknown. This is the default, which is used if you don't specify any other option. By using the `adCmdUnknown` constant, or not specifying any constant at all for the `Options` parameter, you force ADO to make an extra roundtrip to the server to determine the source type. As you would expect, this will decrease your VBA procedure's performance; therefore, you should use `adCmdUnknown` only if you don't know what type of information the `Source` parameter will contain.

Not all options are supported by all data providers. For example, Microsoft Jet OLE DB Provider does not support the `adCmdTableDirect` *cursors.*

In addition to specifying the type of CommandType in the `Options` parameter (see Figure 14.4), you can pass additional information in the `Options` parameter. For example, you can tell ADO how to execute the command by specifying whether ADO should wait while all the records are being retrieved or should continue asynchronously.

Asynchronous Record Fetching

Asynchronous fetching is an ADO feature that allows some records to be downloaded to the client while the remaining records are still being fetched from the database. As soon as the user sees some records, he can begin paging through them. The user does not know that only a few records have been returned. As he pages through the rows backward and forward, a new connection is made to the server and more records are fetched and passed to the client. Once all records have been returned, paging is very quick because all records are on the client. Asynchronous fetching makes it seem to the user that the data retrieval is pretty fast. The downside is that records cannot be sorted until they have all been downloaded.

Additional `Options` parameters are described in the following list. Note that only the first three constants (`adAsyncExecute`, `adAsyncFetch`, and `adAsyncFetchNonBlocking`) can be used with the Recordset's `Open` method. Other constants are used with the Command or Connection object's `Execute` method.

- `adAsyncExecute` (16)—This tells ADO to execute the command asynchronously, meaning that all requested rows are retrieved as soon as they are available. Using `adAsyncExecute` enables the application to perform other tasks while waiting for the cursor to populate.

Note that the `adAsyncExecute` constant cannot be used with `adCmdTableDirect`.

- `adAsyncFetch` (32)—Using this constant requires that you specify a value greater than 1 for the Recordset's CacheSize property. The CacheSize property is used to determine the number of records ADO will hold in local memory. For example, if the cache size is 100, the provider will retrieve the first 100 records after first opening the Recordset object. The `adAsyncFetch` constant tells ADO that the rows remaining after the initial quantity specified in the CacheSize property should be retrieved asynchronously.
- `adAsyncFetchNonBlocking` (64)—This option tells ADO that it should never wait for a row to be fetched. The application will continue execution while records are being continuously extracted from a very large data file.

If the requested recordset row has not been retrieved yet, the current row automatically moves to the end of the file (causing the Recordset's EOF property to become `True`). In other words, the data retrieval process will not block other processes.

Note that `adAsynchFetchNonBlocking` has no effect when the `adCmdTableDirect` option is used to open the recordset. Also, `adAsyncFetchNonBlocking` is not supported with a Server cursor (`adUseServer`) when you use the ODBC provider (MSDASQL).

- `adExecuteNoRecords` (128)—This option tells ADO not to expect any records when the command is executed. Use this option for commands that do not return records, such as INSERT, UPDATE, or DELETE. Use the `adExecuteNoRecords` constant with `adCmdText` to improve the performance of your application. When this option is specified, ADO does not create a Recordset object and does not set any cursor properties.

Note that `adExecuteNoRecords` can only be passed as an optional parameter to the Command or Connection object's `Execute` method and cannot be used when opening a recordset.

- `adExecuteStream` (256)—Indicates that the results of a Command execution should be returned as a stream. The `adExecuteStream` constant can only be passed as an optional parameter to the Command or Connection object's `Execute` method and it cannot be used when opening a recordset.
- `adExecuteRecord` (512)—Indicates that the value of the CommandText property is a command or stored procedure that returns a single row as a Record object (a Record object represents one row of data).
- `adOptionUnspecified` (–1)—Indicates that the command is unspecified. This is the default option.

Note that similar to `adExecuteNoRecords`, `adExecuteStream`, and `adExecuteRecord`, this constant can only be passed as an optional parameter to the Command or Connection object's `Execute` method and cannot be used when opening a recordset.

Opening a Recordset

ADO offers numerous ways of opening a Recordset object. To begin with, you can create ADO Recordsets from scratch without going through any other object. Suppose you want to retrieve all the records from the Employees table. The code you need to write is very simple. Let's try this out in Hands-On 14.8.

Hands-On 14.8. Opening a Recordset (ADO)

1. In the Visual Basic Editor window, choose **Insert | Module**.
2. In the module's Code window, type the following **OpenADORst** procedure:

```
' make sure to set up a reference to
' the Microsoft ActiveX Data Objects 6.1 Library

Sub OpenADORst()
  Dim conn As ADODB.Connection
  Dim rst As ADODB.Recordset

  Set conn = New ADODB.Connection
  With conn
    .Provider = "Microsoft.ACE.OLEDB.12.0"
    .Open "Data Source=" & CurrentProject.Path & _
     "\Northwind 2007.accdb"
  End With

  Set rst = New ADODB.Recordset
  With rst
    .Source = "SELECT * FROM Employees"
    .ActiveConnection = conn
    .Open
    Debug.Print rst.Fields.Count
    .Close
  End With

  Set rst = Nothing
  conn.Close
  Set conn = Nothing
End Sub
```

3. Choose **Run | Run Sub/UserForm** to execute the procedure.

   In the preceding code example, we first define and open a connection to the database. Next, we declare a Recordset object and create a new instance of it. The Recordset object's Source property specifies the data you want to retrieve. The source can be a table, query, stored procedure, view, saved file, or Command object. The SQL `SELECT` statement tells VBA to select all the data from the Employees table. Next, the ActiveConnection property specifies how to connect to the data. We set the ActiveConnection property to the object variable (`conn`) that holds the connection information. Finally, the `Open` method retrieves the specified records into the recordset. Before we close the recordset using the Recordset's `Close` method, we retrieve the number of fields in the open recordset by examining the Recordset's Fields collection and write the result to the Immediate window.

Opening a Recordset Based on a Table or Query

A recordset can be based on a table, view, SQL statement, or command that returns rows. It can be opened via a Connection or Command object's `Execute` method or a Recordset's `Open` method (see the following example procedures).

- Using the `Execute` method of the Connection object:

```
Sub ConnectAndExec()
  Dim conn As ADODB.Connection
  Dim rst As ADODB.Recordset
  Dim fld As Variant

  Set conn = New ADODB.Connection
  conn.Open "Provider=Microsoft.Jet.OLEDB.4.0;" & _
   "Data Source=" & CurrentProject.Path & _
   "\Northwind.mdb"
  Set rst = conn.Execute("SELECT * FROM Employees")
  Debug.Print rst.Source
  Do Until rst.EOF
    Debug.Print "\\\\\\\\\\\\\\\\\\\\\\\\\\\"
    For Each fld In rst.Fields
      Debug.Print fld.Name & "=" & fld.Value
    Next
    'Debug.Print "---new record ---"
    rst.MoveNext
  Loop
  'Debug.Print rst.Fields(1).Value
  rst.Close
  Set rst = Nothing
  conn.Close
  Set conn = Nothing
End Sub
```

Once you open the recordset, you can perform the required operation on its data. In this example, we use the Recordset's Source property to write to the Immediate window the SQL command on which the recordset is based. Next, we loop through the recordset to retrieve the contents of each field in every record. To open the Northwind 2007.accdb database instead of the Northwind.mdb file, change the provider string to `Microsoft.ACE.OLEDB.12.0` *and the name of the database to* `Northwind 2007.accdb`*:*

```
conn.Open "Provider=Microsoft.ACE.OLEDB.12.0;" & _
 "Data Source=" & CurrentProject.Path & _
 "\Northwind 2007.accdb"
```

- Using the `Execute` method of the Command object:

```
Sub CommandAndExec()
  Dim conn As ADODB.Connection
  Dim cmd As ADODB.Command
  Dim rst As ADODB.Recordset

  Set conn = New ADODB.Connection
  With conn
   .ConnectionString = "Provider=Microsoft.Jet.OLEDB.4.0;" & _
     "Data Source=" & CurrentProject.Path & "\Northwind.mdb"
    .Open
  End With

  Set cmd = New ADODB.Command
  With cmd
    .ActiveConnection = conn
    .CommandText = "SELECT * FROM Customers"
  End With

  Set rst = cmd.Execute

  MsgBox rst.Fields(1).Value

  rst.Close
  Set rst = Nothing
  conn.Close
  Set conn = Nothing
End Sub
```

Once you open the recordset, you can perform the required operation on its data. In this example, we display a message with the name of the first customer.

- Using the `Open` method of the Recordset object:

```
Sub RecSetOpen()
  Dim rst As ADODB.Recordset
  Dim strConnection As String

  strConnection = "Provider=Microsoft.Jet.OLEDB.4.0;" & _
   "Data Source=" & CurrentProject.Path & _
   "\Northwind.mdb"

  Set rst = New ADODB.Recordset
  With rst
    .Open "SELECT * FROM Customers", _
     strConnection, adOpenForwardOnly
    .Save CurrentProject.Path & "\MyRst.dat"
    .Close
  End With
  Set rst = Nothing
End Sub
```

Once you open the recordset, you can perform the required operation on its data. In this example, we save the entire recordset to a disk file named MyRst.dat. In Chapter 17, "Using Advanced ADO/DAO Features," you learn how to work with records that have been saved in a file.

The procedure in Hands-On 14.9 illustrates how to open a recordset based on a table or query.

Hands-On 14.9. Opening a Recordset Based on a Table or Query (ADO)

1. In the Visual Basic Editor window, choose **Insert | Module**.
2. In the module's Code window, type the following **OpenRst_TableOrQuery** procedure:

```
Sub OpenRst_TableOrQuery()
  Dim conn As ADODB.Connection
  Dim rst As ADODB.Recordset
```

```
    Set conn = New ADODB.Connection
    With conn
      .Provider = "Microsoft.ACE.OLEDB.12.0"
      .Open "Data Source=" & CurrentProject.Path & _
       "\Northwind 2007.accdb"
    End With

    Set rst = New ADODB.Recordset
    rst.Open "Employees", conn

    Debug.Print "CursorType: " & rst.CursorType & vbCr _
     & "LockType: " & rst.LockType & vbCr _
     & "Cursor Location: " & rst.CursorLocation

    Do Until rst.EOF
      Debug.Print rst.Fields(2)
      rst.MoveNext
    Loop

    rst.Close
    Set rst = Nothing
    conn.Close
    Set conn = Nothing
End Sub
```

3. Choose **Run | Run Sub/UserForm** to execute the procedure.

After opening the recordset, it's a good idea to check what type of recordset was created. Notice that this procedure uses the CursorType, LockType, and Cursor-Location properties to retrieve this information. After the procedure is run, the Immediate window displays the following:

```
CursorType: 0
LockType: 1
Cursor Location: 2
```

Notice that because you did not specify any parameters in the Recordset's `Open` method, you obtained a default recordset. This recordset is forward-only (0), read-only (1), and server-side (2). (For more information, refer to the section titled "Introduction to ADO Recordsets" earlier in this chapter.)

To create a different type of recordset, pass the appropriate parameters to the Recordset's `Open` method. For example, if you open your recordset like this:

```
rst.Open "Employees", conn, adUseClient, adLockReadOnly
```

you will get the static (3), read-only (1), and client-side (3) recordset. In this recordset, you can easily find out the number of records by using the Recordset's RecordCount property:

```
Debug.Print rst.RecordCount
```

Next, this procedure uses the `MoveNext` method to iterate through all the records in the recordset until the end of file (EOF) is reached.

Counting Records

Use the Recordset object's RecordCount property to determine the number of records in a recordset. If the number of records cannot be determined, this property will return –1. The RecordCount property setting depends on the cursor type and the capabilities of the provider. To get the actual count of records, open the recordset with the static (`adOpenStatic`) or dynamic (`adOpenDynamic`) cursor.

To quickly test the contents of the recordset, we write the employees' last names to the Immediate window. Since this recordset contains all the fields in the Employees table, you can add extra code to list the remaining field values.

Is This Recordset Empty?

A recordset may be empty. To check whether your recordset has any records in it, use the Recordset object's BOF and EOF properties. The BOF property stands for "beginning of file," and EOF indicates "end of file."

- If you open a Recordset object that contains no records, the BOF and EOF properties are both set to `True`.
- If you open a Recordset object that contains at least one record, the BOF and EOF properties are `False` and the first record is the current record.

You can use the following conditional statement to test whether there are any records:

```
If rst.BOF and rst.EOF Then
  MsgBox "This recordset contains no records"
End If
```

To open a recordset based on a saved query, replace the table name with your query name.

Opening a Recordset Based on an SQL Statement

The procedure in Hands-On 14.10 demonstrates how to use the Connection object's `Execute` method to open a recordset based on an SQL statement that selects all the employees from the Employees table in the sample Northwind 2007.accdb database. Only the name of the first employee is written to the Immediate window. As in the preceding example, the resulting recordset is forward-only and read-only.

Hands-On 14.10. Opening a Recordset Based on an SQL Statement (ADO)

1. In the Visual Basic Editor window, choose **Insert | Module.**
2. In the module's Code window, type the **CreateRst_WithSQL** procedure shown here:

```
Sub CreateRst_WithSQL()
  Dim conn As ADODB.Connection
  Dim rst As ADODB.Recordset
  Dim strConn As String

  strConn = "Provider = Microsoft.ACE.OLEDB.12.0;" & _
   "Data Source=" & CurrentProject.Path & _
   "\Northwind 2007.accdb"

  Set conn = New ADODB.Connection
  conn.Open strConn

  Set rst = conn.Execute("SELECT * FROM Employees")
  Debug.Print rst("Last Name") & ", " & rst("First Name")

  rst.Close
  Set rst = Nothing
  conn.Close
  Set conn = Nothing
End Sub
```

3. Choose **Run | Run Sub/UserForm** to execute the procedure.

Opening a Recordset Based on Criteria

Instead of retrieving all the records from a specific table or query, you can use the SQL `WHERE` clause to get only those records that meet certain criteria. The procedure in Hands-On 14.11 calls the Recordset's `Open` method to create a forward-only and read-only recordset populated with employees who are sales representatives.

Hands-On 14.11. Opening a Recordset Based on Criteria (ADO)

1. In the Visual Basic Editor window, choose **Insert | Module**.
2. In the module's Code window, type the following **OpenRst_WithCriteria** procedure:

```
Sub OpenRst_WithCriteria()
  Dim conn As ADODB.Connection
  Dim rst As ADODB.Recordset
  Dim strConn As String

  strConn = "Provider=Microsoft.ACE.OLEDB.12.0;" & _
   "Data Source=" & CurrentProject.Path & _
   "\Northwind 2007.accdb"

  Set conn = New ADODB.Connection
  conn.Open strConn

  Set rst = New ADODB.Recordset
  rst.Open "SELECT * FROM Employees WHERE [Job Title] = " & _
   "'Sales Representative'", _
   conn, adOpenForwardOnly, adLockReadOnly

  Do While Not rst.EOF
    Debug.Print rst.Fields(2).Value
    rst.MoveNext
  Loop

  rst.Close
  Set rst = Nothing
  conn.Close
  Set conn = Nothing
End Sub
```

3. Choose **Run | Run Sub/UserForm** to execute the procedure.

Opening a Recordset Directly

If you are planning to open just one recordset from a specific data source, you can take a shortcut and open it directly without first opening a Connection object. This method requires you to specify the source and connection information prior to calling the Recordset object's `Open` method, as shown in Hands-On 14.12.

Hands-On 14.12. Opening a Recordset Directly (ADO)

1. In the Visual Basic Editor window, choose **Insert | Module**.
2. In the module's Code window, type the **OpenRst_Directly** procedure shown here:

```
Sub OpenRst_Directly()
  Dim conn As ADODB.Connection
  Dim rst As ADODB.Recordset

  Set conn = New ADODB.Connection
  With conn
    .Provider = "Microsoft.ACE.OLEDB.12.0"
    .Open "Data Source=" & CurrentProject.Path & _
     "\Northwind 2007.accdb"
  End With

  Set rst = New ADODB.Recordset
  With rst
    .Source = "SELECT * FROM Employees"
    .ActiveConnection = conn
    .Open
  End With
  MsgBox rst.Fields(2)

  rst.Close
  Set rst = Nothing
  conn.Close
  Set conn = Nothing
End Sub
```

3. Choose **Run | Run Sub/UserForm** to execute the procedure.

Moving Around in a Recordset

You can navigate the ADO Recordset by using the following five methods: `MoveFirst`, `MoveLast`, `MoveNext`, `MovePrevious`, and `Move`. The procedure in Hands-On 14.13 demonstrates how to move around in a recordset and retrieve the names of fields and their contents for each record.

Hands-On 14.13. Moving Around in a Recordset (ADO)

1. In the Visual Basic Editor window, choose **Insert | Module**.
2. In the module's Code window, type the following **MoveAround** procedure:

```
Sub MoveAround()
    Dim conn As ADODB.Connection
    Dim rst As ADODB.Recordset
    Dim fld As ADODB.Field
    Dim strConn As String

    strConn = "Provider=Microsoft.Jet.OLEDB.4.0;" & _
      "Data Source=" & CurrentProject.Path & _
      "\Northwind.mdb"

    Set conn = New ADODB.Connection
    conn.Open strConn

    Set rst = New ADODB.Recordset
    rst.Open "SELECT * FROM Customers WHERE " & _
        "ContactTitle = 'Owner'", _
        conn, adOpenForwardOnly, adLockReadOnly
    Do While Not rst.EOF
        Debug.Print "New Record --------------"
        For Each fld In rst.Fields
            Debug.Print fld.Name & " = " & fld.Value
        Next
    rst.MoveNext
    Loop

    rst.Close
    Set rst = Nothing
    conn.Close
    Set conn = Nothing
End Sub
```

3. Choose **Run | Run Sub/UserForm** to execute the procedure.

Finding the Record Position

Use the AbosolutePosition property of the Recordset object to determine the current record number. This property specifies the relative position of a record in an ADO Recordset. The procedure in Hands-On 14.14 opens a recordset filled with employee records from the Employees table in the Northwind database and uses the AbsolutePosition property to return the record number three times during the procedure execution.

Hands-On 14.14. Finding the Record Position (ADO)

1. In the Visual Basic Editor window, choose **Insert | Module**.
2. In the module's Code window, type the following **FindRecordPosition** procedure:

```
Sub FindRecordPosition()
  Dim conn As ADODB.Connection
  Dim rst As ADODB.Recordset
  Dim strConn As String

  strConn = "Provider=Microsoft.Jet.OLEDB.4.0;" & _
   "Data Source=" & CurrentProject.Path & _
   "\Northwind.mdb"

  Set conn = New ADODB.Connection
  conn.Open strConn

  Set rst = New ADODB.Recordset
  With rst
    .Open "SELECT * FROM Employees", conn, adOpenKeyset, _
     adLockOptimistic, adCmdText
    Debug.Print .AbsolutePosition
    .Move 3 ' move forward 3 records
    Debug.Print .AbsolutePosition
    .MoveLast ' move to the last record
    Debug.Print .AbsolutePosition
    Debug.Print .RecordCount
    .Close
  End With
```

```
    Set rst = Nothing
    conn.Close
    Set conn = Nothing
End Sub
```

3. Choose **Run | Run Sub/UserForm** to execute the procedure.

   Notice that at the beginning of the recordset, the record number is 1. Next, the FindRecordPosition procedure uses the `Move` method to move the cursor three rows ahead, after which the AbsolutePosition property returns 4 (1 + 3) as the current record position. Finally, the `MoveLast` method is used to move the cursor to the end of the recordset. The AbsolutePosition property now determines that this is the ninth record (9). The RecordCount property of the Recordset object returns the total number of records (9).

Reading Data from a Field

Use the Fields collection of a Recordset object to retrieve the value of a specific field in an open recordset. The procedure in Hands-On 14.15 uses the `Do...While` loop to iterate through the recordset and prints the names of all the employees to the Immediate window.

Hands-On 14.15. Retrieving Field Values (ADO)

1. In the Visual Basic Editor window, choose **Insert | Module**.
2. In the module's Code window, type the following **ReadField** procedure:

```
Sub ReadField()
  Dim conn As ADODB.Connection
  Dim rst As ADODB.Recordset

  Set conn = New ADODB.Connection
  With conn
    .Provider = "Microsoft.ACE.OLEDB.12.0"
    .Open "Data Source=" & CurrentProject.Path & _
     "\Northwind 2007.accdb"
  End With

  Set rst = New ADODB.Recordset
  rst.Open "SELECT * FROM Employees", conn, adOpenStatic

  Do While Not rst.EOF
    Debug.Print rst.Fields("Last Name").Value
    rst.MoveNext
```

```
  Loop

  rst.Close
  Set rst = Nothing
  conn.Close
  Set conn = Nothing
End Sub
```

3. Choose **Run | Run Sub/UserForm** to execute the procedure.

Returning a Recordset as a String

Instead of using a loop to read the values of fields in all rows of the open recordset, you can use the Recordset object's `GetString` method to get the desired data in one step. The `GetString` method returns a recordset as a string-valued Variant. This method has the following syntax:

```
Variant = Recordset.GetString(StringFormat, NumRows, _
     ColumnDelimiter, RowDelimiter, NullExpr)
```

- The first argument (`StringFormat`) determines the format for representing the recordset as a string. Use the `adAddClipString` constant as the value for this argument.
- The second argument (`NumRows`) specifies the number of recordset rows to return. If blank, `GetString` will return all the rows.
- The third argument (`ColumnDelimiter`) specifies the delimiter for the columns within the row (the default column delimiter is tab (`vbTab`)).
- The fourth argument (`RowDelimiter`) specifies a row delimiter (the default is carriage return (`vbCrLf`)).
- The fifth argument (`NullExpr`) specifies an expression to represent Null values (the default is an empty string (`""`)).

Hands-On 14.16. Converting the Recordset to a String (ADO)

1. In the Visual Basic Editor window, choose **Insert | Module**.
2. In the module's Code window, type the **GetRecords_AsString** procedure shown here:

```
Sub GetRecords_AsString()
  Dim conn As ADODB.Connection
  Dim rst As ADODB.Recordset
  Dim varRst As Variant
```

```
    Dim fso As Object
    Dim myFile As Object

    Set conn = New ADODB.Connection
    With conn
      .Provider = "Microsoft.JET.OLEDB.4.0"
      .Open "Data Source=" & CurrentProject.Path & _
       "\Northwind.mdb"
    End With

    Set rst = New ADODB.Recordset
    rst.Open "SELECT EmployeeId, " & _
     "LastName & "", "" & FirstName AS FullName " & _
     "FROM Employees", _
     conn, adOpenForwardOnly, adLockReadOnly, adCmdText

    If Not rst.EOF Then
      ' Return all rows as a formatted string with
      ' columns delimited by Tabs, and rows
      ' delimited by carriage returns

      varRst = rst.GetString(adClipString, , vbTab, vbCrLf)
      Debug.Print varRst
    End If

    ' save the recordset string to a text file
    Set fso = CreateObject("Scripting.FileSystemObject")
    Set myFile = fso.CreateTextFile(CurrentProject.Path & _
     "\RstString.txt", True)
    myFile.WriteLine varRst
    myFile.Close

    Set fso = Nothing
    rst.Close
    Set rst = Nothing
    conn.Close
    Set conn = Nothing
  End Sub
```

3. Choose **Run | Run Sub/UserForm** to execute the procedure.

The GetRecords_AsString procedure demonstrates how you can transform a recordset into a tab-delimited list of values using the Recordset object's `GetString` method. You can use any characters you want to separate columns and rows. This procedure uses the following statement to convert a recordset to a string:

```
varRst = rst.GetString(adClipString, , vbTab, vbCrLf)
```

Notice that the second argument is omitted. This indicates that we want to obtain all the records. To convert only three records to a string, you could write the following line of code:

```
varRst = rst.GetString(adClipString, 3, vbTab, vbCrLf)
```

The `vbTab` and `vbCrLf` arguments are VBA constants that denote the Tab and carriage return characters.

Because `adClipString`, `vbTab`, and `vbCrLf` are default values for the `GetString` method's arguments, you can skip them altogether. Therefore, to put all of the records in this recordset into a string, you can simply use the `GetString` method without arguments, like this:

```
varRst = rst.GetString
```

Sometimes you may want to save your recordset string to a file. To gain access to a computer's filesystem, the procedure uses the `CreateObject` function to access the FileSystemObject from the Microsoft Scripting Runtime Library. You can easily create a File object by using the `CreateTextFile` method of this object. Notice that the second argument of the `CreateTextFile` method (`True`) indicates that the file should be overwritten if it already exists. Once you have defined your file, you can use the `WriteLine` method of the File object to write the text to the file. In this example, your text is the variable holding the contents of a recordset converted to a string.

Finding Records Using the Find Method

The ADO Object Model provides you with two methods for locating records: `Find` and `Seek`. This section demonstrates how to use the ADO `Find` method to locate all the employee records based on a condition. ADO has a single `Find` method. The search always begins from the current record or an offset from it. The search direction and the offset from the current record are passed as parameters to the `Find` method. The `SearchDirection` parameter can be either `adSearchForward` or `adSearchBackward`.

Hands-On 14.17. Finding Records Using the Find Method (ADO)

1. In the Visual Basic Editor window, choose **Insert | Module**.

2. In the module's Code window, type the following **Find_WithFind** procedure:

```
Sub Find_WithFind()
  Dim conn As ADODB.Connection
  Dim rst As ADODB.Recordset

  Set conn = New ADODB.Connection
  conn.Open "Provider=Microsoft.Jet.OLEDB.4.0;" & _
   "Data Source=" & CurrentProject.Path & _
   "\Northwind.mdb"

  Set rst = New ADODB.Recordset
  rst.Open "Employees", conn, adOpenKeyset, adLockOptimistic

  ' find the first record matching the criteria
  rst.Find "TitleOfCourtesy ='Ms.'"
  Do Until rst.EOF
    Debug.Print rst.Fields("LastName").Value
    ' search forward starting from the next record
    rst.Find "TitleOfCourtesy ='Ms.'", SkipRecords:=1, _
    SearchDirection:=adSearchForward
  Loop

  rst.Close
  Set rst = Nothing
  conn.Close
  Set conn = Nothing
End Sub
```

3. Choose **Run | Run Sub/UserForm** to execute the procedure.

To find the last record, call the `MoveLast` method before using `Find` d. If none of the records meets the criteria, the current record is positioned before the beginning of the recordset (if searching forward) or after the end of the recordset (if searching backward). You can use the EOF or BOF properties of the Recordset object to determine whether a matching record was found.

The ADO `Find` method does not support the `Is` operator. To locate a record that has a Null value, use the equal sign (=). For example:

```
' find records that do not have an entry in the ReportsTo field
rst.Find "ReportsTo = Null"
' find records that have data in the ReportsTo field
rst.Find " ReportsTo <> Null"
```

To find records based on more than one condition, use the Filter property of the Recordset object, as demonstrated in Hands-On 14.19 later in this chapter.

Finding Records Using the Seek Method

You can use the Recordset object's `Seek` method to locate a record based on an index. If you don't specify the index before searching, the primary key will be used. If the record is found, the current row position is changed to that row. The syntax of the `Seek` method looks like this:

```
recordset.Seek KeyValues, SeekOption
```

The first argument of the `Seek` method specifies the key values you want to find. The second argument specifies the type of comparison to be made between the columns of the index and the corresponding `KeyValues`.

The procedure in Hands-On 14.18 uses the `Seek` method to find the first company with an entry in the Region field equal to "SP":

```
rst.Seek "SP", adSeekFirstEQ
```

To find the last record that meets the same condition, use the following statement:

```
rst.Seek "SP", adSeekLastEQ
```

The type of `Seek` to execute is specified by the constants shown in Table 14.6.

TABLE 14.6. Seek method constants

Constant	Value	Description
`adSeekFirstEQ`	1	Seeks the first key equal to `KeyValues`
`adSeekLastEQ`	2	Seeks the last key equal to `KeyValues`
`adSeekAfterEQ`	4	Seeks a key either equal to `KeyValues` or just after where that match would have occurred
`adSeekAfter`	8	Seeks a key just after where a match with `KeyValues` would have occurred
`adSeekBeforeEQ`	16	Seeks a key either equal to `KeyValues` or just before where that match would have occurred
`adSeekBefore`	32	Seeks a key just before where a match with `KeyValues` would have occurred

The `Seek` method is recognized only by the Microsoft Jet 4.0/ACE 12.0 databases. To determine whether the `Seek` method can be used to locate a row in a recordset, use the Recordset object's `Supports` method. This method determines whether a specified Recordset object supports a particular type of feature. The Boolean value of `True` indicates that the feature is supported; `False` indicates that it is not.

```
' find out if the recordset supports the Seek method
MsgBox rst.Supports(adSeek)
```

Hands-On 14.18. Finding Records Using the Seek Method (ADO)

1. In the Visual Basic Editor window, choose **Insert | Module.**
2. In the module's Code window, type the following **Find_WithSeek** procedure:

```
Sub Find_WithSeek()
  Dim conn As ADODB.Connection
  Dim rst As ADODB.Recordset

  Set conn = New ADODB.Connection
  conn.Open "Provider=Microsoft.Jet.OLEDB.4.0;" & _
   "Data Source=" & CurrentProject.Path & _
   "\Northwind.mdb"

  Set rst = New ADODB.Recordset
  With rst
    .Index = "Region"
    .Open "Customers", conn, adOpenKeyset, adLockOptimistic, _
     adCmdTableDirect

    ' find out if this recordset supports the Seek method
    MsgBox rst.Supports(adSeek)
    .Seek "SP", adSeekFirstEQ
  End With

  If Not rst.EOF Then
    Debug.Print rst.Fields("CompanyName").Value
  End If

  rst.Close
  Set rst = Nothing
  conn.Close
  Set conn = Nothing
End Sub
```

3. Choose **Run | Run Sub/UserForm** to execute the procedure.

   If the `Seek` method is based on a multifield index, use the VBA `Array` function to specify values for the `KeyValues` parameter. For example, the Order Details table in the Northwind.mdb database uses a multifield index as the PrimaryKey. This index is a combination of the OrderID and ProductID fields. To find the order in which OrderID = 10295 and ProductID = 56, use the following statement:

```
rst.Seek Array(10295, 56), adSeekFirstEQ
```

Finding a Record Based on Multiple Conditions

ADO's `Find` method does not allow you to find records based on more than one condition. The workaround is using the Recordset object's Filter property to create a view of the recordset that contains only those records that match the specified criteria. The procedure in Hands-On 14.19 uses the Filter property to find the female employees who live in the United States.

Hands-On 14.19. Finding a Record Based on Multiple Criteria (ADO)

1. In the Visual Basic Editor window, choose **Insert | Module**.
2. In the module's Code window, type the **Find_WithFilter** procedure shown here:

```
Sub Find_WithFilter()
  Dim conn As ADODB.Connection
  Dim rst As ADODB.Recordset

  Set conn = New ADODB.Connection
  conn.Open "Provider=Microsoft.Jet.OLEDB.4.0;" & _
   "Data Source=" & CurrentProject.Path & _
   "\Northwind.mdb"

  Set rst = New ADODB.Recordset
  rst.Open "Employees", conn, adOpenKeyset, adLockOptimistic
  rst.Filter = "TitleOfCourtesy ='Ms.' and Country ='USA'"
  Do Until rst.EOF
    Debug.Print rst.Fields("LastName").Value
    rst.MoveNext
  Loop

  rst.Close
  Set rst = Nothing
```

```
    conn.Close
    Set conn = Nothing
End Sub
```

3. Choose **Run | Run Sub/UserForm** to execute the procedure.

Using Bookmarks

When you work with database records, you must keep in mind that the actual number of records in a recordset can change at any time as new records are added or others are deleted. Therefore, you cannot save a record number to return to it later. Because records change all the time, the record numbers cannot be trusted. However, programmers often need to save the position of a record after they've moved to it or found it based on certain criteria. Instead of scrolling through every record in a recordset comparing the values, you can move directly to a specific record by using a bookmark. A *bookmark* is a value that uniquely identifies a row in a recordset.

Use the Bookmark property of the Recordset object to mark the record so you can return to it later. The Bookmark property is read/write, which means that you can get a bookmark for a record or set the current record in a Recordset object to the record identified by a valid bookmark. The Recordset's Bookmark property always represents the current row. Therefore, if you need to mark more than one row for later retrieval, you may want to use an array to store multiple bookmarks (see Hands-On 14.20).

A single bookmark can be stored in a Variant variable. For example, when you get to a particular row in a recordset and decide that you'd like to save its location, store the recordset's bookmark in a variable, like this:

```
varMyBkmrk = rst.Bookmark
```

`varMyBkmrk` is the name of a Variant variable declared with the following statement:

```
Dim varMyBkmrk As Variant
```

To retrieve the bookmark, move to another row, then use the saved bookmark to move back to the original row, like this:

```
rst.Bookmark = varMyBkmrk
```

Because not all ADO Recordsets support the Bookmark property, you should use the `Supports` method to determine if the recordset does. Here's how:

```
If rst.Supports(adBookmark) then
```

```
    MsgBox "Bookmarks are supported."
  Else
    MsgBox "Sorry, can't use bookmarks!"
  End If
```

Recordsets defined with a Static or Keyset cursor always support bookmarks. If you remove the `adOpenKeyset` intrinsic constant from the code used in the next procedure (Hands-On 14.20), the default cursor (`adOpenForwardOnly`) will be used, and you'll get an error because this cursor does not support bookmarks.

Another precaution to keep in mind is that there is no valid bookmark when the current row is positioned at the new row in a recordset. For example, if you add a new record with the following statement:

```
rst.AddNew
```

and then attempt to mark this record with a bookmark:

```
varMyBkmrk = rst.Bookmark
```

you will get an error.

When you close the recordset, bookmarks you've saved become invalid. Also, bookmarks are unique to the recordset in which they were created. This means that you cannot use a bookmark created in one recordset to move to the same record in another recordset. However, if you clone a recordset (that is, you create a duplicate Recordset object), a Bookmark object from one Recordset object will refer to the same record in its clone. (See the section titled "Cloning a Recordset" in Chapter 17.)

Hands-On 14.20. Marking Records with a Bookmark (ADO)

1. In the Visual Basic Editor window, choose **Insert | Module**.
2. In the module's Code window, type the following **TestBookmark** procedure:

```
Sub TestBookmark()
  Dim conn As ADODB.Connection
  Dim rst As ADODB.Recordset
  Dim varMyBkmrk As Variant

  Set conn = New ADODB.Connection
  conn.Open "Provider=Microsoft.Jet.OLEDB.4.0;" & _
   "Data Source=" & CurrentProject.Path & _
   "\Northwind.mdb"

  Set rst = New ADODB.Recordset
```

```
  rst.Open "Employees", conn, adOpenKeyset

  If Not rst.Supports(adBookmark) Then
    MsgBox "This recordset does not support bookmarks!"
    Exit Sub
  End If

  varMyBkmrk = rst.Bookmark
  Debug.Print rst.Fields(1).Value

  ' Move to the 7th row
  rst.AbsolutePosition = 7
  Debug.Print rst.Fields(1).Value

  ' move back to the first row using bookmark
  rst.Bookmark = varMyBkmrk
  Debug.Print rst.Fields(1).Value
  rst.Close
  Set rst = Nothing
End Sub
```

3. Choose **Run | Run Sub/UserForm** to execute the procedure.

   Notice that this procedure uses the AbsolutePosition property of the Recordset object. The absolute position isn't the same as the record number. This property can change if a record with a lower number is deleted.

Using Bookmarks to Filter a Recordset

Bookmarks provide the fastest way of moving through rows. You can also use them to filter a recordset as shown in Hands-On 14.21.

Hands-On 14.21. Using Bookmarks to Filter Records (ADO)

1. In the Visual Basic Editor window, choose **Insert | Module**.
2. In the module's Code window, type the following **Filter_WithBookmark** procedure:

```
Sub Filter_WithBookmark()
  Dim rst As ADODB.Recordset
  Dim varMyBkmrk() As Variant
  Dim strConn As String
  Dim i As Integer
  Dim strCountry As String
```

```
    Dim strCity As String

    i = 0
    strCountry = "France"
    strCity = "Paris"

    strConn = "Provider=Microsoft.Jet.OLEDB.4.0;" & _
     "Data Source=" & CurrentProject.Path & _
     "\Northwind.mdb"

    Set rst = New ADODB.Recordset
    rst.Open "Customers", strConn, adOpenKeyset

    If Not rst.Supports(adBookmark) Then
      MsgBox "This recordset does not support bookmarks!"
      Exit Sub
    End If

    Do While Not rst.EOF
      If rst.Fields("Country") = strCountry And _
       rst.Fields("City") = strCity Then
        ReDim Preserve varMyBkmrk(i)
        varMyBkmrk(i) = rst.Bookmark
        i = i + 1
      End If
      rst.MoveNext
    Loop

    rst.Filter = varMyBkmrk()

    rst.MoveFirst
    Do While Not rst.EOF
      Debug.Print rst("CustomerId") & _
       " - " & rst("CompanyName")
      rst.MoveNext
    Loop
    rst.Close
    Set rst = Nothing
End Sub
```

3. Choose **Run | Run Sub/UserForm** to execute the procedure.

Using the GetRows Method to Fill the Recordset

To retrieve multiple rows from a recordset, use the `GetRows` method, which returns a two-dimensional array. Recall that using arrays in VBA procedures was the main focus of Chapter 7. To find out how many rows were retrieved, use VBA's `UBound` function, as illustrated in Hands-On 14.22. Because arrays are zero-based by default, you must add one (1) to the result of the `UBound` function to get the correct record count.

Hands-On 14.22. Counting the Number of Returned Records (ADO)

1. In the Visual Basic Editor window, choose **Insert | Module**.
2. In the module's Code window, type the following **CountRecords** procedure:

```
Sub CountRecords()
  Dim conn As ADODB.Connection
  Dim rst As ADODB.Recordset
  Dim myarray As Variant
  Dim returnedRows As Integer
  Dim r As Integer ' record counter
  Dim f As Integer ' field counter

  Set conn = New ADODB.Connection
  conn.Open "Provider=Microsoft.Jet.OLEDB.4.0;" & _
   "Data Source=" & CurrentProject.Path & _
   "\Northwind.mdb"

  Set rst = New ADODB.Recordset
  rst.Open "SELECT * FROM Employees", _
   conn, adOpenForwardOnly, adLockReadOnly, adCmdText

  ' Return all rows into array
  myarray = rst.GetRows()
  returnedRows = UBound(myarray, 2) + 1

  MsgBox "Total number of records: " & returnedRows

  ' Find upper bound of second dimension
  For r = 0 To UBound(myarray, 2)
    Debug.Print "Record " & r + 1
    ' Find upper bound of first dimension
    For f = 0 To UBound(myarray, 1)
```

```
      ' Print data from each row in array
      Debug.Print Tab; _
      rst.Fields(f).Name & " = " & myarray(f, r)
    Next f
  Next r

  rst.Close
  Set rst = Nothing
  conn.Close
  Set conn = Nothing
End Sub
```

3. Choose **Run | Run Sub/UserForm** to execute the procedure.

Notice how the CountRecords procedure prints the contents of the array to the Immediate window by using a nested loop.

CHAPTER SUMMARY

In this chapter, you familiarized yourself with various methods of opening DAO and ADO Recordsets, moving around in a recordset, finding, filtering, and bookmarking required records as well as reading the contents of a recordset. You have also learned how to use the Recordset object's properties such as EOF, BOF, and RecordCount. In addition, you found out how to fill the ADO Recordset with the `GetString` and `GetRows` methods.

In the next chapter, you will gain experience performing such important data manipulation tasks as adding, modifying, copying, deleting, and sorting records.

Chapter 15 Working with Records

Now that you've familiarized yourself with various methods of opening, moving around in, and finding records, and reading the contents of a recordset (see Chapter 14), let's look at DAO and ADO techniques for adding, modifying, copying, deleting, and sorting records.

ADDING A NEW RECORD WITH DAO

In the Microsoft Access user interface, before you can add a new record to a table you must first open the appropriate table. In code, you simply open the Recordset object by calling the `OpenRecordset` method. For example, the following statements declare and open the Recordset object based on the Employees table:

```
Dim tblRst As DAO.Recordset
Set tblRst = db.OpenRecordset("Employees")
```

Once the Recordset object is open, use the `AddNew` method to create a blank record. For example:

```
tblRst.AddNew
```

Next, you may set values for all or some of the fields in the new record. You must set the field's value if the Required property of a field is set to `True`. In the Microsoft Access user interface in Table Design view, there will be a Yes entry next to the Required property if the entry in the selected field is required. Here are some examples of setting field values in code:

```
tblRst.Fields("Last Name").Value = "Smith"
tblRst.Fields("Job Title").Value = "Marketing Director"
```

Note that because Value is the default property of a Field object, the use of this keyword is optional and it was omitted in the code of the example procedure in Hands-On 15.1.

After filling in field values, you need to use the `Update` method on the Recordset object to ensure that the newly added record is saved:

```
tblRst.Update
```

Hands-On 15.1 demonstrates how to add a new record to the Employees table and populate some of its fields with values.

Please note files for the "Hands-On" project may be found on the companion CD-ROM.

Hands-On 15.1. Adding a New Record to a Table (DAO)

1. Start Microsoft Access and create a new database named **Chap15.accdb** in your **C:\Access2013_ByExample** folder.
2. In the Access window, press Alt+F11 to switch to the Visual Basic Editor window.
3. In the Visual Basic Editor window, choose **Insert | Module**.

4. In the module's Code window, type the **AddNewRec_DAO** procedure shown here:

```
Sub AddNewRec_DAO()
  Dim db As DAO.Database
  Dim tblRst As DAO.Recordset
  Dim strDb As String

  strDb = "C:\Access2013_ByExample\Northwind 2007.accdb"

  Set db = OpenDatabase(strDb)
  Set tblRst = db.OpenRecordset("Employees")

  With tblRst
    .AddNew
    .Fields("Company") = "Northwind Traders"
    .Fields("Last Name") = "Smith"
    .Fields("First Name") = "Regina"
    .Fields("Job Title") = "Marketing Director"
    .Fields("E-mail Address") = "regina@northwindtraders.com"
    .Update
  End With

  tblRst.Close
  Set tblRst = Nothing
  db.Close
  Set db = Nothing
End Sub
```

5. Choose **Run | Run Sub/UserForm** to execute the procedure.

In a Table-type recordset, the new record is placed in the order identified by the table's index. In a Dynaset-type recordset, the new record is added at the end of the recordset. When you add a new record to a table, the new record does not become the current record. The record that was current prior to adding the new record remains current. In other words, while a new record is being added to the end of the table, the cursor remains in the record that was selected prior to adding a new record. You can, however, make the newly added record current by using the Bookmark and LastModified properties, like this:

```
tblRst.Bookmark = tblRst.LastModified
```

ADDING A NEW RECORD WITH ADO

To add a new record, use the ADO Recordset's `AddNew` method. Use the `Update` method if you are not going to add any more records. In ADO, it is not necessary to call the `Update` method if you are moving to the next record. Calling the `Move` method implicitly calls the `Update` method before moving to the new record. Take a look at the following statements:

```
rst![Last Name] = "Roberts"
rst.MoveNext
```

In this code fragment, the `Update`e method is automatically called when you move to the next record.

The procedure in Hands-On 15.2 demonstrates how to add a new record to the Employees table.

Hands-On 15.2. Adding a New Record to a Table (ADO)

1. In the Visual Basic Editor window, choose **Insert | Module**.
2. In the module's Code window, type the following **AddNewRec_ADO** procedure:

```
' Use the References dialog box
' to set up a reference to
' the Microsoft ActiveX Data 6.1 Object Library

Sub AddNewRec_ADO()
  Dim conn As ADODB.Connection
  Dim rst As ADODB.Recordset
  Dim strConn As String

  strConn = "Provider=Microsoft.ACE.OLEDB.12.0;" & _
   "Data Source=" & CurrentProject.Path & _
   "\Northwind 2007.accdb"

  Set rst = New ADODB.Recordset
  With rst
    .Open "SELECT * FROM Employees", _
     strConn, adOpenKeyset, adLockOptimistic

    ' Add a record and specify some field values
    .AddNew
```

```
        ![Company] = "Northwind Traders"
        ![Last Name] = "Roberts"
        ![First Name] = "Paul"
        ![Job Title] = "Sales Representative"
        ![E-mail Address] = "paul@northwindtraders.com"

        ' Retrieve the Employee ID for the current record
        Debug.Print !ID.Value

        ' Move to the first record
        .MoveFirst
        Debug.Print !ID.Value
        .Close
    End With

    Set rst = Nothing
    Set conn = Nothing
End Sub
```

3. Choose **Run | Run Sub/UserForm** to execute the procedure.

When adding or modifying records, you can set the record's field values in one of the following ways:

```
rst.Fields("First Name").value = "Paul"
```

or

```
rst![First Name] = "Paul"
```

As mentioned earlier, when you use the `AddNew` method to add a new record and then use the `Move` method, the newly added record is automatically saved without explicitly having to call the `Update` method. In the preceding example procedure, we used the `MoveFirst` method to move to the first record; however, you can call any of the other move methods (`Move`, `MoveNext`, `MovePrevious`) to have ADO implicitly call the `Update` method. After calling the `AddNew` method, the new record becomes the current record.

ADDING ATTACHMENTS

In Hands-On 12.8 in Chapter 12, "Creating and Accessing Database Tables and Fields," you learned how to programmatically add an Attachments field to a table. Hands-On 15.3 demonstrates how to use VBA to add external files to records in the Customers table of the Northwind database.

Hands-On 15.3. Using DAO to Add an Attachment to a Table Record

1. In the Visual Basic Editor window, choose **Insert | Module**.
2. In the module's Code window, type the following **AddAttachmentToRecord** procedure:

```
Sub AddAttachmentToRecord()
  Dim db As DAO.Database
  Dim rst As DAO.Recordset2
  Dim rstChild As DAO.Recordset2
  Dim addFlag As Boolean

  Const dirName = "C:\Access2013_HandsOn\External Docs\"
  Const strFile = "California3.jpg"
  Const strDb = "C:\Access2013_ByExample\Northwind 2007.accdb"

  Set db = OpenDatabase(strDb)
  ' Open the recordset for the Customers table
  Set rst = db.OpenRecordset("Customers")
  ' move to the 16th customer (count records from 0)

  rst.Move 15

  ' initialize child recordset
  Set rstChild = rst.Fields("Attachments").Value

  If rstChild.RecordCount > 0 Then
    ' check if the specified file is already attached
    Do Until rstChild.EOF
      If rstChild.Fields("FileName").Value = strFile Then
        addFlag = True
        Exit Do
      End If
    Loop
  End If

  If addFlag Then MsgBox "The specified file " & strFile & _
   " is already attached to this record."

  If Not addFlag Then
    ' put the parent recordset in Edit mode
    rst.Edit
```

```
        ' add a new record to the child recordset
        rstChild.AddNew
        ' load the attachment file
        rstChild.Fields("FileData").LoadFromFile dirName & strFile
        ' update both the child and parent recordsets
        rstChild.Update
        rst.Update
        MsgBox "Successfully attached " & strFile & " to " & _
          rst.Fields(1).Value & " record."
    End If

    Set rstChild = Nothing
    rst.Close
    Set rst = Nothing
    Set db = Nothing
End Sub
```

This procedure adds an attachment to the 16th record in the Customers table. This is a record for Company P. The child recordset holds the records for the Attachment field. Prior to adding a record to this recordset, the procedure checks the RecordCount property of the child recordset to verify that the specified file is not already attached. If RecordCount is greater than zero (0), then the `addFlag` Boolean variable is set to `True` and the user will see a message that the file is already attached. The procedure will then end. If the `addFlag` Boolean variable is `False`, then we know that it is okay to add the file. Note that before adding a new record to the child recordset you must put the parent recordset in Edit mode using the `Edit` method of the Recordset object. Next, call the `AddNew` method of the child recordset to add a new child record, and use the `LoadFromFile` method to load the new attachment file. Be sure to update both the child and parent recordsets.

3. Run the **AddAttachmentToRecord** procedure.
4. Open the **Customers** table in the Northwind 2007 database. Find the 16th record in the table and check out the paper clip column. It should indicate that one record is attached. You can view the attached file by double-clicking the attachment field in the 16th record (see Figure 15.1).
5. Close the **Customers** table and exit the Northwind 2007 database.
6. Return to the Visual Basic Editor window in the Chap15.accdb database and run the AddAttachmentToRecord procedure again to test the condition when the attachment file already exists for the specified record.

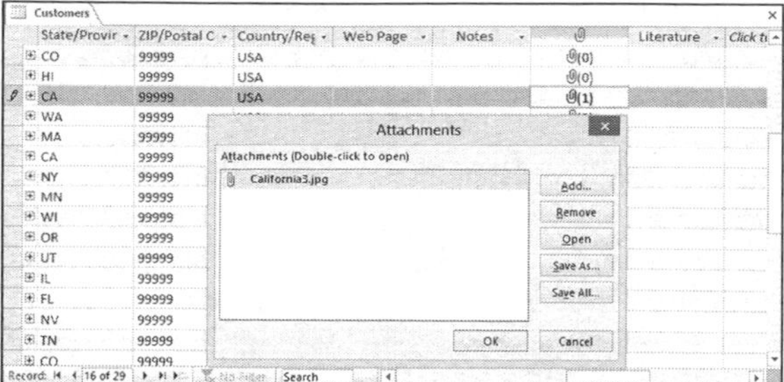

FIGURE 15.1. Attachment files can be added to records in an Access table manually using the Attachments dialog box or via VBA programming.

ADDING VALUES TO MULTIVALUE LOOKUP FIELDS

In Chapter 12 (see Hands-On 12.7), you used DAO to create a multivalue lookup field called Literature. You can add a new value to a multivalue field by modifying its RowSource property when the RowSourceType property is set to `Value List`. The function procedure in the next hands-on exercise adds new values to the Literature multivalue lookup field in the Customers table.

Hands-On 15.4. Using DAO to Add Values to a Multivalue Lookup Field

This hands-on exercise requires the completion of Hands-On 12.7 in Chapter 12.

1. In the Visual Basic Editor window, choose **Insert | Module.**
2. In the module's Code window, type the following **AddToMultiValueList** function procedure:

```
Function AddToMultiValueList(strTblName As String, _
 strMultiFldName As String, strNewVal As String)

  Dim db As DAO.Database
  Dim tdf As DAO.TableDef
  Dim fld As DAO.Field2
  Dim prp As DAO.Property
  Dim strDb As String

  strDb = "C:\Access2013_ByExample\Northwind 2007.accdb"
```

```
  On Error GoTo ErrorHandler

  Set db = OpenDatabase(strDb)
  Set tdf = db.TableDefs(strTblName)
  Set fld = tdf.Fields(strMultiFldName)

  If fld.Properties("RowSourceType").Value = "Value List" Then
    Set prp = fld.Properties("RowSource")
    Debug.Print prp.Value
    If InStr(1, prp.Value, strNewVal) = 0 Then
      prp.Value = prp.Value & Chr(59) & Chr(34) & _
       strNewVal & Chr(34)
      Debug.Print prp.Value
    End If
  End If
ExitHere:
  Set prp = Nothing
  Set fld = Nothing
  Set tdf = Nothing
  Set db = Nothing
  Exit Function
ErrorHandler:
  MsgBox Err.Number & ":" & Err.Description
  GoTo ExitHere
End Function
```

This function procedure takes three arguments: the `strTblName` argument specifies the name of a table where a multivalue lookup field is located; the `strMultiFldName` argument specifies the name of a multivalue lookup field, and the `strNewVal` argument specifies the value you want to add to the list. To work with the specified table, we begin by setting the `tdf` object variable to point to our table:

```
Set tdf = db.TableDefs(strTblName)
```

Recall that the DAO TableDefs collection contains TableDef objects, which are table definitions. Each TableDef object contains a Fields collection. We set up the `fld` object variable to gain access to the specified multivalue lookup field via the Fields collection of the TableDef object:

```
Set fld = tdf.Fields(strMultiFldName)
```

The Field object has a collection of properties. Before we do any work, we check that the RowSourceType property is set to `Value List`. If this test is `True`, we need to get the current value of the RowSource property. We set up the `prp` object variable to point to this property and write the property value to the Immediate window:

```
Set prp = fld.Properties("RowSource")
Debug.Print prp.Value
```

Because we only want to have unique values in the multivalue lookup field, we need to check if the value passed in the `strNewVal` parameter is already in the value list. To do this, you can use the VBA `InStr` function that was introduced in Chapter 12:

```
If InStr(1, prp.Value, strNewVal) = 0 Then
```

Recall that the `InStr` function returns the position of the first occurrence of one string within another. The first parameter is optional. It indicates the character position where the search should start. Obviously, we want to start at the first position so that we can examine the entire value list string. The second parameter is the string to search in. The value of the `prp` variable contains the following string when the function is called:

```
"Product Brochure";"Product Flyer A";"Product Flyer B"
```

The third parameter of the `InStr` function is the string you want to find. We will specify this string when we call the function procedure in the next step. The `InStr` function also has an optional fourth argument that specifies the type of string comparison. When omitted, Access performs a binary comparison where each character matches only itself. This is the default.

The `InStr` function will return a zero (0) when the string you are looking for was not found in the string you searched in. We will then add the new item to the current RowSource value list:

```
prp.Value = prp.Value & Chr(59) & Chr(34) & _
 strNewVal & Chr(34)
```

To add a new value to the list, we use the concatenation character (&). The `Chr(59)` function will give us the required semicolon (;) and the `Chr(34)` is for the double quotes ("). The underscore character (_) simply breaks the long code line into two lines. Notice that the procedure uses the ErrorHandler code to trap errors that may result from entering a nonexistent table or column name.

3. Run the AddToMultiValueList function procedure by typing the following statement in the Immediate window and pressing **Enter** to execute:

```
AddToMultiValueList "Customers", "Literature", "Sales Contract"
```

After you execute the function procedure, the Immediate window should display the original value of the RowSource property and the new updated value:

```
"Product Brochure";"Product Flyer A";"Product Flyer B"
"Product Brochure";"Product Flyer A";"Product Flyer B";"Sales Contract"
```

4. Run the AddToMultiValueList function procedure again by typing the following statement in the Immediate window and pressing **Enter** to execute:

```
AddToMultiValueList "Customers", "Literature", "Dinner Invitation"
```

You should now see in the Immediate window the following two strings:

```
"Product Brochure";"Product Flyer A";"Product Flyer B";"Sales
Contract"
"Product Brochure";"Product Flyer A";"Product Flyer B";"Sales
Contract";"Dinner Invitation"
```

5. Open the **Customers** table in the Northwind 2007 database and take a look at the drop-down list in the **Literature** field. In addition to the values added in Chapter 12, you should see two entries that were added by the VBA code in this hands-on exercise: *Sales Contract* and *Dinner Invitation*.
6. Close the **Customers** table and exit the Northwind 2007 database.

MODIFYING A RECORD WITH DAO

To edit an existing record, use the `OpenRecordset` method to open the Recordset object. Next, locate the record you want to modify. In a Table-type recordset, you can use the `Seek` method and a table index to find a record that meets your criteria. In Dynaset-type and Snapshot-type recordsets, you can use any of the `Find` methods (`FindFirst`, `FindNext`, `FindPrevious`, `FindLast`) to locate the appropriate record. However, recall that you can edit data only in Table-type or Dynaset-type recordsets (Snapshots are used for retrieving data only). Once you've located the record, use the `Edit` method of the Recordset object and proceed to change field values. When you are done with the record modification, invoke the `Update` method for the Recordset object.

The procedure in Hands-On 15.5 demonstrates how to modify a record in the Employees table.

Hands-On 15.5. Modifying a Record in a Table (DAO)

This hands-on exercise requires the completion of Hands-On 15.1.

1. In the Visual Basic Editor window, choose **Insert | Module.**
2. In the module's Code window, type the following **ModifyRecord_DAO** procedure:

```
Sub ModifyRecord_DAO()
  Dim db As DAO.Database
  Dim rst As DAO.Recordset
  Dim strFind As String
  Dim intResult As Integer
  Dim strDb As String

  strDb = "C:\Access2013_ByExample\Northwind 2007.accdb"

  Set db = OpenDatabase(strDb)
  Set rst = db.OpenRecordset("Employees", dbOpenTable)

  rst.MoveFirst
  ' change the Zip/Postal Code of all employees
  ' from 99999 to 99998

  Do While Not rst.EOF
    With rst
      .Edit
      .Fields("Zip/Postal Code") = "99998"
      .Update
      .MoveNext
    End With
  Loop

  ' find the record with the last name of Smith
  ' enter data in Country/Region field
  strFind = "Smith"
  rst.MoveFirst
  rst.Index = "Last Name"

  rst.Seek "=", strFind
  MsgBox rst![Last Name]
  Debug.Print rst.EditMode
```

```
    rst.Edit

    rst![Country/Region] = "USA"
    If rst.EditMode = dbEditInProgress Then
      intResult = MsgBox("Do you want to save the changes " & _
       "to this record?", vbYesNo, _
       "Save or Cancel Changes?")
    End If
    If intResult = 6 Then ' Save changes
      rst.Update
    ElseIf intResult = 7 Then ' Cancel changes
      rst.CancelUpdate
    End If

    rst.Close
    Set rst = Nothing
    db.Close
    Set db = Nothing
  End Sub
```

3. Choose **Run | Run Sub/UserForm** to execute the procedure.

The procedure in Hands-On 15.5 opens a Table-type recordset based on the Employees table and changes the Zip/Postal Code of all employees. Next, the procedure locates a specific employee record. Note that the Index property must be set before using the `Seek` method for searching the Table-type recordset. If you set the Index property to an index that doesn't exist, a runtime error will occur. Once the desired record is located, the procedure displays the employee name in a message box. The exclamation point (!) is used to separate an object's name from the name of the collection of which it is a member. Because the default collection of the Recordset object is the Fields collection, you can omit the default collection name. Next, the procedure places the found employee record into Edit mode and modifies the value of the Country/Region field. The EditMode property of the Recordset object is used to determine if the Edit operation is in progress. The `EditModeEnum` constants, which are shown in Table 15.1, indicate the state of editing for the current record. Before committing the changes to the data, the user is asked to verify if changes should be saved or canceled. If the Yes button is selected in the message box, the Recordset's `Update` method is called; otherwise, the `CancelUpdate` method of the Recordset object will discard the changes to the current record.

TABLE 15.1. EditModeEnum constants used in the EditMode property of the DAO Recordset object

Constant Name	Value	Description
`dbEditNone`	0	`Edit` method not invoked
`dbEditInProgress`	1	`Edit` method invoked
`dbEditAdd`	2	`AddNew` method invoked

At times when working with records you will need to leave the record and discard the changes. To cancel any pending updates to the data, call the `CancelUpdate` *method of the DAO Recordset object. This method aborts any changes you've made to the current row. You can use the* `CancelUpdate` *method to cancel any changes made after the* `Edit` *or* `AddNew` *method was invoked. You can check if there is a pending operation that can be canceled by using the EditMode property of the Recordset object.*

MODIFYING A RECORD WITH ADO

To modify data in a specific field, find the record and set the Value property of the required field to a new value. Always call the `Update` method if you are not planning to edit any more records. If you modify a row and then try to close the recordset without calling the `Update` method first, ADO will trigger a runtime error.

The procedure in Hands-On 15.6 modifies an employee record.

Hands-On 15.6. Modifying a Record (ADO)

This hands-on exercise requires the completion of Hands-On 15.2.

1. In the Visual Basic Editor window, choose **Insert | Module**.
2. In the module's Code window, type the **ModifyRecord_ADO** procedure shown here:

```
Sub ModifyRecord_ADO()
  Dim conn As ADODB.Connection
  Dim rst As ADODB.Recordset
  Dim strConn As String

  strConn = "Provider=Microsoft.ACE.OLEDB.12.0;" & _
   "Data Source=" & CurrentProject.Path & _
```

```
    "\Northwind 2007.accdb"

   Set rst = New ADODB.Recordset

   With rst
     .Open "SELECT * FROM Employees WHERE " _
      & "[Last Name] = 'Roberts'", _
     strConn, adOpenKeyset, adLockOptimistic
     .Fields("City").Value = "Redmond"
     .Fields("State/Province").Value = "WA"
     .Fields("Country/Region").Value = "USA"
     .Update
     .Close
   End With

   Set rst = Nothing
   Set conn = Nothing
End Sub
```

3. Choose **Run | Run Sub/UserForm** to execute the procedure.

This procedure modifies a table record by first accessing the desired fields. You can modify several fields in a specific record by calling the `Update` method and passing it two arrays. The first array should specify the field names, and the second one should list the new values to be entered. For example, the following statement updates the data in the City, State/Province, and Country/Region fields with the corresponding values:

```
rst.Update Array("City", "State/Province", "Country/Region"),
Array("Redmond", "WA", "USA")
```

You can use the same technique with the `AddNew` method.

At times when working with records you will need to leave the record and discard the changes. To prevent ADO from automatically committing that row of data, call the `CancelUpdate` *method of the ADO Recordset object. This method aborts any changes you've made to the current row.*

EDITING MULTIPLE RECORDS WITH ADO

ADO has the ability to perform batch updates. This means that you can edit multiple records and send them to the OLE DB provider in a single operation.

To take advantage of batch updates, you must use the Keyset or Static cursor (see Chapter 14 for more information about cursors).

The procedure in Hands-On 15.7 finds all records in the Employees table where Title is "Sales Representative" and changes it to "Sales Rep." The changes are then committed to the database in a single Update operation.

Hands-On 15.7. Performing Batch Updates (ADO)

1. In the Visual Basic Editor window, choose **Insert | Module**.
2. In the module's Code window, type the following **BatchUpdate_Records_ADO** procedure:

```
Sub BatchUpdate_Records_ADO()
  Dim conn As ADODB.Connection
  Dim rst As ADODB.Recordset
  Dim strConn As String
  Dim strCriteria As String

  strConn = "Provider=Microsoft.ACE.OLEDB.12.0;" & _
   "Data Source=" & CurrentProject.Path & _
   "\Northwind 2007.accdb"

  strCriteria = "[Job Title] = 'Sales Representative'"

  Set conn = New ADODB.Connection
  conn.Open strConn

  Set rst = New ADODB.Recordset

  With rst
    Set .ActiveConnection = conn
    .Source = "Employees"
    .CursorLocation = adUseClient
    .LockType = adLockBatchOptimistic
    .CursorType = adOpenKeyset
    .Open
    .Find strCriteria
    Do While Not .EOF
      .Fields("Job Title") = "Sales Rep"
      .Find strCriteria, 1
    Loop
    .UpdateBatch
```

```
    End With

    rst.Close
    Set rst = Nothing
    conn.Close
    Set conn = Nothing
  End Sub
```

3. Choose **Run | Run Sub/UserForm** to execute the procedure.

   The BatchUpdate_Records_ADO procedure uses the ADO `Find` method to locate all the records that need to be modified. Once the first record is located, it is changed in memory and the find operation goes on to search for the next record and so on until the end of the recordset is reached. Notice that the following statement is issued to search past the current record:

   ```
   .Find strCriteria, 1
   ```

   Once all the records have been located and changed, the changes are all committed to the database in a single operation by issuing the `UpdateBatch` statement.

Updating Data: Differences between ADO and DAO

ADO differs from DAO in the way update and delete operations are performed. In DAO, you are required to use the `Edit` method of the Recordset object prior to making any changes to your data. ADO does not require you to do this; consequently, there is no `Edit` method in ADO. Also, in ADO, your changes are automatically saved when you modify a record. In DAO, leaving a row without first calling the `Update` method of the Recordset object will automatically discard your changes.

DELETING A RECORD WITH DAO

To delete an existing record, open the Recordset object by calling the `OpenRecordset` method, then locate the record you want to delete. In a Table-type recordset, you can use the `Seek` method and a table index to find a record that meets your criteria. In a Dynaset-type recordset, you can use any of the Find methods (`FindFirst`, `FindNext`, `FindPrevious`, `FindLast`) to locate the appropriate record. Next, use the `Delete` method on the Recordset object to perform the deletion. Before using the `Delete` method, it is a good idea to write code to ask the user to confirm or cancel the deletion. Immediately after a record is deleted, there is no current record. Use the `MoveNext` method to move the record pointer to an existing record.

The example procedure in Hands-On 15.8 deletes those employees who have an ID greater than 9.

Hands-On 15.8. Deleting a Record (DAO)

1. In the Visual Basic Editor window, choose **Insert | Module**.
2. In the module's Code window, type the following **DeleteRecord_DAO** procedure:

```
Sub DeleteRecord_DAO()
  Dim db As DAO.Database
  Dim tblRst As DAO.Recordset
  Dim counter As Integer
  Dim strDb As String

  strDb = "C:\Access2013_ByExample\Northwind 2007.accdb"

  Set db = OpenDatabase(strDb)

  ' delete all the employees with ID greater than 9
  Set tblRst = db.OpenRecordset("Employees")
  tblRst.MoveFirst
  Do While Not tblRst.EOF
    Debug.Print tblRst!ID
    If tblRst![ID] > 9 Then
      tblRst.Delete
      counter = counter + 1
    End If
    tblRst.MoveNext
  Loop

  MsgBox "Number of deleted records: " & counter
  tblRst.Close
  Set tblRst = Nothing
  db.Close
  Set db = Nothing
End Sub
```

3. Choose **Run | Run Sub/UserForm** to execute the procedure.

The statement `Do While Not tblRst.EOF` tells Visual Basic to execute the statements inside the loop until the end of file (EOF) is reached. The conditional statement inside the loop checks the value of the ID field and deletes the current record only if the specified condition is `True`. Every time a record is deleted, the `counter` variable's value is increased by 1. The `counter` variable stores the total number of deleted records. After the record is deleted, the `MoveNext` method is called to move the record pointer to the next existing record as long as the end of file has not yet been reached. Even though you can use the `Delete` method and the `While` loop to remove the required records as shown in Hands-On 15.8, it is more efficient to delete records with a Delete query (see Chapter 16).

DELETING A RECORD WITH ADO

To delete a record, find the record you want to delete and call the `Delete` method. After you delete a record, it's still the current record. You must use the `MoveNext` method to move to the next row if you are planning to perform additional operations with your records. An attempt to do anything with the row that has just been deleted will generate a runtime error. The procedure in Hands-On 15.9 deletes a record from the Employees table.

Hands-On 15.9. Deleting a Record (ADO)

This hands-on exercise requires the completion of Hands-On 15.2.

1. In the Visual Basic Editor window, choose **Insert | Module**.
2. In the module's Code window, type the **Delete_Record_ADO** procedure shown here:

```
Sub Delete_Record_ADO()
  Dim conn As ADODB.Connection
  Dim rst As ADODB.Recordset
  Dim strConn As String
  Dim strCriteria As String

  ' call procedure from Hands-On 15.2 to ensure
  ' that we have a record to delete
  AddNewRec_ADO

  strConn = "Provider=Microsoft.ACE.OLEDB.12.0;" & _
```

```
    "Data Source=" & CurrentProject.Path & _
    "\Northwind 2007.accdb"

  Set conn = New ADODB.Connection
  Set rst = New ADODB.Recordset

  With rst
    .Open "SELECT * FROM Employees WHERE " _
     & "[Last Name] ='Roberts'", _
     strConn, adOpenKeyset, adLockOptimistic
    .Delete
    .Close
  End With
  Set rst = Nothing
  Set conn = Nothing
End Sub
```

3. Choose **Run | Run Sub/UserForm** to execute the procedure.

Because we don't want to delete any original rows in the Employees table, the procedure makes a call to the AddNewRec_ADO procedure that we created in Hands-On 15.2 to ensure that we have a custom row to delete.

DELETING ATTACHMENTS

The following hands-on exercise uses the `Delete` method of the Recordset2 object to delete an attachment from a table record.

Hands-On 15.10. Using DAO to Delete an Attachment from a Table Record

This hands-on exercise requires the completion of Hands-On 15.4.

1. In the Visual Basic Editor window, choose **Insert | Module**.
2. In the module's Code window, type the following **RemoveAttachmentFrom-Record** procedure:

```
Sub RemoveAttachmentFromRecord()
  Dim db As DAO.Database
  Dim rst As DAO.Recordset2
  Dim rstChild As DAO.Recordset2
  Dim removeFlag As Boolean
```

```
    Const dirName = "C:\Access2013_HandsOn\External Docs\"
    Const strFile = "California3.jpg"
    Const strDb = "C:\Access2013_ByExample\Northwind 2007.accdb"

    Set db = OpenDatabase(strDb)
    ' Open the recordset for the Customers table
    Set rst = db.OpenRecordset("Customers")
    ' move to the 16th customer
    rst.Move 15
    ' get the child recordset for the Attachment field
    Set rstChild = rst.Fields("Attachments").Value
    ' search for the attachment file and remove it if found

    Do Until rstChild.EOF
      If rstChild.Fields("FileName").Value = strFile Then
        rstChild.Delete
        removeFlag = True
      End If
      rstChild.MoveNext
    Loop

    ' display a message
    If Not removeFlag Then
      MsgBox "The specified file " & strFile & _
       " is not attached to this record.", _
      vbOKOnly + vbInformation, "Nothing to Remove"
    Else
      MsgBox "The specified file " & strFile & _
       " was deleted from this record.", _
      vbOKOnly + vbInformation, "Attachment Removed"
    End If

    ' cleanup code
    rstChild.Close
    Set rstChild = Nothing
    rst.Close
    Set rst = Nothing
    Set db = Nothing
  End Sub
```

3. Run the RemoveAttachmentFromRecord procedure.

4. Open the **Customers** table in the Northwind 2007 database and navigate to the 16th record. The Attachment field in this record should indicate that there are no attachments.
5. Close the **Customers** table and exit the Northwind 2007 database.
6. Run the RemoveAttachmentFromRecord procedure again to test the condition when the attachment file for the specified record does not exist.

COPYING RECORDS TO AN EXCEL WORKSHEET

You can copy the contents of a DAO or ADO Recordset object directly to an Excel worksheet or a worksheet range by using the Workbook Range object's `CopyFromRecordset` method.

- To copy all the records in the Recordset object to a worksheet range starting at cell A1, use the following statement:

  ```
  Set rng = objSheet.Cells(2, 1)
  rng.CopyFromRecordset rst
  ```

  The `rst` following the name of the method is an object variable representing a Recordset object.

- To copy five records to a worksheet range, use the following statement:

  ```
  Set rng = objSheet.Cells(2, 1)
     rng.CopyFromRecordset rst, 5
  ```

- To copy five records and four fields to a worksheet range, use the following statement:

  ```
  Set rng = objSheet.Cells(2, 1)
  rng.CopyFromRecordset rst, 5, 4
  ```

You can also specify the number of records (rows) and fields to be copied using variables:

```
Set rng = objSheet.Cells(2, 1)
rng.CopyFromRecordset rst, myRows, myColumns
```

The procedure in Hands-On 15.11 uses the `CopyFromRecordset` method to copy data from the Employees table to an Excel worksheet (see Figure 15.2).

Hands-On 15.11. Copying Records to an Excel Worksheet (DAO)

1. In the Visual Basic Editor window, choose **Insert | Module**.

2. In the module's Code window, type the following **ExportToExcel_DAO** procedure:

```
Sub ExportToExcel_DAO()
  Dim db As DAO.Database
  Dim rst As DAO.Recordset
  Dim xlApp As Object
  Dim wkb As Object
  Dim objSheet As Object
  Dim rng As Object
  Dim strExcelFile As String
  Dim strDB As String
  Dim strTable As String
  Dim count As Integer
  Dim iCol As Integer
  Dim rowsToReturn As Integer

  strDB = "C:\Access2013_ByExample\Northwind 2007.accdb"
  strTable = "Employees"
  strExcelFile = CurrentProject.Path & "\ExcelFromAccess.xls"

  ' If Excel file already exists, delete it
  If Dir(strExcelFile) <> "" Then Kill strExcelFile

  Set db = OpenDatabase(strDB)
  Set rst = db.OpenRecordset(strTable)

  ' get the number of records from the recordset
  count = rst.RecordCount

  rowsToReturn = CInt(InputBox("How many records to copy?"))
  If rowsToReturn <= count Then

    ' set the reference to Excel and make Excel visible
    Set xlApp = CreateObject("Excel.Application")
    xlApp.Application.Visible = True

    ' set references to the Excel workbook and worksheet
    Set wkb = xlApp.Workbooks.Add
    Set objSheet = xlApp.ActiveWorkbook.sheets(1)
    objSheet.Activate
```

```
      ' write column names to the first worksheet row
      For iCol = 0 To rst.Fields.count - 1
        objSheet.Cells(1, iCol + 1).Value = _
         rst.Fields(iCol).Name
      Next

      ' specify the cell range that will receive the data
      Set rng = objSheet.Cells(2, 1)

      ' copy the specified number of records to the worksheet
      rng.CopyFromRecordset rst, rowsToReturn

      ' autofit the columns to make the data fit
      objSheet.columns.AutoFit

      ' close the workbook
      ' and save it in Excel 97-2003 file format
      wkb.SaveAs fileName:=strExcelFile, FileFormat:=56
      wkb.Close

      ' quit Excel and release object variables
      Set objSheet = Nothing
      Set wkb = Nothing
      xlApp.Quit
      Set xlApp = Nothing
    Else
      MsgBox "Please specify a number less than " _
       & count + 1 & "."
    End If

    db.Close
    Set db = Nothing
  End Sub
```

3. Position the insertion point anywhere within the procedure code and choose **Debug | Step Into** to execute the procedure one line at a time. (Press **F8** to execute each statement.)

   This procedure creates a recordset based on the Employees table and stores the total number of records in the `count` variable. The user is asked to specify the number of records to copy to Excel. If the specified number is less than or equal to the total number of records in the recordset, the code proceeds to copy the records to Excel using the `CopyFromRecordset` method. Notice that the procedure uses the `As Object` clause to declare object variables that will contain references to Excel objects when the procedure is run. When you define an object variable as Object, the variable is late bound. This means that VBA does not know what type of object the variable references until the program is run. To set a reference to Microsoft Excel, it is necessary to use the `CreateObject` function. Once the object is created (Excel.Application), it is referenced with the object variable (`xlApp`). The `CreateObject` function will create a new instance of the Excel application. To use the current instance or to start Excel and load a specific file while Excel is already running, use the `GetObject` function. To view what's going on while the procedure is running, set the Visible property of the Microsoft Excel application to `True`. Then, if you run the ExportToExcel_DAO procedure in step mode, you will be able to check the contents of the Excel window as you execute each statement.

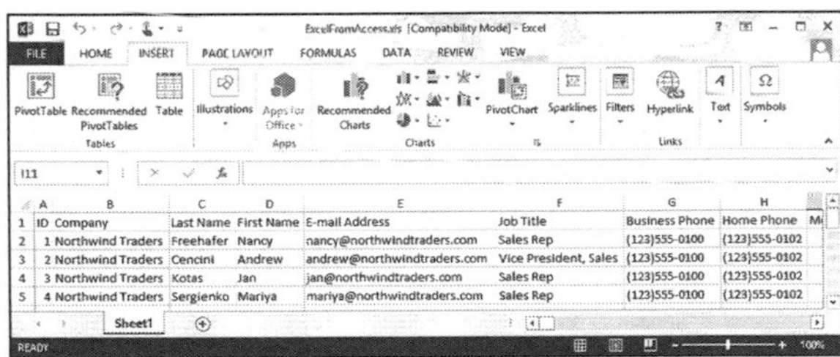

FIGURE 15.2. Access records copied programmatically to Excel.

Before you can copy Access data to the Excel worksheet, you need to set references to the Workbook, Worksheet, and Range objects. Once these references are defined, the procedure uses the `Add` method to add a new Excel workbook and then activates the first worksheet. The Recordset fields' names are written as column names to the first worksheet row. Next, the reference is set to the Range object that will receive the data from the recordset. The `CopyFromRecordset` method is used to copy the specified number of records to the worksheet. Once data is placed in the worksheet, it is fit into the columns with the AutoFit property. The Excel worksheet is then saved in the file format compatible with Excel 97-2003. The Workbook's `SaveAs` method requires the `FileFormat` parameter that specifies the file format for the workbook. The following file formats are used in Excel 2013, 2010 and 2007:

- 50 (`xlExcel12`)—Excel binary workbook with or without macros (.xlsb)
- 52 (`xlOpenXMLWorkbookMacroEnabled`)—.xml file format with or without macros (.xlsm)
- 51 (`xlOpenXMLWorkbook`)—.xml file format without macros (.xlsx)

- 56 (`xlExcel8`)—97-2003 format (.xls)

After saving the workbook, the procedure uses the Workbook `Close` method to close the Excel file. The Excel Application object's `Quit` method is used to close the Excel application.

COPYING RECORDS TO A WORD DOCUMENT

There are several techniques for placing Microsoft Access data in a Microsoft Word document. The procedure in Hands-On 15.12 demonstrates how to use the Recordset's `GetString` method to insert data from the Invoice Data table into a newly created Word document. Hands-On 15.13 shows how to insert data from the Shippers table and format the output using Word's Table object.

Hands-On 15.12. Copying Records to a Word Document (Example 1)

1. In the Visual Basic Editor window, choose **Insert | Module**.
2. Choose **Tools | References** in the Visual Basic Editor window. Scroll down to locate the **Microsoft Word 15 Object Library**, click the checkbox next to it, and then click **OK** to exit.
3. In the module's Code window, type the following **SendToWord_ADO** procedure:

```
' be sure to select Microsoft Word 15 Object Library
' in the References dialog box

Public myWord As Word.Application

Sub SendToWord_ADO()
  Dim conn As ADODB.Connection
  Dim rst As ADODB.Recordset
  Dim doc As Word.Document
  Dim strSQL As String
  Dim varRst As Variant
  Dim f As Variant
  Dim strHead As String

  Set conn = New ADODB.Connection
  Set rst = New ADODB.Recordset

  conn.Provider = "Microsoft.ACE.OLEDB.12.0;" & _
```

```
    "Data Source=" & CurrentProject.Path & _
    "\Northwind 2007.accdb"

  strSQL = "SELECT [Order ID] AS OrderID,"
  strSQL = strSQL & "[Ship Name], "
  strSQL = strSQL & "[Ship City] FROM [Invoice Data]"

  conn.Open
  rst.Open strSQL, conn, adOpenForwardOnly, _
   adLockReadOnly, adCmdText

  ' retrieve data and table headings into variables
  If Not rst.EOF Then
    varRst = rst.GetString(adClipString, , vbTab, vbCrLf)
    For Each f In rst.Fields
      strHead = strHead & f.Name & vbTab
    Next
  End If

  ' notice that Word application is declared
  ' at the top of the module
  Set myWord = New Word.Application

  ' create a new Word document
  Set doc = myWord.Documents.Add
  myWord.Visible = True

  ' paste contents of variables into
  ' Word document
  doc.Paragraphs(1).Range.Text = strHead & vbCrLf
  doc.Paragraphs(2).Range.Text = varRst

  On Error GoTo ErrorHandler
  doc.Close SaveChanges:=wdPromptToSaveChanges
EndProc:
  myWord.Quit
  Set myWord = Nothing
  Exit Sub
ErrorHandler:
  If Err = 4198 Then
    MsgBox "You refused to save this document."
  End If
```

```
  Resume EndProc
End Sub
```

4. Choose **Run | Run Sub/UserForm** to execute the procedure.

This procedure uses the Recordset object's `GetString` method to return recordset data as a string-valued Variant (see "Returning a Recordset as a String" in Chapter 14). Prior to running this procedure you must set a reference to the Microsoft Word 15 Object Library (or its lower version if you do not have the Word 2013 installed on the computer). This reference allows the procedure to access the Word application objects, properties, and methods via its own library. The top of the module contains the declaration of the `myWord` object variable that will point to the Word application. Notice that this variable is declared with the `Public` scope; therefore it can be accessed by other procedures in the current VBA project. (The next hands-on exercise also uses this variable.)

To launch Word and create a new document, we set the Application object to a new instance of `Word.Application` using the `New` keyword:

```
Set myWord = New Word.Application
```

To work with a Word document, the `Add` method of the Word Documents collection is used to create a blank document. We store the reference to this document in the `doc` object variable. To enable the user to see what's going on while the procedure is running, the Visible property of the Word application is set to `True`. Next, the contents of the Recordset and the field names that we previously saved in the string variables are written to the Word document using the Document object's Paragraphs property. The procedure ends by prompting the user to save changes to the Word document. If the user does not opt to save the document, error 4198 is triggered.

Hands-On 15.13. Copying Records to a Word Document (Example 2)

This procedure uses the `myWord` object variable that was declared in Hands-On 15.12 at the top of the module.

1. In the same module's Code window where you entered the procedure in the previous hands-on exercise, type the **SendToWord2** procedure shown here:

```
Sub SendToWord2()
  Dim db As DAO.Database
  Dim doc As Word.Document
  Dim wrdTbl As Word.Table
  Dim rst As DAO.Recordset
```

```
Dim f As Variant
Dim numRows As Integer
Dim numCols As Integer
Dim r As Integer ' row counter
Dim c As Integer ' column counter

Set db = OpenDatabase("C:\Access2013_ByExample\Northwind.mdb")
Set rst = db.OpenRecordset("Shippers")

numRows = rst.RecordCount
numCols = rst.Fields.count

' the myWord application object variable is declared
' at the top of the module
Set myWord = New Word.Application

' create a new Word document
Set doc = myWord.Documents.Add

' insert table
Set wrdTbl = doc.Tables.Add _
 (doc.Range, numRows + 1, numCols)

c = 1
If numRows > 0 Then
  ' Create the column headings in table cells
  For Each f In rst.Fields
    With wrdTbl
      .Cell(1, c).Range.Text = f.Name
      c = c + 1
    End With
  Next f
End If

r = 2
Do While Not rst.EOF
  For c = 1 To numCols
    wrdTbl.Cell(r, c).Range.Text = rst.Fields(c - 1).Value
  Next c
  r = r + 1
  rst.MoveNext
Loop
```

```
    myWord.Visible = True

    rst.Close
    Set rst = Nothing
    Set myWord = Nothing
    db.Close
    Set db = Nothing
End Sub
```

2. Choose **Run | Run Sub/UserForm** to execute the procedure.
3. Close the Word document and exit the Word application after you've looked at the resulting document, shown in Figure 15.3.

FIGURE 15.3. Access records copied programmatically to a Word document.

COPYING RECORDS TO A TEXT FILE

To write records to a text file, save them as a string by using the Recordset object's `GetString` method. Next, create a text file with the `CreateTextFile` method of the FileSystemObject from the Microsoft Scripting Runtime Library.

The procedure in Hands-On 15.14 demonstrates how to write the records from the Order Details table in the Northwind.mdb database to a text file named TestFile. Figures 15.4 and 15.5 show the generated text file after it has been opened in Notepad and in Microsoft Excel, respectively.

Hands-On 15.14. Copying Records to a Text File (ADO)

1. In the Visual Basic Editor window, choose **Insert | Module**.
2. In the module's Code window, type the following **WriteToFile** procedure:

```
Sub WriteToFile()
    Dim conn As ADODB.Connection
    Dim rst As ADODB.Recordset
```

```
    Dim f As ADODB.Field
    Dim fso As Object
    Dim txtfile As Object
    Dim strFileName As String

    Set conn = New ADODB.Connection
    conn.Open "Provider=Microsoft.Jet.OLEDB.4.0;" & _
     "Data Source=" & CurrentProject.Path & _
     "\Northwind.mdb"

    strFileName = CurrentProject.Path & "\TestFile.txt"

    Set fso = CreateObject("Scripting.FileSystemObject")
    Set txtfile = fso.CreateTextFile(strFileName, True)

    Set rst = New ADODB.Recordset
    rst.Open "[Order Details]", conn

    With rst
      For Each f In .Fields
        ' Write field name to the text file
        txtfile.Write (f.Name)
        txtfile.Write Chr(9)
      Next

      ' move to a new line
      txtfile.WriteLine

      ' write out all the records to the text file
      txtfile.Write rst.GetString(adClipString)

      .Close
    End With

    txtfile.Close
    Set rst = Nothing
    conn.Close
    Set conn = Nothing
  End Sub
```

3. Choose **Run | Run Sub/UserForm** to execute the procedure.

This procedure uses the `CreateObject` function to access the FileSystem-Object. The File object is created using the FileSystemObject's `CreateTextFile` method. The first argument of this method specifies the name of the file to create, and the second argument (`True`) indicates that the file should be overwritten if it already exists. Next, the procedure iterates through the recordset based on the Order Details table and writes field names to the text file using the `Write` method of the File object. The data from the recordset is converted into a string using the `GetString` method of the Recordset object and then written to the text file using the File object's `Write` method.

The text file is then closed with the `Close` method.

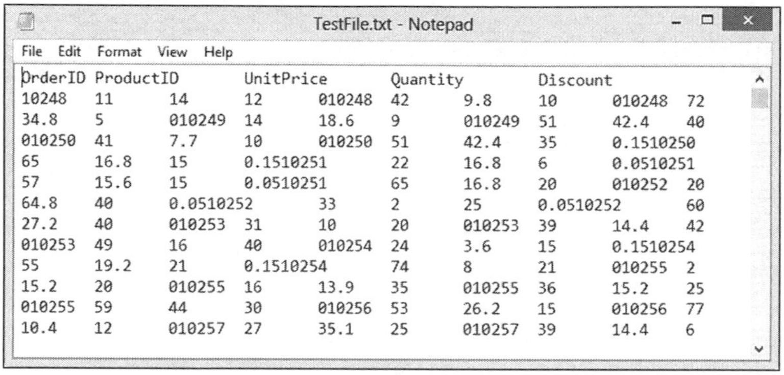

```
TestFile.txt - Notepad
File Edit Format View Help
OrderID ProductID       UnitPrice       Quantity        Discount
10248   11      14      12      010248  42      9.8     10      010248  72
34.8    5       010249  14      18.6    9       010249  51      42.4    40
010250  41      7.7     10      010250  51      42.4    35      0.1510250
65      16.8    15      0.1510251       22      16.8    6       0.0510251
57      15.6    15      0.0510251       65      16.8    20      010252  20
64.8    40      0.0510252       33      2       25      0.0510252       60
27.2    40      010253  31      10      20      010253  39      14.4    42
010253  49      16      40      010254  24      3.6     15      0.1510254
55      19.2    21      0.1510254       74      8       21      010255  2
15.2    20      010255  16      13.9    35      010255  36      15.2    25
010255  59      44      30      010256  53      26.2    15      010256  77
10.4    12      010257  27      35.1    25      010257  39      14.4    6
```

FIGURE 15.4. After running the WriteToFile procedure in Hands-On 15.14, the records from the Order Details table are placed in a text file.

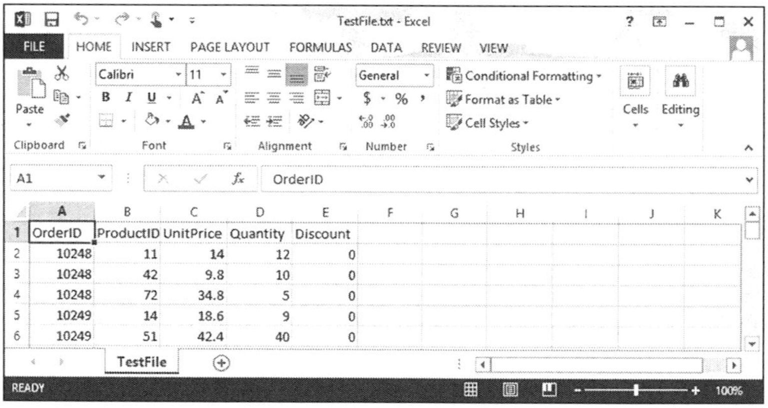

TestFile.txt - Excel

	A	B	C	D	E
1	OrderID	ProductID	UnitPrice	Quantity	Discount
2	10248	11	14	12	0
3	10248	42	9.8	10	0
4	10248	72	34.8	5	0
5	10249	14	18.6	9	0
6	10249	51	42.4	40	0

FIGURE 15.5. The Access-generated text file in Hands-On 15.14 opened in Excel 2013.

FILTERING RECORDS USING THE SQL WHERE CLAUSE

When you want to work only with a certain subset of records, you can filter out those records you don't want to see by using the SQL WHERE clause or the Filter property. You can apply a filter to a Dynaset-type or Snapshot-type Recordset object. The fastest way to filter records is to open a new Recordset object by using an SQL statement that includes a WHERE clause. Hands-On 15.15 provides an example of using the SQL WHERE clause to retrieve product orders with an order quantity greater than 100.

Hands-On 15.15. Filtering Records with the SQL WHERE Clause (DAO)

1. In the Visual Basic Editor window, choose **Insert | Module.**
2. In the module's Code window, type the following **FilterWithSQLWhere_DAO** procedure:

```
Sub FilterWithSQLWhere_DAO()
  Dim db As DAO.Database
  Dim rst As DAO.Recordset
  Dim qdf As DAO.QueryDef
  Dim qryName As String
  Dim mySQL As String
  Dim strDb As String

  strDb = "C:\Access2013_ByExample\Northwind 2007.accdb"

  Set db = OpenDatabase(strDb)

  qryName = "qryOrdersOver100"
  mySQL = "SELECT * FROM " _
   & "[Product Orders] WHERE Quantity > 100;"
  Set qdf = db.CreateQueryDef(qryName)
  qdf.SQL = mySQL
  Set rst = db.OpenRecordset(qryName)
  Debug.Print "There are " & rst.RecordCount & _
   " orders with the order quantity greater than 100."

  rst.Close
  Set rst = Nothing
  Set qdf = Nothing
```

```
    db.Close
    Set db = Nothing
End Sub
```

3. Choose **Run | Run Sub/UserForm** to execute the procedure.

   This procedure creates a simple Select query in the Northwind 2007.accdb database based on the Product Orders table. The SQL `WHERE` clause in the SQL statement specifies that only orders with a quantity greater than 100 should be returned. If the expression contained in the `WHERE` clause is `True`, then the record is selected; otherwise, the record is excluded from the opened set of records.

The procedure in Hands-On 15.16 opens a recordset that contains only records having the value of Null in the Region field or an entry of "Mrs." in the TitleOfCourtesy field.

Hands-On 15.16. Filtering Records with the SQL WHERE Clause (ADO)

1. In the Visual Basic Editor window, choose **Insert | Module**.
2. In the module's Code window, type the following **FilterWithSQLWhere_ADO** procedure:

```
Sub FilterWithSQLWhere_ADO()
  Dim conn As ADODB.Connection
  Dim rst As ADODB.Recordset
  Dim strSQL As String

  strSQL = "SELECT * FROM Employees WHERE IsNull(Region)" & _
   " or TitleOfCourtesy = 'Mrs.' "

  Set conn = New ADODB.Connection
  conn.Open "Provider=Microsoft.Jet.OLEDB.4.0;" & _
   "Data Source=" & CurrentProject.Path & _
   "\Northwind.mdb"

  Set rst = New ADODB.Recordset
  rst.Open strSQL, conn, adOpenKeyset, adLockOptimistic
  MsgBox "Selected " & rst.RecordCount & " records."

  rst.Close
  Set rst = Nothing
```

```
    conn.Close
    Set conn = Nothing
End Sub
```

3. Choose **Run | Run Sub/UserForm** to execute the procedure.

FILTERING RECORDS USING THE FILTER PROPERTY

You can use the DAO or ADO Filter property to obtain a set of records that meet specific criteria.

Hands-On 15.17 uses the Filter property with the DAO Recordset to restrict the subset of records to those in which the employee's city begins with the letter "R."

Hands-On 15.17. Filtering Records Using the Filter Property (DAO)

1. In the Visual Basic Editor window, choose **Insert | Module**.
2. In the module's Code window, type the **FilterRecords_DAO** procedure shown here:

```
Sub FilterRecords_DAO()
    Dim db As DAO.Database
    Dim rst As DAO.Recordset
    Dim FilterRst As DAO.Recordset
    Dim strDb As String

    strDb = "C:\Access2013_ByExample\Northwind 2007.accdb"

    Set db = OpenDatabase(strDb)
    Set rst = db.OpenRecordset("Employees", dbOpenDynaset)
    rst.Filter = "City like 'R*'"
    Set FilterRst = rst.OpenRecordset()

    Do Until FilterRst.EOF
        Debug.Print FilterRst.Fields("Last Name").Value
        FilterRst.MoveNext
    Loop

    FilterRst.Close
    Set FilterRst = Nothing
    rst.Close
    Set rst = Nothing
```

```
    db.Close
    Set db = Nothing
End Sub
```

3. Choose **Run | Run Sub/UserForm** to execute the procedure.

   This procedure begins by opening a Dynaset-type Recordset object based on the Employees table and setting the Filter property on this recordset:

```
rst.Filter = "City like 'R*'"
```

   For the filter to take effect after you set it, you must open a new recordset based on the Recordset object to which the filter was applied:

```
Set FilterRst = rst.OpenRecordset()
```

   Next, the procedure writes to the Immediate window the value of the Last Name field for all of the records in the filtered recordset.

The procedure in Hands-On 15.18 creates a filtered view of customers listed in the Northwind database who are located in Madrid, Spain.

Hands-On 15.18. Filtering Records Using the Filter Property (ADO)

1. In the Visual Basic Editor window, choose **Insert | Module**.
2. In the module's Code window, type the following **FilterRecords_ADO** procedure:

```
Sub FilterRecords_ADO()
    Dim conn As ADODB.Connection
    Dim rst As ADODB.Recordset

    Set conn = New ADODB.Connection
    conn.Open "Provider=Microsoft.Jet.OLEDB.4.0;" & _
     "Data Source=" & CurrentProject.Path & _
     "\Northwind.mdb"

    Set rst = New ADODB.Recordset
    With rst
      .Open "Customers", conn, adOpenKeyset, adLockOptimistic
      .Filter = "City='Madrid' and Country='Spain'"
      MsgBox .RecordCount & " records meet the criteria.", _
       vbInformation, "Customers in Madrid (Spain)"
    End With

    Do Until rst.EOF
```

```
    Debug.Print rst.Fields(1).Value
    rst.MoveNext
  Loop

  rst.Filter = adFilterNone
  MsgBox "Filter was removed. " & vbCr _
   & "The table contains " & rst.RecordCount & " records."

  rst.Close
  Set rst = Nothing
  conn.Close
  Set conn = Nothing
End Sub
```

3. Choose **Run | Run Sub/UserForm** to execute the procedure.

This procedure defines the filter on the Customers table and displays the filtered records. Then the filter is removed by setting the Filter property to `adFilterNone`.

Use the Filter property as a workaround to the ADO `Find` method whenever you need to find records that meet more than one condition. If the specific set of records you want to obtain is located on the SQL Server, you should use stored procedures instead of the Filter property.

SORTING RECORDS

You can use the Recordset object's Sort property to change the order in which records are displayed. The Sort property does not physically rearrange the records; it merely displays the records in the order specified by the index. If you are sorting on non-indexed fields, a temporary index is created for each field specified in the index. This index is removed automatically when you set the Sort property to an empty string. In ADO you can only use Sort on client-side cursors. If you use the server-side cursor, you will receive this error: "The operation requested by the application is not supported by the provider."

The default sort order is ascending. To order a recordset by country in ascending order, then by city in descending order, you would use the following statement:

```
rst.Sort = "Country ASC, City DESC"
```

Although you can use the Sort property to sort your data, you will most likely get better performance by specifying an SQL `ORDER BY` clause in the SQL statement or query used to open the recordset. The procedure in Hands-On 15.19 displays customer records from the Northwind database in ascending order by country.

Hands-On 15.19. Sorting Records (ADO)

1. In the Visual Basic Editor window, choose **Insert | Module.**
2. In the module's Code window, type the following **SortRecords_ADO** procedure:

```
Sub SortRecords_ADO()
  Dim conn As ADODB.Connection
  Dim rst As ADODB.Recordset

  Set conn = New ADODB.Connection
  conn.Open "Provider=Microsoft.Jet.OLEDB.4.0;" & _
   "Data Source=" & CurrentProject.Path & _
   "\Northwind.mdb"

  Set rst = New ADODB.Recordset

  ' sort on nonindexed field
  With rst
    .CursorLocation = adUseClient
    .Open "Customers", conn, adOpenKeyset, adLockOptimistic
    .Sort = "Country"
    Do Until rst.EOF
      Debug.Print rst.Fields("CompanyName").Value & ": " & _
       rst.Fields("Country").Value
    .MoveNext
    Loop

    Debug.Print "------original sort order -----".Sort = ""
    Do Until .EOF
      Debug.Print rst.Fields("CompanyName").Value & ": " & _
       rst.Fields("Country").Value
      .MoveNext
    Loop
    .Close
  End With

  Set rst = Nothing
```

```
    conn.Close
    Set conn = Nothing
End Sub
```

3. Choose **Run | Run Sub/UserForm** to execute the procedure.

   In this procedure, after sorting records in the specified order the Sort property is set to an empty string and records are displayed in the order in which they physically appear in the table.

CHAPTER SUMMARY

This chapter demonstrated several methods of the DAO and ADO Recordset objects you can use for working with records. You learned about the `AddNew`, `Update`, and `Delete` methods for performing such common database tasks as adding, modifying, and deleting records. These methods are suitable for handling a small number of records. Better performance can be achieved by using the SQL `INSERT`, `UPDATE`, and `DELETE` statements, as you will see in Chapters 16 and 25, “Views and Stored Procedures.”

This chapter also showed you how to render your database records into three popular formats: an Excel worksheet, a Word document, and a text file. Because working with large quantities of records can be difficult unless data is properly organized, this chapter also covered methods for filtering and sorting your records.

In the next chapter, you will learn how to create and run Access queries from your VBA procedures.

Chapter 16 Creating and Running Queries with DAO/ADO

Having worked with Microsoft Access for a while, you already know that to retrieve relevant information from your database and perform data-oriented tasks you need to write queries. *Queries* are SQL statements that are saved in the database and can be run at any time. Microsoft Access 2013 supports several types of queries.

The simplest queries allow you to select a set of records from a table. However, when you need to extract information from more than one table at a time, you must write a more complex query by using an SQL `JOIN` statement. Other queries perform specific actions on existing data, such as creating a new table, appending rows to a table, updating the values in a table, or deleting rows from a table. Although Microsoft Access provides a friendly interface—the Query Design view—for creating queries manually, this chapter teaches you how to create and execute the same queries by using DAO and ADO objects as well as SQL Data Manipulation Language (DML) statements in VBA code.

CREATING A SELECT QUERY MANUALLY

Select queries retrieve a set of records from a database table. These queries are easily recognized by the `SELECT` and `FROM` keywords in their syntax. Let's take a look at a couple of examples:

`SELECT LastName FROM Employees`	Selects the LastName field from the Employees table. If there is a space in the field name, enclose the field name in square brackets: [Last Name].
`SELECT FirstName, LastName, PhoneNo FROM Employees`	Selects the FirstName, LastName, and PhoneNo fields from the Employees table.
`SELECT * FROM Employees`	Selects all fields for all records from the Employees table. The asterisk (*) is used to represent all fields.

Often the `WHERE` clause is used with Select queries to specify criteria that determine which records the query will affect. Some examples of using the `WHERE` clause to restrict records are shown in the following table:

`SELECT * FROM Employees WHERE City IN ('Redmond', 'London')`	Selects from the Employees table all fields for all records that have the value Redmond or London in the City field.
`SELECT * FROM Employees WHERE City IN ('Redmond', 'London') AND ReportsTo LIKE 'Buchanan, Steven'`	Selects from the Employees table all fields for all records that have the value Redmond or London in the City field and have a value Buchanan, Steven in the ReportsTo field.
`SELECT * FROM Employees WHERE ((Year([HireDate])<1993) OR (City='Redmond'))`	Selects from the Employees table all fields for all records that have a value less than 1993 in the HireDate field or have the value Redmond in the City field.
`SELECT * FROM Products WHERE UnitPrice BETWEEN 10 AND 25`	Selects from the Products table all fields for all records that have an amount in the UnitPrice field between $10 and $25.
`SELECT * FROM Employees WHERE ReportsTo IS NULL`	Selects from the Employees table all fields for all records that do not have a value in the ReportsTo field.

You can use expressions in `WHERE` clauses to qualify SQL statements. An SQL expression is a string that is used in SQL statements. Expressions can contain literal values, constants, field names, operators, and functions. Several operators that are often used in expressions are shown in Table 16.1.

TABLE 16.1. Operators commonly used in expressions

Operator Name	Description/Usage

`IN`	The `IN` operator is used to determine whether the value of an expression is equal to any of several values in a specified list. If the expression is found in the list of values, the `IN` operator returns True; otherwise, it returns False. You can include the `NOT` logical operator to determine whether the expression is not in the list of values. For example, you can use `NOT IN` to determine which employees don't live in Redmond or London:
	`SELECT * FROM Employees`
	`WHERE City NOT IN ('Redmond', 'London')`

`LIKE`	The `LIKE` operator compares a string expression to a pattern in an SQL expression. For a pattern, you specify the complete value (for example, `LIKE 'Buchanan, Steven'`), or you can use wildcard characters to find a range of values (for example, `LIKE 'B*'`). You can use a number of wildcard characters in the `LIKE` operator pattern (see Table 16.2).

`BETWEEN...AND`	The `BETWEEN...AND` operator is used to determine whether the value of an expression falls within a specified range of values. If the value of the expression is between value1 and value2 (inclusive), the `BETWEEN...AND` operator returns True; otherwise, it returns False. You can include the `NOT` logical operator to evaluate the opposite condition, that is, whether the expression falls outside the range defined by value1 and value2.
	For example, you can select all products with the amount in the UnitPrice field less than $10 and greater than $25:
	`SELECT * FROM Products`
	`WHERE UnitPrice NOT BETWEEN 10 AND 25`

(Contd.)

Operator Name	Description/Usage

`IS NULL`	The `IS NULL` operator is used to determine whether the expression value is equal to the Null value. A Null value indicates missing or unknown data. You can include the `NOT` logical operator to return only records that have values in the specified field.
	For example, you can extract only the employee records that have a value in the ReportsTo field. Records where the ReportsTo field is blank will not be included:
	`SELECT * FROM Employees`
	`WHERE ReportsTo IS NOT NULL`

TABLE 16.2. Wildcard characters used in the LIKE operator patterns

Wildcard	Description
* (asterisk)	Matches any number of characters.
? (question mark)	Matches any single character.
% (percent sign)	Matches any number of characters (used only with the ADO and Jet OLE DB Provider; not in the Access user interface).
_ (underscore)	Matches any single character (used only with the ADO and Jet OLE DB Provider; not in the Access user interface).
# (number sign)	Matches any single digit.
[] (square brackets)	Matches any single character within the list of characters enclosed in brackets.
! (exclamation point)	Matches any single character that is not found in the list enclosed in the square brackets.
- (hyphen)	Matches any one of the range of characters enclosed in the square brackets.

In addition to the `WHERE` clause, you can use predicates to further restrict the set of records to be retrieved. A *predicate* is an SQL statement that qualifies the `SELECT` statement, similar to the `WHERE` clause; however, the predicate must be placed before the column list. Several popular predicates are shown in Table 16.3.

TABLE 16.3. Commonly used predicates in SQL SELECT statements

Predicate Name	Description/Usage

`ALL`	The `ALL` keyword is the default keyword and is used when no predicate is declared in the SQL statement.
	The following two examples are equivalent and return all records from the Employees table:
	`SELECT ALL *`
	`FROM Employees`
	`ORDER BY EmployeeID;`
	
	`SELECT *`
	`FROM Employees`
	`ORDER BY EmployeeID`

`DISTINCT`	The `DISTINCT` keyword eliminates duplicate values from the returned set of records. The values for each field listed in the `SELECT` statement must be unique.
	For example, to return a list of nonduplicate (unique) cities from the Employees table, you can write the following `SELECT` statement:
	`SELECT DISTINCT City`
	`FROM Employees`
	Note: The output of a query that uses `DISTINCT` isn't updatable (it's read-only).

`DISTINCTROW`	While the `DISTINCT` keyword is based on duplicate fields, the `DISTINCTROW` keyword is based on entire rows. It is used only with multiple tables.
	For example, if you join the Customers and Orders tables on the CustomerID field, you can find customers that have at least one order. The Customers table contains no duplicate CustomerID fields, but the Orders table does because each customer can have many orders.
	`SELECT DISTINCTROW CompanyName`
	`FROM Customers, Orders`
	`WHERE Customers.CustomerID =`
	`Orders.CustomerID`
	`ORDER BY CompanyName;`

(Contd.)

Predicate Name	Description/Usage
	Note: If you omit DISTINCTROW, this SELECT statement will produce multiple rows for each company that has more than one order. DISTINCTROW has an effect only when you select fields from some, but not all, of the tables used in the query. DISTINCTROW is ignored if your query includes only one table or if you output fields from all tables.

TOP or PERCENT	The TOP keyword returns a certain number of records that fall at the top or bottom of a range specified by an ORDER BY clause.
	For example, suppose you want to select the five most expensive products:
	`SELECT TOP 5 * FROM Products`
	`ORDER BY UnitPrice DESC`
	The TOP predicate doesn't choose between equal values. If there are equal values present, the TOP keyword will return all rows that have the equal value.
	You can also use the PERCENT keyword to return a percentage of records that fall at the top or bottom of a range specified by an ORDER BY clause.
	For example, to return the lowest 10 percent priced products, you can write the following statement:
	`SELECT TOP 10 PERCENT *`
	`FROM Products`
	`ORDER BY UnitPrice ASC;`
	Note: If you don't include the ORDER BY clause, the SELECT TOP statement will return a random set of rows.

If you'd like to sort records returned by the SELECT statement, use the ORDER BY clause with the ASC (ascending sort) or DESC (descending sort) keywords, as shown in the following example:

`SELECT * FROM`
`Employees`
`ORDER BY Country`
`DESC`
	Select all records from the Employees table and arrange them in descending order based on the Country field. If no order is specified, the order is ascending (ASC) by default.

By default, records are sorted in ascending order. The fields you want to sort by do not need to be enumerated in the SELECT statement's field list. Instead of sorting by field name, you can sort by field position. For example, the statement:

```
SELECT * FROM EMPLOYEES ORDER BY 2
```

will sort the records in ascending order by the second field.

CREATING A SELECT QUERY WITH DAO

In DAO, the QueryDef object represents a saved query in a database. All QueryDef objects are contained in the QueryDefs collection. You can read and set the SQL definition of a Query object using the SQL property. To create a query in code, use the CreateQueryDef method. For example, to create a Select query named myQuery, the following statement is used:

```
Set qdf = db.CreateQueryDef("myQuery", strSQL)
```

When you specify the name for your query, the new QueryDef object is automatically appended to the QueryDefs collection when it is created. The second argument of the CreateQueryDef method is a string variable that holds a valid Access SQL statement. Prior to using this variable, you must assign to it a string expression:

```
strSQL = "SELECT * FROM Employees WHERE TitleOfCourtesy = 'Ms.'"
```

The WHERE clause is used with Select queries to specify criteria that determine which records the query will affect. (See "Creating a Select Query Manually" at the beginning of this chapter.)

The procedure in Hands-On 16.1 selects from the Employees table all records that have a value of "Ms." in the TitleOfCourtesy field. The keyword LIKE can be substituted for the equals sign (=), as in the following:

```
strSQL = "SELECT * FROM Employees WHERE TitleOfCourtesy LIKE 'Ms.'"
```

When creating queries in code, be sure to include an error handler. After all, the query you are trying to create may already exist, or an unexpected error could occur.

Please note files for the "Hands-On" project may be found on the companion CD-ROM.

Hands-On 16.1. Creating a Select Query with DAO

1. Create a new Microsoft Access database named **Chap16.accdb** and save it in your **C:\Access2013_ByExample** folder.
2. In the database window, press **Alt+F11** to switch to the Visual Basic Editor window.
3. In the Visual Basic Editor window, choose **Insert | Module.**
4. In the module's Code window, type the **Create_SelectQuery_DAO** procedure shown here:

```
Sub Create_SelectQuery_DAO()
  Dim db As DAO.Database
  Dim qdf As DAO.QueryDef
  Dim strSQL As String
  Dim strDb As String

  strDb = "C:\Access2013_ByExample\Northwind.mdb"

  On Error GoTo Err_SelectQuery

  strSQL = "SELECT * FROM Employees "
  strSQL = strSQL & "WHERE TitleOfCourtesy = 'Ms.'"
  Set db = OpenDatabase(strDb)
  Set qdf = db.CreateQueryDef("myQuery", strSQL)
ExitHere:
  Set qdf = Nothing
  db.Close
  Set db = Nothing
  Exit Sub
Err_SelectQuery:
  If Err.Number = 3012 Then
    MsgBox "Query with this name already exists."
  Else
    MsgBox Err.Description
  End If
  Resume ExitHere
End Sub
```

5. Choose **Run | Run Sub/UserForm** to execute the procedure.

   When you run the Create_SelectQuery_DAO procedure, the next time you open the Northwind.mdb database you should see the query named myQuery in the list of stored queries in the Access window.

Instead of a query that is saved in the database for future use, it is possible to create a temporary query by setting the QueryDefName property to a zero-length string (""), as in the following example:

```
Set qdf = db.CreateQueryDef("", strSQL)
```

The advantage of temporary queries is that they don't clutter the Access Application window.

CREATING A SELECT QUERY WITH ADO

In ADO, queries, SQL statements, views, and stored procedures are represented by the Command object. This object is part of the ADOX Object Model. The Command object has many properties and methods that will allow you to return records or execute changes to your data (inserts, updates, and deletes). In this chapter, you will become acquainted with the properties of the Command object, including ActiveConnection, CommandText, and CommandType. These properties will be discussed as they appear in the example procedure code. You will also learn how to use the Command object's `Execute` method to run your queries.

The procedure in Hands-On 16.2 demonstrates how to create and save a Select query using ActiveX Data Objects (ADO).

Hands-On 16.2. Creating a Select Query with ADO

1. In the Visual Basic Editor window of the Chap16.accdb database, choose **Insert | Module.**
2. Choose **Tools | References** and select the following object libraries: **Microsoft ADO Ext. 6.0 for DDL and Security Object Library** and **Microsoft ActiveX Data Objects 6.1 Object Library**.
3. In the module's Code window, type the following **Create_SelectQuery_ADO** procedure:

```
Sub Create_SelectQuery_ADO()
  Dim cat As ADOX.Catalog
  Dim cmd As ADODB.Command
  Dim strPath As String
  Dim strSQL As String
  Dim strQryName As String
```

```
    On Error GoTo ErrorHandler

    ' assign values to string variables
    strPath = CurrentProject.Path & "\Northwind 2007.accdb"
    strSQL = "SELECT Employees.* FROM Employees WHERE " _
     & "Employees.City='Redmond';"

    strQryName = "Redmond Employees"

    ' open the Catalog
    Set cat = New ADOX.Catalog
    cat.ActiveConnection = _
     "Provider=Microsoft.ACE.OLEDB.12.0;" & _
     "Data Source=" & strPath

    ' create a query based on the specified
    ' SELECT statement
    Set cmd = New ADODB.Command
    cmd.CommandText = strSQL

    ' add the new query to the database
    cat.Views.Append strQryName, cmd

    MsgBox "The procedure completed successfully.", _
     vbInformation, "Create Select Query"
ExitHere:
    Set cmd = Nothing
    Set cat = Nothing
    Exit Sub

ErrorHandler:
    If InStr(Err.Description, "already exists") Then
      cat.Views.Delete strQryName
      Resume
    Else
      MsgBox Err.Number & ": " & Err.Description
      Resume ExitHere
    End If
End Sub
```

4. Choose **Run | Run Sub/UserForm** to execute the procedure.

The Create_SelectQuery_ADO procedure opens the Catalog object and sets its ActiveConnection property to the Northwind 2007.accdb database:

```
Set cat = New ADOX.Catalog
cat.ActiveConnection="Provider=Microsoft.ACE.OLEDB.12.0;" & _
 "Data Source=" & strPath
```

As you may recall from Chapter 12, the Catalog object represents an entire database. It contains objects that represent all the elements of the database: tables, stored procedures, views, columns of tables, and indexes. The ActiveConnection property of the Catalog object indicates the ADO Connection object the Catalog belongs to. The value of this property can be a reference to the Connection object or a connection string containing the definition for a connection. Next, the procedure defines a Command object and uses its CommandText property to set the SQL statement for the query:

```
Set cmd = New ADODB.Command
cmd.CommandText = strSQL
```

The CommandText property contains the text of a command you want to issue against a provider. In this procedure, we assigned the string variable's value (`strSQL`) to the CommandText property.

The ADO Command object always creates a temporary query. So, to create a stored (saved) query in a database, the procedure must append the Command object to the ADOX Views collection, like this:

```
cat.Views.Append strQryName, cmd
```

When you open the sample Northwind 2007.accdb database after running this procedure, you will find the Redmond Employees query in the Access window.

Row-returning, Nonparameterized Queries

Queries that return records, such as Select queries, are known as row-returning, *nonparameterized* queries.

In ADO, use the View object to work with queries that return records and do not take parameters. All View objects are contained in the Views collection of the ADOX Catalog object. To save these queries in a database, append the ADO Command object to the ADOX Views collection as shown in Hands-On 16.2.

EXECUTING AN EXISTING SELECT QUERY WITH ADO

There's more than one way of executing a row-returning query with ADO. This section demonstrates two procedures that run the Products by Category query located in the Northwind.mdb database.

The procedure in Hands-On 16.3 uses the Command and Recordset objects to perform this task.

Hands-On 16.3. Executing a Select Query

1. In the Visual Basic Editor window, choose **Insert | Module**.
2. In the module's Code window, type the **Execute_SelectQuery_ADO** procedure shown here:

```
Sub Execute_SelectQuery_ADO()
  Dim cmd As ADODB.Command
  Dim rst As ADODB.Recordset
  Dim strPath As String

  strPath = CurrentProject.Path & "\Northwind.mdb"

  Set cmd = New ADODB.Command
  With cmd
    .ActiveConnection = "Provider=Microsoft.Jet.OLEDB.4.0;" & _
     "Data Source=" & strPath
    .CommandText = "[Products by Category]"
    .CommandType = adCmdTable
  End With

  Set rst = New ADODB.Recordset
  Set rst = cmd.Execute

  Debug.Print rst.GetString

  rst.Close
  Set rst = Nothing
  Set cmd = Nothing
  MsgBox "View results in the Immediate window."
End Sub
```

3. Choose **Run | Run Sub/UserForm** to execute the procedure.

In the Execute_Select Query_ADO procedure, the connection to the database is opened by setting the ActiveConnection property of the Command object. Next, the Command object's CommandText property specifies the name of the query you want to run. Notice that you need to place square brackets around the query's name when it contains spaces. The query type is determined by setting the CommandType property of the Command object. Use the `adCmdTable` or `adCmdStoredProc` constants if the query string in the CommandText property is a query name. Finally, the `Execute` method of the Command object executes the query. Notice that the resulting recordset is passed to the Recordset object variable so that you can access the records retrieved by the query. Instead of looping through the records to read the returned records, the procedure uses the Recordset object's `GetString` method to print all the recordset rows to the Immediate window. The `GetString` method returns the recordset as a string (for more information, please see Chapter 14). Figure 16.1 shows the output of the Execute_Select Query_ADO procedure.

```
Immediate
Beverages   Chai    10 boxes x 20 bags  39  0
Beverages   Chang   24 - 12 oz bottles  17  0
Beverages   Chartreuse verte    750 cc per bottle   69  0
Beverages   Côte de Blaye   12 - 75 cl bottles  17  0
Beverages   Ipoh Coffee 16 - 500 g tins 17  0
Beverages   Lakkalikööri    500 ml  57  0
Beverages   Laughing Lumberjack Lager   24 - 12 oz bottles  52  0
Beverages   Outback Lager   24 - 355 ml bottles 15  0
Beverages   Rhönbräu Klosterbier    24 - 0.5 l bottles  125 0
Beverages   Sasquatch Ale   24 - 12 oz bottles  111 0
Beverages   Steeleye Stout  24 - 12 oz bottles  20  0
Condiments  Aniseed Syrup   12 - 550 ml bottles 13  0
```

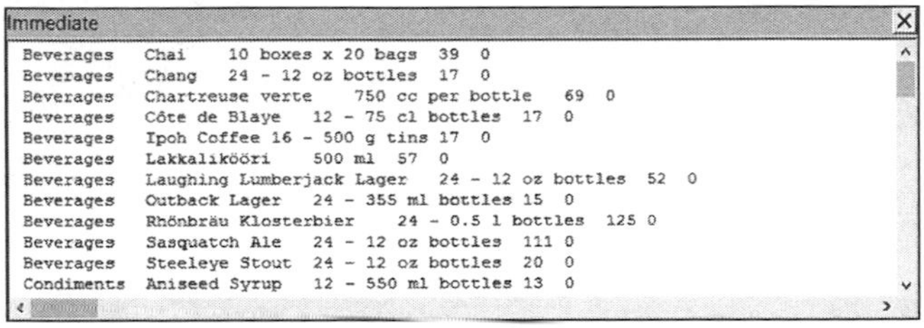

FIGURE 16.1. This is a sample result of records that were generated by executing the Select query in Hands-On 16.3.

The example procedure in Hands-On 16.4 demonstrates another method of running a row-returning query with ADO. Notice that in addition to the ADO Command and Recordset objects, this procedure uses the ADOX Catalog object. The connection to the database is established by setting the ActiveConnection property of the Catalog object and not the Command object, as was the case in Hands-On 16.3.

Hands-On 16.4. Executing a Select Query with an ADO Catalog Object

1. In the Visual Basic Editor window, choose **Insert | Module**.
2. In the module's Code window, type the following **Execute_SelectQuery2_ADO** procedure:

```
Sub Execute_SelectQuery2_ADO()
  Dim cat As ADOX.Catalog
  Dim cmd As ADODB.Command
```

```
    Dim rst As ADODB.Recordset
    Dim strPath As String

    strPath = CurrentProject.Path & "\Northwind.mdb"

    Set cat = New ADOX.Catalog
    cat.ActiveConnection = "Provider=Microsoft.Jet.OLEDB.4.0;" & _
     "Data Source=" & strPath

    Set cmd = New ADODB.Command
    Set cmd = cat.Views("Products by Category").Command

    Set rst = New ADODB.Recordset
    rst.Open cmd, , adOpenStatic, adLockReadOnly, adCmdTable

    Debug.Print rst.GetString
    MsgBox "The query returned " & rst.RecordCount & vbCr & _
     " records to the Immediate window."
    rst.Close
    Set rst = Nothing
    Set cmd = Nothing
    Set cat = Nothing
End Sub
```

3. Choose **Run | Run Sub/UserForm** to execute the procedure.

In this procedure, the following line of code is used to indicate the name of the query to be executed:

```
Set cmd = cat.Views("Products by Category").Command
```

This statement sets the `cmd` object variable to the desired query stored in the Views collection of the ADOX Catalog object. Next, the `Open` method of the Recordset object is used to open the recordset based on the specified query:

```
rst.Open cmd, , adOpenStatic, adLockReadOnly, adCmdTable
```

Notice that several optional arguments of the `Open` method are used to specify the data source: `cmd`, ActiveConnection (a comma appears in this spot because the existing connection is being used), CursorType (`adOpenStatic`), LockType (`adLockReadOnly`), and Options (`adCmdTable`). Refer to Chapter 14 for information about using these ADO constants. Next, the procedure dumps the contents of the records into the Immediate window (just as the procedure in Hands-On 16.3 did) by using the Recordset's `GetString` method. The `MsgBox` function contains a string that includes the information about the number of records retrieved. The RecordCount property of the Recordset object is used to get the record count. To get the correct record count, you must set the CursorType argument of the Recordset's `Open` method to `adOpenStatic`. If you set this argument to `adOpenDynamic` or `adOpenForwardOnly`, the RecordCount property will return –1. To learn more about these constants, refer to the sections in Chapter 14 on working with Recordset objects in ADO.

MODIFYING AN EXISTING QUERY WITH ADO

If you'd like to modify an existing query, follow these steps:

1. Retrieve the query from the Views or Procedures collection of the Catalog object.
2. Set the CommandText property of the Command object to the new SQL statement.
3. Save the changes by setting the Procedure or View object's Command property to the modified Command object.

Earlier in this chapter you learned how to create a Select query named Redmond Employees by using ADO (see Hands-On 16.2). The following hands-on exercise modifies this query so that employee records are ordered by last name.

Hands-On 16.5. Modifying a Select Query with ADO

1. In the Visual Basic Editor window, choose **Insert | Module.**
2. In the module's Code window, type the following **Modify_Query_ADO** procedure:

```
Sub Modify_Query_ADO()
  Dim cat As ADOX.Catalog
  Dim cmd As ADODB.Command
  Dim strPath As String
  Dim newStrSQL As String
  Dim oldStrSQL As String
  Dim strQryName As String

  strPath = CurrentProject.Path & "\Northwind 2007.accdb"

  newStrSQL = "SELECT Employees.* FROM Employees" & _
```

```
     " WHERE Employees.City='Redmond'" & _
     " ORDER BY [Last Name];"

    strQryName = "Redmond Employees"

    Set cat = New ADOX.Catalog
    cat.ActiveConnection = _
     "Provider=Microsoft.ACE.OLEDB.12.0;" & _
     "Data Source=" & strPath

    Set cmd = New ADODB.Command
    Set cmd = cat.Views(strQryName).Command

    ' get the existing SQL statement for this query
    oldStrSQL = cmd.CommandText

    MsgBox oldStrSQL, vbInformation, _
     "Current SQL Statement"

    ' now update the query's SQL statement
    cmd.CommandText = newStrSQL
    MsgBox newStrSQL, vbInformation, _
     "New SQL Statement"

    ' save the modified query
    Set cat.Views(strQryName).Command = cmd

    Set cmd = Nothing
    Set cat = Nothing
End Sub
```

3. Choose **Run | Run Sub/UserForm** to execute the procedure.

When you run this procedure the Redmond Employees query created in Hands-On 16.2 is modified from the following SQL statement:

```
SELECT Employees.*
FROM Employees
WHERE Employees.City='Redmond';666
```

to:

```
SELECT Employees.*
```

```
FROM Employees
WHERE Employees.City='Redmond' ORDER BY [Last Name];
```

CREATING AND RUNNING A PARAMETER QUERY WITH DAO

A special type of a Select query is known as a *Parameter* query. Instead of retrieving the same records each time a query is run, a user can enter the search criteria in a special dialog box at runtime. In DAO, the parameters of a Parameter query are represented by Parameter objects. The QueryDef object contains a Parameters collection. Parameter objects represent existing parameters.

To create a Parameter query, create a query string that includes the PARAMETERS keyword:

```
strSQL = "PARAMETERS [Enter Country] Text;"& _
 "SELECT * FROM CUSTOMERS WHERE Country = [Enter Country];"
```

Before executing an existing Parameter query, assign a value to the parameter, as shown in Hands-On 16.6. Once the parameter value is specified, you need to open a recordset based on the query.

The procedure in Hands-On 16.6 demonstrates how to create and run a Parameter query to retrieve the names of the companies in the user-specified country.

Hands-On 16.6. Creating a Parameter Query with DAO

1. In the Visual Basic Editor window, choose **Insert | Module.**
2. In the module's Code window, type the following **CreateRun_ParameterQuery_DAO** procedure:

```
Sub CreateRun_ParameterQuery_DAO()
  Dim db As DAO.Database
  Dim qdf As DAO.QueryDef
  Dim rst As DAO.Recordset
  Dim strQryName As String
  Dim strSQL As String
  Dim strDb As String

  strDb = "C:\Access2013_ByExample\Northwind.mdb"

  On Error GoTo Err_Handler
```

```
    strQryName = "myParamQuery"
    strSQL = "PARAMETERS [Enter Country] Text; " & _
     "SELECT * FROM Customers WHERE Country = [Enter Country];"

    Set db = OpenDatabase(strDb)
    Set qdf = db.CreateQueryDef(strQryName, strSQL)

RunQuery:
    ' specify the parameter
    qdf.Parameters("Enter Country") = _
     InputBox("Enter the country name:", _
     "Which Country?", "Germany")

    If IsNull(qdf.Parameters("Enter Country").Value) _
     Then GoTo ExitHere

    ' open a recordset based on the specified query
    Set rst = qdf.OpenRecordset(dbOpenDynaset)
    rst.MoveLast
    MsgBox "Number of records: " & rst.RecordCount

    ' write the contents of the second field
    ' to the Immediate window
    rst.MoveFirst
    Do Until rst.EOF
      Debug.Print rst(1)
      rst.MoveNext
    Loop

ExitHere:
    If Not rst Is Nothing Then
      rst.Close
      Set rst = Nothing
    End If
    Set qdf = Nothing
    db.Close
    Set db = Nothing
    Exit Sub

Err_Handler:
    If Err.Number = 3012 Then
      MsgBox "This query already exists."
```

```
      Set qdf = db.QueryDefs(strQryName)
      Resume RunQuery
    Else
      MsgBox Err.Description
    End If
    Resume ExitHere
  End Sub
```

3. Choose **Run | Run Sub/UserForm** to execute the procedure.

This procedure defines a Parameter query that contains one parameter named `Enter Country`. Prior to running this query, the procedure retrieves the name of the country from the user via the VBA `InputBox` method. While the suggested default country name is Germany, the user can supply the name of another country. The supplied value is then used as the value of the `Enter Country` parameter. Next, the recordset is opened based on the specified query, and the number of records for the specified country is retrieved via the RecordCount property of the Recordset object. In order to get the correct record count, we must move to the end of the recordset, using the `MoveLast` method, to access all records. The procedure ends by retrieving to the Immediate window the names of all the companies in the specified country. The procedure contains several labels such as `RunQuery`, `ExitHere`, and `Err_Handler`, which are used in error trapping and ensuring that certain code lines are run only when required. For example, when you execute this procedure again, the statement that attempts to create a query will fail and VBA will generate error 3012. At this point, we want to run the existing query, so we must set the `qdf` object variable with the following statement:

```
Set qdf = db.QueryDefs(strQryName)
```

And then we can safely resume running the code from the label `RunQuery`.

CREATING AND RUNNING A PARAMETER QUERY WITH ADO

In ADO, to create a row-returning, parameterized query, simply add the parameters to the query's SQL string. The parameters must be defined by using the `PARAMETERS` keyword, as in the following:

```
strSQL = "PARAMETERS [Country Name] Text;" & _
 "SELECT Customers.* FROM Customers WHERE " _
 & "Customers.Country=[Type Country Name];"
```

The preceding SQL statement begins by defining one parameter called `Country Name`. This parameter will be able to accept text entries. The second part of the SQL statement selects all the records from the Customers table that have an entry in the Country field equal to the provided parameter value. The complete procedure is shown in Hands-On 16.7.

Hands-On 16.7. Creating a Parameter Query with ADO

1. In the Visual Basic Editor window, choose **Insert | Module.**
2. In the module's Code window, type the following **Create_ParameterQuery_ADO** procedure:

```
Sub Create_ParameterQuery_ADO()
  Dim cat As ADOX.Catalog
  Dim cmd As ADODB.Command
  Dim strPath As String
  Dim strSQL As String
  Dim strQryName As String

  On Error GoTo ErrorHandler

  strPath = CurrentProject.Path & "\Northwind.mdb"

  strSQL = "PARAMETERS [Country Name] Text;" & _
   "SELECT Customers.* FROM Customers WHERE " _
   & "Customers.Country=[Country Name];"

  strQryName = "Customers by Country"

  Set cat = New ADOX.Catalog
  cat.ActiveConnection = _
   "Provider=Microsoft.Jet.OLEDB.4.0;" & _
   "Data Source=" & strPath

  Set cmd = New ADODB.Command
  cmd.CommandText = strSQL

  cat.Procedures.Append strQryName, cmd
  Set cmd = Nothing
  Set cat = Nothing

  MsgBox "The procedure completed successfully.", _
```

```
    vbInformation, "Create Parameter Query"
    Exit Sub

ErrorHandler:
    If InStr(Err.Description, "already exists") Then
      cat.Procedures.Delete strQryName
      Resume
    Else
      MsgBox Err.Number & ": " & Err.Description
    End If
End Sub
```

3. Choose **Run | Run Sub/UserForm** to execute the procedure.

This procedure creates a simple Parameter query with one parameter. Because the ADO Command object always creates a temporary query, you must append the Command object to the ADOX Procedures collection in order to save a parameterized query in a database.

Row-returning, Parameterized Queries

Queries that return records and take parameters are known as row-returning, parameterized queries.

In ADO, use the ADOX Procedure object to work with queries that return records and take parameters. All Procedure objects are contained in the Procedures collection of the ADOX Catalog object. To save these queries in a database, append the ADO Command object to the ADOX Procedures collection.

To execute a Parameter query you must specify the parameter value using the Parameters collection of the Command object, like this:

```
cmd.Parameters("Country Name") = "France"
```

The procedure in Hands-On 16.8 shows how to run the Parameter query created by the procedure in Hands-On 16.7.

Hands-On 16.8. Executing a Parameter Query with ADO

1. In the Visual Basic Editor window, choose **Insert | Module**.
2. In the module's Code window, type the following **Execute_ParamQuery_ADO** procedure:

```
Sub Execute_ParamQuery_ADO(strCountry As String)
```

```
   Dim cat As ADOX.Catalog
   Dim cmd As ADODB.Command
   Dim rst As ADODB.Recordset
   Dim strQryName As String
   Dim strPath As String

   strQryName = "Customers by Country"
   strPath = CurrentProject.Path & "\Northwind.mdb"

   Set cat = New ADOX.Catalog
   cat.ActiveConnection = _
    "Provider=Microsoft.Jet.OLEDB.4.0;" & _
    "Data Source=" & strPath

   Set cmd = New ADODB.Command
   Set cmd = cat.Procedures(strQryName).Command

   ' specify a parameter value
   cmd.Parameters("[Country Name]") = strCountry

   ' use the Execute method of the Command
   ' object to open the recordset
   Set rst = cmd.Execute

   ' return company names to the Immediate window
   Do Until rst.EOF
     Debug.Print rst(1)
     rst.MoveNext
   Loop

   rst.Close
   Set rst = Nothing
   Set cmd = Nothing
   Set cat = Nothing
End Sub
```

3. Execute this procedure from the Immediate window by typing the following statement and pressing **Enter**:

```
Execute_ParamQuery_ADO "Argentina"
```

The Execute_ParamQuery_ADO procedure establishes the connection to the Northwind database. Next, the name of the query is supplied in the following statement:

```
Set cmd = cat.Procedures(strQryName).Command
```

Because this is a Parameter query, the parameter value is specified by using the Parameters collection of the Command object, like this:

```
cmd.Parameters("[Country Name]") = strCountry
```

Then, the Recordset object is opened by using the `Execute` method of the Command object:

```
Set rst = cmd.Execute
```

Finally, the procedure loops through the recordset to retrieve the company names and print them to the Immediate window. After running this procedure, the following lines are returned to the Immediate window for the specified country:

```
Cactus Comidas para llevar
Océano Atlántico Ltda.
Rancho grande
```

Instead of specifying the parameter values before the recordset is open, you can use the `Parameters` argument of the Command object's `Execute` method to pass the parameter value, as follows:

```
Set rst = cmd.Execute(Parameters:=strCountry)
```

CREATING AND RUNNING A MAKE-TABLE QUERY WITH DAO

A Make-Table query creates a new table out of records from one or more tables or queries. Make-Table queries are often used to preserve data as it existed at a particular time or to create a backup copy of a table without backing up the entire database. Use the `SELECT INTO` statement to create a Make-Table query. This statement consists of the following parts:

`SELECT fieldname`	Field name (use * for all fields)
`INTO newTableName`	Name of the new table
`FROM table/queryName`	Name of a table or query from which data is taken
`WHERE condition`	Criteria/limit operation to desired rows (optional)
`ORDER BY fieldname`	Order of the records in the new table (optional)

The procedure in Hands-On 16.9 creates a table of the customers in Brazil.

Hands-On 16.9. Creating and Running a Make-Table Query with DAO

1. In the Visual Basic Editor window, choose **Insert | Module**.
2. In the module's Code window, type the following **MakeATableQuery_DAO** procedure:

```
Sub MakeATableQuery_DAO()
  Dim db As DAO.Database
  Dim qdf As DAO.QueryDef
  Dim strSQL As String
  Dim strDb As String

  strDb = "C:\Access2013_ByExample\Northwind.mdb"

  On Error GoTo Err_Handler

  strSQL = "SELECT * INTO SouthAmericanClients " & _
    "FROM Customers WHERE Country='Brazil';"
  Set db = OpenDatabase(strDb)
  Set qdf = db.CreateQueryDef("", strSQL)
  qdf.Execute
ExitHere:
  Set qdf = Nothing
  db.Close
  Set db = Nothing
  Exit Sub
Err_Handler:
  MsgBox Err.Description
  Resume ExitHere
End Sub
```

3. Choose **Run | Run Sub/UserForm** to execute the procedure.

The `SELECT INTO` statement in the MakeATableQuery_DAO procedure is used to make a new table named SouthAmericanClients containing the names of all Brazilian customers from the Customers table in the Northwind.mdb database. Notice that by not assigning a name to the query, we create a Make-Table query that is temporary (not stored in the Access window):

```
Set qdf = db.CreateQueryDef("", strSQL)
```

CREATING AND RUNNING AN UPDATE QUERY WITH DAO

An Update query is a type of Action query. Update queries are very convenient to use when you want to change fields for a single record or for multiple records in a table. The UPDATE statement consists of the following three parts:

UPDATE	TableName or QueryName
SET	Expression/operation to perform
WHERE	Criteria/limit operation to desired rows

For example, to mark product 10 as discontinued, you would use the following UPDATE statement:

```
UPDATE Products SET Discontinued = True WHERE ProductID = 10
```

The criteria in the WHERE clause is used to determine which rows will be updated. The Update query does not produce a result table. To avoid updating the wrong records, always determine which rows you want to be updated by creating and running a Select query first.

The Execute method of a QueryDef object is used to run any type of Action query. The procedure in Hands-On 16.10 demonstrates how to create and run an Update query with DAO.

Hands-On 16.10. Creating and Running an Update Query with DAO

1. In the Visual Basic Editor window, choose **Insert | Module.**
2. In the module's Code window, type the following **CreateRunUpdateQuery_DAO** procedure:

```
Sub CreateRunUpdateQuery_DAO()
  Dim db As DAO.Database
  Dim qdf As DAO.QueryDef
  Dim strSQL As String
  Dim strDb As String

  strDb = "C:\Access2013_ByExample\Northwind.mdb"

  On Error GoTo Err_Handler

  strSQL = "UPDATE Suppliers INNER JOIN Products ON " & _
   "Suppliers.SupplierID = Products.SupplierID " & _
   "SET Products.UnitPrice = [UnitPrice]+2 " & _
   "WHERE (((Suppliers.CompanyName)='Tokyo Traders'));"
```

```
    Set db = OpenDatabase(strDb)

    Set qdf = db.CreateQueryDef("PriceIncrease", strSQL)
    qdf.Execute
ExitHere:
    Set db = Nothing
    Exit Sub
Err_Handler:
    If Err.Number = 3012 Then
      MsgBox "Query with this name already exists."
    Else
      MsgBox Err.Description
    End If
    Resume ExitHere
End Sub
```

To perform the required update, this procedure needs to join two tables. The Products table is joined with the Suppliers table on the SupplierID field that exists in both tables. Use the `INNER JOIN` statement to combine column values from one row of a table with column values from another row of another (or the same) table to obtain a single row of data. The join condition is specified after the `ON` keyword and determines how the two tables are to be compared to each other to produce the join result. Because the update must occur only for a specific supplier, we also specify the supplier's company name in the `WHERE` clause.

3. Choose **Run | Run Sub/UserForm** to execute the procedure.

After running this procedure, the prices for all products supplied by Tokyo Traders are increased by $2.00.

The following procedure demonstrates how to use the `Execute` method of the DAO Database object to run an existing (previously saved) Update query.

```
Sub UpdateRun_DAO()
    Dim db As DAO.Database
    Dim strDb As String

    strDb = "C:\Access2013_ByExample\Northwind.mdb"

    Set db = OpenDatabase(strDb)
    db.Execute "PriceIncrease"
    db.Close
    Set db = Nothing
End Sub
```

EXECUTING AN UPDATE QUERY WITH ADO

Executing bulk queries that update data is quite easy with ADO. You can use the `Execute` method of the Connection or Command object. The procedure in Hands-On 16.11 uses the Connection object's `Execute` method to update records in the Products table of the Northwind.mdb database where CategoryId is equal to 8. The UnitPrice of the records that match this condition will be increased by one dollar. Note that the number of updated records is returned by the `Execute` method in the `NumOfRec` variable.

Hands-On 16.11. Executing an Update Query with ADO

1. In the Visual Basic Editor window, choose **Insert | Module**.
2. In the module's Code window, type the following **Execute_UpdateQuery_ADO** procedure:

```
Sub Execute_UpdateQuery_ADO()
  Dim conn As ADODB.Connection
  Dim NumOfRec As Integer
  Dim strPath As String

  strPath = CurrentProject.Path & "\Northwind.mdb"

  Set conn = New ADODB.Connection

  conn.Open "Provider=Microsoft.Jet.OLEDB.4.0;" & _
   "Data Source=" & strPath

  conn.Execute "UPDATE Products " & _
   "SET UnitPrice = UnitPrice + 1" & _
   " WHERE CategoryId = 8", _
   NumOfRec, adExecuteNoRecords

  MsgBox NumOfRec & " records were updated."
  conn.Close
  Set conn = Nothing
End Sub
```

3. Choose **Run | Run Sub/UserForm** to execute the procedure.

This procedure uses the Data Manipulation Language (DML) `UPDATE` statement to make a change in the UnitPrice field of the Products table. The `Execute` method of the Connection object allows the provider to return the number of records that were affected via the RecordsAffected parameter. This parameter applies only to Action queries or stored procedures. To get the number of records returned by a result-returning query or stored procedure, you must use the RecordCount property. In the Execute_UpdateQuery_ADO procedure, we store the number of records affected in the string variable `NumOfRec`. Note that when a command does not return a recordset, you should include the constant `adExecuteNoRecords`. The `adExecuteNoRecords` constant can only be passed as an optional parameter to the Command or Connection object's `Execute` method.

The procedure in Hands-On 16.12 demonstrates how to execute an Update query by using the ADO Command object instead of the Connection object used in the preceding example. Running the following example will increase the UnitPrice of all the records in the Products table by 10 percent.

Hands-On 16.12. Executing an Update Query Using the Command Object

1. In the Visual Basic Editor window, choose **Insert | Module**.
2. In the module's Code window, type the **Execute_UpdateQuery2_ADO** procedure shown here:

```
Sub Execute_UpdateQuery2_ADO()
  Dim cmd As ADODB.Command
  Dim NumOfRec As Integer
  Dim strPath As String

  strPath = CurrentProject.Path & "\Northwind.mdb"

  Set cmd = New ADODB.Command
  With cmd
    .ActiveConnection = _
     "Provider=Microsoft.Jet.OLEDB.4.0;" & _
     "Data Source=" & strPath
    .CommandText = "Update Products " & _
     "Set UnitPrice = UnitPrice *1.1"
    .Execute NumOfRec, adExecuteNoRecords
  End With
  MsgBox NumOfRec
  Set cmd = Nothing
End Sub
```

3. Choose **Run | Run Sub/UserForm** to execute the procedure.

Non-row-returning Queries

Queries that do not return records, such as Action queries or Data Definition Language (DDL) queries, are known as non-row-returning queries.

- Action queries are Data Manipulation Language (DML) queries that perform bulk operations on a set of records. They allow you to add, update, or delete records.
- DDL queries are used for creating database objects and altering the structure of a database.
- Use the ADOX Procedure object to work with queries that don't return records. All Procedure objects are contained in the Procedures collection of the ADOX Catalog object. To save these types of queries in a database, append the ADO Command object to the ADOX Procedures collection.

RUNNING AN APPEND QUERY WITH DAO/ADO

Append queries are used for adding records from one or more tables to other tables. You can append records to a table in a current database or another Access or non-Access database. An Append query is an Action query that adds new records to the end of an existing table or query. Append queries don't return records. They are useful for archiving records. Before you can archive the records, you need to create a new table structure to hold the records. To add a record or multiple records to a table, use the `INSERT INTO` statement. This statement has the following parts:

`INSERT INTO target [(Field1, Field2)]`	The name of the table or query to which records are appended. You may indicate the names of the fields to which data is appended.
`SELECT fieldName(s)`	The names of fields from which data is obtained.
`FROM tableName` or `expression`	The name of the table or tables from which records are inserted, the name of a saved query, or a `SELECT` statement.
`WHERE condition`	Criteria/limit operation to desired rows.

The procedure in Hands-On 16.13 demonstrates how to execute an Append query using the `Execute` method of the DAO Database object.

Hands-On 16.13. Running an Append Query with DAO

1. In the Visual Basic Editor window, choose **Insert | Module.**
2. In the module's Code window, type the following **RunAppendQry_DAO** procedure:

```
Sub RunAppendQry_DAO()
  Dim db As DAO.Database
  Dim rst As DAO.Recordset
  Dim strSQL As String
  Dim strDb As String

  strDb = "C:\Access2013_ByExample\Northwind.mdb"

  strSQL = "SELECT * FROM SouthAmericanClients " & _
   "WHERE Country = 'Argentina'"
  Set db = OpenDatabase(strDb)
  Set rst = db.OpenRecordset(strSQL, dbOpenSnapshot)

  If rst.EOF Or rst.BOF Then
    ' Argentina clients not found in destination
    ' table - proceed with insert
    db.Execute "INSERT INTO SouthAmericanClients " & _
     "SELECT * FROM Customers " & _
     "WHERE Country = 'Argentina'"
    MsgBox "Argentina clients have been appended."
  Else
    MsgBox "Clients from Argentina already exist " & _
     "in the destination table."
  End If

  rst.Close
  Set rst = Nothing
  db.Close
  Set db = Nothing
End Sub
```

3. Choose **Run | Run Sub/UserForm** to execute the procedure.

This procedure begins by opening the Northwind database and creating a Snapshot-type recordset based on the supplied SQL query string. Prior to executing the Append query that inserts customers from Argentina into the SouthAmericanClients table, we check the EOF and BOF properties of the DAO Recordset object to determine if the recordset contains any records. If `rst.EOF Or rst.BOF` is True, then there is no current record (the recordset is empty), so we go ahead and use the `Execute` method of the database object to add Argentina customers to the destination table.

The following procedure demonstrates how to execute an Append query using the `Execute` method of the ADO Connection object:

```
Sub RunAppendQry_ADO()
  Dim conn As ADODB.Connection
  Dim strSQL As String
  Dim recAffected As Long

  On Error Resume Next
  Set conn = New ADODB.Connection

  conn.Provider = "Microsoft.Jet.OLEDB.4.0"
  conn.Open "C:\Access2013_ByExample\Northwind.mdb"
  strSQL = "INSERT INTO SouthAmericanClients " & _
   "SELECT * FROM Customers " & _
   "WHERE Country = 'Venezuela'"

  conn.Execute strSQL, recAffected
  If Err <> 0 Then
    Debug.Print "Error Number: " & Err.Number
    Debug.Print "Error Description: " & Err.Description
  Else
   Debug.Print recAffected & " records were inserted."
  End If
  conn.Close
  Set conn = Nothing
End Sub
```

This procedure opens a connection to the Northwind.mdb database. Once the database is open, the `Execute` method of the ADO Connection object is used to execute the specified SQL `INSERT INTO` statement. You can use an optional RecordsAffected parameter (see the `recAffected` variable in the procedure) with the `Execute` method to determine the number of records that the `Execute` method affected. This parameter must be a variable of the Long data type. If the Insert operation was successful, the VBA Err object will return zero (0). The default property of the Err object is Number. Therefore, the statement:

```
If Err <> 0
```

is equivalent to:

```
If Err.Number <> 0
```

If a runtime error occurs, for example, the destination table does not exist, the procedure will print to the Immediate window the error number and its description text. If there were no errors, the Immediate window will contain the number of records that were affected by the Insert operation.

RUNNING A DELETE QUERY WITH DAO

With a Delete query you can delete a single record or multiple records from a database. The `DELETE` statement used to delete rows from a table consists of the following three parts:

DELETE	
FROM	Table name
WHERE	Criteria/limit operation to desired rows

For example, to delete discontinued products from the Products table you would use the following `DELETE` statement:

```
DELETE FROM Products WHERE Discontinued = True
```

To delete all the rows from the Products table, the following statement can be executed:

```
DELETE FROM Products
```

You cannot reverse the operation performed by the `DELETE` statement. Always make a backup copy of your table prior to running a Delete query. It is a good idea to create and run a Select query before using `DELETE` to see which rows will be affected by the Delete operation.

The `Execute` method is used to run Action queries or execute an SQL statement. This method can take optional arguments. For example, in the statement:

```
qdf.Execute dbFailOnError
```

the constant `dbFailOnError` will generate a runtime error and will roll back updates or deletes if an error occurs. Use the RecordsAffected property of the QueryDef object to determine the number of records affected by the most recent `Execute` method. For example, the following statement displays the number of records that were deleted:

```
MsgBox qdf.RecordsAffected & " records were deleted."
```

The procedure in Hands-On 16.14 creates a Delete query and then executes it if the user responds positively to the message shown in Figure 16.2.

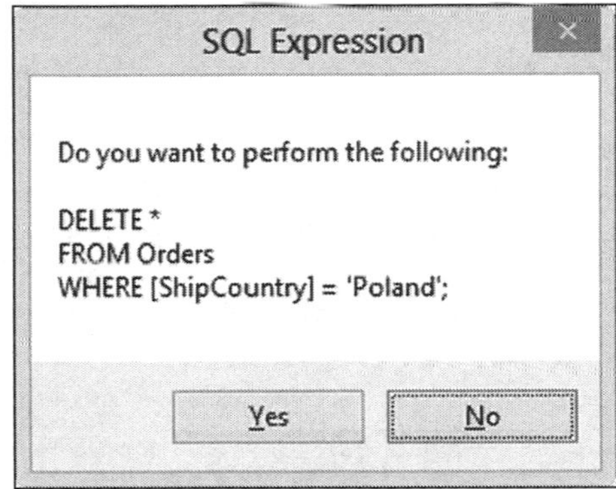

FIGURE 16.2 You can display an SQL statement underlying a query in a message box.

Hands-On 16.14. Running a Delete Query with DAO

1. In the Visual Basic Editor window, choose **Insert | Module**.
2. In the module's Code window, type the following **CreateRunDeleteQuery_DAO** procedure:

```
Sub CreateRunDeleteQuery_DAO()
  Dim db As DAO.Database
  Dim qdf As DAO.QueryDef
  Dim strQryName As String
  Dim strSQL As String
  Dim strDb As String

  strDb = "C:\Access2013_ByExample\Northwind.mdb"

  On Error GoTo ErrorHandler

  strQryName = "DeletePolishOrders"
  strSQL = "DELETE * FROM Orders "
```

```
    strSQL = strSQL & "WHERE [ShipCountry] = 'Poland'"

    Set db = OpenDatabase(strDb)
    Set qdf = db.CreateQueryDef(strQryName, strSQL)

    ' Chr(13) & Chr(13) is a double carriage return
    If (MsgBox("Do you want to perform the following: " & _
     Chr(13) & Chr(13) _
     & qdf.SQL, vbYesNo + vbDefaultButton2, _
     "SQL Expression")) = vbYes Then

    qdf.Execute dbFailOnError
    MsgBox qdf.RecordsAffected & _
     " records were deleted."
    End If
  ExitHere:
    Set qdf = Nothing
    db.Close
    Set db = Nothing
    Exit Sub
  ErrorHandler:
    If Err.Number = 3012 Then
      Set qdf = db.QueryDefs(strQryName)
      Resume Next
    Else
      MsgBox Err.Number & ":" & Err.Description
      Resume ExitHere
    End If
  End Sub
```

3. Choose **Run | Run Sub/UserForm** to execute the procedure.

   This procedure creates a Delete query named DeletePolishOrders in the Northwind database, then runs this query when the user clicks OK to the message. If the specified Delete query already exists in the database, the `qdf` object variable is set to the existing query name and the user is prompted to make a decision about whether the records should be deleted.

CREATING AND RUNNING A PASS-THROUGH QUERY WITH DAO

A *Pass-Through query* works directly with an external ODBC (Open Database Connectivity) data source. Instead of linking to a table that resides on a server, you can send commands directly to the server to retrieve data.

To create a Pass-Through query manually in the Access window, choose Create | Query Design. Close the Show Table dialog box and click Design | Pass-Through. This will bring up a window where you can type a query statement. The SQL statement must be in the format understood by the external data source from which you are retrieving data. Pass-Through queries can also be used in lieu of Action queries when you need to bulk append, update, or delete data in remote databases.

Pass-Through queries can be created and executed programmatically from your VBA procedures. In DAO, use the Connect property to execute an SQL Pass-Through query. If you do not specify a connection string in the Connect property, Access will ask you for the connection information every time you run the Pass-Through query (and this can be very annoying).

The following procedure uses the MaxRecords property to return 15 records from the dbo.entity table located on an SQL server. Notice that the ReturnsRecords property is set to `True`. If your query does not need to return records, set the ReturnsRecords property to `False`.

```
Sub PassThruQry_DAO()
  Dim db As DAO.Database
  Dim qdfPass As DAO.QueryDef

  On Error GoTo err_PassThru

  Set db = CurrentDb
  Set qdfPass = db.CreateQueryDef("GetRecords")

  ' enter your own connect string
  ' supply the server database name you
  ' want to connect to, your User ID,
  ' password, and the Data Source name

  qdfPass.Connect = "ODBC;Database=myDbName; " & _
   "UID=JKO;PWD=tester;DSN=myDataS"
  qdfPass.SQL = "SELECT * FROM dbo.entity"
```

```
    qdfPass.ReturnsRecords = True
    qdfPass.MaxRecords = 15

    DoCmd.OpenQuery "GetRecords"
    Exit Sub
err_PassThru:
    If Err.Number = 3151 Then
      MsgBox Err.Description
      Exit Sub
    End If
    db.QueryDefs.Delete "GetRecords"
    Resume 0
    Exit Sub
End Sub
```

Instead of displaying a datasheet with the records retrieved from the SQL database, the following procedure reads the records to a temporary query and proceeds to open a recordset based on that query. Next, the contents of two fields are printed to the Immediate window.

```
Sub PassThru2()
    Dim db As DAO.Database
    Dim qdfPass As DAO.QueryDef
    Dim rstTemp As DAO.Recordset

    On Error GoTo err_PassThru

    Set db = CurrentDb
    Set qdfPass = db.CreateQueryDef("")

    ' enter your own connect string
    ' supply the server database name you
    ' want to connect to, your User ID,
    ' password, and the Data Source name

    qdfPass.Connect = "ODBC;Database=myDbName;UID=JKO;" & _
     "PWD=tester;DSN=myDataS"
    qdfPass.SQL = "SELECT * FROM dbo.entity"
    qdfPass.ReturnsRecords = True
    qdfPass.MaxRecords = 15
    Set rstTemp = qdfPass.OpenRecordset()
    ' print data from two fields to the Immediate window
```

```
    With rstTemp
      Do While Not .EOF
        Debug.Print .Fields("entity_id"), .Fields("entity_name")
        .MoveNext
      Loop
      .Close
    End With
    SendKeys "^g"
  ExitHere:
    Set db = Nothing
    Exit Sub
  err_PassThru:
    MsgBox Err.Number & ":" & Err.Description
    Resume ExitHere
  End Sub
```

CREATING AND EXECUTING A PASS-THROUGH QUERY WITH ADO

As mentioned earlier, SQL Pass-Through queries are SQL statements that are sent directly to the database server for processing. In earlier versions of Microsoft Access, Pass-Through queries were used with Data Access Objects (DAO) to increase performance when accessing external ODBC data sources. In ADO, you can use the Microsoft OLE DB Provider for SQL Server to directly access the SQL Server. For this reason, you do not need to create Pass-Through queries. However, since it is possible to create a Pass-Through query using ADOX and Microsoft Jet Provider, the next hands-on exercise demonstrates how to do this.

Hands-On 16.15. Creating a Pass-Through Query with ADO

This hands-on exercise requires that you have access to an SQL Server Northwind database and that you make changes in the connection string to point to your server.

1. In the Visual Basic Editor window, choose **Insert | Module**.
2. In the module's Code window, type the **Create_PassThroughQuery** procedure shown here:

```
Sub Create_PassThroughQuery()
  Dim cat As ADOX.Catalog
  Dim cmd As ADODB.Command
```

```
Dim rst As ADODB.Recordset
Dim strPath As String
Dim strSQL As String
Dim strQryName As String
Dim strODBCConnect As String

On Error GoTo ErrorHandler

strSQL = "SELECT Customers.* FROM Customers WHERE " _
 & "Customers.Country='France';"

strQryName = "French Customers"

' modify the following string to connect
' to your SQL Server
strODBCConnect = "ODBC;Driver=SQL Server;" & _
 "Server=PROD15;" & _
 "Database=Northwind;" & _
 "UID=;" & _
 "PWD="

' strODBCConnect = "ODBC;DSN=ODBCNorth;UID=sa;PWD=;"

Set cat = New ADOX.Catalog
cat.ActiveConnection = CurrentProject.Connection

Set cmd = New ADODB.Command
With cmd
  .ActiveConnection = cat.ActiveConnection
  .CommandText = strSQL
  .Properties("Jet OLEDB:ODBC Pass-Through Statement") = True
  .Properties("Jet OLEDB:Pass-Through Query Connect String") _
  = strODBCConnect
End With

cat.Procedures.Append strQryName, cmd

Set cmd = Nothing
Set cat = Nothing
MsgBox "The procedure completed successfully.", _
 vbInformation, "Create Pass-Through Query"
Exit Sub
```

```
ErrorHandler:
  If InStr(Err.Description, "already exists") Then
    cat.Procedures.Delete strQryName
    Resume
  Else
    MsgBox Err.Number & ": " & Err.Description
  End If
End Sub
```

3. Choose **Run | Run Sub/UserForm** to execute the procedure.

This procedure creates a Pass-Through query named French Customers in the current database. Notice that to connect to the SQL Server database, the following string is built and later assigned to the Jet OLEDB:Pass-Through Query Connect String property of the Command object:

```
strODBCConnect = "ODBC;Driver=SQL Server;" & _
 "Server=PROD15;" & _
 "Database=Northwind;" & _
 "UID=;" & _
 "PWD="
```

Needless to say, if you want to try this procedure, you must have access to a remote data source (such as an SQL Server database) and you'll need to modify the preceding string to point to your server. This string allows you to connect via the DSN-less connection. If you prefer, you may build your connection string to the remote data source using the DSN that you define in the Control Panel via Administrative Tools (ODBC). Your connection string could then look like this:

```
strODBCConnect = "ODBC;DSN=myDSN;UID=sa;PWD=;"
```

To create a Pass-Through query, you must also set two provider-specific properties of the Command object: Jet OLEDB:ODBC Pass-Through Statement and Jet OLEDB:Pass-Through Query Connect String.

To permanently store the Pass-Through query in your database, you need to append it to the Catalog object's Procedures collection, like this:

```
cat.Procedures.Append strQryName, cmd
```

After you run the Create_PassThroughQuery procedure, the query can be viewed and accessed from the navigation pane in the Microsoft Access window.

In Hands-On 16.15, you learned how to create a Pass-Through query in VBA with ADO. This query retrieved the list of French customers from the Northwind database located on the SQL Server. The Pass-Through query was named

French Customers and was saved permanently in the Chap16.accdb database. Let's see how you can execute this query from a VBA procedure.

Hands-On 16.16. Executing a Pass-Through Query Saved in Access (ADO)

1. In the Visual Basic Editor window, choose **Insert | Module.**
2. In the module's Code window, type the **Execute_PassThroughQuery_ADO** procedure shown here:

```
Sub Execute_PassThroughQuery_ADO()
  Dim cat As ADOX.Catalog
  Dim cmd As ADODB.Command
  Dim rst As ADODB.Recordset
  Dim strConnect As String

  ' modify the connection string to connect
  ' to your SQL Server Northwind database
  strConnect = "Provider=SQLOLEDB;" & _
   "Data Source=PROD15;" & _
   "Initial Catalog=Northwind;" & _
   "User Id=sa;" & _
   "Password="

  Set cat = New ADOX.Catalog
  cat.ActiveConnection = CurrentProject.Connection

  Set cmd = New ADODB.Command
  Set cmd = cat.Procedures("French Customers").Command
  Set rst = cmd.Execute

  Debug.Print "--French Customers Only--" & vbCrLf _
   & rst.GetString

  Set rst = Nothing
  Set cmd = Nothing
  Set cat = Nothing
End Sub
```

3. Choose **Run | Run Sub/UserForm** to execute the procedure.

The procedure begins by building a connection string to the SQL Server database. This is a standard connection that uses the native OLE DB SQL Server

Provider (SQLOLEDB). This connection requires that you also provide the name of the SQL Server (Data Source), the name of the database from which to retrieve records (Initial Catalog), and the security context with which to log in (User ID, Password). If you connect to your SQL Server database using the NT integrated security, your connection string will look like this:

```
strConnect = "Provider=SQLOLEDB;" & _
 "Data Source=yourServerName;" & _
 "Integrated Security=SSPI;" & _
 "Initial Catalog=Northwind"
```

Because the Pass-Through query you want to execute has been saved in the Access database, you need to open the ADOX Catalog object to access its Procedures collection. The following line of code specifies the name of the query you want to execute and assigns it to the Command object:

```
Set cmd = cat.Procedures("French Customers").Command
```

To execute a Pass-Through query that returns records, you need to use the Recordset object in addition to the Command object. The following statement executes the Pass-Through query:

```
Set rst = cmd.Execute
```

The Pass-Through query executes on the server. To quickly view data on the client machine, we retrieve the contents of the recordset by using the `GetString` method:

```
Debug.Print "--French Customers Only--" & vbCrLf _
 & rst.GetString
```

PERFORMING OTHER OPERATIONS WITH QUERIES

Now that you know how to programmatically create and run various queries using DAO and ADO objects, you may be interested to find out how to use Visual Basic to perform other operations related to queries, such as retrieving a list of queries and their properties, deleting a query, and determining if a query is updatable.

Retrieving Query Properties with DAO

Just like tables and other database objects, queries have properties. To generate a list of properties for a specific query, use the `For Each...Next` looping structure to iterate through the Properties collection of the DAO QueryDef object. The procedure in Hands-On 16.17 demonstrates this.

Hands-On 16.17. Listing Query Properties with DAO

1. In the Visual Basic Editor window, choose **Insert | Module**.
2. In the module's Code window, type the **List_QryProperties_DAO** procedure shown here:

```
Sub List_QryProperties_DAO()
  Dim db As DAO.Database
  Dim prp As DAO.Property
  Dim strDBName As String

  On Error Resume Next

  strDBName = "C:\Access2013_ByExample\Northwind 2007.accdb"
  Set db = OpenDatabase(strDBName)
  For Each prp In db.QueryDefs("Invoice Data").Properties
    Debug.Print prp.Name & "= " & prp.Value
  Next prp
  Set db = Nothing
End Sub
```

3. Choose **Run | Run Sub/UserForm** to execute the procedure.

The following output shows some of the properties printed to the Immediate window by running the code in the preceding List_QryProperties_DAO procedure:

```
Name= Invoice Data
DateCreated= 12/28/2006 5:30:24 PM
LastUpdated= 6/10/2007 6:03:43 PM
Type= 0
SQL= SELECT Orders.[Order ID], Orders.[Ship Name],
Orders.[Ship Address], Orders.[Ship City],
Orders.[Ship State/Province], Orders.[Ship ZIP/Postal Code],
Orders.[Ship Country/Region], Orders.[Customer ID],
Customers.Company AS [Customer Name], Customers.Address,
Customers.City, Customers.[State/Province],
Customers.[ZIP/Postal Code], Customers.[Country/Region],
[Employees Extended].[Employee Name] AS Salesperson,
Orders.[Order Date], Orders.[Shipped Date],
```

```
Shippers.Company AS [Shipper Name], [Order Details].[Product ID], Prod-
ucts.ID AS [Product ID], [Order Details].[Unit Price],
[Order Details].Quantity, [Order Details].Discount,
CCur(Nz([Unit Price]*[Quantity]*(1-[Discount]),0)/100)*100 AS Extended-
Price, Orders.[Shipping Fee], Products.[Product Name]
FROM (Shippers RIGHT JOIN (Customers RIGHT JOIN (Orders LEFT
JOIN [Employees Extended] ON
Orders.[Employee ID]=[Employees Extended].ID) ON Customers.ID=Orders.
[Customer ID]) ON
Shippers.ID=Orders.[Shipper ID]) LEFT JOIN
([Order Details] LEFT JOIN Products ON [Order Details].[Product
ID]=Products.ID) ON Orders.[Order ID]=[Order Details].[Order ID];

Updatable= True
Connect=
ReturnsRecords= True
ODBCTimeout= 60
RecordsAffected= 0
MaxRecords= 0
RecordsetType= 0
Description= (Criteria) Record source for Invoice report.
Based on six tables. Includes expressions that concatenate
first and last employee name and that use the CCur function
to calculate extended price.
OrderByOn= False
Orientation= 0
DefaultView= 2
FilterOnLoad= False
OrderByOnLoad= True
TotalsRow= False
```

Listing All Queries in a Database with DAO/ADO

You can obtain the listing of all queries in a database by using the `For...Each` loop to enumerate the QueryDefs collection of the DAO QueryDef object. The following procedure writes to the Immediate window the names of all queries in the Northwind 2007.accdb database.

```
Sub List_AllQueries_DAO()
  Dim db As DAO.Database
  Dim qdf As DAO.QueryDef
     Dim strDb As String
```

```
    strDb = "C:\Access2013_ByExample\Northwind 2007.accdb"
    Set db = OpenDatabase(strDb)
    For Each qdf In db.QueryDefs
      Debug.Print qdf.Name
    Next qdf

    Set qdf = Nothing
    db.Close
    Set db = Nothing
End Sub
```

The procedure in Hands-On 16.18 retrieves the names of all saved queries in the Northwind.mdb database by iterating through the View objects stored in the ADOX Catalog object's Views collection.

Hands-On 16.18. Listing Queries in a Database with ADO

1. In the Visual Basic Editor window, choose **Insert | Module**.
2. In the module's Code window, type the **List_AllQueries_ADO** procedure shown here:

```
Sub List_AllQueries_ADO()
  Dim cat As New ADOX.Catalog
  Dim v As ADOX.View
  Dim strPath As String

  strPath = CurrentProject.Path & "\Northwind.mdb"
  cat.ActiveConnection = _
   "Provider=Microsoft.Jet.OLEDB.4.0;" & _
   "Data Source= " & strPath

  For Each v In cat.Views
    Debug.Print v.Name
  Next

  Set cat = Nothing
End Sub
```

3. Choose **Run | Run Sub/UserForm** to execute the procedure.
4. After running this procedure, open the Immediate window to view the list of all saved queries in the Northwind.mdb database.

Deleting a Query from a Database with DAO/ADO

To remove a DAO QueryDef object from a QueryDefs collection, use the `Delete` method as shown in Hands-On 16.19. The DeleteAQuery_DAO procedure deletes the query that was created in Hands-On 16.1.

Hands-On 16.19. Deleting a Query from a Database (DAO)

1. In the Visual Basic Editor window, choose **Insert | Module**.
2. In the module's Code window, type the following **DeleteAQuery_DAO** procedure:

```
Sub DeleteAQuery_DAO()
  Dim db As DAO.Database
  Dim qdf As DAO.QueryDef
  Dim strDb As String

  On Error GoTo ErrorHandler

  strDb = "C:\Access2013_ByExample\Northwind.mdb"
  Set db = OpenDatabase(strDb)
  db.QueryDefs.Delete "myQuery"
ExitHere:
  db.Close
  Set db = Nothing
  Exit Sub
ErrorHandler:
  MsgBox Err.Number & ": " & Err.Description
  Resume ExitHere
End Sub
```

3. Choose **Run | Run Sub/UserForm** to execute the procedure.

   After running the procedure in Hands-On 16.19, the query named myQuery is removed from the Northwind.mdb database.

To delete a query in ADO, use the `Delete` method of the Procedures or Views collection. By running the procedure in Hands-On 16.20, you can quickly delete the Redmond Employees query created in Hands-On 16.2.

Hands-On 16.20. Deleting a Query from a Database (ADO)

1. In the Visual Basic Editor window, choose **Insert | Module**.

2. In the module's Code window, type the following **DeleteAQuery_ADO** procedure:

```
Sub DeleteAQuery_ADO()
  Dim cat As New ADOX.Catalog
  Dim strPath As String

  On Error GoTo ErrorHandler

  strPath = CurrentProject.Path & "\Northwind 2007.accdb"
  cat.ActiveConnection = _
   "Provider=Microsoft.ACE.OLEDB.12.0;" & _
   "Data Source= " & strPath

  cat.Views.Delete "Redmond Employees"

ExitHere:
  Set cat = Nothing
  Exit Sub
ErrorHandler:
  If Err.Number = 3265 Then
    MsgBox "Query does not exist."
  Else
    MsgBox Err.Number & ": " & Err.Description
  End If
  Resume ExitHere
End Sub
```

3. Choose **Run | Run Sub/UserForm** to execute the procedure.

After running the procedure in Hands-On 16.20, the query named Redmond Employees is removed from the Northwind 2007.accdb database.

Determining if a Query Is Updatable

When a query is updatable you may edit the values in the result set of records and your changes are automatically reflected in the underlying tables. Microsoft Access's online help lists situations in which query results can or cannot be updated (see Figure 16.3). The DAO QueryDef object has an Updatable property that you can use in your VBA code to find out if the query definition can be updated. However, to determine whether the resulting recordset can be updated, you must use the Updatable property of the DAO Recordset object as demonstrated in Hands-On 16.21. If the Recordset object cannot be edited, the value of the Updatable property is `False`.

The Updatable property of the DAO Snapshot-type and Forward-only-type Recordset objects is always `False`*. The same is true if the Recordset object contains read-only fields. However, when one or more fields are updatable, the property's value is* `True`*. Because a recordset can contain fields that can't be updated, you may want to check the DataUpdatable property of each field in the Fields collection of the Recordset object before attempting to edit a record.*

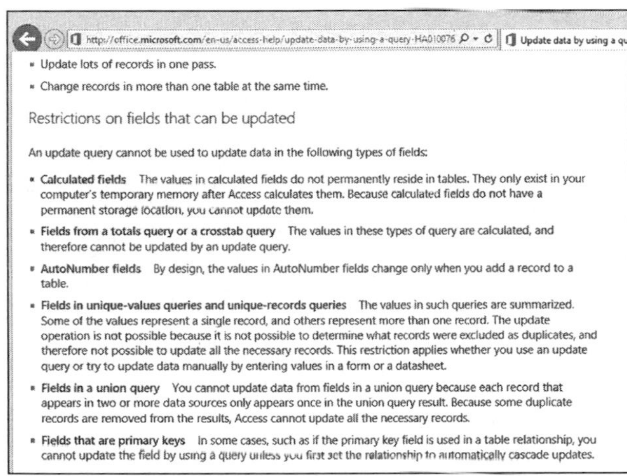

- Update lots of records in one pass.
- Change records in more than one table at the same time.

Restrictions on fields that can be updated

An update query cannot be used to update data in the following types of fields:

- **Calculated fields** The values in calculated fields do not permanently reside in tables. They only exist in your computer's temporary memory after Access calculates them. Because calculated fields do not have a permanent storage location, you cannot update them.
- **Fields from a totals query or a crosstab query** The values in these types of query are calculated, and therefore cannot be updated by an update query.
- **AutoNumber fields** By design, the values in AutoNumber fields change only when you add a record to a table.
- **Fields in unique-values queries and unique-records queries** The values in such queries are summarized. Some of the values represent a single record, and others represent more than one record. The update operation is not possible because it is not possible to determine what records were excluded as duplicates, and therefore not possible to update all the necessary records. This restriction applies whether you use an update query or try to update data manually by entering values in a form or a datasheet.
- **Fields in a union query** You cannot update data from fields in a union query because each record that appears in two or more data sources only appears once in the union query result. Because some duplicate records are removed from the results, Access cannot update all the necessary records.
- **Fields that are primary keys** In some cases, such as if the primary key field is used in a table relationship, you cannot update the field by using a query unless you first set the relationship to automatically cascade updates.

FIGURE 16.3. Records returned by a query may or may not be updatable.

For details, please see: *http://office.microsoft.com/en-us/access-help/update-data-by-using-a-query-HA010076527.aspx.*

The procedure in Hands-On 16.21 checks whether records returned by two queries in the Northwind.mdb database can be edited.

Hands-On 16.21. Determining if Records Returned by a Query Can Be Edited (DAO)

1. In the Visual Basic Editor window, choose **Insert | Module.**
2. In the module's Code window, type the following **IsQryUpdatable_DAO** procedure:

```
Sub IsQryUpdatable_DAO()
  Dim db As DAO.Database
  Dim rst As DAO.Recordset
  Dim fld As DAO.Field
```

```
    Dim strDb As String
    Dim strQryName1 As String
    Dim strQryName2 As String

    strDb = "C:\Access2013_ByExample\Northwind.mdb"
    strQryName1 = "Order Subtotals"
    strQryName2 = "Invoices"

    Set db = OpenDatabase(strDb)

    Set rst = db.OpenRecordset(strQryName1)
    Debug.Print strQryName1 & _
     ": Updatable=" & rst.Updatable
    Set rst = db.OpenRecordset(strQryName2)
    Debug.Print strQryName2 & _
     ": Updatable=" & rst.Updatable
    For Each fld In rst.Fields
      If Not fld.DataUpdatable Then
        Debug.Print fld.Name & " cannot be edited."
      End If
    Next

    rst.Close
    Set rst = Nothing
    db.Close
    Set db = Nothing
  End Sub
```

3. Choose **Run | Run Sub/UserForm** to execute the procedure.
   When you run this procedure, the Updatable property returns True for the Invoices query and False for the Order Subtotals query. The `OpenRecordset` method is used to open each of these queries. The Order Subtotals query is not updatable because its SQL statement contains a `GROUP BY` clause. While the Invoices query is updatable, not all fields in the resulting recordset can be edited (see Figure 16.4).

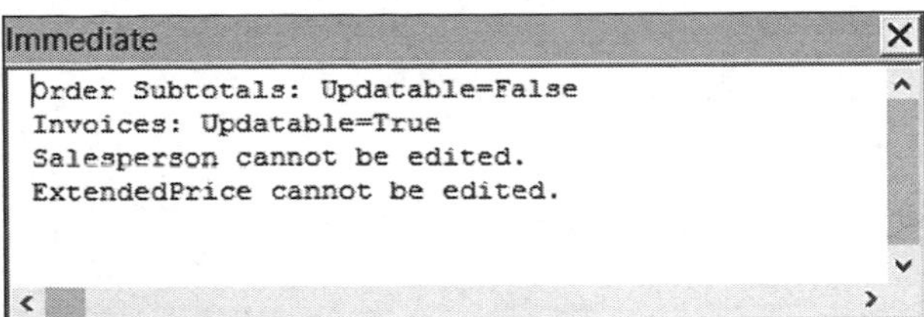

FIGURE 16.4. An updatable query can contain one or more fields that cannot be edited (see Hands-On 16.21).

CHAPTER SUMMARY

Creating and executing queries are the most frequently performed database operations. This chapter has shown you how to create, run, and modify various types of queries using the DAO and ADO code.

In the next chapter, you will learn more about the advanced features of the ADO/DAO Object Model.

Chapter 17 Using Advanced ADO/DAO Features

At this point you should feel pretty comfortable using ADO in most of your Microsoft Access programming endeavors. By using the knowledge you've acquired in the last few chapters, you can switch to any other Office application (Excel, Word, PowerPoint, or Outlook) and start programming. Because you already know the ADO methods of accessing databases and manipulating records, all you need to learn is the object model that the specific application is using. Learning a new type library is not very difficult. Recall that VBA offers the Object Browser that lists all the application's objects, properties, methods, and intrinsic constants that you may need for writing code. However, if you'd like to accomplish more with ADO, this chapter will introduce you to a couple of more advanced ADO features that will set you apart from beginning programmers. You will learn about fabricating, persisting, disconnecting, cloning, and shaping recordsets. You will also learn how to process data modifications and additions by using ADO and DAO transactions.

FABRICATING A RECORDSET

In previous chapters, you worked with recordsets that were created from data that came from a Microsoft Access database, a text or dBASE file, an Excel spreadsheet, or a Word document. You may have also practiced working with a recordset generated from an SQL Server database. In each of these circumstances, to get the necessary data you needed to establish a connection to the appropriate data source. In other words, you worked with recordsets that had a live connection to the data source. These connected recordsets obtained their structure and data from a query to a data source to which they were connected. But what if you need to create a recordset with data that does not come from a data source? As you may recall from Chapter 11, "Data Access Technologies in Microsoft Access," the ADO Object Model allows you to work with both relational and nonrelational data stores.

To store nonrelational data in an ADO Recordset, you can create your recordset from scratch. This recordset will be defined programmatically in memory and will not be connected to any data source. For example, you can easily fabricate a custom recordset that holds nonrelational data, such as the information about the files located in one of your hard drive's directories.

When you create your own recordset from scratch, you define the types of fields in the recordset and then populate the recordset with the information you want. The fields are defined using the Fields collection's `Append` method. You must specify the field name and the data type. The syntax for the `Append` method looks like this:

```
Fields.Append Name, DataType[, FieldSize], [Attribute]
```

Arguments in square brackets are optional. `FieldSize` specifies the size in characters or bytes. `Attribute` specifies characteristics such as whether the field enables Null values or whether it is a primary key or an identity column.

Once you have defined the structure of your recordset, simply open it and populate it with the desired data. You can add data to your custom recordset in the same way you add data to a connected recordset: by using the Recordset object's `AddNew` method.

The procedure in Hands-On 17.1 demonstrates how to create an empty recordset containing three fields (Name, Size, and Modified) and then populate it with files located in a user-specified file folder.

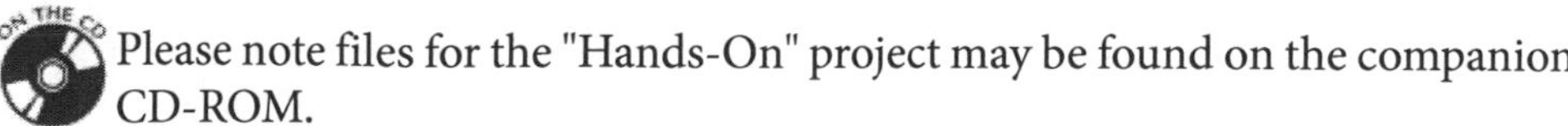
Please note files for the "Hands-On" project may be found on the companion CD-ROM.

Hands-On 17.1. Creating a Custom Recordset (ADO)

1. Start Microsoft Access and create a new database named **Chap17.accdb** in your **C:\Access2013_ByExample** folder.
2. In the database window, press **Alt+F11** to switch to the Visual Basic Editor window.
3. In the Visual Basic Editor window, choose **Tools | References**. In the References dialog box, locate and select **Microsoft ActiveX Data Objects 6.1 Library (or an earlier version)** and then click **OK** to close this dialog box.
4. In the Visual Basic Editor window, choose **Insert | Module**.
5. In the module's Code window, type the following **Custom_Recordset** procedure:

```
Sub Custom_Recordset()
  Dim rst As ADODB.Recordset
  Dim strFile As String
  Dim strPath As String
  Dim strFolder As String

  Const MyFolder = "C:\Access2013_ByExample"
  strPath = InputBox("Enter pathname, e.g., " & MyFolder, _
   "Enter the Folder Name", MyFolder)
  If Right(strPath, 1) <> "\" Then strPath = strPath & "\"
  strFolder = strPath
  strFile = Dir(strPath & "*.*")
  If strFile = "" Then
    MsgBox "This folder does not contain files."
    Exit Sub
  End If
  Set rst = New ADODB.Recordset
  ' Create an empty recordset with 3 fields
  With rst
    Set .ActiveConnection = Nothing
    .CursorLocation = adUseClient
    With .Fields
      .Append "Name", adVarChar, 255
      .Append "Size", adDouble
      .Append "Modified", adDBTimeStamp
    End With
    .Open
    Do While strFile <> ""
```

```
        If strFile = "" Then Exit Do
        ' Add a new record to the recordset
        .AddNew Array("Name", "Size", "Modified"), _
        Array(strFile, FileLen(strFolder & strFile), _
        FileDateTime(strFolder & strFile))
        strFile = Dir
      Loop
      .MoveFirst
      ' Print the contents of the recordset
      ' to the Immediate window
      Do Until .EOF
        Debug.Print !Name & vbTab & !Size & vbTab & !Modified
        .MoveNext
      Loop
      .Close
    End With
    Set rst = Nothing
  End Sub
```

6. Choose **Run | Run Sub/UserForm** to execute the procedure.

In the Custom_Recordset procedure, we start by creating a Recordset object variable. To tell ADO that your recordset is not connected to any database, we set the ActiveConnection property of the Recordset object to `Nothing`. We also set the CursorLocation property to `adUseClient` to indicate that the processing will occur on the client machine as opposed to the database server. Next, we determine what columns the recordset should contain and add these columns to the Recordset's Fields collection by using the `Append` method. Once the structure of your recordset is defined, you can call the `Open` method to actually open your custom recordset. Now you can populate the recordset with the data you want. We obtain the data by looping through the folder the user specified in the input box and reading the information about each file. The VBA `FileLen` function is used to retrieve the size of a file in bytes. Another VBA function, `FileDateTime`, is used to retrieve the date and time a file was last modified. To retrieve the date and time separately, use the `FileDateTime` function as an argument of the `DateValue` or `TimeValue` functions.

Check the following statements in the Immediate window while stepping through the Custom_Recordset procedure:

```
? DateValue(FileDateTime(myFolder & myFile))
? TimeValue(FileDateTime(myFolder & myFile))
```

Now that the recordset is fabricated and populated with the required data, you can display its contents in the Immediate window or send the output to another application. You can also save the recordset to a disk file as explained later in this chapter.

DISCONNECTED RECORDSETS

In the previous section, you learned how to create a recordset from scratch. This recordset had a structure custom-defined by you and was populated with data that did not come from a database. In other words, it was a disconnected recordset that was defined on the fly. A *disconnected recordset* is a recordset that is not connected to a data source. A disconnected recordset can be defined programmatically (as you saw in Hands-On 17.1) or it can get its information from the data source (as shown in Hands-On 17.2).

Using disconnected recordsets allows you to connect to a database, retrieve some records, return the records to the client, and then disconnect from the database. By keeping your connection to a database open just long enough to obtain the required data, you can help conserve valuable server resources. You can work with the disconnected recordset offline and then connect to the database again to add your changes.

To get started using disconnected recordsets, perform Hands-On 17.2. The example procedure retrieves some data from the Orders table in the Northwind database, then disconnects from the database. While disconnected from the database, you can manipulate and examine the content of the retrieved recordset.

Hands-On 17.2. Creating a Disconnected Recordset (ADO)

1. In the Visual Basic Editor window, choose **Insert | Module**.
2. In the module's Code window, type the following **Rst_Disconnected** procedure:

```
Sub Rst_Disconnected()
  Dim conn As ADODB.Connection
  Dim rst As ADODB.Recordset
  Dim strConn As String
  Dim strSQL As String
  Dim strRst As String
  Dim strFilePath As String
```

```
    strFilePath = "C:\Access2013_ByExample\Northwind.mdb"
    strSQL = "SELECT * FROM Orders WHERE CustomerID = 'VINET'"

    strConn = "Provider=Microsoft.Jet.OLEDB.4.0;"
    strConn = strConn & "Data Source = " & strFilePath
    Set conn = New ADODB.Connection
    conn.ConnectionString = strConn
    conn.Open

    Set rst = New ADODB.Recordset
    Set rst.ActiveConnection = conn

    ' retrieve the data
    rst.CursorLocation = adUseClient
    rst.LockType = adLockBatchOptimistic
    rst.CursorType = adOpenStatic
    rst.Open strSQL, , , , adCmdText

    ' disconnect the recordset
    Set rst.ActiveConnection = Nothing

    ' change the CustomerID in the first record to 'OCEAN'
    rst.MoveFirst
    Debug.Print rst.Fields(0) & " was previously: " _
      & rst.Fields(1)
    rst.Fields("CustomerID").Value = "OCEAN"
    rst.Update

    ' stream out the recordset as a comma-delimited string
    strRst = rst.GetString(adClipString, , ",")
    Debug.Print strRst
  End Sub
```

3. Choose **Run | Run Sub/UserForm** to execute the procedure.

Notice that to create a disconnected recordset that gets its data from a data source, you need to set the CursorLocation, LockType, and CursorType properties of the Recordset object. CursorLocation should be set to `adUseClient`. This setting indicates that the cursor will reside on the client computer that is creating the recordset. Set LockType to `adLockBatchOptimistic` to enable multiple records to be updated. Finally, set CursorType to `adOpenStatic` to retrieve the snapshot of the data.

To disconnect a recordset, you must set the Recordset object's ActiveConnection property to `Nothing` after you've called the Recordset's `Open` method.

When the recordset is disconnected from the database, you can freely manipulate its data or pass it to another application or process. In this example procedure, we manipulate our recordset by changing the value of the CustomerID field in the first retrieved record from VINET to OCEAN. Then we create a comma-delimited string using the Recordset object's `GetString` method. The content of the disconnected recordset is then printed out to the Immediate window, as shown here:

```
10274 was previously: VINET
10274,OCEAN,6,8/6/1996,9/3/1996,8/16/1996,1,6.01,
Vins et alcools Chevalier,59 rue de l'Abbaye,Reims,,51100,France
10295,VINET,2,9/2/1996,9/30/1996,9/10/1996,2,1.15,
Vins et alcools Chevalier,59 rue de l'Abbaye,Reims,,51100,France
10737,VINET,2,11/11/1997,12/9/1997,11/18/1997,2,7.79,
Vins et alcools Chevalier,59 rue de l'Abbaye,Reims,,51100,France
10739,VINET,3,11/12/1997,12/10/1997,11/17/1997,3,11.08,
Vins et alcools Chevalier,59 rue de l'Abbaye,Reims,,51100,France
```

SAVING A RECORDSET TO DISK

The ADO has a `Save` method that allows you to save a recordset to disk and work with it from your VBA application. This method takes two parameters. You must specify a filename and one of the following two data formats:

- `adPersistADTG`—Advanced Data TableGram
- `adPersistXML`—Extensible Markup Language

A *saved* (or *persisted*) *recordset* is a recordset that is saved to a file. This file can later be reopened without an active connection.

In this section, you will persist a recordset into a file using the `adPersistADTG` format. You will work with the `adPersistXML` format in Chapter 32, "XML Features in Access 2013."

To save a recordset in a file, you must first open it. When you have applied a filter to a recordset and then decide to save that recordset, only the filtered records will be saved. Using the `Save` method does not close the recordset. You can continue to work with the recordset after it has been saved. However, always remember to close the recordset when you are done working with it.

The procedure in Hands-On 17.3 opens the recordset based on the Customers table. Once the recordset is open, the `Save` method is called to persist the customer records into a file.

Hands-On 17.3. Saving Records to a Disk File (ADO)

1. In the Visual Basic Editor window, choose **Insert | Module.**
2. In the module's Code window, type the following **SaveRecordsToDisk** procedure:

```
Sub SaveRecordsToDisk()
  Dim conn As ADODB.Connection
  Dim rst As ADODB.Recordset
  Dim strFileName As String
  Dim strNorthPath As String

  strFileName = CurrentProject.Path & "\Companies.rst"
  strNorthPath = CurrentProject.Path & "\Northwind.mdb"

  On Error GoTo ErrorHandler

  Set conn = New ADODB.Connection

  With conn
    .Provider = "Microsoft.Jet.OLEDB.4.0"
    .ConnectionString = "Data Source = " & strNorthPath
    .Mode = adModeReadWrite
    .Open
  End With

  Set rst = New ADODB.Recordset
  With rst
    .CursorLocation = adUseClient
    ' Retrieve the data
    .Open "Customers", conn, _
    adOpenKeyset, adLockBatchOptimistic, adCmdTable

    ' Disconnect the recordset
    .ActiveConnection = Nothing

    ' Save the recordset to disk
    .Save strFileName, adPersistADTG
```

```
      .Close
    End With

    MsgBox "Records were saved in " & strFileName & "."
  ExitHere:
    ' Cleanup
    Set rst = Nothing
  Exit Sub
  ErrorHandler:
    If Not IsEmpty(Dir(strFileName)) Then
      Kill strFileName
      Resume
    Else
      MsgBox Err.Number & ": " & Err.Description
      Resume ExitHere
      End If
  End Sub
```

3. Choose **Run | Run Sub/UserForm** to execute the procedure.

   This procedure saves all the data located in the Customers table to a file with an .rst extension. We named this file Companies.rst, but you are free to choose any filename and extension while saving your recordset.

Persisted recordsets are very useful for populating combo boxes or listboxes, especially when the data is located on a server and does not change too often. You can update your data as needed by running a procedure that creates a new dump of the required records and deletes the old disk file. This way, your Access application can display the most recent data in its combo boxes or listboxes without having to connect to a database. Let's look at how you can fill a combo box with a saved recordset by working with Custom Project 17.1.

Custom Project 17.1. Filling a Combo Box with a Disconnected Recordset (ADO)

This custom project requires that you complete Hands-On 17.2.

1. Create an Access form as shown in Figure 17.1. Place a combo box control in the form. Change the Name property of this control to **cboCompany** and set the Caption property of the label control to **Company:**.

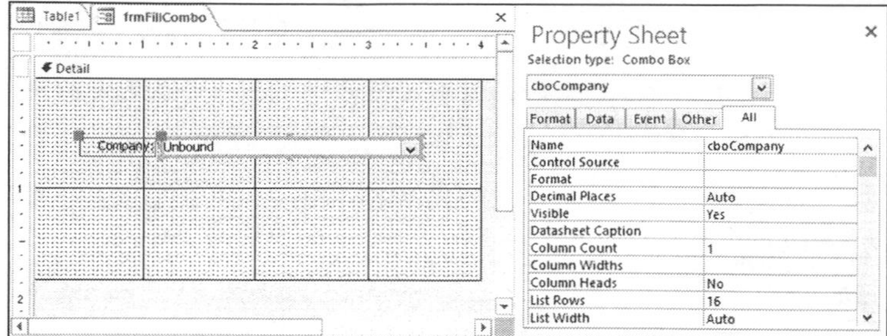

FIGURE 17.1. This custom form is used to demonstrate how you can fill the combo box control with a disconnected recordset.

2. Set the form's Caption property to **Disconnected combo**.
3. Save the form as **frmFillCombo**.
4. In the form's property sheet, activate the **Event** tab and click the button next to the **On Load** event name. In the Choose Builder dialog box, select **Code Builder** and click **OK**.
5. Complete the **Form_Load** procedure as shown here:

```
Private Sub Form_Load()
  Dim rst As ADODB.Recordset
  Dim strRowSource As String
  Dim strName As String

  strName = CurrentProject.Path & "\Companies.rst"

  Set rst = New ADODB.Recordset
  With rst
    .CursorLocation = adUseClient
    .Open strName, , , , adCmdFile
    Do Until .EOF
      strRowSource = strRowSource & rst!CompanyName & ";"
      .MoveNext
    Loop
    With Me.cboCompany
      .RowSourceType = "Value List"
      .RowSource = strRowSource
    End With

    .Close
  End With
  Set rst = Nothing
End Sub
```

6. Open the frmFillCombo form in Form view.
   To populate a combo box with values, the code in the Form_Load procedure changes the RowSourceType property of the combo box control to Value List and sets the RowSource property to the string obtained by iterating though the recordset. When the form opens, its caption is changed to Disconnected combo, as shown in Figure 17.2.
7. Close the Disconnected combo (frmFillCombo) form.

Persisted recordsets are especially handy when you need to support disconnected users or when you want to take data on the road with you. You can save the required set of records to a disk file, send it to your users in remote locations, or take it with you. While disconnected from the database, you or your users can view or modify the records. The next time you connect to the database you can update the original data with your changes using the `BatchUpdate` method. Custom Project 17.2 demonstrates this scenario.

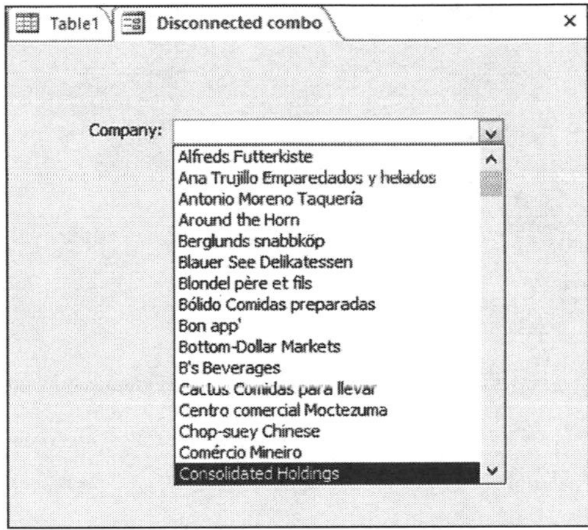

FIGURE 17.2. After opening the form prepared in Custom Project 17.1, the combo box is filled with the names of companies obtained via a persisted recordset.

Custom Project 17.2. Taking Persisted Data on the Road

This custom project requires that you complete Hands-On 17.2.

Part 1: Saving a Recordset to Disk

Before you can take a recordset on the road with you, you must save the records to a disk file. To create the data for this project, prepare and run the procedure in Hands-On 17.3. You should have the Companies.rst file available in your **C:\Access2013_ByExample** folder before you proceed to Part 2.

Part 2: Creating an Unbound Access Form to View and Modify Data

Once you've saved the recordset to a disk file, the recordset becomes portable. You can take the file with you on the road or send it to someone else. Before either one of you can view the data and modify it, however, you need some sort of a user interface. In this part, you will create an unbound Access form that will enable you to work with the file that contains the saved recordset.

1. Create a form as shown in Figure 17.3. Notice that this form contains only a couple of fields from the Customers table. This form serves only as an example. You can use as many fields as you have saved in the disk file.

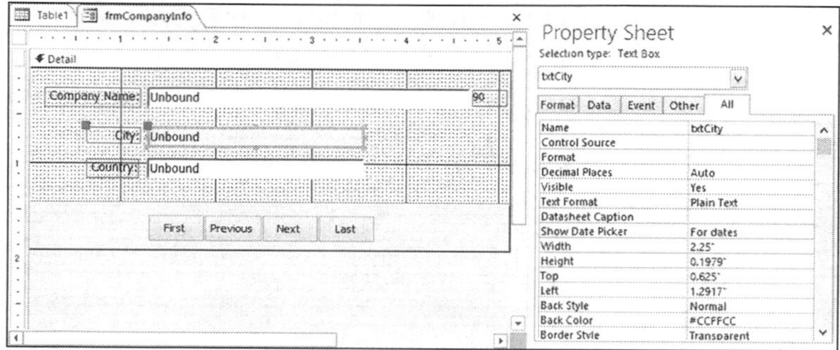

FIGURE 17.3. This custom form is used to demonstrate how you can use the saved recordset in an **unbound form.**

2. Set the following properties for the form's controls:

Object	Property	Setting
Label1	Caption	Company Name:
Text box next to the Company Name label	Name	txtCompany
Label2	Caption	City:
Text box next to the City label	Name Back Color	txtCity Select any color you like
Label3	Caption	Country:
Text box next to the Country label	Name	txtCountry
Label4	Caption Name	90 lbRecordNo
Command button 1	Caption Name	First cmdFirst
Command button 2	Caption Name	Previous cmdPrevious

Object	Property	Setting
Command button 3	Caption Name	Next cmdNext
Command button 4	Caption Name	Last cmdLast

We have set the Back Color property of the txtCity text box in the example application to visually indicate that the user can update only this field's data.

3. To visually match the form in Figure 17.3, draw a rectangle control over the command buttons and set its **Back Color** property to any color you like. Select the rectangle and choose **Arrange | Send to Back** to move the rectangle behind the command buttons.
4. In the property sheet, select **Form** from the drop-down list and activate the **Format** tab. Set the following properties for the form:

Property Name	Setting
Scroll Bars	Neither
Record Selectors	No
Navigation Buttons	No

5. Save the form as **frmCompanyInfo**.

Part 3: Writing Procedures to Control the Form and Its Data

Now that you've designed the form for your data, you need to write a couple of VBA procedures. The first procedure you'll write is an event procedure for the Form_Load event. This procedure will load the form with data from the persisted file. You will start by declaring a module-level Recordset object variable called `rst` and a module-level Integer variable called `counter`. You will also write Click procedures for all the command buttons and a procedure to fill the text boxes with the data from the current record in the recordset. Let's get started!

1. In the form's property sheet, activate the **Event** tab and click the **Build** button next to the **On Load** event name. In the Choose Builder dialog box, select **Code Builder** and click **OK**.
2. Enter the code for the **Form_Load** event procedure as shown here, starting with the declaration of module-level variables:

```
Dim rst As ADODB.Recordset
Dim counter As Integer
```

```
Private Sub Form_Load()
  Dim strFileName As String

  strFileName = CurrentProject.Path & "\Companies.rst"
  On Error GoTo ErrorHandler

  Set rst = New ADODB.Recordset
  With rst
    .CursorLocation = adUseClient
    .Open strFileName, , adOpenKeyset, _
     adLockBatchOptimistic, adCmdFile
  End With

  counter = 1
  Call FillTxtBoxes(rst, Me)

  With Me
 .txtCompany.SetFocus
 .cmdFirst.Enabled = False
    .cmdPrevious.Enabled = False
    .lbRecordNo.Caption = counter
  End With
  ExitHere:
  Exit Sub
ErrorHandler:
  MsgBox Err.Number & ": " & Err.Description
  Resume ExitHere
End Sub
```

The Form_Load event procedure loads Companies.rst from a disk file. To fill the text boxes with the data from the current record in the recordset, you need to write the following code:

```
With Me
  .txtCompany = rst!CompanyName
  .txtCity = rst!City
  .txtCountry = rst!Country
End With
```

Because the preceding code will need to be entered in several procedures in this application, you can save yourself a great deal of typing by placing this code in a subroutine and calling it like this:

```
Call FillTxtBoxes(rst, Me)
```

This statement calls the subroutine named FillTxtBoxes and passes it two arguments: the Recordset object variable and the reference to the current form. The FillTxtBoxes procedure (see Step 3) is entered in a standard module and contains the code shown in the next step.

The `counter` variable, which was declared at the module level, is initialized to the value of 1. We will use this variable to control the display of command buttons on the form. The Form_Load event procedure ends by setting the focus to the first text box (txtCompanyName) and disabling the first two command buttons. These buttons will not be required when the form first opens on the first record.

3. In the Visual Basic Editor Code window, choose **Insert | Module** and type the code of the following **FillTxtBoxes** procedure:

```
Sub FillTxtBoxes(ByVal rst As ADODB.Recordset, frm As Form)
  With frm
    .txtCompany = rst!CompanyName
    .txtCity = rst!City
    .txtCountry = rst!Country
  End With
End Sub
```

This procedure fills the three text boxes on the form with the data from the current record in the recordset. This procedure is called from the Form_Load event procedure and the Click event procedures for each command button.

4. In the Form_frmCompanyInfo Code window, type the following **Click** event procedure for the First command button:

```
Private Sub cmdFirst_Click()
  On Error GoTo Err_cmdFirst_Click

  rst.Update "City", Me.txtCity
  rst.MoveFirst

  Call FillTxtBoxes(rst, Me)

  With Me
    .txtCompany.SetFocus
    .cmdFirst.Enabled = False
    .cmdLast.Enabled = True
    .cmdPrevious.Enabled = False
    .cmdNext.Enabled = True
    counter = 1
```

```
      .lbRecordNo.Caption = counter
    End With
    Exit_cmdFirst_Click:
    Exit Sub
  Err_cmdFirst_Click:
    MsgBox Err.Description
    Resume Exit_cmdFirst_Click
  End Sub
```

5. In the Form_frmCompanyInfo Code window, type the following **Click** event procedure for the Next command button:

```
Private Sub cmdNext_Click()
  On Error GoTo Err_cmdNext_Click

  rst.Update "City", Me.txtCity
  rst.MoveNext
  counter = counter + 1

  Me.cmdFirst.Enabled = True

  Call FillTxtBoxes(rst, Me)

  Me.cmdPrevious.Enabled = True
  Me.lbRecordNo.Caption = counter
  Me.txtCompany.SetFocus
  If counter = rst.RecordCount Then
    Me.cmdNext.Enabled = False
    Me.cmdLast.Enabled = False
  End If

  Exit_cmdNext_Click:
  Exit Sub
Err_cmdNext_Click:
  MsgBox Err.Description
  Resume Exit_cmdNext_Click
End Sub
```

6. In the Form_frmCompanyInfo Code window, type the following **Click** event procedure for the Previous command button:

```
Private Sub cmdPrevious_Click()
  On Error GoTo Err_cmdPrevious_Click
```

```
    rst.Update "City", Me.txtCity
    rst.MovePrevious
    counter = counter - 1

    Call FillTxtBoxes(rst, Me)

    With Me
      .txtCompany.SetFocus
      .cmdLast.Enabled = True
      .cmdNext.Enabled = True
      .lbRecordNo.Caption = counter
    End With
    If counter = 1 Then
      Me.cmdFirst.Enabled = False
      Me.cmdPrevious.Enabled = False
    End If

    Exit_cmdPrevious_Click:
    Exit Sub
  Err_cmdPrevious_Click:
    MsgBox Err.Description
    Resume Exit_cmdPrevious_Click
  End Sub
```

7. In the Form_frmCompanyInfo Code window, type the following **Click** event procedure for the Last command button:

```
Private Sub cmdLast_Click()
  On Error GoTo Err_cmdLast_Click

  rst.Update "City", Me.txtCity
  rst.MoveLast

  Call FillTxtBoxes(rst, Me)

  With Me
    .txtCompany.SetFocus
    .cmdFirst.Enabled = True
    .cmdPrevious.Enabled = True
    .cmdLast.Enabled = False
    .cmdNext.Enabled = False
  End With
  counter = rst.RecordCount
```

```
    Me.lbRecordNo.Caption = counter
    Exit_cmdLast_Click:
    Exit Sub
  Err_cmdLast_Click:
    MsgBox Err.Description
    Resume Exit_cmdLast_Click
  End Sub
```

Notice that all the Click event procedures you prepared in Steps 4–7 contain the following line of code:

```
rst.Update "City", Me.txtCity
```

This statement updates the value of the City field in the recordset with the current value found in the txtCity text box on the form as you move through the records. Although the user can enter data in other text boxes, all modifications are ignored as there is no code in the Click event procedures that will allow changes to fields other than City. Of course, you can easily change this behavior by adding the necessary lines of code. Depending on which button was clicked, certain command buttons are disabled and others are enabled. This gives the user a visual clue of what actions are allowed at a particular moment.
To make the form work, we need to write one more event procedure. Before closing the form, we must make sure that the changes to the City field in the current record are saved and all changes in the City field we made while working with the form data are written back to the disk file. In other words, we must replace the Companies.rst disk file with a new file. This is done in the Form_Unload event procedure as shown in Step 8.

8. In the Form_frmCompanyInfo Code window, type the code of the **Form_Unload** event procedure as shown here:

```
Private Sub Form_Unload(Cancel As Integer)
  If rst.Fields("City").OriginalValue <> Me.txtCity Then
    rst.Update "City", Me.txtCity
  End If
  Kill (CurrentProject.Path & "\Companies.rst")
  rst.Save CurrentProject.Path & "\Companies.rst", _
  adPersistADTG
End Sub
```

ADO Recordsets have a special property called OriginalValue, which is used for storing original values that were retrieved from a database. These original values are left unchanged while you edit the recordset offline. Any changes to the data made locally are recorded using the Value property of the Recordset object. The OriginalValue property is updated with the values changed locally when you reconnect to the database and perform an UpdateBatch operation (see Part 5 in this custom project).

The Form_Unload event occurs when you attempt to close a form but before the form is actually removed from the screen. This is a good place to perform those operations that must be executed before the form is closed. In the Form_Unload procedure, we use the Recordset's OriginalValue property to check whether changes were made to the content of the City field in the current record. If OriginalValue is different from the value found in the current record's txtCity text box, we want to save the record by using the `Update` method of the recordset. Next, we delete the file containing the original recordset and save the current recordset to a file with the same name.

Part 4: Viewing and Editing Data Offline

Now that you've written all the procedures for the custom application, let's begin using the form to view and edit the data.

1. Open the **frmCompanyInfo** form in Form view.
2. In the first record, replace Berlin with **Drezden**.
3. Click the **Last** button, and replace Warszawa with **Opole**.
4. Click the **First** button and notice that the value of City is Drezden, just as you changed it in Step 2.
5. Use the **Next** button to move to the fourth record and replace London with **Dover**.
6. Close the form and then reopen it. Check the values in the City text box in the first, fourth, and last records. You should see Drezden, Dover, and Opole.
7. Close the frmCompanyInfo form.

Part 5: Connecting to a Database to Update the Original Data

After you've made changes to the data by using the custom form, you can send the file with the modified recordset to your database administrator so that he can update the underlying database with your changes. Let's write a procedure that will take care of this task.

The procedure that you are about to write will modify the Customers table in the Northwind database. I recommend that you take a few minutes now and create a copy of this database so that you can restore the original data later if necessary.

1. In the Visual Basic Editor window, choose Insert | Module.

2. In the module's Code window, type the following **UpdateDb** procedure:

```
Sub UpdateDb()
  Dim conn As ADODB.Connection
  Dim rst As ADODB.Recordset
  Dim strNorthPath As String
  Dim strRecStat As String

  On Error GoTo ErrorHandler
  strNorthPath = CurrentProject.Path & "\Northwind.mdb"

  ' Open the connection to the database
  Set conn = New ADODB.Connection
  With conn
    .Provider = "Microsoft.Jet.OLEDB.4.0"
    .ConnectionString = "Data Source = " & strNorthPath
    .Mode = adModeReadWrite
    .Open
  End With

  ' Open the recordset from the local file
  ' that was persisted to the hard drive
  ' and update the data source with the changes
  Set rst = New ADODB.Recordset
  With rst
    .CursorLocation = adUseClient
    .Open CurrentProject.Path & "\Companies.rst", conn, _
    adOpenKeyset, adLockBatchOptimistic, adCmdFile
    .UpdateBatch adAffectAll

    ' Check if there were records with conflicts
    ' during the update
    .Filter = adFilterAffectedRecords
    Do Until .EOF

      strRecStat = strRecStat & " " & rst!City & ":" & rst.Status
      .MoveNext
    Loop
    .Close
    Debug.Print strRecStat
  End With
```

```
  ExitHere:
  Set rst = Nothing
  Set conn = Nothing
  Exit Sub
ErrorHandler:
  MsgBox Err.Number & ": " & Err.Description
  Resume ExitHere
End Sub
```

3. Choose **Run | Run Sub/UserForm** to execute the procedure.

In the UpdateDb procedure, we used the `UpdateBatch` method of the ADO Recordset object to update the underlying database with the changes we made to the data while working with it offline. The `UpdateBatch` method takes an optional parameter that determines how many records will be affected by the update. This parameter can be one of the constants shown in Table 17.1.

TABLE 17.1. Enumerated constants used with the UpdateBatch method

Constant	Value	Description
`adAffectCurrent`	1	Pending changes will be written only for the current record.
`adAffectGroup`	2	Pending changes will be written for the records that satisfy the current filter.
`adAffectAll`	3	Pending changes will be written for all the records in the recordset. This is the default.

When you update the data, your changes are compared with values that are currently in the database. The update will fail if the record was deleted or updated in the underlying database since the recordset was saved to disk. Therefore, after calling the `UpdateBatch` method, you should check the status of the records to locate records with conflicts. To do this, we must filter the recordset to see only the affected records:

```
rst.Filter = adFilterAffectedRecords
```

Next, we loop through the recordset and check the Status property of each record. This property can return different values, as shown in Table 17.2. You can locate these values in the Object Browser by typing RecordStatusEnum in the Search box.

TABLE 17.2. RecordStatusEnum constants returned by the Status property

Constant	Value	Description
adRecCanceled	256	The record was not saved because the operation was canceled.
adRecCantRelease	1024	The new record was not saved because the existing record was locked.
adRecConcurrencyViolation	2048	The record was not saved because optimistic concurrency was in use.
adRecDBDeleted	262144	The record has already been deleted from the data source.
adRecDeleted	4	The record was deleted.
adRecIntegrityViolation	4096	The record was not saved because the user violated integrity constraints.
adRecInvalid	16	The record was not saved because its bookmark is invalid.
adRecMaxChangesExceeded	8192	The record was not saved because there were too many pending changes.
adRecModified	2	The record was modified.
adRecMultipleChanges	64	The record was not saved because it would have affected multiple records.
adRecNew	1	The record is new.
adRecObjectOpen	16384	The record was not saved because of a conflict with an open storage object.
adRecOK	0	The record was successfully updated.
adRecOutOfMemory	32768	The record was not saved because the computer has run out of memory.
adRecPendingChanges	128	The record was not saved because it refers to a pending insert.

(Contd.)

Constant	Value	Description
`adRecPermissionDenied`	65536	The record was not saved because the user has insufficient permissions.
`adRecSchemaViolation`	131072	The record was not saved because it violates the structure of the underlying database.
`adRecUnmodified`	8	The record was not modified.

While iterating through the recordset you can add additional code to resolve any encountered conflicts or check, for example, the original value and the updated value of the fields in updated records. As mentioned earlier, the OriginalValue property returns the field value that existed prior to any changes (since the last `Update` method was called). You can cancel all pending updates by using the `CancelBatch` method.

When you execute the UpdateDb procedure, your changes are written to the database.

4. Open the Northwind database and review the content of the City field in the Customers table. You should see Drezden, Dover, and Opole in the first, fourth, and last records.
5. Close the Northwind database and the Access window in which it was displayed. Do not close the Chap17.accdb database.

This completes Custom Project 17.2 in which you learned how to:

- Save the recordset to disk with the `Save` method
- Create a custom form to view and edit the recordset data in the disk file
- Open the recordset from disk with the `Open` method
- Work with the recordset offline (view and edit data)
- Reopen the connection to the original database and write your changes with the `UpdateBatch` method

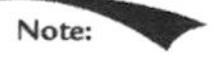

Refer to Chapter 32, "XML Features in Access 2013" to find out how you can save a recordset in XML format using the adPersistXML format.

CLONING A RECORDSET

Sometimes you may want to manipulate a recordset without losing the current position in the recordset. You can do this by *cloning* your original recordset. Use the ADO `Clone` method to create a recordset that is a copy of another recordset. You can create a recordset clone like this:

```
Dim rstOrg As ADODB.Recordset  ' your original recordset
Dim rstClone As ADODB.Recordset      ' cloned recordset
Set rstClone = rstOrg.Clone
```

As you can see from the assignment statement, the `rstClone` object variable contains a reference to the original recordset. After you've used the `Clone` method, you end up with two copies of the recordset that contain the same records but can be filtered and manipulated separately. You can create more than one clone of the original recordset.

Use the `Clone` method when you want to perform an operation on a recordset that requires multiple current records. The Clone object and the original Recordset object each have their own current records; therefore, the record pointers in the original and cloned recordsets can move independently of one another. And, because the clone points to the same set of data as the original, any changes made using either the original recordset or any of its clones will be visible in the original and its clones. However, the original recordset and its clones can get out of sync if you requery the original recordset against the database. When you close the original recordset, the clones remain open until you close them. Closing any of the clones does not close the original recordset.

Because the `Clone` method does not create another copy of the data (it only points to the data), cloning a recordset is faster and more efficient than opening a second recordset based on the same criteria. A recordset created by a method other than cloning will have a different set of bookmarks than the original recordset, even when it is based on the same SQL statement.

You can make a clone read-only by using an optional parameter like this:

```
Set rstClone = rstOrg.Clone(adLockReadOnly)
```

It's worth mentioning that you can only clone bookmarkable recordsets. Use the Recordset object's `Supports` method to find out if the recordset supports bookmarks (see the "Using Bookmarks" section in Chapter 14). If you try to clone a non-bookmarkable recordset, you will receive a runtime error. The clone and the original recordset have the same bookmarks, which you can share. A bookmark reference from one Recordset object refers to the same record in any of its clones.

Custom Project 17.3 demonstrates how the `Clone` method can be used to create a single form for displaying the current and previous records side by side (see Figure 17.4).

Custom Project 17.3. Displaying the Contents of the Current and Previous Record by Using the Clone Method

1. In the Microsoft Access window of the Chap17.accdb database, choose **External Data | Access.** In the File name box of the Get External Data dialog box, enter **C:\Access2013_ByExample\Northwind.mdb**, and then click **OK.** In the Import Objects window, select the **Customers** table and click **OK.** Click **Close** to exit the Get External Data dialog box.
2. Choose **Create | Form Design** and create a form like the one depicted in Figure 17.4. The following steps will help you set up the form and its control properties.

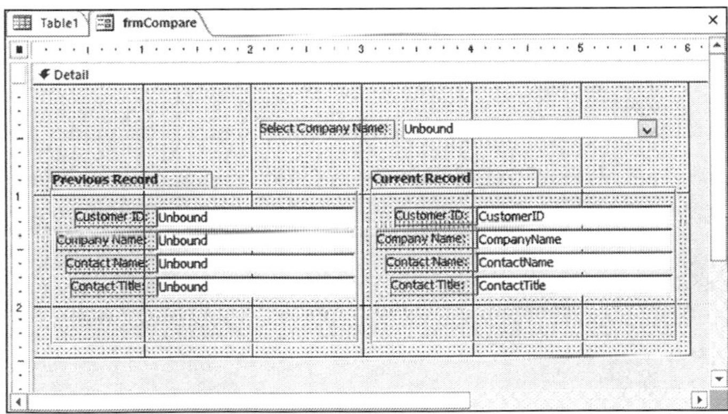

FIGURE 17.4. This custom form is used to demonstrate how you can use recordset cloning to read the contents of the previous record.

3. In the Controls area of the Design tab, click the **Combo Box** control and click inside the form area to position it at the upper right as shown in Figure 17.4. In the Combo Box Wizard's first screen, choose the option button labeled **I want the combo box to look up the values in a table or query**. Click **Next.** Make sure the **Customers** table is selected and click **Next.** The fields available in the Customers table should appear. Move **CustomerID** and **CompanyName** from the Available Fields box to the Selected Fields box, and then click **Next.** Specify **CompanyName** as the Ascending sort order for your combo box, and then click **Next.** In the next wizard dialog, adjust the width of the combo box column to fit the longest company name and click **Finish.** Now you should see the combo box placed on your form.

4. Place the remaining controls on the form and set their properties as shown in the following table (properties of controls that are not listed do not need to be set for this example to work):

Object	Property	Setting
Label1 (in front of the combo box)	Caption Tag	Select Company Name: Cbo
Combo0 (created by the Combo Box Wizard)	Name Tag	CboCompany Cbo
Label2	Caption Tag	Previous Record PrevRec
Label3	Caption	Current Record
Rectangle: Box1	Tag	PrevRec
Label4 (in front of Text box 1)	Caption Tag	Customer ID: PrevRec
Text box 1	Name Tag Control Source	CustIdPrev PrevRec should be blank
Label5 (in front of Text box 2)	Caption Tag	Company Name: PrevRec
Text box 2	Name Tag Control Source	CompanyPrev PrevRec should be blank
Label6 (in front of Text box 3)	Caption Tag	Contact Name: PrevRec
Text box 3	Name Tag Control Source	ContactPrev PrevRec should be blank
Label7 (in front of Text box 4)	Caption Tag	Contact Title: PrevRec
Text box 4	Name Tag Control Source	TitlePrev PrevRec should be blank
Label8 (in front of Text box 5)	Caption	Customer ID:
Text box 5	Name Control Source	CustomerID CustomerID

(Contd.)

Object	Property	Setting
Label9 (in front of Text box 6)	Caption	Company Name:
Text box 6	Name Control Source	CompanyName CompanyName
Label10 (in front of Text box 7)	Caption	Contact Name:
Text box 7	Name Control Source	ContactName ContactName
Label11 (in front of Text box 8)	Caption	Contact Title:
Text box 8	Name Control Source	ContactTitle ContactTitle

5. In the property sheet, select **Form** from the drop-down box and set the following form properties:

Property Name	Setting
Caption	Record Comparison
Scroll Bars	Neither
Record Selectors	No
Navigation Buttons	No

6. Save the form as **frmCompare**.
7. Click the combo box control on the form to select it. Activate the **Event** tab in the property sheet and click to the right of the **AfterUpdate** event name. Select **[Event Procedure]** from the drop-down box, and then click the **Build** button (…) to activate the Code window. Complete the **cboCompany_AfterUpdate** procedure shown here:

```
Private Sub cboCompany_AfterUpdate()
  ' Find the record that matches the control.
  Dim rs As Object
  Dim c As Control

  On Error GoTo ErrHandle

  Set rs = Me.Recordset.Clone
  rs.FindFirst "[CustomerID] = '" & Me![cboCompany] & "'"
  If Not rs.EOF Then Me.Bookmark = rs.Bookmark
    ' Move to the previous record in the clone
    ' so that we can load the previous records'
    ' data in the form's text boxes
```

```
      rs.MovePrevious
      If Not rs.BOF Then
        For Each c In Me.Controls
          c.Visible = True
        Next
        With Me
          .CustIdPrev = rs.Fields(0).Value
          .CompanyPrev = rs.Fields(1)
          .ContactPrev = rs.Fields(2)
          .TitlePrev = rs.Fields(3)
        End With
    Else
      For Each c In Me.Controls
        If c.Tag = "PrevRec" Then
          c.Visible = False
        End If
      Next
    End If
    ExitHere:
    Exit Sub
  ErrHandle:
    MsgBox Err.Number & ":" & Err.Description
    Resume ExitHere
  End Sub
```

Notice that this event procedure begins by creating a clone of the form's recordset. Next, the `FindFirst` method is used to locate the customer record based on the entry selected in the combo box. To ensure that the form's record is in sync with the entry selected in the combo box, the following line of code moves the form's bookmark to the same location as the recordset clone's bookmark as long as we are not at the end of file (EOF):

```
If Not rs.EOF Then Me.Bookmark = rs.Bookmark
```

Next, the procedure ensures that the controls used to display the contents of the previous record are visible whenever the selected record is not the first record. The control's Tag property is used to allow easy selection of controls that need to be hidden or made visible.

8. Press **Ctrl+S** to save the current changes.
9. Test your form by opening it in Form view. Selecting a company name from the combo box should fill the text boxes under the Current Record label

with the selected company's data. The boxes under the Previous Record label should pull company data from the previous record.

Before we start working with this custom project, let's write a Form_Load event procedure to ensure that only the combo box and its label are visible when the form is opened.

10. In the Code window where you have written the cboCompany_AfterUpdate event procedure, select **Form** from the object drop-down box in the top-left corner. Select **Load** from the procedure drop-down box on the right. Complete the code for the **Form_Load** event procedure as shown here:

```
Private Sub Form_Load()
  Dim c As Control

  For Each c In Me.Controls
    If c.Tag <> "cbo" Then
      c.Visible = False
    End If
  Next
End Sub
```

11. Make sure there are no errors in your code by choosing **Debug | Compile Chap17**.
12. Save and close your form.
13. Open the **frmCompare** form in the Form view and test it by choosing various company names from the combo box.
14. Close the form.

Think of ways to improve this form. For example, add a set of controls and write additional code to display the next record.

INTRODUCTION TO DATA SHAPING

Designing database applications often requires that you pull information from multiple tables. For instance, to obtain a listing of customers and their orders, you must link the required tables with SQL JOIN statements as shown here:

```
SELECT Customers.CustomerID AS [Cust Id],
  Customers.CompanyName,
  Orders.OrderDate,
  [Order Details].OrderID,
  Products.ProductName,
```

```
    [Order Details].UnitPrice,
    [Order Details].Discount,
    CCur([Order Details].[UnitPrice]*[Quantity]*(1-[Discount])/100)*100
       AS [Extended Price]
FROM Products
    INNER JOIN ((Customers
    INNER JOIN Orders ON Customers.CustomerID = Orders.CustomerID)
    INNER JOIN [Order Details]
    ON Orders.OrderID = [Order Details].OrderID)
    ON Products.ProductID = [Order Details].ProductID
    ORDER BY Customers.CustomerID, Orders.OrderDate DESC;
```

When you execute this SQL statement in the Northwind.mdb database, your output will match Figure 17.5.

Query1

Customer I	Company Name	Order Dat	Order I	Product Name	Unit Pric	Discou	Extended Pr
ALFKI	Alfreds Futterkiste	09-Apr-1998	11011	Fløtemysost	$21.50	0%	$430.00
ALFKI	Alfreds Futterkiste	09-Apr-1998	11011	Escargots de Bourgogne	$13.25	5%	$503.50
ALFKI	Alfreds Futterkiste	16-Mar-1998	10952	Grandma's Boysenberry Spread	$25.00	5%	$380.00
ALFKI	Alfreds Futterkiste	16-Mar-1998	10952	Rössle Sauerkraut	$45.60	0%	$91.20
ALFKI	Alfreds Futterkiste	15-Jan-1998	10835	Original Frankfurter grüne Soße	$13.00	20%	$20.80
ALFKI	Alfreds Futterkiste	15-Jan-1998	10835	Raclette Courdavault	$55.00	0%	$825.00
ALFKI	Alfreds Futterkiste	13-Oct-1997	10702	Lakkalikööri	$18.00	0%	$270.00
ALFKI	Alfreds Futterkiste	13-Oct-1997	10702	Aniseed Syrup	$10.00	0%	$60.00
ALFKI	Alfreds Futterkiste	03-Oct-1997	10692	Vegie-spread	$43.90	0%	$878.00
ALFKI	Alfreds Futterkiste	25-Aug-1997	10643	Rössle Sauerkraut	$45.60	25%	$513.00
ALFKI	Alfreds Futterkiste	25-Aug-1997	10643	Chartreuse verte	$18.00	25%	$283.50
ALFKI	Alfreds Futterkiste	25-Aug-1997	10643	Spegesild	$12.00	25%	$18.00
ANATR	Ana Trujillo Emparedados y helados	04-Mar-1998	10926	Mozzarella di Giovanni	$34.80	0%	$348.00

Record: 1 of 2155 No Filter Search

FIGURE 17.5. When you use SQL JOIN statements you get a flat recordset with a lot of duplicate information.

When you output your data in a standard way by using the SQL `JOIN` syntax, you get a lot of duplicate information. You can eliminate this redundant information by using an advanced feature of ADO known as a shaped (or hierarchical) recordset.

Data shaping allows you to create recordsets within recordsets with a single ADO object. This sort of hierarchical data arrangement is often seen as a *parent-child relationship*. The parent recordset contains the child recordset. A child recordset can contain another child recordset, which is a grandchild of the original recordset. A parent-child relationship can be placed in an easy-to-read tree structure. You will produce such a structure in Custom Project 17.4 later in this chapter. For now, let's focus on learning some new concepts that will enable you to present your data in a format that's easy to view and navigate.

Writing a Simple SHAPE Statement

You can easily create a hierarchy of data by using a data shaping language. All you need to know is how to use the following three commands: SHAPE, APPEND, and RELATE. The basic syntax looks like this:

```
SHAPE {parent-command}
APPEND ({child-command} [[AS] table-alias]
RELATE (parent-column TO child-column)
```

parent-command and child-command are often SQL SELECT statements that pull the data from the required tables. Let's look at the following example that uses the preceding syntax:

```
SHAPE {SELECT CustomerID AS [Cust Id], CompanyName AS (Company)
Customers}
 APPEND ({SELECT CustomerId, OrderDate, OrderId, Freight FROM Orders}
 AS custOrders
 RELATE (CustomerID TO CustomerID)
```

The preceding statement is a shaped recordset. This statement selects two fields from the Customers table and four fields from the Orders table. By using this SHAPE statement, you can list all orders for each of the customers in the Customers table without returning any redundant information.

Notice that there are two SELECT statements in this recordset:

- The first SELECT statement is the parent recordset. This recordset retrieves the data from the Customers table. Notice this SELECT statement is surrounded by curly braces and preceded by the SHAPE command, which defines a recordset.
- The second SELECT statement is the child recordset. It gets the data from the Orders table. Notice that this SELECT statement is also surrounded by curly braces; however, it is preceded by the APPEND clause and an opening parenthesis. The APPEND clause will add the child recordset to the parent.

When you append a child recordset to the parent recordset, a new field (column) is created in a parent recordset. This field is called a chapter column and has a data type called adChapter. You can use the AS *clause to assign a name to the chapter column. If the appended column has no chapter alias, a name will be generated for it automatically. In our example, the chapter column is called custOrders. Always specify an alias for your child recordset if you are planning to refer to it later in your code.*

After specifying the SELECT statement for the child recordset, you must indicate how you want the two recordsets to be linked. You do this with the RELATE clause. The column (CustomerID) from the parent recordset is related to the column (CustomerID) of the child recordset. Notice that you don't have to specify table names in the RELATE clause. Always specify the name of the parent column first.

The fields you use to relate parent and child recordsets must be in both recordsets. For example, you could not relate both recordsets if you did not select CustomerID from the Orders table.

Finally, remember to place a closing parenthesis at the end of the statement.

Working with Data Shaping

To work with data shaping in your VBA procedure, you need two providers: one for the data shaping functionality and the other for the data itself. Therefore, before you can create shaped (hierarchical) recordsets in your programs, you will need to specify:

- The name of a service provider

  The data shaping functionality is provided by the data shaping service for OLE DB. The name of this service provider is MSDataShape and it is specified as the value of the Connection object's Provider property like this:

  ```
  conn.Provider = "MSDataShape"
  ```

  or it can be a connection string like this:

  ```
  "Provider=MSDataShape"
  ```

- The name of a data provider

  Because a shaped recordset needs to be populated with rows of data, you must specify the name of a data provider as the value of the DataProvider property of the Connection object:

  ```
  conn.DataProvider = "Microsoft.Jet.OLEDB.4.0;"
  ```

  or in the connection string like this:

  ```
  "Data Provider=Microsoft.Jet.OLEDB.4.0;"
  ```

The following is a code fragment from the procedure in Hands-On 17.4 that demonstrates how to specify the names of the data and service providers:

```
' define database connection string
' using the OLE DB provider
```

```
' and Northwind database as Data Source
strConn = "Data Provider=Microsoft.Jet.OLEDB.4.0;"
strConn = strConn & "Data Source = " & _
  "C:\Access2013 ByExample\Northwind.mdb"
' specify Data Shaping provider
' and open connection to the database
Set conn = New ADODB.Connection
With conn
  .ConnectionString = strConn
  .Provider = "MSDataShape"
  .Open
End With
```

Data Shaping with Other Databases

The data shaping service creates a shaped (hierarchical) recordset from any data supplied by a data provider. In order to provide shaped data from a database other than Microsoft Access, let's say, an SQL Server database, a connection string might look like this:

```
Dim conn As ADODB.Connection
Set conn = New ADODB.Connection
conn.Open = "Provider = MSDataShape;" & _
 "Data Provider = SQLOLEDB;" & _
 "Server=myServerName;" & _
 "Initial Catalog = Northwind;" & _
 "User ID = myId; Password="
```

or like this:

```
Dim conn As ADODB.Connection
Set conn = New ADODB.Connection
conn.Provider = "MSDataShape"
conn.Open "Data Provider=SQLOLEDB; Integrated Security=SSPI;" & _
"Database=Northwind"
```

In Hands-On 17.4, you learn how to create a shaped recordset in a VBA procedure and display hierarchical data in the Immediate window (see Figure 17.6).

Hands-On 17.4. Creating a Shaped Recordset (ADO)

1. In the Visual Basic Editor window of the Chap17 database, choose **Insert | Module.**
2. In the module's Code window, enter the **ShapeDemo** procedure shown here:

```
Sub ShapeDemo()
  Dim conn As ADODB.Connection
  Dim rst As ADODB.Recordset
  Dim rstChapter As Variant
  Dim strConn As String
  Dim shpCmd As String

  ' define database connection string
  ' using the OLE DB provider
  ' and Northwind database as Data Source
  strConn = "Data Provider=Microsoft.Jet.OLEDB.4.0;"
  strConn = strConn & "Data Source = " & _
   "C:\Access2013_ByExample\Northwind.mdb"

  ' specify Data Shaping provider
  ' and open connection to the database
  Set conn = New ADODB.Connection
  With conn
    .ConnectionString = strConn
    .Provider = "MSDataShape"
    .Open
  End With
  ' define the SHAPE command for the shaped recordset
  shpCmd = "SHAPE {SELECT CustomerID AS [Cust Id], " & _
   " CompanyName AS Company FROM Customers}" & _
   " APPEND ({SELECT CustomerID, OrderDate," & _
   " OrderID, Freight FROM Orders} AS custOrders" & _
   " RELATE [Cust Id] TO CustomerID)"

  ' create and open the parent recordset
  ' using the open connection
  Set rst = New ADODB.Recordset
  rst.Open shpCmd, conn

  ' output data from the parent recordset
```

```
    Do While Not rst.EOF
      Debug.Print rst("Cust Id"); _
      Tab; rst("Company")
      rstChapter = rst("custOrders")
      ' write out column headings
      ' for the child recordset
      Debug.Print Tab; _
       "OrderDate", "Order #", "Freight"
      ' output data from the child recordset
      Do While Not rstChapter.EOF
        Debug.Print Tab; _
         rstChapter("OrderDate"), _
         rstChapter("OrderID"), _
         Format(rstChapter("Freight"), "$ #.##")
        rstChapter.MoveNext
      Loop
      rst.MoveNext
    Loop

    ' Cleanup
    rst.Close
    Set rst = Nothing
    Set conn = Nothing
  End Sub
```

3. Choose **Run | Run Sub/UserForm** to execute the procedure.

This procedure begins by specifying the data provider and data source name in the `strConn` variable. Next, we define a new ADO Connection object and set the ConnectionString property of this object to the `strConn` variable. Now that we have the data provider name and also know which database we need to pull the data from, we specify the data shaping service provider. This is done by using the Provider property of the Connection object. We set this property to MSDataShape, which is the name of the service provider for the hierarchical recordsets. Now we are ready to actually open a connection to the database. Before we can pull the required data from the database, we define the shaped recordset statement and store it in the `shpCmd` String variable. Next, we create a new Recordset object and open it using the open database connection. Then, we populate it with the content of the `shpCmd` variable like this:

```
Set rst = New ADODB.Recordset
rst.Open shpCmd, conn
```

Now that we have filled the hierarchical recordset, we begin to loop through the parent recordset. The first statement in the loop:

```
Debug.Print rst("Cust Id"); Tab; rst("Company")
```

will write out the customer ID (Cust Id) and the company name (Company) to the Immediate window.

In the second statement in the loop:

```
rstChapter = rst("custOrders")
```

we create a Recordset object variable based on the value of the custOrders field. As you recall from an earlier discussion, custOrders is an alias for the child recordset. The object variable (`rstChapter`) can be any name you like as long as it's not a VBA keyword.

Because a child recordset is simply a field in a parent recordset, when you retrieve the value of that field you will get the entire recordset filtered to include only the related records.

Before iterating through the child recordset, the column headings are output to the Immediate window for the fields we want to display. This way it is much easier to understand the meaning of the data in the child recordset. The next block of code loops through the child recordset and dumps the data to the Immediate window under the appropriate column heading. Once the data is retrieved for each parent record, we can close the recordset and release the memory.

```
Immediate
TORTU          Tortuga Restaurante
               OrderDate      Order #        Freight
               8/8/1996        10276         $ 13.84
               8/29/1996       10293         $ 21.18
               9/12/1996       10304         $ 63.79
               10/2/1996       10319         $ 64.5
               4/25/1997       10518         $ 218.15
               6/23/1997       10576         $ 18.56
               9/22/1997       10676         $ 2.01
               1/20/1998       10842         $ 54.42
               2/27/1998       10915         $ 3.51
               5/4/1998        11069         $ 15.67
TRADH          Tradição Hipermercados
               OrderDate      Order #        Freight
               7/5/1996        10249         $ 11.61
               8/28/1996       10292         $ 1.35
               4/4/1997        10496         $ 46.77
               7/22/1997       10606         $ 79.4
               1/13/1998       10830         $ 81.83
               1/15/1998       10834         $ 29.78
               1/19/1998       10839         $ 35.43
TRAIH          Trail's Head Gourmet Provisioners
               OrderDate      Order #        Freight
               6/19/1997       10574         $ 37.6
```

FIGURE 17.6. After running the ShapeDemo procedure in Hands-On 17.4, you can see the contents of the hierarchical recordset in the Immediate window.

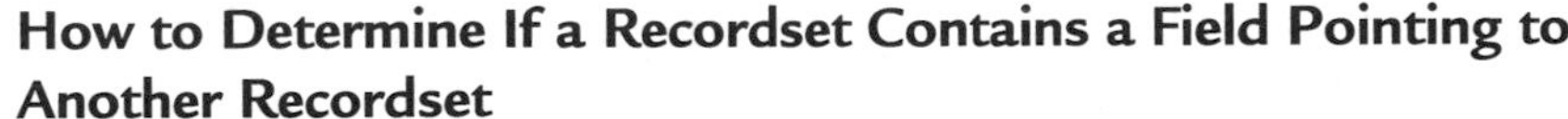

How to Determine If a Recordset Contains a Field Pointing to Another Recordset

To find out if a certain recordset contains another recordset, you can use the following conditional statement:

```
Dim rst as New ADODB.Recordset
If rst.Fields("custOrders").Type = adChapter then
    Debug.Print "This is a child recordset"
End If
```

custOrders is the chapter column alias you created with the AS *clause while appending a child recordset to the parent.*

Writing a Complex SHAPE Statement

In the previous section, you worked with a simple SHAPE statement that displayed order information for each customer in the Northwind.mdb database in the Immediate window. You learned how to nest a child recordset within a parent recordset and access the fields in both. In the following sections, you will learn how to write more complex SHAPE statements that include multiple child and grandchild recordsets.

Shaped Recordsets with Multiple Children

Data shaping does not limit you to having just one child recordset within a parent recordset. You can specify as many children as you want. For example, to display a parent with two children, use the following syntax:

```
SHAPE {SELECT * FROM Parent}
APPEND ({SELECT * FROM Child1}
RELATE parent-column TO child1-column) AS child1-alias,
({SELECT * FROM Child2}
RELATE parent-column TO child2-column) AS child2-alias
```

Notice that additional children (siblings) are added to the end of the APPEND clause.

Suppose you want to display both the orders and products for a customer in the Northwind database. Using the syntax provided earlier, you can shape your hierarchical recordset as demonstrated in the ShapeMultiChildren procedure shown in Hands-On 17.5.

Hands-On 17.5. Creating a Shaped Recordset with Multiple Children (ADO)

1. In the Visual Basic Editor window of the Chap17 database, choose **Insert | Module**.
2. In the module's Code window, enter the **ShapeMultiChildren** procedure shown here:

```
Sub ShapeMultiChildren()
  Dim conn As ADODB.Connection
  Dim rst As ADODB.Recordset
  Dim rstChapter1 As Variant
  Dim rstChapter2 As Variant
  Dim strConn As String
  Dim shpCmd As String
  Dim strParent As String
  Dim strChild1 As String
  Dim strChild2 As String
  Dim strLink As String
  Dim str1stChildName As String
  Dim str2ndChildName As String

  ' define database connection string
  ' using the OLE DB provider
  ' and Northwind database as Data Source
  strConn = "Data Provider=Microsoft.Jet.OLEDB.4.0;"
  strConn = strConn & "Data Source = " & _
   "C:\Access2013_ByExample\Northwind.mdb"

  ' specify Data Shaping provider
  ' and open connection to the database
  Set conn = New ADODB.Connection
  With conn
    .ConnectionString = strConn
    .Provider = "MSDataShape"
    .Open
  End With

  ' define the SHAPE command for the shaped recordset

  strParent = "SELECT CustomerID AS [Cust Id], " & _
   "CompanyName AS Company FROM Customers"
```

```
strChild1 = "SELECT CustomerID, OrderDate," & _
 "OrderID, Freight FROM Orders"

strChild2 = "SELECT Customers.CustomerID," & _
 "Products.ProductName FROM Products " & _
 "INNER JOIN ((Customers INNER JOIN Orders ON " & _
 "Customers.CustomerID = Orders.CustomerID) " & _
 "INNER JOIN [Order Details] ON " & _
 "Orders.OrderID = [Order Details].OrderID) ON " & _
 "Products.ProductID = [Order Details].ProductID " & _
 "Order By Products.ProductName"

str1stChildName = "custOrders"
str2ndChildName = "custProducts"

strLink = "RELATE [Cust Id] TO CustomerID"

shpCmd = "SHAPE {"
shpCmd = shpCmd & strParent
shpCmd = shpCmd & "}"
shpCmd = shpCmd & " APPEND ({"
shpCmd = shpCmd & strChild1
shpCmd = shpCmd & "}"
shpCmd = shpCmd & strLink
shpCmd = shpCmd & ")"
shpCmd = shpCmd & " AS " & str1stChildName
shpCmd = shpCmd & ", ({"
shpCmd = shpCmd & strChild2
shpCmd = shpCmd & "} "
shpCmd = shpCmd & strLink
shpCmd = shpCmd & ")"
shpCmd = shpCmd & " AS " & str2ndChildName

' create and open the parent recordset
' using the open connection
Set rst = New ADODB.Recordset
rst.Open shpCmd, conn

' output data from the parent recordset
Do While Not rst.EOF
  Debug.Print rst("Cust Id"); Tab; rst("Company")
  rstChapter1 = rst("custOrders")
```

```
    ' write out column headings
    ' for the 1st child recordset
    Debug.Print Tab(4); " (" & rst("Cust Id") & " Orders)"
    Debug.Print Tab; "OrderDate", "Order #", "Freight"

    ' output data from the 1st child recordset
    Do While Not rstChapter1.EOF
      Debug.Print Tab; _
      rstChapter1("OrderDate"), _
      rstChapter1("OrderID"), _
      Format(rstChapter1("Freight"), "$ #,#0.00")
      rstChapter1.MoveNext
    Loop

    rstChapter2 = rst("custProducts")
    ' write out column headings
    ' for the 2nd child recordset
    Debug.Print Tab(4); " (" & rst("Cust Id") & " Products)"

    ' output data from the 2nd child recordset
    Do While Not rstChapter2.EOF
      Debug.Print Tab; _
      rstChapter2("ProductName")
      rstChapter2.MoveNext
    Loop
    rst.MoveNext
  Loop

  ' Cleanup
  rst.Close
  Set rst = Nothing
  Set conn = Nothing
  End Sub
```

3. Choose **Run | Run Sub/UserForm** to execute the procedure.

   The SHAPE statement in this procedure has been specially formatted so that you can easily create any shaped recordset containing multiple children by replacing SELECT statements with your own. This procedure produces the output in the Immediate window as shown in Figure 17.7. Notice that each customer has two child records: Orders and Products.

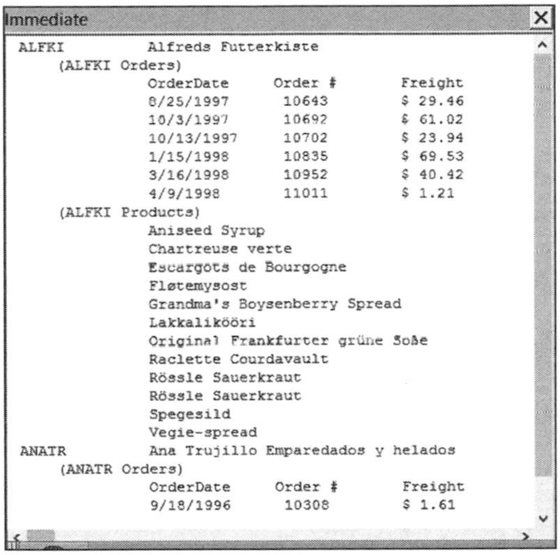

```
Immediate
ALFKI          Alfreds Futterkiste
    (ALFKI Orders)
               OrderDate      Order #        Freight
               8/25/1997       10643         $ 29.46
               10/3/1997       10692         $ 61.02
               10/13/1997      10702         $ 23.94
               1/15/1998       10835         $ 69.53
               3/16/1998       10952         $ 40.42
               4/9/1998        11011         $ 1.21
    (ALFKI Products)
               Aniseed Syrup
               Chartreuse verte
               Escargots de Bourgogne
               Fløtemysost
               Grandma's Boysenberry Spread
               Lakkalikööri
               Original Frankfurter grüne Soße
               Raclette Courdavault
               Rössle Sauerkraut
               Rössle Sauerkraut
               Spegesild
               Vegie-spread
ANATR          Ana Trujillo Emparedados y helados
    (ANATR Orders)
               OrderDate      Order #        Freight
               9/18/1996       10308         $ 1.61
```

FIGURE 17.7. After running the ShapeMultiChildren procedure in Hands-On 17.5, you can see the output of the hierarchical recordset with multiple children in the Immediate window.

Shaped Recordsets with Grandchildren

In addition to the parent recordset having multiple children, the child recordset can contain a child of its own. Simply put, your hierarchical recordset can contain grandchildren. Creating such a hierarchy is a bit harder, but it can be tackled in no time if you take a step-by-step approach. The `SHAPE` syntax that includes grandchildren looks like this:

```
SHAPE {SELECT * FROM Parent}
APPEND ((SHAPE {SELECT * FROM Child}
APPEND ({SELECT * FROM Grandchild}
RELATE child-column TO grandchild-column) AS grandchild-alias)
RELATE parent-column TO child-column) as child-alias
```

Notice that when grandchildren are present, the child recordset is appended with another `SHAPE` command.

Although you can have as many children or grandchildren as you want, it will be more difficult to write a `SHAPE` statement that uses more than three or four levels.

In Custom Project 17.4, you create a shaped recordset that contains both children and grandchildren. Next, you display this recordset on the Access form in the ActiveX TreeView control (see Figure 17.11. for the final output). This project will also introduce you to using aggregate functions within your shaped recordsets.

Custom Project 17.4. Using Hierarchical Recordsets

Part 1: Creating a Form with a TreeView Control

1. In the Access window of the Chap17.accdb database, choose **Create | Form Design**. The Form design window opens.
2. In the Controls area of the Design tab, click the More button in the Scroll area, and choose **ActiveX Controls** (see Figures 17.8 and 17.9).

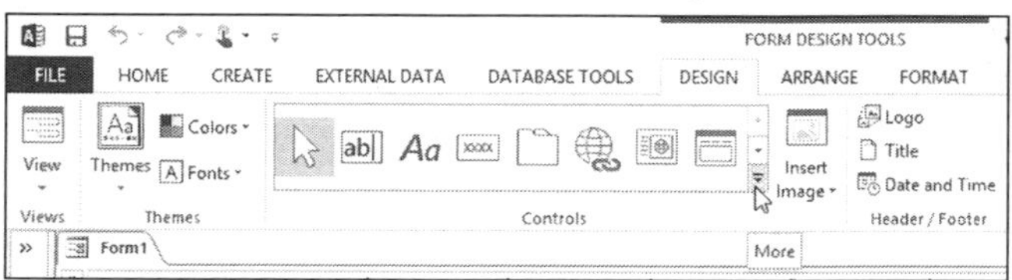

FIGURE 17.8. Adding an ActiveX control to an Access form (Step 1).

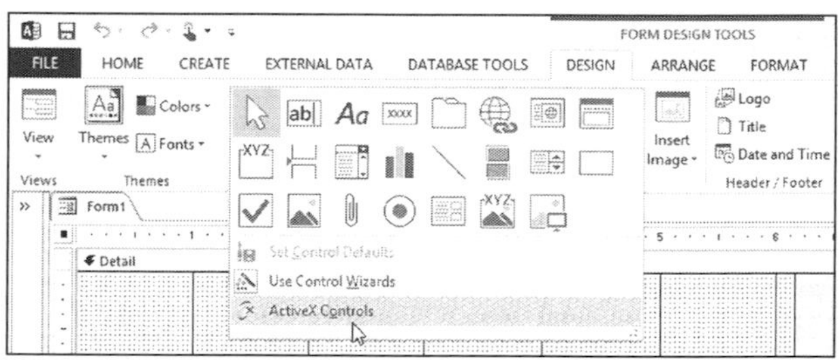

FIGURE 17.9. Adding an ActiveX control to an Access form (Step 2).

3. In the Insert ActiveX Control window, choose **Microsoft TreeView Control, version 6.0** as shown in Figure 17.10, and click **OK** to place a TreeView control on the form.

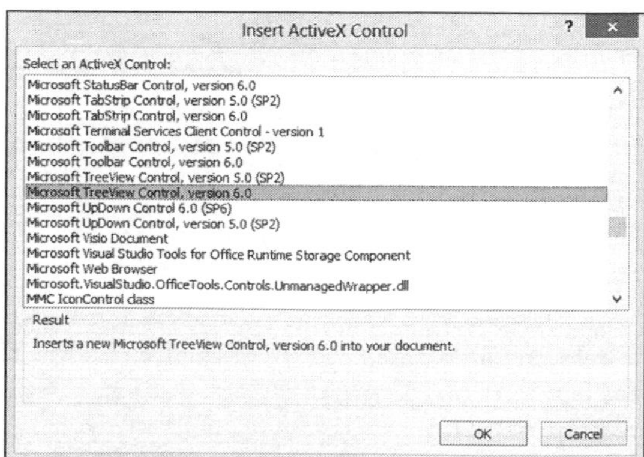

FIGURE 17.10. The Microsoft TreeView control provides an excellent way to display shaped recordsets in an Access form.

4. Resize the TreeView control and the form to match Figure 17.11.
5. Click the TreeView control to select it. In the property sheet, change the Name property of the TreeView control from TreeView0 to **myTreeCtrl**.

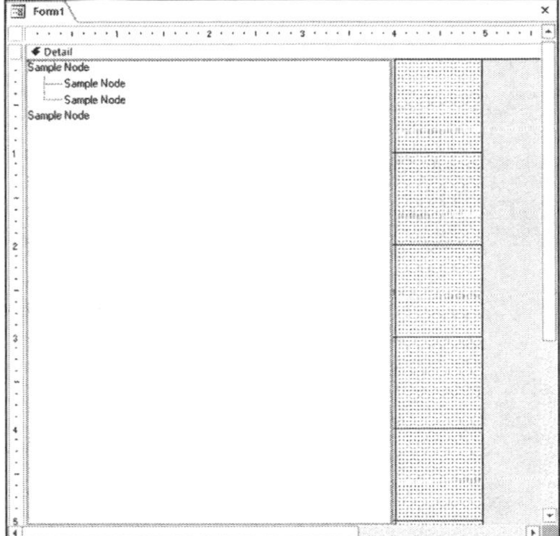

FIGURE 17.11. A TreeView control after being placed and resized on the Access form.

6. Right-click the TreeView control in Design view and choose **TreeCtrl_Object | Properties.** Adjust the custom properties of the TreeView control as listed on the General tab in Figure 17.12.

In addition to the properties listed in the property sheet, the ActiveX TreeView control exposes a number of custom properties that can be adjusted via the TreeCtrl Properties dialog box, as shown in Figure 17.12.

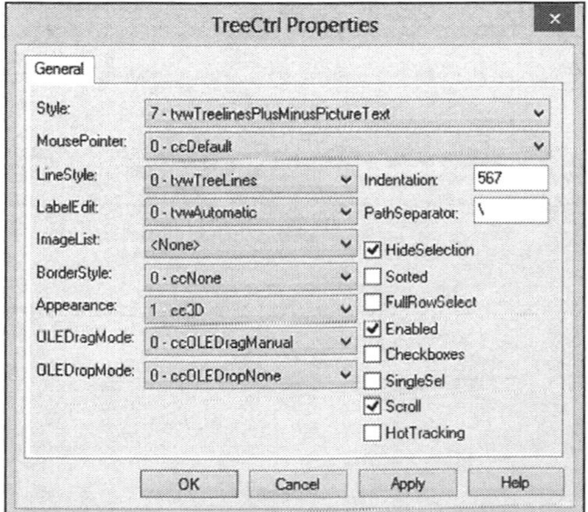

FIGURE 17.12. You can set custom properties of the TreeView control in the TreeCtrl Properties dialog box.

7. Save the form as **frmOrders**.

Part 2: Writing an Event Procedure for the Form Load Event

1. In the property sheet, select **Form** from the drop-down box and click the **Event** tab for the selected form.
2. Click the **Build** button (…) next to the **On Load** event name to display the Choose builder dialog box.
3. In the Choose Builder dialog box, select **Code Builder** and click **OK**. The form module window appears with the following Form_Load event procedure stub:

```
Private Sub Form_Load()
End Sub
```

4. Type the code for the **Form_Load** event procedure shown here, or copy the procedure code from **C:\Access2013_HandsOn\Chap17.txt**:

```
Private Sub Form_Load()
  Dim conn As ADODB.Connection
  Dim rstCustomers As ADODB.Recordset
  Dim rstOrders As ADODB.Recordset
  Dim rstOrderDetails As ADODB.Recordset
  Dim fld As Field
  Dim objNode1 As Node
  Dim objNode2 As Node
  Dim strConn As String
  Dim strSQL As String
  Dim strSQLCustomers As String
  Dim strSQLOrders As String
  Dim strSQLOrderDetails As String
  Dim strSQLRelParentToChild As String
  Dim strSQLRelGParentToParent As String

  ' Create the ADO Connection object
  Set conn = New ADODB.Connection

  ' Specify a valid connection string
  strConn = "Data Provider=Microsoft.Jet.OLEDB.4.0;"
  strConn = strConn & "Data Source = " & _
    "C:\Access2013_ByExample\Northwind.mdb"
    conn.ConnectionString = strConn
```

```
' Specify the Data Shaping provider
conn.Provider = "MSDataShape"

' Open the connection
conn.Open

' Specify SELECT statement for the Grandparent
strSQLCustomers = "SELECT CustomerID AS [Cust #]," & _
  "CompanyName AS [Customer] " & _
  "FROM Customers"
' Specify SELECT statement for the Parent
strSQLOrders = "SELECT OrderID AS [Order #]," & _
  "OrderDate AS [Order Date]," & _
  "Orders.CustomerID AS [Cust #] " & _
  "FROM Orders ORDER BY OrderDate DESC"

' Specify SELECT statement for the Child
strSQLOrderDetails = "SELECT od.OrderID AS [Order #]," & _
  "p.CategoryId AS [Category]," & _
  "p.ProductName AS [Product]," & _
  "od.Quantity," & _
  "od.ProductId," & _
  "od.UnitPrice AS [Unit Price]," & _
  "(od.UnitPrice * od.Quantity) " & _
  "AS [Extended Price] " & _
  "FROM [Order Details] od " & _
  "INNER JOIN Products p " & _
  "ON od.ProductID = p.ProductID " & _
  "ORDER BY p.CategoryId, p.ProductName"

' Specify RELATE clause to link Parent to Child
strSQLRelParentToChild = "RELATE [Order #] TO [Order #]"

' Specify RELATE clause to link Grandparent to Parent
strSQLRelGParentToParent = "RELATE [Cust #] TO [Cust #]"

' Build complete SQL statement for the shaped recordset
' adding aggregate functions for the Grandparent and Parent
strSQL = "SHAPE(SHAPE{" & strSQLCustomers & "}"
strSQL = strSQL & "APPEND((SHAPE{" & strSQLOrders & "} "
strSQL = strSQL & "APPEND({" & strSQLOrderDetails & "} "
strSQL = strSQL & strSQLRelParentToChild & ") AS rstOrderDetails,"
```

```
strSQL = strSQL & "COUNT(rstOrderDetails.Product) "
strSQL = strSQL & "        AS [Items On Order],"
strSQL = strSQL & "SUM(rstOrderDetails.[Extended Price]) "
strSQL = strSQL & "        AS [Order Total])"
strSQL = strSQL & strSQLRelGParentToParent & ") AS [rstOrders],"
strSQL = strSQL & "SUM(rstOrders.[Order Total]) "
strSQL = strSQL & "        AS [Cust Grand Total]"
strSQL = strSQL & ") AS rstCustomers"

' Create and open the Grandparent recordset
Set rstCustomers = New ADODB.Recordset
rstCustomers.Open strSQL, conn

' Fill the TreeView control
Do While Not rstCustomers.EOF
  Set objNode1 = myTreeCtrl.Nodes.Add _
   (Text:=rstCustomers.Fields(0) & _
   "   " & rstCustomers.Fields(1) & _
   "   ($ " & rstCustomers.Fields(3) & ")")
   Set rstOrders = rstCustomers.Fields("rstOrders").Value
     Do While Not rstOrders.EOF
        Set objNode2 = myTreeCtrl.Nodes.Add _
           (relative:=objNode1.Index, _
           relationship:=tvwChild, _
           Text:=rstOrders.Fields(0) & _
           "  " & rstOrders.Fields(1) & _
           "  " & rstOrders.Fields(4) & " (items)" & _
           "    $" & rstOrders.Fields(5) & _
           " (Order Total)")
           Set rstOrderDetails = _
           rstOrders.Fields("rstOrderDetails").Value
           Do While Not rstOrderDetails.EOF
             myTreeCtrl.Nodes.Add _
             relative:=objNode2.Index, _
             relationship:=tvwChild, _
             Text:=rstOrderDetails.Fields(3) & _
             "  " & rstOrderDetails.Fields(2) & _
             "  $" & rstOrderDetails.Fields(6) & _
             "  (" & rstOrderDetails.Fields(3) & _
             " x $" & rstOrderDetails.Fields(5) & ")"
             rstOrderDetails.MoveNext
           Loop
```

```
                rstOrders.MoveNext
            Loop
            rstCustomers.MoveNext
        Loop

        ' Cleanup
        rstCustomers.Close
        Set rstCustomers = Nothing
        Set conn = Nothing
    End Sub
```

5. Choose **Tools | References** and set the reference to the **Microsoft Windows Common Controls 6.0 (SP6)**. If this reference is not listed in the Available References list box, click the **Browse** button. In the Add a Reference window, in the System32 folder, select **ActiveX Controls (*.ocx)** in the files of type drop down box, and scroll down to locate and select **MSCOMCTL.OCX**. Click the **Open** button to confirm your selection, and then **OK** to exit the References window.
6. Press Ctr+F11 to return to the Access window and open **frmOrders** in Form view.

When you open the frmOrders form, the Form_Load procedure populates the TreeView control with the data from the Northwind.mdb database. As you can see in Figure 17.13, the results are quite impressive. Double-clicking on the nodes in the TreeView control expands and collapses the details underneath those nodes.

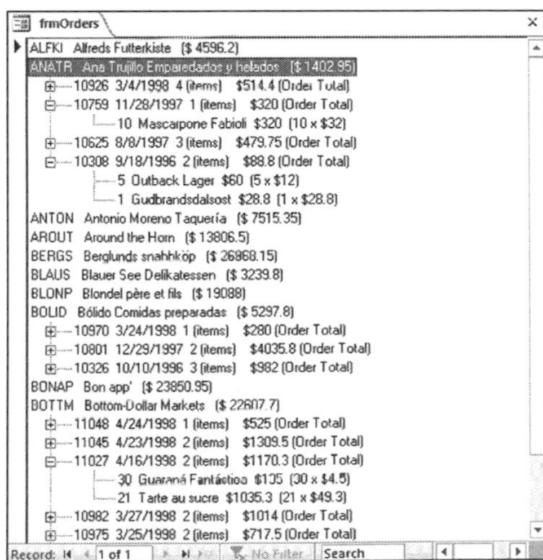

FIGURE 17.13. The TreeView control is filled with the data from the Northwind database when the user opens the form.

Prior to populating the TreeView control with the data, we connect to the database and enlist the services of the Data Shaping provider:

```
conn.Provider = "MSDataShape"
```

Because a TreeView control displays data as a hierarchy, we need to build a complex SQL statement using the SHAPE syntax we learned in preceding sections. To make things easier for ourselves, we start by defining SQL statements with fields we want to display for parent, child, and grandchild recordsets. Notice that we renamed some fields using the AS clause. We also defined separate statements to allow us to link grandparent to parent and parent to child. The structure we need to create can be illustrated like this:

```
Grandparent
Parent
Child
```

Now that we've defined the relationship and the fields for our data hierarchy, we use the SHAPE commands to build the complete SHAPE statement:

```
strSQL = "SHAPE(SHAPE{" & strSQLCustomers & "}"
strSQL = strSQL & "APPEND((SHAPE{" & strSQLOrders & "}"
strSQL = strSQL & "APPEND({" & strSQLOrderDetails & "}"
strSQL = strSQL & strSQLRelParentToChild & ") AS rstOrderDetails,"
strSQL = strSQL & "COUNT(rstOrderDetails.Product)"
strSQL = strSQL & " AS [Items On Order],"
strSQL = strSQL & "SUM(rstOrderDetails.[Extended Price])"
strSQL = strSQL & " AS [Order Total])"
strSQL = strSQL & strSQLRelGParentToParent & ") AS [rstOrders],"
strSQL = strSQL & "SUM(rstOrders.[Order Total])"
strSQL = strSQL & " AS [Cust Grand Total]"
strSQL = strSQL & ") AS rstCustomers"
```

While creating the SHAPE statement, we added additional calculated fields using the aggregate functions. For instance, in the parent recordset (rstOrders) we calculated the number of items ordered using the COUNT function:

```
COUNT(rstOrderDetails.Product) AS [Items On Order]
```

We also used the SUM function to obtain the total amount of the order:

```
SUM(rstOrderDetails.[Extended Price]) AS [Order Total]
```

In the grandparent recordset (rstCustomers), we used the SUM function to calculate the total amount owed by a customer.

When expanded, the complete SHAPE statement will look as follows:

```
strSQL = "SHAPE(SHAPE{"
strSQL = strSQL & "SELECT CustomerID AS [Cust #],"
strSQL = strSQL & "CompanyName AS [Customer]"
strSQL = strSQL & "FROM Customers"
strSQL = strSQL & "}"
strSQL = strSQL & "APPEND((SHAPE{"
strSQL = strSQL & "SELECT OrderID AS [Order #],"
strSQL = strSQL & "OrderDate AS [Order Date],"
strSQL = strSQL & "Orders.CustomerID AS [Cust #]"
strSQL = strSQL & "FROM Orders "
strSQL = strSQL & "ORDER BY OrderDate DESC"
strSQL = strSQL & "}"
strSQL = strSQL & "APPEND({"
strSQL = strSQL & "SELECT od.OrderID AS [Order #],"
strSQL = strSQL & "p.CategoryId AS [Category],"
strSQL = strSQL & "p.ProductName AS [Product],"
strSQL = strSQL & "od.Quantity,"
strSQL = strSQL & "od.ProductID,"
strSQL = strSQL & "od.UnitPrice AS [Unit Price],"
strSQL = strSQL & "(od.UnitPrice * od.Quantity) "
strSQL = strSQL & "AS [Extended Price] "
strSQL = strSQL & "FROM [Order Details] od INNER JOIN Products p"
strSQL = strSQL & " ON od.ProductID = p.ProductID "
strSQL = strSQL & "ORDER BY p.CategoryId, p.ProductName"
strSQL = strSQL & "}"
strSQL = strSQL & "RELATE [Order #] TO [Order #]"
strSQL = strSQL & ")"
strSQL = strSQL & "AS rstOrderDetails,"
strSQL = strSQL & "COUNT(rstOrderDetails.Product) "
strSQL = strSQL & "AS [Items On Order],"
strSQL = strSQL & "SUM(rstOrderDetails.[Extended Price]) "
strSQL = strSQL & "AS [Order Total])"
strSQL = strSQL & "RELATE [Cust #] TO [Cust #]"
strSQL = strSQL & ") "
strSQL = strSQL & "AS [rstOrders],"
strSQL = strSQL & "SUM(rstOrders.[Order Total]) "
strSQL = strSQL & "AS [Cust Grand Total]) AS rstCustomers"
```

Notice that the SHAPE statement we built contains standard fields pulled from the database tables and child recordsets (rstOrders, rstOrderDetails), as well as calculated columns. The rstOrders recordset is a field in the rstCustomers recordset. This field contains order information for a customer. rstOrderDetails is a field within the rstOrders recordset. This field contains the order details information for a customer's order.

Now that we've completed the `SHAPE` statement, we can open the grandparent recordset and begin populating the TreeView control with our data.

A TreeView control consists of Node objects, which you can expand or collapse to display or hide child nodes. Nodes that have child nodes are referred to as *parent* nodes. The nodes located at the top of the tree control are referred to as *root* nodes. Root nodes can have *sibling* nodes that are located on the same level. For example, customer ALFKI (see Figure 17.13) is a root node, and so is the customer ANATR, ANTON, and so on. They are also siblings of one another.

To populate a TreeView control, we use the `Add` method of the Nodes collection like this:

```
Set objNode1 = myTreeCtrl.Nodes.Add
```

`objNode1` is an object variable representing the Node object. The first node added to a TreeView is a root node. The `Add` method of the Nodes collection uses the following syntax:

```
object.Add([relative,] [relationship,] [key], text[, image,]
[selectedimage])
```

The only required arguments in the syntax are `object` and `text`. The object is the object variable (`myTreeCtrl`) representing the TreeView control. The text is a string that appears in the node. The following complete statement:

```
Set objNode1 = myTreeCtrl.Nodes.Add _
 (Text:=rstCustomers.Fields(0) & _
 "   " & rstCustomers.Fields(1) & _
 "   ($ " & rstCustomers.Fields(3) & ")")
```

creates a root node to display the following information:

```
Cust # (rstCustomers.Fields(0))
Customer (rstCustomers.Fields(1))
Cust Grand Total (rstCustomers.Fields(3))
```

Because the preceding statement appears inside a looping structure, the TreeView control will display all the customers at their root level.

Now that we've taken care of the root node, we go on to add children and grandchildren. A child node has a relationship to a parent node that has already been added. To define a child node, in addition to the required `text` argument, we will use two optional arguments of the `Add` method as follows:

- `relative`—This is the index number or key of a preexisting Node object. In our example, we used the index of the parent node that we just created (`relative:=objNode1.Index`).

When a Node object is created, it is automatically assigned an index number. This number is stored in the Node object's Index property.

- `relationship`—Specifies the type of relationship you are creating. Use the `tvwChild` setting to create a child node of the node named in the `relative` argument. The statement that creates a child node looks like this:

```
Set objNode2 = myTreeCtrl.Nodes.Add _
 (relative:=objNode1.Index, _
 relationship:=tvwChild, _
 Text:=rstOrders.Fields(0) & _
 "  " & rstOrders.Fields(1) & _
 "  " & rstOrders.Fields(4) & " (items)" & _
 "  $" & rstOrders.Fields(5) & _
 " (Order Total)")
```

The preceding statement displays order information for a customer. The child node `text` argument is set to display:

```
Order # (rstOrders.Fields(0))
Order Date (rstOrders.Fields(1))
Items On Order (rstOrders.Fields(4))
Order Total (rstOrders.Fields(5))
```

Because this statement appears inside a looping structure, the TreeView control will display the order information for each customer. Finally, we add grandchildren using the following statement:

```
myTreeCtrl.Nodes.Add _
 relative:=objNode2.Index, _
 relationship:=tvwChild, _
 Text:=rstOrderDetails.Fields(3) & _
 "  " & rstOrderDetails.Fields(2) & _
 "  $" & rstOrderDetails.Fields(6) & _
 "  (" & rstOrderDetails.Fields(3) & _
 " x $" & rstOrderDetails.Fields(5) & ")"
```

This statement displays order details for a customer's order. Notice that this Node object references the index number of the child object that has just been added (`relative:=objNode2.Index`).

The grandchild node `text` argument is set to display:

```
Quantity (rstOrderDetails.Fields(3))
Product (rstOrderDetails.Fields(2))
Extended Price (rstOrderDetails.Fields(6))
Quantity x Unit Price (rstOrderDetails.Fields(3) & _
"x $" & rstOrderDetails.Fields(5))
```

The looping structure ensures that these order details are listed for all customers' orders.

Now that you are done with this custom project, you should be able to provide your own hierarchical data in a pretty neat user interface.

TRANSACTION PROCESSING

To improve your application's performance and to ensure that database activities can be recovered in case an unexpected hardware or software error occurs, consider grouping sets of database activities into a transaction. A *transaction* is a set of operations that are performed together as a single unit. If you use an automatic teller machine (ATM), you are already familiar with transaction processing. When you go to the bank to get cash, your account must be debited. In other words, the cash withdrawal must be deducted from your savings or checking account. A transaction is a two-sided operation. If anything goes wrong during the transaction, the entire transaction is canceled. If both operations succeed, that is, you get the cash and the bank debits your account, the transaction's work is saved (or committed).

Database transactions often involve modifications and additions of one or more records in a single table or in several tables. When a transaction has to be undone or canceled, the transaction is rolled back. Often, when you perform batch updates to database tables and an error occurs, updates to all tables must be canceled or the database could be left in an inconsistent state.

Transactions are extremely important for maintaining data integrity and consistency. If you don't use transactions for operations that should be performed together, the database could be left in an inconsistent state, resulting not only in loss of important information but also in a number of other headaches.

In ADO, the Connection object offers three methods (`BeginTrans`, `CommitTrans`, and `RollbackTrans`) for managing transaction processing. You should use these methods to save or cancel a series of changes made to the data as a single unit.

- `BeginTrans`—Begins a new transaction
- `CommitTrans`—Saves any changes and ends the current transaction
- `RollbackTrans`—Cancels any changes made during the current transaction and ends the transaction

Please note that in ADO a transaction is limited to one database because the Connection object can only point to one database.

To work with transaction processing in DAO, use the transaction methods of the Workspace or DBEngine object: `BeginTrans`, `CommitTrans`, and `Rollback`. Within a Workspace transaction you can perform operations on more than one connection or database.

Creating a Transaction with ADO

Use the `BeginTrans` method to specify the beginning of a transaction, and the `CommitTrans` method to save the changes. `BeginTrans` and `CommitTrans` are used in pairs. The data-modifying instructions you place between these keywords are stored in memory until VBA encounters the `CommitTrans` statement. After reaching `CommitTrans`, Access writes to the disk the changes that have occurred since the `BeginTrans` statement; therefore, any changes you've made in the tables become permanent.

If an error is generated during the transaction process, the `RollbackTrans` statement placed further down in your procedure will undo all changes made since the `BeginTrans` statement. The rollback ensures that the data is returned to the state it was in before you started the transaction.

Using transaction processing helps improve database performance because the operations carried out during a transaction are run in memory. If the transaction succeeds, the results are written to the disk in a single operation. If any operation included in a transaction fails, the transaction is simply aborted and no changes are written to the database. If you don't use transactions, the results of each operation must be written to the disk separately—a process that consumes more database resources.

The procedure in Hands-On 17.6 assumes that you want to enter an order for a new customer. Because this customer does not exist in the database, you will use a transaction to ensure that the new order is entered only after the customer record has been created in the Customers table. The result is shown in Figure 17.14.

Hands-On 17.6. Using a Database Transaction to Insert Records (ADO)

1. In the Visual Basic Editor window, choose **Insert | Module**.
2. In the module's Code window, enter the **Create_Transaction_ADO** procedure as shown here:

```
Sub Create_Transaction_ADO()
  Dim conn As ADODB.Connection

  On Error GoTo ErrorHandler

  Set conn = New ADODB.Connection

  With conn
    .Provider = "Microsoft.Jet.OLEDB.4.0"
    .ConnectionString = "Data Source = " & _
     "C:\Access2013_ByExample\Northwind.mdb"
    .Open
    .BeginTrans

    ' insert a new customer record
    .Execute "INSERT INTO Customers " & _
     "Values ('GWIPO','Gwiazda Polarna'," & _
     "'Marcin Garnia', 'Sales Manager'," & _
     "'ul.Majewskiego 10', 'Warszawa', Null, '02-106'," & _
     "'Poland', ,0114822230445', Null)"

    , insert the order for that customer
    .Execute "INSERT INTO Orders " & _
     " (CustomerId, EmployeeId, OrderDate, RequiredDate) " & _
     " Values (,GWIPO', 1, Date(), Date()+5)"
    .CommitTrans
    .Close
   MsgBox "Both inserts completed."
  End With

  ExitHere:
  Set conn = Nothing
  Exit Sub
ErrorHandler:
  If Err.Number = -2147467259 Then
```

```
      MsgBox Err.Description
      Resume ExitHere
    Else
      MsgBox Err.Description
      With conn
        .RollbackTrans
        .Close
      End With
      Resume ExitHere
    End If
  End Sub
```

3. Choose **Run | Run Sub/UserForm** to execute the procedure.

   The first SQL `INSERT INTO` statement inserts the customer data into the Customers table in the Northwind.mdb database. Before the customer can actually order specific products, a record must be added to the Orders table. The second SQL `INSERT INTO` statement takes care of this task. Because both inserts must occur prior to filling in order details, they are treated as a single transaction. If an error occurs anywhere (for example, the Orders table is open in Design view), the entire transaction is rolled back. Notice how the `INSERT INTO` statement is used in this procedure. If you do not specify the field names, you will need to include values for each field in the table.

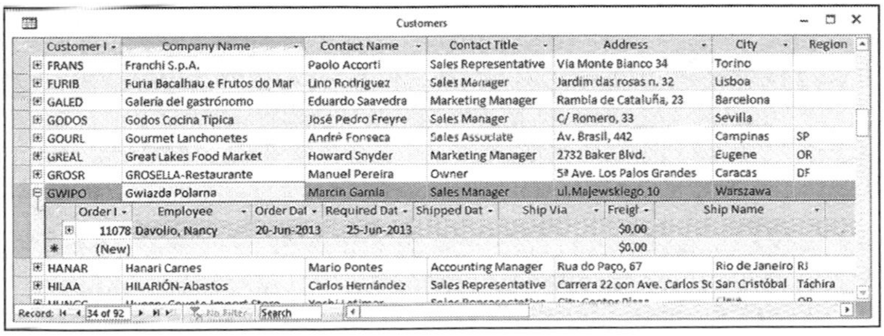

Customers

Customer I	Company Name	Contact Name	Contact Title	Address	City	Region
FRANS	Franchi S.p.A.	Paolo Accorti	Sales Representative	Via Monte Bianco 34	Torino	
FURIB	Furia Bacalhau e Frutos do Mar	Lino Rodriguez	Sales Manager	Jardim das rosas n. 32	Lisboa	
GALED	Galería del gastrónomo	Eduardo Saavedra	Marketing Manager	Rambla de Cataluña, 23	Barcelona	
GODOS	Godos Cocina Típica	José Pedro Freyre	Sales Manager	C/ Romero, 33	Sevilla	
GOURL	Gourmet Lanchonetes	André Fonseca	Sales Associate	Av. Brasil, 442	Campinas	SP
GREAL	Great Lakes Food Market	Howard Snyder	Marketing Manager	2732 Baker Blvd.	Eugene	OR
GROSR	GROSELLA-Restaurante	Manuel Pereira	Owner	5ª Ave. Los Palos Grandes	Caracas	DF
GWIPO	Gwiazda Polarna	Marcin Garnia	Sales Manager	ul.Majewskiego 10	Warszawa	
HANAR	Hanari Carnes	Mario Pontes	Accounting Manager	Rua do Paço, 67	Rio de Janeiro	RJ
HILAA	HILARIÓN-Abastos	Carlos Hernández	Sales Representative	Carrera 22 con Ave. Carlos Sc	San Cristóbal	Táchira

Order I	Employee	Order Dat	Required Dat	Shipped Dat	Ship Via	Freigt	Ship Name
11078	Davolio, Nancy	20-Jun-2013	25-Jun-2013			$0.00	
(New)						$0.00	

Record: 34 of 92 No Filter Search

FIGURE 17.14. After running the procedure in Hands-On 17.6, a record for a new customer, "GWIPO," is added to the Customers and Orders tables.

Creating a Transaction with DAO

The DAO Object Model supports transactions through the `BeginTrans`, `CommitTrans`, and `Rollback` methods of the Workspace and DBEngine objects. When you use these methods with the DBEngine object, the transaction is applied to the default workspace—DBEngine.Workspaces(0). If you need to manage transactions or connections to multiple databases, use the Workspace object.

A Workspace object represents a user's session. A transaction on a workspace will affect all data modifications made within the workspace. You can manage transactions independently across Database objects by creating additional Workspace objects.

As in ADO, use the `BeginTrans` method to specify the beginning of a transaction, the `CommitTrans` method to save the changes, and `Rollback` to cancel the transaction. `BeginTrans` and `CommitTrans` are used in pairs. The data-modifying instructions you place between these keywords are stored in memory until VBA encounters the `CommitTrans` statement. After reaching `CommitTrans`, Access writes to the disk the changes that have occurred since the `BeginTrans` statement; therefore, any changes you've made in the tables become permanent.

If an error is generated during the transaction process, the `Rollback` statement placed further down in your procedure will undo all changes made since the `BeginTrans` statement, which ensures that the data is returned to the state it was in before you started the transaction.

Transaction processing should be used for archiving historical data. For instance, the procedure in Hands-On 17.7 selects all orders placed in 1996 and appends them to an archive table in another database (Chap11.accdb). Then the records are deleted from the source table.

Hands-On 17.7. Using a Database Transaction to Archive Records (DAO)

1. This Hands-On requires the Chap11.accdb database that was created in Chapter 11.
2. In the Microsoft Access window of the Chap17.accdb database, choose **External Data | Access**. In the File name box of the Get External Data dialog box, enter **C:\Access2013_ByExample\Northwind.mdb**, and then click **OK**. In the Import Objects window, select the **Orders** table and click **OK**. Click **Close** to exit the Get External Data dialog box.
3. In the Visual Basic Editor window of the Chap17 database, choose **Insert | Module**.
4. In the module's Code window, enter the **OrdersArchive1996_DAO** procedure shown here:

```
Sub OrdersArchive1996_DAO()
  Dim db As DAO.Database
  Dim blnTrans As Boolean
  Dim strSQL As String
```

```
On Error GoTo ErrorHandler

'begin transaction
DBEngine.BeginTrans
blnTrans = True

Set db = CurrentDb()

' create an archive table on the fly
' and fill it with records
strSQL = "SELECT * INTO OrdersArchive1996 IN " & _
 """C:\Access2013_ByExample\Chap11.accdb""" & _
 " FROM Orders WHERE Orders.OrderDate " & _
 "BETWEEN #1/1/1996# AND #12/31/1996#;"

db.Execute strSQL, dbFailOnError

' delete records from the source table
If db.RecordsAffected <> 0 Then
  strSQL = "DELETE FROM Orders " & _
   "WHERE Orders.OrderDate " & _
   "BETWEEN #1/1/1996# AND #12/31/1996#;"

  db.Execute strSQL, dbFailOnError

  ' ask user if OK to commit changes
  If MsgBox("Click OK if you want to archive " _
   & db.RecordsAffected & " records.", vbOKCancel + _
   vbQuestion + vbDefaultButton2, "Proceed?") = vbOK Then
    DBEngine.CommitTrans
  Else
    If blnTrans Then DBEngine.Rollback
  End If
Else
  DBEngine.Rollback
  MsgBox "There are no records to archive " & _
   "with the specified criteria.", _
   vbInformation + vbOKOnly, "Records not found"
End If
Cleanup:
Set db = Nothing
Exit Sub
```

```
ErrorHandler:
  If Err.Number = 3010 Then
    strSQL = "INSERT INTO OrdersArchive1996 IN " & _
     """C:\Access2013_ByExample\Chap11.accdb""" & _
     " SELECT * FROM Orders WHERE Orders.OrderDate " & _
     "BETWEEN #1/1/1996# AND #12/31/1996#;"
    Resume 0
  Else
    If blnTrans Then DBEngine.Rollback
      MsgBox Err.Description
      Resume Cleanup
  End If
End Sub
```

5. Choose **Run | Run Sub/UserForm** to execute the procedure.

In this procedure, we start a transaction with the DBEngine object's `BeginTrans` method and set the transaction flag to `True` (`blnTrans`) to indicate that the transaction is active. We also initiate the Database object variable to point to the current database. The first data operation in this transaction requires that we create a table in another Access database to store the selected records from the Orders table in the current database. In the Access user interface, we would simply create a Make-Table query; in VBA programming, we can use the SQL `SELECT...INTO` statement. The first part of this statement specifies the fields we want to select; in this case we use a wildcard (*) to denote that all fields should be copied into the new table. This is followed by the `INTO` clause and the name of the table to be created.

If the table already exists, then the `SELECT...INTO` statement will fail and VBA will respond with error 3010. We must set an error trap (see the ErrorHandler code). To add the data to the existing table, we must use the SQL `INSERT INTO` statement. The name of the table in the `SELECT...INTO` statement is followed by the `IN` clause and the name of the external database into which data is to be inserted. You need to specify the full path to the database file.

The name of the external database is followed by the `FROM` clause and the name of the existing table from which records are selected. You may select data from more than one table. You may also specify selection criteria following the `WHERE` clause. After creating the SQL statement, we execute it using the `Execute` method of the Database object.

Notice the use of the `dbFailOnError` option with the `Execute` method. If the statement fails, `dbFailOnError` will generate an error message we can trap. Without it, you are not notified of any errors, and the entire procedure may not produce the intended results. You can see how the error trap works by running the procedure more than once. If the `Execute` statement succeeds, we proceed to delete records from the source table. However, we don't want to execute the delete code if the `SELECT` statement returned no records. After the `Execute` command is run, we use the RecordsAffected property of the Database object to obtain the number of records affected by the most recent `Execute` command.

If we have more than one record, we specify the records to delete using the SQL `DELETE` statement, and then carry out the delete operation by calling the `Execute` method of the Database object. If `dbFailOnError` did not notify us of any errors, we assume that the `Execute` statement succeeded and we can commit the transaction. Before carrying out this operation, we ask the user to confirm or cancel the transaction. If the user chooses not to go ahead with the changes, we roll back the transaction. We also withdraw changes to the records if there were no records to archive.

It is important to keep in mind that in case of an error you must roll back the transaction. Always check if the transaction is still active by using a flag. Rolling back the transaction will ensure that the transaction doesn't stay active after your VBA procedure has ended.

6. Run the procedure once again in step mode (using **F8**) to walk through the error code.

CHAPTER SUMMARY

This chapter covered quite a bit of advanced ADO and DAO material you will find useful in developing professional applications in Microsoft Access. You started by creating your own recordset from scratch and using it for storing nonrelational data. Next, you learned how to disconnect a recordset from a database and work with it offline. You also learned that a recordset can be saved to a disk file and later reopened without an active connection to the database. Next, you discovered how you can use the `Clone` method of the Recordset object to create a recordset that is a copy of another recordset. Finally, you familiarized yourself with the concepts of data shaping and learned statements that make it possible to create impressive hierarchical views of your data. You also learned how transactions are used to ensure that certain database operations are always performed as a single unit.

In the next chapter, you will learn how to write VBA procedures that handle database security.

Chapter 18 Implementing Database Security

The .accdb file format does not support user-level security. This means that you cannot create user and group accounts or assign object permissions in Access ACCDB databases. This chapter focuses on implementing database security in Access databases created in the .mdb file format.

In the course of this chapter, you will learn how to:

- Use the Users and Groups collections of the ADOX Catalog object to create and manage security user accounts.
- Use the `GetPermissions` and `SetPermissions` methods of the ADOX User and Group objects to retrieve and set permissions on database objects.
- Use the `ChangePassword` method of the ADOX User object to change the user's password.
- Use the `CompactDatabase` method of the JRO JetEngine object to set a database password.

To use ADOX and JRO in your VBA procedures, you must set a reference to the Microsoft ADO Ext. 6.0 for DDL and Security Object Library and Microsoft Jet and Replication Objects (JRO) Library (choose Tools | References in the Visual Basic Editor window to open the References dialog box).

TWO TYPES OF SECURITY IN MICROSOFT ACCESS

Depending on your requirements, Microsoft Access allows you to implement share-level or user-level security to protect and secure your Access database. As mentioned earlier, user-level security can only be implemented in Access databases created in the .mdb file format.

Share-Level Security (in Access .accdb and .mdb File Formats)

Using passwords to secure the database or objects in the database is known as *share-level security*. When you set a password on the database, users are required to enter a password in order to gain access to the data and database objects. Anyone with the password has unrestricted access to all Access data and database objects.

To manually change the database password:

- For an Access database in the .accdb file format, choose File | Info | Encrypt with Password.
- For an Access database in the .mdb file format, choose File | Info | Set Database Password.

Refer to the sections titled "Setting a Database Password Using the CompactDatabase Method" and "Setting a Database Password Using the NewPassword Method" later in this chapter to set a database password from within a VBA procedure

User-Level Security

User-level security is a relatively complex process that secures the code and objects in your database so that users can't accidentally modify or change them. With this type of security you can provide the most restrictive access to the database and the objects it contains. When you use user-level security, a *workgroup information file* is used to determine who can open a database and what objects are available to them.

The workgroup information file holds group and user information, including passwords. The information contained in this file determines not only who can open the database, but also the permissions users and groups have on the objects in the database. The workgroup information file contains built-in groups (Admins and Users) and a generic user account (Admin) with unlimited privileges on the database and the objects it contains.

When an Access .mdb database file is open in Access 2013, you can manually implement user-level security by choosing File | Info | Users and Permissions. You can also define user and group accounts and their passwords from your VBA procedures by using ADO code, as demonstrated later in this chapter.

UNDERSTANDING WORKGROUP INFORMATION FILES

To successfully run the procedures in this chapter, you need to know the location of the workgroup information file on your computer. This file, also known as *system database* (System.mdw), is created automatically on your computer (see Table 18.1).

Please note that the Application Data folder (used for storing System.mdw file for an Access database in the .mdb file format) is a hidden folder. To browse this folder, perform the following steps:

In Windows 7: activate Windows Explorer and choose Tools | Folder Options. In the Folder Options window, click the View tab, click the option button next to Show hidden files and folders, and click OK.

In Windows 8: activate Windows Explorer and click the View tab. Check the Hidden Items in the View/Hide section of the ribbon.

Now you should be able to access the path, where <username> is the name of your user profile. Take a few minutes right now to locate the System.mdw file on your machine.

TABLE 18.1. The workgroup information file in different versions of Access

Access Version	Default Workgroup Information Filename	Workgroup Information File Location
2.0	System.mda	C:\Access
95	System.mdw	C:\MSOffice\Access
97	System.mdw	C:\Windows\System
2000	System.mdw	C:\Program Files\Common Files\System
2002-2003	System.mdw	C:\Documents and Settings\<username>\Application Data\Microsoft\Access
2007-2010	System.mdw	Windows XP: C:\Documents and Settings\<username>\Application Data\Microsoft\Access Windows Vista /Windows 7: C:\Users\<username>\AppData\Roaming\Microsoft\Access\System.mdw
2013	System.mdw	Windows 7/ Windows 8: C:\Users\<username>\AppData\Roaming\Microsoft\Access\System.mdw

You can also find the location and name of the workgroup information file by starting Microsoft Access and opening any MDB database. Switch to the Visual Basic Editor window and activate the Immediate window. Type the following statement on one line (beginning with a question mark) and press Enter to execute:

```
? CurrentProject.Connection.Properties(
"Jet OLEDB:System Database").Value
```

When you press Enter, Access displays the full path of the workgroup information file that the currently open database uses for its security information. Jet OLEDB:System Database is a provider-specific property of the Microsoft OLE DB Provider for Jet in the ADO Properties collection of the Connection object.

Access uses the workgroup information file to store the following information:

- The name of each user and group

- The list of users who belong to each group
- The encrypted logon password for each workgroup user
- The Security Identifier (SID) of each user and group in a binary format

Once you add user and group accounts to your database, the workgroup information file will contain vital security information. *YOU DON'T WANT TO LOSE THIS INFORMATION.* Always take time to make a backup copy of the System.mdw file and store it in a safe location. This way, if the original file gets corrupted, you'll be able to quickly restore your backup file and avoid having to recreate user and group accounts.

The workgroup information file is like any other Access database file except that it contains hidden system tables with information regarding user and group accounts and their actual permissions. However, you cannot change the security information by opening this file directly. All the security data stored in hidden system tables is encrypted and protected. Changes to the workgroup information file are done automatically by the JetEngine when you use the built-in Access commands to manage security or execute ADO/JRO code in your VBA procedures.

You can use the same workgroup information file for more than one database or you can create a separate workgroup information file for each database you are securing. You can also give this file a name other than the default System.mdw. Most people find it best to use the same name as the database file. For example, if your secured database file is named Assets.mdb, you could create a workgroup information file called Assets.mdw and put it in the same folder as the database file. This way, you'd know right away that these two files are associated with one another even after many weeks or months have passed since you created them. Keeping track of which workgroup information file goes with which database can be quite challenging, especially if you are managing more than a couple of secured Access databases.

If you try to open a secured database while another workgroup information file is active, Access displays the following message:

You do not have the necessary permissions to use the <name> object. Have your system administrator or the person who created this object establish the appropriate permissions for you.

If you receive the preceding message while opening an Access database in the .mdb file format, you should look for the accompanying workgroup information file and perform one of the following:

- **Set Up a Shortcut**

Set up a shortcut to the database file that uses the /WRKGRP command-line switch to load the specified workgroup information file when the database is opened (see Custom Project 18.1).

- **Use the Workgroup Administrator Tool in Microsoft Access 2013**

1. Start Access and open any Access database.
2. Press **Alt+F11** to switch to the Visual Basic Editor window.
3. Choose **View | Immediate Window**.
4. In the Immediate window, type the following statement and press **Enter** to execute:

```
DoCmd.RunCommand acCmdWorkgroupAdministrator
```

5. In the Workgroup Administrator dialog box, click **Join**, then click **Browse**.
6. Locate the workgroup information file and then click **Open**. See Table 18.1 for the .mdw filenames used with various versions of Access.
7. In the Workgroup Administrator dialog box, click **OK**, then click **Exit**.

Creating and Joining Workgroup Information Files

When you open a database, Microsoft Access reads the workgroup information file to find out who is allowed to access the database. If security was put into place, you will be prompted for the user ID and password. Custom Project 18.1 walks you through the steps required to create and join a new workgroup information file. Once you join the workgroup, you create a new Access database and set up a password for the Admin user. This information is saved in the workgroup information file that you just joined. The workgroup information file is created using the User-Level Security Wizard. This option is available by choosing File | Info | Users and Permissions.

Securing a database boils down to creating a new workgroup information file, adding a new member to the Admins group, and removing the default Admin user from that group. You also need to remove permissions from the Admin user and from the Users group, and assign permissions to your own groups that you create. Don't be discouraged if you need to go over the security steps more than once. Access security is complex and can be approached from many different angles. Books of several hundred pages have been written to explain its inner workings. The approach presented here simply provides us with a secured Access database file we use to perform the programming exercises in this chapter. Although you could learn how to use the ADOX commands for managing security using the currently open unsecured Access database, this particular approach gives you a better set of skills to build from. So let's begin.

Please note files for the "Hands-On" project may be found on the companion CD-ROM.

Custom Project 18.1. Securing a Microsoft Access MDB Database

You must complete this project in order to work with the hands-on exercises in this chapter.

1. Start Microsoft Access 2013 and create a new blank database called **SpecialDb.mdb** in your **C:\Access2013_ByExample** folder. Be sure to select **Microsoft Access Databases (2002-2003) (*.mdb)** file format.

   You will use the built-in User-Level Security Wizard to secure the blank Access database (**SpecialDb.mdb**) you just created.
2. Choose **File | Info | User and Group Permissions | User-Level Security Wizard**.
3. Click **Yes** in response to the message that the database should be opened in shared mode to run the Security Wizard. Microsoft Access closes the database and reopens it in shared mode.
4. Microsoft Access then automatically activates the Security Wizard (see Figure 18.1). Click **Next** to continue.

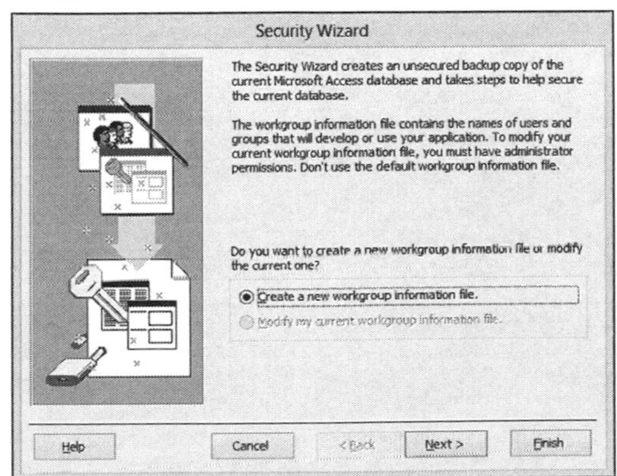

FIGURE 18.1. Security Wizard (screen 1).

Another Security Wizard window appears (see Figure 18.2). Do not make any changes in this screen. Click **Next** to continue.

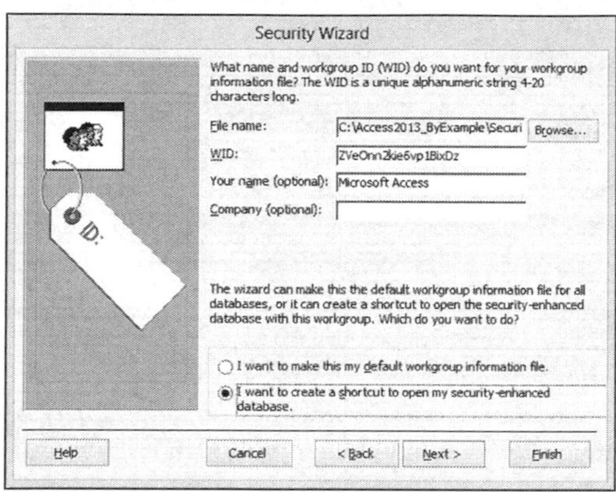

FIGURE 18.2. Security Wizard (screen 2). The workgroup information file named Security1.mdb stores user and group account information for the SpecialDb database.

5. The Security Wizard window now shows an empty tabbed screen that normally displays database objects (Figure 18.3). Because our database does not contain any tables, queries, reports, etc., there's nothing you can do in this screen. Click **Next** to continue.

FIGURE 18.3. Security Wizard (screen 3).

6. The Security Wizard window now displays a list of optional security accounts that you could include in your new workgroup information file (Figure 18.4). Because we will define our accounts in programming code later in this chapter, do not make any selections in this screen. Click **Next** to continue to the next screen.

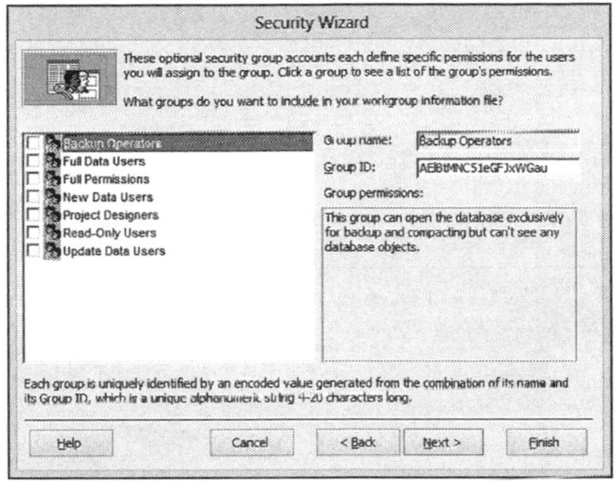

FIGURE 18.4. Security Wizard (screen 4).

7. Now the Security Wizard asks whether you want to grant permissions to the Users group (Figure 18.5). The Users group will have no permissions, so do not make any changes in this screen. We will work with permissions in our VBA procedures later. Click **Next** to continue.

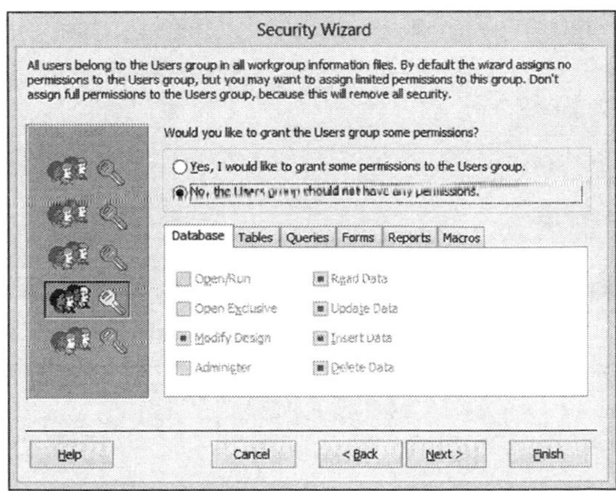

FIGURE 18.5. Security Wizard (screen 5).

8. Now the Security Wizard shows a screen (Figure 18.6) where you finally can do a little bit of work. You need to define a new user in your database. This user will function as a new Admin. Let's call this user **Developer** and allow him to log into the database using **chapter18** as a password. Fill in the User name and the Password boxes as shown in Figure 18.6 and click the **Add This User to the List** button. **Developer** should now appear in the users list (see Figure 18.7). Do not leave this screen yet.

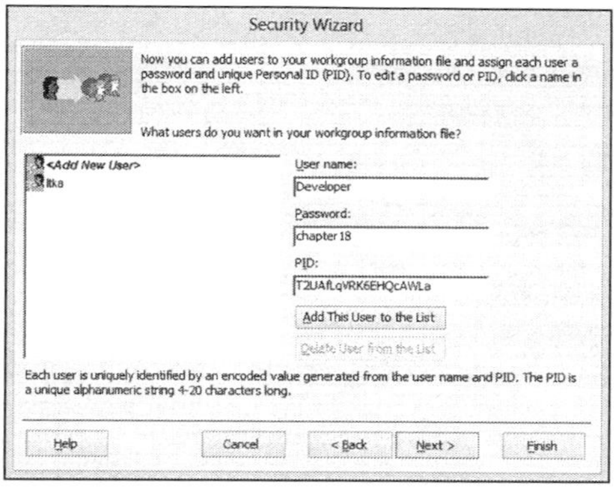

FIGURE 18.6. Security Wizard (screen 6a).

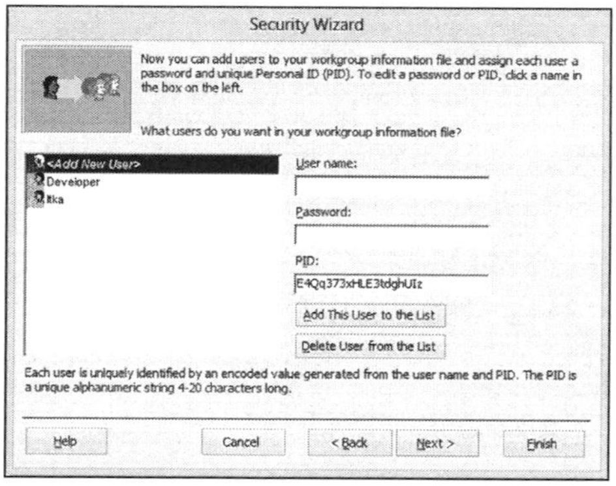

FIGURE 18.7. Security Wizard (screen 6b).

9. Now remove the user account you used to log into Access. In the list of users, select the username you logged in with and click the **Delete User from the List** button. Now **Developer** is the only user in our database, as shown in Figure 18.8. Click **Next** to continue.

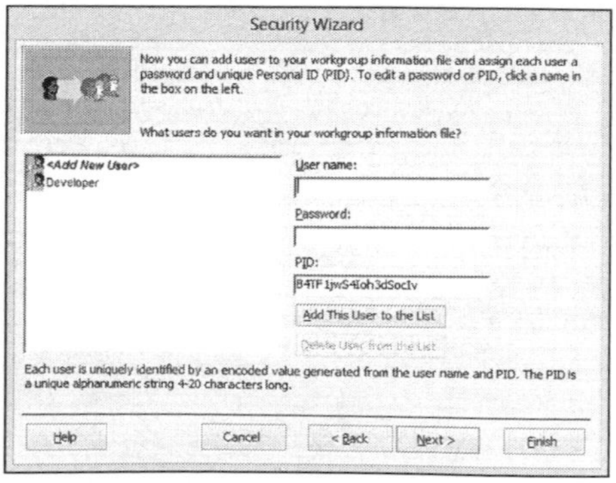

FIGURE 18.8. Security Wizard (screen 6c).

10. The Security Wizard shows the screen where you can assign users to groups in the workgroup information file (Figure 18.9). Notice that the user (**Developer**) you created in Step 8 is a member of the Admins group. Click **Next** to continue.

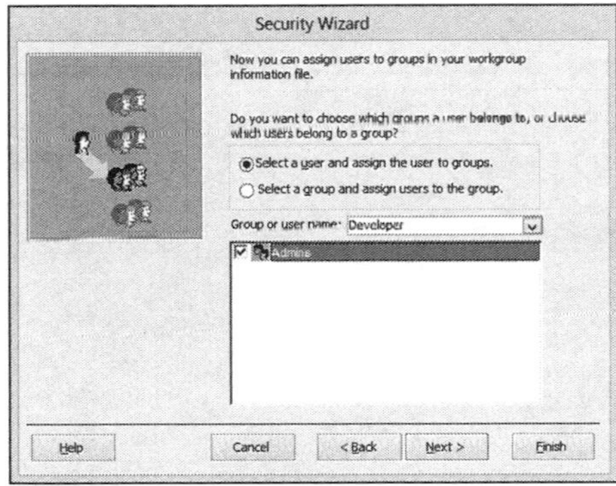

FIGURE 18.9. Security Wizard (screen 7).

11. The Security Wizard has now collected all the required information. As a final step, it suggests the name for the backup copy of the unsecured database (Figure 18.10). The unsecured database in this case is the blank database you started with. When you are done with this final step, your database will still be blank; however, it will be secured. Do not make any changes in this screen. Click **Finish**.

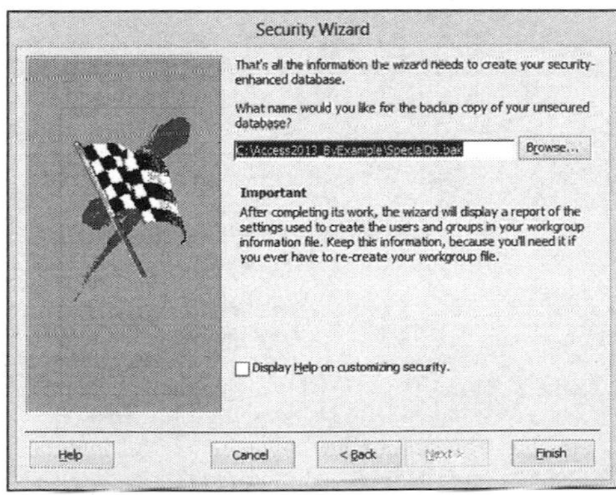

FIGURE 18.10. Security Wizard (screen 8).

12. Access performs its final tasks of securing your database and displays the Security Wizard report (Figure 18.11). If you are connected to a printer, it's a good idea to take a minute now to print this report. You can also magnify the report to read it on screen. When you are done, close the Security Wizard report window.

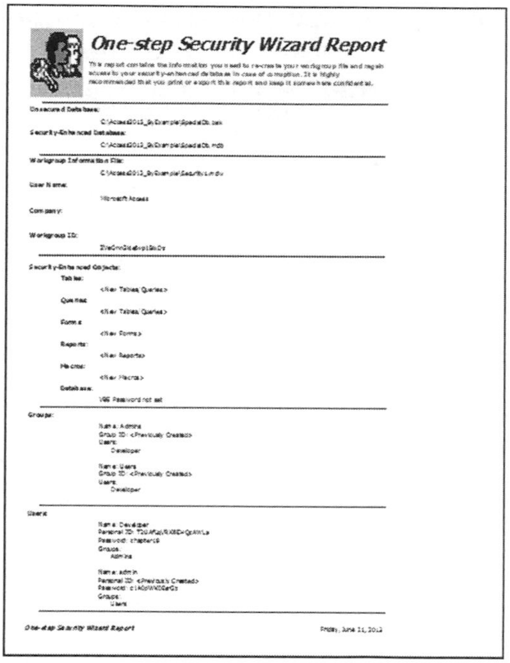

FIGURE 18.11. Security Wizard (screen 9).

13. When you close the window, the Security Wizard displays a warning message that asks whether you would like to save the report as a Snapshot (.snp) file that you can view later. For this exercise, click **No**. You should see the confirmation message that the Security Wizard has encoded your database and to reopen the database you must use the new workgroup file you created by closing Access and reopening it. You'll do as suggested in the next section. Click **OK** to this message.
14. Close the Microsoft Access window.

OPENING A SECURED MDB DATABASE

The following files were added to your **C:\Access2013_ByExample** folder when you completed Custom Project 18.1:

- A database file named *SpecialDb.mdb*
- A workgroup information file named *Security1.mdb* that stores user and group account information for the SpecialDb database
- A backup copy of the SpecialDb database named *SpecialDb.bak*

Also, there is a shortcut on your desktop (created by the Security Wizard) that allows you to quickly start the SpecialDb database using the new workgroup information file (Security1.mdb). If you right-click that desktop shortcut and choose Properties, you will see in the Target box the following path:

```
"C:\Program Files\Microsoft Office 15\root\office15\MSACCESS.EXE"
"C:\Access2013_ByExample\SpecialDb.mdb" /WRKGRP
"C:\Access2013_ByExample\Security1.mdw"
```

Because this path is very long it's shown on three lines. Notice that the first part of this path is the location of the Microsoft Access executable program on your disk enclosed by quotation marks. The path to the MSAccess.exe file is followed by a space and the full path of the database file (also in quotation marks). Because this database file is secured, we must also include a space and a command-line switch, /WRKGRP, followed by a space and the name of the accompanying workgroup information file (also in quotation marks).

The /WRKGRP command-line switch tells Access that you want to start a database with a specific workgroup. If you know which user account you want to log on with, you can use the /User and /Pwd command-line switches to avoid being prompted by Access for the username and password:

```
"C:\Program Files\Microsoft Office 15\root\office15\MSACCESS.EXE"
"C:\Access2013_ByExample\SpecialDb.mdb" /WRKGRP
"C:\Access2013_ByExample\Security1.mdw
    /User "Developer" /Pwd "chapter18"
```

The information about the username and password follows the name of the workgroup information file and a single space.

Now that you know how the path to a secured database is built, you can create similar shortcuts to other secured databases if they use different workgroup information files.

Hands-On 18.1. Opening a Secured MDB Database

This hands-on exercise requires prior completion of Custom Project 18.1.

1. On your desktop, double-click the shortcut to **SpecialDb.mdb** to open this database. Because this database is protected, a logon box appears. Enter **Developer** in the Name box and **chapter18** in the Password box and click **OK**.
2. Now that your secured database file is open, let's take a look at the changes the Security Wizard has made in the Users and Groups accounts. Choose **File | Info | Users and Permissions | User and Group Accounts**. Notice that the

Admin user is a member of the Users group (see Figure 18.12). The Security Wizard removed the Admin account from the Admins group. If you open the Name drop-down list in the User area of this screen and select Developer, you will see that Developer is a member of two groups: Admins and Users. Click **Cancel** to exit the User and Group Accounts window.

FIGURE 18.12. In Custom Project 18.1, you removed the default Admin user from the Admins group while running the built-in User-Level Security Wizard.

3. Having checked the Users and Groups accounts, you can also examine the changes made by the Security Wizard in the group permissions. Choose **File | Info | Users and Permissions | User and Group Permissions**. The users Developer and Admin don't have permissions on any new objects (see Figure 18.13). To view group permissions, click the **Groups** option button. The Admins group has all the necessary permissions to administer the database while the Users group has no permissions at all. You will learn how to grant and revoke permissions to database objects in the example procedures in this chapter. Now click **Cancel** to exit the User and Group Permissions window.

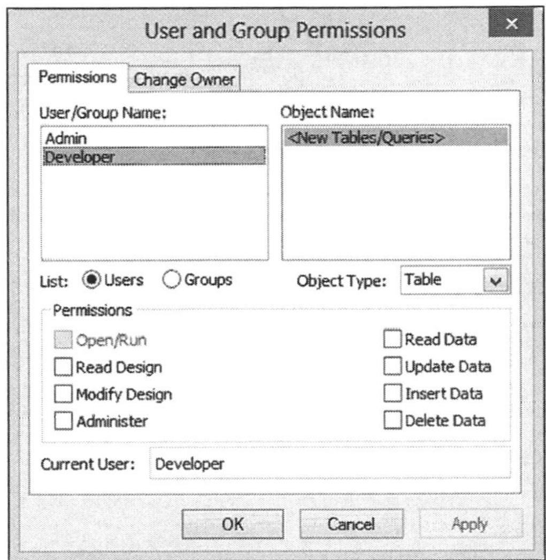

FIGURE 18.13. Use the User and Group Permissions window to check current permissions for the users Admin and Developer after running the User-Level Security Wizard in Custom Project 18.1.

4. Now let's import a couple of objects into this database. We will need them for our tests later in this chapter when we learn to handle permissions for database objects. In the Access window, choose **External Data | Access**. In the Get External Data dialog box, enter **C:\Access2013_ByExample\ Northwind.mdb** in the File name box and click **OK**.
5. In the Import Objects window, click **Select All** to select all the tables. Click the **Queries** tab, then choose **Select All** to select all the queries. Finally, click **OK** to begin importing. When the import operation is completed, click the **Close** button.
6. The objects you selected in Step 5 have now been added to your database. Close the SpecialDb database and close the Access application.

CREATING AND MANAGING GROUP AND USER ACCOUNTS

To create a new group account from a VBA procedure using ADO, open the ADOX Catalog object by specifying the connection to the appropriate database and use the `Append` method of the Catalog object's Groups collection to add a new group account.

To create a new user account, pass the name and password to the `Append` method of the Users collection. Specifying a password at this time is optional. You can assign a password later with the User object's `ChangePassword` method.

The procedure in Hands-On 18.2 illustrates how to create two group accounts and a user account in the secured database (SpecialDb.mdb) that you created in Custom Project 18.1.

Hands-On 18.2. Creating User and Group Accounts (ADO)

This hands-on exercise requires that you have completed Custom Project 18.1.

1. Start Microsoft Access and create a new database named **Chap18.mdb** in your **C:\Access2013_ByExample** folder. Make sure you select Microsoft Access Databases (2002-2003) (*.mdb) file format.
2. Press **Alt+F11** to switch to the Visual Basic Editor window and choose **Insert | Module**.
3. Choose **Tools | References** and click the checkbox next to the following three object libraries: **Microsoft ActiveX Data Objects 6.1 Library, Microsoft ADO Ext. 6.0 for DDL and Security Object Library**, and **Microsoft Jet and Replication Objects 2.6 Library**. After making these selections, click **OK** to exit the References dialog box.
4. Activate the Immediate window by choosing **View | Immediate Window**. Type the following statement in the Immediate window and press **Enter**:

```
DoCmd.RunCommand acCmdWorkgroupAdministrator
```

When you press Enter, Access loads the Workgroup Administrator tool, which lets you check the path to the workgroup information file that is currently being used. In Access 2013 there is no command in the user interface to access this tool. You must enter the preceding code to use the Workgroup Administrator tool.

Perform one of the following steps:

a. If System.mdw appears in the Workgroup path, click **OK** to exit the Workgroup Administrator dialog box and proceed with step 5.

b. If the Workgroup path includes the Security1.mdw file that was created in Custom Project 18.1, click the **Join** button to join another workgroup. Use the **Browse** button in the Workgroup Information File dialog box to select and open **System.mdw**. Refer to the beginning of this chapter for information on the default location of this file. Once you select the correct file, the dialog box should display its full path. Click **OK** to exit this dialog box. Access will display a message box saying you successfully joined the workgroup defined by the selected information file. Click **OK** to the message and click **OK** in the Workgroup Administrator dialog box to exit. Proceed to Step 5.

5. In the module's Code window, enter the following **Create_UserAnd-Group_ADO** procedure:

```
Sub Create_UserAndGroup_ADO()
Dim cat As ADOX.Catalog
Dim conn As ADODB.Connection
Dim strDB As String
Dim strSysDB As String
Dim strGrpName1 As String
Dim strGrpName2 As String
Dim strUsrName As String

On Error GoTo ErrorHandler

strDB = "C:\Access2013_ByExample\SpecialDb.mdb"
strSysDB = "C:\Access2013_ByExample\Security1.mdw"
strGrpName1 = "Masters"
strGrpName2 = "Elite"
strUsrName = "PowerUser"
' open connection to the database
' using the specified system database
Set conn = New ADODB.Connection
With conn
  .Provider = "Microsoft.Jet.OLEDB.4.0"
  .Properties("Jet OLEDB:System Database") = strSysDB
  .Properties("User ID") = "Developer"
  .Properties("Password") = "chapter18"
  .Open strDB
End With

' Open the catalog
Set cat = New ADOX.Catalog
With cat
  .ActiveConnection = conn
  ' create group accounts
  .Groups.Append strGrpName1
  .Groups.Append strGrpName2
  Debug.Print "Successfully created group accounts."
  ' create a user account
  .Users.Append strUsrName, "star"
  Debug.Print "Successfully created user account."
  ' Add user to the group
```

```
      .Users(strUsrName).Groups.Append strGrpName2
      Debug.Print strUsrName & " is a member of the " & _
       strGrpName2 & " group account."
    End With

  ExitHere:
    Set cat = Nothing
    conn.Close
    Set conn = Nothing
    Exit Sub
  ErrorHandler:
    MsgBox Err.Description
    Resume ExitHere
  End Sub
```

6. Choose **Run | Run Sub/UserForm** to execute the Create_UserAndGroup_ADO procedure.

   Upon executing this procedure, two new group accounts named Masters and Elite are established in the secured SpecialDb database you created in Custom Project 18.1. A new user account named PowerUser is added and made a member of the Elite group account. Notice that before opening the database we need to set the Jet OLEDB:System Database property in the Properties collection of the ADO Connection object to specify the path and name of the workgroup information file that should be active when the database is opened. We also set the User ID and Password properties to log onto the database. After opening the database, we open the Catalog object and use the `Append` method of the Catalog's Groups collection to add new group accounts. The Groups collection contains all groups in the specified workgroup information file. The `Append` method of the Catalog's Users collection is used to create a new user account. This user account is then appended to the Groups collection and made a member of a particular group (Elite). You can verify that the accounts were indeed created by opening the SpecialDb.mdb file.
7. If you'd like to take a moment now, open the SpecialDb database using the shortcut on your desktop. Once the database is open, choose **File | Info | Users and Permissions | User and Group Accounts**. Notice that the database now contains the Masters and Elite groups in addition to the default Admins and Users groups (see Figure 18.14).

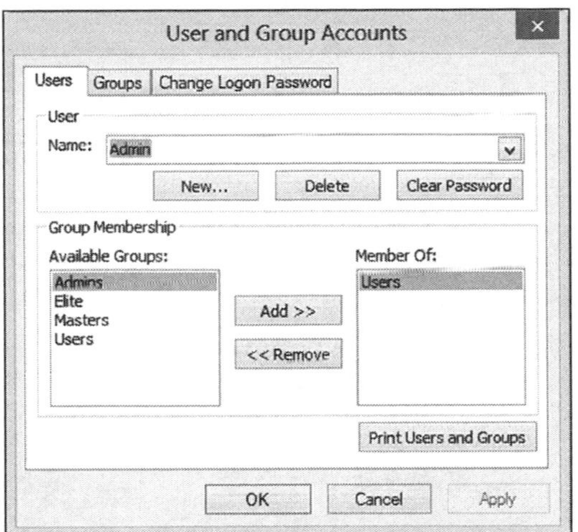

FIGURE 18.14. The Elite and Masters group accounts are created by running the procedure in Hands-On 18.2.

8. Close the SpecialDb database and the Access window in which it was displayed. Be careful not to close the Chap18.mdb database you are working with.

Deleting User and Group Accounts

Use the `Delete` method of the Catalog object's Users collection to delete a user account. Use the `Delete` method of the Catalog object's Groups collection to delete a group account.

The procedure in Hands-On 18.3 deletes the user account named PowerUser and the group account named Masters that were created in Hands-On 18.2.

Hands-On 18.3. Deleting User and Group Accounts (ADO)

This hands-on exercise requires the prior completion of Custom Project 18.1 and Hands-On 18.2.

1. In the Visual Basic Editor window of Chap18.mdb, choose **Insert | Module.**
2. In the module's Code window, enter the following **Delete_UserAndGroup** procedure:

```
Sub Delete_UserAndGroup(UserName As String, _
   GroupName As String)
  Dim cat As ADOX.Catalog
  Dim conn As ADODB.Connection
  Dim strDB As String
  Dim strSysDB As String
```

```
    On Error GoTo ErrorHandler

    strDB = "C:\Access2013_ByExample\SpecialDb.mdb"
    strSysDB = "C:\Access2013_ByExample\Security1.mdw"

    ' Open connection to the database using
    ' the specified system database
    Set conn = New ADODB.Connection
    With conn
      .Provider = "Microsoft.Jet.OLEDB.4.0"
      .Properties("Jet OLEDB:System Database") = strSysDB
      .Properties("User ID") = "Developer"
      .Properties("Password") = "chapter18"
      .Open strDB
    End With

    ' Open the catalog
    Set cat = New ADOX.Catalog
    With cat
      .ActiveConnection = conn
      ' Delete user
      .Users.Delete UserName
      ' Delete group
      .Groups.Delete GroupName
    End With

  ExitHere:
    Set cat = Nothing
    conn.Close
    Set conn = Nothing
    Exit Sub
  ErrorHandler:
    MsgBox Err.Description
    Resume ExitHere
  End Sub
```

3. To run this procedure, enter the following statement in the Immediate window and press **Enter** to execute it:

```
Delete_UserAndGroup "PowerUser", "Masters"
```

After running the Delete_UserAndGroup procedure, the Masters group account and the PowerUser user account are removed from the SpecialDb database.

Listing User and Group Accounts

The procedure in Hands-On 18.4 demonstrates how to retrieve the names of all defined group and user accounts from the Groups and Users collections of the Catalog object (see Figure 18.15).

Hands-On 18.4. Listing Group and User Accounts (ADO)

1. In the Visual Basic Editor window of Chap18.mdb, choose **Insert | Module.**
2. In the module's Code window, enter the **List_GroupsAndUsers_ADO** procedure as shown here:

```
Sub List_GroupsAndUsers_ADO()
  Dim conn As ADODB.Connection
  Dim cat As ADOX.Catalog
  Dim grp As New ADOX.Group
  Dim usr As New ADOX.User
  Dim strDB As String
  Dim strSysDB As String

  strDB = "C:\Access2013_ByExample\SpecialDb.mdb"
  strSysDB = "C:\Access2013_ByExample\Security1.mdw"

  ' Open connection to the database using
  ' the specified system database
  Set conn = New ADODB.Connection
  With conn
    .Provider = "Microsoft.Jet.OLEDB.4.0"
    .Properties("Jet OLEDB:System Database") = strSysDB
    .Properties("User ID") = "Developer"
    .Properties("Password") = "chapter18"
    .Open strDB
  End With

  ' Open the catalog
  Set cat = New ADOX.Catalog
  cat.ActiveConnection = conn
  ' list group and user accounts
  For Each grp In cat.Groups
    Debug.Print "Group: " & grp.Name
  Next
```

```
    For Each usr In cat.Users
      Debug.Print "User: " & usr.Name
    Next

    Set cat = Nothing
    conn.Close
    Set conn = Nothing

    MsgBox "Groups and users are listed in the Immediate window."
  End Sub
```

3. Choose **Run | Run Sub/UserForm** to execute the procedure.

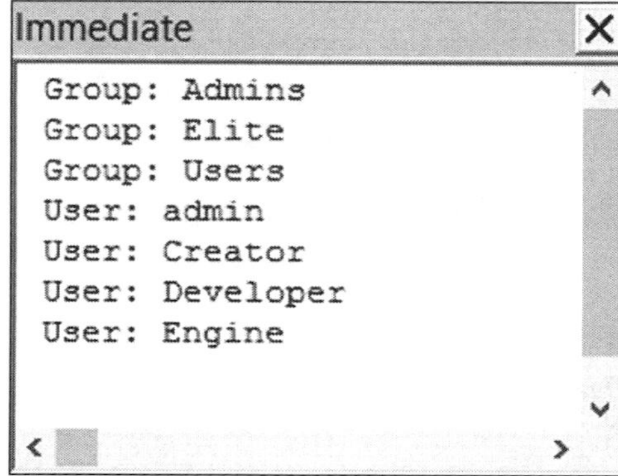

FIGURE 18.15. The names of existing security group and user accounts are written to the Immediate window by the procedure in Hands-On 18.4.

Notice that in addition to the user accounts you have defined, Access reveals the names of its two built-in users: Creator and Engine. To keep these built-in users from showing up in your users listing, use the following conditional statement:

```
If usr.Name <> "Creator" And usr.Name <> "Engine" Then
  Debug.Print "User:" & usr.Name
End If
```

Listing Users in Groups

Sometimes you will need to know which users belong to which groups. The procedure in Hands-On 18.5 demonstrates how to obtain such a list, which is shown in Figure 18.16.

Hands-On 18.5. Listing Users in Groups (ADO)

1. In the Visual Basic Editor window of Chap18.mdb, choose **Insert | Module**.
2. In the module's Code window, enter the **List_UsersInGroups** procedure as shown here:

```
Sub List_UsersInGroups()
  Dim conn As ADODB.Connection
  Dim cat As ADOX.Catalog
  Dim grp As New ADOX.Group
  Dim usr As New ADOX.User
  Dim strDB As String
  Dim strSysDB As String

  strDB = "C:\Access2013_ByExample\SpecialDb.mdb"
  strSysDB = "C:\Access2013_ByExample\Security1.mdw"

  ' Open connection to the database using
  ' the specified system database
  Set conn = New ADODB.Connection
  With conn
    .Provider = "Microsoft.Jet.OLEDB.4.0"
    .Properties("Jet OLEDB:System Database") = strSysDB
    .Properties("User ID") = "Developer"
    .Properties("Password") = "chapter18"
    .Open strDB
  End With

  ' Open the catalog
  Set cat = New ADOX.Catalog
  cat.ActiveConnection = conn
  For Each grp In cat.Groups
    Debug.Print "Group: " & grp.Name
    If cat.Groups(grp.Name).Users.Count = 0 Then
      Debug.Print vbTab & "There are no users in the " & _
       grp & " group."
    End If
    For Each usr In cat.Groups(grp.Name).Users
      Debug.Print vbTab & "User: " & usr.Name
    Next usr
  Next grp
```

```
    Set cat = Nothing
    conn.Close
    Set conn = Nothing
    MsgBox "Groups and Users are listed in the Immediate window."
End Sub
```

3. Choose **Run | Run Sub/UserForm** to execute the procedure.

```
Immediate
Group: Admins
    User: Developer
Group: Elite
    There are no users in the Elite group.
Group: Users
    User: admin
    User: Developer
```

FIGURE 18.16. After running the procedure in Hands-On 18.5, security group account names and the corresponding user accounts are listed in the Immediate window.

SETTING AND RETRIEVING USER AND GROUP PERMISSIONS

Users and groups of users can be granted specific permissions to database objects. For example, a user or an entire group of users can be authorized to only read an object's contents, while other users or groups can have less restrictive access to a database, allowing them to modify or delete objects.

It is important to understand that when you set permissions for a group, every user in that group automatically inherits those permissions. Also, keep in mind that while the user and group accounts are stored in the workgroup information file, the permissions that those users and groups have to specific objects are stored in system tables in your database.

The following sections of this chapter will familiarize you with using ADOX to retrieve, list, and set permissions for various database objects.

Determining the Object Owner

The database, and every object in the database, has an owner. The owner is the user who created that particular object. The object owner has special privileges, including the ability to assign or revoke permissions for that object. To retrieve the name of the object owner, use the `GetObjectOwner` method of a Catalog object. This method takes two parameters: the object's name and the object's type. For example, to determine the owner of a table, use the following syntax:

```
cat.GetObjectOwner(myObjName, adPermObjTable)
```

where `cat` is an object variable representing the ADOX Catalog object, `myObjName` is the name of a database table, and `adPermObjTable` is a built-in ADOX constant specifying the type of object. The constants for the Type parameter can be looked up in the Object Browser, as shown in Figure 18.17.

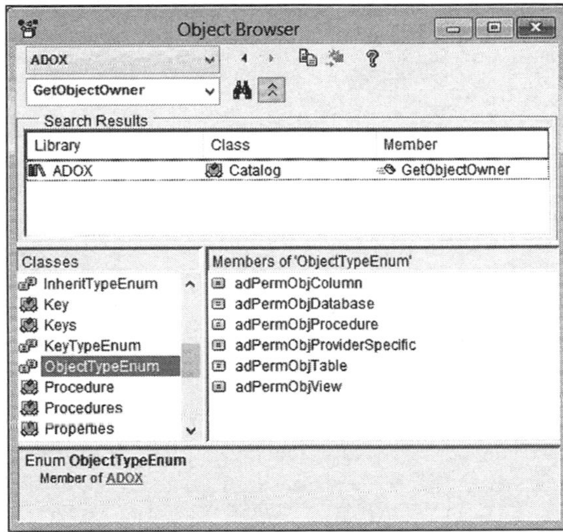

FIGURE 18.17. The Object Browser displays the available constants for the Type parameter of the GetObjectOwner method.

Hands-On 18.6. Retrieving the Name of the Object Owner (ADO)

1. In the Visual Basic Editor window of Chap18.mdb, choose **Insert | Module.**
2. In the module's Code window, enter the **Get_ObjectOwner** procedure as shown here:

```
Sub Get_ObjectOwner()
  Dim conn As ADODB.Connection
  Dim cat As ADOX.Catalog
  Dim strObjName As Variant
  Dim strDB As String
  Dim strSysDB As String
```

```
    strDB = "C:\Access2013_ByExample\SpecialDb.mdb"
    strSysDB = "C:\Access2013_ByExample\Security1.mdw"
    strObjName = "Customers"

    ' Open connection to the database using
    ' the specified system database
    Set conn = New ADODB.Connection
    With conn
      .Provider = "Microsoft.Jet.OLEDB.4.0"
      .Properties("Jet OLEDB:System Database") = strSysDB
      .Properties("User ID") = "Developer"
      .Properties("Password") = "chapter18"
      .Open strDB
    End With

    ' Open the catalog
    Set cat = New ADOX.Catalog
    cat.ActiveConnection = conn

    ' Display the name of the table owner
    MsgBox "The owner of the " & strObjName & _
     " table is " & vbCr _
     & cat.GetObjectOwner(strObjName, adPermObjTable) & "."

    Set cat = Nothing
    conn.Close
    Set conn = Nothing
End Sub
```

3. Choose **Run | Run Sub/UserForm** to execute the procedure.

To set the ownership of an object with ADOX, use the `SetObjectOwner` method of the Catalog object like this:

```
cat.SetObjectOwner("Customers", adPermObjTable, "PowerUser")
```

The preceding statement says that the ownership of the Customers table is to be transferred to the user named PowerUser. Note that currently there is no such user in the SpecialDb database. Recall that we created the PowerUser user account in Hands-On 18.2 and deleted it in Hands-On 18.3. If you want to experiment with changing object ownership, you need to make appropriate changes in the example procedure using the information you have already learned.

Setting User Permissions for an Object

With ADOX, you set permissions on an object by using the `SetPermissions` method. User-level security can be easier to manage if you set permissions only for groups, and then assign users to the appropriate groups. Recall that permissions set for the group are automatically inherited by all users in that group. The `SetPermissions` method, which can be used for setting both user and group permissions, has the following syntax:

```
GroupOrUser.SetPermissions(Name, ObjectType, Action, Rights[, Inherit] [,ObjectTypeId])
```

- `Name`—The name of the object to set permissions on.
- `ObjectType`—The type of object the permissions are set for. (See Figure 18.17 for the names of the ADOX built-in constants that can be used to specify the `Type` parameter.)
- `Action`—The type of action to perform when setting permissions. Use the `adAccessSet` constant for Microsoft Access databases to specify that the group of users will have exactly the requested permissions.
- `Rights`—A Long value containing a bitmask indicating the permissions to set. The `Rights` argument can consist of a single permissions constant or several constants combined with the OR operator. See Figure 18.18 for the names of the ADOX built-in constants that can be used in the `Rights` argument to specify the type of permissions to set.

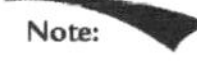

A bitmask is a numeric value intended for a bit-by-bit value comparison with other numeric values, usually to flag options in parameters or return values. In Visual Basic, this comparison is done with bitwise logical operators, such as AND and OR. The ADOX `GetPermissions` *and* `SetPermissions` *methods use the bitwise logical operator OR to retrieve the bitmask for the existing permissions and to add new permissions to the bitmask.*

The last two arguments (those in square brackets) are optional:

- `Inherit`—A Long value that specifies how objects will inherit these permissions. The default value is `adInheritNone`.
- `ObjectTypeId`—A Variant value that specifies the GUID (global unique identifier) for a provider object type not defined by OLE DB. This parameter is required if `ObjectType` is set to `adPermObjProviderSpecific` (which is used for setting permissions for forms, reports, and macros); otherwise, it is not used. See Table 18.2 for available GUIDs.

TABLE 18.2 GUIDs for provider objects

Object	GUID
Form	{c49c842e-9dcb-11d1-9f0a-00c04fc2c2e0}
Report	{c49c8430-9dcb-11d1-9f0a-00c04fc2c2e0}
Macro	{c49c842f-9dcb-11d1-9f0a-00c04fc2c2e0}

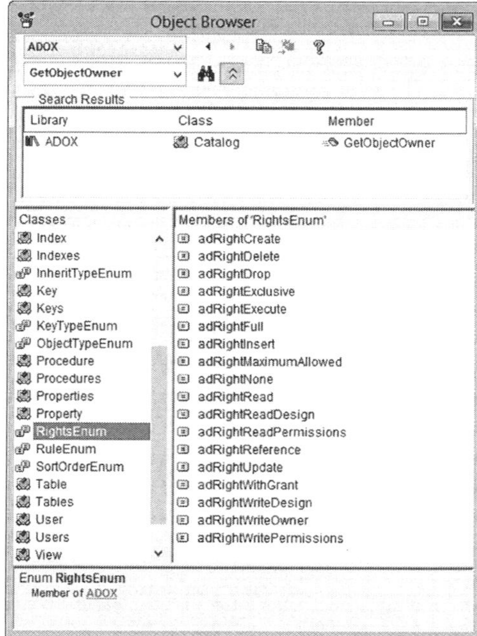

FIGURE 18.18. In ADOX, you can use many security constants for setting permissions to database objects.

The example procedure in Hands-On 18.7 grants a user the permission to read (`adRightRead`), insert (`adRightInsert`), update (`adRightUpdate`), and delete (`adRightDelete`) records.

Hands-On 18.7. Setting User Permissions for an Object (ADO)

1. In the Visual Basic Editor window of Chap18.mdb, choose **Insert | Module.**
2. In the module's Code window, enter the **Set_UserObjectPermissions** procedure as shown here:

```
Sub Set_UserObjectPermissions()
  Dim conn As ADODB.Connection
  Dim cat As ADOX.Catalog
```

```
  Dim strDB As String
  Dim strSysDB As String

  On Error GoTo ErrorHandler

  strDB = "C:\Access2013_ByExample\SpecialDb.mdb"
  strSysDB = "C:\Access2013_ByExample\Security1.mdw"

  ' Open connection to the database using
  ' the specified system database
  Set conn = New ADODB.Connection

  With conn
    .Provider = "Microsoft.Jet.OLEDB.4.0"
    .Properties("Jet OLEDB:System Database") = strSysDB
    .Properties("User ID") = "Developer"
    .Properties("Password") = "chapter18"
    .Open strDB
  End With

  ' Open the catalog
  Set cat = New ADOX.Catalog
  cat.ActiveConnection = conn
  ' add a user account
  cat.Users.Append "PowerUser", "star"

  ' Set permissions for PowerUser on the Customers table
  cat.Users("PowerUser").SetPermissions "Customers", _
   adPermObjTable, _
   adAccessSet, _
   adRightRead Or _
   adRightInsert Or _
   adRightUpdate Or _
   adRightDelete
  MsgBox "Read, Insert, Update and Delete permissions " & _
   vbCrLf & " were set on Customers table " & _
   "for PowerUser."
ExitHere:
  Set cat = Nothing
  conn.Close
  Set conn = Nothing
  Exit Sub
```

```
ErrorHandler:
  If Err.Number = -2147467259 Then
    MsgBox "PowerUser user already exists."
    Resume Next
  Else
    MsgBox Err.Description
    Resume ExitHere
  End If
End Sub
```

3. Choose **Run | Run Sub/UserForm** to execute the procedure.

Setting User Permissions for a Database

To specify permissions for the database, specify an empty string (“”) as the name of the database:

```
cat.Users("PowerUser").SetPermissions "", _
adPermObjDatabase, _
adAccessSet, adRightExclusive
```

This statement gives the user named PowerUser the right to open the database exclusively.

Figure 18.19 displays the permissions for the SpecialDb database that are set when the example procedure in Hands-On 18.8 is run.

Hands-On 18.8. Setting User Permissions for a Database (ADO)

1. In the Visual Basic Editor window of Chap18.mdb, choose **Insert | Module**.
2. In the module's Code window, enter the **Set_UserDbPermissions_ADO** procedure as shown here:

```
Sub Set_UserDbPermissions_ADO()
  Dim conn As ADODB.Connection
  Dim cat As ADOX.Catalog
  Dim strDB As String
  Dim strSysDB As String

  On Error GoTo ErrorHandler

  strDB = "C:\Access2013_ByExample\SpecialDb.mdb"
  strSysDB = "C:\Access2013_ByExample\Security1.mdw"

  ' Open connection to the database using
```

```
    ' the specified system database
    Set conn = New ADODB.Connection
    With conn
        .Provider = "Microsoft.Jet.OLEDB.4.0"
        .Properties("Jet OLEDB:System Database") = strSysDB
        .Properties("User ID") = "Developer"
        .Properties("Password") = "chapter18"
        .Open strDB
    End With

    ' Open the catalog
    Set cat = New ADOX.Catalog
    cat.ActiveConnection = conn

    ' add a user account
    cat.Users.Append "PowerUser", "star"

    ' Set permissions for PowerUser
    cat.Users("PowerUser").SetPermissions "", _
      adPermObjDatabase, adAccessSet, adRightExclusive
    MsgBox "PowerUser has been granted " & vbCrLf & _
      "permission to open the database exclusively."
  ExitHere:
    Set cat = Nothing
    conn.Close
    Set conn = Nothing
    Exit Sub
  ErrorHandler:
    If Err.Number = -2147467259 Then
        ' because PowerUser user already exists
        ' we ignore this statement
        Resume Next
    Else
        MsgBox Err.Description
        Resume ExitHere
    End If
  End Sub
```

3. Choose **Run | Run Sub/UserForm** to execute the procedure.

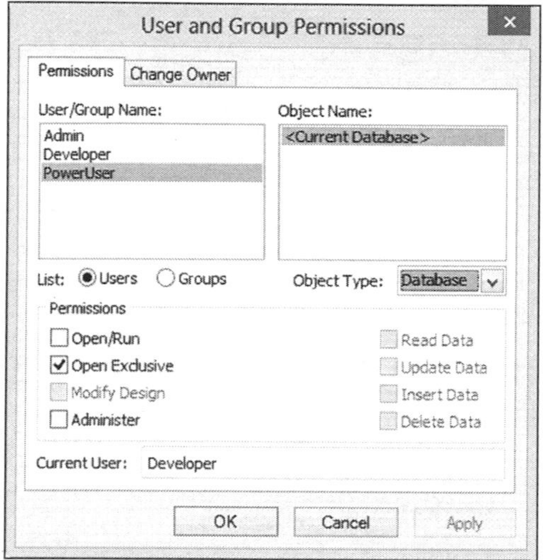

FIGURE 18.19. The settings shown here are found in the User and Group Permissions window for the SpecialDb database after running the Set_UserDbPermissions_ADO procedure in Hands-On 18.8.

Setting User Permissions for Containers

Now that you've learned how to grant permissions to a user for a specific object such as a table or query, you may want to know how to specify permissions for an entire set of objects such as tables, queries, forms, reports, and macros.

Each Database object has a Containers collection consisting of built-in Container objects. A Container object groups together similar types of Document objects. You can use the Containers collection to set security for all Document objects of a given type. You can set the permissions that users and groups will receive by default on all newly created objects in a database by passing in Null for the object name argument of the ADOX `SetPermissions` method, as shown in the example procedure in Hands-On 18.9.

Hands-On 18.9. Setting User Permissions for Containers (ADO)

1. In the Visual Basic Editor window of Chap18.mdb, choose **Insert | Module.**
2. In the module's Code window, enter the **Set_UserContainerPermissions_ ADO** procedure as shown here:

```
Sub Set_UserContainerPermissions_ADO()
  Dim conn As ADODB.Connection
  Dim cat As ADOX.Catalog
  Dim strDB As String
```

```
   Dim strSysDB As String

   On Error GoTo ErrorHandler

   strDB = "C:\Access2013_ByExample\SpecialDb.mdb"
   strSysDB = "C:\Access2013_ByExample\Security1.mdw"

   ' Open connection to the database using
   ' the specified system database
   Set conn = New ADODB.Connection
   With conn
     .Provider = "Microsoft.Jet.OLEDB.4.0"
     .Properties("Jet OLEDB:System Database") = strSysDB
     .Properties("User ID") = "Developer"
     .Properties("Password") = "chapter18"
     .Open strDB
   End With

   ' Open the catalog
   Set cat = New ADOX.Catalog
   cat.ActiveConnection = conn

   ' add a user account
   cat.Users.Append "PowerUser", "star"

   ' Set permissions for PowerUser on the Tables Container
   cat.Users("PowerUser").SetPermissions Null, _
    adPermObjTable, _
    adAccessSet, _
    adRightRead Or _
    adRightInsert Or _
    adRightUpdate Or _
    adRightDelete, adInheritNone
   MsgBox "You have successfully granted " & vbCrLf & _
    "permissions to PowerUser on the Tables Container."
ExitHere:
   Set cat = Nothing
   conn.Close
   Set conn = Nothing
   Exit Sub
ErrorHandler:
   If Err.Number = -2147467259 Then
```

```
      ' because PowerUser user already exists
      ' we ignore this statement
      Resume Next
    Else
      MsgBox Err.Description
      Resume ExitHere
    End If
  End Sub
```

3. Choose **Run | Run Sub/UserForm** to execute the procedure.

   This procedure gives the PowerUser account the permission to design, read, update, insert, and delete data for all newly created tables and queries. Notice that Null is passed as the first argument of the `SetPermissions` method to indicate that permissions are to be set only on new objects of the type specified by the second argument of this method.

   After executing this procedure, the user account PowerUser has the permissions listed in Figure 18.20 on all newly created Table and Query objects.

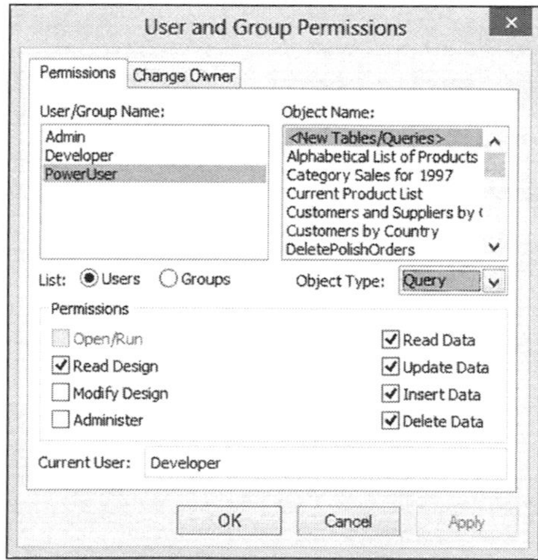

FIGURE 18.20. The settings shown here are found in the User and Group Permissions window after running the Set_UserContainerPermissions_ADO procedure in Hands-On 18.9.

Checking Permissions for Objects

You can retrieve the permissions for a particular user or group on a particular object with the ADOX `GetPermissions` method. Because this method returns a numeric permission value for the specified object, you must write more code to decipher the returned value if you want to display the names of constants representing permissions. The procedure in Hands-On 18.10 demonstrates how to retrieve the permissions set for PowerUser on the Customers table in a sample database (Figure 18.21).

Hands-On 18.10. Checking Permissions for a Specific Object (ADO)

1. In the Visual Basic Editor window of Chap18.mdb, choose **Insert | Module.**
2. In the module's Code window, enter the **GetObjectPermissions_ADO** procedure as shown here:

```
Sub GetObjectPermissions_ADO(strUserName As String, _
  varObjName As Variant, _
  lngObjType As ADOX.ObjectTypeEnum)

  Dim conn As ADODB.Connection
  Dim cat As ADOX.Catalog
  Dim strDB As String
  Dim strSysDB As String
  Dim listPerms As Long
  Dim strPermsTypes As String

  On Error GoTo ErrorHandler

  strDB = "C:\Access2013_ByExample\SpecialDb.mdb"
  strSysDB = "C:\Access2013_ByExample\Security1.mdw"

  ' Open connection to the database using
  ' the specified system database
  Set conn = New ADODB.Connection
  With conn
      .Provider = "Microsoft.Jet.OLEDB.4.0"
      .Properties("Jet OLEDB:System Database") = strSysDB
      .Properties("User ID") = "Developer"
      .Properties("Password") = "chapter18"
      .Open strDB
  End With

  ' Open the catalog
  Set cat = New ADOX.Catalog
  cat.ActiveConnection = conn
  ' add a user account
  cat.Users.Append "PowerUser", "star"
```

```
    listPerms = cat.Users(strUserName) _
      .GetPermissions(varObjName, lngObjType)
    Debug.Print listPerms

    If (listPerms And ADOX.RightsEnum.adRightCreate) = _
      adRightCreate Then
        strPermsTypes = strPermsTypes & "adRightCreate" & vbCr
    End If
    If (listPerms And RightsEnum.adRightRead) = _
      adRightRead Then
        strPermsTypes = strPermsTypes & "adRightRead" & vbCr
    End If
    If (listPerms And RightsEnum.adRightUpdate) = _
      adRightUpdate Then
        strPermsTypes = strPermsTypes & "adRightUpdate" & vbCr
    End If
    If (listPerms And RightsEnum.adRightDelete) = _
      adRightDelete Then
        strPermsTypes = strPermsTypes & "adRightDelete" & vbCr
    End If
    If (listPerms And RightsEnum.adRightInsert) = _
      adRightInsert Then
        strPermsTypes = strPermsTypes & "adRightInsert" & vbCr
    End If
    If (listPerms And RightsEnum.adRightReadDesign) = _
      adRightReadDesign Then
        strPermsTypes = strPermsTypes & "adRightReadDesign" & vbCr
    End If

    Debug.Print strPermsTypes
    MsgBox "Permissions are listed in the Immediate Window."
ExitHere:
    Set cat = Nothing
    conn.Close
    Set conn = Nothing
    Exit Sub
ErrorHandler:
    If Err.Number = -2147467259 Then
        ' because PowerUser user already exists
        ' we ignore this statement
        Resume Next
    Else
```

```
        MsgBox Err.Description
        Resume ExitHere
    End If
End Sub
```

3. To run the GetObjectPermissions_ADO procedure, type the following statement in the Immediate window and press **Enter** to execute it:

```
GetObjectPermissions_ADO "PowerUser", "Customers", adPermObjTable
```

```
Immediate
GetObjectPermissions_ADO "PowerUser", "Customers", adPermObjTable
-1073642496
adRightRead
adRightUpdate
adRightDelete
adRightInsert
adRightReadDesign
```

FIGURE 18.21. The procedure in Hands-On 18.10 writes the permissions found for PowerUser in the Customers table to the Immediate window.

Setting a Database Password Using the CompactDatabase Method

You can implement share-level security by setting a database password. When you set a database password, the password dialog box will appear when you open the database. Only users with a valid password can open the database.

You cannot use ADOX objects to set a database password. Instead, you have to use the objects from the Microsoft Jet and Replication Objects (JRO) Library. Use the `CompactDatabase` method of the JRO JetEngine object and specify the `Password` parameter. Remember that passwords are case sensitive.

The procedure in Hands-On 18.11 sets the sample database password to "welcome." Before running the procedure in Hands-On 18.11, you must set a reference to the Microsoft Jet and Replication Objects Library. To do this, choose Tools | References in the Visual Basic Editor window and select the required library in the list of Available References.

Hands-On 18.11. Setting a Database Password (JRO)

1. Create a new Access database named **PasswordTest.mdb** in your **C:\Access2013_ByExample** folder. Close this database before proceeding to step 2.
2. In the Visual Basic Editor window of Chap18.mdb, choose **Insert | Module**.

3. In the module's Code window, enter the **Change_DBPassword** procedure as shown here:

```
Sub Change_DBPassword()
  Dim jetEng As JRO.JetEngine
  Dim strCompactFrom As String
  Dim strCompactTo As String
  Dim strPath As String

  On Error GoTo ErrHandler
  strPath = CurrentProject.Path & "\"

  strCompactFrom = "PasswordTest.mdb"
  strCompactTo = "PasswordTest_Compact.mdb"

  Set jetEng = New JRO.JetEngine
  ' Compact the database specifying the new database password
  jetEng.CompactDatabase "Data Source=" & strPath & _
    strCompactFrom & ";", "Data Source=" & strPath & _
    strCompactTo & ";" & "Jet OLEDB:Database Password=welcome"

ExitHere:
  Set jetEng = Nothing
  Exit Sub
ErrHandler:
  If Err.Number = -2147217897 Then
      Kill strPath & strCompactTo
      Resume
  Else
      MsgBox Err.Number & ": " & Err.Description
      Resume ExitHere
  End If
End Sub
```

4. Choose **Run | Run Sub/UserForm** to execute the procedure.
5. After you run this procedure, open the **PasswordTest_Compact.mdb** database file. You should be prompted for the password. Type **welcome** to log in.
6. Close the **PasswordTest_Compact.mdb** file.

Setting a Database Password Using the NewPassword Method

The DAO Object Model has a `NewPassword` method of the Database object that you can use to change the password of an existing Microsoft Access database in

.accdb or .mdb file format. The `NewPassword` method requires two parameters. The first one specifies the old password, and the second one provides the new password. Both passwords can be up to 20 characters long and can include any characters except the ASCII character 0 (Null). Use a zero-length string ("") for the old password if the database does not have a password. Use a zero-length string ("") for the new password to clear the password. Password operations require that the database is open in exclusive mode. Remember that passwords are case sensitive.

The procedure in Hands-On 18.12 sets the password for the Chap01.accdb database you created at the beginning of this book.

Hands-On 18.12. Setting a Database Password (DAO)

1. In the Visual Basic Editor window of Chap18.mdb, choose **Insert | Module.**
2. In the module's Code window, enter the **Set_DBPassword_DAO** procedure as shown here:

```
Sub Set_DBPassword_DAO()
  Dim db As DAO.Database
  Dim strDB As String

  strDB = CurrentProject.Path & "\Chap01.accdb"

  Set db = OpenDatabase(strDB, True)

  db.NewPassword "", "chapter1"
  db.Close
  Set db = Nothing
End Sub
```

3. Choose **Run | Run Sub/UserForm** to execute the procedure.
   The second parameter (`True`) in the `OpenDatabase` method tells VBA to open the database in exclusive mode.
4. Open the **Chap01.accdb** database and notice that you are now prompted to enter a password. Type **chapter1** for the password and click **OK**. When the database opens, close it and close the Access window in which it was opened. Do not close the Chap18.mdb file.

You can unset the password on the Chap01.accdb database by running the following procedure:

```
Sub Unset_DBPassword_DAO()
  Dim db As DAO.Database
```

```
    Dim strDB As String

    strDB = CurrentProject.Path & "\Chap01.accdb"
    Set db = OpenDatabase(strDB, True, False, ";pwd=chapter1")

    db.NewPassword "chapter1", ""
    db.Close
    Set db = Nothing
End Sub
```

Notice the parameters of the `OpenDatabase` method. `True` specifies that the database is to be opened in exclusive mode, and `False` indicates that the database should be opened in read/write mode. Because the Chap01.accdb database has been protected with a password by the Set_DBPassword_DAO procedure in Hands-On 18.12, we also had to specify the password in the connect parameter. After you run the preceding procedure, you will not be prompted for a password when you open the Chap01.accdb database.

Changing a User Password

User passwords are stored in the workgroup information file. To change a user's password in VBA code, use the ADOX User object's `ChangePassword` method. This method takes as parameters the user's current password and the new password. If a user does not yet have a password, use an empty string ("") for the user's current password.

The procedure in Hands-On 18.13 demonstrates how to change a password for the Admin user. Recall that Admin is the default user account that has a blank password. In an unsecured Access database, all users are automatically logged on using the Admin account. When establishing user-level security, you should start by changing the password for the Admin user. Changing an Admin password activates the Logon dialog box the next time you start Microsoft Access. Only users with a valid username and password will be able to log onto the database. Although users are permitted to change their own passwords, only a user who belongs to the Admins group can clear a password that another user has forgotten.

Hands-On 18.13. Changing a User Password (ADO)

1. Create a new Access database named **AdminPwd.mdb** in your **C:\Access2013_ByExample** folder. Close this database before proceeding to Step 2.
2. In the Visual Basic Editor window of Chap18.mdb, choose **Insert | Module**.

3. In the module's Code window, enter the **Change_UserPassword_ADO** procedure as shown here:

```
Sub Change_UserPassword_ADO()
  Dim cat As ADOX.Catalog
  Dim strDB As String
  Dim strSysDB As String

  On Error GoTo ErrorHandler

  strDB = CurrentProject.Path & "\AdminPwd.mdb"
  ' change the path to use the default workgroup
  ' information file on your computer
  strSysDB = "C:\Users\Itka\" & _
    "Application Data\Microsoft\Access\System.mdw"

  'strSysDB = "C:\Access2013_ByExample\Security1.mdw"

  ' Open the catalog, specifying the system database to use
  Set cat = New ADOX.Catalog
  With cat
    .ActiveConnection = "Provider='Microsoft.Jet.OLEDB.4.0';" & _
    "Data Source='" & strDB & "';" & _
    "Jet OLEDB:System Database='" & strSysDB & "';" & _
    "User Id=Admin;Password=;"

    ' Change the password for the Admin user
    .Users("Admin").ChangePassword "", "secret"
  End With

ExitHere:
  Set cat = Nothing
  Exit Sub
ErrorHandler:
  MsgBox Err.Description
  GoTo ExitHere
End Sub
```

4. Choose **Run | Run Sub/UserForm** to execute the procedure.
5. When you open the AdminPwd.mdb database after running the procedure, a Logon dialog box will appear, as shown in Figure 18.22. Enter **Admin** in the Name text box and **secret** in the Password text box, and click **OK**.

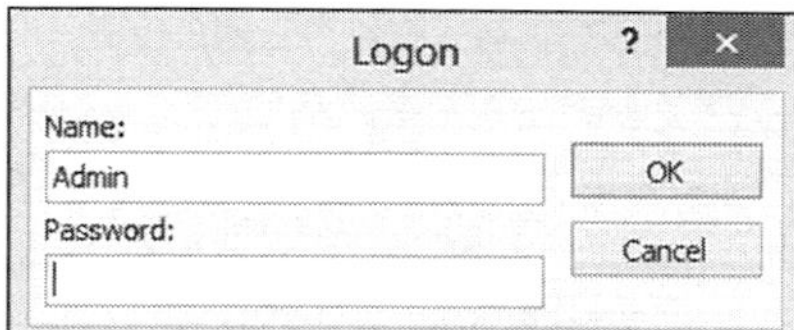

FIGURE 18.22. A Logon dialog box requests the username and password for the Admin user.

6. Remove the Admin password by choosing **File | Info | Users and Permissions | User and Group Accounts**. Click the **Change Logon Password** tab and type the old password (secret). Click the **Apply** button, and click **OK** to exit the User and Group Accounts window (Figure 18.23).

FIGURE 18.23. You can remove the Admin password via the Change Logon Password tab in the User and Group Accounts window or by modifying the VBA code shown in Hands-On 18.13.

7. After removing the Admin password, reopen the **AdminPwd.mdb** file. The database should now open without prompting you to enter a password.
8. Close the **AdminPwd.mdb** database and exit the Access window in which the file was opened.

ENCRYPTING A SECURED MDB DATABASE

To achieve a higher level of security and protect your database from unauthorized access, you can encrypt it. Prior to encrypting, secure your database by setting user and group permissions on database objects. To encrypt a database, you must be the owner or the creator of the database, or a member of the Admins group in the workgroup information file (System.mdw) that was in use when the database was created.

Use the `CompactDatabase` method of the Microsoft Jet and Replication Objects (JRO) JetEngine object to encrypt or decrypt a database. To use the JRO JetEngine object, you must first set a reference to the Microsoft Jet and Replication Objects Library. To encrypt the database, set the Jet OLEDB:Encrypt Database property to `True` in the connection string destination argument of the `CompactDatabase` method.

After a database has been encrypted, it cannot be read or written to directly by using any utility program or word processor. The procedure in Hands-On 18.14 creates an encrypted version of the SpecialDb.mdb database that you created in Custom Project 18.1.

Hands-On 18.14. Encrypting a Database (JRO)

The procedure in this hands-on exercise should be run after you have initially secured your database by creating the necessary user and group accounts and assigned user and group permissions on database objects.

1. In the Visual Basic Editor window of Chap18.mdb, choose **Insert | Module.**
2. In the module's Code window, enter the **EncryptDb** procedure as shown here:

```
Sub EncryptDb()
  Dim jetEng As JRO.JetEngine
  Dim strCompactFrom As String
  Dim strCompactTo As String
  Dim strSource As String
  Dim strDest As String
  Dim strSysDB As String

  strCompactFrom = CurrentProject.Path & "\SpecialDb.mdb"
  strCompactTo = CurrentProject.Path & "\SpecialDb_Enc.mdb"
  strSysDB = CurrentProject.Path & "\Security1.mdw"

  On Error GoTo HandleErr

  ' Use the CompactDatabase method to create
  ' a new, encrypted version of the database
  Set jetEng = New JRO.JetEngine

  strSource = "Data Source=" & strCompactFrom & ";" & _
   "Jet OLEDB:System Database=" & strSysDB & ";" & _
   "User ID=Developer" & ";" & _
   "Password=chapter18"
```

```
    strDest = "Data Source=" & strCompactTo & ";" & _
     "Jet OLEDB:Engine Type=5;" & _
     "Jet OLEDB:Encrypt Database=True"

    jetEng.CompactDatabase strSource, strDest
  ExitHere:
    Set jetEng = Nothing
    Exit Sub
  HandleErr:
    MsgBox Err.Number & ": " & Err.Description
    Resume ExitHere
  End Sub
```

3. Choose **Run | Run Sub/UserForm** to execute the procedure.

To open the secured SpecialDb_Enc database file, you must provide the name of the workgroup information file as shown here:

```
"C:\Program Files\Microsoft Office 15\root\office15\MSACCESS.EXE"
"C:\Access2013_ByExample\SpecialDb_Enc.mdb" /WRKGRP
"C:\Access2013_ByExample\Security1.mdw"
/User "Developer" /Pwd "chapter18"
```

Use the preceding command to set up a shortcut on your desktop similar to the one that was set up by the Access Security Wizard when you completed Custom Project 18.1. You should log in as Developer using "chapter18" as the password.

CHAPTER SUMMARY

In this chapter, you worked with VBA procedures that implemented share-level and user-level security in Microsoft Access databases. You found out that in Access 2013, user-level security can only be used in Access databases created in the .mdb file format. You learned to work with workgroup information files and practiced creating and modifying user and group accounts, and setting user permissions to a database and its objects. You also learned how to set a database password on Access databases in the .mdb and .accdb file formats, and how to write a procedure that encrypts a secured .mdb file.

This chapter concludes Part V of this book in which we focused on performing important database tasks using DAO and ADO.

The next chapter will show you how to use the Data Definition Language to work with tables and fields.

Part 6 Programming with the Jet Data Definition Language

Data Definition Language (DDL) is a component of Structured Query Language (SQL), which is used for defining database objects (tables, views, stored procedures, primary keys, indexes, and constraints) and managing database security. In this part of the book, you will learn how to use DDL with Jet databases, ADO, and the Jet 4.0/ACE OLE DB Provider.

Chapter 19 Creating, Modifying, and Deleting Tables and Fields

In Part V of this book, you tried out different methods that are available in Microsoft Access 2013 for creating and manipulating databases via VBA programming code using the DAO and ADO object models. In particular, you learned how to create new databases from scratch, add tables and indexes, set up relationships between tables, secure a database with a password, define user and group security accounts, and handle object permissions. In addition to using DAO and ADO, you can perform many of the mentioned database tasks by using Data Definition Language (DDL), which is a component of Structured Query Language (SQL).

SQL is a widely used language for data retrieval and manipulation in databases. The SQL specification (known as ANSI SQL-89) was first published in 1989 by the American National Standards Institute (ANSI). The ANSI SQL standard was revised in 1992; this version is referred to as ANSI SQL-92 or SQL-2. This revised specification is supported by the major database vendors, many of whom have created their own extensions of the SQL language. Microsoft Access 2013 supports both SQL specifications and refers to them as *ANSI SQL query modes*.

While the ANSI-89 SQL query mode (also called Microsoft Jet SQL and ANSI SQL) uses the traditional Jet SQL syntax, the ANSI-92 SQL mode uses syntax that is more compliant with SQL-92 and Microsoft SQL Server. For example, ANSI-92 uses the percent sign (%) and the underscore character (_) for its wildcards instead of the asterisk (*) and the question mark (?), which are commonly used in VBA. Microsoft Access Jet Engine does not implement the complete ANSI SQL-92 standard and provides its own Jet 4.0 ANSI SQL-92 extensions to support new features of Access. You can use the ANSI-92 syntax in your VBA procedures with the Microsoft OLE DB Provider for Jet or with the Data Definition Language, which we cover in this part of the book. ANSI-89 is the default setting for a new Microsoft Access database in Access 2002-2003 and 2000 file formats. Because the two ANSI SQL query modes are not compatible, you must decide which query mode you are going to use for the current database. This can easily be done in the Microsoft Access user interface as outlined in Hands-On 19.1.

Please note files for the "Hands-On" project may be found on the companion CD-ROM.

Hands-On 19.1. Setting the ANSI SQL Query Mode

1. Start Microsoft Access 2013 and create a new database named **Chap19.accdb** in your **C:\Access2013_ByExample** folder.
2. Click the **File** tab and select the **Options** button.
3. In the left pane of the Access Options window, select **Object Designers**.
4. In the right pane, in the Query design section, look for the SQL Server Compatible Syntax (ANSI 92) area (see Figure 19.1). Set the query mode to ANSI-92 SQL by clicking the **This database** checkbox. (You can set the query mode to ANSI-89 SQL by clearing the **This database** checkbox.)

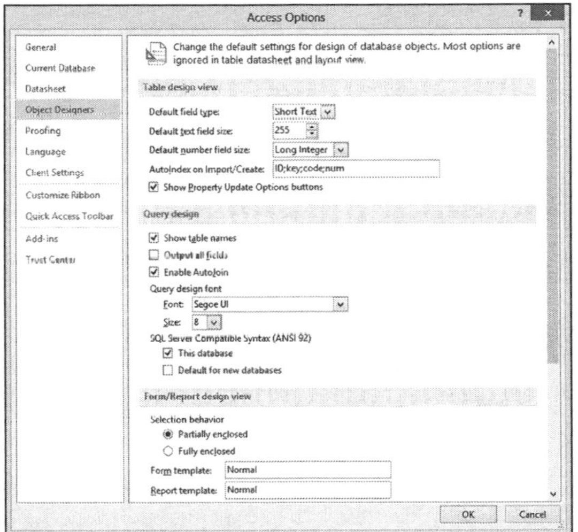

FIGURE 19.1. Use the Access Options window to set the ANSI SQL query mode for the current database or all new databases.

5. Click **OK** to exit the Access Options window. Microsoft Access displays a message as shown in Figure 19.2.
6. Click **OK** to accept the message. The Microsoft Access database will close and reopen with the new settings in effect.

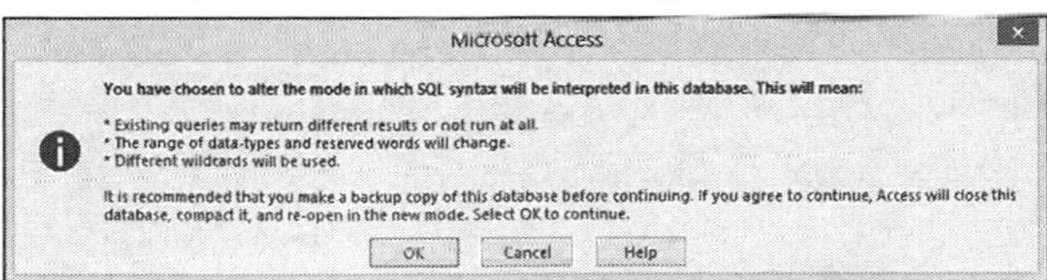

FIGURE 19.2. When you change the query mode to ANSI-92, Microsoft Access displays an informational message alerting you to possible problems.

There are two areas of Microsoft Access SQL:

- Data Definition Language (DDL) offers a number of SQL statements to manage database security and to create and alter database components (such as tables, indexes, relationships, views, and stored procedures). These statements are: `CREATE TABLE`, `DROP TABLE`, `ALTER TABLE`, `CREATE INDEX`, `DROP INDEX`, `CHECK CONSTRAINT`, `CREATE VIEW`, `DROP VIEW`, `CREATE PROCEDURE`, `DROP PROCEDURE`, `EXECUTE`, `ALTER DATABASE`, `ADD USER`, `ALTER USER`, `CREATE USER`, `CREATE GROUP`, `DROP GROUP`, `DROP USER`, `GRANT`, and `REVOKE`.
- Data Manipulation Language (DML) offers SQL statements that allow you to retrieve and manipulate data contained in the database tables as

well as perform transactions. These statements are: `SELECT`, `UNION`, `UPDATE`, `DELETE`, `INSERT INTO`, `SELECT INTO`, `INNER JOIN`, `LEFT JOIN`, `RIGHT JOIN`, `TRANSFORM`, `PARAMETERS`, `BEGIN TRANSACTION`, `COMMIT`, and `ROLLBACK`.

This chapter and the remaining chapters of Part VI focus on using the DDL language for creating and changing the underlying structure of a database. To get the most out of these chapters, you should be familiar with using DAO and ADO, discussed in Part V.

CREATING TABLES

Using the Microsoft Access SQL `CREATE TABLE` statement and the `Execute` method of either the DAO Database object or the ADO Connection object, you can define a new table, its fields, and field constraints. The `CREATE TABLE` statement can only be used with Microsoft Jet and Microsoft Access engine databases. The two examples that follow illustrate how to create a table named tblSchools in the currently open database and in a new database using ADO.

Hands-On 19.2. Creating a Table in the Current Database (DDL with ADO)

1. In the **Chap19.accdb** database that you created in Hands-On 19.1, switch to the Visual Basic Editor window and choose **Tools | References**. In the References dialog box, scroll down to locate **Microsoft ActiveX Data Objects 6.1 Library**. Click the checkbox to the left of this library name to set a reference to it and click **OK** to exit the dialog box.
2. Choose **Insert | Module** to add a new module to the current VBA project.
3. In the module's Code window, type the following **CreateTable** procedure:

```
Sub CreateTable()
  ' you must set up a reference to
  ' the Microsoft ActiveX Data Objects Library
  ' in the References dialog box
  Dim conn As ADODB.Connection
  Dim strTable As String

  On Error GoTo ErrorHandler

  Set conn = CurrentProject.Connection
```

```
    strTable = "tblSchools"
    conn.Execute "CREATE TABLE " & strTable & _
     "(SchoolID AUTOINCREMENT(100, 5), " & _
     "SchoolName CHAR," & _
     "City CHAR (25), District CHAR (35), " & _
     "YearEstablished DATE);"

    Application.RefreshDatabaseWindow
  ExitHere:
    conn.Close
    Set conn = Nothing
    Exit Sub
  ErrorHandler:
    MsgBox Err.Number & ":" & Err.Description
    Resume ExitHere
  End Sub
```

4. Position the insertion point anywhere within the code of the CreateTable procedure and press **F5** or choose **Run | Run Sub/UserForm** to execute the procedure.

This procedure uses ADO to establish a connection to the current database (Chap19.accdb). The ADO Connection object's `Execute` method is used to execute the Data Definition Language `CREATE TABLE` statement that defines a new table and its fields. The first field is named SchoolID and its data type is defined as AutoNumber.

The seed and increment values of AutoNumber columns are specified using the following syntax:

```
Column_name AUTOINCREMENT (seed, increment)
```

The table, tblSchools, has an AutoNumber column with a seed of 100 and an increment of 5:

```
SchoolID AUTOINCREMENT(100, 5)
```

When you switch to the database window and open this table in Datasheet view, the SchoolID for the first record will be 100, the second will be 105, the third 110, and so on.

Three fields are defined as Text fields and one field as a Date/Time field. The Text fields are defined using the CHAR data type (see Table 19.1). To specify the size of the Text field, put the appropriate value between parentheses. If the size of the Text field is not specified, it is assumed to be 255 characters long.

When you examine the code of the CreateTable procedure and compare the resultant table in Figure 19.3, you will notice that Access SQL uses different data types than those available in the Table Design window. Table 19.1 shows the equivalent SQL data types.

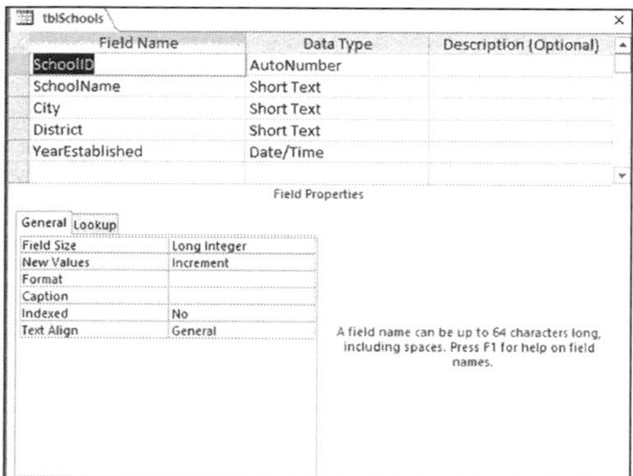

FIGURE 19.3. The tblSchools table was generated by the CreateTable procedure in Hands-On 19.2 using the Microsoft Access SQL statement CREATE TABLE.

TABLE 19.1. Table design data types and their Access SQL equivalents

Table Design Data Types	Access SQL Data Types
Text	TEXT, ALPHANUMERIC, CHAR, CHARACTER, STRING, or VARCHAR
Memo	LONGTEXT, MEMO, LONGCHAR, or NOTE
Number (Field Size = Byte)	BYTE or INTEGER1
Number (Field Size = Integer)	SHORT, INTEGER2, or SMALLINT
Number (Field Size = Long Integer)	COUNTER, INTEGER, INT, or AUTOINCREMENT
Number (Field Size = Single)	SINGLE, FLOAT4, or REAL
Number (Field Size = Double)	DOUBLE, FLOAT, or NUMBER
Date/Time	DATETIME, DATE, TIME, or TIMESTAMP
Currency	CURRENCY or MONEY
AutoNumber (Field Size = Long Integer)	AUTOINCREMENT or COUNTER
AutoNumber (Field Size = Replication Id)	GUID
Yes/No	BOOLEAN, BIT, LOGICAL, LOGICAL1, or YESNO
OLE Object	LONGBINARY, OLEOBJECT, or GENERAL

The `RefreshDatabaseWindow` method of the Application object ensures that the database window is updated after the creation of the new table object. The error-handling code will alert you if an error is encountered. Try to run this procedure again in step mode (F8) to see what happens. Notice that the procedure uses two labels to mark appropriate sections in the procedure. The `On Error GoTo ErrorHandler` statement will transfer the procedure execution to the line labeled `ErrorHandler` when an error is triggered. Statements following this label will be executed until the `Resume` statement is encountered. This statement will direct the code execution to the line labeled `ExitHere`. The `Exit Sub` statement in the `ExitHere` block of code will allow us to exit the procedure whether or not an error is encountered.

Additional Code on CD: see CreateTable2_Chap19.txt

Sometimes you may be required to create a new database and a new table in one procedure. Hands-On 19.3 demonstrates how to create a table in a brand-new database.

Hands-On 19.3. Creating a Table in a New Database (DDL with ADO/ADOX)

1. In the Visual Basic Editor window, choose **Tools | References**. In the References dialog box, scroll down to locate **Microsoft ADO Ext. 6.0 for DDL and Security Object Library**. Click the checkbox to the left of the library name to set a reference to it. Also, make sure that the Microsoft ActiveX Data Objects 6.1 Library is selected. Click **OK** to exit the dialog box.
2. In the module's Code window, enter the following **CreateTableInNewDB** procedure:

```
Sub CreateTableInNewDB()
  ' use the References dialog box to set up a reference to
  ' Microsoft ADO Ext. 6.0 for DDL and Security Object Library
  ' and Microsoft ActiveX Data Objects 6.1 Library
  Dim cat As ADOX.Catalog
  Dim conn As ADODB.Connection
  Dim strDB As String
  Dim strTable As String
  Dim strConnect As String

  On Error GoTo ErrorHandler
```

```
    Set cat = New ADOX.Catalog
    strDB = CurrentProject.Path & "\Sites.mdb"
    strConnect = "Provider=Microsoft.Jet.OLEDB.4.0;" & _
     "Data Source=" & strDB

    ' create a new database file
    cat.Create strConnect
    MsgBox "The database was created (" & strDB & ")."

    ' set connection to currently open catalog
    Set conn = cat.ActiveConnection

    strTable = "tblSchools"
    conn.Execute "CREATE TABLE " & strTable & _
     "(SchoolID AUTOINCREMENT(100, 5), " & _
     "SchoolName CHAR," & _
     "City CHAR (25), District CHAR (35), " & _
     "YearEstablished DATE);"
  ExitHere:
    Set cat = Nothing
    Set conn = Nothing
    Exit Sub
  ErrorHandler:
    If Err.Number = -2147217897 Then
      ' delete the database file if it exists
      Kill strDB
      ' start from statement that caused this error
      Resume 0
    Else
      MsgBox Err.Number & ": " & Err.Description
      GoTo ExitHere
    End If
  End Sub
```

3. Position the insertion point anywhere within the CreateTableInNewDB procedure and press **F5** or choose **Run | Run Sub/UserForm** to execute the procedure.

The CreateTableInNewDB procedure shown here creates a new database named Sites.mdb in the current folder. As you may recall from Chapter 11, you can create a Microsoft Access database by using the `Create` method of the ADOX Catalog object. Before creating a table in the new database, set the `conn` object variable to the currently open Catalog, like this:

```
Set conn = cat.ActiveConnection
```

Use the Connection object's `Execute` method to create a new table named tblSchools in the Sites.mdb file. Like other procedure examples in this section, this table contains an AutoNumber field with a sequence starting at 100 that will be incremented by 5 as new columns are added. Notice that the error-handling code demonstrated in this procedure is slightly different from previous examples. If you know the type of error that is most likely to occur, you can check for the error number in the error handler and execute the appropriate statement when the condition is met. If the database already exists, it will be deleted using the VBA `Kill` statement (don't do this in the production environment unless you are absolutely certain this is what you want to do). The statement `Resume 0` in the error-handling code will return the code execution to the line that caused the error. If other errors are encountered, error information will appear in a message box and the code execution will continue from the line following the `ExitHere` label.

DELETING TABLES

It's time to remove some of our test data by using the `DROP TABLE` statement to delete an existing table from a database. Note that a table must be closed before it can be deleted. The procedure in Hands-On 19.4 will delete the tblSchool table that was created in Hands-On 19.2.

Hands-On 19.4. Deleting a Table

This hands-on exercise requires the prior completion of Hands-On 19.2.

1. In the Visual Basic Editor window, choose **Insert | Module.**
2. In the module's Code window, enter the following **DeleteTable** procedure:

```
Sub DeleteTable()
  Dim conn As ADODB.Connection
  Dim strTable As String

  On Error GoTo ErrorHandler
  Set conn = CurrentProject.Connection

  strTable = "tblSchools"
  conn.Execute "DROP TABLE " & strTable
  Application.RefreshDatabaseWindow
ExitHere:
```

```
    conn.Close
    Set conn = Nothing
    Exit Sub
  ErrorHandler:
    If Err.Number = -2147217900 Then
      DoCmd.Close acTable, strTable, acSavePrompt
      Resume 0
    Else
      MsgBox Err.Number & ":" & Err.Description
      Resume ExitHere
    End If
  End Sub
```

3. Position the insertion point anywhere within the code of the DeleteTable procedure and press **F5** or choose **Run | Run Sub/UserForm** to execute the procedure.

You can also execute the `DROP TABLE` statement directly in the Microsoft Access user interface's Data Definition Query window by following these steps:

4. Choose **Create | QueryDesign**.
5. Click the **Close** button in the Show Table dialog box.
6. Choose **Design | SQL | Data Definition**.
7. Enter the following statement in the Query window:

```
DROP TABLE tblSchools2;
```

8. Choose **Design | Run**.

MODIFYING TABLES WITH DDL

You can modify a table definition by altering, adding, or dropping columns and constraints. *Constraints* allow you to enforce integrity by creating rules for a table. The procedures in the following sections illustrate how to use Microsoft Access SQL DDL statements to:

- Add new columns to a table
- Change the column's data type
- Change the size of a Text column
- Delete a field from a table
- Add a primary key to an existing table

- Add a unique, multiple-field index to an existing table
- Delete an index
- Set a default value for a column in a table
- Change the seed and increment values of AutoNumber columns

Adding New Fields to a Table

Use the `ALTER TABLE` statement followed by a table name to modify the design of a table after it has been created with the `CREATE TABLE` statement. Prior to modifying the structure of an existing table, it's recommended that you make a backup copy of the table.

The `ALTER TABLE` statement can be used with the `ADD COLUMN` clause to add a new field to the table. For example, the procedure in Hands-On 19.5 adds a Currency field called Budget2014 to the tblSchools table using the following statement:

```
ALTER TABLE tblSchools ADD COLUMN Budget2014 MONEY
```

When you add a new field to a table, you should specify the name of the field, its data type and, for Text and Binary fields, the size of the field.

Hands-On 19.5. Adding a New Field to an Existing Table

1. Run the procedure in Hands-On 19.2 to create the tblSchools table in the current database if you deleted the table in Hands-On 19.4.
2. In the Visual Basic Editor window, choose **Insert | Module**.
3. In the module's Code window, enter the following **AddNewField** procedure:

```
Sub AddNewField()
  Dim conn As ADODB.Connection
  Dim strTable As String
  Dim strCol As String

  On Error GoTo ErrorHandler
  Set conn = CurrentProject.Connection

  strTable = "tblSchools"
  strCol = "Budget2014"

  conn.Execute "ALTER TABLE " & strTable & _
```

```
    " ADD COLUMN " & strCol & " MONEY;"
ExitHere:
  conn.Close
  Set conn = Nothing
  Exit Sub
ErrorHandler:
  MsgBox Err.Number & ":" & Err.Description
  Resume ExitHere
End Sub
```

4. Position the insertion point anywhere within the code of the AddNewField procedure and press **F5** or choose **Run | Run Sub/UserForm** to execute the procedure.

Changing the Data Type of a Table Column

You can use the `ALTER COLUMN` clause in the `ALTER TABLE` statement to change the data type of a table column. You must specify the name of the field, the desired data type, and the size of the field, if required.

The procedure in Hands-On 19.6 changes the data type of the SchoolID field in the tblSchools table from AutoNumber to a 15-character Text field.

Hands-On 19.6. Changing the Field Data Type

This hands-on exercise uses the tblSchools table created in Hands-On 19.2 and recreated in Hands-On 19.5.

1. In the same module where you entered the procedure in Hands-On 19.5, enter the following **ChangeFieldType** procedure:

```
Sub ChangeFieldType()
  Dim conn As ADODB.Connection
  Dim strTable As String
  Dim strCol As String

  On Error GoTo ErrorHandler
  Set conn = CurrentProject.Connection
  strTable = "tblSchools"
  strCol = "SchoolID"
  conn.Execute "ALTER TABLE " & strTable & _
   " ALTER COLUMN " & strCol & " CHAR(15);"
ExitHere:
  conn.Close
```

```
    Set conn = Nothing
    Exit Sub
  ErrorHandler:
    MsgBox Err.Number & ":" & Err.Description
    Resume ExitHere
  End Sub
```

2. Position the insertion point anywhere within the code of the ChangeFieldType procedure and press **F5** or choose **Run | Run Sub/UserForm** to execute the procedure.

   This procedure modifies the SchoolID data type to store text data. You can double-check the procedure changes by switching to the Microsoft Access window and opening the tblSchools table in Design view.

   When done, make sure you close the tblSchools table.

Changing the Size of a Text Column

It's easy to increase or decrease the size of a Text column. Simply use the ALTER TABLE statement followed by the name of the table, and the ALTER COLUMN clause followed by the name of the column whose size you want to modify. Then specify the data type of the column and the new column size.

Hands-On 19.7 modifies the size of the SchoolName field from the default 255 characters to 40.

Hands-On 19.7. Changing the Size of a Field

This hands-on exercise uses the tblSchools table created in Hands-On 19.2.

1. In the same module where you entered the procedure from the previous hands-on exercise, enter the following **ChangeFieldSize** procedure:

```
Sub ChangeFieldSize()
  Dim conn As ADODB.Connection
  Dim strTable As String
  Dim strCol As String

  On Error GoTo ErrorHandler
  Set conn = CurrentProject.Connection
  strTable = "tblSchools"
  strCol = "SchoolName"

conn.Execute "ALTER TABLE " & strTable & _
 " ALTER COLUMN " & strCol & " CHAR(40);"
```

```
ExitHere:
  conn.Close
  Set conn = Nothing
  Exit Sub
ErrorHandler:
  MsgBox Err.Number & ":" & Err.Description
  Resume ExitHere
End Sub
```

2. Position the insertion point anywhere within the code of the ChangeFieldSize procedure and press **F5** or choose **Run | Run Sub/UserForm** to execute the procedure.

   This procedure sets the size of the SchoolName field to 40 characters. You can double-check the procedure changes by switching to the Microsoft Access window and opening the tblSchools table in Design view.

   When done, be sure to close the tblSchools table.

Deleting a Column from a Table

Use the `DROP COLUMN` clause in the `ALTER TABLE` statement to delete a column from a table. You only need to specify the name of the field you want to remove.

The example procedure in Hands-On 19.8 deletes the Budget2014 column from the tblSchools table.

Hands-On 19.8. Deleting a Field from a Table

This hands-on exercise uses the tblSchools table created in Hands-On 19.2. Make sure this table contains the Budget2014 column, which was added in Hands-On 19.5.

1. In the same module where you entered previous hands-on exercises, enter the following **DeleteField** procedure:

```
Sub DeleteField()
  Dim conn As ADODB.Connection
  Dim strTable As String
  Dim strCol As String

  On Error GoTo ErrorHandler
  Set conn = CurrentProject.Connection

  strTable = "tblSchools"
  strCol = "Budget2014"
```

```
conn.Execute "ALTER TABLE " & strTable & _
 " DROP COLUMN " & strCol & ";"
ExitHere:
  conn.Close
  Set conn = Nothing
  Exit Sub
ErrorHandler:
  MsgBox Err.Number & ":" & Err.Description
  Resume ExitHere
End Sub
```

2. Position the insertion point anywhere within the code of the DeleteField procedure and press **F5** or choose **Run | Run Sub/UserForm** to execute the procedure.

   This procedure removes the Budget2014 field from the tblSchools table.

Adding a Primary Key to a Table

You can use the `ADD CONSTRAINT` clause in the `ALTER TABLE` statement to define one or more columns as a primary key. The primary key is defined using the `PRIMARY KEY` keywords.

Hands-On 19.9 defines a primary key for the tblSchools table created in Hands-On 19.2. The result is shown in Figure 19.4.

Hands-On 19.9. Adding a Primary Key to a Table

This hands-on exercise uses the tblSchools table created in Hands-On 19.2.

1. In the same module where you entered previous hands-on exercises, enter the following **AddPrimaryKey** procedure:

```
Sub AddPrimaryKey()
  Dim conn As ADODB.Connection
  Dim strTable As String
  Dim strCol As String

  On Error GoTo ErrorHandler
  Set conn = CurrentProject.Connection

  strTable = "tblSchools"
  strCol = "SchoolID"

  conn.Execute "ALTER TABLE " & strTable & _
```

```
    " ADD CONSTRAINT pKey PRIMARY KEY " & _
    "(" & strCol & ");"
ExitHere:
  conn.Close
  Set conn = Nothing
  Exit Sub
ErrorHandler:
  MsgBox Err.Number & ":" & Err.Description
  Resume ExitHere
End Sub
```

2. Position the insertion point anywhere within the code of the AddPrimaryKey procedure and press **F5** or choose **Run | Run Sub/UserForm** to execute the procedure.

Adding a Multiple-Field Index to a Table

Use the `ADD CONSTRAINT` clause and the `UNIQUE` keyword in the `ALTER TABLE` statement to add a multiple-field index. The `UNIQUE` keyword prevents duplicate values in the index.

Hands-On 19.10. Adding a Unique Index Based on Two Fields to an Existing Table

This hands-on exercise uses the tblSchools table created in Hands-On 19.2.

1. In the same module where you entered previous hands-on exercises, enter the following **AddMulti_UniqueIndex** procedure:

```
Sub AddMulti_UniqueIndex()
  Dim conn As ADODB.Connection
  Dim strTable As String
  Dim strCol As String

  On Error GoTo ErrorHandler
  Set conn = CurrentProject.Connection

  strTable = "tblSchools"
  strCol = "SchoolID, District"

  conn.Execute "ALTER TABLE " & strTable & _
   " ADD CONSTRAINT multiIdx UNIQUE " & _
   "(" & strCol & ");"
ExitHere:
  conn.Close
```

```
    Set conn = Nothing
    Exit Sub
ErrorHandler:
    MsgBox Err.Number & ":" & Err.Description
    Resume ExitHere
End Sub
```

2. Position the insertion point anywhere within the code of the AddMulti_UniqueIndex procedure and press **F5** or choose **Run | Run Sub/UserForm** to execute the procedure. Figure 19.4 shows the result.

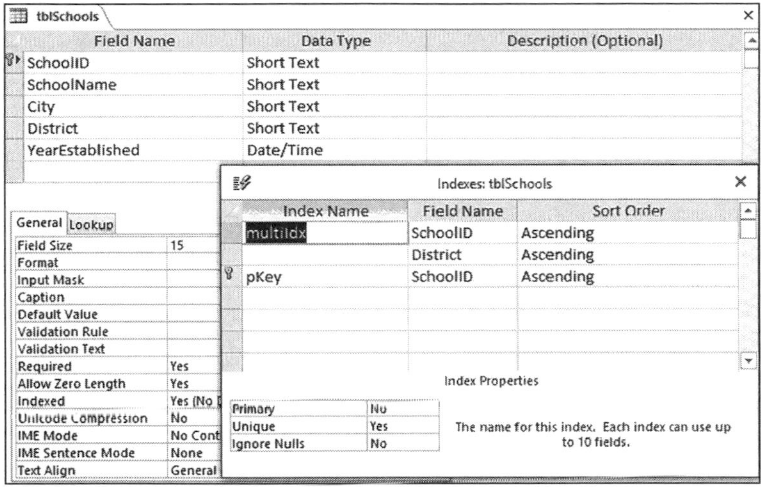

FIGURE 19.4. After running the procedures in Hands-On 19.9 and 19.10, the tblSchools table contains a primary key and a unique index based on two fields.

Deleting an Indexed Column

Deleting an index field is a two-step process:

- Use the `DROP CONSTRAINT` clause to delete an index. You must specify the index name.
- Use the `DROP COLUMN` clause to delete the desired column. You must specify the column name.

Both clauses must be used in the `ALTER TABLE` statement.

The procedure in Hands-On 19.11 deletes the District column from the tblSchools table. Recall that the procedure in Hands-On 19.10 added a multiple-field index based on the SchoolID and District columns.

Hands-On 19.11. Deleting a Field that Is Part of an Index

This hands-on exercise uses the tblSchools table created in Hands-On 19.2. You must perform Hands-On 19.10 prior to running this procedure.

1. In the same module where you entered previous hands-on exercises, enter the following **DeleteIdxField** procedure:

```
Sub DeleteIdxField()
  Dim conn As ADODB.Connection
  Dim strTable As String
  Dim strCol As String
  Dim strIdx As String

  On Error GoTo ErrorHandler
  Set conn = CurrentProject.Connection

  strTable = "tblSchools"
  strCol = "District"
  strIdx = "multiIdx"

  conn.Execute "ALTER TABLE " & strTable & _
   " DROP CONSTRAINT " & strIdx & ";"

  conn.Execute "ALTER TABLE " & strTable & _
   " DROP COLUMN " & strCol & ";"

ExitHere:
  conn.Close
  Set conn = Nothing
  Exit Sub
ErrorHandler:
  MsgBox Err.Number & ":" & Err.Description
  Resume ExitHere
End Sub
```

2. Position the insertion point anywhere within the code of the DeleteIdxField procedure and press **F5** or choose **Run | Run Sub/UserForm** to execute the procedure.

Deleting an Index

Use the `DROP CONSTRAINT` clause to delete an index. You must specify the index name.

The procedure in Hands-On 19.12 deletes a primary key index from the tblSchools table.

Hands-On 19.12. Deleting an Index

This hands-on exercise uses the tblSchools table created in Hands-On 19.2. You must perform Hands-On 19.10 prior to running this procedure.

1. In the same module where you entered previous hands-on exercises, enter the following **RemovePrimaryKeyIndex** procedure:

```
Sub RemovePrimaryKeyIndex()
  Dim conn As ADODB.Connection
  Dim strTable As String
  Dim strIdx As String

  On Error GoTo ErrorHandler
  Set conn = CurrentProject.Connection

  strTable = "tblSchools"
  strIdx = "pKey"

  conn.Execute "ALTER TABLE " & strTable & _
   " DROP CONSTRAINT " & strIdx & ";"

ExitHere:
  conn.Close
  Set conn = Nothing
  Exit Sub
ErrorHandler:
  MsgBox Err.Number & ":" & Err.Description
  Resume ExitHere
End Sub
```

2. Position the insertion point anywhere within the code of the RemovePrimaryKeyIndex procedure and press **F5** or choose **Run | Run Sub/UserForm** to execute the procedure.

After running the procedures in Hands-On 19.11 and 19.12, the Indexes window (see Figure 19.4. earlier) should be empty.

Setting a Default Value for a Table Column

Specifying a default value for a field automatically enters that value in the field each time a new record is added to a table unless the user provides a value for the field. Using DDL, you can add a default value for an existing column with the `SET DEFAULT` clause. The required syntax is as follows:

```
ALTER TABLE table_name ALTER [COLUMN] column_name
SET DEFAULT default-value;
```

The `[COLUMN]` in the syntax is optional.

Hands-On 19.13. Setting a Default Value for a Field

This hands-on exercise uses the tblSchools table created in Hands-On 19.2.

1. In the same module where you entered previous hands-on exercises, enter the following **SetDefaultFieldValue** procedure:

```
Sub SetDefaultFieldValue()
  Dim conn As ADODB.Connection
  Dim strTable As String
  Dim strCol As String
  Dim strDefVal As String
  Dim strSQL As String

  On Error GoTo ErrorHandler
  Set conn = CurrentProject.Connection

  strTable = "tblSchools"
  strCol = "City"
  strDefVal = "Boston"
  strSQL = "ALTER TABLE " & strTable & _
   " ALTER " & strCol & " SET DEFAULT " & strDefVal

  conn.Execute strSQL
ExitHere:
  conn.Close
  Set conn = Nothing
  Exit Sub
ErrorHandler:
  MsgBox Err.Number & ":" & Err.Description
  Resume ExitHere
End Sub
```

2. Position the insertion point anywhere within the code of the SetDefaultField-Value procedure and press **F5** or choose **Run | Run Sub/UserForm** to execute the procedure. Figure 19.5 shows that the Default Value property of the City field has been set to Boston.

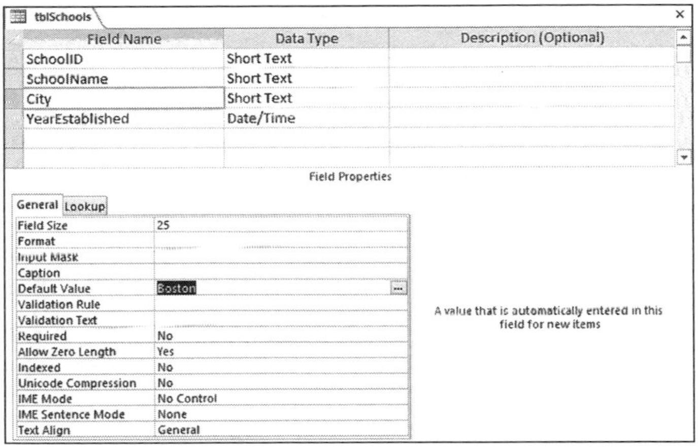

FIGURE 19.5. After running the procedure in Hands-On 19.13, the Default Value property of the City field is set to Boston.

Changing the Seed and Increment Values of AutoNumber Columns

When a table contains a field with an AutoNumber data type, you can set a seed value and an increment value. The seed value is the initial value for the column, and the increment value is the number added to the seed value to obtain a new counter value for the next record. If not specified, both seed and increment values default to 1. You can use DDL to change the seed and increment values of AutoNumber columns by using one of the following three statements:

```
ALTER TABLE Table_name
ALTER COLUMN Column_name AUTOINCREMENT (seed, increment)
ALTER TABLE Table_name
ALTER COLUMN Column_name COUNTER (seed, increment)
ALTER TABLE Table_name
ALTER COLUMN Column_name IDENTITY (seed, increment)
```

The example procedure in Hands-On 19.14 modifies the seed value of the existing AutoNumber column in the SchoolID column to start at 1000. Because we changed the SchoolID column's data type to the Text data type in one of the earlier hands-on exercises, you will modify the SchoolID column in the Sites.mdb file you created in Hands-On 19.3 earlier in this chapter.

Hands-On 19.14. Changing the Start (Seed) Value of the AutoNumber Field

This hands-on exercise uses the Sites.mdb database file and tblSchools table created in Hands-On 19.3.

1. In the same module where you entered previous hands-on exercises, enter the following **ChangeAutoNumber** procedure:

```
Sub ChangeAutoNumber()
  Dim conn As ADODB.Connection
  Dim strDb As String
  Dim strConnect As String
  Dim strTable As String
  Dim strCol As String
  Dim intSeed As Integer

  On Error GoTo ErrorHandler

  strDb = CurrentProject.Path & "\" & "Sites.mdb"
  strConnect = "Provider=Microsoft.Jet.OLEDB.4.0;" & _
   "Data Source=" & strDb

  strTable = "tblSchools"
  strCol = "SchoolID"
  intSeed = 1000

  Set conn = New ADODB.Connection
  conn.Open strConnect
  conn.Execute "ALTER TABLE " & strTable & _
   " ALTER COLUMN " & strCol & _
   " COUNTER (" & intSeed & ");"
ExitHere:
  conn.Close
  Set conn = Nothing
  Exit Sub
ErrorHandler:
  If Err.Number = -2147467259 Then
    MsgBox "The database file cannot be located.", _
    vbCritical, strDb
    Exit Sub
  Else
    MsgBox Err.Number & ":" & Err.Description
```

```
      Resume ExitHere
    End If
  End Sub
```

2. Position the insertion point anywhere within the code of the ChangeAutoNumber procedure and press **F5** or choose **Run | Run Sub/UserForm** to execute the procedure.
3. Launch Microsoft Access with the Sites.mdb database and open the **tblSchools** table.
4. Enter a couple of new records in this table. In the YearEstablished field, enter the date in the format **mm/dd/yyyy**. Note that the first new record is numbered 1000, the second 1001, the third 1002, and so on.
5. Close the **Sites.mdb** database file.

CHAPTER SUMMARY

In this chapter, you learned various Data Definition Language (DDL) commands for creating a new Access database, as well as creating, modifying, and deleting tables. You also learned how to add, modify, and delete fields and indexes, how to change the seed and increment values for AutoNumber fields, and how to change a field's data type. You also practiced assigning default values to table fields.

In the next chapter, you will learn about several DDL commands used for establishing relationships between tables and controlling referential integrity.

Chapter 20 ENFORCING DATA INTEGRITY AND RELATIONSHIPS BETWEEN TABLES

When creating tables in a database, you often need to define rules regarding the values allowed in columns (fields). As mentioned in Chapter 19, constraints allow you to enforce integrity by creating rules for a table. The five types of constraints are listed in Table 20.1.

TABLE 20.1. Table constraints

Constraint Name	Usage
PRIMARY KEY	Identifies the column or set of columns whose values uniquely identify a row in a table.
FOREIGN KEY	Defines the relationship between tables and maintains data integrity when records are being added, changed, or deleted in a table.
UNIQUE	Ensures that no duplicate values are entered in a specific column or combination of columns that is not a table's primary key.
NOT NULL	Specifies that a column cannot contain a Null value. Primary key columns are automatically defined as NOT NULL. Note: *A Null value is not the same as zero (0), blank, or a zero-length character string (""). A Null value indicates that no entry has been made. You can determine if a field contains a Null value by using the* `IsNull` *function.*
CHECK	Enforces integrity by limiting the values that can be placed in a column.

When constraints are added, all existing data is verified for constraint violations.

USING CHECK CONSTRAINTS

Tables and columns can contain multiple CHECK constraints. A *CHECK constraint* can validate a column value against a logical expression or another column in the same or another table. What you can't do with the CHECK constraint is to specify the custom validation message, as is possible to do in the Access user interface.

The procedure in Hands-On 20.1 uses a PRIMARY KEY constraint explicitly named PrimaryKey to identify the ID column as a primary key. The CHECK constraint used in this procedure ensures that only numbers within the specified range are entered in the YearsWorked column. You can apply CHECK constraints to a single column or to multiple columns. When a table is deleted, CHECK constraints are also dropped.

Please note files for the "Hands-On" project may be found on the companion CD-ROM.

Hands-On 20.1. Using a CHECK Constraint to Specify a Condition for All Values Entered for the Column

1. Start **Microsoft Access 2013** and create a new database named **Chap20.accdb** in your **C:\Access2013_ByExample** folder.
2. Switch to the Visual Basic Editor window and choose **Tools | References**. In the References dialog box, scroll down to locate **Microsoft ActiveX Data Objects 6.1 Library**. Click the check box to the left of this library name to set a reference to it and click OK to exit the dialog box.
3. Choose **Insert | Module** to add a new module to the current VBA project.
4. In the module's Code window, type the **CheckColumnValue** procedure shown below:

```
Sub CheckColumnValue()
    ' you must set up a reference to
    ' the Microsoft ActiveX Data Objects Library
    ' in the References dialog box
    Dim conn As ADODB.Connection
    Dim strTable As String
    On Error GoTo ErrorHandler
    Set conn = CurrentProject.Connection
    strTable = "tblAwards"

    conn.Execute "CREATE TABLE " & strTable & _
    "(ID AUTOINCREMENT CONSTRAINT PrimaryKey PRIMARY KEY," & _
    "YearsWorked INT, CONSTRAINT FromTo " & _
    "CHECK (YearsWorked BETWEEN 1 AND 30));"

ExitHere:
    conn.Close
    Set conn = Nothing
    Exit Sub
ErrorHandler:
    MsgBox Err.Number & ":" & Err.Description
    Resume ExitHere
End Sub
```

5. Position the insertion point anywhere within the code of the **Check Column-Value** procedure and press F5 or choose **Run | Run Sub/UserForm** to execute the procedure.

The CheckColumnValue procedure creates the tblAwards table with the CHECK constraint.

6. Open the **tblAwards** table and enter a value in the **YearsWorked** column that does not fall between **1** and **30**. You should receive the message shown in Figure 20.1.

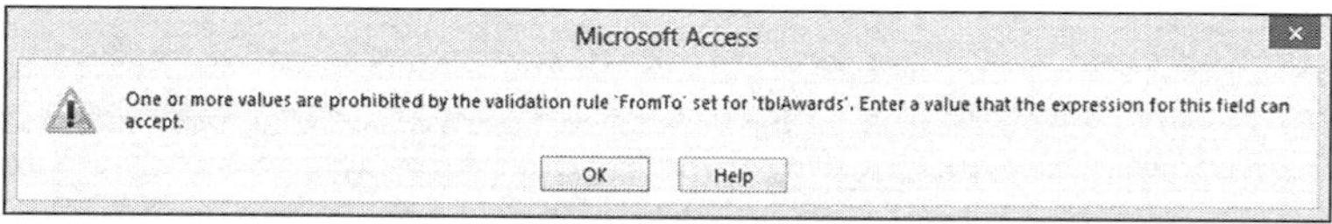

FIGURE 20.1. This message appears when you attempt to enter a value in the YearsWorked column that is not within the range of values specified by the FromTo constraint.

Hands-On 20.2 demonstrates how to create a CHECK constraint to ensure that the value of the Items column in the tblBookOrders table is less than the value of the MaxUnits column in the tblSupplies table for the specified ISBN number. This hands-on exercise also illustrates how to use the SQL Data Manipulation Language (DML) statements `INSERT INTO`, `BEGIN TRANSACTION`, `COMMIT TRANSACTION`, and `ROLLBACK TRANSACTION`.

Hands-On 20.2. Creating a Table with a Validation Rule Referencing a Column in Another Table

1. In the same module where you entered the procedure in Hands-On 20.1, enter the **ValidateAgainstCol_InAnotherTbl** procedure shown here:

```
Sub ValidateAgainstCol_InAnotherTbl()
  Dim conn As ADODB.Connection
  Dim strTable1 As String
  Dim strTable2 As String
  Dim InTrans As Boolean

  On Error GoTo ErrorHandler

  Set conn = CurrentProject.Connection
  strTable1 = "tblSupplies"
  strTable2 = "tblBookOrders"

  conn.Execute "BEGIN TRANSACTION"
  InTrans = True
  conn.Execute "CREATE TABLE " & strTable1 & _
   "(ISBN CHAR CONSTRAINT " & _
```

```
    "PrimaryKey PRIMARY KEY, " & _
    "MaxUnits LONG);", adExecuteNoRecords

   conn.Execute "INSERT INTO " & strTable1 & _
    " (ISBN, MaxUnits) " & _
    " Values ('158-76609-09', 5);", _
    adExecuteNoRecords

   conn.Execute "INSERT INTO " & strTable1 & _
    " (ISBN, MaxUnits) " & _
    " Values ('167-23455-69', 7);", _
    adExecuteNoRecords

   conn.Execute "CREATE TABLE " & strTable2 & _
    "(OrderNo AUTOINCREMENT CONSTRAINT " & _
    "PrimaryKey PRIMARY KEY, " & _
    "ISBN CHAR, Items LONG, " & _
    "CONSTRAINT OnHandConstr CHECK " & _
    "(Items <(SELECT MaxUnits FROM " & strTable1 & _
    " WHERE ISBN = " & strTable2 & ".ISBN)));", _
    adExecuteNoRecords

conn.Execute "COMMIT TRANSACTION"
  InTrans = False
  Application.RefreshDatabaseWindow
ExitHere:
  conn.Close
  Set conn = Nothing
  Exit Sub
ErrorHandler:
  If InTrans Then
    conn.Execute "ROLLBACK TRANSACTION"
    Resume ExitHere
  Else
    MsgBox Err.Number & ":" & Err.Description
    Exit Sub
  End If
End Sub
```

2. Position the insertion point anywhere within the code of the ValidateAgainstCol_InAnotherTbl procedure and press **F5** or choose **Run | Run Sub/UserForm** to execute the procedure.

This procedure creates two tables. Because the Items column in the tblBookOrders table needs to be validated against the contents of the MaxUnits column in the tblSupplies table, we wrapped the process of creating these tables and entering data in the tblSupplies table into a transaction. To trap errors that could occur during the procedure execution, we declared a Boolean variable named `InTrans` to help us determine whether an error occurred during the transaction; if the value of `InTrans` is True, we will cancel the transaction. Notice that in Access SQL syntax we use the `BEGIN TRANSACTION` statement to start the transaction, the `COMMIT TRANSACTION` statement to save the results of the transaction, and the `ROLLBACK TRANSACTION` statement to cancel any changes. These transaction statements can only be used through the Jet OLE DB Provider and ADO. They will cause an error when used with the Access user interface or DAO.

In this example procedure, we used the `adExecuteNoRecords` option to specify that no rows should be returned. You can use this setting with the Connection or Command object's `Execute` method to improve performance when no rows are returned or when you don't plan to access the returned rows in your procedure code. If you omit this setting, your ADO code will still execute successfully, but the ADO will unnecessarily create a Recordset object as the return value for the `Execute` method. Using the `adExecuteNoRecords` setting is one of several techniques for optimizing data access using ADO.

3. Open the **tblBookOrders** table and enter the record shown at the top of Figure 20.2.

   When you try to save this record or move to the next data row, Access will display a message informing you that the value you are trying to enter is prohibited.
4. Enter the value of **4** in the **Items** column. This time Access approves of the entry and no error message is displayed.
5. Close the **tblBookOrders** table.

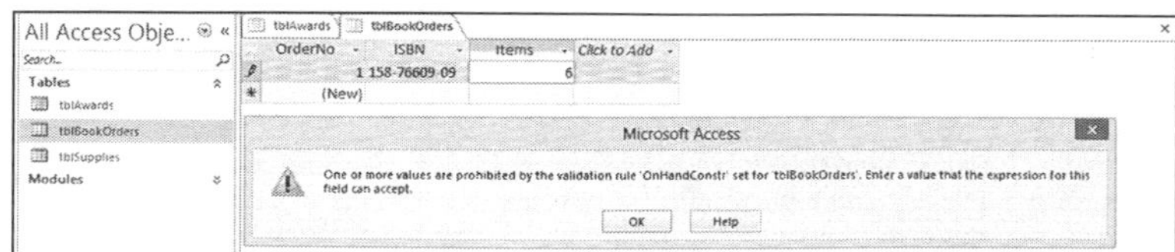

FIGURE 20.2. When you attempt to enter a value that does not meet the validation rule, Microsoft Access displays an error message.

6. Click **OK** to dismiss the message box, then press **Esc** to cancel the data entry.
7. Right-click the **tblBookOrders** tab and choose **Close** from the shortcut menu.

8. In the object Navigation pane on the left side of the database window, right-click the **tblBookOrders** table and choose **Delete**. Click **Yes** to confirm the deletion. Access will respond with the error message shown in Figure 20.3.

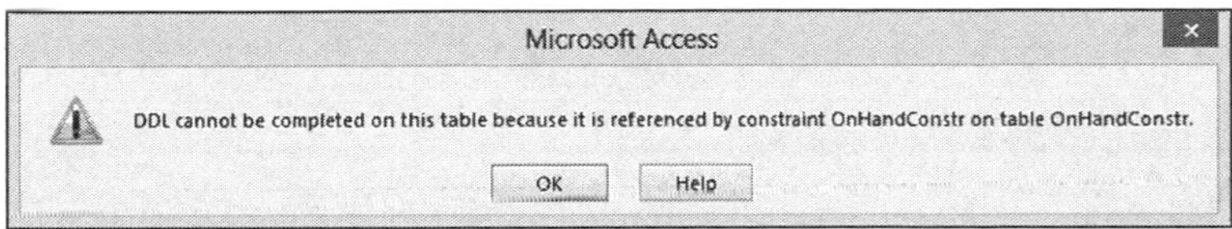

FIGURE 20.3. If you try to manually delete a table referenced by the CHECK constraint, Microsoft Access will display an error message.

Now, let's see how you can use the Access user interface to issue commands that delete tables and CHECK constraints.

Hands-On 20.3. Deleting Tables and Constraints Using the Access User Interface

This hands-on exercise requires that you have created the tblBookOrders and tblAwards tables in Hands-On 20.1 and 20.2.

1. In the database window, choose **Create | Query Design**.
2. In the Show Table dialog box, click the **Close** button.
3. Choose **Design | Data Definition**.
4. In the Data Definition Query window, enter the statement shown in Figure 20.4.

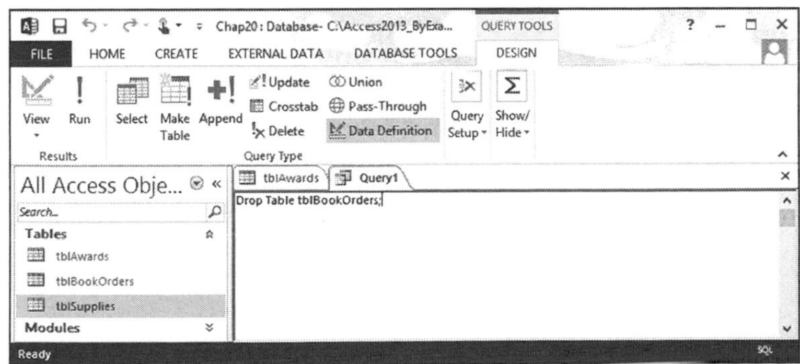

FIGURE 20.4. To delete a table that contains a CHECK constraint, type the DROP TABLE statement in the Data Definition Query window, then click Run.

5. To run the Data Definition query, click the **Run** button in the Ribbon.
   Note that a table must be closed before it can be deleted. If you don't want to delete a table but need to remove a constraint from a table, use the following syntax:

```
ALTER TABLE table_name DROP CONSTRAINT constraint_name
```

To remove a constraint, you must know its name.

6. To delete the constraint from the tblAwards table, type the statement shown in Figure 20.5 in the Data Definition Query window. Make sure that the tblAwards is closed.

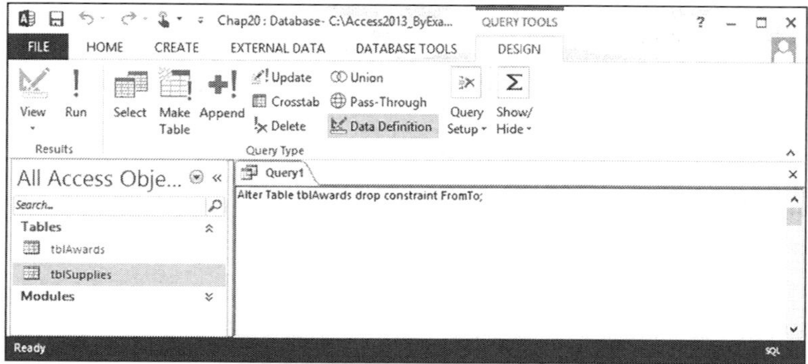

FIGURE 20.5. To remove a table constraint, use the DROP CONSTRAINT statement with ALTER TABLE.

Before using `ALTER TABLE`*, it is a good idea to make a backup copy of the table.*

7. Click the **Run** button on the Ribbon to execute the statement that will delete the constraint.
8. On your own, delete the tblSupplies table using the Data Definition Query window.

ESTABLISHING RELATIONSHIPS BETWEEN TABLES

To establish a link between the data in two tables, add one or more columns that hold one table's primary key values to the other table. This column becomes a *foreign key* in the second table. In SQL DDL, you can use a FOREIGN KEY constraint to reference another table. Foreign keys can be single- or multicolumn.

A *FOREIGN KEY constraint* enforces referential integrity by ensuring that changes made to data in the primary key table do not break the link to data in the foreign key table. For example, you cannot delete a record in a primary key table or change a primary key value if the deleted or changed primary key value corresponds to a value in the FOREIGN KEY constraint of another table. The `REFERENCES` clause identifies the parent table of the relation.

To create a brand-new table and relate it to an existing table, the following steps are required:

1. Use the CREATE TABLE statement followed by a table name.

```
CREATE TABLE tblOrder_Details
```

2. Follow the table name with one or more column definitions. A *column definition* consists of ColumnName followed by the data type and column size (if required).

```
InvoiceID CHAR, ProductId CHAR, Units LONG, Price MONEY
```

3. To designate a primary key, use the CONSTRAINT clause followed by the constraint name, the PRIMARY KEY clause, and the name of the column or columns to be designated as the primary key.

```
CONSTRAINT PrimaryKey PRIMARY KEY (InvoiceId, ProductId)
```

4. To designate a foreign key, use the CONSTRAINT clause followed by the constraint name, the FOREIGN KEY clause, and the name of the column to be designated as the foreign key.

```
CONSTRAINT fkInvoiceId FOREIGN KEY (InvoiceId)
```

5. Use the REFERENCES clause to specify the parent table to which a relationship is established.

```
REFERENCES tblProduct_Orders
```

6. If required, specify ON UPDATE CASCADE and/or ON DELETE CASCADE to enable referential integrity rules with cascading updates or deletes.

```
ON UPDATE CASCADE ON DELETE CASCADE
```

You may choose not to enforce referential integrity rules by specifying ON UPDATE NO ACTION *or* ON DELETE NO ACTION, *or skipping the* ON UPDATE *or* ON DELETE *keywords. If you choose this path, you will not be able to change the value of a primary key if matching records exist in the foreign table.*

Refer to the procedure in Hands-On 20.4 to find out how to correctly combine the preceding example statements into a single SQL statement.

Hands-On 20.4. Relating Two Tables and Setting up Cascading Referential Integrity Rules

1. In the Visual Basic Editor window, choose **Insert | Module**.

2. In the module's Code window, enter the **RelateTables** procedure shown here:

```
Sub RelateTables()
  Dim conn As ADODB.Connection
  Dim strPrimaryTbl As String
  Dim strForeignTbl As String

  On Error GoTo ErrorHandler

  Set conn = CurrentProject.Connection
  strPrimaryTbl = "tblProduct_Orders"
  strForeignTbl = "tblOrder_Details"

  conn.Execute "CREATE TABLE " & strPrimaryTbl & _
    "(InvoiceID CHAR(15), PaymentType CHAR(20), " & _
    " PaymentTerms CHAR(25), Discount LONG, " & _
    " CONSTRAINT PrimaryKey PRIMARY KEY (InvoiceID));", _
    adExecuteNoRecords

  conn.Execute "CREATE TABLE " & strForeignTbl & _
    "(InvoiceID CHAR(15), ProductID CHAR(15), " & _
    " Units LONG, Price MONEY, " & _
    "CONSTRAINT PrimaryKey PRIMARY KEY " & _
    "(InvoiceID, ProductID), " & _
    "CONSTRAINT fkInvoiceID FOREIGN KEY (InvoiceID) " & _
    "REFERENCES " & strPrimaryTbl & _
    " ON UPDATE CASCADE ON DELETE CASCADE);", _
    adExecuteNoRecords

  Application.RefreshDatabaseWindow
ExitHere:
  conn.Close
  Set conn = Nothing
  Exit Sub
ErrorHandler:
  MsgBox Err.Number & ":" & Err.Description
  Resume ExitHere
End Sub
```

3. Position the insertion point anywhere within the code of the RelateTables procedure and press **F5** or choose **Run | Run Sub/UserForm** to execute the procedure.

The RelateTables procedure creates and joins two tables. A primary key table named tblProduct_Orders is created with a primary key on the InvoiceID field. The foreign key table named tblOrder_Details is created with a multifield primary key index based on the ProductID and InvoiceID fields. The `REFERENCES` clause specifies the tblProduct_Orders table as the parent table. The created relationship has the referential integrity rules enforced via the `ON UPDATE CASCADE` and `ON DELETE CASCADE` statements.

The outcome of the RelateTables procedure is illustrated in the following figures. Figure 20.6. displays the one-to-many relationship between tblProduct_Orders and tblOrder_Details. Figure 20.7. presents the Edit Relationships window in which both cascading updates and deletes are selected.

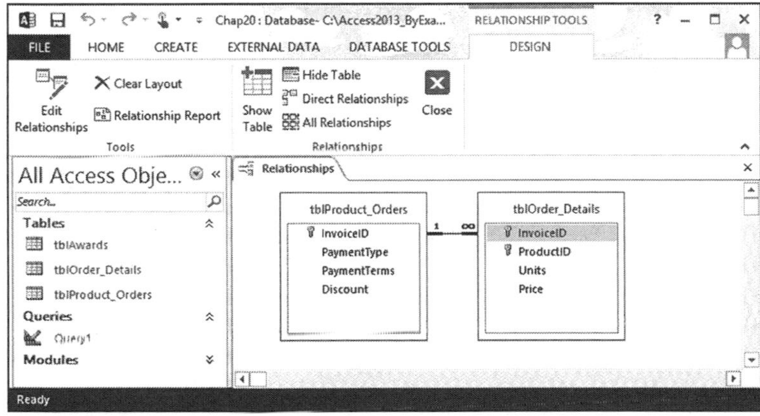

FIGURE 20.6. To access the Relationships window, choose Database Tools | Relationships.

Edit Relationships

Table/Query: tblProduct_Orders | Related Table/Query: tblOrder_Details

InvoiceID | InvoiceID

OK | Cancel | Join Type.. | Create New..

☑ Enforce Referential Integrity

☑ Cascade Update Related Fields

☑ Cascade Delete Related Records

Relationship Type: One-To-Many

FIGURE 20.7. To access the Edit Relationships window, choose Design | Edit Relationship.

USING THE DATA DEFINITION QUERY WINDOW

To enhance your understanding of creating tables and relationships with Data Definition Language, perform Hands-On 20.5 using the Data Definition Query window.

Hands-On 20.5. Running DDL Statements in the Microsoft Access User Interface

Each of the statements in this hands-on exercise can be executed by choosing Design | Run.

1. In the database window, choose **Create | Query Design**.
2. In the Show Table dialog box, click the **Close** button.
3. Choose **Design | Data Definition**.
4. In the Data Definition Query window that appears, enter the statement shown earlier in Figure 20.4 and run the query.
5. To create a table on the primary (one) side of the relationship, type the following statement on one line in the query window and run the query:

```
CREATE TABLE myPrimaryTbl(ID COUNTER CONSTRAINT pKey
 PRIMARY KEY, COUNTRY TEXT(15));
```

6. To create a table on the foreign (many) side of the relationship, delete the preceding statement, then type the following statement, and run the query:

```
CREATE TABLE myForeignTbl(ID LONG, Region TEXT (15));
```

7. To create a one-to-many relationship between myPrimaryTbl and myForeignTbl, delete the preceding statement, then type the following statement on one line in the query window, and run the query:

```
ALTER TABLE myForeignTbl ADD CONSTRAINT Rel FOREIGN KEY(ID)
 REFERENCES myPrimaryTbl (ID);
```

8. In the database window, choose **Database Tools | Relationships**.
9. In the Relationships window, choose **Design | All Relationships**. This will add both tables (myPrimaryTbl and myForeignTbl) to the Relationships window (see Figure 20.8).

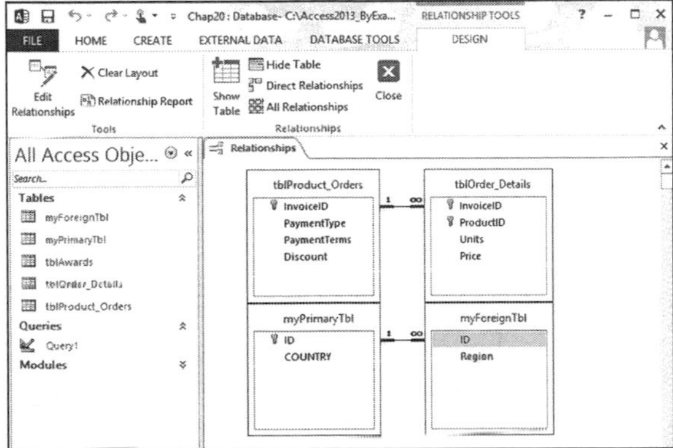

FIGURE 20.8. Notice that the tables you created by running the DDL statements in Steps 5 and 6 are joined on the ID column (see Step 7).

10. Right-click the line that joins the myPrimaryTbl and myForeignTbl tables and choose **Edit Relationship** to open the Edit Relationships dialog box, as shown in Figure 20.9. You can also double-click the line to open this dialog box.

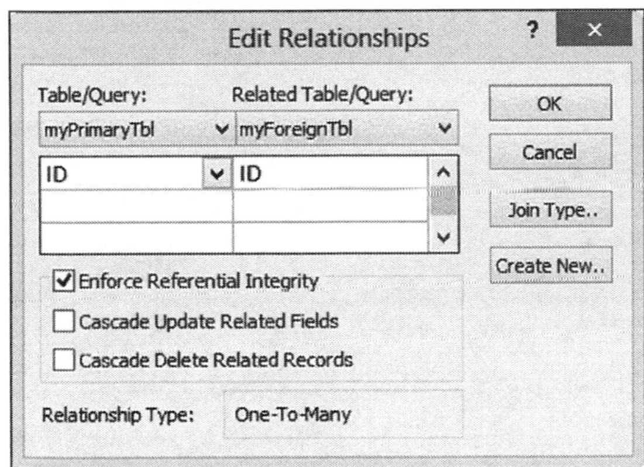

FIGURE 20.9. You can edit relationships between tables via the Edit Relationships window.

11. Click **Cancel** to exit the Edit Relationships window.
12. To delete the relationship between the tables, type the following statement in the Data Definition Query window (overwriting the previously entered statement), and run the query:

```
ALTER TABLE myForeignTbl DROP CONSTRAINT Rel;
```

13. To delete the table on the one side (myPrimaryTbl), type the following statement in the Data Definition Query window (overwriting the preceding statement), and run the query:

```
DROP TABLE myPrimaryTbl;
```

14. To delete the table on the many side (myForeignTbl), type the following statement in the Data Definition Query window (overwriting the preceding statement), and run the query:

```
DROP TABLE myForeignTbl;
```

CHAPTER SUMMARY

In this short chapter, you learned how to enforce data integrity by creating rules for tables with constraints. You learned how to validate data against another column in the same table or a column located in another table. You also learned how to use the Access Data Definition Query window to delete tables that have constraints and remove constraints from a table. Finally, you saw how you can establish relationships between tables using DDL commands inside a VBA procedure.

The next chapter focuses on ways to use DDL for defining and removing indexes and primary keys.

Chapter 21

Defining Indexes and Primary Keys

Indexes speed the processes of finding and sorting records. You should create indexes for fields that are frequently used in searches and in sorting. You can create an index on a new or existing table. An index can be made of one or more fields. This chapter presents a number of procedures that use Data Definition Language statements to define indexes and primary keys.

CREATING TABLES WITH INDEXES

You can create an index while creating a table by using the CONSTRAINT clause with the CREATE TABLE statement. The procedure in Hands-On 21.1 creates a new table called Supplier1 with a unique index called idxSupplierName based on the SupplierName field.

Please note files for the "Hands-On" project may be found on the companion CD-ROM.

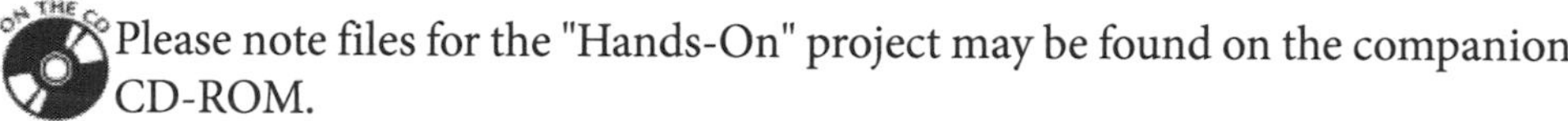

Hands-On 21.1. Creating a Table with a Single-Field Index

1. Start Microsoft Access 2013 and create a new database named **Chap21.accdb** in your **C:\Access2013_ByExample** folder.
2. Switch to the Visual Basic Editor window and choose **Tools | References**. In the References dialog box, scroll down to locate **Microsoft ActiveX Data Objects 6.1 Library**. Click the checkbox to the left of this library name to set a reference to it and click **OK** to exit the dialog box.
3. Choose **Insert | Module** to add a new module to the current VBA project.
4. In the module's Code window, type the following **SingleField_Index** procedure:

```
Sub SingleField_Index()
  Dim conn As ADODB.Connection
  Dim strTable As String

  On Error GoTo ErrorHandler

  Set conn = CurrentProject.Connection

  strTable = "Supplier1"

  conn.Execute "CREATE TABLE " & strTable _
   & "(SupplierID INTEGER, " _
   & "SupplierName CHAR (30), " _
   & "SupplierPhone CHAR (12), " _
   & "SupplierCity CHAR (19), " _
   & "CONSTRAINT idxSupplierName UNIQUE " _
   & "(SupplierName));"
  Application.RefreshDatabaseWindow
ExitHere:
```

```
    conn.Close
    Set conn = Nothing
    Exit Sub
  ErrorHandler:
    MsgBox Err.Number & ":" & Err.Description
    Resume ExitHere
  End Sub
```

5. Position the insertion point anywhere within the code of the SingleField_Index procedure and press **F5** or choose **Run | Run Sub/UserForm** to execute the procedure.

When you run this procedure then switch to the Access database window, you will notice a table named Supplier1 that contains four fields. There is a unique index on the SupplierName field, as shown in Figure 21.1.

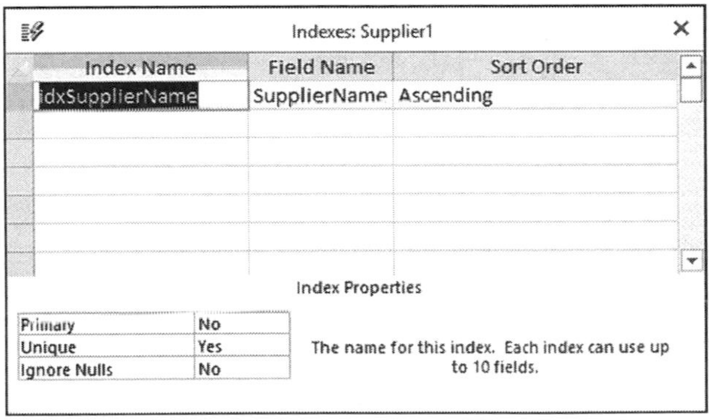

FIGURE 21.1. The idxSupplierName index was created by running the procedure in Hands On 21.1.

ADDING AN INDEX TO AN EXISTING TABLE

To add an index to an existing table, use the `CREATE INDEX` statement. You can add an index based on one or more fields. The procedure in Hands-On 21.2 demonstrates how to add an index to the Supplier1 table you created in Hands-On 21.1.

Hands-On 21.2. Adding a Single-Field Index to an Existing Table

1. In the same module where you entered the procedure in Hands-On 21.1, enter the **SingleField_Index2** procedure shown here:

```
Sub SingleField_Index2()
  Dim conn As ADODB.Connection
```

```
  Dim strTable As String

  On Error GoTo ErrorHandler

  Set conn = CurrentProject.Connection
  strTable = "Supplier1"

  conn.Execute "CREATE INDEX idxCity ON " & strTable & _
   "(SupplierCity);"
ExitHere:
  conn.Close
  Set conn = Nothing
  Exit Sub
ErrorHandler:
  MsgBox Err.Number & ":" & Err.Description
  Resume ExitHere
End Sub
```

2. Position the insertion point anywhere within the code of the SingleField_Index2 procedure and press **F5** or choose **Run | Run Sub/UserForm** to execute the procedure.

The preceding procedure adds a single-field index named idxCity to the Supplier1 table. The following procedure will add a multiple-field index named idxSupplierNameCity to the Supplier2 table.

```
Sub MultiField_Index()
  Dim conn As ADODB.Connection
  Dim strTable As String

  On Error GoTo ErrorHandler
  Set conn = CurrentProject.Connection

  strTable = "Supplier2"

  conn.Execute "CREATE TABLE " & strTable _
   & "(SupplierID INTEGER, " _
   & "SupplierName CHAR(30), " _
   & "SupplierPhone CHAR(12), " _
   & "SupplierCity CHAR(19), " _
   & "CONSTRAINT idxSupplierNameCity UNIQUE " _
   & "(SupplierName, SupplierCity));"

  Application.RefreshDatabaseWindow
```

```
ExitHere:
  conn.Close
  Set conn = Nothing
  Exit Sub
ErrorHandler:
  MsgBox Err.Number & ":" & Err.Description
  Resume ExitHere
End Sub
```

CREATING A TABLE WITH A PRIMARY KEY

When you create a database table, you should define a primary key to uniquely identify rows within the table. A *primary key* allows you to relate a particular table with other tables in the database (for procedure examples, refer to the previous chapter). A table can have only one primary key; however, a primary key can consist of more than one column.

To create a table with a primary key, use the CONSTRAINT clause with the CREATE TABLE statement. The procedure in Hands-On 21.3 uses the following CONSTRAINT clause to create a single-field primary key based on the SupplierID field:

```
CONSTRAINT idxPrimary PRIMARY KEY(SupplierID)
```

To create a table with a primary key based on two or more columns, specify column names in parentheses following the PRIMARY KEY keywords. For example, the following CONSTRAINT clause will create a primary key based on the SupplierID and SupplierName columns:

```
CONSTRAINT idxPrimary PRIMARY KEY (SupplierID, SupplierName)
```

Hands-On 21.3. Creating a Single-Field Primary Key

1. Switch to the Visual Basic Editor window and insert a new module.
2. In the module's Code window, enter the **SingleField_PKey** procedure shown here:

```
Sub SingleField_PKey()
  Dim conn As ADODB.Connection

  Dim strTable As String

  On Error GoTo ErrorHandler

  Set conn = CurrentProject.Connection
  strTable = "Supplier3"
```

```
    conn.Execute "CREATE TABLE " & strTable _
     & "(SupplierID INTEGER, " _
     & "SupplierName CHAR(30), " _
     & "SupplierPhone CHAR(12), " _
     & "SupplierCity CHAR(19), " _
     & "CONSTRAINT idxPrimary PRIMARY KEY " _
     & "(SupplierID));"
    Application.RefreshDatabaseWindow
  ExitHere:
    conn.Close
    Set conn = Nothing
    Exit Sub
  ErrorHandler:
    MsgBox Err.Number & ":" & Err.Description
    Resume ExitHere
  End Sub
```

3. Position the insertion point anywhere within the code of the SingleField_PKey procedure and press **F5** or choose **Run | Run Sub/UserForm** to execute the procedure. After running this procedure, you will have a primary key index named idxPrimary based on the SupplierID column, as shown in Figure 21.2.

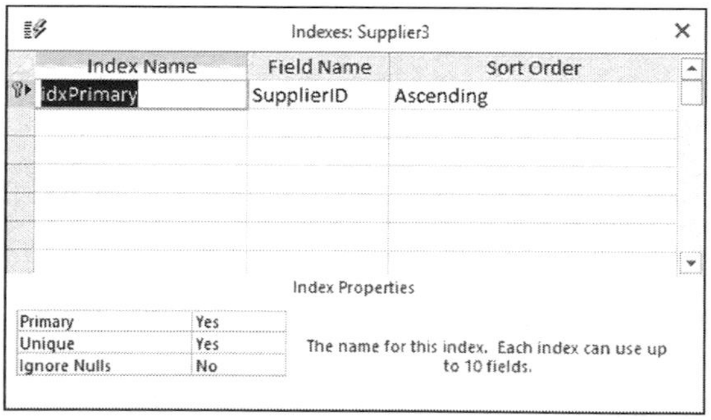

Indexes: Supplier3

Index Name	Field Name	Sort Order
idxPrimary	SupplierID	Ascending

Index Properties

Primary	Yes
Unique	Yes
Ignore Nulls	No

The name for this index. Each index can use up to 10 fields.

FIGURE 21.2. The result of running the SingleField_PKey procedure in Hands-On 21.3 is a primary key index named idxPrimary based on the SupplierID column.

CREATING INDEXES WITH RESTRICTIONS

You can use the `CREATE INDEX` statement to add an index to an existing table. The `CREATE INDEX` statement can be used with the following options:

- `PRIMARY` option—Creates a primary key index that does not allow duplicate values in the key.

- `DISALLOW NULL` option—Creates an index that does not allow adding records with Null values in the indexed field.
- `IGNORE NULL` option—Creates an index that does not include records with Null values in the key.

Use the `WITH` keyword to declare the preceding index options.

The procedure in Hands-On 21.4 designates the SupplierID field as the primary key by using the `PRIMARY` option (see Figure 21.3).

Hands-On 21.4. Creating a Primary Key Index with Restrictions

1. Switch to the Visual Basic Editor window and insert a new module.
2. In the module's Code window, enter the following **Index_WithPrimaryOption** procedure:

```
Sub Index_WithPrimaryOption()
  Dim conn As ADODB.Connection
  Dim strTable As String

  On Error GoTo ErrorHandler

  Set conn = CurrentProject.Connection
  strTable = "Supplier1"
  conn.Execute "CREATE INDEX idxPrimary1 ON " & strTable _
   & "(SupplierID) WITH PRIMARY;"
ExitHere:
  conn.Close
  Set conn = Nothing
  Exit Sub
ErrorHandler:
  MsgBox Err.Number & ":" & Err.Description
  Resume ExitHere
End Sub
```

3. Position the insertion point anywhere within the code of the Index_WithPrimaryOption procedure and press **F5** or choose **Run | Run Sub/UserForm** to execute the procedure.

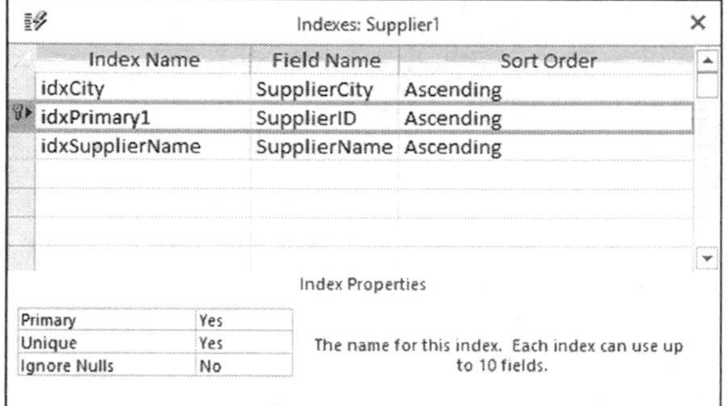

FIGURE 21.3. The index created by the procedure in Hands-On 21.4 has the Primary and Unique properties set to Yes, which means that this index is a primary key and every value in this index must be unique.

Primary key indexes are automatically created as unique indexes.

You can prohibit the entry of Null values in the indexed fields by using the `DISALLOW NULL` option as shown in the example procedure in Hands-On 21.5. The result of running this procedure is an index called idxSupplierCity that does not allow Null values, as shown in Figure 21.4.

Hands-On 21.5. Creating an Index that Disallows Null Values in the Key

1. Switch to the Visual Basic Editor window and insert a new module.
2. In the module's Code window, enter the following **Index_WithDisallowNullOption** procedure:.

```
Sub Index_WithDisallowNullOption()
  Dim conn As ADODB.Connection
  Dim strTable As String

  On Error GoTo ErrorHandler
  Set conn = CurrentProject.Connection

  strTable = "Supplier3"
  conn.Execute "CREATE INDEX idxSupplierCity ON " & strTable _
   & "(SupplierCity) WITH DISALLOW NULL;"
ExitHere:
  conn.Close
  Set conn = Nothing
  Exit Sub
ErrorHandler:
```

```
    MsgBox Err.Number & ":" & Err.Description
    Resume ExitHere
  End Sub
```

3. Position the insertion point anywhere within the code of the Index_WithDisallowNullOption procedure and press **F5** or choose **Run | Run Sub/UserForm** to execute the procedure.

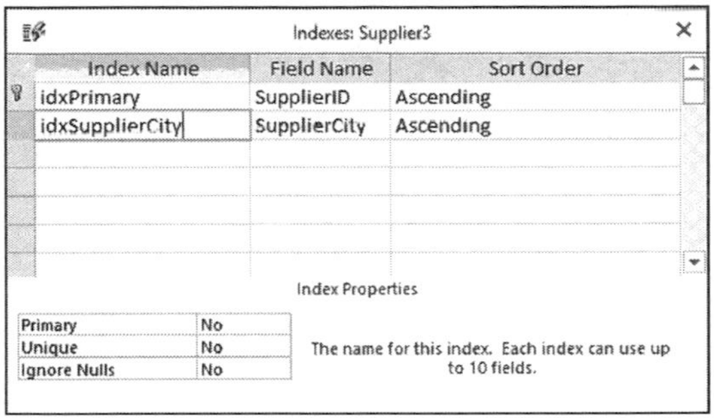

FIGURE 21.4. The result of running the procedure in Hands-On 21.5 is an index called idxSupplierCity that does not allow Null values.

You can prevent records with Null values in the indexed fields from being included in the index by using the `IGNORE NULL` option, as illustrated in Hands-On 21.6. Figure 21.5 shows the result of this procedure.

Hands-On 21.6. Creating an Index with the Ignore Null Option

1. Switch to the Visual Basic Editor window and insert a new module.
2. In the module's Code window, enter the following **Index_WithIgnoreNullOption** procedure:

```
Sub Index_WithIgnoreNullOption()
  Dim conn As ADODB.Connection
  Dim strTable As String

  On Error GoTo ErrorHandler

  Set conn = CurrentProject.Connection
  strTable = "Supplier3"

  conn.Execute "CREATE INDEX idxSupplierPhone ON " & strTable _
   & "(SupplierPhone) WITH IGNORE NULL;"
ExitHere:
  conn.Close
```

```
    Set conn = Nothing
    Exit Sub
  ErrorHandler:
    MsgBox Err.Number & ":" & Err.Description
    Resume ExitHere
  End Sub
```

3. Position the insertion point anywhere within the code of the Index_WithIgnoreNullOption procedure and press **F5** or choose **Run | Run Sub/UserForm** to execute the procedure.

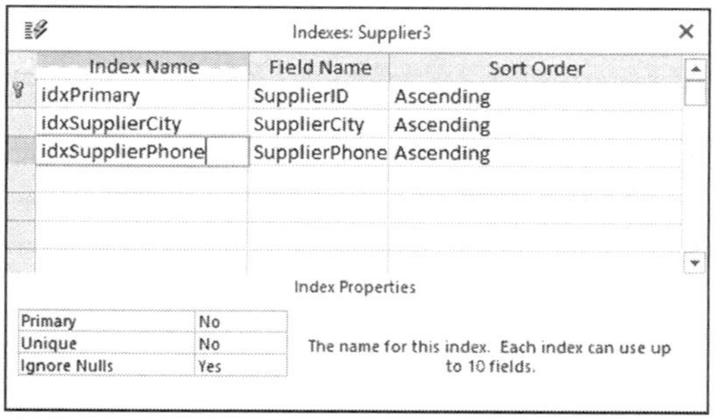

Indexes: Supplier3

Index Name	Field Name	Sort Order
idxPrimary	SupplierID	Ascending
idxSupplierCity	SupplierCity	Ascending
idxSupplierPhone	SupplierPhone	Ascending

Index Properties

Primary	No
Unique	No
Ignore Nulls	Yes

The name for this index. Each index can use up to 10 fields.

FIGURE 21.5. The result of running the procedure in Hands-On 21.6 is an index called idxSupplierPhone that allows Null values in the key. However, records containing Null values will be excluded from any searches that use that index.

DELETING INDEXES

Use the `DROP INDEX` statement to remove an index. Anytime you want to delete a column that is part of an index, you must first remove the index using the `DROP CONSTRAINT` or `DROP INDEX` statement. Before removing the index, make sure that the table containing the index is closed. The procedure in Hands-On 21.7 deletes the index named idxSupplierName from the Supplier1 table.

Hands-On 21.7. Deleting an Index

1. Switch to the Visual Basic Editor window and insert a new module.
2. In the module's Code window, enter the following **DeleteIndex** procedure:

```
Sub DeleteIndex()
  Dim conn As ADODB.Connection
  Dim strTable As String

  On Error GoTo ErrorHandler
```

```
    Set conn = CurrentProject.Connection
    strTable = "Supplier1"

    conn.Execute "DROP INDEX idxSupplierName ON " & strTable & ";"
ExitHere:
    conn.Close
    Set conn = Nothing
    Exit Sub
ErrorHandler:
    MsgBox Err.Number & ":" & Err.Description
    Resume ExitHere
End Sub
```

3. Position the insertion point anywhere within the code of the DeleteIndex procedure and press **F5** or choose **Run | Run Sub/UserForm** to execute the procedure.
4. On your own, write a procedure to remove other indexes created in this chapter's procedures.

CHAPTER SUMMARY

This chapter introduced you to using DDL statements for creating indexes. Columns that are frequently used in database queries should be indexed to allow for faster access to the information. However, if you frequently add, delete, and update rows, you might want to limit the number of indexes, as they take up disk space and slow data operations. You also learned that a primary key is a special type of index that allows you to uniquely identify rows in a table as well as create a relationship between two tables, as demonstrated in the previous chapter.

The next chapter introduces you to the DDL statements that are used to manage database security.

Chapter 22 Database Security

The procedures in this chapter demonstrate how to use simple Data Definition Language statements to easily manage database and user passwords, create or delete user and group accounts, and grant or delete permissions for database objects.

SETTING THE DATABASE PASSWORD

Database security can be handled at share level or user level. Share-level security is the easiest to implement, as it only requires that you set a password on the database. As mentioned earlier in this book, user-level security can only be used with Access databases created in the .mdb file format.

To set a new database password or change an existing password, use the `ALTER DATABASE` PASSWORD statement in the following format:

```
ALTER DATABASE PASSWORD newPassword oldPassword
```

When setting the password for the first time, use Null for the old password (see the example procedure in Hands-On 22.1). The Access database must be opened in exclusive mode to perform password operations. Therefore, when using ADO, set the ADO Connection object's Mode property to `adModeShareExclusive` before opening a database.

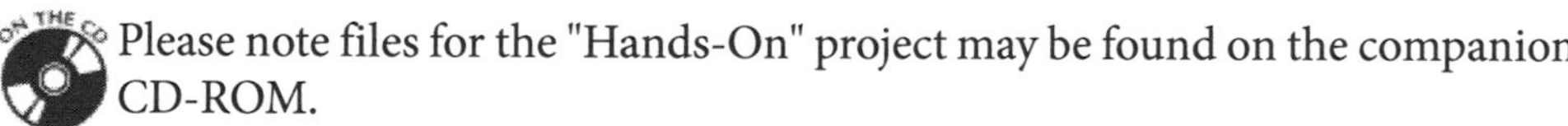

Please note files for the "Hands-On" project may be found on the companion CD-ROM.

Hands-On 22.1. Setting a Database Password

This hands-on exercise sets a database password on the Chap21.accdb database file.

1. Start Microsoft Access 2013 and create a new database named **Chap22.accdb** in your **C:\Access2013_ByExample** folder.
2. Switch to the Visual Basic Editor window and choose **Tools | References**. In the References dialog box, scroll down to locate **Microsoft ActiveX Data Objects 6.1 Library**. Click the checkbox to the left of this library name to set a reference to it and click **OK** to exit the dialog box.
3. Choose **Insert | Module** to add a new module to the current VBA project.
4. In the module's Code window, type the following **SetDBPassword** function:

```
Function SetDBPassword(strFullFilePath)
  Dim conn As ADODB.Connection
  On Error GoTo ErrorHandler
  Set conn = New ADODB.Connection
  With conn
    .Mode = adModeShareExclusive
```

```
      .Open "Provider = Microsoft.ACE.OLEDB.12.0;" & _
       "Data Source=" & strFullFilePath & ";"
      .Execute "ALTER DATABASE PASSWORD secret null "
    End With
ExitHere:
    If Not conn Is Nothing Then
      If conn.State = adStateOpen Then conn.Close
    End If
    Set conn = Nothing
    Exit Function
ErrorHandler:
    MsgBox Err.Number & ":" & Err.Description
    Resume ExitHere
End Function
```

5. Execute the SetDBPassword function from the Immediate window by typing the following statement and pressing **Enter**:

```
SetDBPassword "C:\Access2013_ByExample\Chap21.accdb"
```

While you may supply the name of a different Access database to the SetDB-Password function, the database must be in the new .accdb file format because the function opens a connection to the database file using the ACE OLE DB Provider.

After opening a database in exclusive mode, this function procedure changes the database password from Null to "secret." Notice that the new password is listed first, followed by the old password. Notice also how the function uses the State property of the ADO Connection object to determine whether the connection to the database is open. State returns `adStateOpen` if the Connection object is open and `adStateClosed` if it is not.

REMOVING THE DATABASE PASSWORD

To remove a database password, replace the existing password with Null. The password can be removed by using the `ALTER DATABASE PASSWORD` statement, as illustrated in the preceding section. When the database is secured with a password, you will need to use the Jet/ACE OLEDB:Database Password property to specify the password to open the database. This is a Microsoft Jet 4.0/ACE OLE DB Provider-specific property of the Connection object. The following procedure shows how to remove the password "secret" from the Chap21.accdb database that was set by the SetDBPassword function in Hands-On 22.1.

Hands-On 22.2. Deleting a Database Password

This procedure requires prior completion of Hands-On 22.1.

1. In the same module where you entered the SetDBPassword function (Hands-On 22.1), enter the **ResetDBPassword** function shown here:

```
Function ResetDBPassword(strFullFilePath, _
  strNewPwd, strOldPwd)

  Dim conn As ADODB.Connection

  On Error GoTo ErrorHandler
  Set conn = New ADODB.Connection

  With conn
  .Mode = adModeShareExclusive
  .Open "Provider = Microsoft.ACE.OLEDB.12.0;" & _
    "Data Source=" & strFullFilePath & _
    "; Jet OLEDB:Database Password=" & strOldPwd & ";"
    .Execute "ALTER DATABASE PASSWORD " & strNewPwd & " " & _
    strOldPwd
  End With
ExitHere:
  If Not conn Is Nothing Then
    If conn.State = adStateOpen Then conn.Close
  End If
  Set conn = Nothing
  Exit Function
ErrorHandler:
  MsgBox Err.Number & ":" & Err.Description
  Resume ExitHere
End Function
```

2. Execute the ResetDBPassword function from the Immediate window by typing the following statement on one line and pressing **Enter**:

```
ResetDBPassword "C:\Access2013_ByExample\Chap21.accdb", "null",
 "secret"
```

3. Close the **Chap22.accdb** database file.

CREATING A USER ACCOUNT

Establishing database security at a user level is more involved than setting a database password. It requires that you create group and user accounts and assign permissions to groups and users to perform operations on various database objects. Use the `CREATE USER` statement to create a new user account. Specify the username to log in to the account followed by the required password and a personal identifier (PID) to make the account unique. The syntax of creating a user account looks like this:

```
CREATE USER userLoginName password PID
```

You can create more than one user account at a time by separating the usernames with a comma.

The procedure in Hands-On 22.3 sets up a new user account for GeorgeM with "fisherman" as the login password and "0302" as the PID. While this example procedure uses a simple PID number, the PID number you choose for a production environment should be from 4 to 20 characters long (preferably a combination of numbers and uppercase and lowercase letters that will be difficult for someone to guess).

Hands-On 22.3. Creating a User Account

1. Create a new Microsoft Access 2013 database named **Chap22.mdb** in your **C:\Access2013_ByExample** folder. Be sure to select Microsoft Access databases (2002-2003)(*.mdb) file format.
2. Switch to the Visual Basic Editor window and choose **Tools | References**. In the References dialog box, scroll down to locate **Microsoft ActiveX Data Objects 6.1 Library**. Click the checkbox to the left of this library name to set a reference to it and click **OK** to exit the dialog box.
3. Choose **Insert | Module** to add a new module to the current VBA project.
4. In the module's Code window, enter the **CreateUserAccount** procedure shown here:

```
Sub CreateUserAccount()
  Dim conn As ADODB.Connection

  On Error GoTo ErrorHandler
  Set conn = CurrentProject.Connection

  conn.Execute "CREATE USER GeorgeM fisherman 0302"
ExitHere:
```

```
    If Not conn Is Nothing Then
      If conn.State = adStateOpen Then conn.Close
    End If
    Set conn = Nothing
    Exit Sub
  ErrorHandler:
    MsgBox Err.Number & ":" & Err.Description
    Resume ExitHere
  End Sub
```

5. Run the CreateUserAccount procedure.
6. Press **Alt+F11** to switch to the Microsoft Access application window. Choose **File | Info | Users and Permissions | User and Group Accounts.**

After running the CreateUserAccount procedure in Hands-On 22.3, you will see a listing for the GeorgeM user account in the User and Group Accounts window, as shown in Figure 22.1.

FIGURE 22.1. The User and Group Accounts window.

7. Click **Cancel** to close the User and Group Accounts window.

CHANGING A USER PASSWORD

A user account password can be changed by using the `ALTER USER` statement in the following form:

```
ALTER USER userAccountName PASSWORD newPassword oldPassword
```

The procedure in Hands-On 22.4 changes the GeorgeM account's user password from "fisherman" to "primate."

Hands-On 22.4. Changing a User Password

This hands-on exercise requires prior completion of Hands-On 22.3.

1. In the same module where you entered the CreateUserAccount procedure in Hands-On 22.3, enter the following **ChangeUserPassword** procedure as:

```
Sub ChangeUserPassword()

  Dim conn As ADODB.Connection
  On Error GoTo ErrorHandler

  Set conn = CurrentProject.Connection

  conn.Execute "ALTER USER GeorgeM PASSWORD primate fisherman"
ExitHere:
  If Not conn Is Nothing Then
    If conn.State = adStateOpen Then conn.Close
  End If
  Set conn = Nothing
  Exit Sub
ErrorHandler:
  MsgBox Err.Number & ":" & Err.Description
  Resume ExitHere
End Sub
```

2. Run the ChangeUserPassword procedure.

CREATING A GROUP ACCOUNT

Use the `CREATE GROUP` statement to create a new group account. You must specify the group name followed by a unique PID (personal identifier):

```
CREATE GROUP groupName PID
```

You can create more than one group at a time by separating the group names with a comma. The procedure in Hands-On 22.5 creates a new group account called Mozart with “2012Best” as the PID.

Hands-On 22.5. Creating a Group Account

1. In the same module where you entered the ChangeUserPassword procedure in Hands-On 22.4, enter the **CreateGroupAccount** procedure shown here:

```
Sub CreateGroupAccount()
  Dim conn As ADODB.Connection

  On Error GoTo ErrorHandler
  Set conn = CurrentProject.Connection

  conn.Execute "CREATE GROUP Mozart 2012Best"
ExitHere:
  If Not conn Is Nothing Then
    If conn.State = adStateOpen Then conn.Close
  End If
  Set conn = Nothing
  Exit Sub
ErrorHandler:
  MsgBox Err.Number & ":" & Err.Description
  Resume ExitHere
End Sub
```

2. Run the CreateGroupAccount procedure.

   The Groups tab in the User and Group Accounts window (see Figure 22.2) will now list the name of the newly created Mozart user group.

FIGURE 22.2. The User and Group Accounts window shows the Mozart group after running the CreateGroupAccount procedure in Hands-On 22.5.

ADDING USERS TO GROUPS

Use the ADD USER statement to make a user account a member of a group. Specify the user account name followed by the TO keyword and a group name:

```
ADD USER userAccountName TO groupName
```

Hands-On 22.6. Making a User Account a Member of a Group

This hands-on exercise requires prior completion of Hands-On 22.3 and 22.5.

1. In the same module where you entered the procedure in Hands-On 22.5, enter the following **AddUserToGroup** procedure:

```
Sub AddUserToGroup()
  Dim conn As ADODB.Connection

  On Error GoTo ErrorHandler
  Set conn = CurrentProject.Connection

  conn.Execute "ADD USER GeorgeM TO Mozart"
ExitHere:
  If Not conn Is Nothing Then
    If conn.State = adStateOpen Then conn.Close
  End If
  Set conn = Nothing
  Exit Sub
ErrorHandler:
  MsgBox Err.Number & ":" & Err.Description
  Resume ExitHere
End Sub
```

2. Run the AddUserToGroup procedure.

The user account GeorgeM is now a member of the Mozart group account. This can be easily verified by opening the User and Group Accounts window in the Access application window (see Step 6 in Hands-On 22.3) and selecting GeorgeM from the Name drop-down.

REMOVING A USER FROM A GROUP

To delete a user from a group, use the `DROP USER` statement followed by the username, the `FROM` keyword, and the group name. For example, to delete the GeorgeM account from the Mozart group, use the following statement:

```
DROP USER GeorgeM FROM Mozart
```

Hands-On 22.7. Removing a User Account from a Group

This hands-on exercise requires prior completion of Hands-On 22.5 and 22.6.

1. In the same module where you entered the procedure in Hands-On 22.6, enter the **RemoveUserFromGroup** procedure shown here:

```
Sub RemoveUserFromGroup()
  Dim conn As ADODB.Connection

  On Error GoTo ErrorHandler

  Set conn = CurrentProject.Connection

  conn.Execute "DROP USER GeorgeM FROM Mozart"

ExitHere:
  If Not conn Is Nothing Then
    If conn.State = adStateOpen Then conn.Close
  End If
  Set conn = Nothing
  Exit Sub
ErrorHandler:
  MsgBox Err.Number & ":" & Err.Description
  Resume ExitHere
End Sub
```

2. Run the RemoveUserFromGroup procedure to remove the GeorgeM user account from the Mozart group.

DELETING A USER ACCOUNT

To delete a user account, use the `DROP USER` statement followed by the user account name, as demonstrated by the DeleteUserAccount procedure in Hands-On 22.8.

Hands-On 22.8. Deleting a User Account

This procedure requires prior completion of Hands-On 22.3.

1. In the same module where you entered the procedures in the previous hands-on exercises, enter the following **DeleteUserAccount** procedure:

```
Sub DeleteUserAccount()
  Dim conn As ADODB.Connection

  On Error GoTo ErrorHandler

  Set conn = CurrentProject.Connection

  conn.Execute "DROP USER GeorgeM"
ExitHere:
  If Not conn Is Nothing Then
    If conn.State = adStateOpen Then conn.Close
  End If
  Set conn = Nothing
  Exit Sub
ErrorHandler:
  MsgBox Err.Number & ":" & Err.Description
  Resume ExitHere
End Sub
```

2. Run the DeleteUserAccount procedure to delete the user account named GeorgeM.

GRANTING PERMISSIONS FOR AN OBJECT

Use the GRANT statement to assign security permissions for an object in a database to an existing user or group account. The procedure in Hands-On 22.9 grants the SELECT, DELETE, INSERT, and UPDATE permissions on all tables to the Mozart group.

The GRANT statement requires the following:

- A list of privileges to be granted
- The keyword ON followed by the name of a table, a nontable object, or an object container (e.g., Tables, Forms, Reports, Modules, Scripts)

- The keyword TO followed by the user or group name

```
GRANT listOfPermissions ON tableName | objectName |
 containerName TO accountName
```

Please note that in addition to tables, the Tables container contains queries, views, and procedures, and the Scripts container includes macros.

Hands-On 22.9. Granting Permissions for Tables to an Existing Group

This hands-on exercise requires prior completion of Hands-On 22.5.

1. In the same module where you entered the procedure in Hands-On 22.8, enter the following **SetTblPermissions** procedure:

```
Sub SetTblPermissions()
  Dim conn As ADODB.Connection

  On Error GoTo ErrorHandler

  Set conn = CurrentProject.Connection

  conn.Execute "GRANT SELECT, DELETE, INSERT, " _
   & "UPDATE ON CONTAINER TABLES TO Mozart"
ExitHere:
  If Not conn Is Nothing Then
    If conn.State = adStateOpen Then conn.Close
  End If
  Set conn = Nothing
  Exit Sub
ErrorHandler:
  MsgBox Err.Number & ":" & Err.Description
  Resume ExitHere
End Sub
```

2. Run the SetTblPermissions procedure.
   After running the SetTblPermissions procedure, you can open the User and Group Permissions window (choose **File | Info | Users and Permissions | User and Group Permissions**) to check out the privileges granted to the members of the Mozart group, as shown in Figure 22.3.

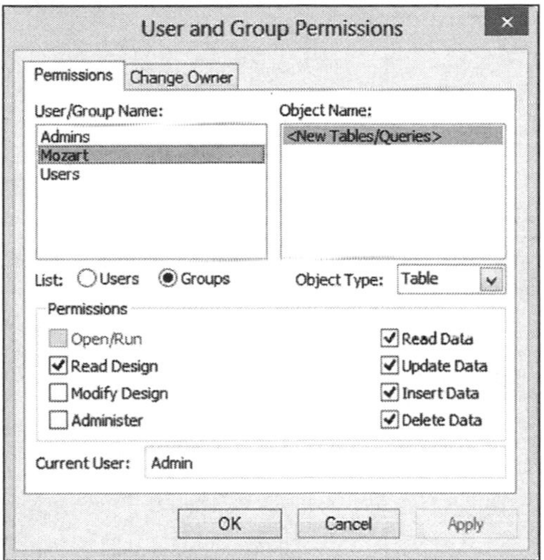

FIGURE 22.3. Verifying the group permissions to database objects.

REVOKING SECURITY PERMISSIONS

Use the `REVOKE` statement to revoke security permissions for an object from an existing user or group account. This statement has the following form:

```
REVOKE listOfPermissions ON tableName | objectName | containerName
FROM accountName
```

The procedure in Hands-On 22.10 removes the privilege of deleting tables from the members of the Mozart group (see Figure 22.4).

Hands-On 22.10. Revoking Security Permissions

This hands-on exercise requires prior completion of Hands-On 22.5 and 22.9.

1. In the same module where you entered the procedure in Hands-On 22.9, enter the **RevokePermission** procedure shown here:

```
Sub RevokePermission()
  Dim conn As ADODB.Connection

  On Error GoTo ErrorHandler

  Set conn = CurrentProject.Connection
```

```
    conn.Execute "REVOKE DELETE ON CONTAINER TABLES FROM Mozart"
  ExitHere:
    If Not conn Is Nothing Then
      If conn.State = adStateOpen Then conn.Close
    End If
    Set conn = Nothing
    Exit Sub
  ErrorHandler:
    MsgBox Err.Number & ":" & Err.Description
    Resume ExitHere
  End Sub
```

2. Run the RevokePermission procedure.

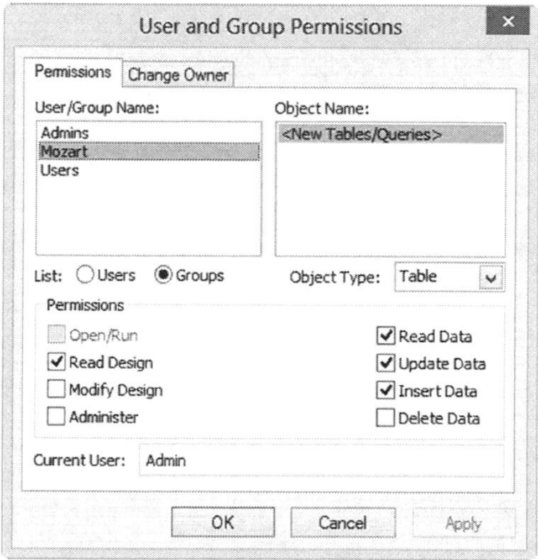

FIGURE 22.4. After running the procedure in Hands-On 22.10, the Delete Data permission on new tables and queries for the members of the Mozart group is turned off.

DELETING A GROUP ACCOUNT

Use the `DROP GROUP` statement to delete a group account. You only need to specify the name of the group account you want to delete. To delete more than one account, separate each group name with a comma.

Hands-On 22.11. Deleting a Group Account

This hands-on exercise requires prior completion of Hands-On 22.5.

1. In the same module where you entered the procedure in Hands-On 22.9, enter the following **DeleteGroupAccount** procedure:

```
Sub DeleteGroupAccount()
  Dim conn As ADODB.Connection

  On Error GoTo ErrorHandler

  Set conn = CurrentProject.Connection

  conn.Execute "DROP GROUP Mozart"
ExitHere:
  If Not conn Is Nothing Then
    If conn.State = adStateOpen Then conn.Close
  End If
  Set conn = Nothing
  Exit Sub
ErrorHandler:
  MsgBox Err.Number & ":" & Err.Description
  Resume ExitHere
End Sub
```

2. Run the DeleteGroupAccount procedure to delete the Mozart group account.

CHAPTER SUMMARY

This chapter demonstrated the use of Data Definition Language (DDL) statements in VBA procedures for managing security in a Microsoft Access database. You used the `ALTER DATABASE PASSWORD` statement to create, modify, and remove the database password. You managed user-level accounts with the `CREATE`, `ADD`, `ALTER`, and `DROP` statements. You also learned how to use the `GRANT` and `REVOKE` statements to establish and remove permissions on database objects for user and group accounts in an Access MDB database created in the 2002-2003 file format.

In the next chapter, you will learn how to organize your data using structures known as views and how to use stored procedures in lieu of Access Action and Parameter queries.

Chapter 23 Views and Stored Procedures

In this chapter, we will work with advanced Data Definition Language statements that are used for creating, altering, and deleting two special database objects known as views and stored procedures. These objects are used to perform various query operations. *Views* are like Access Select queries; however, you can't use the ORDER BY clause to sort your data or use parameters to filter records. *Stored procedures* perform the same operations as Access Action and Parameter queries. They can also be used for creating sorted Select queries. Stored procedures are saved precompiled so that at runtime the procedure executes much faster than a standard SQL statement. Learning how to create and use views and stored procedures will give you more control over your database.

CREATING A VIEW

If you want users to view and update data in a table or set of tables, but you do not want them to open the underlying tables directly, you can create a view. An SQL *view* is like a virtual table. Similar to an Access Select query, a view can display data from one or more tables. Instead of providing all the available data in your tables, you decide exactly what fields you'd like to include for viewing.

To create a view, use a SELECT statement to select the columns you want to include in the view and the FROM keyword to specify the table. Next, associate the SELECT statement with a CREATE VIEW statement. The syntax looks like this:

```
CREATE VIEW viewName [(columnNames)]
AS
SELECT (columnNames)
FROM tableName;
```

Views must have unique names in the database. The name of the view cannot be the same as the name of an existing table. Specifying the names of columns following the name of the view is optional (note the square brackets in the preceding syntax). Column names must be specified in the SELECT statement. Use the asterisk (*) to select all columns.

Let's put more meaning into the preceding syntax. The following example statement creates a view that lists only orders with a Freight amount less than $20.

```
CREATE VIEW cheapFreight
AS
SELECT Orders.OrderID,
Orders.[Shipping Fee],
Orders.[ShipCountry/Region]
FROM Orders
WHERE Orders.[Shipping Fee] < 20;
```

The SELECT statement that defines the view cannot contain any parameters and cannot be typed directly in the SQL pane of the Query window. It must be used through the ADO Connection object's Execute method after establishing the connection to a database, as illustrated here:

```
Sub Create_View_CheapFreight()
  Dim conn As ADODB.Connection
  Set conn = CurrentProject.Connection
```

```
    conn.Execute "CREATE VIEW CheapFreight AS " & _
     "SELECT Orders.[Order ID], Orders.[Shipping Fee], " & _
     "Orders.[Ship Country/Region] " & _
     "FROM Orders WHERE Orders.[Shipping Fee] < 20;"
    Application.RefreshDatabaseWindow
    conn.Close
    Set conn = Nothing
End Sub
```

The `Application.RefreshDatabaseWindow` statement ensures that after the view is created it is immediately listed in the Navigation pane in the Access application window. If you omit this statement, you will need to refresh the Navigation pane manually by selecting any object in it and pressing Shift+F5.

To return data from the CheapFreight view, simply double-click its name in the Navigation pane.

A view can be used as if it were a table. The following statement can be used to return all records from the CheapFreight view:

```
SELECT * FROM CheapFreight;
```

Remember that a view never stores any data; it simply returns the data as stated in the `SELECT` statement used in the view definition. Because a view is like a Select query, you can use the `OpenQuery` method of the Access DoCmd object to open it from your VBA code:

```
Sub OpenView()
  DoCmd.OpenQuery "CheapFreight", acViewNormal
End Sub
```

The `OpenQuery` method is used to carry out the OpenQuery action in Visual Basic.

To get working experience with the views, let's proceed to the hands-on section. We will start by creating a view called vw_Employees. This view is based on the Employees and Orders tables, and contains four columns (Employee ID, Full Name, Job Title, and Order ID).

Please note files for the "Hands-On" project may be found on the companion CD-ROM.

Hands-On 23.1. Creating a View Based on a Table

1. Start Microsoft Access 2013 and create a new database named **Chap23.accdb** in your **C:\Access2013_ByExample** folder.

2. Choose **External Data | Access.**
3. In the File name box of the Get External Data dialog box, enter **C:\Access2013_ByExample\Northwind 2007.accdb** and click **OK.**
4. In the Import Objects dialog box, select the **Employees, Orders**, and **Shippers** tables and click **OK.**
5. Click **Close** to exit the Get External Data dialog box.
   The Employees, Orders and Shippers tables are now listed in the Navigation pane.
6. Switch to the Visual Basic Editor window and choose **Tools | References.** In the References dialog box, scroll down to locate **Microsoft ActiveX Data Objects 6.1 Library**. Click the checkbox to the left of this library name to set a reference to it and click **OK** to exit the dialog box.
7. Choose **Insert | Module** to add a new module to the current VBA project.
8. In the module's Code window, type the following **Create_View** procedure:

```
' Don't forget to set up a reference to the
' Microsoft ActiveX Data Objects 6.1 Library
' in the References dialog box

Sub Create_View()
  Dim conn As ADODB.Connection

  Set conn = CurrentProject.Connection

  On Error GoTo ErrorHandler
  conn.Execute "CREATE VIEW vw_Employees AS " & _
   "SELECT Employees.ID AS [Employee ID], " & _
   "[First Name] & chr(32) & [Last Name] AS [Full Name], " & _
   "[Job Title], Orders.[Order ID] AS [Order ID] " & _
   "FROM Employees " & _
   "INNER JOIN Orders ON " & _
   "Orders.[Employee ID] = Employees.ID;"
   Application.RefreshDatabaseWindow
ExitHere:
  If Not conn Is Nothing Then
    If conn.State = adStateOpen Then conn.Close
  End If
  Set conn = Nothing
  Exit Sub
ErrorHandler:
  If Err.Number = -2147217900 Then
```

```
      conn.Execute "DROP VIEW vw_Employees"
      Resume
    Else
      MsgBox Err.Number & ":" & Err.Description
      Resume ExitHere
    End If
  End Sub
```

9. Run the Create_View procedure.

   This procedure creates a view named vw_Employees. If the view already exists, it will be deleted using the `DROP VIEW` statement. The `chr(32)` statement will insert a space between the first and last name.

   Notice that views don't differ much from saved queries. When you open the view created by the Create_View procedure in Design view, you will notice that this view is simply a Select query. Because the query defined by the `SELECT` statement is updatable, the vw_Employees view is also updatable. If the query were not updatable, the view would be read-only.

   Views cannot contain the `ORDER BY` clause. To return the records in a specific order, you might want to use the view in a stored procedure, as discussed later in this chapter.

ENUMERATING VIEWS

You can find out the names of the views that have been created by iterating through the Views collection of the ADOX Catalog object, as illustrated in Hands-On 23.2.

Hands-On 23.2. Generating a List of Saved Views

1. In the Visual Basic Editor window, choose **Tools | References**. In the References dialog box, scroll down to locate **Microsoft ADO Ext. 6.0 for DDL and Security Object Library**. Click the checkbox to the left of this library name to set a reference to it and click **OK** to exit the dialog box.
2. Choose **Insert | Module** to add a new module to the current VBA project.
3. In the module's Code window, enter the **List_Views** procedure shown here:

```
' Don't forget to set up a reference to the
' Microsoft ADO Ext. 2.8 for DDL and Security

Sub List_Views()
```

```
    Dim cat As New ADOX.Catalog
    Dim myView As ADOX.View

    cat.ActiveConnection = CurrentProject.Connection

    For Each myView In cat.Views
      Debug.Print myView.Name
    Next myView
  End Sub
```

4. Run the List_Views procedure.

   The List_Views procedure writes the names of the existing views to the Immediate window.

DELETING A VIEW

Use the `DROP VIEW` statement to delete a particular view from the database. You must specify the name of the view you want to delete. The following example procedure deletes a view named vw_Employees that was created by the procedure in Hands-On 23.1.

Note that both the `CREATE VIEW` and `DROP VIEW` statements can only be executed using the `Execute` method of the ADO Connection object.

Hands-On 23.3. Deleting a View

1. In the same module where you entered the procedure in Hands-On 23.2, enter the **Delete_View** procedure shown here:

```
Sub Delete_View()
  Dim conn As ADODB.Connection

  Set conn = CurrentProject.Connection

  On Error GoTo ErrorHandler
  conn.Execute "DROP VIEW vw_Employees"
ExitHere:
  If Not conn Is Nothing Then
    If conn.State = adStateOpen Then conn.Close
  End If
  Set conn = Nothing
```

```
    Exit Sub
ErrorHandler:
    If Err.Number = -2147217865 Then
      MsgBox "The view was already deleted."
      Exit Sub
    Else
      MsgBox Err.Number & ":" & Err.Description
      Resume ExitHere
    End If
End Sub
```

2. Run the Delete_View procedure.
3. Run the List_Views procedure from Hands-On 23.2 to ensure that the vw_Employees view was deleted.

CREATING A STORED PROCEDURE

Stored procedures allow you to perform bulk operations that delete, update, or append records. Unlike views, stored procedures allow the ORDER BY clause and parameters. Use the CREATE PROCEDURE (or CREATE PROC) statement to create a stored procedure. You must specify the name of the stored procedure and the AS keyword followed by the desired SQL statement that performs the required database operation. The syntax is as follows:

```
CREATE PROC[EDURE] procName
[(param1 datatype1 [, param2 datatype2 [, ...] ])]
AS
  sqlStatement;
```

The name of the stored procedure cannot be the same as the name of an existing table. To pass values to a stored procedure, include parameters after the procedure name. Parameter names are followed by a data type and are separated by commas. The parameters are listed in parentheses (see Hands-On 23.4 in the next section). Up to 255 parameters can be specified in the parameter list. If the stored procedure does not require parameters, the AS keyword immediately follows the name of the stored procedure.

The SQL statement for the stored procedure can be prepared in the Access Query Design window and then copied to the VBA procedure from the SQL view and appropriately formatted.

To return the employee records from the vw_Employees view (see Hands-On 23.1) ordered by Full Name, the following stored procedure can be written:

```
CREATE PROCEDURE usp_EmpByFullName
AS
  SELECT * FROM vw_Employees
  ORDER BY [Full Name];
```

This stored procedure selects all columns that exist in the vw_Employees view and orders the returned data by the Full Name field. Notice that this procedure does not require any parameters. You might want to precede the name of the stored procedure with a prefix indicating the type of stored procedure. The "usp" prefix is often used to indicate a user-defined stored procedure.

Like views, stored procedures are created via the ADO Connection object's `Execute` method after establishing a connection to the database. Therefore, you can use the following VBA code to create the usp_EmpByFullName stored procedure:

```
Sub Create_StoredProc()
  Dim conn As ADODB.Connection

  Set conn = CurrentProject.Connection
  conn.Execute "CREATE PROCEDURE usp_EmpByFullName AS " & _
   "SELECT * FROM vw_Employees " & _
   "ORDER BY [Full Name];"
  Application.RefreshDatabaseWindow
  conn.Close
  Set conn = Nothing
End Sub
```

Once created, stored procedures can be executed in the Access user interface by double-clicking the stored procedure name in the Navigation pane, or from VBA code by calling the `EXECUTE` statement with the ADO Connection object's `Execute` method (see Hands-On 23.5).

CREATING A PARAMETERIZED STORED PROCEDURE

Most advanced stored procedures require one or more parameters. The parameters are then used as part of the SQL statement, usually the `WHERE` clause. When creating a parameterized stored procedure, Access allows you to specify up to 255 parameters in the parameters list. The stored procedure parameters must be separated by commas and enclosed in parentheses.

The procedure in Hands-On 23.4 creates a stored procedure that allows you to insert a new record into the Shippers table on the fly by supplying the required parameter values. Note that the SQL Data Manipulation Language (DML) INSERT INTO statement is used for adding new records to a table.

Hands-On 23.4. Creating a Stored Procedure that Accepts Parameters

1. Switch to the Visual Basic Editor window and insert a new module.
2. In the module's Code window, enter the following **Create_SpWithParam** procedure:

```
Sub Create_SpWithParam()
  Dim conn As ADODB.Connection

  On Error GoTo ErrorHandler

  Set conn = CurrentProject.Connection

  conn.Execute "CREATE PROCEDURE usp_procEnterData " & _
   "(@Company TEXT (50), " & _
   "@Tel TEXT (25)) AS " & _
   "INSERT INTO Shippers (Company, [Business Phone]) " & _
   "VALUES (@Company, @Tel);"
  Application.RefreshDatabaseWindow
ExitHere:
  If Not conn Is Nothing Then
    If conn.State = adStateOpen Then conn.Close
  End If
  Set conn = Nothing
  Exit Sub
ErrorHandler:
  If InStr(1, Err.Description, "procEnterData") Then
     conn.Execute "DROP PROC procEnterData"
    Resume
  Else
    MsgBox Err.Number & ":" & Err.Description
    Resume ExitHere
  End If
End Sub
```

3. Run the Create_SpWithParam procedure.

   The preceding stored procedure requires two values to be entered at runtime. The first value is passed by the @Company parameter and the second value by the @Tel parameter. In this example, the names of the parameters have been preceded with the @ sign for easy migration of the stored procedure into the SQL Server environment. If you omit the @ sign, the procedure will still execute correctly in Microsoft Access. If the procedure already exists, it will be dropped using the `DROP PROC` statement.

Like views, stored procedures appear in the Navigation pane in the Access application window. Because we used the SQL `INSERT INTO` statement, Microsoft Access treats this stored procedure as a parameterized Append query.

4. Run the stored procedure named **usp_procEnterData** by double-clicking its name in the Navigation pane of the Access application window. Figures 23.1 through 23.4 outline the process of running this stored procedure, and Figure 23.5 shows the result.

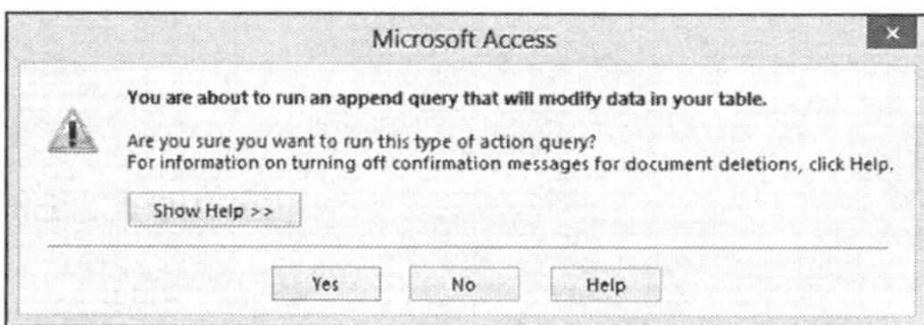

FIGURE 23.1. When you double-click a stored procedure name in the Navigation pane of the Access database window, Access displays this message when the stored procedure expects parameters and its SQL statement attempts to insert data into a table.

Enter Parameter Value ? ×
Company
Orient Express
OK Cancel

FIGURE 23.2. Because the stored procedure expects some input, you are prompted for the first parameter value.

Enter Parameter Value ? ×
Tel
800-234-0747
OK Cancel

FIGURE 23.3. Here you are prompted to enter the phone number for the second stored procedure parameter.

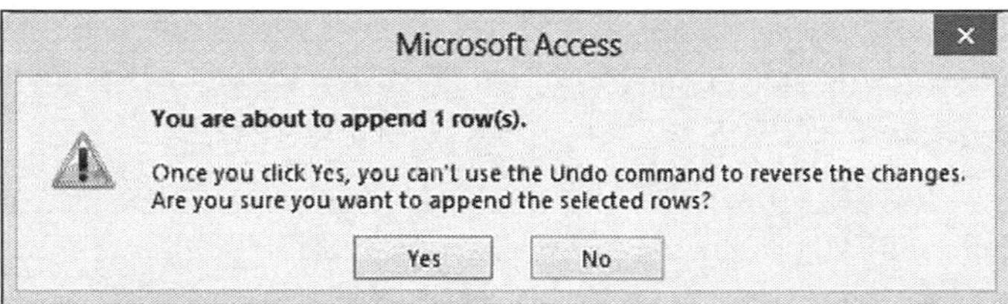

FIGURE 23.4. Once all input has been gathered via the parameters, Access informs you about the action that is to be performed. Click Yes to execute the stored procedure or No to cancel.

ID	Company	Last Name	First Name	E-mail Addres	Job Title	Business Phone
1	Shipping Company A					
2	Shipping Company B					
3	Shipping Company C					
4	Orient Express					800-234-0747
###						

Record: 4 of 4 No Filter Search

FIGURE 23.5. After you click Yes, Access runs the Append query. To view the result of this operation, double-click the Shippers table in the Navigation pane. Notice that a new record (Orient Express) was added to the Shippers table.

EXAMINING THE CONTENTS OF A STORED PROCEDURE

You can examine the contents of the stored procedure created in Hands-On 23.4 by right-clicking on the usp_procEnterData procedure in the Navigation pane and choosing Design View. Figure 23.6 displays the Design view of the Append query. Other stored procedures that you create may be presented as different Action queries.

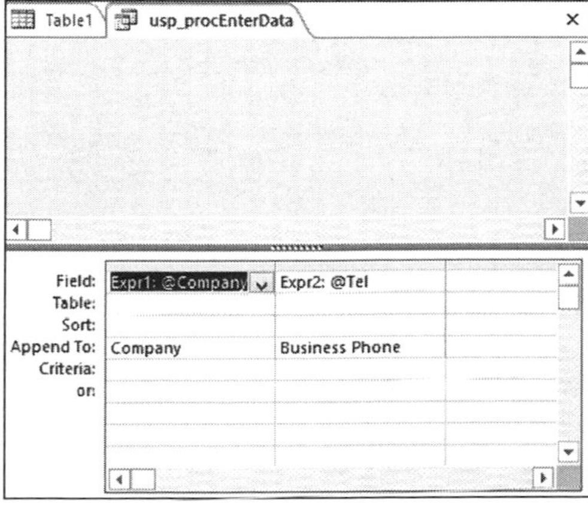

FIGURE 23.6. To view or modify the contents of a stored procedure, open it in Design view.

You can examine the SQL statements used by Access to execute your stored procedure by switching to the SQL view (click Design | View and select SQL View), as shown in Figure 23.7.

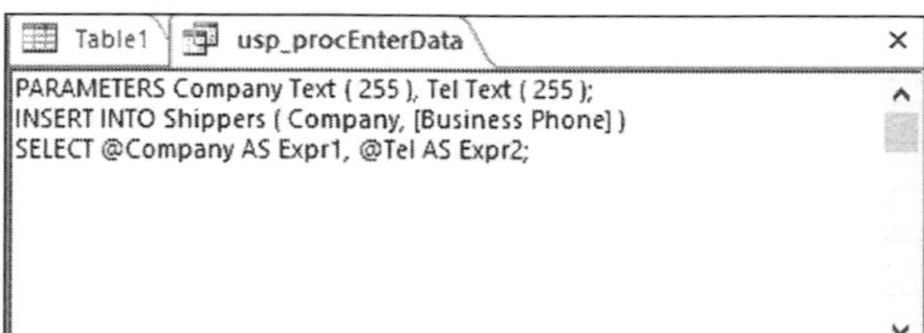
Table1
usp_procEnterData
PARAMETERS Company Text (255), Tel Text (255);
INSERT INTO Shippers (Company, [Business Phone])
SELECT @Company AS Expr1, @Tel AS Expr2;

FIGURE 23.7. The SQL view of the Query window displays the SQL statement that Access will execute when you run the stored procedure created in Hands-On 23.4.

EXECUTING A PARAMETERIZED STORED PROCEDURE

In the preceding section, you learned how to run a parameterized stored procedure from the Access user interface. To execute an existing stored procedure from VBA code, use the `Execute` method of the ADO Connection or Command object. Here's how:

- With the `Execute` method of the Connection object:

  ```
  conn.Execute "usp_procEnterData"
  ```

- With the `Execute` method of the Command object:

  ```
  cmd.CommandText = "usp_procEnterData"
  cmd.CommandType = adCmdStoredProc
  cmd.Execute
  rst. Open cmd
  ```

If the stored procedure requires parameters, parameter values follow the procedure name as a comma-separated list. Here's an example procedure that executes the usp_procEnterData stored procedure and contains the values for its two parameters:

```
Sub RunProc_WithParam()
  Dim conn As ADODB.Connection

  Set conn = CurrentProject.Connection
  conn.Execute "usp_procEnterData ""My Company2"", ""(234) 334-3344"""
  conn.Close
  Set conn = Nothing
End Sub
```

Instead of surrounding parameters with sets of double quotes, you can use single quotes like this:

```
conn.Execute "procEnterData 'My Company2', '(234) 334-3344'"
```

The procedure in Hands-On 23.5 runs the stored procedure named usp_procEnterData created in Hands-On 23.4. Notice how this procedure uses the `InputBox` function to obtain the parameter values from the user instead of hard-coding them in the `Execute` method of the Connection object (as shown in the preceding example). Still another way of providing parameter values to a stored procedure would be via an Access form. This is left for you to try on your own.

Hands-On 23.5. Executing a Parameterized Stored Procedure

1. Switch to the Visual Basic Editor window and insert a new module.
2. In the module's Code window, enter the following **Execute_StoredProcWith-Param** procedure:

```
Sub Execute_StoredProcWithParam()
  Dim conn As ADODB.Connection
  Dim strCompany As String
  Dim strPhone As String

  On Error GoTo ErrorHandler

  Set conn = CurrentProject.Connection
  strCompany = InputBox("Please enter company name:", "Company")
  strPhone = InputBox("Please enter the phone number:", "Phone")
  If strCompany <> "" And strPhone <> "" Then
    conn.Execute "usp_procEnterData " & strCompany & _
          ", " & strPhone
  End If
ExitHere:
  If Not conn Is Nothing Then
    If conn.State = adStateOpen Then conn.Close
  End If
  Set conn = Nothing
  Exit Sub
ErrorHandler:
  MsgBox Err.Number & ":" & Err.Description
  Resume ExitHere
End Sub
```

3. Run the Execute_StoredProcWithParam procedure.

When you run the parameterized stored procedure in Hands-On 23.5, Access displays an input box for each parameter. After you have supplied values for both required parameters, a new record is entered into the Shippers table.

DELETING A STORED PROCEDURE

Use the `DROP PROCEDURE` (or `DROP PROC`) statement to delete a stored procedure. The syntax looks like this:

```
DROP PROC[EDURE] procedureName
```

The following example procedure deletes the stored procedure named usp_procEnterData from the current database.

Hands-On 23.6. Deleting a Stored Procedure

1. Switch to the Visual Basic Editor window and insert a new module.
2. In the module's Code window, enter the **Delete_StoredProc** procedure shown here:

```
Sub Delete_StoredProc()
  Dim conn As ADODB.Connection

  On Error GoTo ErrorHandler
  Set conn = CurrentProject.Connection

  conn.Execute "DROP PROCEDURE usp_procEnterData; "
ExitHere:
  If Not conn Is Nothing Then
    If conn.State = adStateOpen Then conn.Close
  End If
  Set conn = Nothing
  Exit Sub
ErrorHandler:
  If InStr(1, Err.Description, "cannot find") Then
    MsgBox "The procedure you want to delete " & _
     "does not exist.", _
     vbDefaultButton1 + vbInformation, "Request failed"
  Else
    MsgBox Err.Number & ":" & Err.Description
```

```
    End If
    Resume ExitHere
End Sub
```

3. Run the Delete_StoredProc to remove the usp_procEnterData procedure from the database.

CHANGING DATABASE RECORDS WITH STORED PROCEDURES

Stored procedures can perform various actions similar to what Access Action queries and Select queries with parameters can do. For example, here's how you would write a statement to create a stored procedure that, when executed, deletes a record from the Shippers table:

```
conn.Execute "CREATE PROCEDURE usp DeleteRec " & _
 "(ID Integer) " & _
 "AS " & _
 "DELETE * FROM Shippers WHERE ID = ID;"
```

To update a phone number in a specified record in the Shippers table, you may want to create a stored procedure that performs the specified record update with the following statement:

```
conn.Execute "CREATE PROCEDURE usp_UpdatePhone " & _
 "(ID Integer, tel text (25)) " & _
 "AS " & _
 "UPDATE Shippers SET [Business Phone] = tel " & _
 "WHERE ID = ID;"
```

CHAPTER SUMMARY

This chapter introduced you to two powerful database objects you can use in Access: views and stored procedures. You learned how views are used as virtual tables to make specific rows and columns from one or more tables available to your Access users. Remember that views are similar to SELECT statements, except they cannot contain the ORDER BY clause to sort the data and they do not allow parameters. Views can be used in queries to hide from users the complexity of joins between the tables. Converting your Access queries into views and stored procedures will help with migration of your Access applications to the SQL Server environment in the future.

This chapter concludes Part VI of this book, which presented numerous examples of using SQL DDL statements inside VBA procedures. In particular, you learned how DDL statements are used to create tables, views, stored procedures, primary keys, indexes, and constraints that define the database. You also learned some advanced Data Manipulation Language (DML) statements. Although there is more to Access SQL than this part of the book has covered, the information presented here should be quite sufficient to get you started using SQL in your own Access database applications.

Part 7 ENHANCING THE USER EXPERIENCE

The behavior of Microsoft Access objects such as forms, reports, and controls can be modified by writing programming code known as an event procedure or an event handler. In this part of the book, you will learn how you can design more effective and visually appealing forms and reports, and make your forms, reports, and controls perform useful tasks by writing event procedures in class modules. You also learn how to use VBA, macros, and XML to customizc the user interface in Access 2013.

Chapter 24 Enhancing Access Forms

Access 2013 offers users a great number of features in the form design area. For example, the Layout view gives form interface a true WYSIWYG: you can see the live data as you design your form without the need to constantly switch between the Design and Form views. The form features include various methods of creating forms, the Split Form, Bound Image controls, the Attachments control, styles and AutoFormats, rich text support as well as various ways of grouping controls by using the layouts. Many features are available in datasheets, including the Date Picker, alternating row colors, the Totals row, truncated number displays, sorting and filtering, and an easy way to add new list items to combo boxes.

The Web Browser control extends the capabilities of Access forms by enabling users to view and interact with Web data directly from Access. The Navigator control allows the creation of modern-looking tab-style navigation forms that replace the old-fashioned switchboard style form used in earlier versions of Access. With the Navigation control, it is possible to create richer user interfaces with "parent" and "child" navigation forms. You can further enhance your Access forms by linking subreports to the form and making it possible for users to view information related to the record as they navigate the form. You can publish your application to the Web using SharePoint. With the Web Form Designer you can quickly and easily create forms in tabular format that are properly rendered on SharePoint via Access Services.

CREATING ACCESS FORMS

The form buttons on the Ribbon's Create tab (see Figure 24.1) enable users to create both simple and advanced forms.

The classic Access form with a columnar layout is generated automatically by selecting a record source (a table or query) in the Navigation pane and clicking the Form button. The next button, called Form Design, is used to create a blank form in Design view. This form is not bound to any data source. Instead, you are presented with a list of tables and queries on which you can base your form. The Blank Form button on the Ribbon can be used to create a custom form from scratch in Layout view. As in earlier versions of Access, this form is not connected to any data source and can be used to create any form you want. The Form Wizard button allows you to create simple customizable forms. When you use the Form Wizard, Access allows you to specify which fields to include from which table or query and makes it easy to choose from a variety of AutoFormats. Using the wizard lets you choose only the fields you want so you don't have to spend extra time deleting the fields you don't want and repositioning the remaining fields.

The Navigation button is a gallery control that allows you to create forms that browse to other forms and reports. When you click the Navigation button, it will present a selection of different layouts (see Figure 24.1).

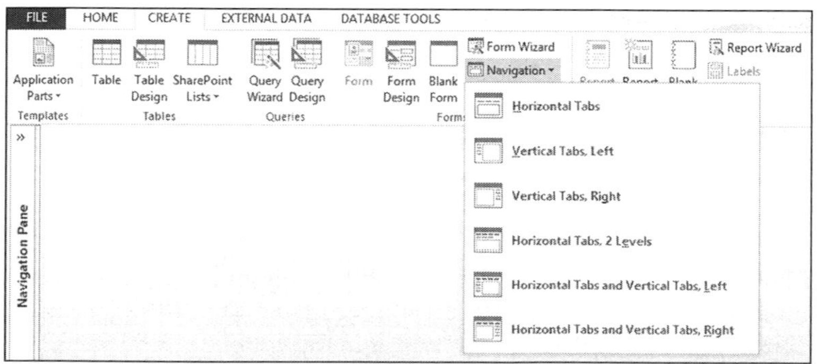

FIGURE 24.1. Use the Navigation control button on the Create tab (Forms group) to create customizable navigation forms.

The More Forms button (see Figure 24.2) provides additional types of forms users can create in Access 2013: Multiple Items, Datasheet, Split Form, and Modal Dialog. Unlike previous versions, Access 2013 no longer provides options for creating PivotChart, and PivotTable forms.

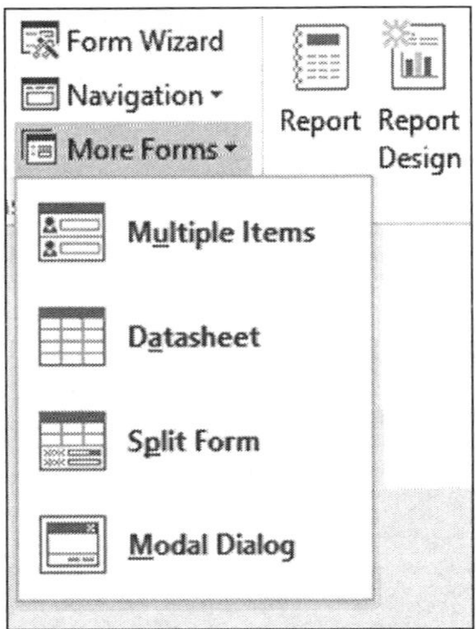

FIGURE 24.2. The More Forms button provides many types of forms that you can create in Access 2013.

The Multiple Items form is a standard continuous form used in earlier versions of Access. This form (see Figure 24.3) displays multiple records in a datasheet, with one record per row, and allows you to arrange the controls any way you want.

Employees

ID	Company	Last Name	First Name	E-mail Address	Job Title
1	Northwind Traders	Freehafer	Nancy	nancy@northwindtraders.com	Sales Rep
2	Northwind Traders	Cencini	Andrew	andrew@northwindtraders.com	Vice President, Sales
3	Northwind Traders	Kotas	Jan	jan@northwindtraders.com	Sales Rep
4	Northwind Traders	Sergienko	Mariya	mariya@northwindtraders.com	Sales Rep
5	Northwind Traders	Thorpe	Steven	steven@northwindtraders.com	Sales Manager
6	Northwind Traders	Neipper	Michael	michael@northwindtraders.com	Sales Rep

Record: 1 of 9 No Filter Search

FIGURE 24.3. The Multiple Items form in Access 2013.

The Datasheet form (see Figure 24.4) organizes data like the Multiple Items form but looks more like an Excel worksheet with one record per row.

The Split Form contains a datasheet and a standard Access form (see Figure 24.5). The datasheet displays multiple records. Simply click on the record in the datasheet and the form will change to show the details for this record, which you can edit. Access can create these types of forms with ease without asking you a single question.

The Modal Dialog form opens a form that functions like a modal window. This means that the user will not be able to activate any other object before closing that form. Modal Dialog forms are very useful when you need to gather specific information from users before allowing them to perform other actions.

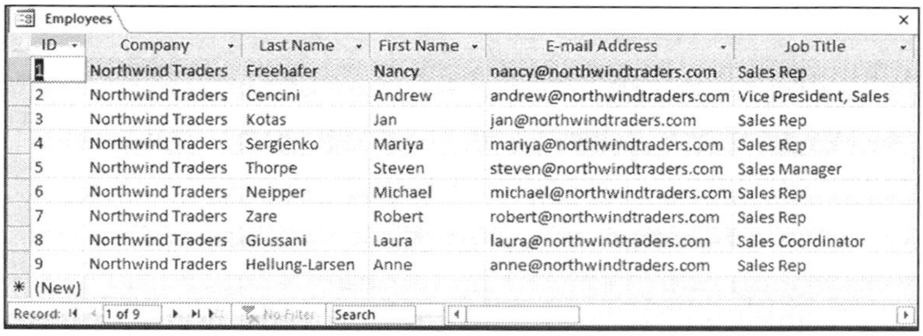

FIGURE 24.4. The Datasheet form resembles an Excel worksheet.

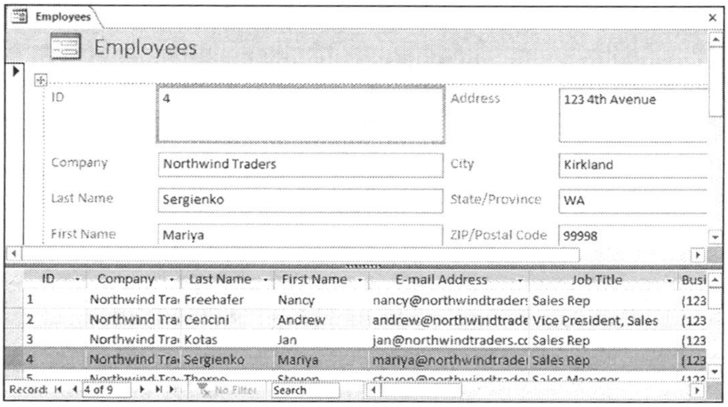

FIGURE 24.5. The Split Form is a combination of a standard form in an upper section and a datasheet in a lower section. It allows easy browsing through the records and entering or editing data for the selected record in the standard form.

GROUPING CONTROLS USING LAYOUTS

In Access you can group controls through a feature known as Layouts. Layout view enables you to work with entire groups of controls without having to guess whether the controls are properly sized and positioned. Take a look at Figure 24.6 and notice the Selector widget. When you click on the widget, you will see which controls are included in that layout. Using the Layout view you can easily move controls around in the form and resize them. To control the layouts, use the buttons in the Control Layout section of the Ribbon's Arrange tab (see Figure 24.7).

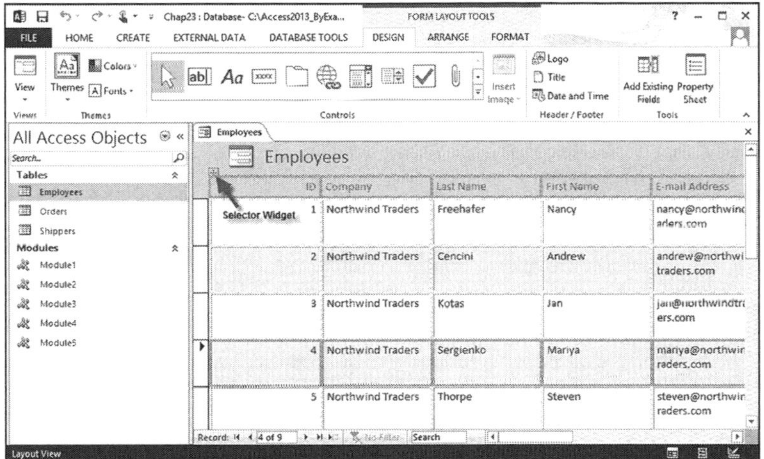

FIGURE 24.6. A group of controls on the form can be moved easily by using the anchor point (Selector widget).

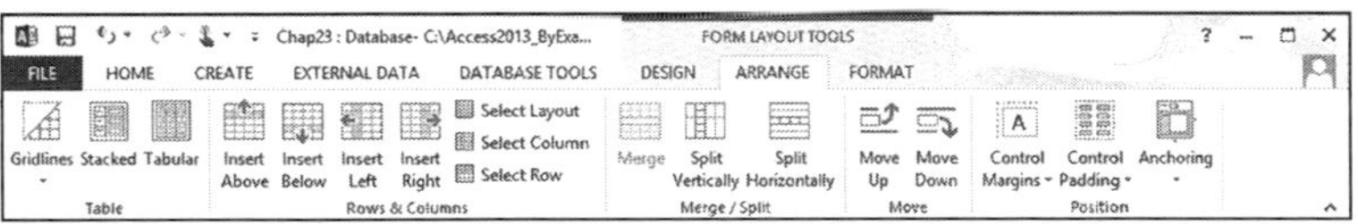

FIGURE 24.7. The Form Layout Tools section of the Ribbon provides numerous options for controlling the layout of the controls placed on the form.

Groups of controls can be moved to a new layout in one step. The tabular layout makes it easy to group the controls similar to a spreadsheet, with labels positioned across the top and data displayed in columns below the labels. To do this quickly, select the control group you want to reposition by clicking the anchor (Selector widget), then click the Tabular button on the Ribbon. The Stacked button can be used to create a layout similar to a paper form with labels to the left of the data. Removing entire groups of controls is also easily done by clicking the Selector widget and pressing the Delete key.

You can use anchoring to tie a control or a group of controls to a section or another control so it moves into place in accordance with the parent. The Anchoring options in Figure 24.8 show various positions where controls can be moved and ways they can be stretched to maximize the use of the space available on the form.

To keep users from making changes to the form, you can disallow the Layout view in the property sheet for the form.

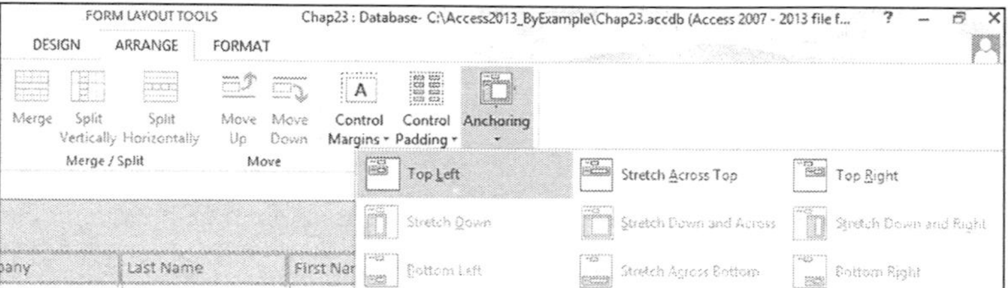

FIGURE 24.8. The Anchoring button reveal several options for positioning and stretching form controls.

RICH TEXT SUPPORT IN FORMS

In Chapter 12, you learned how to enable rich text formatting in memo fields. You can also use rich text formatting on Access forms via the Text Box control. For an unbound text box, all you need to do is open the property sheet for the text box and set its Text Format property to Rich Text (see Figure 24.9). This change will enable text formatting tools on the Ribbon. At runtime, users will be able to change the font style and color; add bold, italics, and highlighting; and apply other formatting. If the text box is bound to a field in a table, you must first change the Text Format property of the table field before changing the property of the control. If you forget to do this, Access will display the warning message shown in Figure 24.10.

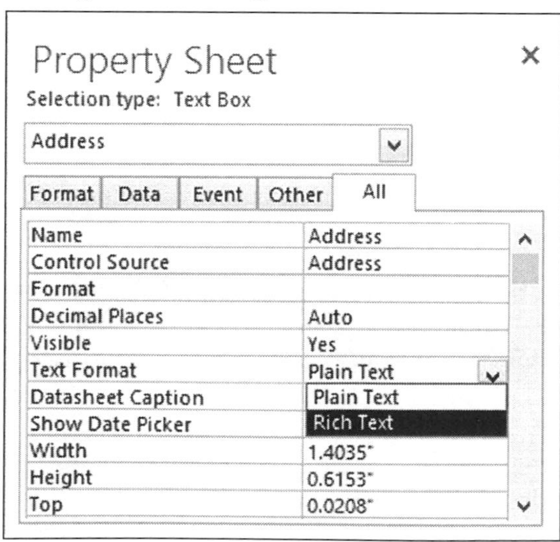

FIGURE 24.9. Access allows Rich Text format to be used in text box controls placed on the form.

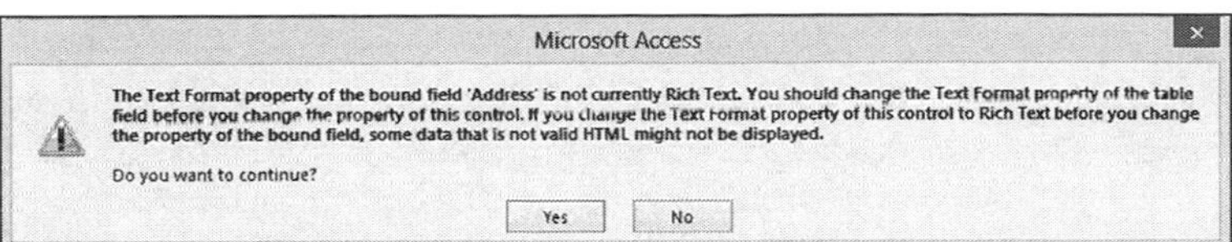

FIGURE 24.10. Before changing the Text Format of a bound text box control to Rich Text, be sure to change the Text Format property of the table field.

USING BUILT IN FORMATTING TOOLS

Access provides a gallery of themes you can use to give your forms a pleasing and consistent look. To preview the available designs, switch to the Form Layout view and click on the Themes button in the Themes section of the Design tab (see Figure 24.11).

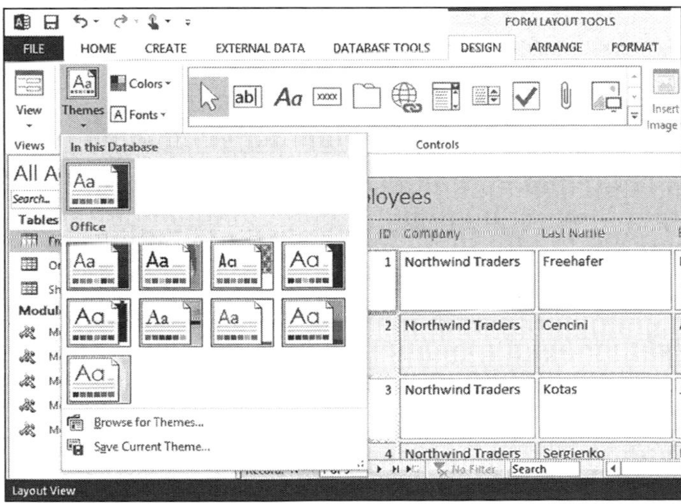

FIGURE 24.11. The Themes button provides a gallery of quick formats in Microsoft Access.

USING IMAGES IN ACCESS FORMS

Access 2007 came with better image support. Earlier versions of Access converted images from their native format and stored them as bitmaps (.bmp). This format caused a significant increase in the size of the database because bitmap files are not compressed. Also, any image transparency features were lost during the conversion process. If you use the Image control in Access 2007/2010/2013 and specify the image in the Picture property, Access will store the image in its native format with no conversion. Images with transparency work just fine. You can see examples of transparent buttons and pictures in the Northwind 2007 database's forms (see Figure 24.12).

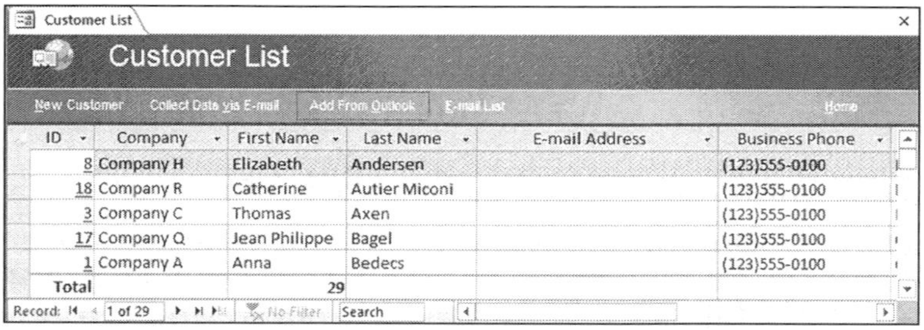

FIGURE 24.12. Access form with transparent images and buttons.

The older MDB databases do not support saving images in the native format, so there is a special database property that lets you choose whether the images should be converted to DIB (device-independent bitmap) or stored in their native format. The default setting is to store images in their native format and convert them to bitmaps for MDB databases. If you want your images to be displayed in previous versions of Access, choose the second option button under the Picture Property Storage Format setting (see Figure 24.13).

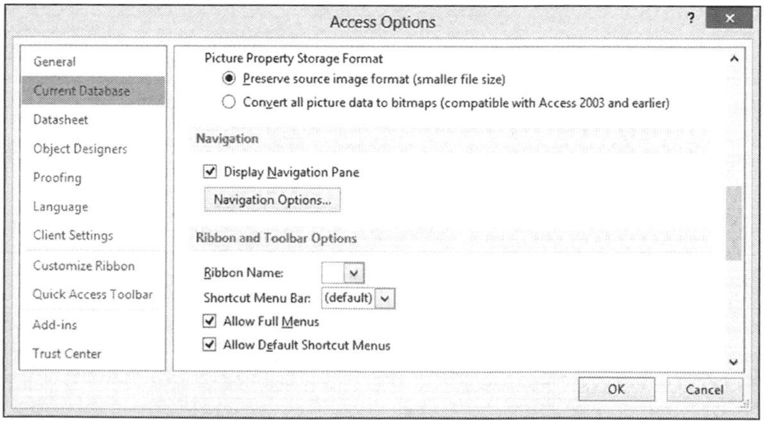

FIGURE 24.13. You can tell Access how to store images in older versions of Access (2003 and earlier) by using the options under Picture Property Storage Format. The Access Options window can be accessed by clicking File | Options.

Access 2007 introduced a bound Image control. Access 2003 and earlier needed lots of VBA code to display images on forms and reports when the images were stored in the directories on disk. The Image control can be bound to the image path. To add an image to your form, place the form in Design view and click the Image button ( ) in the Controls group of the Ribbon's Design tab. When you click the form grid, Access displays the Insert Picture dialog box where you can browse to your picture location on disk. When you click OK, the selected picture is placed in the Image control on the form. If you activate the property sheet for the Image control, you will see that Access has placed the filename in the Picture property (see Figure 24.14). If you don't like the picture you've chosen, you can simply click the ellipsis button (...) next to the Picture property and choose another image.

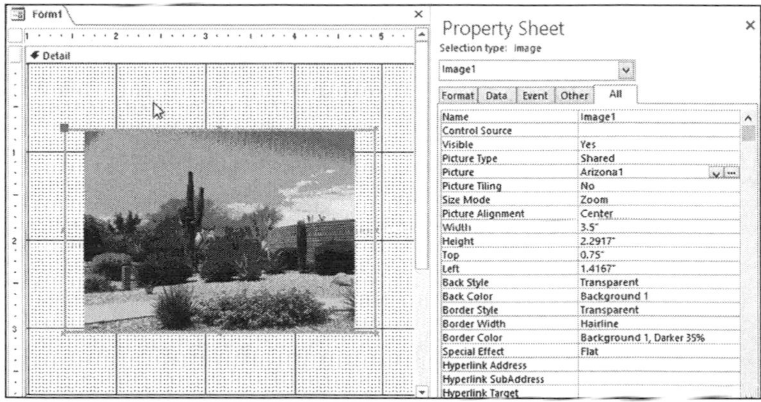

FIGURE 24.14. The picture is shown here using the Image control placed on an Access form.

USING THE ATTACHMENTS CONTROL

In the Controls group of the Form Design tab, you will find an Attachments control ( ) that enables you to attach a file or a collection of files to any database record. When you click on the Attachments field on the form, Access displays a small toolbar with three buttons (see Figure 24.15). The Forward and Backward buttons allow you to move through the attached files, and the third button opens the Attachments dialog box. You can also right-click the Attachments field and choose the same options from the shortcut menu.

Recall that in Chapter 15 (see Hands-On 15.3), you wrote a VBA procedure that added an attachment to a customer record in the Northwind 2007 database. Let's now see how you can work with the attachments in an Access form.

Please note files for the "Hands-On" project may be found on the companion CD-ROM.

Custom Project 24.1. Working with the Attachments Control

1. Start Microsoft Access 2013 and create a new database named **Chap24.accdb** in your **C:\Access2013_ByExample** folder.
2. Choose **External Data | Access** to import the Customers table. In the File name box, type **C:\Access2013_ByExample\Northwind 2007.accdb** and click **OK**. In the Import Objects window, activate the **Tables** tab and select the **Customers** table, then click **OK**. Click **Close** to exit the dialog boxes when the import process has completed.
3. In the Navigation pane, double-click the **Customers** table.
4. Double-click on the paper clip icon for the third record. You should see an empty Attachments dialog box. Click the **Add** button, select the **California1.jpg** and **California2.jpg** images from your **C:\Access2013_HandsOn\External Docs** folder, and click **Open**. The names of the selected files are now listed in the Attachments dialog box.
5. Click **OK** to close the dialog box and press **Ctrl+S** to save the record. Notice that the paper clip column now displays the number of attached files in parentheses next to the paper clip icon for the record.
6. Close the Customers table.
7. Highlight the **Customers** table in the Navigation pane, then click the **Form** button in the Forms group on the Ribbon's Create tab.

Access will display a form as shown in Figure 24.15.

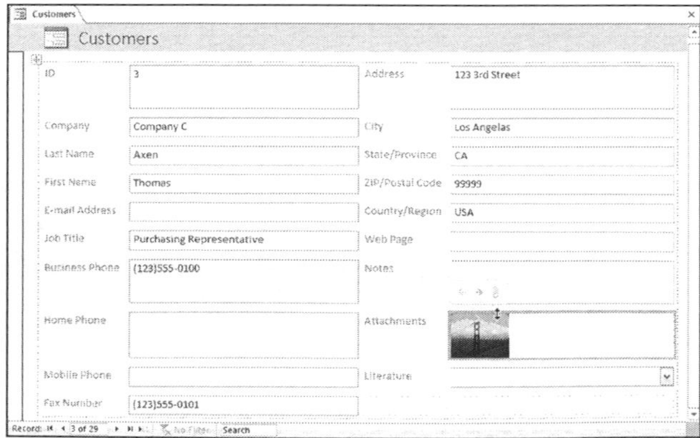

FIGURE 24.15. The Access form uses the Attachments control to show images attached to a record.

8. Activate the record for the third customer.
9. On the Customers form, click the **Attachments** control next to the Attachments label; notice a small toolbar with three buttons. Scroll through the attached files by clicking the Forward and Backward buttons.

Let's modify the form to display additional information about the attachments.

10. Switch to the Design view of the Customers form and use the **Text Box control** in the Controls group of the Design tab to add a text box to the form as shown in Figure 24.16. Change the default label of the text box control to **Current File** as shown.

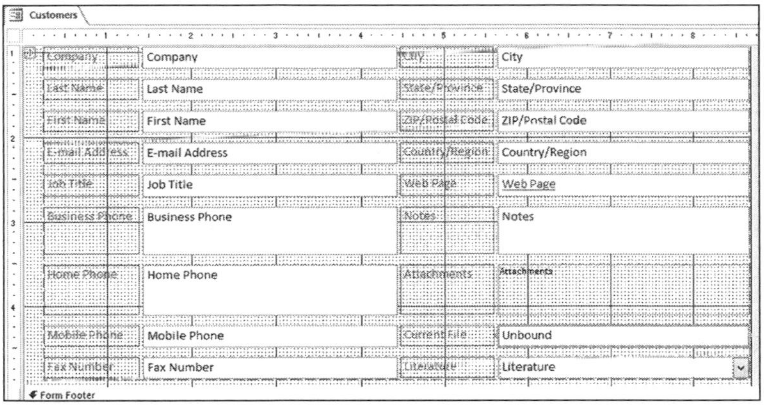

FIGURE 24.16. Placing unbound text box controls on the form.

11. In the form grid, click the unbound text box next to Current File. In the property sheet for this text box, click the **All** tab and type **txtCurrentFileName** in the Name property. Click the **Format** tab and change the Back Color property to any color you like.
12. In the form grid, click the **Current File** label. In the property sheet for the selected label control, click the **All** tab and type **lblCurrentFile** in the Name property.
13. In the form grid, click the **Attachments** label. In the property sheet for this label, click the **All** tab and type **lblTotalFiles** in the Name property.

    Now let's write an event procedure to display information about the attached file.
14. In the form grid, click the **Attachments** control. In the property sheet for this control, click the **Event** tab, then click the **Browse** button next to the **On Attachment Current** property. In the Choose Builder dialog box, select **Code Builder** and click **OK**.

    Access will write the stub of the **Attachments_AttachmentCurrent** event procedure.
15. Complete the code of the **Attachments_AttachmentCurrent** procedure as shown in Figure 24.17.

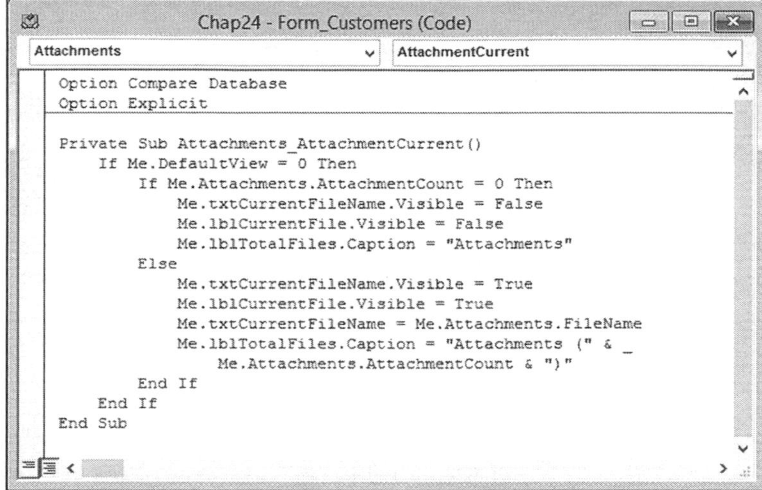

FIGURE 24.17. Use the AttachmentCurrent event procedure for the Attachments control to retrieve information about attachments and load it into your form's controls.

The Attachments control comes with special properties that apply to working with the Attachment data type. The FileName property returns the name of the attached file. If you need to display the file extension, use the FileType property. The AttachmentCount property returns the number of attachments stored for the record.

The Attachments control has a special event called AttachmentCurrent. This event is similar to the form's OnCurrent event. It is triggered when you move the focus from one attachment to another. The code shown in Figure 24.17 begins by checking whether the form's default view is set to Single Form (0). If DefaultView is set to display other types of Access forms, the code in the event procedure will not run. The procedure hides the txtCurrentFileName text box control and its label lblTotalFiles for all records that do not have any attachments. This is done by setting the Visible property of the text box control and label control to False. Next, the procedure fills in the text boxes with the values retrieved from the AttachmentCount and FileName properties. Notice how the procedure manipulates the Attachments label control to display the total number of attachments for records that have them.

16. Press **Alt+F11** to return to the Microsoft Access window and activate the **Customers** form in Form view.
17. Scroll to the third customer record.
18. Notice that the Attachments label now shows the number of attached files in parentheses. There is also a text box below the attachment listing the cur-

rent filename (see Figure 24.18). To scroll through the available files, select the Attachments field and click the Forward button in the tiny pop-up toolbar. Notice that when the new file loads into the Attachments control, the Current File box displays the name of the file being viewed.

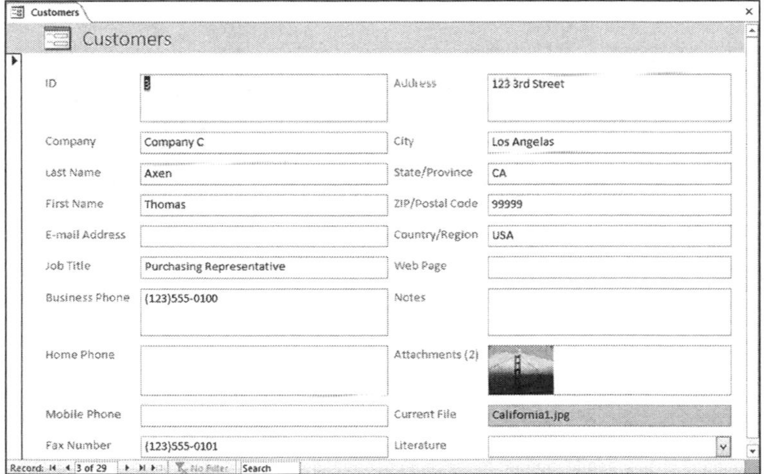

FIGURE 24.18. The Current File text box control added to the form provides information about the attachment filename currently displayed in the Attachments control. The Attachments label has been modified to include the total number of attached files for records that contain attachments.

19. **Press Ctrl+S** to save changes to the Customers form, and then close this form.

CHAPTER SUMMARY

This chapter presented a quick overview of types of forms you can create with Access 2013 and types of formatting you can apply to make your forms more attractive.

You learned how you can group form controls using the layouts, implement rich formatting in form controls, professionally format your forms using built-in themes, and enhance forms with images.

The chapter's main project focused on using the Attachments control in an Access form and showed you how to write an event procedure to display additional information about the attachments. You may want to treat it as a "warm-up" exercise for the next chapter, which gives you a complete overview and working knowledge of event procedures you can write for Access forms to change or enhance their default behavior.

Chapter 25 Using Form Events

Chapter 1 provided a quick introduction to events, event properties, and event procedures as well as an example event procedure that changed the background color of a text box control placed on a form. Now is a good time to go back to the beginning of this book and review these topics. Here's a rundown of the terms you need to be familiar with:

- **Event**—Events are things that happen to an object. Events occur when you move a mouse, press a key, make changes to data, open a form, or add, modify or delete a record, etc. An event can be triggered by the user or by the operating system.
- **Event property**—Forms, reports, and controls have various event properties you can use to trigger desired actions. When an event occurs, Microsoft Access runs a procedure assigned to an event property. Event properties are listed in the Event tab of the object's property sheet. The name of the event property begins with "On" and is followed by the event's name. Therefore, the On Click event property corresponds to the Click event, and the On Got Focus event property is used for responding to the GotFocus event.
- **Event procedure**—This is programming code you write to specify how a form, report, or control should respond to a particular event. By writing event procedures you can modify the application's built-in response to an event.

- **Event trapping**—When you assign programming code to an event property, you set an event trap. When you trap an event, you interrupt the default processing that Access would normally carry out in response to the user's keypress or mouseclick.
- **Sequence of events**—Events occur in a predefined order. For example, the Click event occurs before the DoubleClick event. When you perform an action, several events occur, one after the other. For instance, the following form and control events occur when you open a form:

Open → Load → Resize → Activate → Current → Enter (control) → GotFocus (control)

Closing the form triggers the following control and form events:

Exit (control) → LostFocus (control) → Unload → Deactivate → Close

To find out whether a particular event is triggered in response to a user action, you may want to place the `MsgBox` statement inside the event procedure for the event you want to test. Microsoft Access forms, reports, and controls recognize various events.

Events can be organized by object (form, report, control) or by cause (what caused the event to happen). This chapter contains numerous examples of event procedures you can write to make your forms and reports dynamic. You can also experiment with various events in the data entry/lookup application (AssetsDataEntry.mdb) located on the companion CD-ROM.

Microsoft Access forms can respond to a variety of events. These events allow you to manage entire records and respond to changes in the data. You can determine what happens when records are added, changed, or deleted, or when a different record becomes current. You can decide how the form appears to the user when it is first displayed on the screen and what happens when the form is closed. You can also manage problems that occur when the data is unavailable. As you design your custom forms, you will find that some form events are used more frequently than others. The following sections provide hands-on examples of event procedures you can write for Access forms.

DATA EVENTS

Data events occur when you change the data in a control or record placed on a form, or when you move the focus from one record to another.

Current

The Current event occurs when the form is opened or requeried and when the focus moves to a different record. Use the Current event to synchronize data among forms or move focus to a specific control.

The event procedure in Hands-On 25.1 sets the BackColor property of the form's header (Section 1) to red (255) for each discontinued product. The Form_Current event will occur each time you move to a new record if the specified condition is true.

Please note files for the "Hands-On" project may be found on the companion CD-ROM.

Hands-On 25.1. Writing the Form_Current Event Procedure

1. Start Microsoft Access 2013 and create a new database named **Chap25.accdb** in your **C:\Access2013_ByExample** folder.
2. Import all the tables, queries, forms, reports, macros, and modules from the Northwind.mdb sample database to your Chap25.accdb database. To do this, in the Access window, choose **External Data | Access**. In the File name box, type **C:\Access2013_ByExample\Northwind.mdb** and click **OK**. In the Import Objects window, select the **Tables** tab and click the **Select All** button. This will highlight all the tables. Select the **Queries** tab and click the **Select All** button. Select the **Forms** tab and click the **Select All** button. Select the **Reports** tab and click the **Select All** button. Do the same for macros and modules. After selecting all the objects on the specified tabs, click **OK** to begin importing. Click the **Close** button when done.
3. In the Access window of the Chap25.accdb database, right-click on the **Products** form and choose **Design View**. Make sure the form's property sheet is visible and the Selection Type is set to Form. To activate the property sheet, choose **Property Sheet** in the Tools section of the Design tab.
4. In the form's property sheet, click the **Event** tab. Click next to the **On Current** event property and choose **[Event Procedure]** from the drop-down box. Click the **Build** button (…).

   Access opens the Visual Basic Editor window and writes the stub of the Form_Current event procedure.
5. Complete the code of the **Form_Current** event procedure as shown here:

```
Private Sub Form_Current()
  If Discontinued = True Then
```

```
      Me.Section(1).BackColor = 255
      Me.Picture = ""
    Else
      Me.Picture = "C:\Access2013_HandsOn\External Docs\" & _
       "Pinelumb.jpg"
    End If
  End Sub
```

6. To test this event procedure, activate the **Products** form that is currently open in Design view. You can quickly switch to the selected form from Visual Basic by clicking the View Object button in the Project Explorer window. Next, in the Access window, click the **View** button on the Ribbon to display the form in Form view. Use the record selectors to move to record 5. Because this record is marked as Discontinued, the code in the Form_Current event will change the form header section's color to red (see Figure 25.1). The records that are not discontinued will appear with the background image specified in the `Else` clause.
7. Close the Products form and save all the changes when prompted.

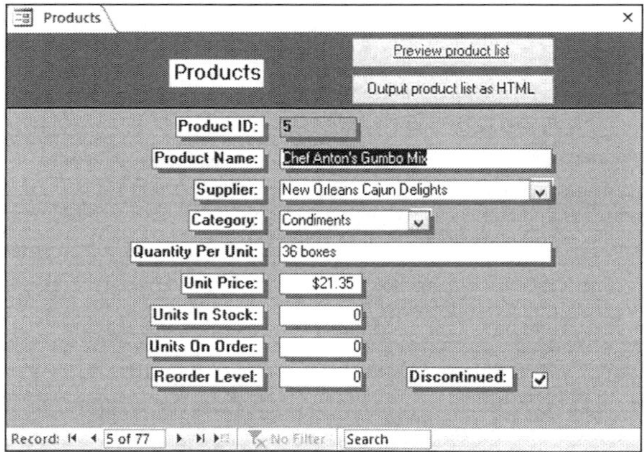

FIGURE 25.1. The Products form displays a red header background when a product is marked as Discontinued.

BeforeInsert

The BeforeInsert event occurs when the first character is typed in a new record but before the new record is created. Use the BeforeInsert event to verify that the data is valid or to display information about data being added. This event is quite useful for placing default values in the fields at runtime. The BeforeInsert event can be canceled if the data being added does not meet specific criteria.

The event procedure in Hands-On 25.2 demonstrates how to enter a default value in the Country field when a user begins to enter data in the form.

Hands-On 25.2. Writing the Form_BeforeInsert Event Procedure

For this hands-on exercise, we will create a new form based on the Customers table.

1. Highlight the **Customers** table in the left pane of the Access window. Choose **Create | Form Wizard.**
2. Select the following fields: **CustomerID**, **CompanyName**, **Address**, **City**, **Region**, **PostalCode**, and **Country**. Step through the Form Wizard screens, pressing the **Next** button until you get to the screen where you are asked for the form's title. Type **New Customers** for the form's title, select the **Modify the form's design** option button, and click **Finish.**
   Access opens the New Customers form in Design view.
3. In the property sheet, select **Form** from the drop-down box, and click the **Data** tab. Set the Data Entry property to **Yes.**
4. In the form's property sheet, click the **Event** tab. Click next to the **Before Insert** event property and choose **[Event Procedure]** from the drop-down box. Click the **Build** button (...).
   Access opens the Visual Basic Editor window and writes the stub of the Form_BeforeInsert event procedure.
5. Complete the code of the **Form_BeforeInsert** event procedure as shown here:

```
Private Sub Form_BeforeInsert(Cancel As Integer)
  Me.Country = "USA"
End Sub
```

6. To test this event procedure, activate the **New Customers** form in Form view.
7. Type **JANIT** in the CustomerID field. Notice that as soon as you start filling in the form's text boxes, the text "USA" appears in the Country field.
8. Click the **Esc** key twice to undo the changes to the form.
9. Close the New Customers form and save all the changes when prompted.

AfterInsert

The AfterInsert event occurs when a new record has been inserted. Use this event to requery the recordset when a new record is added or to display other information. The event procedure in Hands-On 25.3 retrieves the total number of records in the Customers table after a new record has been inserted.

Hands-On 25.3. Writing the Form_AfterInsert Event Procedure

This hands-on exercise uses the New Customers form created in Hands-On 25.2.

1. In the Visual Basic Editor's Project Explorer window, double-click **Form_New Customers**.
2. In the Code window, you will see the Form_BeforeInsert event procedure prepared in Hands-On 25.2. Below this procedure code, enter the **Form_AfterInsert** event procedure as shown here:

```
Private Sub Form_AfterInsert()
  Dim db As DAO.Database
  Dim rst As DAO.Recordset

  Set db = CurrentDb()
  Set rst = db.OpenRecordset("Customers")

  MsgBox "Total Number of Records: " & _
   rst.RecordCount & "."

  rst.Close
  Set rst = Nothing
  Set db = Nothing
End Sub
```

3. To test this event procedure, open the **New Customers** form in Form view. Type **TRYIT** in the Customer ID text box and **Test Events** in the Company Name text box. Now use the record selector to move to the next record. Access executes the code in the Form_AfterInsert event procedure and displays the total number of records.
4. Close the New Customers form and save all the changes if prompted.

BeforeUpdate

The BeforeUpdate event occurs after a record has been edited but before it is written to the table. This event is triggered by moving to another record or attempting to save the current record. The BeforeUpdate event takes place after the BeforeInsert event. Use this event to validate the entire record and display a message to confirm the change. The BeforeUpdate event can be canceled if the record cannot be accepted. The event procedure in Hands-On 25.4 will supply the value for the CustomerID field before the newly entered record is saved.

Hands-On 25.4. Writing the Form_BeforeUpdate Event Procedure

This hands-on exercise uses the New Customers form created in Hands-On 25.2.

1. In the Visual Basic Editor's Project Explorer window, double-click **Form_New Customers**.
2. In the Code window, other event procedures prepared in Hands-On 25.2 and 25.3 will be listed. Enter the following **Form_BeforeUpdate** event procedure below the code of the last procedure:

```
Private Sub Form_BeforeUpdate(Cancel As Integer)
  If Not IsNull(Me.CompanyName) Then
    Me.CustomerID = Left(CompanyName, 3) & _
     Right(CompanyName, 2)
    MsgBox "You just added Customer ID: " & _
     Me.CustomerID
  Else
    MsgBox "Please enter Company Name.", _
     vbOKOnly, "Missing Data"
    Me.CompanyName.SetFocus
    Cancel = True
  End If
End Sub
```

3. To test this event procedure, open the **New Customers** form in Form view. Type **Event Enterprises** in the Company Name box. Click the record selector to move to the next record. The BeforeUpdate event procedure code will run at this point and you will see a message box with the custom-generated Customer ID. Click **OK** to the message. Another message will appear with the number of total records. This message box is generated by the AfterInsert event procedure that was prepared in Hands-On 25.3. Click **OK** to this message.
4. Close the New Customers form and save changes to the form if prompted.

AfterUpdate

The AfterUpdate event occurs after the record changes have been saved in the database. It is also invoked when a control loses focus and after the data in the control has changed. Use the AfterUpdate event to update data in other controls on the form or to move the focus to a different record or control. The event procedure in Hands-On 25.5 creates an audit trail for all newly added records, as illustrated in Figure 25.2.

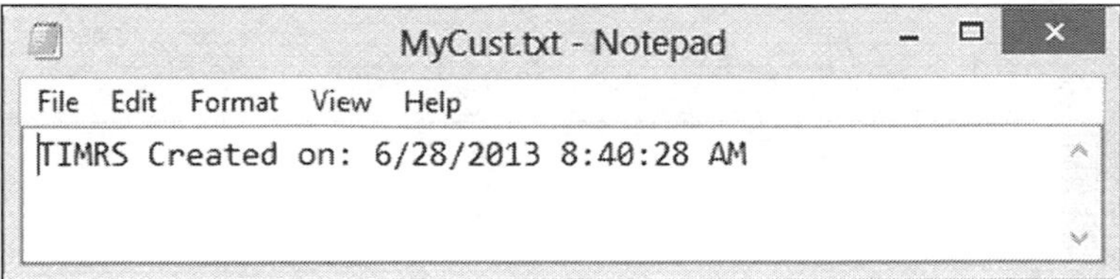

FIGURE 25.2. The Form_AfterUpdate event procedure is used here to store information about newly added records in a text file.

Hands-On 25.5. Writing the Form_AfterUpdate Event Procedure

This hands-on exercise requires the New Customers form that was created in Hands-On 25.2.

1. In the Visual Basic Editor window, choose **Tools | References**. Locate and select **Microsoft Scripting Runtime** in the Available References list and click **OK**.
2. In the Project Explorer window, double-click **Form_New Customers**.
3. Other procedures that were prepared in Hands-On 25.2, 25.3, and 25.4 will be listed in the Code window. Enter the following **Form_AfterUpdate** event procedure below the code of the last procedure:

```
Private Sub Form_AfterUpdate()
  Dim fso As FileSystemObject
  Dim objFile As Object
  Dim strFileName As String

  On Error Resume Next
  strFileName = "C:\Access2013_ByExample\MyCust.txt"
  Set fso = New FileSystemObject
  Set objFile = fso.GetFile(strFileName)

  If Err.Number = 0 Then
    ' open text file
    Set objFile = fso.OpenTextFile(strFileName, 8)
  Else
    ' create a text file
    Set objFile = fso.CreateTextFile(strFileName)
  End If

  objFile.writeLine UCase(Me.CustomerID) & _
   " Created on: " & Date & " " & Time
  objFile.Close
```

```
    Set fso = Nothing
    MsgBox "This record was logged in: " & strFileName
  End Sub
```

This event procedure first checks whether the specified text file exists on your computer. If the file is found, then the `Err.Number` statement returns zero. At this point you want to open the file. The "8" represents the open mode for appending. Use "2" if you want to replace the contents of a file with the new data.

4. To test the event procedure, open the **New Customers** form in Form view. Type **Time Organizers** in the Company Name box. Click the record selector to move to the next record. The BeforeUpdate event procedure code you prepared in Hands-On 25.4 will run at this point and you should see a message box that displays the custom-generated Customer ID. Click **OK** to the message. The next message box notifies you about the location of the audit trail (the result of the AfterUpdate event procedure prepared in this exercise). Click **OK** to the message. Another message will appear with the number of total records. This message box is generated by the AfterInsert event procedure that was prepared in Hands-On 25.3. Click **OK** to this message.

   As you enter more customer records using the New Customers form, events are executed in the following order:

   BeforeInsert (Hands-On 25.2)
   BeforeUpdate (Hands-On 25.4)
   AfterUpdate (Hands-On 25.5)
   AfterInsert (Hands-On 25.3)

5. Close the New Customers form and save changes to the form if prompted.

Dirty

The Dirty event occurs when the contents of a form or the text portion of a combo box changes. This event will be triggered by an attempt to enter a character directly in the form's text box or combo box. Use this event to determine if the record can be changed. The event procedure in Hands-On 25.6 disallows changes to form data when the CategoryID is less than or equal to 4.

Hands-On 25.6. Writing the Form_Dirty Event Procedure

1. Highlight the **Categories** table in the left pane of the Access window. Choose **Create | Form Wizard**.
2. Add all the fields as listed in the Categories table. Step through the Form Wizard screens, clicking the **Next** button until you get to the screen where you are

asked for the form's title. Type **Product Categories** for the form's title, select the **Modify the form's design** option button, and click **Finish**.

Access opens the Product Categories form in Design view.

3. In the property sheet, select **Form** from the drop-down box, and click the **Event** tab. Click next to the **On Dirty** event property and choose **[Event Procedure]** from the drop-down box. Click the **Build** button (...).

Access opens the Visual Basic Editor window and writes the stub of the Form_Dirty event procedure.

4. Complete the code of the **Form_Dirty** event procedure as shown here:

```
Private Sub Form_Dirty(Cancel As Integer)
  If CategoryID <= 4 Then
    MsgBox "You cannot make changes in this record."
    Cancel = True
  End If
End Sub
```

5. To test this event procedure, open the **Product Categories** form in Form view. Try to make any changes to the original records. You will not be able to make changes to the data if the product's CategoryID is less than or equal to 4.
6. Close the Product Categories form and save changes to the form when prompted.

OnUndo

The OnUndo event occurs when the user undoes a change to a combo box control, form, or text box control. By setting the `Cancel` argument to `True`, you can cancel the undo operation and leave the control or form in its edited state. The Undo event for forms is triggered when the user clicks the Undo button, presses the Esc key, or calls the `Undo` method.

Delete

The Delete event occurs when you select one or more records for deletion and before the records are actually removed from the table. Use this event to place restrictions on the data that can be deleted. When deleting multiple records, the Delete event occurs for each record. This enables you to confirm or cancel each deletion in your event procedure code. You can cancel the deletion in the Delete or BeforeDelConfirm events by setting the `Cancel` argument to `True`.

The event procedure in Hands-On 25.7 demonstrates how to disallow deletion of records when CategoryID is less than or equal to 8 and ask the user to confirm the deletion for other records.

Hands-On 25.7. Writing the Form_Delete Event Procedure

This hands-on exercise uses the Product Categories form created in Hands-On 25.6.

1. In the Visual Basic Editor's Project Explorer window, double-click **Form_Product** Categories, which was created in Hands-On 25.6.
2. In the Code window, you will see the Form_Dirty event procedure that was prepared in Hands-On 25.6. Below this procedure code, enter the **Form_Delete** event procedure as shown here:

```
Private Sub Form_Delete(Cancel As Integer)
  If CategoryID <= 8 Then
    MsgBox "You can't delete the original categories."
    Cancel = True
  Else
    If MsgBox("Do you really want to delete " & _
     "this record?", vbOKCancel, _
     "Delete Verification") = vbCancel Then
      Cancel = True
    End If
  End If
End Sub
```

3. To test this event procedure, open the **Product Categories** form in Form view. Click on the record selector to the left of the first record and press the **Delete** key. At this point Access will execute the code of the Form_Delete event procedure. You should see the message that you cannot delete original product categories.
4. Click the **New** button on the Ribbon to add a new record to the form. Enter a new category named **Organic Food** and save the record. Now press the **Delete** button on the Ribbon to delete this record. If there is no code in the Form_BeforeDelConfirm event procedure (see Hands-On 25.8), you will be prompted twice to confirm the deletion. Go ahead with the deletion by clicking **OK** to the first message and **Yes** to the second.
5. Close the Product Categories form and save changes to the form when prompted.

BeforeDelConfirm

The BeforeDelConfirm event occurs after the Delete event but before the Delete Confirm message box is displayed. If you don't write your own BeforeDelConfirm event, Access will display a standard delete confirmation message as described in Hands-On 25.7. You can use this event to write a custom deletion confirmation message. The event procedure in Hands-On 25.8 demonstrates how to suppress the default message.

Hands-On 25.8. Writing the Form_BeforeDelConfirm Event Procedure

This hands-on exercise uses the Product Categories form created in Hands-On 25.6.

1. In the Visual Basic Editor's Project Explorer window, double-click **Form_ Product Categories**, which was created in Hands-On 25.6 and modified in Hands-On 25.7.
2. In the Code window, two event procedures are shown that were prepared in Hands-On 25.6 and 25.7. Enter the following **Form_BeforeDelConfirm** event procedure below the code of the last procedure:

```
Private Sub Form_BeforeDelConfirm(Cancel _
 As Integer, Response As Integer)
  Response = acDataErrContinue
End Sub
```

   In this procedure code, the statement `Response = acDataErrContinue` will suppress the default message box that Microsoft Access normally displays when you attempt to delete a record.
3. To test this event procedure, open the **Product Categories** form in Form view. Click the **New** button on the Ribbon to add a new record, save it, and then delete it. The Form_Delete event procedure prepared in Hands-On 25.7 will be executed at this point, and you will see a dialog with your custom prompt to confirm the deletion. Click **Yes**. Notice that Access does not display its default message asking you to confirm the deletion of the specified number of records.
4. Close the Product Categories form and save changes to the form when prompted.

Instead of writing your custom confirmation message in the Form_ Delete event procedure, you can place it in the Form_BeforeDelConfirm event procedure as shown here:

```
Private Sub Form_BeforeDelConfirm(Cancel As Integer, _
  Response As Integer)
  ' remove the default Access message box
  ' that prompts to confirm deletion
  Response = acDataErrContinue
  If MsgBox("Do you really want to delete this record?", _
   vbOKCancel) = vbCancel Then
    Cancel = True
  End If
End Sub
```

AfterDelConfirm

The AfterDelConfirm event occurs after the record is actually deleted or after deletion is canceled in the BeforeDelConfirm event procedure. Use the AfterDelConfirm event to move to another record or to display a message indicating whether the deletion was successful. The `Status` argument allows you to check whether deletion progressed normally or was canceled by the user or Visual Basic. The following constants can be used for the `Status` argument in the AfterDelConfirm event procedure: `acDelete` (6), `acDeleteCancel` (1), `acDeleteOK` (0), or `acDeleteUserCancel` (2).

The event procedure in Hands-On 25.9 displays a message when a record is successfully deleted.

Hands-On 25.9. Writing the Form_AfterDelConfirm Event Procedure

This hands-on exercise uses the Product Categories form created in Hands-On 25.6.

1. In the Visual Basic Editor's Project Explorer window, double-click **Form_Product Categories.**
2. The Code window appears with several event procedures that were prepared in previous hands-on exercises. Enter the following **Form_AfterDelConfirm** event procedure below the code of the last procedure:

```
Private Sub Form_AfterDelConfirm(Status As Integer)
  MsgBox "The selected record was deleted."
  Debug.Print "Status = " & Status
End Sub
```

3. To test this event procedure, open the **Product Categories** form in Form view. Add a new record, save it, and delete it. Access will execute the code in the Form_Delete event procedure (Hands-On 25.7) that displays a message box asking you whether you want to delete the record. Click **Yes**. Access will then check the code in Form_BeforeDelConfirm (Hands-On 25.8). The statement `Response = acDataErrContinue` will cause Access to suppress its default Delete Confirm message box and you will not be prompted again to reconfirm the deletion. Finally, Form_AfterDelConfirm will run and you will see a message about the successful deletion.
4. Close the Product Categories form and save changes to the form when prompted.

FOCUS EVENTS

Focus events occur when a form becomes active or inactive and when a form or form control loses or gains the focus.

Activate

The Activate event occurs whenever the form gains the focus and becomes the active window. This situation occurs when the form is first opened and when the user activates the form again by clicking on the form or one of its controls. Use this event to display or hide supporting forms.

The event procedure in Hands-On 25.10 will hide the tab labeled Personal Information when the Employees form is displayed. Notice that the tabs are numbered beginning with 0, hence the second tab in the tab control placed on the form has an index value of 1.

Hands-On 25.10. Writing the Form_Activate Event Procedure

1. In the Visual Basic Editor's Project Explorer window, double-click **Form_Employees**.
2. The Code window contains a number of event procedures and functions already written for this form. Enter the following **Form_Activate** event procedure below the code of the last procedure:

```
Private Sub Form_Activate()
  Me.TabCtl0.Pages(1).Visible = False
End Sub
```

3. To test this event procedure, open the **Employees** form in Form view. Notice that only the tab labeled Company Info is shown.
4. Close the Employees form and save changes to the form when prompted.

Deactivate

The Deactivate event occurs when the user switches to another form or closes the form. Use this event to display or hide supporting forms. The event procedure in Hands-On 25.11 will display a message when the focus moves to a different form.

Hands-On 25.11. Writing the Form_Deactivate Event Procedure

1. In the Visual Basic Editor's Project Explorer window, double-click **Form_Employees**.

2. The Code window contains a number of event procedures and functions already written for this form. Enter the following **Form_Deactivate** event procedure below the code of the last procedure:

```
Private Sub Form_Deactivate()
  MsgBox "You are leaving the " & Me.Name & " form."
  If Me.Dirty Then
    DoCmd.Save acForm, Me.Name
    MsgBox "Your changes have been saved."
  End If
End Sub
```

3. To test this event procedure, open the **Products** form in Form view. Next, activate the **Employees** form in Form view and change the phone extension in the first employee record. Now go back to the Products form. You should get two messages as programmed in the Form_Deactivate event procedure. Click **OK** to each message.
4. Close the Employees and the Products forms.

GotFocus

The GotFocus event happens when a form receives the focus, provided that there are no visible or enabled controls on the form. The GotFocus event is frequently used for controls placed on the form and rarely used for the form itself.

LostFocus

The LostFocus event happens when a form loses focus, provided there are no visible or enabled controls on the form. This event is frequently used for controls placed on the form and rarely used for the form itself.

MOUSE EVENTS

Mouse events occur when you move a mouse or click any of the available mouse buttons.

Click

The Click event occurs when you click a mouse button on a blank area of a form, a form's record selector, or a control placed on the form.

The event procedure in Hands-On 25.12 will cause a text box control to move one inch to the right when you click the record selector.

Hands-On 25.12. Writing the Form_Click Event Procedure

1. Create a new form with two text boxes. Position both text boxes starting at 1 inch on the horizontal ruler. Save the form as **Mouse Test**.
2. In the form's property sheet, make sure **Form** is selected and click the **Event** tab. Click next to the **On Click** event property and choose [**Event Procedure**] from the drop-down box. Click the **Build** button (…).

   Access opens the Visual Basic Editor window and writes the stub of the Form_Click event procedure.
3. Complete the code of the **Form_Click** event procedure as shown here:

```
Private Sub Form_Click()
  MsgBox "Form Click Event Occurred."
  Me.Text0.Left = Text0.Left + 1440
End Sub
```

   The first text box control placed on the form is automatically named Text0. The Left property is used to specify an object's location on a form or report. This procedure moves a text box control one inch to the right. Screen measurements are expressed in units called twips, and there are 1440 twips per inch. Thus, to calculate the new position of the text box, you must add 1440 to the current position.
4. To test this event procedure, open the **Mouse Test** form in Form view. Click on the record selector (a bar to the left of a record). This will cause the Form_Click event procedure code to execute and you will see a message box. After clicking **OK** in response to the message, the first text box control will move one inch to the right as illustrated in Figure 25.3.
5. Close the Mouse Test form and save changes to the form when prompted.

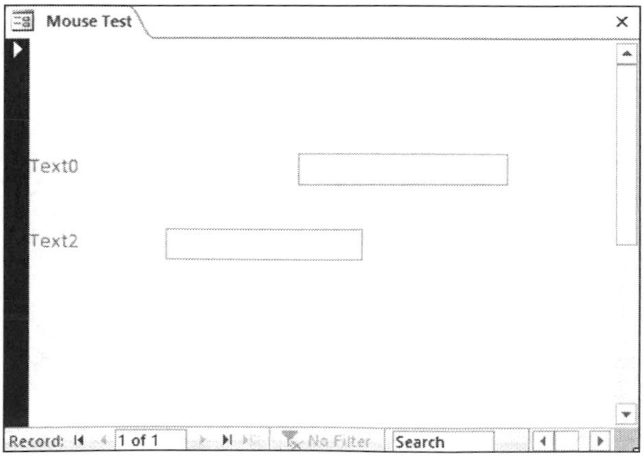

FIGURE 25.3. The Form_Click event procedure has moved the first text box to the right.

DblClick

The DblClick event occurs when you double-click on a blank area of the form, the form's record selector, or a control placed on the form.

MouseDown

The MouseDown event occurs when you click and hold on a blank area of the form, the form's record selector, or a control placed on the form. This event occurs before the Click event. The MouseDown event has four arguments:

- **Button**—Identifies the state of the mouse buttons. Use `acLeftButton` to check for the left mouse button, `acRightButton` to check for the right mouse button, and `acMiddleButton` to check for the middle mouse button.
- **Shift**—Specifies the state of the Shift, Ctrl, and Alt keys when the button specified by the `Button` argument was pressed or released. Use `acShiftMask` (`1`) to test for the Shift key, `acCtrlMask` (`2`) to test for the Ctrl key, and `acAltMask` (`4`) to test for the Alt key. You can test for any combination of buttons. For example, to specify that Ctrl and Alt were pressed, use the value of `6` (2+4) as the Shift argument.
- **X**—Specifies the horizontal (x) position from the left edge of the form or control.
- **Y**—Specifies the vertical (y) position from the top edge of the form or control.

The event procedure in Hands-On 25.13 displays two messages when the form's MouseDown event is fired. The first message tells whether you pressed the Alt, Ctrl, or Shift key, and the second one announces which mouse button was used.

Hands-On 25.13. Writing the Form_MouseDown Event Procedure

1. Create a new form based on the Products table adding all the available fields to the form. Save this form as **Products Test**.
2. In the form's property sheet, make sure **Form** is selected and click the **Event** tab. Click next to the **On Mouse Down** event property and choose **[Event Procedure]** from the drop-down box. Click the **Build** button (...).
3. Enter the following code in the **Form_MouseDown** event:

```
Private Sub Form_MouseDown(Button As Integer, _
  Shift As Integer, _
```

```
     X As Single, _
     Y As Single)
    Debug.Print "Mouse Down"

    Select Case Shift
      Case 0
        MsgBox "You did not press a key."
      Case 1 ' or acShiftMask
        MsgBox "You pressed SHIFT."
      Case 2 ' or acCtrlMask
        MsgBox "You pressed CTRL."
      Case 3
        MsgBox "You pressed CTRL and SHIFT."
      Case 4 ' or acAltMask
        MsgBox "You pressed ALT."
      Case 5
        MsgBox "You pressed ALT and SHIFT."
      Case 6
        MsgBox "You pressed CTRL and ALT."
      Case 7
        MsgBox "You pressed CTRL, ALT, and SHIFT."
    End Select

    If Button = 1 Then ' acLeftButton
      MsgBox "You pressed the left button."
    ElseIf Button = 2 Then ' acRightButton
      MsgBox "You pressed the right button."
    ElseIf Button = 4 Then ' acMiddleButton
      MsgBox "You pressed the middle button."
    End If
  End Sub
```

4. To test this event procedure, switch to the **Products Test** form and open it in Form view. Click on the record selector while holding down any mouse button and pressing the Shift, Ctrl, or Alt keys or combinations of these keys.
5. Close the Products Test form and save changes to the form when prompted.

MouseMove

The MouseMove event occurs when you move the mouse over a blank area of the form, the form's record selector, or a control placed on the form. The MouseMove event occurs before the Click event and has the same arguments as the MouseDown event.

MouseUp

The MouseUp event occurs when you release the mouse button. It occurs before the Click event and uses the same arguments as the MouseDown and MouseMove events.

MouseWheel

The MouseWheel event occurs in Form view or Datasheet view when the user rotates the mouse wheel on a mouse device that has a wheel. This event takes the following two arguments:

- **Page**—Returns True if the page was changed.
- **Count**—Specifies the number of lines that were scrolled with the mouse wheel.

Because there is no Cancel argument, you cannot use the MouseWheel event to prevent users from using the mouse wheel to scroll through records on a form.

KEYBOARD EVENTS

Keyboard events occur when you hold down, press, or release a key on the keyboard or send a keystroke by using the `SendKeys` statement in Visual Basic or the SendKeys action in a macro.

The keyboard events occur in the following sequence:

KeyDown → KeyPress → KeyUp

If the form's KeyPreview property is set to Yes, all keyboard events occur first for the form, and then for the control that has the focus. When you press and hold down the key, the KeyDown and KeyPress events occur repeatedly. When you release the key, the KeyUp event occurs.

KeyDown

The KeyDown event occurs when you press a key while a form or control has the focus. This event is also triggered by using the `SendKeys` statement in Visual Basic or the SendKeys action in a macro. If the form's KeyPreview property is set to `Yes`, all keyboard events occur first for the form, and then for the control that has the focus.

The KeyDown event takes the following two arguments:

- **KeyCode**—Determines which key was pressed. To specify keycodes, use members of the KeyCodeConstants class in the VBA Object Library in the Object Browser. To prevent an object from receiving the keystroke, set KeyCode to zero (0).
- **Shift**—Determines if the Shift, Ctrl, or Alt key was pressed. Use acShift-Mask(1) to test for the Shift key, acCtrlMask(2) to test for the Ctrl key, and acAltMask(4) to test for the Alt key. You can test for any combination of buttons. For example, to specify that Ctrl and Alt were pressed, use the value of 6 (2+4) as the `Shift` argument.

The event procedure in Hands-On 25.14 displays a message when you press one of the following keys: F1, Home, Tab, Shift, Ctrl, Alt, or Delete.

Hands-On 25.14. Writing the Form_KeyDown Event Procedure

1. Open the **Products** form in Design view. In the form's property sheet, make sure **Form** is selected and click the **Event** tab. Set the **Key Preview** property to **Yes**.
2. Save the **Products** form.
3. Click next to the **On Key Down** event property and choose **[Event Procedure]** from the drop-down box. Click the **Build** button (…).
   In the Code window there are a couple of event procedures already written for this form.
4. Enter the following **Form_KeyDown** event procedure below the last procedure code.

```
Private Sub Form_KeyDown(KeyCode As Integer, Shift As Integer)
  Select Case KeyCode
    Case vbKeyF1
      MsgBox "You pressed the F1 key."
    Case vbKeyHome
      MsgBox "You pressed the Home key."
    Case vbKeyTab
      MsgBox "You pressed the Tab key."
  End Select

  Select Case Shift
    Case acShiftMask
      MsgBox "You pressed the SHIFT key."
    Case acCtrlMask
```

```
        MsgBox "You pressed the CTRL key."
      Case acAltMask
        MsgBox "You pressed the ALT key."
    End Select
    If KeyCode = vbKeyDelete Then
      MsgBox "Delete Key is not allowed."
      KeyCode = 0
    End If
  End Sub
```

5. To test this event procedure, open the **Products** form in Form view. Press one of the following keys: **F1**, **Home**, **Tab**, **Shift**, **Ctrl**, **Alt**, or **Delete**. Click **OK** to the message.
6. Close the Products form and save changes to the form when prompted.

KeyPress

The KeyPress event occurs when you press and release a key or a key combination. This event is also triggered by using the `SendKeys` statement in Visual Basic or the SendKeys action in a macro. If the form's KeyPreview property is set to `Yes`, all keyboard events occur first for the form, and then for the control that has the focus.

The KeyPress event responds only to the ANSI characters generated by the keyboard, the Ctrl key combined with a character from the standard alphabet or a special character, and the Enter or Backspace key. Other keystrokes are handled by the KeyDown and KeyUp event procedures. `KeyAscii` is a read/write argument that specifies which ANSI key was pressed. To cancel the keystroke in the KeyPress event, set the `KeyAscii` argument to `0`. The KeyPress event treats uppercase and lowercase letters as different characters.

The event procedure in Hands-On 25.15 prints the ASCII code and the value of the pressed key to the Immediate window. Upon pressing the Escape key (`KeyAscii=27`), the user is prompted to save changes. Clicking Yes to the message will cause the form to be closed. All other keystrokes are ignored.

Hands-On 25.15. Writing the Form_KeyPress Event Procedure

1. Open the **Suppliers** form in Design view. In the property sheet, make sure that **Form** is selected and click the **Event** tab. Set the **Key Preview** property to **Yes**. Click next to the **On keypress** event property and choose **[Event Procedure]** from the drop-down box. Click the **Build** button (...).

   In the Code window there are a couple of event procedures already written for this form.

2. Enter the following **Form_KeyPress** event procedure below the last procedure code.

```
Private Sub Form_KeyPress(KeyAscii As Integer)
  Debug.Print "keypress: KeyAscii = " & KeyAscii & _
    Space(1) & "= " & Chr(KeyAscii)
  If KeyAscii = 27 Then
    If MsgBox("Save changes to this form?", _
     vbYesNo) = vbYes Then
      DoCmd.Close acForm, Me.Name, acSaveYes
    Else
       KeyAscii = 0
    End If
  Else
    KeyAscii = 0
  End If
End Sub
```

The statement `KeyAscii = 0` will disable any input to all the controls on the form. Recall that a form's Key Preview property determines whether form keyboard events are invoked before control keyboard events. To prevent keystrokes from going to the form's controls, the KeyPreview property must be set to Yes.

Note that the KeyPress event is not triggered by the Delete key. You can delete any data on this form as long as there is no custom VBA code written in the KeyDown or KeyUp event procedure that blocks the use of this key.

3. To test this event procedure, open the **Suppliers** form in Form view. Try to edit a field by typing some text. Because the input to all the controls on the form has been disabled by the Form_KeyPress event procedure, you cannot see any input. However, when you switch to the Immediate window, you will see the complete listing of keys that you pressed. Switch back to the Suppliers form and press the **Escape** key. If you agree to save changes to this form, the form will be closed.

KeyUp

The KeyUp event occurs when you release a key while a form or control has the focus. This event is also triggered by using the `SendKeys` statement in Visual Basic or the SendKeys action in a macro. If the form's KeyPreview property is set to Yes, all keyboard events occur first for the form, and then for the control that has the focus.

The KeyUp event takes the following two arguments:

- **KeyCode**—Determines which key was pressed. To specify keycodes, use members of the KeyCodeConstants class in the VBA Object Library in the Object Browser. To prevent an object from receiving the keystroke, set `KeyCode` to zero (`0`).
- **Shift**—Determines if the Shift, Ctrl, or Alt key was pressed. Use `acShiftMask` (1) to test for the Shift key, `acCtrlMask` (`2`) to test for the Ctrl key, and `acAltMask` (`4`) to test for the Alt key. You can test for any combination of buttons. For example, to specify that Ctrl and Alt were pressed, use the value of `6` (2+4) as the `Shift` argument.

The event procedure in Hands-On 25.16 will print to the Immediate window the keycode and the value of the key that was released. Also, the information about `KeyCode` and the state of the Shift key will be shown in the form's caption.

Hands-On 25.16. Writing the Form_KeyUp Event Procedure

1. Open the **Suppliers** form in Design view. In the form's property sheet, make sure **Form** is selected and click the **Event** tab. Make sure the **Key Preview** property is set to **Yes**.
2. Click next to the **On Key Up** event property and choose **[Event Procedure]** from the drop-down box. Click the **Build** button (…).

   The Code window contains a couple of event procedures already written for this form.
3. Enter the following **Form_KeyUp** event procedure below the code of the last procedure:

```
Private Sub Form_KeyUp(KeyCode As Integer, _
  Shift As Integer)
  Debug.Print "Key up : " & Chr(KeyCode) & _
   "(" & KeyCode & ") " & _
   Shift
  Me.Caption = Me.Name
  Me.Caption = Me.Caption & ": KeyCode=" & _
   KeyCode & " " & "Shift=" & Shift
End Sub
```

4. To test this event procedure, open the **Suppliers** form in Form view. Press various keys on the keyboard and notice the key information in the form's caption.
5. Switch to the Visual Basic Editor window and activate the Immediate window.

You should see a listing of the keys that were pressed and released while performing Step 4.

6. Close the Suppliers form and save changes to the form when prompted.

ERROR EVENTS

The Error event is triggered by runtime errors generated either in the Microsoft Access interface or by the Microsoft Jet/ACE database engine. The Error event does not trap VBA errors.

Error

The Error event occurs when there is a problem accessing data for the form. Use this event to suppress the standard error messages and display a custom error message instead.

The Error event takes the following two arguments:

- **DataErr**—Contains the number of the Microsoft Access error that occurred.
- **Response**—Determines whether or not error messages should be displayed. It may be one of the following constants:
  - **acDataErrContinue**—Ignore the error and continue without displaying the default Microsoft Access error message.
  - **acDataErrDisplay**—Display the default Microsoft Access error message. This is the default.

The event procedure in Hands-On 25.17 displays a custom message when an attempt is made to add a new record with a customer ID that already exists. The standard Microsoft Access error message is not displayed.

Hands-On 25.17. Writing the Form_Error Event Procedure

1. Create a new form based on the Customers table. Add all the fields from the Customers table and save the new form as **Customers Data Entry**.
2. Activate the **Customers Data Entry** form in Design view. In the property sheet, make sure **Form** is selected and click the **Data** tab. Set the form's **DataEntry** property to **Yes**.
3. Click the **Event** tab, set the **On Error** property to **[Event Procedure]**, and press the **Build** button (…).

Access will create the event procedure stub.

4. Enter the following **Form_Error** event procedure:

```
Private Sub Form_Error(DataErr As Integer, _
  Response As Integer)

  Dim strMsg As String
  Dim custId As String

  Const conDuplicateKey = 3022
  custId = Me.CustomerID

  If DataErr = conDuplicateKey Then
    ' Don't show built-in error messages
    Response = acDataErrContinue
    strMsg = "Customer " & custId & " already exists."
    ' Show a custom error message
    MsgBox strMsg, vbCritical, "Duplicate Value"
  End If
End Sub
```

5. Open the **Customers Data Entry** form in Form view. Enter **ALFKI** in the CustomerID field and **Alfred Fiki** in the Company Name field. Click the **Save** button. When you try to save this record, the Form_Error event procedure code will cause a message box to appear, saying that the customer already exists. Click **OK** to the message. Press **Esc** to cancel the changes to this record.
6. Close the Customers Data Entry form and save the changes to the form.

FILTER EVENTS

Filter events are triggered by opening or closing a filter window or when you are applying or removing a filter.

Filter

The Filter event occurs when you design a filter to limit the form's records to those matching specified criteria. This event takes place when you select the Filter by Form or Advanced Filter/Sort options. Use this event to remove the filter that was previously set, to enter initial settings for the filter, or to call your own custom filter dialog box. To cancel the filtering command, set the `Cancel` argument for the event procedure to `True`.

The event procedure in Hands-On 25.18 allows the use of the Filter by Form option but disallows the use of the Advanced Filter/Sort option.

Hands-On 25.18. Writing the Form_Filter Event Procedure

This hands-on exercise uses the Product Categories form created in Hands-On 25.6.

1. In the Visual Basic Editor's Project Explorer window, double-click **Form_ Product Categories.**
2. The Code window shows other event procedures already written for this form. Enter the following **Form_Filter** event procedure below the code of the last procedure:

```
Private Sub Form_Filter(Cancel As Integer, _
 FilterType As Integer)
  Select Case FilterType
    Case acFilterByForm
      MsgBox "You selected to filter records " & _
       "by form.", vbOKOnly + vbInformation, _
       "Filter By Form"
      Me.CategoryName.SetFocus
      Me.CategoryID.Enabled = False
    Case acFilterAdvanced
      MsgBox "You are not authorized to use " & _
       " Advanced Filter/Sort.", _
       vbOKOnly + vbInformation, _
       "Advanced Filter By Form"
      Cancel = True
  End Select
End Sub
```

3. To test this event procedure, open the **Product Categories** form in Form view.
4. In the Sort & Filter area of the Ribbon, choose **Advanced | Filter by Form.** The code in the Form_Filter event procedure runs and you will see a message box. Click **OK.** The Filter by Form dialog box appears with the Category ID text box disabled. You can disable certain controls on the form if you don't want the user to filter by them.
5. Filter the form to display only records for **Seafood or Meat/Poultry**. Be sure to click **Toggle Filter** in the Sort & Filter area of the Ribbon after setting up filter criteria.
6. Now, remove the filter by clicking **Toggle Filter** again.

7. Choose **Advanced | Advanced Filter/Sort**. You will not be able to use the advanced filter for this form because the form's Filter event has disabled this action.
8. Close the Product Categories form and save changes to the form when prompted.

ApplyFilter

The ApplyFilter event occurs when you apply the filter to restrict the records. This event takes place when you select the Apply Filter/Sort, Filter by Selection, or Remove Filter/Sort options. Use this event to change the form display before the filter is applied or undo any changes made when the Filter event occurred.

The `ApplyType` argument can be one of the predefined constants shown in Table 25.1.

***TABLE 25.1.* ApplyType argument constants**

Constant Name	Constant Value
`acShowAllRecords`	0
`acApplyFilter`	1
`acCloseFilterWindow`	2
`acApplyServerFilter`	3
`acCloseServerFilterWindow`	4

The event procedure in Hands-On 25.19 displays a different message depending on whether or not the user has made a selection in the Filter by Form dialog box.

Hands-On 25.19. Writing the Form_ApplyFilter Event Procedure

This hands-on exercise uses the Product Categories form created in Hands-On 25.6.

1. In the Visual Basic Editor's Project Explorer window, double-click **Form_ Product Categories.**
2. The Code window shows other event procedures already written for this form. Enter the following **Form_ApplyFilter** event procedure below the code of the last procedure:

```
Private Sub Form_ApplyFilter(Cancel As Integer, _
 ApplyType As Integer)
```

```
    Dim Response As Integer

    If ApplyType = acApplyFilter Then
      If Me.Filter = "" Then
        MsgBox "You did not select any criteria.", _
         vbOKOnly + vbCritical, "No Selection"
        GoTo ExitHere
      End If
      Response = MsgBox("The selected criteria " & _
       "is as follows:" & vbCrLf & _
       Me.Filter, vbOKCancel + vbQuestion, _
       "Filter Criteria")
    End If

    If Response = vbCancel Then
      Cancel = True
    End If
    If ApplyType = acShowAllRecords Then
      Me.Filter = ""
      MsgBox "Filter was removed."
    End If
    If ApplyType = acCloseFilterWindow Then
      Response = MsgBox("Are you sure you " & _
       "want to close the Filter window?", vbYesNo)
      If Response = vbNo Then
        Cancel = True
      End If
    End If
  ExitHere:
    With Me.CategoryID
      .Enabled = True
      .SetFocus
    End With
  End Sub
```

3. To test this event procedure, open the **Product Categories** form in Form view.
4. From the Sort & Filter area of the Ribbon, choose **Advanced | Filter by Form**. The Form_Filter event will be triggered (see Hands-On 25.18). Click **OK** to the message box.
5. Select a category from the **Category Name** combo box and click **Toggle Filter** on the Ribbon. This action will trigger the Form_ApplyFilter event procedure. Experiment with the form filter, testing other situations such as click-

ing Toggle Filter when the filtering criteria were not specified or closing the Filter by Form dialog box.

6. Close the Product Categories form and save changes to the form when prompted.

TIMING EVENTS

Timing events occur in response to a specified amount of time passing.

Timer

The Timer event occurs when the form is opened. The duration of this event is determined by the value (milliseconds) entered in the TimerInterval property located on the Event tab of the form's property sheet. Use this event to display a splash screen when the database is opened. The Timer event is helpful in limiting the time the record remains locked in multiuser applications.

The event procedure in Hands-On 25.20 will flash the button's text, "Preview Product List" (or the entire button if you use the commented code instead). For the code to work, you must start the timer by changing the TimerInterval property from 0 (stopped) to the desired interval. A timer interval of 1,000 will invoke a timer event every second. The form's Load event procedure sets the form's TimerInterval property to 250, so the button text (or the entire button) is toggled once every quarter second. You may change the timer interval manually by typing the value next to the form's TimerInterval property in the property sheet or by placing the following statement in the Form_Load event:

```
Me.TimerInterval = 250
```

Hands-On 25.20. Writing the Form_Timer Event Procedure

1. In the Visual Basic Editor's Project Explorer window, double-click the **Products** form.
2. The Code window shows other event procedures already written for this form. Enter the following **Form_Timer** event procedure below the code of the last procedure:

```
Private Sub Form_Timer()
  Static OnOff As Integer

  If OnOff Then
```

```
      Me.PreviewReport.Caption = "Preview Product List"
      ' Me.PreviewReport.Visible = True
    Else
      Me.PreviewReport.Caption = ""
      ' Me.PreviewReport.Visible = False
    End If
    OnOff = Not OnOff
  End Sub
```

3. Activate the **Products** form in Design view. In the property sheet, make sure **Form** is selected and click the **Event** tab. Enter **250** for the TimerInterval property.
4. Switch the form to Form view. Notice the flashing effect of the Preview Product List button's text.
5. Close the Products form and save changes to the form when prompted.

To make the entire button flash, uncomment the commented lines of code and comment the original lines. Next, open the Products form in Form view and notice that the entire button is now flashing.

EVENTS RECOGNIZED BY FORM SECTIONS

In addition to trapping events for the entire form, you can write event procedures for the following form sections: Detail, FormHeader, FormFooter, PageHeader, and PageFooter. Form sections respond to the following events: Click, DblClick, MouseDown, MouseUp, and MouseMove.

DblClick (Form Section Event)

The DblClick event occurs when you double-click inside the form's header or footer section.

The example procedure in Hands-On 25.21 demonstrates how to randomly change the background color for each of the form's sections every time you double-click anywhere within the form's Detail section.

Hands-On 25.21. Writing the Detail_DblClick Event Procedure

1. In the Navigation pane of the Chap25.accdb database, open the **Product Categories** form in Design view. Recall that you created this form in Hands-On 25.6.
2. Increase the size of the header and footer so that they are visible when you run the form.

3. In the property sheet, choose **Detail** from the drop-down box. Click the **Event** tab and select **[Event Procedure]** next to the **DblClick** property name. Click the **Build** button (...).

4. In the Code window, you should have the stub of the Detail_DblClick event procedure already written for you. Complete this procedure as shown here:

```
Private Sub Detail_DblClick(Cancel As Integer)
  With Me
    .Section(acHeader).BackColor = _
     RGB(Rnd * 128, _
     Rnd * 256, _
     Rnd * 255)
    .Section(acDetail).BackColor = _
     RGB(Rnd * 128, _
     Rnd * 256, _
     Rnd * 255)
    .Section(acFooter).BackColor - _
     RGB(Rnd * 128, _
     Rnd * 256, _
     Rnd * 255)
  End With
End Sub
```

5. To test this event procedure, open the **Product Categories** form in Form view. Double-click anywhere in the Detail section of the form and see the colors of the Detail, Header, and Footer sections change.

UNDERSTANDING AND USING THE OPENARGS PROPERTY

It's been over a decade since Microsoft introduced in Access an extremely useful property of the Form and Report objects called `OpenArgs`. Using the `OpenArgs` property you can pass parameters to the form or report when you open it with the `DoCmd` command. The `OpenArgs` property also comes in handy when:

- You want to pass values from one form to another,
- You want to move the focus to a specific record when the form opens,
- You want to automatically populate a control on the form,
- You want to restrict access to certain forms.

Note: *To use the `OpenArgs` property with the Access reports, turn to Chapter 27.*

The `OpenArgs` property is a string expression. It can be used both in macros and in VBA code. Only one `OpenArgs` string can be used in the `OpenForm` or `OpenReport` command, however, by combining values into one string separated by a unique character, and using the `Split` function, you can overcome this limitation. Before we delve into a practical example, let's take a look at the complete syntax of the `OpenForm` method:

```
DoCmd.OpenForm FormName, View, FilterName, WhereCondition, DataMode, WindowMode, OpenArgs
```

The parameter definitions are listed in Table 25.2.

TABLE 25.2. Parameters used with the OpenForm method of the DoCmd object.

Parameter Name	Data Type	Description
`FormName` (This parameter is required.)	`Variant`	A string expression containing the name of a form in the current database.
`View`	`acFormView`	The `acFormView` constant specifies the view in which form should open. The default is `acNormal`.
`FilterName`	`Variant`	A string expression containing the name of a query in the current database.
`WhereCondition`	`Variant`	A string expression containing the `SQL WHERE` clause without the word WHERE.
`DataMode`	`acFormOpenDataMode`	A `acFormOpenDataMode` constant specifies the data entry mode for the form and applies only to forms open in the Form view or Datasheet view. The default is `acFormPropertySettings`.
`WindowMode`	`acWindowMode`	A `acWindowMode` constant specifies the window mode in which the form opens. The default is `acWindowNormal`.
`OpenArgs`	`Variant`	A string expression used to set the form's `OpenArgs` property in a VBA code or in a macro.

The Hands-On 25.22 shows you how to use the `OpenArgs` property to pass values from a custom form (frmOpenArgs) to the Northwind 2007 database built-in form (Employee List).

Hands-On 25.22. Passing values to a form using the OpenArgs property

1. Open the **Northwind 2007_Revised.accdb** database from your Access2013_HandsOn folder. Cancel the login dialog box upon loading of the database.
2. In the Navigation pane on the left, double click the **frmOpenArgs** to open it in Form view (see Figure 25.4).
3. Select the last value from the drop-down box, and click the Execute button. Access displays the Employee List form as shown in Figure 25.5.
4. Switch to the Visual Basic Editor window and analyze the VBA code in the Form_Employee List form class module, Form_frmEmpOpenArgs form class module, and in the OpenArgs_Demo module. Use the debugging techniques that you acquired earlier in this book to step line by line through the code sections.

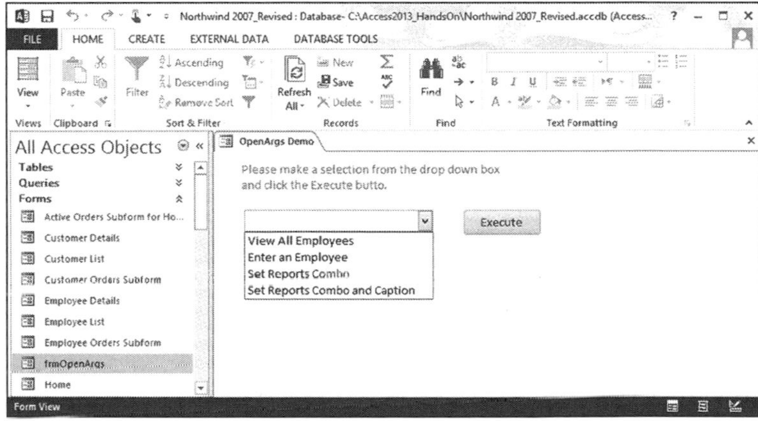

FIGURE 25.4. Working with the OpenArgs Demo (frmOpenArgs form).

Notice that the example form contains a combo box control with four items. Every time you select an item from the combo box and click the Execute button, an Employee List form is loaded with a slightly different effect. The code attached to the click event of the Execute button is shown below:

```
Private Sub cmdOpenEmpList_Click()
On Error GoTo Err_cmdOpenEmpList_Click

Dim strFormToOpen As String
Dim strUserSelection As String
strFormToOpen = "Employee List"

If IsOpenForm(strFormToOpen) Then
```

```
        DoCmd.Close acForm, strFormToOpen
        DoEvents
    End If

    If Not IsNull(cboSelection) Then
       strUserSelection = cboSelection.Value

       Select Case cboSelection
          Case "View All Employees"
            DoCmd.OpenForm FormName:=strFormToOpen, _
                View:=acNormal, WindowMode:=acWindowNormal, _
                OpenArgs:=strUserSelection
          Case "Enter an Employee"
            DoCmd.OpenForm FormName:=strFormToOpen, _
                View:=acNormal, DataMode:=acFormAdd, _
                OpenArgs:=strUserSelection
          Case "Set Reports Combo"
            DoCmd.OpenForm strFormToOpen, acNormal, _
            , , , acWindowNormal, "Customer Address Book"
          Case "Set Reports Combo and Caption"
            DoCmd.OpenForm strFormToOpen, acNormal, _
            , , , acWindowNormal, _
            Me.Name & "|" & "Customer Phone Book"

       End Select
    Else
        MsgBox "Please make a selection from the combo box."
    End If
    Exit_cmdOpenEmpList_Click:
        Exit Sub
    Err_cmdOpenEmpList_Click:
        MsgBox Err.Description
        Resume Exit_cmdOpenEmpList_Click
    End Sub
```

Notice how this event procedure uses the `OpenArgs` property of the form to send different values to the Employee List form. To open a form, we simply use the `OpenForm` method of the `DoCmd` object and pass the name of the form as well as other parameters that define the type of view, data mode, window mode, and the `OpenArgs`. The parameters used with the `DoCmd` object are listed in Table 25.2. These parameters can be passed by name (as shown in the first two `Select Case` statements, or in line (as shown in the last two `Select Case` statements).

The Form_Load event procedure of the Employee List form reads the values placed in the OpenArgs property and makes changes to the specified form controls:

```
Private Sub Form_Load()
   Dim aArgs() As String
   Dim counter As Integer

    If Not IsNull(Me.OpenArgs) Then
        If Me.OpenArgs = "Customer Address Book" Then
            Me.cboReports = Me.OpenArgs
            Me.cboReports.Width = 2800
            Exit Sub
        End If

        If DelimFound(Me.OpenArgs, "|") Then
            MsgBox "Passing multiple values."
            aArgs() = Split(Me.OpenArgs, "|")
            For counter = 0 To UBound(aArgs)
                If aArgs(counter) = "frmOpenArgs" Then
                    Me.Auto_Title0.Caption = _
                        Me.Auto_Title0.Caption & _
                            " called from " & aArgs(counter)
                End If
                If aArgs(counter) = "Customer Phone Book" Then
                    Me.cboReports = aArgs(counter)
                    Me.cboReports.Width = 2800
                End If
                Debug.Print counter & ":" & aArgs(counter)
            Next counter
        Else
            Me.Auto_Title0.Caption = Me.OpenArgs
        End If

    End If
End Sub
```

This procedure begins by checking whether the OpenArgs property contains any values. If the property is not Null, Access will run the remaining code prior to loading the form. Notice that to determine whether the OpenArgs property is passing more than one value, we make a call to the custom DelimFound function (see the second code excerpt below). We pass two values to the DelimFound function. The first value is the contents of the `OpenArgs` property; the second value is the delimiter. In this example, we are using the Pipe character (|) as the delimiter. If the delimiter is found, we need to extract the values from the `OpenArgs` property by using the `Split` function:

```
aArgs() = Split(Me.OpenArgs, "|")
```

The extracted values are stored in the `aArgs` array variable. The For...Next loop is then used to iterate through the array and assign the values to the form controls. In this process we assign corresponding values to the `Auto_Title0` and the `cboReports` controls.

The supplemental function procedures that the `Click` event procedure and the `Load` event procedure call are placed in a standard module called OpenArgs_Demo.

The `IsOpenForm` function returns `true` if the Employee List form is open and `false` if it is closed. If the form is open, the `cmdOpenEmpList_Click` event procedure will close it prior to executing the remaining code.

```
Function IsOpenForm(strFormName As String) _
    As Boolean
    IsOpenForm = Application.CurrentProject. _
        AllForms(strFormName).IsLoaded
End Function
```

The `DelimFound` function checks if the specified delimiter can be found in the string passed in the `OpenArgs` property. This is done by using the built-in `InStr` function.

```
Function DelimFound(strOpenArgs As String, _
    strDelim As String) As Boolean

    If InStr(1, strOpenArgs, strDelim) Then
        DelimFound = True
    Else
        DelimFound = False
    End If

End Function
```

FIGURE 25.5. When you select the last value from the OpenArgs Demo drop-down list (see Figure 25.4) Access displays the Employee List form with changes made to the form caption and the Reports drop down list.

CHAPTER SUMMARY

In this chapter, you learned that numerous events can occur on a Microsoft Access form and that you can react to a specific form event by writing an event procedure. If you don't write your own code to handle a particular form event, Access will use its default handler for the event. You have also learned how to use the Form's OpenArgs property to pass values from one form to another.

After trying out numerous hands-on exercises presented in this chapter, you should have a good understanding of how to write event procedures for an Access form. You should also be able to recognize the importance of form events in an Access application.

For more hands-on experience with event programming, proceed to the next chapter, which discusses the events recognized by controls placed on an Access form.

Chapter 26 Events Recognized by Controls

In addition to the events for forms introduced in Chapter 25, you can control a great many events that occur for labels, text boxes, combo boxes, list boxes, option buttons, checkboxes, and other controls installed by default with an Access application. These events make it possible to manage what happens on a field level.

The best way to learn about form, report, or control events is to develop an application that addresses specific needs. For example, the AssetsDataEntry.mdb database keeps track of computer assets in various companies. We will use this database to further experiment with event programming. The main data entry form is divided into four easy-to-maintain sections as illustrated in Figure 26.1.

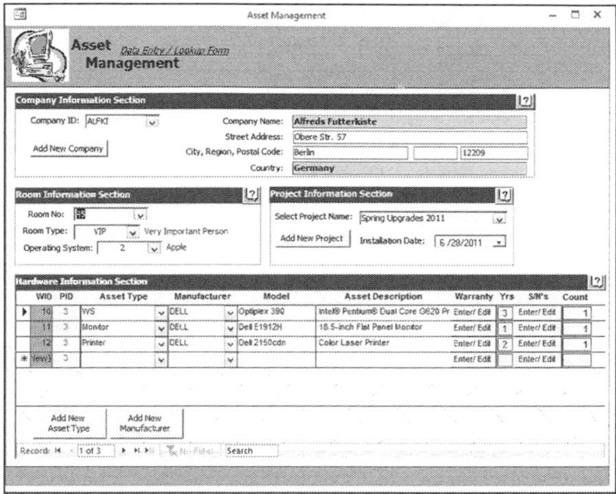

FIGURE 26.1. Custom data entry form.

Please note files for the "Hands-On" project may be found on the companion CD-ROM.

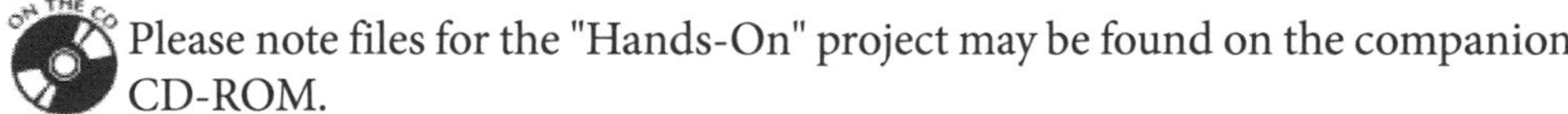

Hands-On 26.1. Launching the Custom Access Application

1. Copy the **AssetsDataEntry.mdb** file from the **C:\Access2013_HandsOn** folder to your **C:\Access2013_ByExample** folder.
2. To access the database source code, double-click the **C:\Access2013_ByExample\AssetsDataEntry.mdb** file while holding down the **Shift** key.
3. When Access loads the database, locate the **frmDataEntryMain** form in the Navigation pane and open it in Design view.

Now that the main data entry form is open, let's proceed to examine the events that this form's controls respond to.

ENTER (CONTROL)

The Enter event occurs before a control actually receives the focus from another control on the same form. The Enter event applies to text boxes, combo boxes, list boxes, option buttons, checkboxes, option groups, command buttons, toggle buttons, bound and unbound object frames, and subform and subreport controls. You can use the Enter event to display a message directing the user to first fill in another control on the form.

For example, when a user attempts to make a selection from the combo box controls located in the Room Information and Project Information sections of the Asset Management form (Figure 26.1) without first specifying the Company ID, the Enter event procedures may be triggered for: cboRooms_Enter, cboRoomType_Enter, cboOS_Enter, and cboProject_Enter.

Hands-On 26.2. Using the Enter Event Procedure for the Combo Box Control

1. Open the **frmDataEntryMain** form in form view.
2. Click inside the combo box control located to the right of **Room No**. This action will fire the following Enter event procedure:

```
Private Sub cboRooms_Enter()
  If Me.cboCompanyID = "" Or IsNull(Me.cboCompanyID) Then
    MsgBox "Please select Company ID.", _
     vbInformation + vbOKOnly, _
     "Missing Company ID"
    Me.cboCompanyID.SetFocus
    Exit Sub
  End If
End Sub
```

3. Click **OK** to the information message generated by the cboRooms_Enter event procedure. Notice that the cursor has been positioned inside the combo box control containing Company IDs. Don't make any selections from the Company ID combo box at this time.
4. Click on the combo box control next to **Room Type**. This action will fire the following Enter event procedure:

```
Private Sub cboRoomType_Enter()
  If Me.cboCompanyID = "" Or IsNull(Me.cboCompanyID) Then
    MsgBox "Please select Company ID.", _
     vbInformation + vbOKOnly, "Missing Company ID"
    Me.cboCompanyID.SetFocus
    Exit Sub
  End If
  If Me.cboRooms = "" Or IsNull(Me.cboRooms) Then
    MsgBox "Please specify or select Room number.", _
     vbInformation + vbOKOnly, "Missing Room Number"
    Me.cboRooms.SetFocus
    Exit Sub
  End If
End Sub
```

When you click the cboRoomType combo box control, the Enter event checks whether the cboCompanyID combo box control or cboRooms combo box control is empty. If no selection has been made in these controls, a message box is displayed and the focus is moved to the appropriate combo box control.

5. Click **OK** to the information message generated by the cboRoomType_Enter event procedure and notice that the cursor has again been positioned inside the Company ID combo box control.

BEFOREUPDATE (CONTROL)

The BeforeUpdate event occurs when you attempt to save the record or leave the control after making changes. This event applies to text boxes, combo boxes, list boxes, option buttons, checkboxes, and bound object frames. Use this event to validate the entry.

For example, the combo box control in the Company Information section of the Asset Management form causes Access to display a custom message if the value of the cboCompanyID combo box control is Null. To cancel the Update event, the `Cancel` argument has been set to `True`.

Hands-On 26.3. Using the BeforeUpdate Event Procedure for the Combo Box Control

1. Press **Alt+F11** to switch to the Visual Basic Editor window.
2. In the Project Explorer window, double-click the **frmDataEntryMain** form.
3. From the Object drop-down box at the top-left side of the Code window, select **cboCompanyID**. In the Procedure drop-down box at the top-right side of the Code window, select **BeforeUpdate**.
4. The Code window should display this event procedure:

```
Private Sub cboCompanyID_BeforeUpdate(Cancel As Integer)
  Dim strMsg As String, strTitle As String
  Dim intStyle As Integer

  If IsNull(Me!cboCompanyID) Or Me!cboCompanyID = "" Then
    strMsg = "You must pick a value from the Company ID list."
    strTitle = "Company ID Required"
    intStyle = vbOKOnly
    MsgBox strMsg, intStyle, strTitle
    Cancel = True
  End If
End Sub
```

5. Position the cursor on the line with the `If` statement, then press **F9** or choose **Debug | Toggle Breakpoint**.
6. Activate the **frmDataEntryMain** form in Form view and make a selection from the Company ID combo box.

   When you make your selection, the BeforeUpdate event procedure is fired and the Code window appears in break mode. Press **F8** to step through the code line by line. Because you have not set up more breakpoints, you cannot see that two other events (cboCompanyID_AfterUpdate and cboRooms_Enter) were triggered when you made a selection from the Company ID combo box.
7. When the procedure finishes executing, activate the **frmDataEntryMain** form. You should see the text boxes filled with a company name and address and the cursor positioned inside the Room No combo box and ready for the next selection or data entry.

AFTERUPDATE (CONTROL)

The AfterUpdate event occurs after the data in the control has been modified. It applies to text boxes, combo boxes, list boxes, option buttons, checkboxes, and bound object frames. Unlike the BeforeUpdate event, the AfterUpdate event cannot be canceled. Use this event to fill in other controls on the form based on the newly entered or selected value.

For example, after updating the cboCompanyID combo box in the Company Information section of the Asset Management form, the following event procedure is executed:

```
Private Sub cboCompanyID_AfterUpdate()
  With Me
    .txtCompanyName = Me.[cboCompanyID].Column(1)
    .txtAddress = Me.cboCompanyID.Column(2)
    .txtCity = Me.cboCompanyID.Column(3)
    .txtRegion = Me.cboCompanyID.Column(4)
    .txtPostalCode = Me.cboCompanyID.Column(5)
    .txtCountry = Me.cboCompanyID.Column(6)
    .cboRooms.Value = vbNullString
    .cboRooms.Requery
    .txtRoomDescription = vbNullString
    .cboRoomType = vbNullString
    .cboOS = vbNullString
    .txtOperatingSystem = vbNullString
```

```
      .cboProject = vbNullString
      .txtPID = vbNullString
    End With
    If Me.cboRooms.ListCount = 0 Then
      'do not display column headings
      Me.cboRooms.ColumnHeads = False
    Else
      Me.cboRooms.ColumnHeads = True
    End If
    Me.cboRooms.SetFocus
  End Sub
```

In the preceding procedure, the company address information is filled in based on the contents of the cboCompanyID columns. For example, to fill in the street address, you can read the value of the Columns() property of the cbo-CompanyID control, even though this column is not visible when you view the combo box:

```
Me.txtAddress = Me.cboCompanyID.Column(2)
```

Note that because the combo box column numbering begins with zero (0), this statement actually reads the contents of the third column. Next, the combo box labeled Room No is requeried and a number of other controls on the form are cleared.

Also, note how the intrinsic constant named `vbNullString` is used here instead of an empty string ("") to clear text boxes or combo boxes on a form. The final procedure code segment contains the `If...Then...Else` statement that sets the ColumnHeads property of the cboRooms control to `False` if there are no rooms associated with the selected Company ID.

The last line of the code

```
Me.cboRooms.SetFocus
```

moves the focus to the combo box control with the room numbers. When this code is executed, the cboRooms_Enter event procedure will be triggered.

Hands-On 26.4. Using the AfterUpdate Event Procedure for the Combo Box Control

1. In the Code window of the **frmDataEntryMain** form, ensure that the Object drop-down box displays **cboCompanyID**.
2. Choose the **AfterUpdate** event from the Procedure drop-down box, then set a breakpoint on the first line of this procedure.

3. Switch to Form view for the frmDataEntryMain form, then make another selection from the **Company ID** combo box. When the Code window appears in break mode, step through the code line by line by pressing **F8**. Notice that the following three event procedures are run:

```
cboCompanyID_BeforeUpdate
cboCompanyID_AfterUpdate
cboRooms_Enter
```

4. Choose **Debug | Clear All Breakpoints** to remove the breakpoint you set in this and the previous hands-on exercise.
5. When the procedure finishes executing, activate the **frmDataEntryMain** form.

NOTINLIST (CONTROL)

The NotInList event is triggered if the user enters a value that is not in the list when the LimitToList property of a combo box control is set to `True`. The NotIn-List event procedure can take the following two arguments:

- **NewData**—A string that Access uses for passing the user-entered text to the event procedure.
- **Response**—An integer specifying what Access should do after the procedure executes. This argument can be set to one of the following constants:
  - `acDataErrAdded`—Set the `Response` argument to `acDataErrAdded` if the event procedure enters a new value in the combo box. This constant tells Access to requery the combo box, adding the new value to the list.
  - `acDataErrDisplay`—Set the `Response` argument to `acDataErrDisplay` if you want Access to display the default error message when a user attempts to add a new value to the combo box. The default Access message requires the user to enter a valid value from the list.
  - `acDataErrContinue`—Set the `Response` argument to `acDataErrContinue` if you display your own message in the event procedure. Access will not display its default error message.

The NotInList event applies only to combo boxes. Use this event to display a custom warning message or to trigger a custom function that allows the user to add a new item to the list. For example, after attempting to enter a nonexistent value in the combo box labeled Room Type in the Room Information section of the Asset Management form, this event procedure is executed:

```
Private Sub cboRoomType_NotInList(NewData As String, _
 Response As Integer)

  MsgBox "Please select a value from the list.", _
   vbInformation + vbOKOnly, "Invalid entry"
  ' Continue without displaying default error message.
  Response = acDataErrContinue
End Sub
```

The cboRoomType_NotInList code displays a custom message if a user attempts to type an invalid entry in the cboRoomType combo box control on the form.

Hands-On 26.5. Using the NotInList Event Procedure for the Combo Box Control

1. In the Code window of the **frmDataEntryMain** form, enter the code of the **cboRoomType_NotInList** procedure as shown in the previous section.
2. Open the **frmDataEntryMain** form in Form vew.
3. Select a company ID from the **Company ID** combo box.
4. Select a room number from the **Room No** combo box, or type a value in this box.
5. Type a new value in the **Room Type** combo box, then click on the **Operating System** combo box. This will trigger the cboRoomType_NotInList event procedure code to run. Your custom error message should appear. Click **OK** to the message box. Notice that Access does not display its own default message because we set the `Response` argument to `acDataErrContinue`.
6. Select a value from the **Room Type** combo box.

CLICK (CONTROL)

The Click event occurs when the user clicks a control with the left mouse button or presses an Enter key when a command button placed on a form has its Default property set to `Yes`. The Click event applies only to forms, form sections, and controls on a form. The Asset Management data entry form contains several command buttons that allow the user to add new values to appropriate combo box selections. For example, when the user clicks the button labeled Add New Company, the following Click event procedure is triggered:

```
Private Sub cmdNewCompany_Click()
  On Error GoTo Err_cmdNewCompany_Click

  Dim stDocName As String
  Dim stLinkCriteria As String

  stDocName = "frmAddCompany"
  DoCmd.OpenForm stDocName, , , stLinkCriteria

Exit_cmdNewCompany_Click:
  Exit Sub
Err_cmdNewCompany_Click:
  MsgBox Err.Description
  Resume Exit_cmdNewCompany_Click
End Sub
```

This event procedure opens a window titled New Company Data Entry Screen (Figure 26.2), where the user can enter new company information.

New Company Data Entry Screen

Enter New Company Information

Company ID:

Company Name:

Street Address:

City: Region: Postal Code:

Country:

Save Cancel

FIGURE 26.2. This data entry form is used for adding new companies to the database.

When the user clicks the Save button on the New Company Data Entry Screen window, the Click event procedure attached to this button ensures that:

- All text boxes have been filled in
- The Company ID does not contain more than five characters
- The Postal Code text box contains a five-digit zip code for the United States
- The Company ID does not already exist in the table

Notice that the New Company Data Entry form is unbound (it isn't connected to a record source such as a table, query, or SQL statement). After successful data validation, the procedure uses the `AddNew` method of the ADO Recordset object to create a new record. This record is added to the tblCompanies table that provides the record source for the Company ID combo box control on the Asset Management data entry form. Next, the cboCompanyID combo box control on the Asset Management form is requeried so that the new Company ID can be accessed from the drop-down list when the user returns to the form.

```
Private Sub cmdSaveCompanyInfo_Click()
  Dim conn As ADODB.Connection
  Dim rst As New ADODB.Recordset
  Dim ctrl As Control
  Dim count As Integer

  On Error GoTo Err_cmdSaveCompanyInfo_Click

  'validate data prior to save

  For Each ctrl In Me.Controls
    If ctrl.ControlType = acTextBox And IsNull(ctrl) _
     Or IsEmpty(ctrl) Then
      count = count + 1
      If count > 0 Then
        MsgBox "All text fields must be filled in.", _
         vbInformation + vbOKOnly, _
         "Missing Data"
        ctrl.SetFocus
        Exit Sub
      End If
    End If
  Next

  If Len(Me.txtAddCompanyID) <> 5 Then
    MsgBox "The Company ID requires 5 characters"
    Me.txtAddCompanyID.SetFocus
    Exit Sub
  End If

  'check the zipcode field
  If Len(Me.txtAddPostalCode) <> 5 And _
   UCase(Me.txtAddCountry) = "USA" Then
    MsgBox "Please enter a five-digit zip code " & _
     "for the United States.", _
```

```
        vbInformation + vbOKOnly, "Invalid Zip Code"
      Me.txtAddPostalCode.SetFocus
      Exit Sub
    End If

    'are any alphabetic characters in zip code?
    If Not IsNumeric(Me.txtAddPostalCode) And _
     UCase(Me.txtAddCountry) = "USA" Then
      MsgBox "You can't have letters in Zip Code.", _
        vbInformation + vbOKOnly, "Invalid Zip Code"
      Me.txtAddPostalCode.SetFocus
      Exit Sub
    End If

    'save the data
    Set conn = CurrentProject.Connection
    With rst
      .Open "SELECT * FROM tblCompanies", _
        conn, adOpenKeyset, adLockOptimistic
      'check if the CompanyID is not a duplicate
      .Find "CompanyID='" & Me.txtAddCompanyID & "'"
      'if Company already exists then get out

      If Not rst.EOF Then
        MsgBox "This Company is already in the list : " _
          & rst("CompanyID"), _
        vbInformation + vbOKOnly, "Duplicate Company ID"
        Me.txtAddCompanyID.SetFocus
        Exit Sub
      End If

      .AddNew
      !CompanyID = Me.txtAddCompanyID
      !CompanyName = Me.txtAddCompanyName
      !Address = Me.txtAddAddress
      !City = Me.txtAddCity
      !Region = Me.txtAddRegion
      !PostalCode = Me.txtAddPostalCode
      !Country = Me.txtAddCountry
      .Update
      .Close
    End With
```

```
    Set rst = Nothing
    conn.Close
    Set conn = Nothing

    'requery the combo box on the main form
    Forms!frmDataEntryMain.cboCompanyID.Requery
    'close the form
    DoCmd.Close

  Exit_cmdSaveCompanyInfo_Click:
    Exit Sub
  Err_cmdSaveCompanyInfo_Click:
    MsgBox Err.Description
    Resume Exit_cmdSaveCompanyInfo_Click
  End Sub
```

Hands-On 26.6. Using the Click Event Procedure for the Command Button Control

1. Open the **frmDataEntryMain** form in Form view.
2. Click the **Add New Company** command button.
3. When the New Company Data Entry Screen window appears, enter the information shown in Figure 26.3.

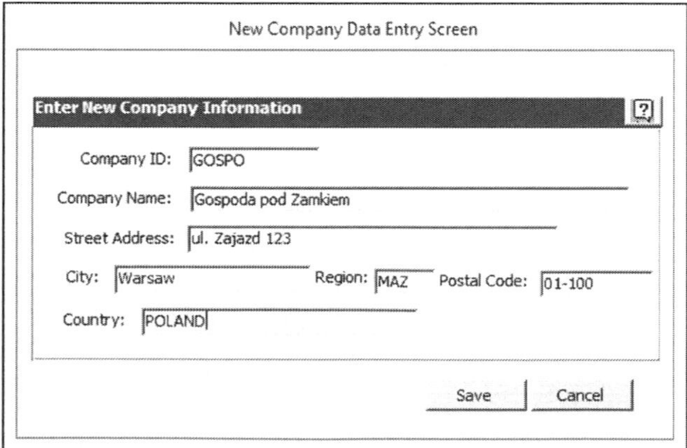

FIGURE 26.3. After saving the new company information in this window, the Company ID will appear in the Company ID combo box on the main form.

4. Click the **Save** button to save the company information. Access will run the cmdSaveCompanyInfo_Click event procedure, as shown earlier. If you have not entered data according to the criteria listed in this event procedure, Access will not allow you to save data until you correct the problem.

5. Back on the main form, select the newly added company (**GOSPO**) from the Company ID combo box.

Notice how the data entry form displays a number of icons with a question mark. Each icon is actually a command button with a Click event attached to it. When you click on the question mark button, a simple form will appear with help information pertaining to the form's section or the data entry screen.

For example, the following Click event procedure is executed upon clicking the question mark button in the Room Information section on the Asset Management data entry form:

```
Private Sub cmdRoomInfoSec_Click()
  Dim stDocName As String
  Dim stLinkCriteria As String

  On Error GoTo Err_cmdRoomInfoSec_Click

  stDocName = "frmHelpMe"
  stLinkCriteria = "HelpId = 2"
  DoCmd.OpenForm stDocName, , , stLinkCriteria

Exit_cmdRoomInfoSec_Click:
  Exit Sub
Err_cmdRoomInfoSec_Click:
  MsgBox Err.Description
  Resume Exit_cmdRoomInfoSec_Click
End Sub
```

This procedure loads the appropriate help topic into the text box control, as illustrated in Figure 26.4.

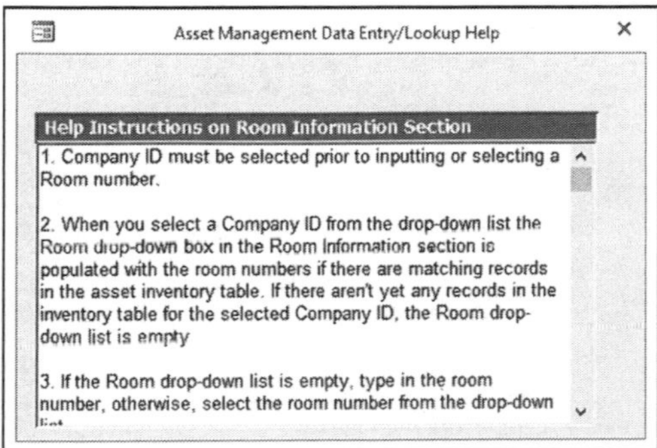

FIGURE 26.4. By clicking the question mark button in each section of the data entry form, users can get detailed guidelines on how to work with the form section.

DBLCLICK (CONTROL)

The DblClick event occurs when the user double-clicks the form or control. This event applies only to forms, form sections, and controls on a form, not controls on a report. Hands-On 26.7 demonstrates how the user of the Asset Management application can delete an asset by double-clicking on its name.

Hands-On 26.7. Using the DblClick Event Procedure for the Listbox Control

1. Open the **frmDataEntryMain** form in Form view.
2. Make appropriate selections on the Asset Management data entry form.
3. Click the **Add New Asset Type** button in the Hardware Information section. If this button cannot be clicked, you have not made all the necessary selections in the upper part of the form.
4. The Add New Asset Type Data Entry Screen window will appear, as shown in Figure 26.5.

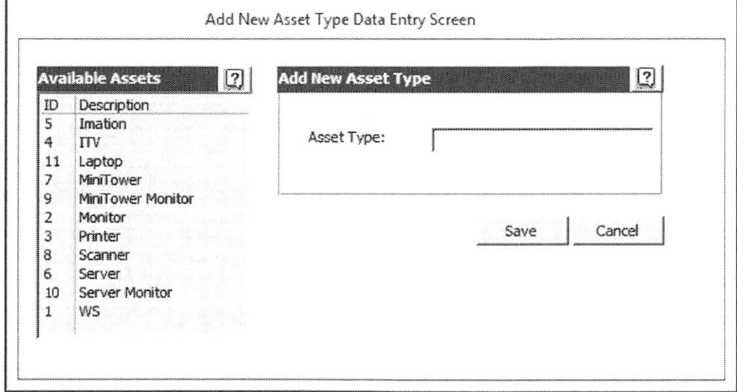

FIGURE 26.5. This form allows the user to add a new entry to the Asset Type column in the Hardware Information section of the Asset Management form or delete the asset entry by double-clicking on the entry in the Available Assets listbox.

5. In the Add New Asset Type Data Entry Screen window, enter **iPad** in the Asset Type text box and click the **Save** button.
6. In the main form, open the combo box in the **Asset Type** column and scroll down to view the newly added asset type—iPad. Do not make any selection in this combo box.
7. Click the **Add New Asset Type** button in the Hardware Information section to return to the Add New Asset Type Data Entry Screen window.

The left side of the data entry screen (see Figure 26.6) displays a listbox with the currently available assets. When the user double-clicks any item in the list, the following DblClick event procedure will determine whether or not the item can be deleted:

```
Private Sub lboxCategories_DblClick(Cancel As Integer)
    Dim conn As ADODB.Connection
    Dim myAsset  As String
    Dim myAssetDesc As String
    Dim Response As String
    Dim strSQL As String

    myAsset = Me.lboxCategories.Value
    myAssetDesc = Me.lboxCategories.Column(1)

      If myAsset >= 1 And myAsset <= 11 Then
        MsgBox "Cannot Delete - This item is being used.", _
         vbOKOnly + vbCritical, "Asset Type: " & myAsset
        Exit Sub
      End If

      If (Not IsNull(DLookup("[AssetType]", "tblProjectDetails", _
         "[AssetType] = " & myAsset))) Or _
        Not IsNull(DLookup("[EquipCategoryID]", _
         "tblEquipInventory", _
         "[EquipCategoryID] = " & myAsset)) Then

        MsgBox "This item cannot be delcted.", _
            vbOKOnly + vbCritical, "Asset Type: " & myAsset
      Else
        Response = MsgBox("Do you want to delete this Asset?", _
           vbYesNo, "Delete - " & myAssetDesc & " ?")
         If Response = 6 Then
           Set conn = CurrentProject.Connection
           strSQL = "DELETE * FROM tblEquipCategories "
           strSQL = strSQL & "Where EquipCategoryID = "
           conn.Execute (strSQL & myAsset)
           conn.Close
           Set conn = Nothing
           Me.lboxCategories.Requery
         End If
      End If
      DoCmd.Close
```

```
    'requery the combo box on the subform
    Forms!frmDataEntryMain.frmSubProjectDetails.Form.EquipCatId. _
     Requery
End Sub
```

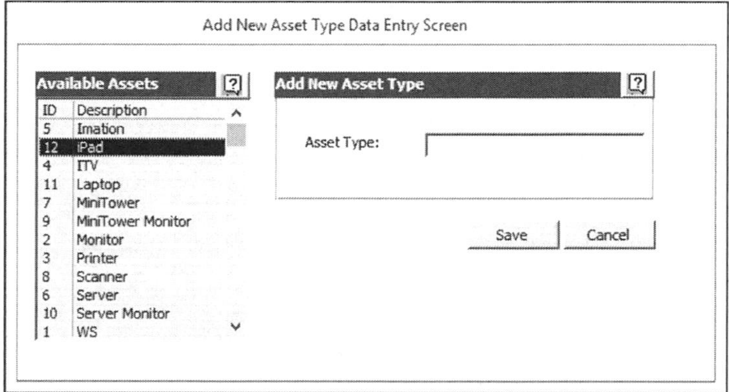

FIGURE 26.6. You can delete an item from the Available Assets list only if the item has not yet been used during the data entry.

8. Double-click on **iPad** in the Available Assets listbox. The DblClick event procedure attached to the listbox will ask you whether you want to delete this asset. Click **Yes** to the message.

   Notice that the iPad entry disappears from the Available Assets listbox.
9. Click the **Cancel** button to exit the Add New Asset Type Data Entry Screen window.

CHAPTER SUMMARY

In this chapter, you worked with a custom Microsoft Access application and examined event procedures for various controls placed on an Access form. There are other event procedures not discussed here that control how the Asset Management form and its controls respond to the user's actions. As you explore this application on your own, you will start noticing the areas where writing additional event handlers would prove beneficial to the application's users. So get to it! Tear this sample application apart. Rebuild it. Add new features. Change the user interface if you want. Learn how to handle whatever event may come your way. Be prepared, because events happen frequently in an Access application, and sooner or later you'll need to respond to them.

The next chapter focuses on working with Access reports and controlling report behavior with event programming.

Chapter 27

Enhancing Access Reports and Using Report Events

Reports have always been a very popular and widely used feature in Access. Access reporting is very interactive thanks to a view called Report view. With the Report view you can easily perform data searches, sorting and filtering. You can also copy the data. Many of the Access form features are also available for reports. For example, to make long tabular reports easier to read, you can apply alternating row shading just by changing the Alternate Back Color property in the report's Detail section. Like forms, reports can utilize bound Image controls, rich text formatting, and filtering and sorting features. The Layout view makes it easier to design reports; because you are working with the live data in Layout view, you do not need to switch between Design and Report views to see how the final report will look. The Layout view allows formatting report sections and controls, adding new fields, applying AutoFormats, grouping and sorting data, and changing many of the report's properties. As with the Layouts feature in forms, layouts can be used in reports for resizing and moving groups of controls together, or adding grid lines that grow or shrink with the data. Designing objects in Layout view has become very easy. You can drop any control anywhere within the layout. Your controls can span multiple rows and columns. Like forms, reports can be enhanced by applying a consistent style using Office themes. The reports distribution is easy with the portable document format (.pdf) and XML Paper Specification (.xps) format. The feature most appreciated by all Access users is the ability to view a report as a subform on a form. Additionally, you can specify the name of a report by using the SourceObject property of a subform control on a form.

CREATING ACCESS REPORTS

The Access 2013 Report Wizard will walk you through the report creation process by presenting various options to choose from, such as selecting a data source for your report, determining grouping and sorting criteria, and offering formatting options (layout, orientation, and style). Using the wizard makes it easy to create a report based on multiple tables.

If you need more control over creating a report, you may want to try Report Designer. You can bind your report to a table, query, or SQL statement, and add VBA code behind a report as demonstrated later in this chapter. When using Report Designer, you will not be able to see the actual data from tables and views at design time. You need to switch to Print Preview to view the entire report. To overcome this limitation, try working with the Layout view. In Layout view, you can add different types of controls, as well as functionality for sorting, grouping and calculating totals. You can also apply different formatting to the Layout view while viewing the actual data from your tables and queries.

In addition to creating Access reports via these built-in tools, you can create an Access report programmatically by using the `CreateReport` method of the Application object.

USING REPORT EVENTS

When an Access report is run, a number of events can occur. The following examples demonstrate how to control what happens not only when the report is opened, activated, deactivated, or closed, but also when there are no records for the report to display or the report record source simply does not exist.

Open

The Open event for a report occurs when the report is opened. Use this event to display support forms or custom buttons, or to change the record source for the report. The event procedure in Hands-On 27.1 demonstrates how to change a report's record source on the fly.

Please note files for the "Hands-On" project may be found on the companion CD-ROM.

Hands-On 27.1. Writing the Report_Open Event Procedure

1. Start Microsoft Access 2013 and create a new database named **Chap27.accdb** in your **C:\Access2013_ByExample** folder.
2. Import all the tables, queries, forms, reports, macros, and modules from the **Northwind.mdb** sample database to your **Chap27.accdb** database.
   To do this, in the Access window, choose **External Data | Access**. In the File name box, type **C:\Access2013_HandsOn\Northwind.mdb** and click **OK**. In the Import Objects window, select the **Tables** tab and click the **Select All** button. This will highlight all the tables. Select the **Queries** tab and click the **Select All** button. Select the **Forms** tab and click the **Select All** button. Select the **Reports** tab and click the **Select All** button. Do the same for macros and modules. After selecting all the objects on the specified tabs, click **OK** to begin importing. Click the **Close** button when done.
3. In the Access window's Navigation pane, select the **Customers** table and choose **Create | Report Wizard**. Select all the fields for the report and click the **Next** button. Continue clicking **Next** until you get to the Report Wizard screen where you can specify the title for your report. Type **rptCustomers** for the title and click **Finish**. Access opens the report in Print Preview.
4. Right-click the report tab and choose **Design view** from the context menu.
5. In the Report Header area, click the report title (label control) to select it. Resize the control to allow for longer text that will be entered dynamically by the event procedure in Step 7 below. In the property sheet for the selected label control, click the **All** tab and enter **lblCustomers** as the Name property and enter **Customers** as the Caption property.
6. In the property sheet, select **Report** from the drop-down box and click the **Event** tab. Click next to the **On Open** event property and choose **[Event Procedure]** from the drop-down box. Click the **Build** button (…).
7. Access opens the Visual Basic Editor window and writes the stub of the Report_Open event procedure. Complete the code of the following **Report_Open** event procedure:

```
Private Sub Report_Open(Cancel As Integer)
  Dim strCustName As String
```

```
    Dim strSQL As String
    Dim strWHERE As String

    On Error GoTo ErrHandler
    strSQL = "SELECT * FROM Customers"

    strCustName = InputBox("Type the first letter of " & _
     " the Company Name or type an asterisk (*) to view" & _
     " all companies.", "Show All /Or Filter")

    If strCustName = "" Then
      Cancel = True
    ElseIf strCustName = "*" Then
      Me.RecordSource = strSQL
      Me.lblCustomers.Caption = "All Customers"
    Else
      strCustName = "'" & Trim(strCustName) & "*'"
      strWHERE = " WHERE CompanyName Like " _
       & strCustName & ""
      Debug.Print strSQL
      Debug.Print strWHERE
      Me.RecordSource = strSQL & strWHERE
      Me.lblCustomers.Caption = "Selected Customers" & _
        " (" & UCase(strCustName) & ")"
    End If
    Exit Sub
  ErrHandler:
    MsgBox Err.Description
  End Sub
```

8. Switch to the rptCustomers report's Design view and choose **Home | View | Print Preview**. A message box will appear where you can enter an asterisk (*) to view all customers or the first letter of a company name if you'd like to limit your records. To cancel the report, click **Cancel** or press the **Esc** key.

Close

The Close event occurs when you close the report. Use this event to close supporting forms or to perform other cleanup operations. You cannot cancel the Close event. Figure 27.1 shows the Report_Close event procedure for the Report_Sales by Year report. This report opens via the Sales by Year Dialog form where the user can specify the report beginning and ending dates. This form remains open while the report is open and is closed during the Report_Close event.

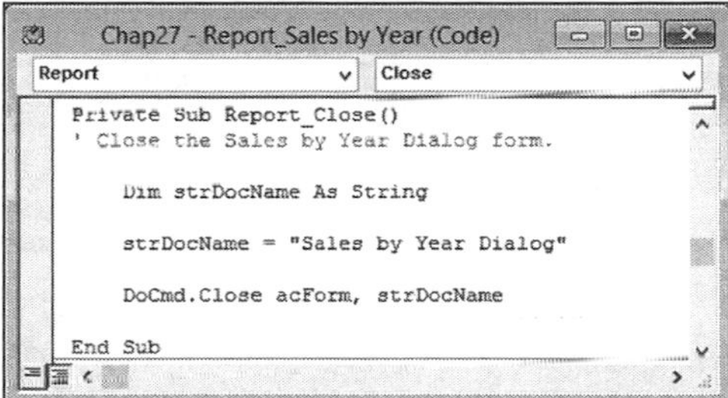

FIGURE 27.1. The Report_Close event procedure is often used to close supporting forms.

Activate

The Activate event occurs when the report is opened right after the Open event but before the event for the first section of the report. The procedure in Hands-On 27.2 displays a message when the report is open in Print Preview and returns the name of the default printer to the Immediate window.

Hands-On 27.2. Writing the Report_Activate Event Procedure

This hands-on exercise uses the rptCustomers report created in Hands-On 27.1.

1. In the Visual Basic Editor's Project Explorer window, double-click the **rptCustomers** report. In the Code window, enter the following Report_Activate event procedure:

```
Private Sub Report_Activate()
  If Me.CurrentView = acCurViewPreview Then
    MsgBox "Activating Print Preview of " _
     & Me.Name & " report."
    Debug.Print "Default Printer: " & _
     Application.Printer.DeviceName
  End If
End Sub
```

Notice how the CurrentView property is used to determine the current view of an object. Table 27.1 lists the CurrentView property constants.

TABLE 27.1. CurrentView property names and values

CurrentView Property Name	Value	Description
`acCurViewDesign`	0	The object is in Design view.
`acCurViewFormBrowse`	1	The object is in Form view.

(Contd.)

CurrentView Property Name	Value	Description
acCurViewDatasheet	2	The object is in Datasheet view.
acCurViewPivotTable	3	The object is in PivotTable view.
acCurViewPivotChart	4	The object is in PivotChart view.
acCurViewPreview	5	The object is in Print Preview.
acCurViewReportBrowse	6	The object is in Report view.
acCurViewLayout	7	The object is in Layout view.

2. In the Navigation pane of the Access window, right-click the **rptCustomers** report and choose **Print Preview**. Enter your report criteria when prompted. Upon activation of the report, the Report_Activate event will fire and a message will be displayed. Click **OK** to the message, and then switch to the Immediate window to check out the name of the default printer.
3. Close the rptCustomers report and save changes to the report when prompted.

Deactivate

The Deactivate event occurs when a report loses the focus to a table, query, form, report, macro, module, or database window. This event occurs before the Close event for the report.

NoData

The NoData event occurs when the record source for the report contains no records. This event allows you to cancel the report when no records are available. The event procedure in Hands-On 27.3 displays a message when the user enters criteria that are not met.

Hands-On 27.3. Writing the Report_NoData Event Procedure

This hands-on exercise uses the rptCustomers report created in Hands-On 27.1.

1. In the Visual Basic Editor's Project Explorer window, double-click the **Report_rptCustomers** report. In the Code window, enter the following **Report_NoData** event procedure:

```
Private Sub Report_NoData(Cancel As Integer)
  MsgBox "There is no data for the criteria " & _
   "you entered."
  Cancel = True
End Sub
```

2. Switch to the Access window and open the **rptCustomers** report. Request to see customers with a company name starting with the letter "X." Because there aren't any company names beginning with "X," a message box will be displayed, saying that there is no data for the criteria entered, and the report will be canceled.

Page

The Page event occurs after a page is formatted but before it is printed. Use the Page event to customize the appearance of your printed reports by adding lines, circles, and graphics. The event procedure in Hands-On 27.4 will draw a red border around the report pages.

Hands-On 27.4. Drawing a Page Border Using the Report_Page Event Procedure

This hands-on exercise uses the rptCustomers report created in Hands-On 27.1.

1. In the Visual Basic Editor's Project Explorer window, double-click the **rptCustomers** report. In the Code window, enter the following **Report_Page** event procedure:

```
Private Sub Report_Page()
  Me.DrawWidth = 15 ' pixels
  Me.Line (0, 0)-(Me.ScaleWidth, Me.ScaleHeight), vbRed, B
End Sub
```

Notice that the `DrawWidth` method specifies the thickness of the line and the `Line` method draws a line with the upper-left corner at (0, 0) and the lower-right corner at (Me.ScaleWidth, Me.ScaleHeight). The ScaleWidth and ScaleHeight properties specify the width and height of the report.

2. Switch to the Access window and open the **rptCustomers** report in Print Preview. Notice that when the report appears on the screen, a red border surrounds the pages (see Figure 27.2).
3. Close the rptCustomers report and save changes when prompted.

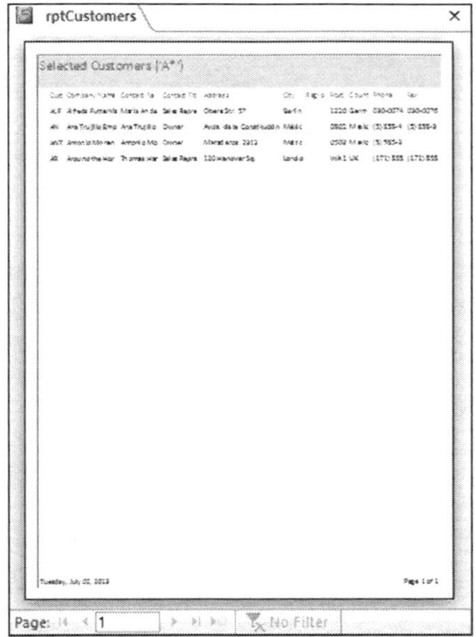

FIGURE 27.2. You can frame your Access report pages with a red line by implementing the Report_Page event procedure shown in Hands-On 27.4.

Error

The Error event is triggered by errors in accessing the data for the report. Use this event to replace the default error message with your custom message. The Error event takes the following two arguments:

- **DataErr**—Contains the number of the Microsoft Access error that occurred.
- **Response**—Determines whether or not error messages should be displayed. It may be one of the following constants:
  - **acDataErrContinue**—Ignore the error and continue without displaying the default Microsoft Access error message.
  - **acDataErrDisplay**—Display the default Microsoft Access error message. This is the default.

The Report_Error event procedure in Hands-On 27.5 illustrates how to use the value of the `DataErr` argument together with the `AccessError` method to determine the error number and its descriptive string.

The statement

```
Response = acDataErrContinue
```

will prevent the standard Microsoft Access error message from appearing. The Error event for reports works the same as the Error event for forms—but only Microsoft Access ACE or Jet Engine errors can be trapped here.

To trap errors in your VBA code, use the `On Error GoTo` statement to direct the procedure flow to the location of the error-handling statements in your procedure.

Hands-On 27.5. Writing the Report_Error Event Procedure

This hands-on exercise uses the rptCustomers report created in Hands-On 27.1.

1. In the Navigation pane, rename the Customers table **Customers2.**
2. In the Visual Basic Editor's Project Explorer window, double-click the **rptCustomers** report. In the Code window, enter the following **Report_Error** event procedure:

```
Private Sub Report_Error(DataErr As Integer, _
 Response As Integer)
  ' obtain information about the error
  MsgBox Application.AccessError(DataErr), _
   vbOKOnly, "Error Number: " & DataErr
  If DataErr = 3078 Then
    Response = acDataErrContinue
    MsgBox "Your custom error message goes here."
  End If
End Sub
```

3. Switch to the Access window and open the **rptCustomers** report. When the input box appears prompting you for the criteria, type any letter and press **OK.** At this point the Report_Error event will fire because the underlying data for the rptCustomers report does not exist. Because you renamed the Customers table that this report uses for its data source, Microsoft Access cannot locate the data and generates the error.
4. In the Navigation pane, change the **Customers2** table's name back to **Customers** and open the **rptCustomers** report to ensure that it does not produce unexpected errors.
5. Close the report when finished and save the changes when prompted.

EVENTS RECOGNIZED BY REPORT SECTIONS

An Access report can contain various sections such as Report Header/Footer, Page Header/Footer, the Detail section, and Group Headers/Footers. All report sections can respond to the Format and Print events. These events occur when you print or preview a report. In addition, the Report Header/Footer and the Detail sections recognize the Retreat event that occurs when Access returns to a previous section during report formatting.

Format (Report Section Event)

A Format event occurs for each section in a report before Microsoft Access formats the section for previewing or printing. This event takes the following two arguments:

- **Cancel**—Determines if the formatting of the section occurs. To cancel the section formatting, set this argument to `True`.
- **FormatCount**—Is an integer that specifies whether the Format event has occurred more than once for a section. If a section does not fit on one page and the rest of the section needs to be moved to the next page of the report, the `FormatCount` argument is set to `2`.

Use the Format event in the appropriate report section for changes that affect page layout, as described in Table 27.2. For changes that don't affect page layout, use the Print event for the report section.

TABLE 27.2. Effect of the Format event on report sections

Report Sections	Description of Event
Detail	The Format event occurs for each record in the section just before Microsoft Access formats the data in the record. You can access the data in the current record using the event procedure.
Group Headers	The Format event occurs for each new group. You can access the data in the Group Header and the data in the first record in the Detail section using the event procedure.
Group Footers	The Format event occurs for each new group. You can access the data in the Group Footer and the data in the last record in the Detail section via an event procedure.

The event procedure in Hands-On 27.6 demonstrates how to make reports easier to read by shading alternate rows.

Hands-On 27.6. Shading Alternate Rows Using the Detail_Format Event Procedure

This hands-on exercise uses the rptCustomers report created in Hands-On 27.1.

1. In the Visual Basic Editor's Project Explorer window, double-click the **rptCustomers** report. In the Code window, enter the following **Detail_Format** event procedure. Do not type the `Option Compare Database` and `Option Explicit` statements if they are already present at the top of the Code window.

```
Option Compare Database
Option Explicit

Dim shaded As Boolean

Private Sub Detail_Format(Cancel As Integer, _
 FormatCount As Integer)
  If shaded Then
    Me.Detail.BackColor = vbYellow
  Else
    Me.Detail.BackColor = vbWhite
  End If
  shaded = Not shaded
End Sub
```

Notice that at the top of the module sheet (in the module's Declarations area) we have placed the following statement:

```
Dim shaded As Boolean
```

This statement declares the global variable of the Boolean type to keep track of the alternate rows.

When you run the report, upon printing the Detail section, Access will check the value of the `shaded` variable. If the value is True, it will change the background of the formatted row to yellow (which produces a light gray background when printed on a noncolor printer). The shaded value will then be set to False for the next row by using the following statement:

```
shaded = Not shaded
```

This statement works as a toggle. If shaded was True, it will be False now, and so on.

2. Modify the **Report_Open** event procedure as follows:

3. Add the following statement just below the other variable declarations that are already present inside this procedure:

```
Dim ctrl As TextBox
```

4. Enter the following code before the Exit Sub statement:

```
For Each ctrl In Me.Detail.Controls
    If ctrl.BackStyle = 1 Then ctrl.BackStyle = 0
Next
```

The For Each loop will iterate through the controls in the detail section of the rptCustomers report and set the Back Style property of each text box control to 0 (Transparent).

5. Switch to the Access window and open the **rptCustomers** report in Print Preview. When the input box prompts you for the criteria, type an asterisk (*) and press **OK**. Figure 27.3 shows the result of this procedure.

In Access 2013/2010/2007 you can shade alternate rows by setting the AlternateBackColor property for the Detail section instead of writing VBA code for the Format event of the Detail section.

6. Close the rptCustomers report, saving changes when prompted.

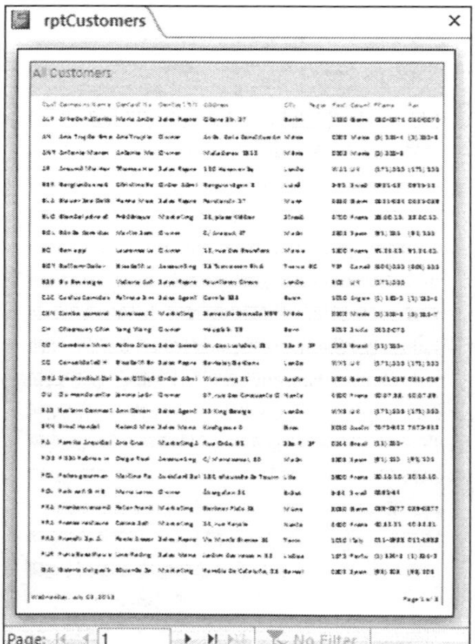

FIGURE 27.3. You can shade alternate rows in an Access report by setting the AlternateBackColor property for the Detail section or by writing VBA code for the Format event of the Detail section.

The next hands-on exercise demonstrates how to suppress the Page Footer on the first page of your report by placing code in the PageFooterSection_Format event procedure.

Hands-On 27.7. Suppressing the Page Footer Using the PageFooterSection_Format Event Procedure

1. Using the Report Wizard, create a report called **rptProducts** based on the Products table. Choose the following fields for this report: **ProductID**, **ProductName**, **UnitPrice**, and **UnitsInStock**. On the last page of the Report Wizard, select the **Modify the report's design** option button.
2. In the Design view of the rptProducts report, select **PageFooterSection** in the property sheet. Click the **Event** tab. Click next to the **On Format** property and select [**Event Procedure**] from the drop-down list. Click the **Build** button (…) to activate the Code window.
3. In the Code window for rptProducts, enter the following **PageFooterSection_Format** event procedure:

```
Private Sub PageFooterSection_Format(Cancel As Integer, _
 FormatCount As Integer)
  Dim ctrl As Control

  For Each ctrl In Me.PageFooterSection.Controls
    If Me.Page = 1 Then
      ctrl.Visible = False
    Else
      ctrl.Visible = True
    End If
  Next ctrl
End Sub
```

4. Switch to the Access window and open the **rptProducts** report in Print Preview. Notice that the Footer section does not appear on the first page of the report.
5. Close the rptProducts report and save changes to the report when prompted.

Print (Report Section Event)

The Print event occurs after the data in a report section has been formatted but before the data is printed. The Print event occurs only for sections that are actually printed, as described in Table 27.3. To access data from sections that are not printed, use the Format event.

You can use the `PrintCount` argument to check whether the Print event has occurred more than once for a record. If part of a record is printed on one page and the rest is printed on the next page, the Print event will occur twice, and the `PrintCount` argument will be set to 2. You can use the `Cancel` argument to cancel the printing of a section.

TABLE 27.3. Effect of the Print event on report sections

Report Section	Description of Event
Detail	The Print event occurs for each record in the Detail section just before Microsoft Access prints the data in the record.
Group Headers	The Print event occurs for each new group.
Group Footers	The Print event occurs for each new group.

The event procedure in Hands-On 27.8 demonstrates how to print a record range indicator in the report's Footer. This indicator will display the range of records printed on each page. You can easily modify this example procedure to print the first and last customer ID on the page (see the discussion that follows this hands-on exercise).

Hands-On 27.8. Displaying a Record Range in the Report's Footer Using the Detail_Print Event Procedure

This hands-on exercise uses the rptCustomers report you created in Hands-On 27.1.

1. Open the **rptCustomers** report in Design view and place two unbound text boxes in the report's Page Footer section.
2. Change the **Name** property of the first box to **txtPage** and set its **Visible** property to **No.** Delete the label control in front of this text box.
3. Name the second text box **txtRange** and set the **Caption** property of its label control to **Records.**
4. In the property sheet for the **txtRange** and **Records** controls, set the **Display When** property to **Print Only** (see the note at the end of this exercise).
5. In the property sheet, select **Detail** from the drop-down box and click the **Event** tab. Set the **On Print** property of the Detail section to **[Event Procedure]** and write the code for the **Detail_Print** event as shown here:

```
Private Sub Detail_Print(Cancel As Integer, _
 PrintCount As Integer)
  Static rCount As Integer
  Static start As Integer
```

```
    Static firstID As String
    Static lastID As String

    If Me.Page <> Me.txtPage Then
      start = Me.CurrentRecord
      firstID = CustomerID
      Me.txtPage = Me.Page
      rCount = 0
    End If
      rCount = rCount + 1
      lastID = CustomerID
    If start <= rCount Then
      Me.txtRange = start & "-" & rCount
      ' Me.txtRange = UCase(firstID) & "-" & UCase(lastID)
    Else
      rCount = Me.CurrentRecord
      lastID = CustomerID
    End If
  End Sub
```

The Detail_Print event procedure is triggered for each record. It uses the `start` and `rCount` variables to keep track of the first and last items on the page.

6. In the Code window, enter the **PageHeaderSection_Print** event procedure shown here:

```
  Private Sub PageHeaderSection_Print(Cancel As Integer, _
   PrintCount As Integer)
    Me.txtPage = 0
  End Sub
```

7. To test this event procedure, switch to the Access window and open the **rpt-Customers** report in Print Preview displaying all customers. Notice the record range indicator at the bottom of the report page, as shown in Figure 27.4.
8. Close the rptCustomers report and save changes when prompted.

The PageHeaderSection_Print event procedure will set the value of the unbound txtPage text box to zero (0) whenever the Print event occurs for a new page.

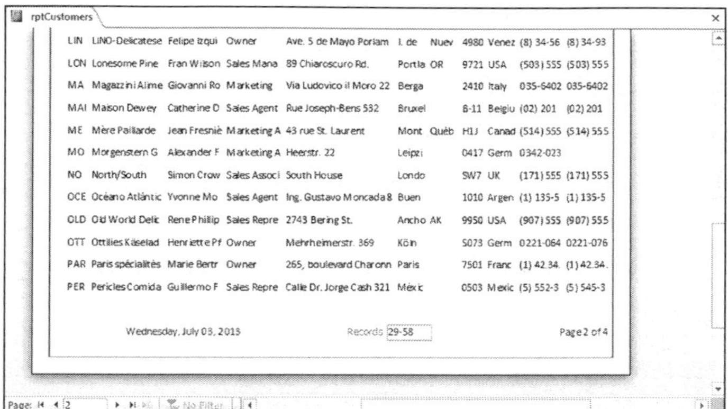

FIGURE 27.4. This report displays the record range indicator at the bottom of the page (see Hands-On 27.8).

You can modify the event procedure in Hands-On 27.8 to print the first and last customer IDs on the page as shown in Figure 27.5. Simply replace the following statement in the Detail_Print event procedure

```
Me.txtRange = start & "-" & rCount
```

with the following line of code:

```
Me.txtRange = UCase(firstID) & "-" & UCase(lastID)
```

FIGURE 27.5. This report displays the first and last Customer ID for a specific page at the bottom of each printed page.

When you open the rptCustomers report in Report view instead of in Print Preview, you will notice that there is no calculated value in the Records text box at the bottom of the page. The reason for this is that in Report view there aren't any pages. The entire report is one big continuous page. Since there aren't any pages Access cannot calculate any values that depend on the Page or Pages properties. Also, it's important to remember that the Print event for report sections does not fire in Report view. You can tell Access to display certain controls only in Print Preview by changing the Display When property of the control to Print Only in the property sheet.

Retreat (Report Section Event)

The Retreat event occurs when Microsoft Access returns to previous sections of the report during report formatting. For example, after formatting a report section, if Access discovers that the data will not fit on the page, it will go back to the necessary location in the report to ensure that the section can properly begin on the next page.

The Retreat event occurs after the Format event but before the Print event. This event applies to all report sections except Page Headers and Footers. The Retreat event occurs for Group Headers and Footers whose KeepTogether property has been set to Whole Group or With First Detail. This event is also triggered in subreports whose CanGrow or CanShrink properies have been set to True.

The Retreat event makes it possible to undo any changes made during the Format event for the section. The Retreat event is demonstrated in the sample Northwind.mdb database's Sales by Year report, as shown in Figure 27.6.

```
Private Sub GroupFooter1_Retreat()
' If ShowDetails check box on Sales by Year Dialog form is checked,
' set value of Show text box to True so that page header will print on
' next page.

    If Forms![Sales by Year Dialog]!ShowDetails Then Me!Show.Value = True

End Sub
```

FIGURE 27.6. The Sales by Year report in the Northwind.mdb database uses the GroupFooter1_Retreat event procedure to control printing of a page header.

USING THE REPORT VIEW

Reports have an interactive view called Report view, as shown in Figure 27.7. This is the default view for all new reports created in Access 2007-2013. In this view you can easily copy data by selecting it and then clicking the Copy button in the Clipboard group of the Home tab or pressing Ctrl+C. If you need to find particular data in the report, use the Find button in the Find group of the Home tab or press Ctrl+F. Access will pop up the standard Find dialog box in which you can enter your search criteria. Filtering and sorting is also enabled for the Report view via the buttons located in the Filter & Sort section of the Home tab. A report open in Report view isn't divided into pages; it is a single big page. If you have any calculations that depend on the Page or Pages properties of the report, they may not return the correct results. Certain report events, such as Print and Format, will not be triggered when the report is displayed in the Report view. The Report view has its own new event called Paint that is used with sections in Report view. This event fires whenever a section needs to be drawn on the screen. Use this event to conditionally format controls in that view, as shown in Hands-On 27.9.

The Paint event fires multiple times for each section of the report because Access paints various elements of the given section separately at different times. The calculated controls and the items that require a change in background or foreground colors are each painted separately.

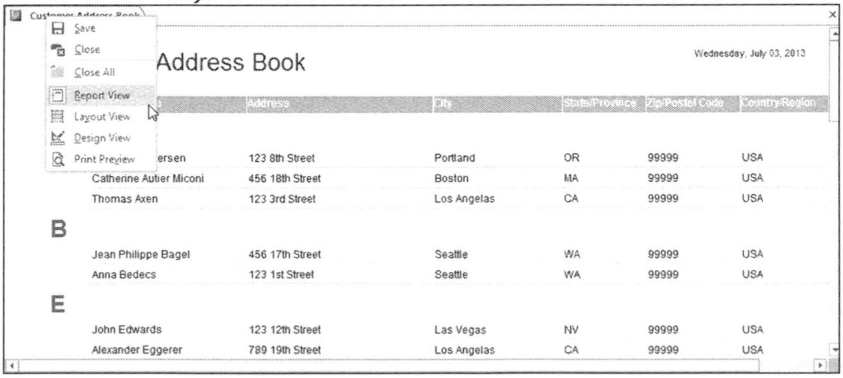

FIGURE 27.7. Access 2013 reports can be displayed using four different views: Report View, Layout View, Design View, and Print Preview.

Hands-On 27.9. Conditionally Formatting a Control in Report View

1. Open the **Northwind 2007.accdb** database in your **C:\Access2013_ByExample** folder. Close the Login box when prompted to log in.
2. In the Navigation pane, right-click the **Customer Address Book** report and choose **Design View**.

3. Click the **Detail** section and activate the property sheet.
4. In the property sheet of the Detail section, click the **Event** tab, select **Event Procedure** from the drop-down box next to the **On Paint** property, then click the **Ellipsis** button (...).

   Access activates the Code window and writes the stub of the Detail_Paint event procedure.
5. Complete the code of the **Detail_Paint** procedure as shown here:

```
Private Sub Detail_Paint()
  If Me.City.Value = "Chicago" Then
    Me.City.ForeColor = vbBlue
  Else
    Me.City.ForeColor = vbBlack
  End If
End Sub
```

   This event procedure will set the ForeColor property for a control called City to blue when the city name is Chicago and display the names of all other cities in black. This procedure will be triggered when you open the report in Report view.
6. Press **Ctrl+S** to save the changes in the Code window.
7. Press **Ctrl+F11** to return to the Microsoft Access window.
8. Click the **View** button in the Views group of the Design tab.
9. Access displays the report in its default Report view. Press **Ctrl+F** to activate the Find dialog box. Enter **Chicago** and click **Find Next**. Access locates the first customer who lives in Chicago. Click **Find Next** again to locate the next customer. Notice that all the occurrences of "Chicago" are shown in blue.
10. Close the Customer Address Book report.

SORTING AND GROUPING DATA

Access offers users a convenient interface for grouping data, adding totals, and filtering. These features are available from a separate Group, Sort, and Total pane as shown in Figure 27.8. To work with this pane, open the report in Layout view, click the Design tab, and select the Group & Sort button in the Grouping & Totals section. When you click on the Add a Group or Add a Sort buttons in the pane at the bottom of the report, Access will walk you through the steps required to create new report groups, add totals, or sort (Figure 27.9).

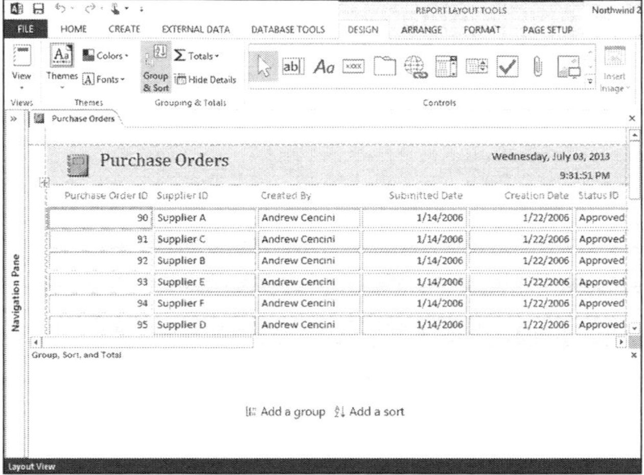

FIGURE 27.8. The Group, Sort, and Total pane provides a quick way to group and sort data, and add calculations in Access reports.

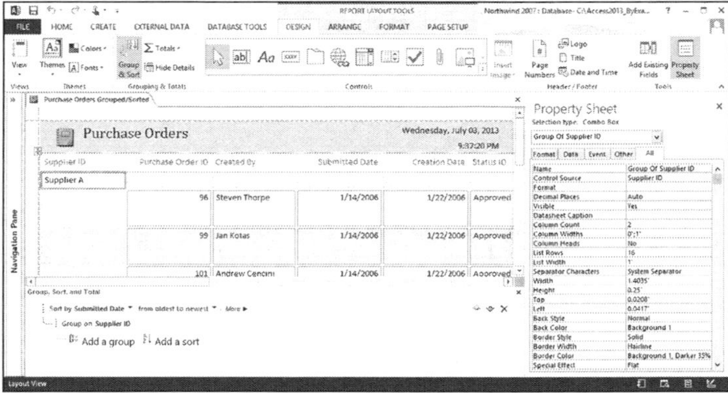

FIGURE 27.9. The Group, Sort, and Total pane indicates that the report is sorted by the Submitted Date field and grouped by the Supplier ID field.

SAVING REPORTS IN .PDF OR .XPS FILE FORMAT

Access reports can be saved to the .pdf or .xps format, as shown in Figure 27.10. The Portable Document Format (.pdf) preserves document formatting and makes files easy to distribute and print. Reports distributed as .pdf files retain their format and are protected so that the data may not be copied or changed. Another format that you can use for your report distribution is the XML Paper Specification (.xps) format, which also retains the format of the original document.

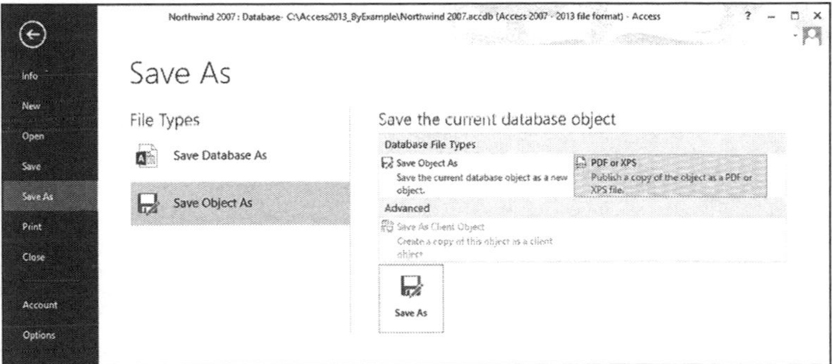

FIGURE 27.10. To save your report to .pdf or .xps file format, click File | Save As. Select Save Object As, select PDF or XPS, and click the Save As button. Access will display the Publish As PDF or XPS dialog box as shown in Figure 27.11.

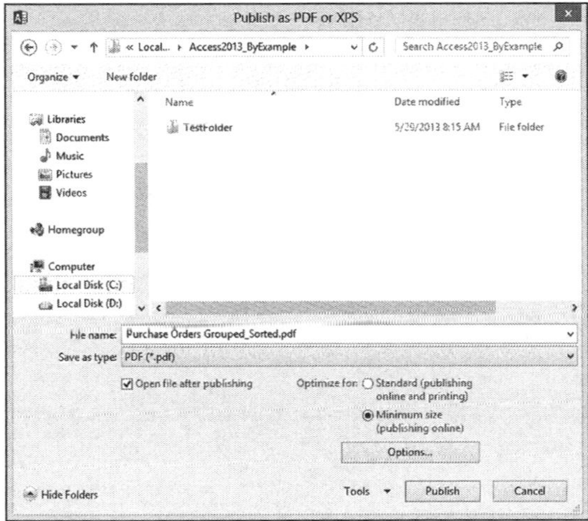

FIGURE 27.11. You can select the required file format for saving the Access report in the Publish As PDF or XPS dialog box. Figure 27.10 shows how to activate this dialog.

USING THE OPENARGS PROPERTY OF THE REPORT OBJECT

Like forms, Access reports have a very useful property called `OpenArgs` that you can use from a VBA code or a macro to pass a value to a report as the report is opened. Use the `OpenReport` method of the `DoCmd` object in the following form:

```
DoCmd.OpenReport (reportname, view, filtername, wherecondition, window-
mode, OpenArgs)
```

The `OpenArgs` argument is a string expression of the Variant data type. As demonstrated in Chapter 25, you can pass multiple values in the `OpenArgs` argument by concatenating your values.

The `OpenArgs` property can be used to set a report format or to determine what data the report should display. With the `OpenArgs` property, you can reuse the same report, instead of creating a new report for a similar requirement.

The Hands-On 27.10 demonstrates how to filter a report with the help of the `OpenArgs` property.

Hands-On. 27.10. Using the OpenArgs property to filter an Access report

1. Open the **Northwind 2007_Revised.accdb** database in your **C:\Access 2013_HandsOn** folder. Cancel out of the Login dialog box.
2. In the Navigation pane on the left, locate and double-click the **frmEmployee-Address**.

   You should see the form as shown in Figure 27.12.

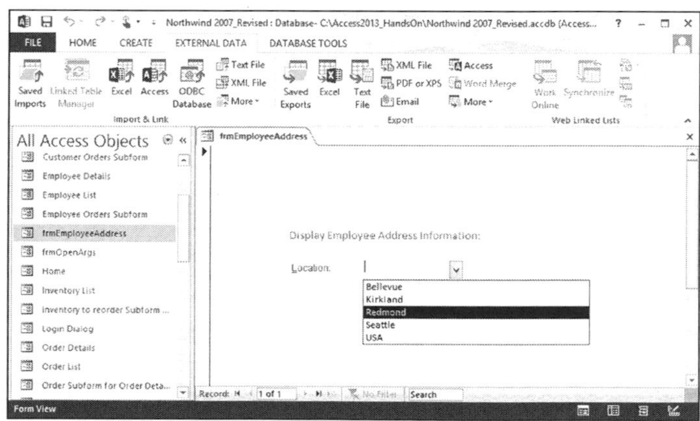

FIGURE 27.12. The form used to filter the Employee Address Book report by City or Country/Region.

3. Choose **Redmond** from the combo box.

   Access executes the following code in the `cboReports_AfterUpdate` event procedure:

```
Private Sub cboReports_AfterUpdate()
    Dim strFilterBy As Variant
    Dim strRpt As String

    strRpt = "Employee Address Book"
```

```
        If SysCmd(acSysCmdGetObjectState, acReport, _
            strRpt) <> 0 Then
            DoCmd.Close acReport, strRpt
        End If

        strFilterBy = Me.cboReports.Value
        DoCmd.OpenReport ReportName:=strRpt, _
            View:=acViewReport, _
            OpenArgs:=strFilterBy
    End Sub
```

This event procedure closes the Employee Address Book report if it is open. The `SysCmd` method is used here to return the state of a specified database object. Use this method to find out whether the object is open, is a new object, or has been changed but not saved. For more information on using this method in your VBA procedures, see the online help.

Next, the procedure stores the selected value in the `strFilterBy` variable. This variable is then referenced in the `OpenArgs` property when the report is opened with the `OpenReport` method. The `Report_Load` event procedure of the Report_Employee Address book (see the code below) then checks the `OpenArgs` property for the Null value. If the property is not `Null`, the strFilter variable is set to contain filtering criteria for the City or Country/Region fields. If you selected Redmond from the form's combo box, the `strFilter` will be set to City = 'Redmond'. The statement `Me.OpenArgs` returns the value stored in the `OpenArgs` property. With the filtering expression set, all you need to do is tell Access to turn the filter on by using the `FilterOn` property and set the `Filter` property to the strFilter variable.

```
Private Sub Report_Load()
    Dim strFilter As String

    If IsNull(Me.OpenArgs) Then
        Exit Sub
    Else
        If Me.OpenArgs = "USA" Then
            strFilter = "[Country/Region] = '" & Me.OpenArgs & "'"
        Else
            strFilter = "City = '" & Me.OpenArgs & "'"
        End If
        Me.FilterOn = True
            Me.Filter = strFilter
```

```
    End If
End Sub
```

After the procedure finishes executing its code, you should see the Employee Address Book filtered by Redmond or whatever item you specified in the form's combo box (Figure 27.13).

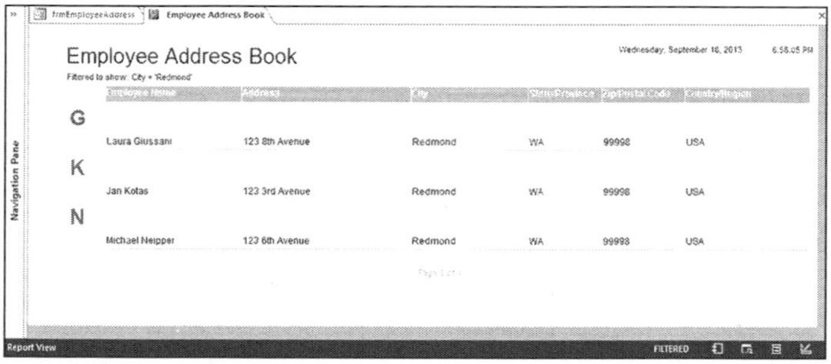

FIGURE 27.13. This report was filtered by using the value passed in the OpenArgs property.

To filter the report again, make another selection from the form's combo box.

CHAPTER SUMMARY

In this chapter, you discovered various ways of creating reports in Access 2013 and learned how you can extend your reports by incorporating some VBA code. You worked with several events that fire when the Access report is run. By writing your own event procedures you can specify what happens when the report is opened, activated, deactivated, or closed. You can also display a custom message when an error occurs or the report does not contain any data, or you can make last-minute changes to the report format before it is printed or previewed. In the last section of this chapter, you were introduced to the OpenArgs property of the Report object and learned how to use it to filter a report.

This chapter barely scratched the surface of what is possible and doable with reports. There are numerous templates in Access 2013 that you can study to gain more insight into designing very appealing, informative, and interactive reports.

In the next chapter, you learn how to handle an object's events from standalone class modules, as well as how to program and raise your own events.

Chapter 28 Advanced Event Programming

So far in this book you've worked with event procedures that executed from the form or report class module when a certain event occurred for a form, report, or control. You have probably noticed that event programming, as you've seen it implemented in the form and report class modules, requires that you copy and paste your existing event code into new form or report events in order to obtain exactly the same functionality. For instance, say you added certain features to a text box on one form and now you'd like to have a text box on other forms behave in the same way. You could react to the text box's events in the same way on all your forms by entering the same event procedure code in a form class module for each form, or you could save keystrokes by learning how to centralize and reuse your event code.

You can avoid typing the same event procedure code again and again by using classes. Recall that we've already used classes in this book; in Chapter 8, you learned how you can design your own objects in VBA by writing code in a standalone class module. You worked with property procedures that allowed data to be read or written to the object. You also learned how to create functions in a class module that worked as object methods. In this chapter, you learn how to react to an object's events from a standalone class module.

You will need to become familiar with the following VBA terms:

- **Event sink**—A class that implements an event. Only classes can sink events.
- **Event source**—An object that raises events. An event source can have multiple event sinks. Note that source and sink terminology is derived from electronics. A device that outputs current when active is said to be sourcing current. A device that draws current into it when active is said to be sinking current.
- **WithEvents**—A keyword that allows you to handle an object's events inside classes other than form or report classes. The variable that you declare for the `WithEvents` keyword is used to handle an object's events.
- **Event**—A statement used to declare a user-defined event. The `Event` declaration must appear in a class module.
- **RaiseEvent**—A statement used to call a custom event. The custom event must first be declared using the `Event` statement.

SINKING EVENTS IN STANDALONE CLASS MODULES

Instead of writing your event procedures in the form and report class modules, you can make the maintenance of your Microsoft Access applications much simpler by writing the event code in standalone class modules.

Recall that a *standalone class module* is a special type of class module that is not associated with any particular form or report. This class module can be inserted in the Visual Basic Editor window by choosing Insert | Class Module. In addition to creating custom objects (see examples in Chapter 8), standalone class modules can implement object events.

The process of listening to an object's events is called *sinking the event*. To sink (handle) events in a standalone class module, you must use the `WithEvents` keyword. This keyword will tell the class that you want to sink some or all of the object's events in the class module. You determine which events you want to sink by writing appropriate event code (see Custom Project 28.1). Only classes can sink events. Therefore, the `WithEvents` keyword can only be used in classes. You can use the `WithEvents` keyword to declare as many individual variables as you need; however, you cannot create arrays using `WithEvents`.

An object that generates events is called an *event source*. The process of broadcasting an event is called *sourcing the event*. To handle events raised by an event source, you must declare an object variable using the `WithEvents` keyword. For example, to react to form events in a standalone class module, you would need to enter the following module-level variable declaration in the class module:

```
Private WithEvents m_frm As Access.Form
```

In this statement, `m_frm` is the name of the object variable that references the Form object. While you can use any variable name you want, this variable cannot be a generic object. That means you cannot declare it as Object. If the variable were declared as Object, Visual Basic wouldn't know what type library should be used. Therefore, it would not be able to provide you with the names of events for which you can write code.

Now, let's walk through these new concepts step by step. Custom Project 28.1 demonstrates how to create a record logger class that handles a form's AfterUpdate event. Each time the AfterUpdate event occurs, this class will enter information about the newly created record into a text file.

Please note files for the "Hands-On" project may be found on the companion CD-ROM.

Custom Project 28.1. Sinking Events in a Standalone Class Module

Part 1: File Preparation

1. Start Microsoft Access and create a new database named **Chap28.accdb** in your **C:\Access2013_ByExample** folder.
2. Import the Customers, Products, Suppliers, and Categories tables from the sample Northwind.mdb database. To do this, in the Access window, choose **External Data | Access**. In the File name box, type **C:\Access2013_HandsOn\Northwind.mdb** and click **OK**. In the Import Objects window, on the **Tables** tab, select the **Customers**, **Products**, **Suppliers**, and **Categories** tables and click **OK** to begin importing. Click the **Close** button when done.

3. In the Navigation pane of the Access application window, select the **Customers** table and choose **Create | Form Wizard** to create a new form based on the Customers table. Select all the fields from the Customers table, choose **Columnar** layout, and specify **frmCustomers** as the form's title. After you click **Finish**, the newly designed frmCustomers form will appear in the Form view as shown in Figure 28.1.

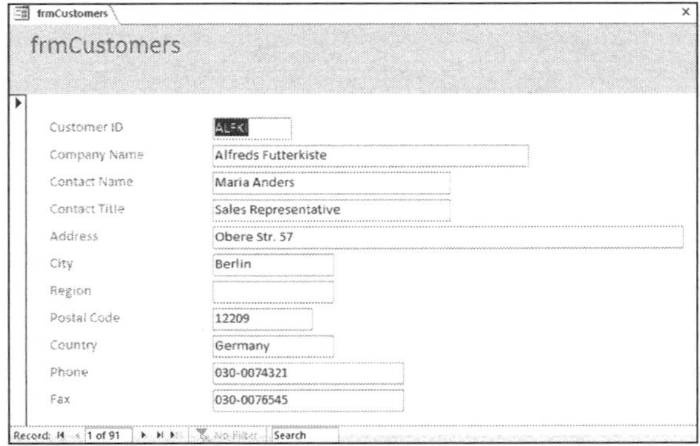

FIGURE 28.1. The frmCustomers form is used in Custom Project 28.1 to demonstrate how an object's event can be handled outside of the form class module.

4. Close the frmCustomers form created in Step 3.

Part 2: Creating the cRecordLogger Class

1. Press **Alt+F11** to activate the Visual Basic Editor window.
2. In the Project Explorer window, select **Chap28 (Chap28)** and choose **Insert | Class Module**. A new class called Class1 will appear in the Project Explorer window. Use the Name property in the Properties window to change the name of the class to **cRecordLogger** (see Figure 28.2).

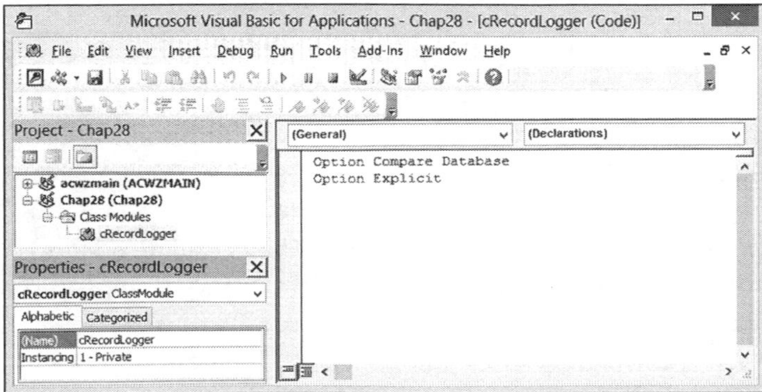

FIGURE 28.2. Use the Name property in the Properties window to change the name of the class module.

3. In the cRecordLogger class module's Code window, enter the following module-level variable declaration just below the `Option Compare Database` and `Option Explicit` statements:

```
Private WithEvents m_frm As Access.Form
```

After declaring the object variable using `WithEvents`, the variable name `m_frm` appears in the Object box in your class module (see Figure 28.3). When you select this variable from the drop-down list, the valid events for that object will appear in the Procedure box (see Figure 28.4). By choosing an event from the Procedure drop-down list, an empty procedure stub will be added to the class module where you can write your code for handling the selected event. By default, Access adds the Load event procedure stub after an object is selected from the Object drop-down list.

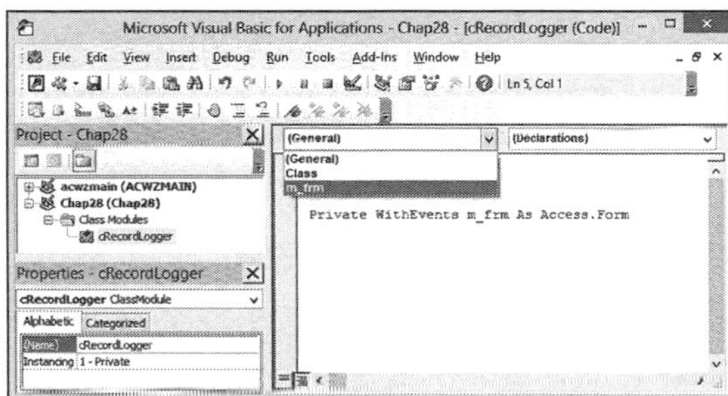

FIGURE 28.3. The Object drop-down list in the cRecordLogger's Code window lists the m_frm object variable that was declared using the WithEvents keyword.

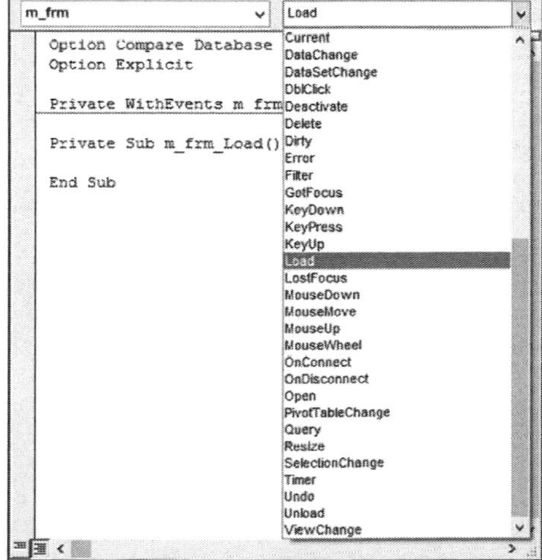

FIGURE 28.4. The Procedure drop-down list in the cRecordLogger's Code window lists the valid events for the object declared with the WithEvents keyword.

4. In the cRecordLogger class module's Code window, enter the following Property procedure just below the variable declaration:

```
Public Property Set Form(cur_frm As Form)
  Set m_frm = cur_frm
  m_frm.AfterUpdate = "[Event Procedure]"
End Property
```

In order to sink events in the class module, you must tell the class which specific form's events the class should be responding to. You do this by writing the Property Set procedure. Recall from Chapter 8 that Property Set procedures are used to assign a reference to an object. Therefore, the statement:

```
Set m_frm = cur_frm
```

will assign the current form (passed in the `cur_frm` variable) to the `m_frm` object variable declared in Step 3. Pointing the object variable (`m_frm`) at the object (`cur_frm`) isn't enough. Access will not raise the event unless the object's Event property is set to [Event Procedure]. Therefore, the second statement in the preceding procedure:

```
m_frm.AfterUpdate = "[Event Procedure]"
```

will ensure that Access knows that it must raise the form's AfterUpdate event.

5. Choose **Tools | References** and add the **Microsoft Scripting Runtime Library** to the class module. You will need this library to gain access to the FileSystemObject in the next step.
6. In the cRecordLogger class module's Code window, enter the following **m_frm_AfterUpdate** event procedure:

```
Private Sub m_frm_AfterUpdate()
  Dim fso As FileSystemObject
  Dim myFile As Object
  Dim strFileN As String
  Dim ctrl As Control

  On Error Resume Next

  Set fso = New FileSystemObject
  strFileN = "C:\Access2013_ByExample\MyCust.txt"
  Set myFile = fso.GetFile(strFileN)

 If Err.Number = 0 Then
    ' open text file
```

```
    Set myFile = fso.OpenTextFile(strFileN, 8)
  Else
    ' create a text file
    Set myFile = fso.CreateTextFile(strFileN)
  End If

  For Each ctrl In m_frm.Controls
    If ctrl.ControlType = acTextBox And _
      InStr(1, ctrl.Name, "ID") Then
      myFile.WriteLine "ID:" & ctrl.Value & _
       " Created on: " & Date & " " & Time & _
       " (Form: " & m_frm.Name & ")"
      MsgBox "See the audit trail in " & strFileN & "."
      Exit For
    End If
  Next
  myFile.Close
  Set fso = Nothing
End Sub
```

The code inside the m_frm_AfterUpdate event procedure will be executed after Access finds that the form's AfterUpdate property is set to [Event Procedure]. This code tells Access to open or create a text file named **C:\Access2013_ByExample\MyCust.txt** and write a line consisting of the value of the ID control on the form, the date and time the record was inserted or modified, and the name of the form. Notice how the `InStr` function is used to locate the control whose name contains the "ID" string. The first argument of the `InStr` function determines the character position where the search should begin, the second argument is the string being searched, and the third argument is the string expression being sought within the string specified in the second argument.

7. Save the code that you wrote in the class module by clicking the **Save** button on the toolbar or choosing **File | Save**. When the Save As dialog box appears with cRecordLogger in the text box, click **OK**.

For the events to actually fire now that you've written the code to handle the event in the standalone class module, you need to instantiate the class and pass it the object whose events you want to track. This requires that you write a couple of lines of code in your form's class module.

Part 3: Creating an Instance of the Custom Class in the Form's Class Module

1. Switch to the Access window by pressing **Alt+F11**. In the Navigation pane, right-click the **frmCustomers** form you created in Step 3 of Part 1. Select **Design View** from the shortcut menu.
2. In the property sheet, make sure **Form** is selected in the drop-down list, and activate the **Event** tab. Click next to the **On Open** property and select **[Event Procedure]**. Click the **Build** (...) button. Access activates the Code window and writes the procedure stub for the Form_Open event. Complete the code of this event procedure and the Form_Close event procedure, as shown here:

```
Private clsRecordLogger As cRecordLogger
Private Sub Form_Open(Cancel As Integer)
  Set clsRecordLogger = New cRecordLogger
  Set clsRecordLogger.Form = Me
End Sub
Private Sub Form_Close()
  Set clsRecordLogger = Nothing
End Sub
```

To instantiate a custom class module, we begin by declaring a module-level object variable, `clsRecordLogger`, as the name of our custom class, cRecordLogger. You can choose any name you wish for your variable name.
Next, we instantiate the class in the Form_Open event procedure by using the following `Set` statement:

```
Set clsRecordLogger = New cRecordLogger
```

Notice that you must use the `New` keyword to create a new object of a particular class. By setting the reference to an actual instance of the object when the form first opens, we ensure that the object refers to an actual object by the time the event is first fired. The second statement in the Form_Open event procedure:

```
Set clsRecordLogger.Form = Me
```

sets the Form property defined by the Property procedure in the class module (see Step 4 of Part 2) to the Form object whose events we want to sink. The `Me` keyword represents the current instance of the Form class.
When you are done pointing the object variable to the instance of the custom class, it is a good idea to release the variable reference. We've done this by setting the object variable `clsRecordLogger` to `Nothing` in the Form_Close event procedure. The complete code entered in the frmCustomers form class module is shown in Figure 28.5.

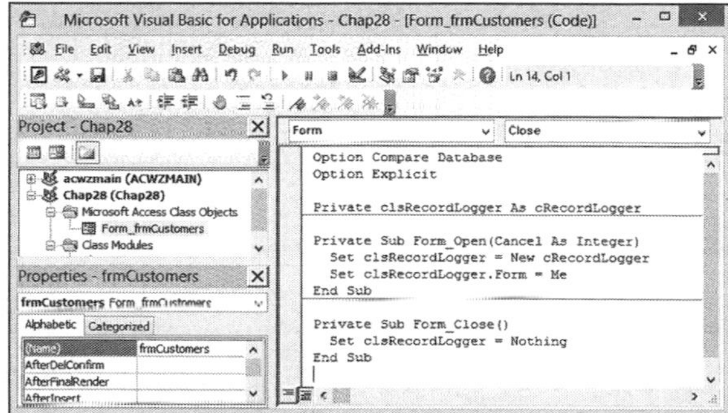

FIGURE 28.5. To sink form events in a custom class module, you must enter some code in the form class module.

3. Press **Ctrl+S** to save the code you entered in Step 2 or click the **Save** button on the toolbar.
4. Close the frmCustomers form.

Now that all the code has been written in the standalone class module and in the form class module, it's time to test our project.

Part 4: Testing the cRecordLogger Custom Class

1. Switch to the Access window and open the **frmCustomers** form in Form view.
2. Choose **Home | New**.
3. Enter **MARSK** in the Customer ID text box and **Marski Enterprises** in the Company Name text box (see Figure 28.6). Press the record selector on the left side of the form to save the record.

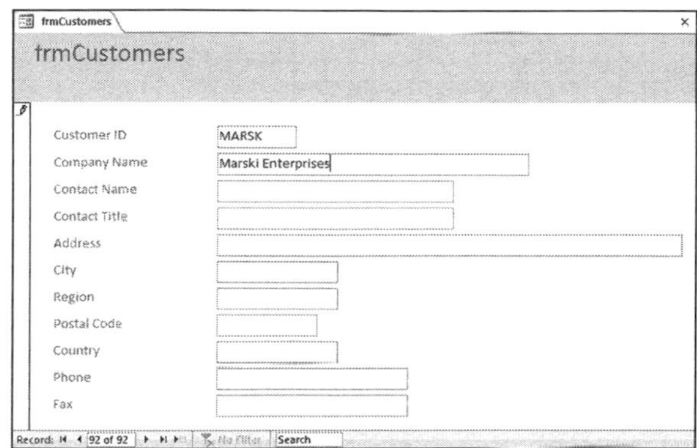

FIGURE 28.6. The frmCustomers form in the Data Entry mode is used for testing out the custom cRecordLogger class.

When you save the newly entered record, a message box appears with the text "See the audit trail in C:\Access2013_ByExample\MyCust.txt." Recall that this message was programmed inside the m_frm_AfterUpdate() event procedure in the cRecordLogger class module. It looks like our custom class has successfully sunk the AfterUpdate event. The form's AfterUpdate event was propagated to the custom class module.

4. Click **OK** to close the message box.
5. Activate Windows Explorer and open the **C:\Access2013_ByExample\MyCust.txt** file.

   The MyCust.txt file (see Figure 28.7) displays the record log. You may want to revise the m_frm_AfterUpdate() event procedure so that you can track whether a record was created or modified.

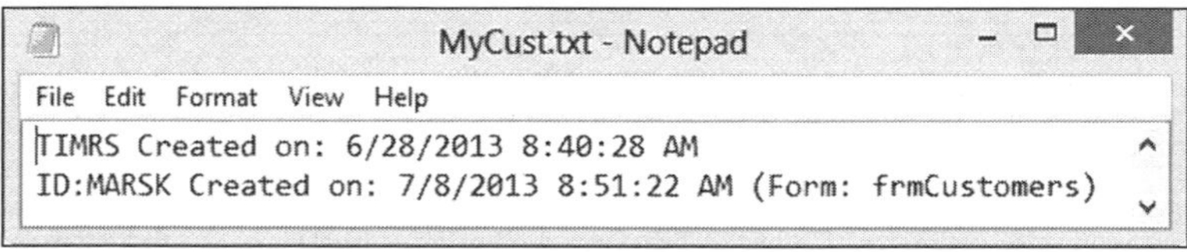

FIGURE 28.7. The MyCust.txt file is used by the cRecordLogger custom class for tracking record additions.

6. Close the **C:\Access2013_ByExample\MyCust.txt file.**
7. Add a few more records to the frmCustomers form and check out the **C:\Access2013_ByExample\MyCust.txt file.**
8. Close the frmCustomers form.

Now that you know how to sink the form's AfterUpdate event outside the form class module, you can use the same idea to sink other form events in a class module and make your code easier to implement and maintenance free. Just remember that if you want to sink events in a standalone class module, you must write code in two places: in your class module and in your form or report class module. The class module must contain a module-level `WithEvents` variable declaration, and you must set the reference to an actual instance of the object in the form or report module.

Part 5: Using the cRecordLogger Custom Class with another Form

The code you've written so far in this project is ready for reuse in another Microsoft Access form. In the remaining steps, we will hook it up to the frmProducts form. Let's begin by creating this form.

1. In the Navigation pane of the Access window, select the **Products** table and choose **Create | Form**. Access creates a form and displays it in the Form Layout view.

2. Press **Ctrl+S** or click the **Save** button on the toolbar to save the form. In the Save As dialog box, type **frmProducts** for the form name and click **OK.**
3. Activate the **frmProducts** form in Design view. In the property sheet, make sure **Form** is selected from the drop-down box and click the **Other** tab. Set the **Has Module** property of the form to **Yes**. Save the changes to the form by pressing **Ctrl+S.**
4. Switch to the Visual Basic Editor window, and double-click the **Form_frmProducts** object in the Project Explorer window. Access opens the Code module for the form.
5. Copy the code from the frmCustomers Code window to the frmProducts Code window. The code in the frmProducts Code window should match Figure 28.5 shown earlier.
6. In the frmProducts Code window, replace all the references to the object variable `clsRecordLogger` with **clsRecordLogger2.**

To quickly perform this operation, position the cursor inside the first `clsRecordLogger` *variable name and choose Edit | Replace. The Find What text box should automatically display the name of the variable you want to replace. Type clsRecordLogger2 in the Replace With text box and click the Replace All button. Click OK to confirm the replacement of four instances of variable names. Click Cancel to exit the Replace dialog box.*

7. Save and close the frmProducts form.
8. Open the **frmProducts** form in Form view and choose **Home | New.**
9. Enter **Delicious Raisins** in the Product Name text box and press the record selector on the left side of the form to save the record.

   At this point you should receive the custom message about the audit trail that you defined in the AfterUpdate event procedure within the custom cRecordLogger class module. This indicates that the AfterUpdate event that was raised by the form when you saved the newly entered record was successfully propagated to the custom class module.
10. Click **OK** to close the message box.
11. Close the frmProducts form.
12. Open the **C:\Access2013_ByExample\MyCust.txt** file to view the record log. Close this file when you are finished.

You may want to choose a more generic name for your record log text file if it will be used for tracking various types of information.

WRITING EVENT PROCEDURE CODE IN TWO PLACES

If you write event procedure code for the same event both in the form module and in the class module, the code defined in the form class module will run first, followed by the code in the custom class module. You can easily test this by entering the following Form_AfterUpdate event procedure code in the form class module of the frmCustomers or frmProducts forms prepared in Custom Project 28.1:

```
Private Sub Form_AfterUpdate()
  MsgBox "Transferring control to the custom class."
  ' when you click OK to this message, the code
  ' inside the AfterUpdate procedure in the custom
  ' class module will run
End Sub
```

When you open the form and add and save a new record, the Form_After-Update event will fire and you will see the message about transferring control to the custom class. Next, the AfterUpdate event procedure will run in the custom class, and you will see a message informing you that you can view the audit trail in the specified text file.

RESPONDING TO CONTROL EVENTS IN A CLASS

Everyone designing Microsoft Access forms sooner or later realizes that it takes a long time to customize some of the controls placed on the form. It's no wonder then that once the control is working correctly, there is a tendency toward copying the control and its event procedures to a new form that requires a control with the same functionality. If you followed this chapter carefully, you already know a better (and a neater) solution. By using the `WithEvents` keyword you can create an object variable that points to the control raising the events. Instead of responding to control events in the form module, you will react to these events in a different location: a standalone class module. This lets you write centralized code that is easy to implement in other form controls of the same type.

Suppose you need a text box that converts lowercase letters to uppercase and disallows numbers. Hands-On 28.1 demonstrates how to create a text box with these features and hook it up with any Microsoft Access form.

Hands-On 28.1. Responding to Control Events in a Class

This hands-on exercise requires prior completion of Custom Project 28.1.

1. Activate the Visual Basic Editor window and choose **Insert | Class Module**. A new class named Class1 will appear in the Project Explorer window.
2. In the Properties window, click the **(Name)** property and type **UCaseBox** as the new name of Class1. Click again on the **(Name)** property to save the new name. You should see the UCaseBox entry under the Class Modules folder in the Project Explorer.
3. In the UCaseBox class module's Code window, enter the following code:

```
Private WithEvents txtBox As Access.TextBox

Public Function InitializeMe(myTxt As TextBox)
  Set txtBox = myTxt
  txtBox.OnKeyPress = "[Event Procedure]"
End Function

Private Sub txtBox_KeyPress(KeyAscii As Integer)
  Select Case KeyAscii
    Case 48 To 57
      MsgBox "Numbers are not allowed!"
      KeyAscii = 0
    Case Else
      ' convert to uppercase
      KeyAscii = Asc(UCase(Chr(KeyAscii)))
  End Select

  txtBox.FontBold = True
  txtBox.FontItalic = True
  txtBox.BackColor = vbYellow
End Sub
```

Notice that to respond to a control's events in a class module you start by declaring a module-level object variable using the `WithEvents` keyword. In our text box example, we declared the object variable `txtBox` as an Access text box control.

Because the form can contain more than one text box control, we should tell the class which text box it needs to respond to. We do this by creating a Property Set procedure (similar to the one created in Custom Project 28.1) or a function procedure like the one shown here. We called this function `InitializeMe`, but you can use any name you wish. Recall from Chapter 8 that a function entered in a class module serves as an object's method. We will call the `InitializeMe` method later from a form class module and pass it the actual control we want it to respond to (see Step 7). The `InitializeMe` method will assign the passed control to the `WithEvents` object variable like this:

```
Set txtBox = myTxt
```

Next, we set the text box KeyPress property to [Event Procedure] to tell the class that we are interested in tracking this particular event.

Finally, we write the event procedure code for the text box control's KeyPress event. This code begins by checking the value of the key that was pressed by the user. If a number was entered, the user is advised that numbers aren't allowed and the digit is removed from the text box by setting the value of `KeyAscii` to zero (0). Otherwise, if the user typed a lowercase letter, the character is converted to uppercase using the following statement:

```
KeyAscii = Asc(UCase(Chr(KeyAscii)))
```

`KeyAscii` is an integer that returns a numerical ANSI keycode. To convert the `KeyAscii` argument into a character, we use the `Chr` function:

```
Chr(KeyAscii)
```

Once we've converted a key into a character, we use the `UCase` function to convert it to uppercase:

```
UCase(Chr(KeyAscii))
```

Finally, we translate the character back to an ANSI number by using the `Asc` function:

```
Asc(UCase(Chr(KeyAscii)))
```

The txtBox_KeyPress event procedure ends by adding some visual enhancements to the text box. The text entered in it will appear in bold italic type on a yellow background.

4. Save the code you entered in the UCaseBox class module's Code window by pressing the **Save** button on the toolbar.
5. In the Project Explorer window, double-click the **Form_frmProducts**. The Form_frmProducts class module's Code window should already contain code you entered while working with Part 3 of Custom Project 28.1 earlier in this chapter. To connect the UCaseBox class module with the actual text box on any Access form, you would need to enter the following code in a form's class module (do not enter it yet):

```
' module-level variable declaration
Private clsTextBox1 As UCaseBox
Private Sub Form_Open(Cancel As Integer)
  Set clsTextBox1 = New UCaseBox
  clsTextBox1.InitializeMe Me.Controls("ProductName")
End Sub
Private Sub Form_Close()
  Set clsTextBox1 = Nothing
End Sub
```

Since the frmProducts form already contains a call to the cRecordLogger class created earlier, all of the procedures we need are already in place; therefore, we will simply add the appropriate lines of code to the existing procedures. So let's proceed.

6. In the Form_frmProducts Code window, enter the following module-level variable declaration just above the Form_Open event procedure (see Figure 28.8):

```
Private clsTextBox1 As uCaseBox
```

This statement declares the `clsTextBox1` class variable. This variable is used in instantiating the UCaseBox object and connecting it with the actual text box control on the form (see the next step).

7. Enter the following lines of code before the `End Sub` statement of the Form_Open event procedure (see Figure 28.8):

```
Set clsTextBox1 = New uCaseBox
clsTextBox1.InitializeMe Me.Controls("ProductName")
```

Before our UCaseBox class can respond to a text box's events, you need these two lines of code; the first one sets the class variable `clsTxtBox1` to a new instance of the UCaseBox class, and the second one calls the class `InitializeMe` method and supplies it with the name of the text box control.
Enter the following line of code before the `End Sub` statement of the Form_Close event procedure (see Figure 28.8):

```
Set clsTextBox1 = Nothing
```

When we are done with the object variable, we set it to `Nothing` to release the resources that have been assigned to it.

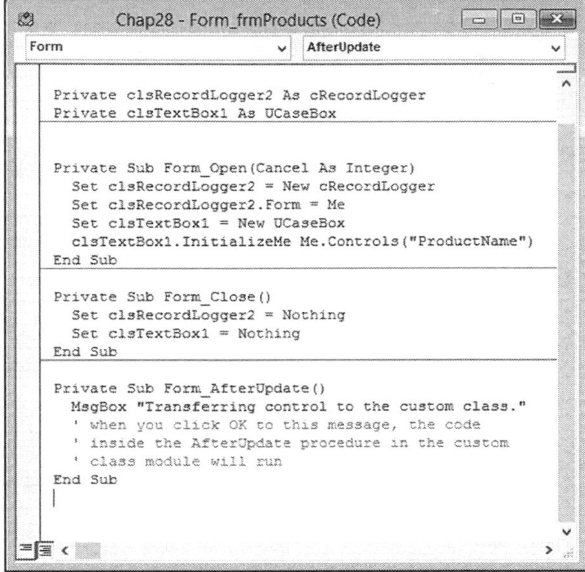
```
Private clsRecordLogger2 As cRecordLogger
Private clsTextBox1 As UCaseBox

Private Sub Form_Open(Cancel As Integer)
  Set clsRecordLogger2 = New cRecordLogger
  Set clsRecordLogger2.Form = Me
  Set clsTextBox1 = New UCaseBox
  clsTextBox1.InitializeMe Me.Controls("ProductName")
End Sub

Private Sub Form_Close()
  Set clsRecordLogger2 = Nothing
  Set clsTextBox1 = Nothing
End Sub

Private Sub Form_AfterUpdate()
  MsgBox "Transferring control to the custom class."
  ' when you click OK to this message, the code
  ' inside the AfterUpdate procedure in the custom
  ' class module will run
End Sub
```

FIGURE 28.8. The form class module shows code that instantiates and hooks up objects created in the cRecordLogger and UCaseBox class modules with the form and text box control.

8. Save the changes made in the Code window by clicking the **Save** button on the toolbar.
9. Open the **frmProducts** form in Form view and choose **Home | New**.
10. Enter **prune butter** in the Product Name text box. Notice that as you type, the characters you enter are converted to uppercase. They are also made bold and italic, and appear on a yellow background. If you happen to press a number key, which is disallowed by your custom KeyPress event, you receive an error message. Click on the record selector to save the record. Because this form also responds to the AfterUpdate event that we programmed in Custom Project 28.1, you should see two message boxes when you save this form.
11. Close the frmProducts form.

DECLARING AND RAISING EVENTS

Standalone class modules automatically support two events: Initialize and Terminate. Use the Initialize event to give the variables in your classes initial values. The Initialize event is called when you make a new instance of a class. The Terminate event is called when you set the instance to `Nothing`. In addition to these default events, you can define custom events for your class module.

To create a custom event, use the `Event` statement in the declaration section of a class module. For example, the following statement declares an event named SendFlowers that requires two arguments:

```
Public Event SendFlowers(ByVal strName As String, cancel As Boolean)
```

The `Event` statement declares a user-defined event. This statement is followed by the name of the event and any arguments that will be passed to the event procedure. Arguments are separated by commas. An event can have `ByVal` and `ByRef` arguments. Recall that when passing the variable `ByRef`, you are actually passing the memory location of the variable. If you pass a variable `ByVal`, you are sending a copy of the variable.

When declaring events with arguments, bear in mind that events cannot have named arguments, optional arguments, or `ParamArray` arguments. The `Public` keyword is optional as events are public by default.

Use the `RaiseEvent` statement to fire the event. This is usually done by creating a method in a class module. For example, here's how you could trigger the SendFlowers event:

```
Public Sub Dispatch(ByVal toWhom As String, cancel As Boolean)
   RaiseEvent SendFlowers(toWhom, True)
End Sub
```

Events can only be raised in the module in which they are declared using the `Event` statement. After declaring the event and writing the method that will be used for raising the event, you need to switch to the form class module and perform the following tasks:

- Declare a module-level variable of the class type using the `WithEvents` keyword
- Assign an instance of the class containing the event to the object defined using the `WithEvents` statement
- Write a procedure that calls the class method
- Write the event handler code

The next hands-on exercise demonstrates how a user-defined event can be used in a class. We will learn how to raise the SendFlowers event from a Microsoft Access form.

Hands-On 28.2. Declaring and Raising Events

1. Activate the Visual Basic Editor window and choose **Insert | Class Module**. A new class named Class1 will appear in the Project Explorer window.
2. In the Properties window, click the **(Name)** property and type **cDispatch** as the new name of Class1. Click again on the **(Name)** property to save the new name. You should see the cDispatch entry under the Class Modules folder in the Project Explorer.
3. In the cDispatch class module's Code window, enter the following code:

```
Public Event SendFlowers(ByVal strName As String, _
  cancel As Boolean)

Sub Dispatch(ByVal ToWhom As String, cancel As Boolean)
  If ToWhom = "Julitta" Then
    cancel = True
    MsgBox "Dispatch to " & ToWhom & " was cancelled.", _
     vbInformation + vbOKOnly, "Reason Unknown"
  Else
    RaiseEvent SendFlowers(ToWhom, True)
  End If
End Sub
```

   The first statement in the preceding code declares a custom event called SendFlowers. This event will accept two arguments: the name of the person to whom flowers should be sent and a Boolean value of `True` or `False` that will allow you to cancel the event if necessary.

   Next, the Dispatch procedure is used as a class method. The code states that the flowers should be sent to the person whose name is passed in the `ToWhom` argument as long as the person's name is not "Julitta." The `RaiseEvent` statement will call the event handler that we will write in a form module in a later step.

4. Create a new form as shown in Figure 28.9. This form isn't bound to any data source. Use the property sheet to set the **Name** property of the text box control to **Recipient** and the **Caption** property of the accompanying label control to **Recipient Name:**. Set the `Name` property of the command button to **cmdFlowers** and its `Caption` property to **Send Flowers**. Save this form as **frmFlowers**.

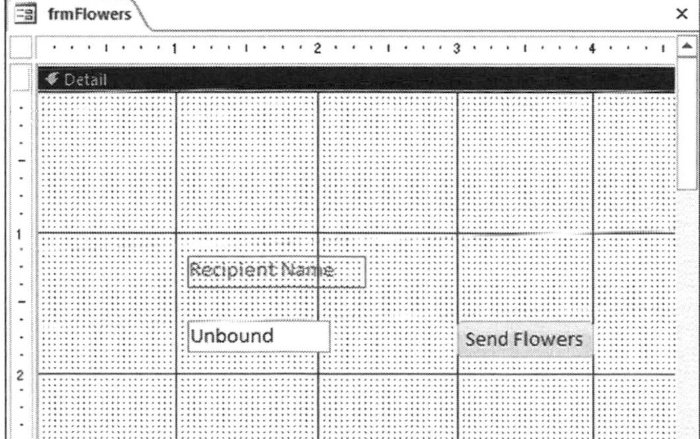

FIGURE 28.9. The frmFlowers form is used in Hands-On 28.2 to demonstrate the process of raising and handling custom events.

5. While the frmFlowers form is displayed in Design view, click the **View Code** button in the Tools area of the Design tab.
6. Enter the following code in the **Form_frmFlowers** Code window:

```
Private WithEvents clsDispatch As cDispatch
Private Sub Form_Load()
  Set clsDispatch = New cDispatch
End Sub

Private Sub Form_Close()
  Set clsDispatch = Nothing
End Sub
```

Our form class can respond to events from an object only if it has a reference to that object. Therefore, at the top of the form class module we declare the object variable `clsDispatch` by using the `WithEvents` keyword. This means that from now on the instance of the cDispatch class is associated with events.
The next step involves setting the object variable to an object. In the Form_Load event procedure, we create a class object with the `Set` statement and the `New` keyword. When the object variable is no longer needed, we release the reference to the object by setting the object variable to `Nothing` (see the preceding Form_Close event procedure).

Now that we are done with declaring, setting, and resetting the object variable, let's proceed to write some code that will allow us to raise the SendFlowers event when we click on the Send Flowers button.

7. In the Form_frmFlowers Code window, enter the following **Click** event procedure for the cmdFlowers command button that you placed on the frmFlowers form:

```
Private Sub cmdFlowers_Click()
  If Len(Me.Recipient) > 0 Then
    clsDispatch.Dispatch Me.Recipient, False
  Else
    MsgBox "Please specify the recipient name."
    Me.Recipient.SetFocus
    Exit Sub
  End If
End Sub
```

   Notice that this event procedure begins by checking whether the user has entered data in the Recipient text box. If the data exists, the `Dispatch` method is called; otherwise, the user is asked to enter data in the text box. When calling the `Dispatch` method we must provide two arguments that this method expects: the name of the recipient and the value for the Boolean variable `Cancel`. Recall that the `Dispatch` method has the necessary code that raises the SendFlowers event (see Step 3). Now what's left to do is to write an event handler for the SendFlowers event.

8. Select the **clsDispatch** variable from the Object drop-down list in the upper-left corner of the Form_frmFlowers Code window. As you make this selection, a template of the event procedure will be inserted into the Code window as shown here:

```
Private Sub clsDispatch_SendFlowers(ByVal strName As String, _
 cancel As Boolean)
End Sub
```

   The code that you write within this procedure stub will be executed when the event is generated by the object.

9. Enter the following statement inside the clsDispatch_SendFlowers procedure stub:

```
MsgBox "Flowers will be sent to " & strName & ".", , _
 "Order taken"
```

   Our custom event is not overly exciting but should give you an understanding of how custom events are declared and raised in a standalone class module and how they are consumed in a client application (form class module). The complete code entered in the form class module is shown in Figure 28.10.

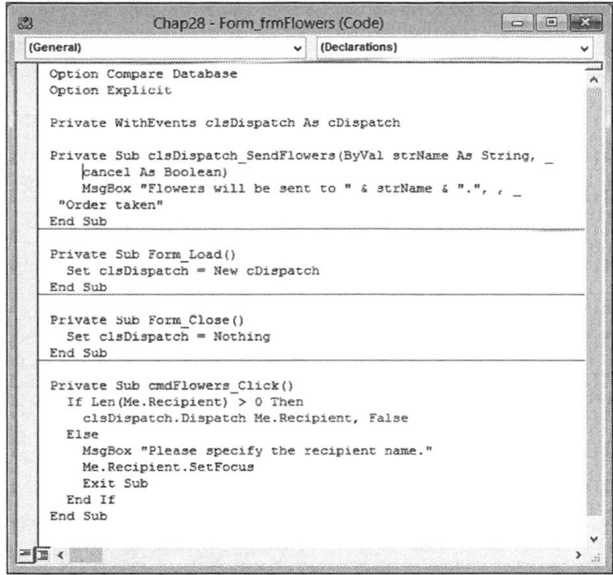

Chap28 - Form_frmFlowers (Code)
(General)
(Declarations)
Option Compare Database
Option Explicit
Private WithEvents clsDispatch As cDispatch
Private Sub clsDispatch_SendFlowers(ByVal strName As String, _
cancel As Boolean)
MsgBox "Flowers will be sent to " & strName & ".", , _
"Order taken"
End Sub
Private Sub Form_Load()
Set clsDispatch = New cDispatch
End Sub
Private Sub Form_Close()
Set clsDispatch = Nothing
End Sub
Private Sub cmdFlowers_Click()
If Len(Me.Recipient) > 0 Then
clsDispatch.Dispatch Me.Recipient, False
Else
MsgBox "Please specify the recipient name."
Me.Recipient.SetFocus
Exit Sub
End If
End Sub

FIGURE 28.10. The form class module shows code that uses a custom object with its events.

10. Save the changes you made in the Code window by clicking the **Save** button on the toolbar.
11. To test the code, open the **frmFlowers** form in Form view, type any name in the Recipient text box, and click the **Send Flowers** button. You should see the message generated by the SendFlowers custom event. Also see what happens when you type Julitta in the text box.

Be sure to try out the example provided in the online help for the `RaiseEvent` *statement topic. The quickest way to find this example is by positioning the cursor in the* `RaiseEvent` *statement (located in the cDispatch class module) and pressing F1.*

CHAPTER SUMMARY

In this chapter, you were introduced to advanced concepts in event-driven programming. You learned how you can make your code more manageable and portable to other objects by responding to events in class modules other than form modules. This chapter has also shown you the process of creating your own events for a class and raising them from a public method by calling the `RaiseEvent` statement with the arguments defined for the event. The important thing to understand is that while events happen all the time whether or not you respond to them, you are the one to decide where to respond to the built-in events. And, if you ever find yourself short of an event, you can always create one that does exactly what you need by using the knowledge acquired in this chapter.

In the next chapter, you learn how to use VBA, macros, and XML to customize the user interface in Access 2013.

Chapter 29 Programming the User Interface

If you have used two previous versions of Microsoft Access, you are already familiar with the Fluent Ribbon user interface that replaced the menus and toolbars in earlier versions of Access. If you are new to Access, there is nothing you need to unlearn to take full advantage of this chapter. This chapter provides an overview of the programing elements available in the Ribbon and shows how you can customize the user interface (UI) in your Access database applications.

THE INITIAL MICROSOFT ACCESS 2013 WINDOW

When you launch Access, you are presented with several Backstage View buttons (Custom web app and Blank Database plus a selection of prebuilt templates) so that you can get a head start on your Access database project (Figure 29.1). The templates come with ready-to-use tables, forms, reports, queries, relationships, and macros that can be modified as needed. When you click on the type of template you want to create, Access displays a dialog box that prompts you for the file name and the location of your new database along with a button to create the file.

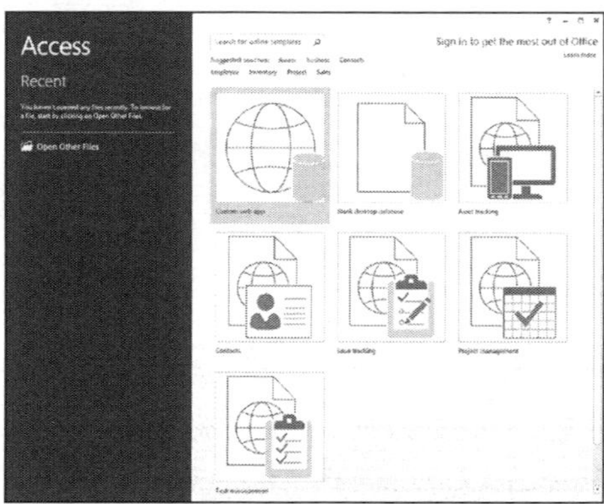

FIGURE 29.1. The Access 2013 startup screen Backstage View.

The Quick Access Toolbar

Once you open an existing Access database or create a new one, you will notice above the File tab is a special tool area known as the Quick Access toolbar where you can quickly access the most frequently used commands. The Quick Access toolbar will expand to accommodate as many commands as you wish to add. You can add other Access built-in commands to the Quick Access toolbar by using any of the following methods:

- Click on the drop-down arrow in the Quick Access toolbar and select a command you want to add or click More Commands.
- Right-click on the Quick Access toolbar and choose Customize Quick Access Toolbar.
- Click File | Options and choose Quick Access Toolbar.

Together with the titlebar and the tabs, the Quick Access toolbar belongs to a large rectangular area called the Ribbon. This area is positioned at the top of the UI window. The Quick Access toolbar was designed for the convenience of end users. Developers should not alter this toolbar. However, if you have a valid reason to hide the contents of this toolbar or add other buttons to it, you can apply your own customizations as described later in this chapter.

CUSTOMIZING THE NAVIGATION PANE

When an Access database is open, you can easily access all of your objects via the Navigation pane on the left side of the window (see Figure 29.2). If you need more screen real estate, you can hide the Navigation pane by clicking on the Shutter Bar Open/Close button (the double arrow at the top of the pane) or by pressing F11.

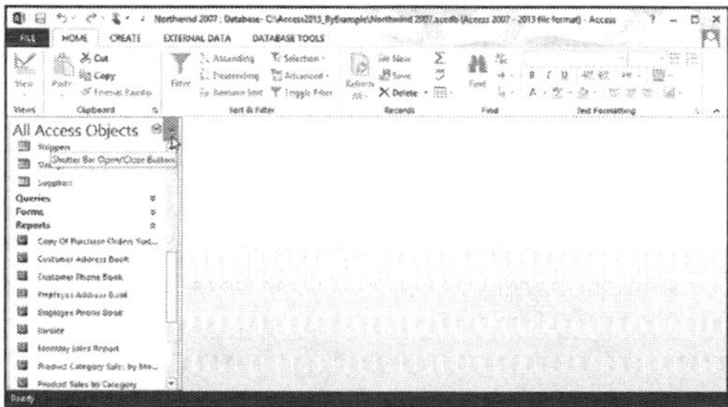

FIGURE 29.2. The Navigation pane in Access 2013.

Use the Navigation pane to organize your objects by type, date created or modified, or related table, or create your own custom groups of objects. By clicking on the down arrow button at the top of the Navigation pane, you can define how you view and manage database objects (see Figure 29.3).

To sort and filter your database objects, activate the Search Bar, or access the navigation options, right-click the top bar of the Navigation pane (Figure 29.4).

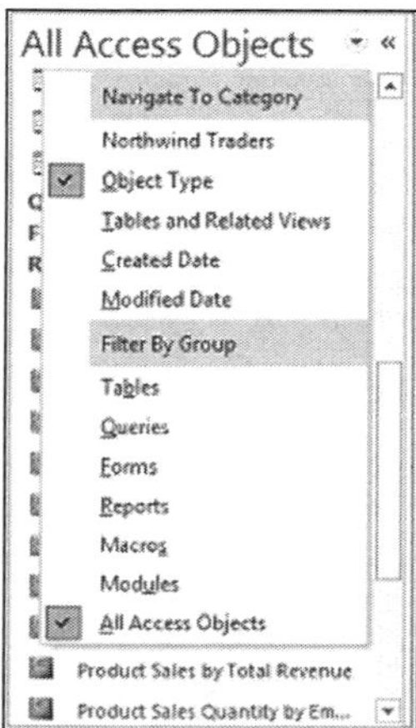

FIGURE 29.3. Grouping options in the Navigation pane.

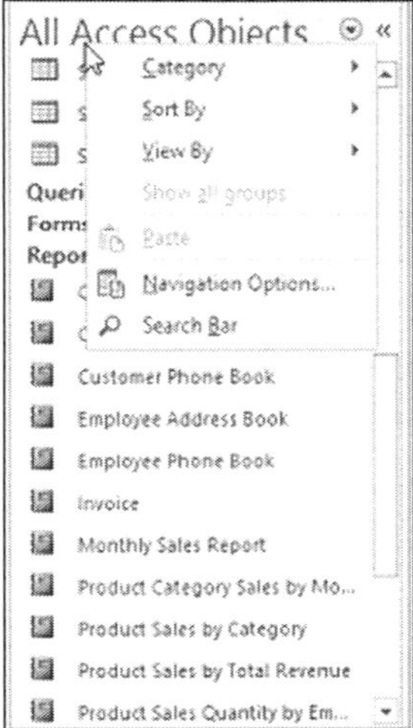

FIGURE 29.4. Objects in the Navigation pane can be easily categorized, sorted, and filtered. Use the Search Bar to locate a hard-to-find object. Use the Navigation Options tool to create custom groups of objects.

The Navigation Options dialog box (see Figure 29.5) allows you to create any number of custom groups for organizing your objects according to specific database needs. The exercise in Hands-On 29.1 will walk you through the process of creating a custom group to track your development efforts.

Please note files for the "Hands-On" project may be found on the companion CD-ROM.

Hands-On 29.1. Adding a Custom Group to the Navigation Pane

1. Open the **C:\Access2013_ByExample\Northwind2007.accdb** database. Cancel the login box.
2. Right-click the top bar of the Navigation pane and choose **Navigation Options**.
3. Click the **Add Item** button and type a new name for the category: **Objects in Development**.
4. While the new Objects in Development category is selected, click the **Add Group** button and enter **Dev Tables** for the new group name.
5. Click the **Add Group** button again to add another group named **Dev Queries**.
6. Add two more groups under Objects in Development named **Dev Forms** and **Dev Reports**. See Figure 29.5 for the final output.
7. Click **OK** to close the Navigation Options dialog box.

8. Click on the down arrow button at the top of the Navigation pane and choose **Objects in Development** (Figure 29.6).

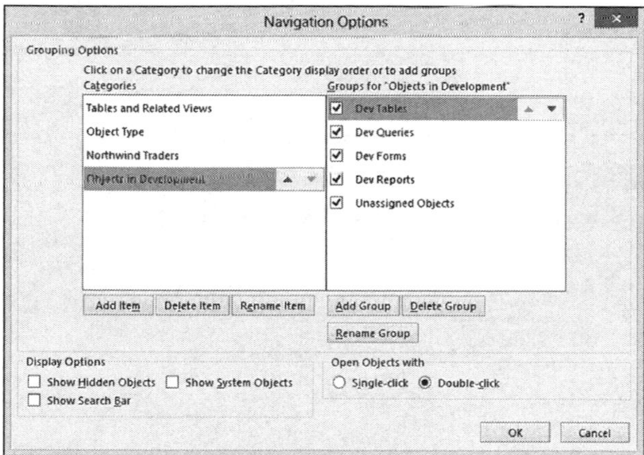

FIGURE 29.5. The Navigation Options dialog box.

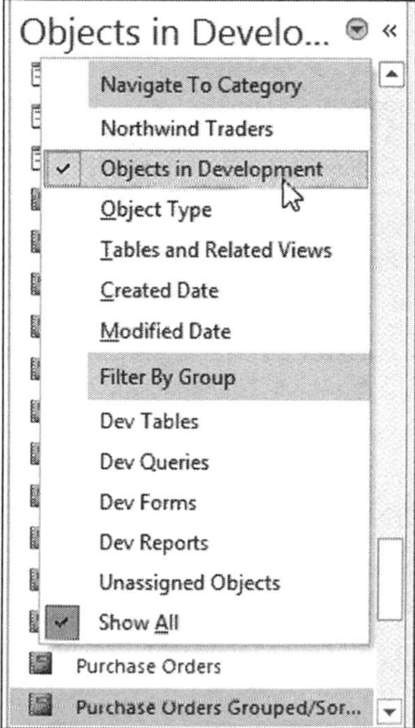

FIGURE 29.6. Displaying a custom group in the Navigation pane.

The next logical step is placing some database objects into your custom groups. Hands-On 29.2 requires prior completion of Hands-On 29.1.

Hands-On 29.2. Assigning Objects to Custom Groups in the Navigation Pane

1. In the Northwind 2007.accdb database you opened in Hands-On 29.1, right-click the **Customers** table and choose **Copy**.
2. Right-click anywhere in the Navigation pane and choose **Paste**. In the Paste Table As dialog box, enter **Companies** for the new name of the table and select the **Structure Only** option button. Click **OK** to exit the dialog box.
3. Drag the **Companies** table from the Unassigned Objects group to the **Dev Tables** group.

   When you place a database object into a custom group in the Navigation pane, Access creates a shortcut to this object (see Figure 29.7). You can rename your shortcut by right-clicking its name and choosing Rename Shortcut. You can also hide the shortcut in the group or remove it from the group, provided the Navigation pane has not been locked (this is discussed in the next section). In addition to the Hidden attribute, each shortcut has a Disable Design View shortcuts attribute you can set to prevent users from switching to Design view when using the shortcut. You must restart your Access database for this property change to take effect. To display the shortcut properties as shown in Figure 29.7, right-click on the Companies shortcut under the Dev Tables group and choose Table Properties. You can drag any object listed in the Unassigned Objects group into your custom groups.

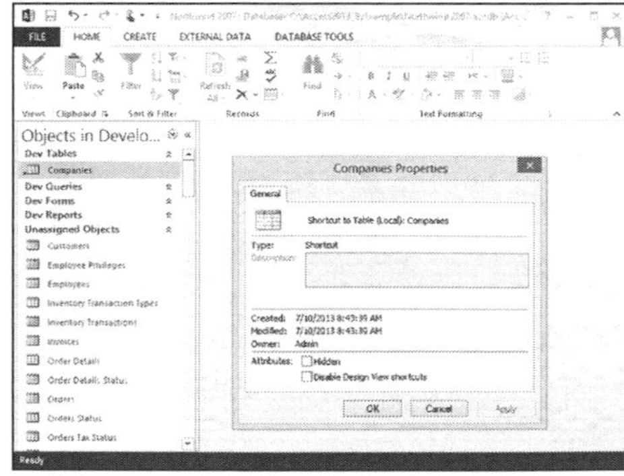

FIGURE 29.7. Navigation pane with custom groupings.

USING VBA TO CUSTOMIZE THE NAVIGATION PANE

You can lock down and customize the Navigation pane programmatically by using the following methods of the DoCmd object: `LockNavigationPane`, `NavigateTo`, and `SetDisplayedCategories`. There are also two methods of the Application object (`ExportNavigationPane` and `ImportNavigationPane`) that enable you to quickly apply the same Navigation pane customizations to any other Access database.

Locking the Navigation Pane

To prevent users from deleting database objects that are displayed in the Navigation pane, use the following statement:

```
DoCmd.LockNavigationPane True
```

The `LockNavigationPane` method of the DoCmd object requires a `Lock` argument. Use the Boolean value of `True` to lock the Navigation pane and `False` to unlock it.

Controlling the Display of Database Objects

To automatically navigate to a specific category in the Navigation pane upon startup of your Access database or to display only certain objects in the category, use the `NavigateTo` method of the DoCmd object. This method takes two arguments: `Category` (required) and `Group` (optional). The `Category` argument specifies the category you want to navigate to. This argument can be the name of your custom category, such as the Objects in Development category you created in Hands-On 29.1, or a constant representation of the Object Type, Tables and Views, Created Date, and Modified Date categories. The `Group` argument is optional. If you omit it, the Navigation pane will display all database objects arranged by the criteria specified in the `Category` argument. See Table 29.1 for valid `Group` arguments for the various `Category` arguments.

TABLE 29.1. Category and Group arguments used in the NavigateTo method

Category Argument	Category Argument Constant	Group Argument	Group Argument Constant
Object Type	acNavigationCategoryObjectType	Tables Forms Reports Queries Pages Macros Modules	acNavigationGroupTables acNavigationGroupForms acNavigationGroupReports acNavigationGroupQueries acNavigationGroupPages acNavigationGroupMacros acNavigationGroupModules
Tables and Views	acNavigationCategoryTablesAndViews	Name of a specific table or view in your database.	

(Contd.)

Category Argument	Category Argument Constant	Group Argument	Group Argument Constant
Modified Date	acNavigationCategory-ModifiedDate	Today Yesterday Last Month Older	acNavigationGroupToday acNavigationGroupYesterday acNavigationGroupLastMonth acNavigationGroupOlder
Created Date	acNavigationCategoryCre-atedDate	Today Yesterday Last Month Older	acNavigationGroupToday acNavigationGroupYesterday acNavigationGroupLastMonth acNavigationGroupOlder
Custom	Name of your custom category.	Name of one of the custom groups you have created for the specified custom category.	

Let's get some practice with the `NavigateTo` method.

Hands-On 29.3. Using the NavigateTo Method to Control the Display of Database Objects in the Navigation Pane

1. In the Northwind 2007.accdb database, press **Alt+F11** to switch to the Visual Basic Editor window. Press **Ctrl+G** (or choose **View | Immediate Window**) to open the Immediate window.
2. In the Immediate window, type each of the following `DoCmd.NavigateTo` statement examples, pressing **Enter** to execute each statement. After the execution of each statement, check the resulting display in the Navigation pane of the Access main window (see Figure 29.8).

```
DoCmd.NavigateTo "acNavigationCategoryCreatedDate"
```

This statement will navigate to the Created Date category and display all database objects.

```
DoCmd.NavigateTo "acNavigationCategoryObjectType",
"acNavigationGroupForms"
```

This statement will navigate to the Object Type category and select the Forms group.

```
DoCmd.NavigateTo "acNavigationCategoryTablesAndViews",
"Invoices"
```

This statement will navigate to the Invoices table in the Tables and Views category.

```
DoCmd.NavigateTo "Objects in Development", "Dev Forms"
```

This statement will navigate to the Dev Forms group objects in the Objects in Development category created in Hands-On 29.1.

```
DoCmd.NavigateTo "acNavigationCategoryModifiedDate",
"acNavigationGroupOlder"
```

This statement will navigate to the Modified Date category and display all the database objects beginning with a date earlier than the previous month.

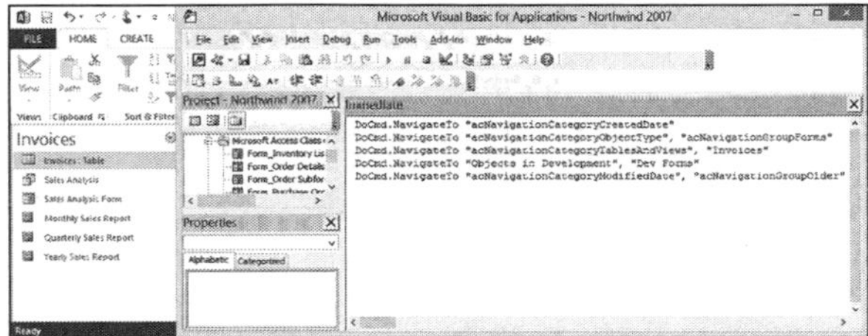

FIGURE 29.8. Testing out the NavigateTo method of the DoCmd object in the Immediate window. Notice how the main Access window and the Microsoft Visual Basic Editor window are positioned so that the Navigation pane stays in view as you execute each statement.

3. In the VBE window, press **Ctrl+G**, **Ctrl+A**, then the **Delete** key to remove the contents of the Immediate window.

Setting Displayed Categories

The `SetDisplayedCategories` method of the DoCmd object is used to specify which categories should be displayed under Navigate To Category in the titlebar of the Navigation pane. Use this method to show and hide groups from the top bar of the Navigation pane. For example, the following statement will remove the custom category Objects in Development from the Navigation pane titlebar's drop-down list:

```
DoCmd.SetDisplayedCategories False, "Objects in Development"
```

Notice that the `SetDisplayedCategories` method uses two arguments. The first argument specifies whether to show or hide the category. Use the Boolean value of `False` to hide the category specified in the second argument of this method or `True` to show the category. The second argument denotes the name of the category you want to show or hide. Do not specify this argument if you want to show or hide all categories.

Saving and Loading the Configuration of the Navigation Pane

The configuration of the Navigation pane can be saved at any time with the `ExportNavigationPane` method of the Application object. This method requires one argument—the path and the name of the XML file where you want to save the configuration of the Navigation pane. For example, the following statement will save the current configuration of the Navigation pane to North2007NavConfig.xml in the C:\Access2013_ByExample folder:

```
Application.ExportNavigationPane
"C:\Access2013_ByExample\North2007NavConfig.xml"
```

To load a saved Navigation pane configuration from the XML file, use the `ImportNavigationPane` method of the Application object:

```
Application.ImportNavigationPane
"C:\Access2013_ByExample\North2007NavConfig.xml", False
```

Notice that the `ImportNavigationPane` method used in the previous statement has two arguments. The first one specifies the path and name of the XML file that contains the Navigation pane configuration to load. The second argument is optional. When set to `True`, the imported categories will be appended to the existing categories. The default value is `False`.

Hands-On 29.4 demonstrates how to save the current configuration of the Navigation pane and then load it into another Access database.

Hands-On 29.4. Saving and Loading the Configuration of the Navigation Pane

1. In the VBE window of the Northwind 2007.accdb database, type the following statement on one line in the Immediate window and press **Enter** to execute:

   ```
   Application.ExportNavigationPane
   "C:\Access2013_ByExample\North2007NavConfig.xml"
   ```

2. Switch to Windows Explorer and check that the North2007NavConfig.xml file is in the Access2013_ByExample folder.
3. Double-click the filename to open it in the browser. Figure 29.9 displays the partial content of the configuration file.

   The XML file contains the objects and structure of the Access Navigation pane. This file includes information about the contents of the Navigation pane system tables: MSysNavPaneGroupCategories, MSysNavPaneGroups, MSysNavPaneGroupToObjects, and MSysNavPaneObjectIDs.

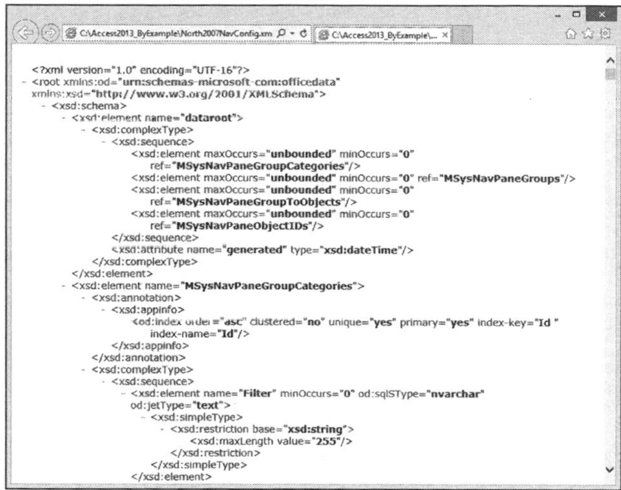

```
<?xml version="1.0" encoding="UTF-16"?>
- <root xmlns:od="urn:schemas-microsoft-com:officedata"
  xmlns:xsd="http://www.w3.org/2001/XMLSchema">
  - <xsd:schema>
    - <xsd:element name="dataroot">
      - <xsd:complexType>
        - <xsd:sequence>
            <xsd:element maxOccurs="unbounded" minOccurs="0"
              ref="MSysNavPaneGroupCategories"/>
            <xsd:element maxOccurs="unbounded" minOccurs="0" ref="MSysNavPaneGroups"/>
            <xsd:element maxOccurs="unbounded" minOccurs="0"
              ref="MSysNavPaneGroupToObjects"/>
            <xsd:element maxOccurs="unbounded" minOccurs="0"
              ref="MSysNavPaneObjectIDs"/>
          </xsd:sequence>
          <xsd:attribute name="generated" type="xsd:dateTime"/>
        </xsd:complexType>
      </xsd:element>
    - <xsd:element name="MSysNavPaneGroupCategories">
      - <xsd:annotation>
        - <xsd:appinfo>
            <od:index order="asc" clustered="no" unique="yes" primary="yes" index-key="Id "
              index-name="Id"/>
          </xsd:appinfo>
        </xsd:annotation>
      - <xsd:complexType>
        - <xsd:sequence>
          - <xsd:element name="Filter" minOccurs="0" od:sqlSType="nvarchar"
            od:jetType="text">
            - <xsd:simpleType>
              - <xsd:restriction base="xsd:string">
                  <xsd:maxLength value="255"/>
                </xsd:restriction>
              </xsd:simpleType>
            </xsd:element>
```

FIGURE 29.9. The current configuration of the Navigation pane is saved in this XML file.

4. Close the Explorer window.
5. Close the Northwind 2007.accdb database.
6. Create a new Access database named **Load_North2007NavConfig.accdb** in your **C:\Access2013_ByExample** folder.
7. Click the top bar of the Navigation pane and view the Navigation pane titlebar's drop-down list before proceeding to import the saved configuration file.
8. Press **Alt+F11**, then press **Ctrl+G** to activate the Immediate window. Type the following statement on one line and press **Enter** to execute:

```
Application.ImportNavigationPane "C:\Access2013_ByExample\
North2007NavConfig.xml", False
```

9. Press **Alt+F11** to switch back to the Access application window.
10. Click the top bar of the Navigation pane and display the Navigation pane titlebar's drop-down list again (see Figure 29.10).

Notice the additional entries in the drop-down list: Northwind Traders and Objects in Development.

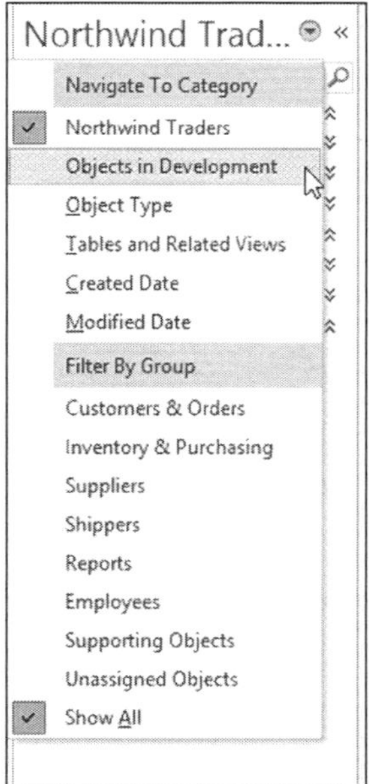

FIGURE 29.10. The Navigation pane can be easily modified using the external XML file containing the custom configuration settings.

Now that you know how to manually and programmatically control the Navigation pane, you should find it easy to provide users with the needed customization of the Access database navigation system. The next section will expand your knowledge of the Access user interface by giving you a quick overview of the Ribbon.

A QUICK OVERVIEW OF THE ACCESS 2013 RIBBON INTERFACE

All Microsoft Access program commands can be accessed from the Ribbon (see Figure 29.11). Beginning with the release of Access 2007, the Ribbon replaced the system of menus and toolbars found in earlier versions of Access. The Ribbon contains the titlebar, the File tab (for access to the Backstage View), the Quick Access toolbar, and a number of other tabs listing various commands. Each tab on the Ribbon provides access to features and commands related to a particular database task. For example, you can use the Create tab to quickly create new tables, forms, reports, queries, macros, modules, and Microsoft Windows SharePoint Services lists (see Figure 29.12). Related commands within a tab are organized into groups. For example, the Create tab divides its commands into six groups: Templates, Tables, Queries, Forms, Reports, and Macros & Code. This type of organization makes it easy to locate a particular command.

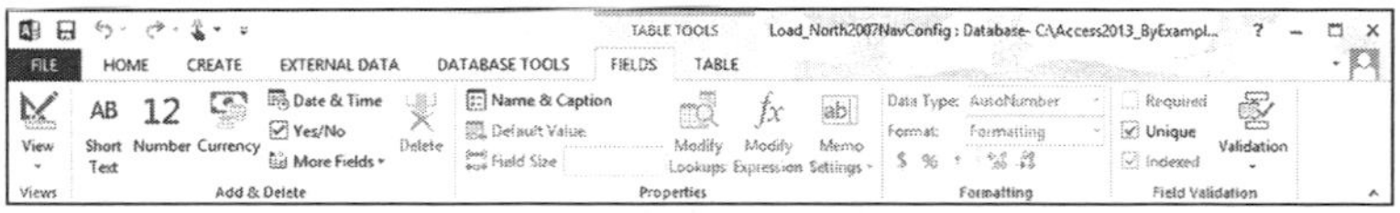

FIGURE 29.11. The rectangular area at the top of the Access 2013 window is called the Ribbon.

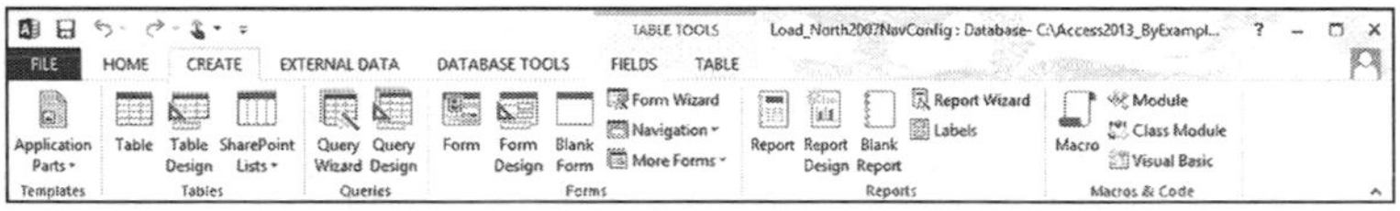

FIGURE 29.12. All Access commands related to creating various database objects are grouped on the Create tab.

Various program commands are displayed as large or small buttons. Large buttons denote frequently used commands, while small buttons show specific features of the main commands. For example, in the Forms group there is a large Form button and a small Form Wizard button. Some large and small command buttons include drop-down lists of other specialized commands. For example, the small More Forms button drop-down contains additional methods for creating a form: Multiple Items, Datasheet, Split Form, and Modal Dialog (Figure 29.13).

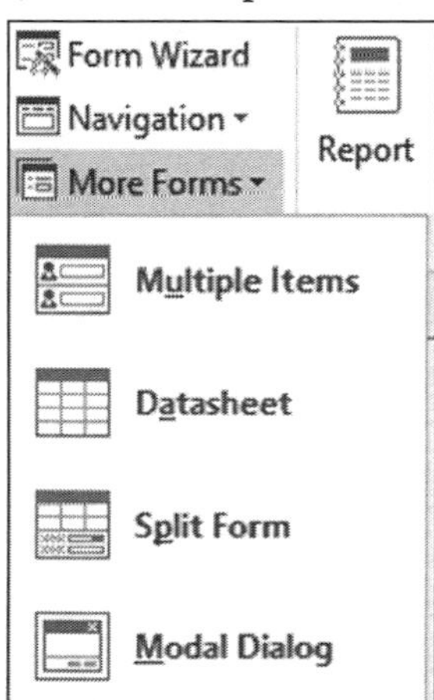

FIGURE 29.13. Additional commands can be accessed by clicking on the down arrow to the right of the button control.

Some controls that you find on the Ribbon do not display commands. Instead, they provide a visual clue of the output you might expect when a specific option is selected. These types of controls are known as *galleries*. Gallery controls are often used to present various formatting options, such as the margin settings shown in Figure 29.14.

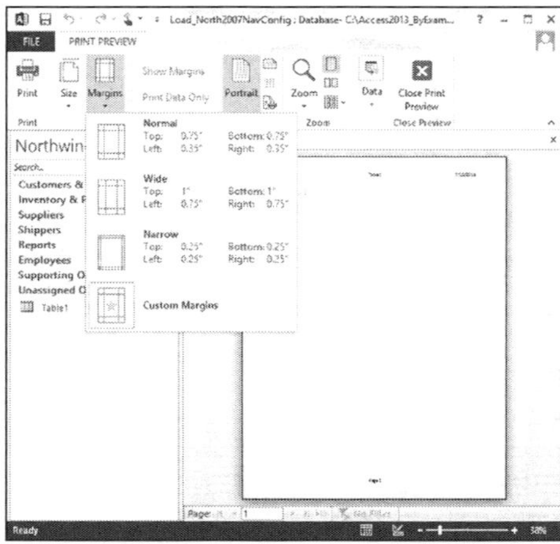

FIGURE 29.14. Clicking on the Margins button on the Print Preview tab displays a gallery of different margin settings.

Some tab groups have dialog box launchers in the bottom-right corner (see Figure 29.15) that display a dialog box in which you can set several advanced options at once.

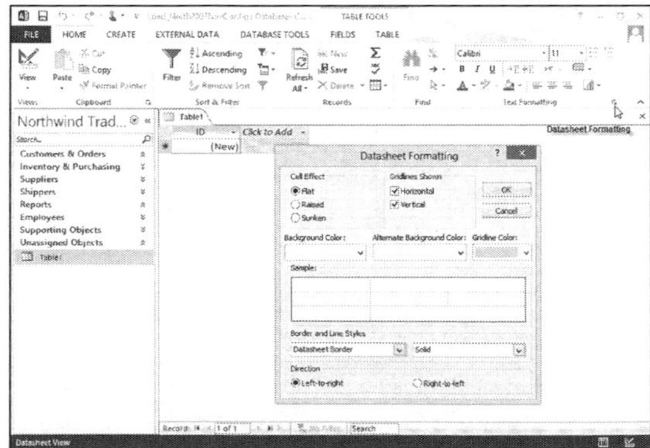

FIGURE 29.15. The dialog box launcher button in the bottom-right corner of the Text Formatting group on the Home tab will display the Datasheet Formatting dialog box.

In addition to the main Ribbon tabs, there are also contextual tabs that contain commands that apply to what you are doing. When a particular object is selected, the Ribbon displays a contextual tab that provides commands for working with that object. For example, when a table is open in Datasheet view, the Ribbon displays a contextual tab called Table Tools. Clicking on the Table Tools tab activates the Fields tab (see Figure 29.16). The contextual tab disappears when you cancel the selection of the object. In other words, close the datasheet and the Table Tools tab will be gone.

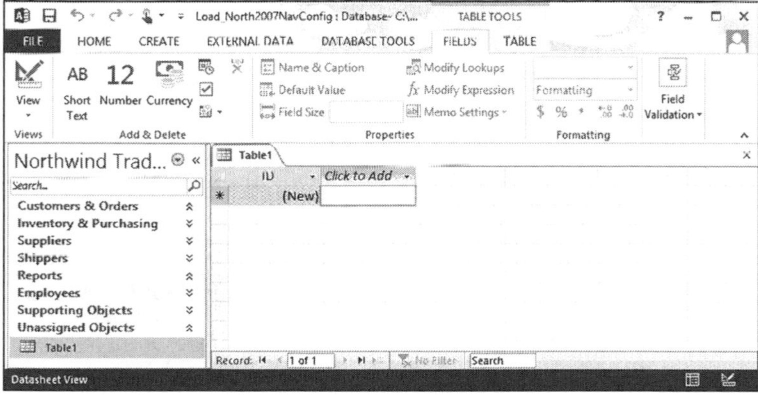

FIGURE 29.16. A contextual tab (Table Tools) in the Ribbon.

Now that you've reviewed the main features of the Ribbon interface, let's look at how you can extend it with your own tabs and controls. The next section introduces you to Ribbon programming.

RIBBON PROGRAMMING WITH XML, VBA, AND MACROS

The components of the Ribbon user interface can be manipulated programmatically using Extensible Markup Language (XML) or other programming languages. Refer to Chapter 32 ("XML Features in Access 2013") for an introduction to using XML with Access.

All Office 2013 applications use the Ribbon and rely on the programming model known as Ribbon extensibility, or *RibbonX*.

This section introduces you to customizing Access 2013 Ribbons by using XML markup. No special tools are required to perform these customizations. XML is plain text; therefore, you can use any text editor to create your customization files. In the examples that follow, we'll be using the simple Windows Notepad.

Your Ribbon customizations can be stored in a special Access table, in a VBA procedure, or in another Access database, or they can be linked to from an Excel spreadsheet. When storing your customizations in a location other than the Access table, you must call the `LoadCustomUI` method of the Application object to load your XML markup manually and then set the Ribbon name in your program at runtime. Ribbon customizations can be applied to the entire application or to specific forms and reports.

Simple text editors such as Notepad do not provide tools for validating your XML markup. You must be extra careful to write well-formed XML or your code will fail (see Chapter 32 for an introduction to XML terms and markup). If you have access to a copy of Visual Studio® 2005 or later, you can use its editor to perform the XML validation based on the XSD schema file. You can download this file (customUI.xsd) from the following Website:

http://www.microsoft.com/download/en/details.aspx?displaylang=en&id=1574

If you'd like to work with the XML in a tree-based format, you may want to download a free copy of XML Notepad 2007 from:

http://www.microsoft.com/download/en/details.aspx?id=7973

None of these advanced tools are necessary for the completion of this chapter's Ribbon customizations. Each new tool, especially an advanced one, requires that you first familiarize yourself with its interface. To get started with XML programming without further delays, the built-in Windows Notepad will do.

Creating the Ribbon Customization XML Markup

To make custom changes to the Ribbon user interface in Access 2013, you need to prepare an XML markup file that specifies all your customizations. The XML markup file that we will use in the hands-on exercise in this section is shown in Figure 29.17. You can see the resulting output in Figure 29.18.

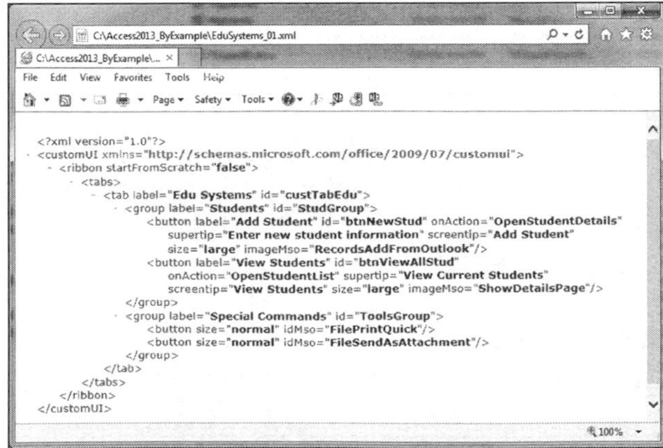


```
<?xml version="1.0"?>
- <customUI xmlns="http://schemas.microsoft.com/office/2009/07/customui">
  - <ribbon startFromScratch="false">
    - <tabs>
      - <tab label="Edu Systems" id="custTabEdu">
        - <group label="Students" id="StudGroup">
            <button label="Add Student" id="btnNewStud" onAction="OpenStudentDetails"
              supertip="Enter new student information" screentip="Add Student"
              size="large" imageMso="RecordsAddFromOutlook"/>
            <button label="View Students" id="btnViewAllStud"
              onAction="OpenStudentList" supertip="View Current Students"
              screentip="View Students" size="large" imageMso="ShowDetailsPage"/>
          </group>
        - <group label="Special Commands" id="ToolsGroup">
            <button size="normal" idMso="FilePrintQuick"/>
            <button size="normal" idMso="FileSendAsAttachment"/>
          </group>
        </tab>
      </tabs>
    </ribbon>
</customUI>
```

FIGURE 29.17. This XML file defines a new tab with two groups for the existing Access 2013 Ribbon. See the output this file produces in Figure 29.18.

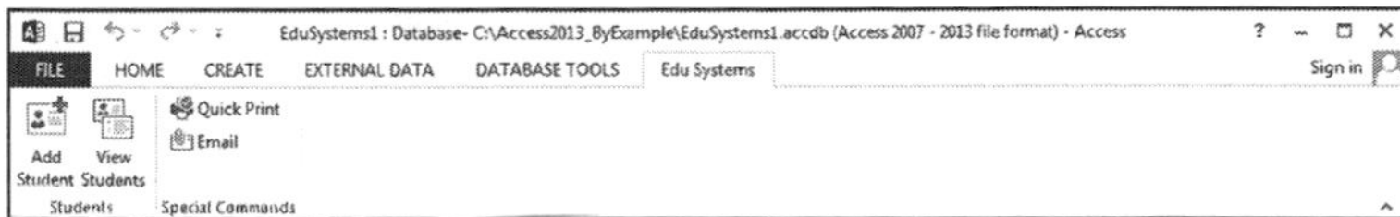

FIGURE 29.18. The custom Edu Systems tab is based on the XML markup file shown in Figure 29.17.

Hands-On 29.5. Creating an XML Document with Ribbon Customizations

1. Open **Windows Notepad** and type the following XML markup, or copy the code from **C:\Access2013_HandsOn\EduSystems_01.txt:**

```
<customUI xmlns="http://schemas.microsoft.com/office/2009/07/customui">
  <ribbon startFromScratch="false">
  <tabs>
  <tab id="custTabEdu" label="Edu Systems">
  <group id="StudGroup" label="Students">
  <button id="btnNewStud" imageMso="RecordsAddFromOutlook"
  size="large" label="Add Student" screentip="Add Student"
  supertip="Enter new student information"
  onAction="OpenStudentDetails" />
  <button id="btnViewAllStud" imageMso="ShowDetailsPage"
  size="large" label="View Students" screentip="View Students"
  supertip="View Current Students" onAction="OpenStudentList" />
  </group>
  <group id="ToolsGroup" label="Special Commands">
  <button idMso="FilePrintQuick" size="normal" />
  <button idMso="FileSendAsAttachment" size="normal" />
  </group>
  </tab>
  </tabs>
  </ribbon>
</customUI>
```

Let's go over the contents of this file. In Chapter 32, you will learn that every XML document consists of a number of elements, called *nodes*. In any XML document there must be a root node, or a top-level element. In the Ribbon customization file, the root tag is <customUI>. The root's purpose is to specify the Office 2013 RibbonX XML namespace:

```
<customUI xmlns="http://schemas.microsoft.com/office/2009/07/customui">
```

Namespaces are used to uniquely identify elements in the XML documents and avoid name collisions when elements with the same name are combined in the same document. If you were to customize the Office 2007 Ribbon, you would use the following namespace instead:

```
<customUI xmlns="http://schemas.microsoft.com/office/2006/01/customui">
```

The xmlns attribute of the <customUI> tag holds the name of the default namespace to be used in the Ribbon customization. Notice that the root element encloses all other elements of this XML document: ribbon, tabs, tab, group, and button. Each element consists of a beginning and ending tag. For example, <customUI> is the name of the beginning tag and </customUI> is the ending tag.
The actual Ribbon definition is contained within the <ribbon> tag:

```
<ribbon startFromScratch="false">
[Include xml tags to specify the required ribbon customization]
</ribbon>
```

The startFromScratch attribute of the <ribbon> tag defines whether you want to replace the built-in Ribbon with your own (true) or add a new tab to the existing Ribbon (false).

Hiding the Elements of the Access User Interface

Setting `startFromScratch="true"` in the <ribbon> tag will hide the default Ribbon as well as the contents of the Quick Access toolbar.
The File tab will be left with only three commands: New, Open, and SaveAs, and Close Database.

To create a new tab set in the Ribbon, use the <tabs> tag. Each tab element is defined with the <tab> tag. The label attribute of the tab element specifies the name of your custom tab. The name in the id attribute is used to identify your custom tab:

```
<tabs>
<tab id="custTabEdu" label="Edu Systems">
```

Ribbon tabs contain controls organized in groups. You can define a group for the controls on your tab with the <group> tag. The example XML markup file defines the following two groups for the Edu Systems tab:

```
<group id="StudGroup" label="Students">
<group id="ToolsGroup" label="Special Commands">
```

Similar to the tab node, the group nodes of the XML document contain the id and label attributes. Placing controls in groups is easy. The group labeled Students has two custom button controls, identified by the <button> elements. The group labeled Special Commands also contains two buttons; however, unlike the Students group, the buttons placed here are built-in Office system controls rather than custom controls. You can quickly determine this by looking at the id attribute for the control. Any attribute that ends with "Mso" refers to a built-in Office item:

```
<button idMso="FilePrintQuick" size="normal" />
```

You can download control IDs for built-in controls in all applications that use the Office Fluent user interface from the following Website:

http://www.microsoft.com/download/en/details.aspx?id=6627

As mentioned earlier in this chapter, buttons placed on the Ribbon can be large or small. You can define the size of the button with the size attribute set to "large" or "normal." Buttons can have additional attributes:

```
<button id="btnNewStud" imageMso="RecordsAddFromOutlook"
size="large" label="Add Student"
screentip="Add Student" supertip="Enter new student information"
onAction="OpenStudentDetails" />
```

The imageMso attribute denotes the name of the existing Office icon. You can use images provided by any Office application. To provide your own image, you must use the getImage attribute in the XML markup (see more information in the section "Using Images in Ribbon Customizations" later in this chapter).

The screentip and supertip attributes allow you to specify the short and longer text that should appear when the mouse pointer is positioned over the button.

The controls that you specify in the XML markup perform their designated actions via *callback procedures*. For example, the onAction attribute of a button control contains the name of the callback procedure that is executed when the button is clicked. When that procedure completes, it calls back the Ribbon to provide the status or modify the Ribbon. You will write the callback procedures for the onAction attribute in the next section (see Custom Project 29.1).

Buttons borrowed from the Office system do not require the onAction attribute. When clicked, these buttons will perform their default built-in action. Before finishing off the XML Ribbon customization document, always make sure that you have included all the required ending tags:

```
</tab>
</tabs>
</ribbon>
</customUI>
```

2. Save the file as **C:\Access2013_ByExample\EduSystems_01.xml**. By entering the XML extension, the text file is saved as an XML document.
3. To ensure that this XML document is well formed (it follows the formatting rules for XML), open it in a browser. If the browser can read the document, then its output should match Figure 29.17. If the browser finds problems with the document, it will show you the incorrect statement. It is up to you to figure out what correction is required. Open the file in Notepad, correct the erroneous code, save the file, and test it again by loading it in the browser.
4. Close the browser.

   At this point, you should have a well-formed XML document with Ribbon customizations.

Now that you know how to structure an XML document for Ribbon customizations, you should find it straightforward to add other features to the Access Ribbon as they are discussed in this chapter.

Loading Ribbon Customizations from an External XML Document

Since your first Ribbon customization is already in an external XML document, we will go ahead and load it into Access using the combination of VBA and macros. In a later section of this chapter, you will learn how to load the same XML markup into a local Access table and have Access take care of the Ribbon modifications at startup.

Custom Project 29.1 walks you through the steps required to integrate Ribbon customizations into your database application.

Custom Project 29.1. Applying Ribbon Customizations from an External XML File

This custom project depends on the XML document prepared in Hands-On 29.5.

Part 1: Setting Access Options

1. Copy **EduSystems1.accdb** database from the **C:\Access2013_HandsOn folder** to your **C:\Access2013_ByExample** folder.
2. Open the EduSystems1 database and click the **File | Options**, then **Client Settings**.

3. In the General section, enable the **Show add-in user interface errors** option (Figure 29.19).

   When you enable this option, you will be able to see error messages if errors are encountered when you load your Ribbon customizations.

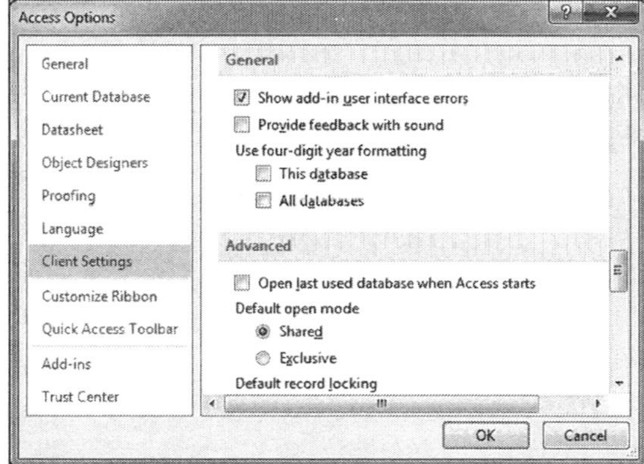

FIGURE 29.19. When you check Show add-in user interface errors, Access will notify you about any problems in the Ribbon XML.

4. Click **OK** to close the Access Options dialog box.

Part 2: Setting Up the Programming Environment

1. Choose Database Tools | Visual Basic to switch to the Visual Basic Editor window.
2. Choose **Tools | References**. In the References dialog box, add references to the following two libraries: **Microsoft Office 15.0 Object Library** and **Microsoft Scripting Runtime** (Figure 29.20).

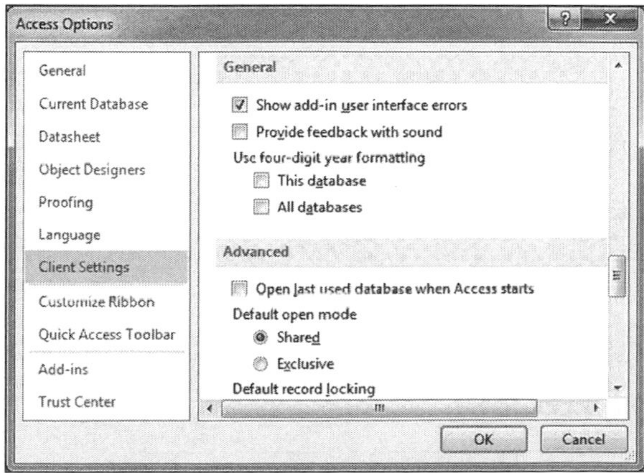

FIGURE 29.20. To avoid compile errors, you must set library references as shown here.

3. Click **OK** to close the **References** dialog box.
4. Choose **Insert | Module**.
5. In the Properties window, change the **Name** property of the module to **RibbonModification**.

Part 3: Writing VBA Code

1. In the RibbonModification Code window, enter the following VBA procedures, or copy and paste the VBA code from **C:\Access2013_HandsOn\RibbonVBA.txt**:

```
Sub OpenStudentDetails(ByVal control As IRibbonControl)
  DoCmd.OpenForm "Student Details", acNormal, , , acFormAdd
End Sub
Sub OpenStudentList(ByVal control As IRibbonControl)
  DoCmd.OpenForm "Student List", acNormal
End Sub
```

Notice that both of the preceding procedures open the specified Access forms. In addition, the OpenStudentDetails procedure opens the Student Details form in Add mode. You may recall that these procedures (OpenStudentDetails and OpenStudentList) are the names of the callback procedures that were specified in the onAction attribute of the button XML:

```
<button id="btnNewStud" imageMso="RecordsAddFromOutlook"
  size="large" label="Add Student"
  screentip="Add Student" supertip="Enter new student information"
  onAction="OpenStudentDetails" />
<button id="btnViewAllStud" imageMso="ShowDetailsPage"
  size="large" label="View Students"
  screentip="View Students"
  supertip="View Current Students"
  onAction="OpenStudentList" />
```

A callback procedure executes some action and then notifies the Ribbon that the task has been completed. The onAction callback can be handled by a VBA procedure, a macro, or an expression. When using VBA, the callback must include the IRibbonControl parameter and return type, as shown here:

```
Sub OpenStudentDetails(ByVal control As IRibbonControl)
Sub OpenStudentList(ByVal control As IRibbonControl)
```

The IRibbonControl parameter is the control that was clicked. This control is passed to your VBA code by the Ribbon. For VBA to recognize this parameter, we added a reference to the Microsoft Office 15.0 Object Library in Part 2 of this custom project.

The IRibbonControl Properties

You can view the properties (Context, Id, and Tag) of the IRibbonControl object in the Object Browser. The Context property returns the active window that contains the Ribbon interface, in this case Microsoft Access. The Id property contains the ID of the control that was clicked. The Tag property can be used to store additional information with the control. To use this property, you need to add the tag attribute to the XML markup. You can write a more generic procedure to handle the callbacks by using the Tag property. For example, instead of writing a separate procedure to open the Student Details and Student List forms as we did in Custom Project 29.1, you could write a single procedure like this:

```
Sub OpenFrm(ByVal control AS IRibbonControl)
  Select Case control.Id
    Case "btnNewStud"
      DoCmd.OpenForm "Student Details", acNormal, , , acFormAdd
    Case "btnViewAllStud"
      DoCmd.OpenForm "Student List", acNormal
  End Select
End Sub
```

Next, you would need to add the tag attribute to the XML markup and change the onAction callback to the OpenFrm procedure name:

```
<button id="btnNewStud" imageMso="RecordsAddFromOutlook"
  size="large" label="Add Student"
  screentip="Add Student" supertip="Enter new student information"
  onAction="OpenFrm" tag="Student Details" />
<button id="btnViewAllStud" imageMso="ShowDetailsPage"
  size="large" label="View Students"
  screentip="View Students"
  supertip="View Current Students"
  onAction="OpenFrm" tag="Student List" />
```

You can see the implementation of the preceding technique in the EduSystems2.accdb database and EduSystems_02.xml document located in the Access2013_HandsOn folder.

2. In the RibbonModification Code module, enter the following **LoadRibbon** function procedure, or copy and paste the procedure code from **C:\Access2013_HandsOn\RibbonVBA.txt**:

```
Public Function LoadRibbon()
  Dim strXML As String
  Dim oFso As New FileSystemObject
  Dim oTStream As TextStream

  ' Open the file containing the Ribbon customizations
  ' and return a TextStream object that will be used
  ' for reading from the file
  Set oTStream = oFso.OpenTextFile("C:\Access2013_ByExample" & _
   "\EduSystems_01.XML", ForReading)

  ' Read the entire stream into a string variable
  strXML = oTStream.ReadAll

  ' Close the TextStream object
  oTStream.Close

  ' Free up resources
  Set oTStream = Nothing
  Set oFso = Nothing

  ' load XML markup that represents a customized Ribbon
  Application.LoadCustomUI "EduTabR", strXML
End Function
```

This procedure uses the `LoadCustomUI` method of the Application object to load into Access the XML markup that contains your Ribbon customizations. To use this method you must pass the name of the Ribbon and the XML code that defines the customized Ribbon. In this example, EduTabR is the name of our customized Ribbon. You can name your Ribbon anything you want. The `strXML` variable contains the XML markup. Because the XML markup must be passed as a text string, the procedure begins by accessing the FileSystemObject from the Microsoft Scripting Runtime (see Part 2) and reading the contents of the XML file using the `ReadAll` method of the TextStream object.

In Part 4 of this custom project, you will call the LoadRibbon function from the AutoExec macro to make the custom Ribbon available to the database application on startup.

3. Click the **Save** button to save changes in the RibbonModification module.

4. Choose **Debug | Compile EduSystems1** to ensure that the VBA code does not contain spelling or other errors. If errors are found, correct them before proceeding to the next step.
5. Press **Alt+Q** to close the Visual Basic Editor window and return to Microsoft Access.

Part 4: Calling the LoadRibbon function from an Autoexec Macro

1. In the Access window's Navigation pane, right-click the AutoExec macro name and select **Design View**.
2. In the macro design window, select **RunCode** from the Add New Action drop-down list. Enter **LoadRibbon()** in the Function Name text box (see Figure 29.21). The function name should automatically appear when you start typing its name.
3. Press **Ctrl+S** or click the **Save** icon on the Quick Access toolbar.
4. An Access macro named AutoExec runs automatically each time you open the database. If you need to open a database without running this macro or to bypass other startup options, hold down the Shift key when opening the database. For more information about using macros in Access 2013, refer to Chapter 30 ("Macros and Templates.").
5. To run the AutoExec macro right now, click the **Run** button on the Design tab. If you clicked the Run button and received no error, the macro has run successfully and your Ribbon customization (EduTabR) has been loaded. For the changes in the Ribbon to become visible, you must complete the steps in Part 5 of this custom project.
6. Close the Macro Designer window.

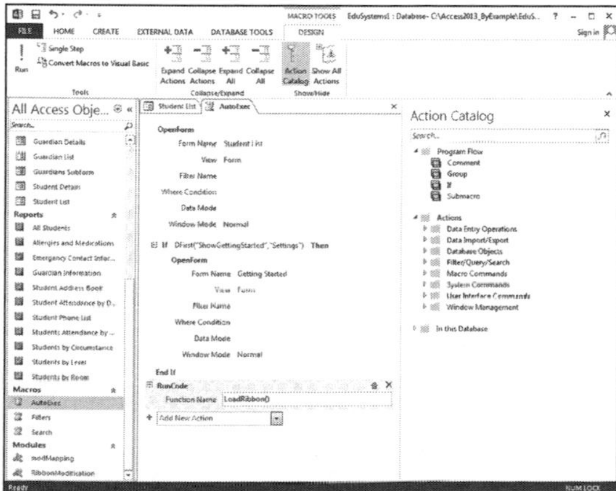

FIGURE 29.21. To load a Ribbon customization in your Access database, enter the custom LoadRibbon() function in the AutoExec macro.

Part 5: Applying the Customized Ribbon

7. Click the **File tab**, then click **Options**.
8. Click the **Current Database** option. In the Ribbon and Toolbar Options section, choose **EduTabR** from the Ribbon Name list (Figure 29.22).

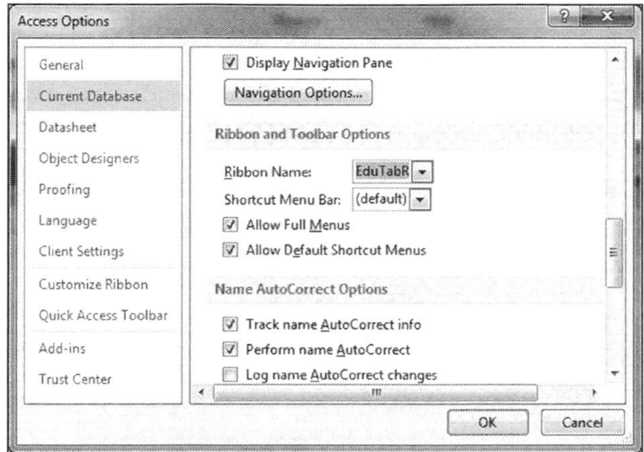

FIGURE 29.22. Enabling a customized Ribbon in the current database.

9. Click **OK** to close the Access Options window.
   Microsoft Access displays a message informing you that you must close and reopen the current database for the specified option to take effect.
10. Click **OK** to the message. Then close and restart the **EduSystems1** database.
    When the database reopens, you should see in the default database Ribbon your custom tab named Edu Systems (see Figure 29.18 earlier in this chapter). Before going on to the next section, allow some time for testing the controls placed in this custom tab.
11. Close the **EduSystems1.accdb** database.

Embedding Ribbon XML Markup in a VBA Procedure

In Custom Project 29.1, you saw how to load a Ribbon customization from an external XML document. Because the name and path of this document are hard-coded in the LoadRibbon function, prior to loading the database you must make sure that the XML markup file exists in the specified folder or Access will greet you with one or more error messages. If you don't want to worry about the location of the XML markup file, you can place the XML markup directly inside the VBA function procedure that loads the Ribbon, as shown in Figure 29.23. While the formatting of the XML string is more time consuming than referencing the file directly, it will ensure that your Ribbon markup travels with the database. Placing Ribbon XML markup inside a VBA procedure is not recommended if you plan on using the same Ribbon customizations in more than one database. If the Ribbon needs to be modified, you would need to make changes in several places, which can become confusing.

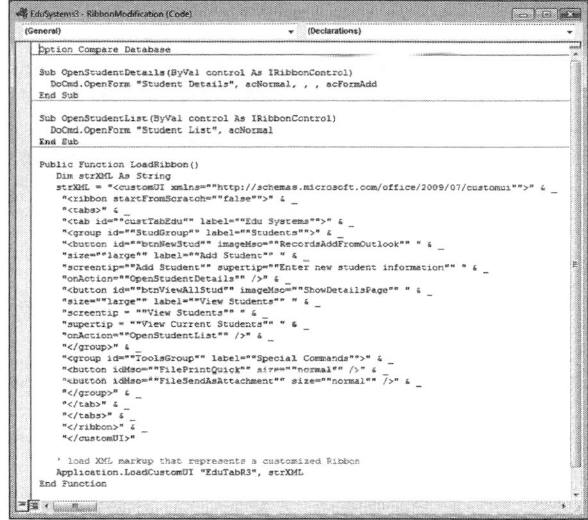

```
Option Compare Database

Sub OpenStudentDetails(ByVal control As IRibbonControl)
  DoCmd.OpenForm "Student Details", acNormal, , , acFormAdd
End Sub

Sub OpenStudentList(ByVal control As IRibbonControl)
  DoCmd.OpenForm "Student List", acNormal
End Sub

Public Function LoadRibbon()
   Dim strXML As String
   strXML = "<customUI xmlns=""http://schemas.microsoft.com/office/2009/07/customui"">" & _
    "<ribbon startFromScratch=""false"">" & _
    "<tabs>" & _
    "<tab id=""custTabEdu"" label=""Edu Systems"">" & _
    "<group id=""StudGroup"" label=""Students"">" & _
    "<button id=""btnNewStud"" imageMso=""RecordsAddFromOutlook"" " & _
    "size=""large"" label=""Add Student"" " & _
    "screentip=""Add Student"" supertip=""Enter new student information"" " & _
    "onAction=""OpenStudentDetails"" />" & _
    "<button id=""btnViewAllStud"" imageMso=""ShowDetailsPage"" " & _
    "size=""large"" label=""View Students"" " & _
    "screentip = ""View Students"" " & _
    "supertip = ""View Current Students"" " & _
    "onAction=""OpenStudentList"" />" & _
    "</group>" & _
    "<group id=""ToolsGroup"" label=""Special Commands"">" & _
    "<button idMso=""FilePrintQuick"" size=""normal"" />" & _
    "<button idMso=""FileSendAsAttachment"" size=""normal"" />" & _
    "</group>" & _
    "</tab>" & _
    "</tabs>" & _
    "</ribbon>" & _
    "</customUI>"

   ' load XML markup that represents a customized Ribbon
   Application.LoadCustomUI "EduTabR3", strXML
End Function
```

FIGURE 29.23. Ribbon XML markup can be embedded inside the VBA procedure (see the EduSystems3.accdb database in the Access2013_HandsOn folder on the CD-ROM).

Storing Ribbon Customization XML Markup in a Table

If you store your Ribbon customization XML markup in a local database table, your XML code will be loaded automatically at startup and you won't need to write a special VBA function to load your markup as you did in Custom Project 29.1. To store your XML in a table, you must create a system table named USysRibbons. This table must include two fields: a text field named RibbonName and a memo field named RibbonXML. Access expects these specific column names and data types to read your Ribbon customizations. Any additional fields in this table will be ignored. In the RibbonName field, enter a unique name that identifies your custom Ribbon. The RibbonXML field must contain the XML customization markup to be applied to the Ribbon. Please note that the USysRibbons table is a hidden system table. To show this table in the Navigation pane, you must tell Access to show hidden objects (see the next hands-on exercise). You can define multiple Ribbons in your database application by adding a new record to the USysRibbons table.

Let's now proceed to Hands-On 29.6 in which you create the USysRibbons table to store the Ribbon customization prepared earlier in this chapter.

Hands-On 29.6. Creating a Local System Table to Store Ribbon Customization

This hands-on exercise requires the XML document prepared in Hands-On 29.5.

1. Copy the Access database named **EduSystems_Local.accdb** from your **C:\Access2013_HandsOn** folder to your **C:\Access2013_ByExample** folder.
2. Open the **EduSystems_Local.accdb** database and click **Create | Table Design**.
3. In Table Design view, enter the table structure as shown in Figure 29.24. To make the RibbonName field the primary key, select this field and click the **Primary Key** button in the Tools group of the Design tab.
4. Save the table as **USysRibbons**. (Press **Ctrl+S** or click the **Save** button on the Quick Access toolbar to open the Save As dialog box.)

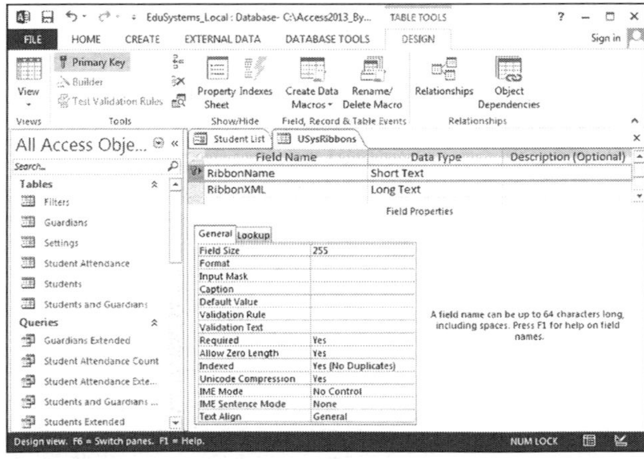

FIGURE 29.24. USysRibbons is a special system table used for storing Ribbon customizations.

5. Open the **C:\Access2013_ByExample\EduSystems_01.xml** in Windows Notepad. Press Ctrl+A to select all text and Ctrl+C to copy it to the clipboard, then close Notepad.
6. Back in Microsoft Access, in the Design tab, click the **View** button and open the table in the Datasheet view.
7. In the RibbonName field, enter **TestRibbonTab**. Press Ctrl+V to paste the entire contents of the **C:\Access2013_ByExample\EduSystems_01.xml** document into the RibbonXML field. Expand the row width so that the entire XML markup is visible. Make changes in the onAction attribute of the buttons as

shown in Figure 29.25. In the onAction attribute for btnNewStud, enter **RibbonLib.OpenStudentDetails.** In the onAction attribute for btnViewAllStud, enter RibbonLib.OpenStudentList.

FIGURE 29.25. The USysRibbons table with a record defining Ribbon customization. To define multiple Ribbons in your application, simply add a new record to this table.

8. Save and close the **USysRibbons** table.

Showing System Objects in the Navigation Pane

By default, system tables do not show in the Navigation pane. If you need to open the USysRibbons table to correct any errors or add a new record, you must enable the system objects in the Navigation Options dialog box, as follows:

1. Click the **File** tab, then click **Options**.
2. Click **Current Database**, then in the Navigation section click the **Navigation Options** button.
3. Select **Show System Objects** and click **OK** to close the Navigation Options dialog box.

Click **OK** to exit the Access Options dialog box.

The next step is to enter the callbacks that are needed for the button actions. In Custom Project 29.1, you wrote VBA callback procedures for the btnNewStud and btnViewAllStud buttons. Instead of a VBA callback, the onAction attribute of the button control can invoke a macro. Macro callbacks do not require that you return a value to the Ribbon. Also, your Ribbon customization can be functional even in safe mode (when code is not enabled for the database). It is up to you to decide whether to write VBA callbacks or to create simple macro actions for your custom Ribbon controls.

Hands-On 29.7 demonstrates the implementation of macros in the onAction attribute for controls. This hands-on exercise also introduces you to submacros. *Submacros* are like subroutines. Instead of cluttering the Navigation pane with a large number of small macros that perform a specific task, you can define a series of actions in one place as a submacro and then call that submacro whenever it's needed.

Hands-On 29.7. Using Macros Instead of VBA Callbacks

1. In the database window, click the **Create** tab, then in the Macros & Code group, click the **Macro** button.
2. In the macro design window, select **Submacro** from the Add New Action drop-down list.
3. In the Submacro name text box, enter **OpenStudentDetails**.
4. Specify the form settings as shown in Figure 29.26:
   a. Select **OpenForm** from the Add New Action drop-down list.
   b. Select **StudentDetails** from the Form Name drop-down list.
   c. Select **Add** from the Data Mode drop-down list.
5. Select **Submacro** from the Add New Action drop-down list located below End Submacro.
6. In the Submacro name text box, enter **OpenStudentList**.
7. Specify the form settings as shown in Figure 29.27:
   a. Select **OpenForm** from the Add New Action drop-down list.
   b. Select **StudentList** from the Form Name drop-down list.

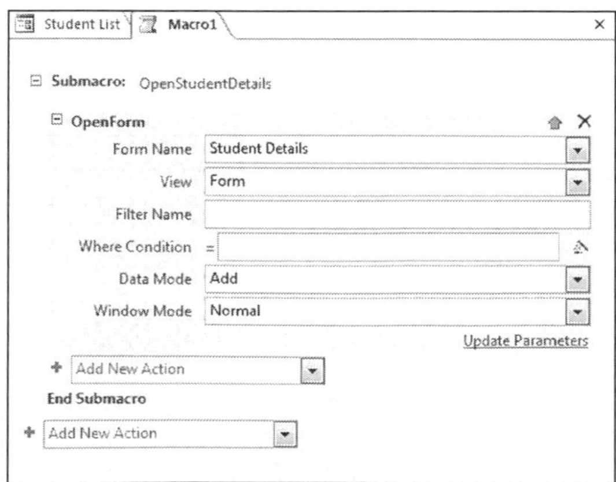

FIGURE 29.26. Creating the OpenStudentDetails submacro.

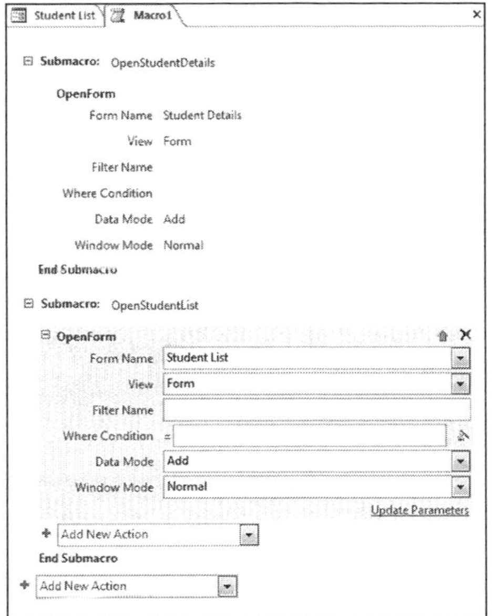

FIGURE 29.27. Creating the OpenStudentList submacro.

8. Press **Ctrl+S** to invoke a Save As dialog box. Enter **RibbonLib** for your macro name. This macro contains the two submacros created in earlier steps.
9. Close the Macro Designer window.

   Now that your macro callbacks are ready, you must tell Access to read your Ribbon definition from the USysRibbons system table. To do this, you must close and restart the application.
10. Close and then reload the **EduSystems_Local.accdb** database.

    When the application starts, Access looks for the USysRibbons system table. If the table exists, Access proceeds to read the data. If any errors are encountered in the Ribbon definition and you have set the option to Show add-in user interface errors (see Figure 29.20 earlier), you will see error messages similar to the one shown in Figure 29.28. Needless to say, you must correct all the errors before Access can display your customization in the Ribbon.

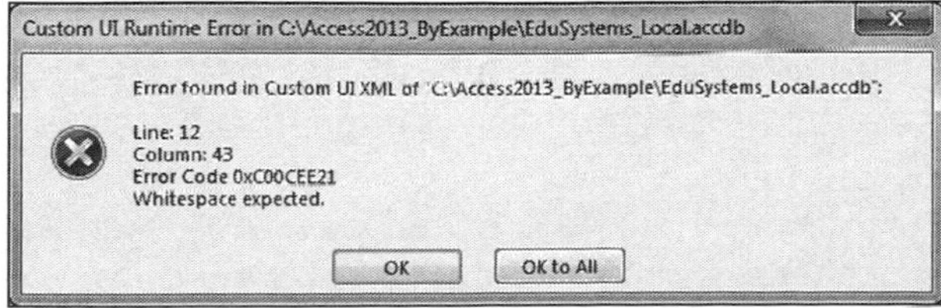

FIGURE 29.28. Upon loading the database, Access displays an error message if errors are found in the Ribbon customization markup.

If there are no errors, Access loads your customization; however, before you can see the Ribbon you need to tell Access to apply your Ribbon customization when the application is started.

11. Click the **File** tab, then click **Options**.
12. Click **Current Database**. In the Ribbon and Toolbar Options section, choose the name of your customized Ribbon from the Ribbon Name list: **TestRibbonTab**.
13. Click **OK** to close the Access Options window.
    Access will advise you that you must close and restart the application before the changes take effect.
14. Close and restart the database.
    When the EduSystems_Local database is reloaded, it should see your custom Ribbon tab named Edu Systems. Take the time to test the controls placed on this tab to make sure that the macro actions are invoked correctly.
15. Close the **EduSytems_Local.accdb** database.

If you don't want Access to automatically load Ribbon customizations from the USysRibbons table, simply rename this table.

Assigning Ribbon Customizations to Forms and Reports

In addition to customizing the main database Ribbon, Access allows you to create Ribbons that are associated with a particular form or report. To display Ribbon content for forms and reports, you can use a contextual tabset called AccessFormReportExtensibility. This tabset is hidden by default; however, it will become visible when it has controls to display. You will insert some commands into this contextual tabset in Custom Project 29.2. Because the contextual tabset takes focus when the form or report is first opened, your users will be able to see right away the special controls you've made available for them. These controls can include built-in icons from other Access tabs or your own custom buttons and other types of controls as discussed later in this chapter.

Keep in mind that Ribbon customizations for forms and reports are only displayed when a form or report is loaded or activated, and they are removed when the object is closed or deactivated. While a specific form or report is in use, you may also hide other built-in Ribbon items. You can do this by setting the visible attribute of a Ribbon item to `False`. This will prevent users from using features of the program that you don't want to be available.

To assign a custom Ribbon to a form or report, you need to open a form or report in Design or Layout view. On the Other tab of the property sheet, choose the Ribbon you want to apply from the Ribbon Name list.

Custom Project 29.2. Creating and Assigning Ribbon Customization to a Report

This custom project requires access to the **EduSystems_Local.accdb** database and the USysRibbons table that was created in Hands-On 29.6.

Part 1: Creating Ribbon Customization for a Report Using a Local System Table

1. Open the **C:\Access2013_ByExample\EduSystems_Local.accdb** database. This database will display a custom tab named Edu Systems. Recall that the XML markup for this customization is stored in the local system table named USysRibbons. In this exercise, you will add another record to this table to specify a Ribbon customization for an Access report. Before you proceed to the next step, make sure that the USysRibbons table is displayed in the Navigation pane. To unhide the table, follow the steps outlined in the sidebar "Showing System Objects in the Navigation Pane" earlier.
2. Open the USysRibbons table and enter a new record for the Ribbon named **AlergMedRpt** as shown in Figure 29.29. You can copy the XML markup file from **C:\Access2013_HandsOn\EduSystems_04.xml.**

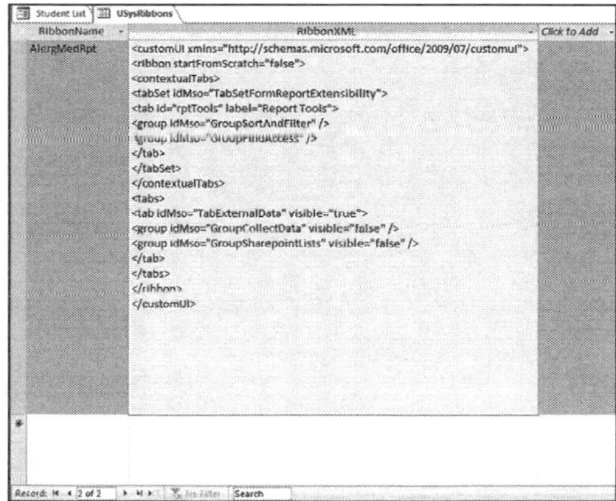

FIGURE 29.29. Entering Ribbon customization for a report.

The RibbonXML field contains the XML markup you want to apply to a report. The RibbonName can be any name you want to use to identify this customization. To have Access use the special contextual tabset available for forms and reports, you must use the <contextualTabs> XML tag. Within this tag, use the <tabSet> tag. Because this tabset is defined by Access, you must specify `TabSetFormReportExtensibility` in the idMso attribute:

```
<contextualTabs>
<tabSet idMso="TabSetFormReportExtensibility">
```

In the next statement, assign a custom ID and a name to the tab that will contain your customization:

```
<tab id="rptTools" label="Report Tools">
```

The preceding XML statement tells Access to place the focus on the Report Tools tab when the report is opened.

The next two XML statements define the controls you want to display:

```
<group idMso="GroupSortAndFilter" />
<group idMso="GroupFindAccess" />
</tab>
```

In this example, you are telling Access to simply add the Sort and Filter and Find groups from its library of built-in controls. As mentioned earlier, you can download the list of control IDs from the Microsoft Website. Since at this time you are not defining other customizations to appear on this tab, you need to close this XML group by including the closing tags:

```
</tabSet>
</contextualTabs>
```

When the report is loaded, you also want to disable certain built-in features such as controls that collect data and use SharePoint lists. This can be done by setting the visible attribute of the named built-in control groups to `false`:

```
<tabs>
<tab idMso="TabExternalData" visible="true">
<group idMso="GroupCollectData" visible="false" />
<group idMso="GroupSharepointLists" visible="false" />
</tab>
</tabs>
```

To finish off the customization markup, you need to include the ending tags:

```
</ribbon>
</customUI>
```

3. Press **Ctrl+S** to save changes to the USysRibbons table.
4. Close the **USysRibbons** table.

Part 2: Making Access Aware of the New Customization

Remember that the Ribbon customization cannot be displayed until you close and reopen the database.

1. Close and reopen the **EduSystems_Local.accdb** database.

When Access loads, it will read the Ribbon customizations from the USysRibbons table. Now is the time to tell Access to load the customized Ribbon for a specific report.

You should follow the same steps for creating and assigning Ribbon customizations for a form. Of course, your XML markup for a form ought to include the features related to forms and not reports.

Part 3: Assigning a Ribbon Customization to a Report

1. In the Navigation pane, right-click the **Allergies and Medications** report and choose **Design View**.
2. If the property sheet is not displayed, press **Alt+Enter** to display it. Make sure **Report** is selected in the selection list at the top of the property sheet.
3. In the property sheet, click the **Other** tab, click the down arrow next to the **Ribbon Name** property, and choose **AlergMedRpt** from the drop-down list (see Figure 29.30).

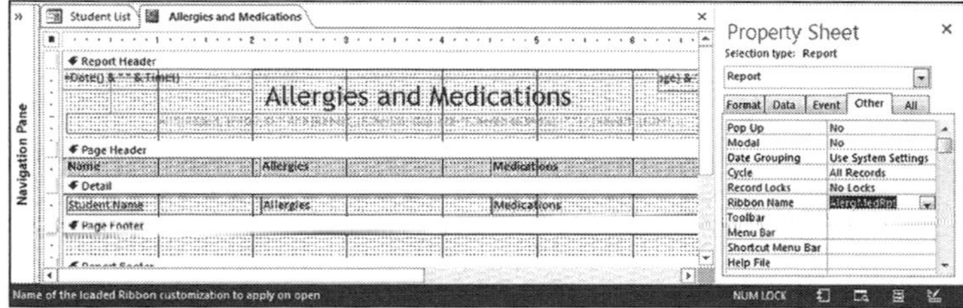

FIGURE 29.30. Use the Ribbon Name property of the report to assign your Ribbon customization to the active report.

4. Press **Ctrl+S** to save the changes.
5. Close the **Allergies and Medications** report, then reopen it.

   Notice that when the report opens, the focus is on your custom Ribbon tab named Report Tools (Figure 29.31).

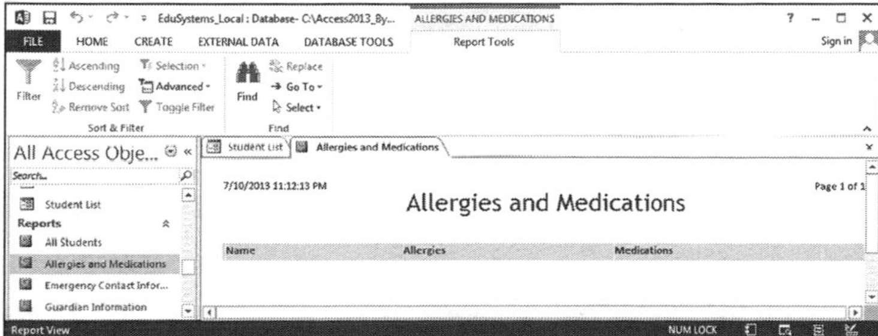

FIGURE 29.31. The custom Report Tools tab appears in the Access Ribbon when the Allergies and Medications report is opened.

6. Click the **External Data** tab and notice that only two control groups are shown: Import & Link and Export. The Collect Data tab that normally appears for reports is removed from the Ribbon. This report group is made invisible when the Allergies and Medications report is active, and appears on the External Data tab when any other report is open.
7. Close the **Allergies and Medications** report when you are finished viewing Ribbon customizations.
8. Close the **EduSystems_Local.accdb** database.

USING IMAGES IN RIBBON CUSTOMIZATIONS

The images you have learned to use so far in your Ribbon customizations are images provided by any Office application that implements the Ribbon. You already know that to reuse an Office icon you must use the imageMso attribute of a control. However, instead of using built-in Office images you can also use your own BMP, GIF, and JPEG image files. These images can be stored in a directory on your computer or a network drive, or in an Access table, then passed to your Ribbon controls via the loadImage callback for the Ribbon or the getImage callback for a control.

Requesting Images via the loadImage Callback

You can specify the name of a custom image file to be loaded for a specific control on the Ribbon by using the image attribute. When you request the image via the image attribute, the loadImage callback is called. To load images dynamically with one procedure call, define the callback procedure name in the loadImage attribute of the customUI node. Here's a fragment of the XML markup file that we'll use in Custom Project 29.3 to implement this method of loading images:

1. In the first line of your Ribbon customization markup (inside the <customUI> tag), use the loadImage attribute and specify the name of the callback procedure:

```
<customUI xmlns="http://schemas.microsoft.com/office/2009/07/customui"
loadImage="OnLoadImage">
```

2. When defining your Ribbon controls, use the image attribute and specify the name of the image file:

```
<group id="ImagesGroup" label ="Special Features">
<button id="btnNotes" label="Open Notepad"
image="Note.gif" size="large" onAction="OpenNotepad" />
```

```
<button id="btnComputer" label="Computer folder"
image="MyFolder.gif" size="normal" />
</group>
```

3. Write the loadImage callback procedure (OnLoadImage) in a VBA module:

```
Public Sub OnLoadImage(imgName As String, ByRef image)
  Dim strImgFileName As String
  strImgFileName = "C:\Access2013_ByExample\images\" & imgName
  Set image = LoadPicture(strImgFileName)
End Sub
```

Notice that to load a picture from a file, you need to use the `LoadPicture` function. This function is a member of the stdole.StdFunctions library. The library file, which is called stdole2.tlb, is installed in the System or System32 folder on your computer and is available to your VBA procedures without setting additional references. The `LoadPicture` function returns an object of type IPictureDisp that represents the image. You can view objects, methods, and properties available in the stdole library by activating the Object Browser in the Visual Basic Editor window.

4. Write the callback procedure for the button labeled Open Notepad:

```
Public Sub OpenNotepad(ctl As IRibbonControl)
  Shell "Notepad.exe", vbNormalFocus
End Sub
```

The OpenNotepad procedure tells Access to use the `Shell` function to open Windows Notepad. Notice that the name of the program's executable file is in double quotes. The second argument of the `Shell` function is optional. This argument specifies the window style, that is, how the program will appear once it is launched. The `vbNormalFocus` constant will open Notepad in a normal size window with focus. If the window style is not specified, the program will be minimized with focus (`vbMinimizedFocus`).

Let's proceed to Custom Project 29.3, which adds two new buttons with custom images to the Ribbon.

Custom Project 29.3. Loading Custom Images Using the loadImage Callback

This project requires access to the **EduSystems_Local.accdb** database and the USysRibbons table that was created in Hands-On 29.6. To use custom images, copy the **Images** folder from the **C:\Access2013_HandsOn** folder to your **C:\Access2013_ByExample** folder.

Part 1: Creating Ribbon Customization for Loading Custom Images

1. Open the **C:\Access2013_ByExample\EduSystems_Local.accdb** database.
2. In the Navigation pane, double-click the **USysRibbons** table to open it. If you cannot find this table, refer to Part 1 in Custom Project 29.2.
3. Enter a new record for the Ribbon named **CustomImage1** as shown in Figure 29.32. You can copy the XML markup from **C:\Access2013_HandsOn\EduSystems_05.txt**.

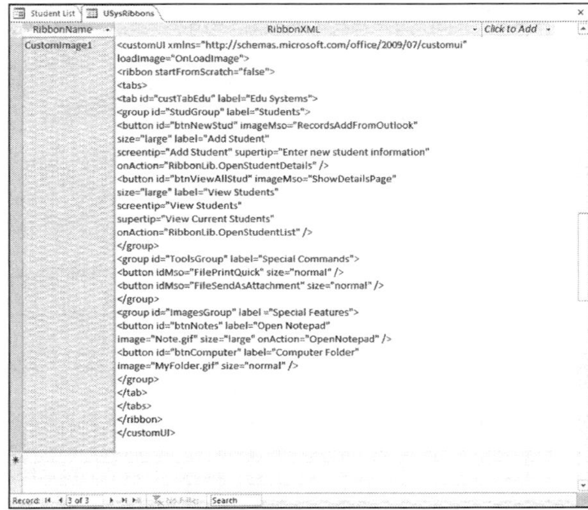

FIGURE 29.32. Entering Ribbon customization for loading custom images.

In the first line of the Ribbon customization markup (inside the <customUI> tag), notice that we've added the loadImage attribute. This attribute specifies the name of the callback procedure, OnLoadImage, that will handle loading the custom images included in the Special Features group. The Special Features group contains two images to be loaded from the C:\Access2013_ByExample\Images folder. Notice that the names of these images are specified in the image attribute of each button control in this group. You do not need to specify the file path; the OnLoadImage procedure will contain this information. For the button to perform some action you need to include the onAction attribute with the name of the macro, VBA procedure, or expression to be executed. This example does not define the onAction callback for the button named Computer Folder. To test your skills, you can add your own action for this button when you have completed this project.

4. Press **Ctrl+S** to save changes to the USysRibbons table.
5. Close the **USysRibbons** table.

Part 2: Setting Up the Programming Environment

1. Press **Alt+F11** to switch to the Visual Basic Editor window.
2. Choose **Tools | References**. In the References dialog box, add a reference to the following library: **Microsoft Office 15.0 Object Library**.
3. Click **OK** to close the References dialog box.

Part 3: Writing the VBA Callback Procedures

1. Choose **Insert | Module.**
2. In the module Code window, enter the following VBA procedures:

```
Public Sub OnLoadImage(imgName As String, ByRef image)
  Dim strImgFileName As String
  strImgFileName = "C:\Access2013_ByExample\images\" & imgName

  Set image = LoadPicture(strImgFileName)
End Sub

Public Sub OpenNotepad(ctl As IRibbonControl)
  Shell "Notepad.exe", vbNormalFocus
End Sub
```

For the explanations of these procedures, please refer to the beginning of this section

3. Press **Ctrl+S** to save changes in the Code window. When asked to name your module, enter any name you want.
4. Choose **File | Close and Return to Microsoft Access.**

Part 4: Making Access Aware of the New Customization

Remember that the Ribbon customization cannot be displayed until you close and reopen the database.

1. Close and reopen the **EduSystems_Local.accdb** database.
   When Access loads, it will read the Ribbon customizations from the USysRibbons table.
2. Click the **File** tab, then click **Options**.
3. Click the **Current Database** option. In the Ribbon and Toolbar Options section, choose **CustomImage1** from the Ribbon Name list.
4. Click **OK** to close the Access Options window.
   Microsoft Access displays a message informing you that you must close and reopen the current database for the specified option to take effect.

5. Click **OK** to the message. Then close and restart the **EduSystems_Local** database.

   When the database reopens, you should see in the default database Ribbon your custom tab named Edu Systems (Figure 29.33).

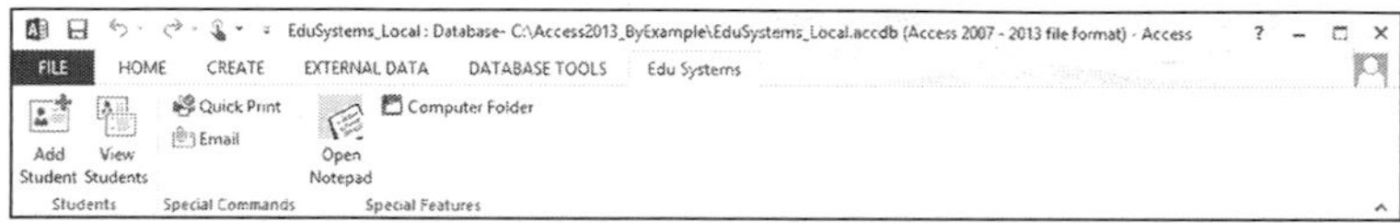

FIGURE 29.33. The Ribbon customization as defined in Custom Project 29.3.

6. Try out one of the buttons by clicking the **Open Notepad** button to open Windows Notepad. Then close Notepad.

Before going on to the next section, take time to modify the Ribbon XML to include the onAction callback for the button labeled Computer Folder and write your own custom VBA procedure to execute when this button is clicked. For example, you can make this button display a dialog box asking the user for the name of the folder to create, then use the VBA built-in function `MkDir` to create it. Use the Object Browser to locate this function. Remember that you will have to close and reopen the database for Access to recognize your Ribbon changes.

Requesting Images via the getImage Callback

Custom images can also be loaded to the Ribbon using the getImage attribute of a control. The procedure you specify in this attribute will retrieve the correct image from the specified location using the same `LoadPicture` function you worked with in the previous section. The following XML markup adds two new controls with custom images to the Special Features group that was defined in Custom Project 29.3:

```
<customUI xmlns="http://schemas.microsoft.com/office/2009/07/
loadImage="OnLoadImage">
  <ribbon startFromScratch="false">
  <tabs>
  <tab id="custTabEdu" label="Edu Systems">
  <group id="StudGroup" label="Students">
  <button id="btnNewStud" imageMso="RecordsAddFromOutlook"
  size="large" label="Add Student"
  screentip="Add Student" supertip="Enter new student information"
  onAction="RibbonLib.OpenStudentDetails" />
  <button id="btnViewAllStud" imageMso="ShowDetailsPage"
  size="large" label="View Students"
  screentip="View Students"
```

```
supertip="View Current Students"
onAction="RibbonLib.OpenStudentList" />
</group>
<group id="ToolsGroup" label="Special Commands">
<button idMso="FilePrintQuick" size="normal" />
<button idMso="FileSendAsAttachment" size="normal" />
</group>
<group id="ImagesGroup" label="Special Features">
<button id="btnNotes" label="Open Notepad"
image="Note.gif" size="large"
onAction="OpenNotepad" />
<button id="btnComputer" label="Computer Folder"
image="MyFolder.gif" size="normal" />
<button id="btnRedStar" label="Honor Student"
getImage="OnGetImage" size="large" />
<gallery id="glHolidays" label="Holidays" columns="3" rows="4"
getImage="OnGetImage" getItemCount="OnGetItemCount"
getItemLabel="OnGetItemLabel" getItemImage="OnGetItemImage"
getItemID="onGetItemID" onAction="onSelectedItem" />
</group>
</tab>
</tabs>
</ribbon>
</customUI>
```

In the preceding Ribbon customization markup, we are using all the controls that have been added thus far in this chapter's hands-on exercises and projects. In addition, the Special Features group now includes a new button labeled Honor Student and a gallery control labeled Holidays:

```
<button id="btnRedStar" label="Honor Student"
getImage="OnGetImage" size="large" />
<gallery id="glHolidays" label="Holidays" columns="3" rows="4"
getImage="OnGetImage" getItemCount="OnGetItemCount"
getItemLabel="OnGetItemLabel" getItemImage="OnGetItemImage"
getItemID="onGetItemID" onAction="onSelectedItem" />
```

In this XML markup, the gallery control will perform the action specified in the onSelectedItem callback procedure. To specify your own callback procedure for the Honor Student button, you must add the onAction attribute to this button, then write the appropriate VBA code. Notice that the gallery control has many attributes that contain static text or define callbacks. We will discuss them later. Right now, let's focus on the image loading process. Both the button and the gallery controls use the getImage attribute with the OnGetImage callback procedure. This procedure will tell Access to load the appropriate image to the Ribbon for each of these controls:

```
Public Sub OnGetImage(ctl As IRibbonControl, ByRef image)
 Select Case ctl.id
   Case "btnRedStar"
     Set image = LoadPicture("C:\Access2013_ByExample\images\redstar.gif")
   Case "glHolidays"
     Set image = LoadPicture("C:\Access2013_ByExample\images\Square0.gif")
 End Select
End Sub
```

Notice that the decision as to which image should be loaded is based on the ID of the control in the `Select Case` statement. The gallery control also uses the OnGetItemImage callback procedure (defined in the getItemImage attribute) to load custom images for its drop-down selection list (see Figure 29.34).

Use the columns and rows attributes to specify the number of columns and rows in the gallery when it is opened. If you need to define the height and width of images in the gallery, use the itemHeight and itemWidth attributes (not used in this example due to the simplicity of the utilized images). The getItemCount and getItemLabel attributes contain callback procedures that provide information to the Ribbon on how many items should appear in the drop-down list and the names of those items. The getItemImage attribute contains a callback procedure that specifies the images to be displayed next to each gallery item. The getItemID attribute specifies the onGetItemID callback procedure that will provide a unique ID for each of the gallery items.

Now that we've discussed the Ribbon customization markup, let's go over the VBA callbacks that are referenced in it. The following procedures need to be added to the VBA module for the preceding XML markup to work:

```
Public Sub OnGetItemCount(ctl As IRibbonControl, ByRef count)
  count = 12
End Sub
```

In this procedure, we use the `count` parameter to return to the Ribbon the number of items we want to have in the gallery control.

```
Public Sub OnGetItemLabel(ctl As IRibbonControl, _
  index As Integer, ByRef label)
  label = MonthName(index + 1)
End Sub
```

This procedure will label each of the gallery items. The VBA `MonthName` function is used to retrieve the name of the month based on the value of the index. The initial value of the index is zero (0). Therefore, `index + 1` will return February To display the month's name abbreviated (Jan, Feb, etc.), specify `True` as the second parameter to this function:

```
label = MonthName(index + 1, True)
```

If you are using a localized version of Microsoft Office (French, Spanish, etc.), the `MonthName` function will return the name of the month in the specified interface language.

The next callback procedure shows how to load images for each gallery item:

```
Public Sub OnGetItemImage(ctl As IRibbonControl, _
  index As Integer, ByRef image)

  Dim imgPath As String

  imgPath = "C:\Access2013_ByExample\images\square"
  Set image = LoadPicture(imgPath & index + 1 & ".gif")
End Sub
```

Each item in the gallery must have a unique ID, so the onGetItemID callback uses the `MonthName` function to specify the ID:

```
Public Sub onGetItemID(ctl As IRibbonControl, _
 index As Integer, ByRef id)

  id = MonthName(index + 1)


End Sub
```

The last procedure you need to write for the gallery control should define the actions to be performed when an item in the gallery is clicked. This is done via the following onSelectedItem callback that was specified in the onAction attribute of the XML markup:

```
Public Sub onSelectedItem(ctl As IRibbonControl, _
 selectedId As String, _
 selectedIndex As Integer)

  Select Case selectedIndex
    Case 6
```

```
        MsgBox "Holiday 1: Independence Day, July 4th", _
         vbInformation + vbOKOnly, _
         selectedId & " Holidays"
      Case 11
        MsgBox "Holiday 1: Christmas Day, December 25th", _
         vbInformation + vbOKOnly, _
         selectedId & " Holidays"
      Case Else
        MsgBox "Please program holidays for " & selectedId & ".", _
         vbInformation + vbOKOnly, _
         " Under Construction"
    End Select
  End Sub
```

In the preceding callback procedure, the `selectedId` parameter returns the name that was assigned to the label, while the selectedIndex parameter is the position of the item in the list. The first item in the list (January) is indexed with zero (0), the second with 1, and so forth. In this procedure, we have just coded two holidays: one for the month of July (`selectedIndex=6`) and one for December (`selectedIndex=11`). The `Case Else` clause in the `Select Case` statement provides a message when other months are selected.

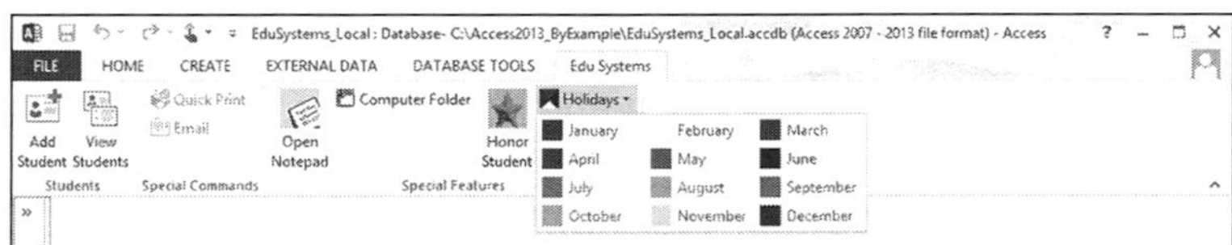

FIGURE 29.34. Customized Ribbon with the gallery control.

To implement the Ribbon customization shown in Figure 29.34, follow the steps outlined in Hands-On 29.8.

Hands-On 29.8. Loading Custom Images Using the getImage Callback

This hands-on exercise requires access to the EduSystems_Local.accdb database and the USysRibbons table that was created in Hands-On 29.6. This exercise assumes that you have also completed Custom Project 29.3, which presented a method of loading images via the loadImage callback. By now you should be very familiar with the Ribbon customization process, and thus this exercise outlines only the main steps you need to take to complete it.

For a detailed explanation of the process, refer to the previous exercises and projects. The images used in this example are located in the C:\Access2013_ByExample\Images folder.

1. In the **USysRibbons** table of the **EduSystems_Local** database, add a new record. In the RibbonName field, enter **CustomImage2** for the name of the new Ribbon customization. In the RibbonXML field, paste the XML markup from **C:\Access2013_HandsOn\EduSystems_06.txt**. Press **Ctrl+S** to save the changes, then close the **USysRibbons** table.
2. Press **Alt+F11** to switch to the Visual Basic Editor window. You should see one module with VBA procedures that were added in Custom Project 29.3. You do not need to create a new module for this customization. Simply enter in the existing module Code window the VBA procedures discussed earlier in this section (OnGetImage, OnGetItemCount, OnGetItemLabel, OnGetItemImage, onGetItemID, and onSelectedItem). Press **Ctrl+S** to save the changes in your module and exit Visual Basic Editor.
3. Close and restart the **EduSystems_Local** database. When the database is reloaded, click the **File** tab and select **Options.** In the Access Options window, click **Current Database**, and select your new Ribbon (**CustomImage2**) from the Ribbon Name list in the Ribbon and Toolbar Options section. Click **OK** to close the Access Options window. Microsoft Access will display a message informing you that you must close and reopen the current database for the specified option to take effect. Click **OK** to the message. Then close and restart the **EduSystems_Local** database.

The customized Ribbon should appear as shown earlier in Figure 29.34. Test the gallery control by clicking on some of the month items.

Instead of loading custom images from a computer folder, you can create an Access table to store your images and then use the Recordset object in the getImage callback to read the images from the table. This table should contain at least two fields: the ControlID field with the name of the control and the ImageFileName field specifying the name of the image file for the control. Custom images can also be stored and loaded from an Attachment field, which is available in Access databases created in the .accdb file format.

Understanding Attributes and Callbacks

Ribbon controls have properties defined by attributes, such as id, label, enabled, screen tip, and so on. By using a specific attribute you can modify the appearance of a control either at design time or at runtime. To define a control attribute at runtime, simply set it to the allowable value right in the Ribbon customization XML markup. For example, you can provide the name for your control in the label attribute. The control label can contain up to 1,024 characters.

If the attribute value is unknown at design time, add the prefix "get" to the design-time attribute name and specify the name of the callback procedure or macro as the attribute value. For example, if the control's label needs to be defined at runtime, use the getLabel attribute and specify the name of the callback procedure:

```
<group id="Today's Events" getLabel="getEventDate">
```

When the Ribbon is loaded, the procedure in the getLabel attribute will run and provide the actual value of the attribute:

```
Public Sub getEventDate(ctl As IRibbonControl, _
  ByRef ReturnValue As Variant)

  ReturnValue = "Events for " & Format(Now(), "mm/dd/yyyy")
End Sub
```

This procedure will display the current date in the name of the group label. Although many times you will see the callback procedure name prefixed by "get" or "onGet," keep in mind that you do not have to give the callback procedure the same name as the attribute it is used with. Use any name that makes sense to you. The only requirement is that the callback procedure matches a particular *signature*, which is the declaration of the procedure, the parameters, and return types. For example, the callback for the onAction attribute of a button control has the following signature:

```
Public Sub NameOfCallback(control As IRibbonControl)
```

IRibbonControl is the control that was clicked. This control is passed to your procedure by the Ribbon. You can specify your own name for the control parameter. For example:

```
Public Sub NameOfCallback(ctl As IRibbonControl)
```

Before using the IRibbonControl, you need to add a reference to the Microsoft Office 15.0 Object Library in your VBA project. The onAction attribute is a special type of attribute that does not need to be prefixed by the word "get" to point to a callback procedure.

USING VARIOUS CONTROLS IN RIBBON CUSTOMIZATIONS

Now that you know how to go about creating the XML markup for your Ribbon customizations and loading and applying the custom Ribbon to a database, form, or report, let's look at other types of controls you can show in the Ribbon to give your database application a more polished and professional look. You can reuse the EduSystems_Local database used in the earlier examples to create additional Ribbon customizations that utilize the controls discussed in this section.

Creating Toggle Buttons

A *toggle button* is a button that alternates between two states. Many formatting features such as Bold, Italic, or Format Painter are implemented as toggle buttons. When you click a toggle button, the button stays down until you click it again. To create a toggle button, use the <toggleButton> XML tag as shown here:

```
<toggleButton id="tglNewStudent" label="New Student Questionnaire"
size="normal" getPressed="OnGetPressed" onAction="ShowHideQ" />
```

You can add a built-in image to the toggle button with the imageMso attribute, or use a custom image as discussed earlier in this chapter. To find out whether the toggle button is pressed, include the getPressed attribute in your XML markup. The getPressed callback procedure provides two arguments: the control that was clicked and the pressed state of the toggle button:

```
Sub OnGetPressed(control As IRibbonControl, _
 ByRef pressed)

  If control.id="tglNewStudent" then
    pressed = False
  End If
End Sub
```

The preceding callback routine will ensure that the specified toggle button is not pressed when the Ribbon is loaded.

To perform an action when the toggle button is clicked, set the onAction attribute to the name of your custom callback procedure. This callback also provides two arguments: the control that was clicked and the state of the toggle button:

```
Sub ShowHideQ(control As IRibbonControl, pressed As Boolean)
  If pressed Then
    MsgBox "The toggle button is pressed."
  Else
    MsgBox "The toggle button is not pressed."
  End If
End Sub
```

If the toggle button is pressed, the value of the pressed argument will be True; otherwise, it will be False. The toggle button created in this section is shown in Figure 29.35.

FIGURE 29.35. The custom toggle button Student Questionnaire will become highlighted when pressed and will return to its normal state when clicked again.

Creating Split Buttons, Menus, and Submenus

A *split button* is a combination of a button or toggle button and a menu. Clicking the button performs one default action, and clicking the drop-down arrow opens a menu with a list of related options to select from. To create the split button, use the <splitButton> tag. Within this tag, you need to define a <button> or a <toggleButton> control and the <menu> control, as shown in the following XML markup:

```
<group id="OtherControlsGroup" label="Other Controls" >
<splitButton id="btnSplit1" size="large" >
<button id="btnImport" label="Import More"
imageMso="ImportAccess" />
<menu id="mnuImport" label="More Import Formats"
itemSize="normal" >
<menuSeparator id="mnuDiv1" title="Other Databases" />
<button id="btnImportODBC" label="ODBC database"
imageMso="ImportOdbcDatabase" />
<button id="btnImportDbase" label="Dbase file"
imageMso="ImportDBase" />
<button id="btnImportParadox" label="Paradox file"
imageMso="ImportParadox" />
<menuSeparator id="mnuDiv2" title="Spreadsheet Files" />
<menu id="mnuExcel" label="Excel File Formats"
imageMso="ImportExcel" itemSize="normal" >
<checkBox id="xlsFormat" label="xls file" />
<checkBox id="xlsxFormat" label="xlsx file" />
</menu>
<button id="btnImportLotus" label="Lotus 1-2-3 file"
imageMso="ImportLotus" />
<menuSeparator id="mnuDiv3" title="Other Files" />
```

```
<button id="btnText" label="Text file"
imageMso="ImportTextFile" />
<button id="btnXML" label="XML file" imageMso="ImportXmlFile" />
<button id="btnHTML" label="HTML file"
imageMso="ImportHtmlDocument" />
<button id="btnOutlook" label="Outlook folder"
imageMso="ImportOutlook" />
<button id="btnSharepoint" label="SharePoint List"
imageMso="ImportSharePointList" />
</menu>
</splitButton>
</group>
```

The <checkBox> tag used in the preceding example XML is discussed in detail in the next section.

You can specify the size of the items in the menu using the itemSize attribute. To display a description for each menu item below the item label, set the itemSize attribute to large (`itemSize="large"`) and use the description attribute to specify the text. The <menuSeparator> tag can be used inside the menu node to break the menu into sections. Each menu segment can then be titled using the title attribute, as shown in the preceding example. You can add the onAction attribute to each menu button to specify the callback procedure or macro to execute when the menu item is clicked. In addition to button controls, menus can contain toggle buttons, checkboxes, gallery controls, split buttons, nested menus, and dynamic menus. Figure 29.36 displays the Ribbon with split buttons, menus, and submenus created in this section.

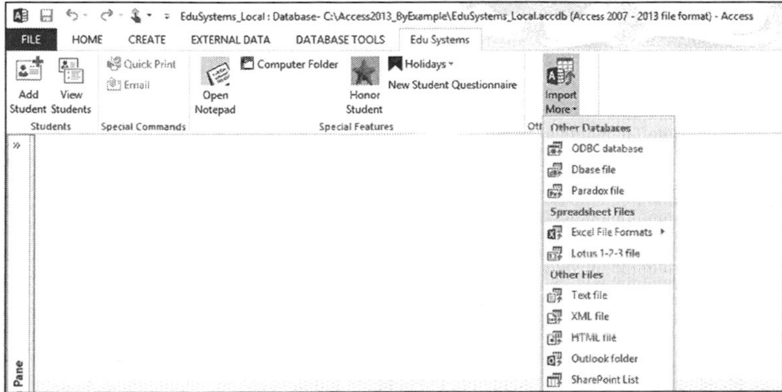

FIGURE 29.36. Custom split button controls can use the built-in Office images. They can also contain menus and submenus consisting of checkboxes.

Creating Checkboxes

The checkbox control is used to provide an option, such as true/false or on/off. It can be included inside a menu control as was demonstrated in the previous section or used as a separate control on the Ribbon. To create a checkbox, use the <checkBox> tag, as shown in the following XML:

```
<separator id="OtherControlsDiv1" />
<labelControl id="TitleForBox1" label="Areas of Interest (please
  check below)" />
<box id="boxLayout1">
<checkBox id="chkSafety" label="School Safety"
  enabled="true" visible="true"
  onAction="DoSomething" />
<checkBox id="chkHealth" label="Health" enabled="false" />
<checkBox id="chkSportsMusic" getLabel="onGetLabel" />
</box>
```

In the preceding XML markup, the <separator> tag will produce the vertical bar that visually separates controls within the same Ribbon group (see Figure 29.37). The <labelControl> tag can be used to display static text anywhere in the Ribbon. In this example, we use it to place a header over a set of controls. To control the layout of various controls (to display them horizontally instead of vertically), use the <box> tag. You can define whether a checkbox should be visible or hidden by setting the visible attribute to true or false. To disable a checkbox, set the enabled attribute to false; this will cause the checkbox to appear grayed out. Notice that the checkbox labeled Health is not active (it is grayed out).

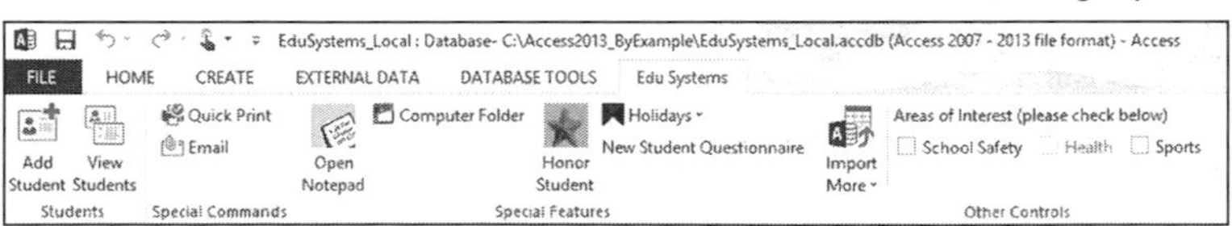

FIGURE 29.37. These checkbox controls are laid out horizontally.

Similar to other controls, labels for checkboxes can contain static text in the label attribute, or they can be assigned dynamically using the callback procedure in the getLabel attribute:

```
<checkBox id="chkSportsMusic" getLabel="onGetLabel" />
```

The getLabel attribute points to the onGetLabel callback procedure, which needs to be added to your VBA module:

```
Public Sub onGetLabel(ctl As IRibbonControl, ByRef label)
  If ctl.id = "chkSportsMusic" And _
```

```
    Weekday(Now(), vbWednesday) Then
      label = "Sports"
    Else
      label = "Music"
    End If
End Sub
```

This procedure will run automatically when the Ribbon loads. If today happens to be Wednesday, you will see a checkbox for Sports; otherwise, it will be Music.

The action of the checkbox control is handled by the callback procedure in the onAction attribute:

```
<checkBox id="chkSafety" label="School Safety"
  enabled="true" visible="true"
  onAction="DoSomething" />
```

The DoSomething procedure needs to be added to the VBA module for the School Safety checkbox to respond to a user's click:

```
Public Sub DoSomething(ctl As IRibbonControl, _
 pressed As Boolean)
  If ctl.id = "chkSafety" And pressed Then
    MsgBox "Safety is our number one concern."
  Else
    MsgBox "Sorry to hear that safety is not your concern."
  End If
End Sub
```

To get the checked state for a checkbox, point to your callback procedure in the getPressed attribute, similar to what we've done earlier with the toggle button. The default VBA syntax for this callback is as follows:

```
Sub GetPressed(control As IRibbonControl, ByRef return)
```

As mentioned earlier, callback procedures don't need to be named the same as the attribute they are used with. Also, you may change the callback's argument names as desired.

Creating Edit Boxes

Use the <editBox> tag to provide an area on the Ribbon where users can type text or numbers:

```
<editBox id="txtFullName" label="First and Last Name:"
sizeString="AAAAAAAAAAAAAAAA" maxLength="25"
onChange="onFullNameChange" />
```

Figure 29.38 shows the result of the preceding XML markup.

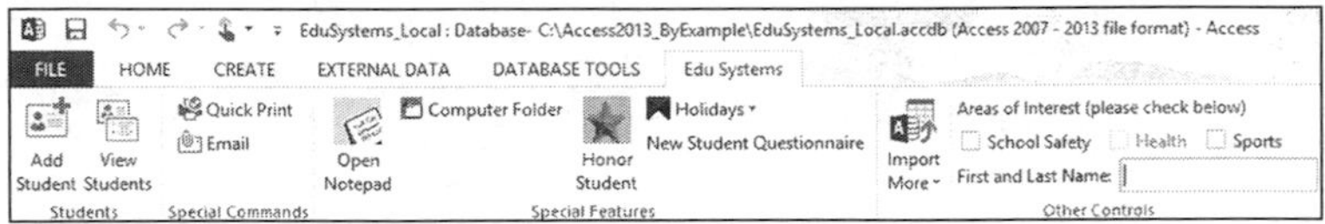

FIGURE 29.38. An edit box control allows data entry directly on the Ribbon.

The sizeString attribute specifies the width of the edit box. Set it to a string that will give you the width you want. The maxLength attribute allows you to limit the number of characters and/or digits that can be typed in the edit box. If the text entered exceeds the specified number of characters (25 in this case), Access automatically displays a balloon message on the Ribbon: "The entry may contain no more than 25 characters."

When the entry is updated in an edit box control, the callback procedure specified in the onChange attribute is called:

```
Public Sub onFullNameChange(ctl As IRibbonControl, _
 text As String)
  If text <> "" Then
    MsgBox "Is '" & text & _
     "' your real name?"
  End If
End Sub
```

When the user enters text in the edit box, the procedure will display a message box.

Creating Combo Boxes and Drop Downs

There are three types of drop-down controls that can be placed on the Ribbon: combo box, drop down, and gallery.

These controls can be dynamically populated at runtime by writing callbacks for their getItemCount, getItemID, getItemLabel, getItemImage, getItemScreentip, or getItemSupertip attributes. The combo box and drop-down controls can also be made static by defining their drop-down content using the <item> tag, as shown here:

```
<separator id="OtherControlsDiv2" />
<comboBox id="cmbLang" label="Languages"
supertip="Select Language Guide"
  onChange="OnChangeLang" >
  <item id="English" label="English" />
  <item id="Spanish" label="Spanish" />
```

```
    <item id="French" label="French" />
    <item id="German" label="German" />
    <item id="Russian" label="Russian" />
  </comboBox>
```

To separate the combo box control from other controls in the same Ribbon group, this example uses the <separator> tag. Notice that each <item> tag specifies a new drop-down row.

A combo box is a combination of a drop-down list and a single-line edit box, allowing the user to either type a value directly into the control or choose from the list of predefined options. Use the sizeString attribute to define the width of the edit box.

The combo box control does not have the onAction attribute. It uses the onChange attribute that specifies the callback to execute when the item selection changes:

```
Public Sub OnChangeLang(ctl As IRibbonControl, _
 text As String)

  MsgBox "You selected the " & text & " language guide."
End Sub
```

Notice that the onChange callback provides only the text of the selected item; it does not give you access to the selected index. If you need the index of the selection, use the dropdown control instead, as shown here:

```
<dropDown id="drpBoro" label="City Borough"
  supertip="Select School Borough"
  onAction="OnActionBoro" >
  <item id="M" label="Manhattan" />
  <item id="B" label="Brooklyn" />
  <item id="Q" label="Queens" />
  <item id="I" label="Staten Island" />
  <item id="X" label="Bronx" />
</dropDown>
```

The onAction callback of the drop-down control will give you both the selected item's ID and its index:

```
Public Sub OnActionBoro(ctl As IRibbonControl, _
 ByRef SelectedID As String, _
 ByRef SelectedIndex As Integer)
```

```
    MsgBox "Index=" & SelectedIndex & " ID=" & SelectedID
End Sub
```

Figure 29.39 shows the combo box and drop-down controls created in this section.

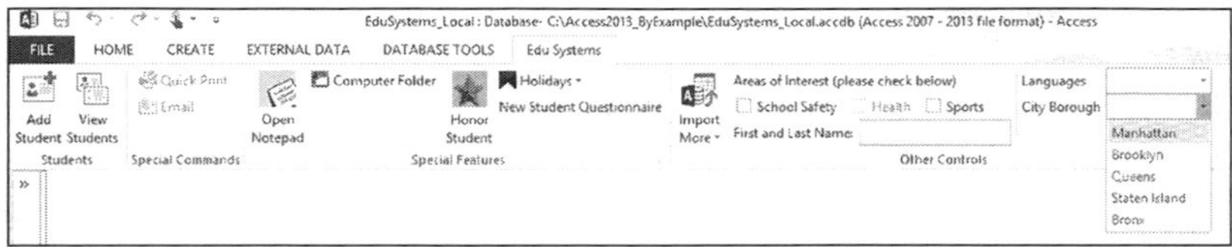

FIGURE 29.39. The Languages combo box and City Borough drop-down controls look the same on the Ribbon.

The gallery control was introduced earlier in this chapter in the section titled "Requesting Images via the getImage Callback." This control cannot be static; it must be dynamically populated at runtime.

Creating a Dialog Box Launcher

Some Ribbon tabs have a small dialog launcher button at the bottom-right corner of a group (see Figure 29.15 earlier). You can use this button to open a special form that allows the user to set up many options at once, or you can display a form that contains specific information. To add a custom dialog launcher button to the Ribbon, use the <dialogBoxLauncher> tag, as shown here:

```
<dialogBoxLauncher>
  <button id="Launch1"
  screentip="Show Product Key"
  onAction="OnActionLaunch" />
</dialogBoxLauncher>
```

The dialog box launcher control must contain a button. The OnAction attribute for the button contains the callback procedure that will execute when the button is clicked:

```
Public Sub OnActionLaunch(ctl As IRibbonControl)
  ' open the About Microsoft Office Access box
  DoCmd.RunCommand acCmdAboutMicrosoftAccess
End Sub
```

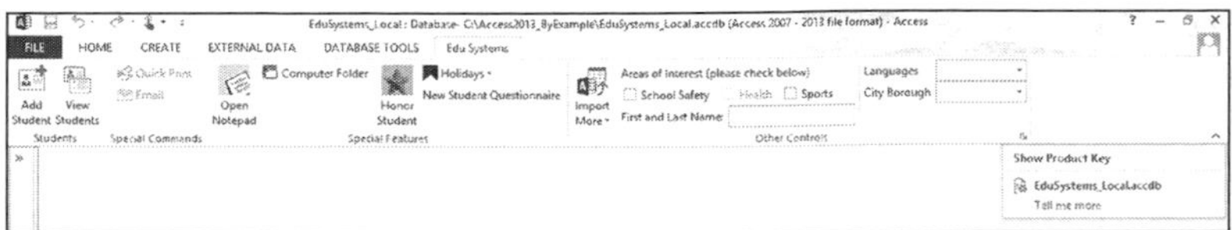

FIGURE 29.40. A dialog box launcher control on the Ribbon.

The dialog box launcher control must appear as the last element within the containing group element in the XML markup. The entire definition of the custom Edu Systems Ribbon tab created in this chapter and depicted in Figure 29.40 is available in the EduSystems_12_withDialogLauncher.txt file in your C:\Access2013_HandsOn folder.

Disabling a Control

You can disable a built-in or custom Ribbon control by using the enabled or getEnabled attribute. Here's how we disabled our custom check box control earlier by using the enabled attribute:

```
<checkBox id="chkHealth" label="Health" enabled="false" />
```

Use the getEnabled attribute to disable a control based on some conditions or simply display a "not authorized" message. The following XML code shows how to disable the built-in Relationships button on the Ribbon's Database Tools tab:

```
<!-- Built-in commands section -->
<commands>
<command idMso="DatabaseRelationships" onAction="DisableRelations" />
</commands>
```

To make your XML code more readable, you can include comments between the <!-- and --> characters. The <command> tag can be used to refer to any built-in command. This tag must appear in the <commands> section of the XML code. To see the exact position of the above XML markup in the Ribbon Customization, open the EduSystems_13_DisableAndRepurpose.txt file in your C:\Access2013_HandsOn folder. Notice the built-in command section just before the line:

```
<ribbon startFromScratch="false">
```

The onAction attribute contains the following callback procedure that will display a message when the Relationships button is clicked:

```
Sub DisableRelations(ctl As IRibbonControl, _
 ByRef cancelDefault)
  MsgBox "You are not authorized to use this function."
  cancelDefault = True
End Sub
```

You can add more code to this procedure if you need to cancel the control's default behavior only when certain conditions have been satisfied.

Repurposing a Built-in Control

It is possible to change the purpose of a built-in Ribbon button. For example, when the user clicks the DatabaseDocumentor button (Database Tools | Analyze Group) while the Student List form is open, you could display a Database Properties dialog box instead of the default Documentor dialog box:

```
<command idMso="DatabaseDocumentor" onAction="ShowDbProperties" />

Public Sub ShowDbProperties(ctl As IRibbonControl, _
  ByRef cancelDefault)

  If CurrentProject.AllForms("Student List").IsLoaded Then
    ' display Database Properties dialog box instead
    DoCmd.RunCommand acCmdDatabaseProperties
  Else
    cancelDefault = False
  End If
End Sub
```

Only simple buttons that perform an action when clicked can be repurposed. You cannot repurpose advanced controls such as combo boxes, drop downs, or galleries.

Refreshing the Ribbon

So far in this chapter you've seen how to use callback procedures to specify the values of control attributes at runtime. But what if you need to update your custom Ribbon or the controls placed in the Ribbon based on what the user is doing in your application? The good news is that you can change the attribute values at any time by using the `InvalidateControl` method of the IRibbonUI object. To use this object, start by adding the onLoad attribute to the customUI element in your Ribbon customization XML:

```
<customUI xmlns="http://schemas.microsoft.com/office/2009/07/customui"
loadImage="OnLoadImage" onLoad="RefreshMe" >
```

The onLoad attribute points to the callback procedure that will give you a copy of the Ribbon that you can use to refresh anytime you want. In this example, the onLoad callback procedure name is RefreshMe.

Let's say you have a checkbox that is disabled when the Ribbon is first loaded and you would like to enable it when the user enters text in an edit box. Also, upon entry you want the text of the edit box to appear in uppercase. To implement the onLoad callback, start by declaring a Public module-level variable of type IRibbonUI:

```
Public objRibbon As IribbonUI
```

The preceding statement should appear in the declaration section at the top of the VBA module. To keep track of the state of the two Ribbon controls we are interested in, declare two Private module-level variables:

```
Private strUserTxt As String
Private isCtlEnabled As Boolean
```

Next, enter the callback procedure that will store a copy of the Ribbon in the objRibbon variable and assign an initial value to the `isCtlEnable` variable:

```
' callback for the onLoad attribute of customUI
Public Sub RefreshMe(ribbon As IRibbonUI)
  Set objRibbon = ribbon
  isCtlEnabled = False
End Sub
```

When the Ribbon loads, the checkbox control will be disabled. You will also have a copy of the IRibbonUI object saved for later use. Now, let's take a look at the XML markup used in this scenario:

```
<checkBox id="chkHealth" label="Health"
getEnabled="onGetEnabled_Health" />
<editBox id="txtFullName" label="First and Last Name:"
  sizeString="AAAAAAAAAAAAAAAAAA" maxLength="25"
  getText="getEditBoxText" onChange="onFullNameChangeToUcase" />
```

These checkbox and edit box controls were introduced earlier in this chapter (see Figure 29.39). In order to change the enabled state of the checkbox control based on the user action, the getEnabled attribute must be used. The callback procedure for this attribute is as follows:

```
Public Sub onGetEnabled_Health(control As IRibbonControl, _
  ByRef enabled)
  enabled = isCtlEnabled
End Sub
```

When the Ribbon is loaded, the onGetEnabled_Health procedure will provide the value for the getEnabled attribute. The Health checkbox will be displayed in the Ribbon in its disabled mode because we have set the value of the `isCtlEnabled` variable to `False` in the RefreshMe procedure.

The edit box control contains two attributes that require callback procedures. The getText attribute points to the following callback:

```
Public Sub getEditBoxText(control As IRibbonControl, _
  ByRef text)
  text = UCase(strUserTxt)
End Sub
```

The preceding callback uses the VBA built-in `UCase` function to change the text that the user entered in the edit box to uppercase letters. When text is updated in the edit box, the procedure in the onChange attribute is called:

```
Public Sub onFullNameChangeToUcase(ByVal control As IRibbonControl, _
 text As String)
  If text <> "" Then
    strUserTxt = text
    objRibbon.InvalidateControl "txtFullName"
    isCtlEnabled = True
  Else
    isCtlEnabled = False
  End If
  objRibbon.InvalidateControl "chkHealth"
End Sub
```

The preceding callback begins by checking the value of the `text` parameter provided by the Ribbon. If this parameter contains a value other than an empty string (" "), the text the user entered is stored in the `strUserTxt` variable. Before a change can occur in the Ribbon control, you need to mark the control as invalid. This is done by calling the `InvalidateControl` method of the IRibbonUI object that we have stored in the `objRibbon` variable:

```
objRibbon.InvalidateControl "txtFullName"
```

This statement will tell the txtFullName control to refresh itself the next time it is displayed. When the control is invalidated, it will automatically call its callback functions. The getEditBoxText callback procedure in the onChange attribute will execute, causing the text entered in the txtFullName edit box control to appear in uppercase letters.

The second action that we want to perform is to enable the chkHealth checkbox control when the user enters text in the edit box control and keep this button disabled when the edit box control is empty. This is done by setting the `isCtlEnabled` Boolean variable to `True` or `False` and invalidating the chkHealth checkbox control. When the chkHealth control is marked as invalid, it will call its callback functions. The onGetEnabled_Health callback procedure in the getEnabled attribute will execute, causing the control to appear in the enabled state if the txtFullName edit box control contains any text.

Figure 29.41 shows the Ribbon after it has been refreshed.

FIGURE 29.41. The Ribbon controls are shown here after the Ribbon refresh. The Health checkbox is enabled upon entry of text in the First and Last Name edit box, and disabled when the entry is deleted.

The XML markup for the final Ribbon customization is contained in the C:\Access2013_HandsOn\ EduSystems_14_WithRefresh.txt file. You will find the completed database will all Ribbon customizations demonstrated in this chapter in the C:\Access2013_ByExample\EduSystems_Local.accdb file.

Figure 29.42 shows the names of all custom Ribbons that where tried out in this chapter.

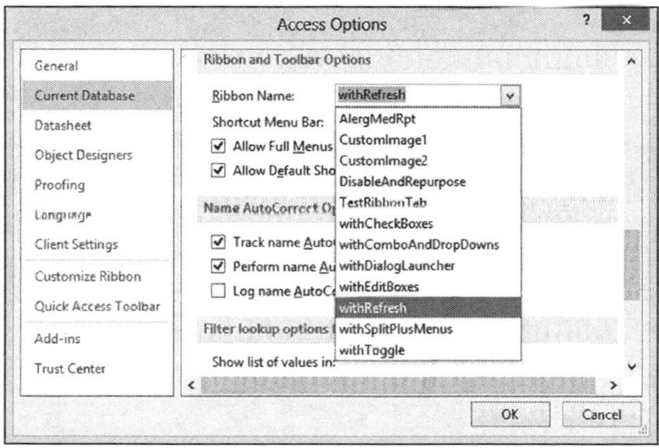

FIGURE 29.42. Each time you apply a different Ribbon customization you need to close and reopen the Access database.

The IRibbonUI object has only two methods: `InvalidateControl` *and* `Invalidate`*. Use the* `InvalidateControl` *method to refresh an individual control. Use the* `Invalidate` *method to refresh all controls in the Ribbon.*

THE COMMANDBARS OBJECT AND THE RIBBON

You can make your custom Ribbon button match any built-in button by using the CommandBars object. This object has been extended with several get methods that expose the state information for the built-in controls: GetEnabledMso, GetImageMso, GetLabelMso, GetPressedMso, GetScreentipMso, GetSupertipMso, and GetVisibleMso. Use these methods in your callbacks to check the built-in control's properties. For example, the following statement will return False if the Ribbon's built-in Cut button is currently disabled (grayed out), and True if it is enabled (ready to use):

```
MsgBox Application.CommandBars.GetEnabledMso("Cut")
```

Notice that the GetEnabledMso method requires that you provide the name of the built-in control. To see the result of the preceding statement, simply type it in the Immediate window and press Enter.

The GetImageMso method is very useful if you'd like to reuse any of the built-in button images in your own controls. This method allows you to get the bitmap for any imageMso tag. For example, to retrieve the bitmap associated with the Cut button on the Ribbon, enter the following statement in the Immediate window:

```
MsgBox Application.CommandBars.GetImageMso("Cut", 16, 16)
```

The preceding GetImageMso method uses three arguments: the name of the built-in control, and the width and height of the bitmap image in pixels. Because this method returns the IPictureDisp object, it is very easy to place the retrieved bitmap onto your own custom Ribbon control by writing a simple VBA callback for your control's getImage attribute.

In addition to the methods that provide information about the properties of the built-in controls, the CommandBars object also includes a handy ExecuteMso method that can be used to trigger the built-in control's default action. This method is quite useful when you want to perform a click operation for the user from within a VBA procedure or want to conditionally run a built-in feature.

Let's take a look at the example implementation of the GetImageMso and ExecuteMso methods. Here's the XML definition for a custom Ribbon button (see Figure 29.43):

```
<button id="btnRptWizard" label="Use Report Wizard" size="normal"
  getImage="onGetBitmap" onAction="DoDefaultPlus" />
```

The preceding XML code can be added to any of the custom Ribbon definitions you've already defined in the USysRibbons table. Now let's look at the VBA part. You want the button to use the same image as the built-in button labeled Report Wizard. When the button is clicked, you'd like to display the built-in Report Wizard dialog box only when a certain condition is true. Here is the code you need to add to your VBA module:

```
Sub onGetBitmap(ctl As IRibbonControl, ByRef image)
  Set image = Application.CommandBars. _
  GetImageMso("CreateReportFromWizard", 16, 16)
End Sub
```

When the Ribbon is loaded, the onGetBitmap callback automatically retrieves the image bitmap from the Report Wizard button's imageMso attribute and assigns it to the getImage attribute of your button. When your button is clicked and the Student List form is open, the Report Wizard dialog box will pop up; if the specified object is not open, the user will see a message box:

```
Sub DoDefaultPlus(ctl As IRibbonControl)
  If Application.CurrentObjectName = "Student List" Then
    Application.CommandBars.ExecuteMso "CreateReportFromWizard"
  Else
    MsgBox "To run this Wizard you need to open " & _
     " the Student List Form", _
    vbOKOnly + vbInformation, "Action Required"
  End If
End Sub
```

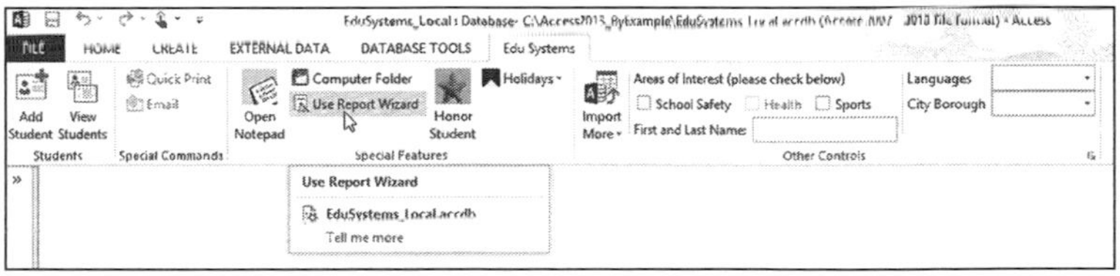

FIGURE 29.43. The custom Use Report Wizard button in the Special Features group of the Edu Systems tab uses a built-in image and runs a built-in Access feature based on the condition specified in the callback assigned to its onAction attribute.

You will find the XML markup discussed in this section in the EduSystems_15_withCommandBars.txt file in your C:\Access2013_HandsOn folder.

TAB ACTIVATION AND GROUP AUTO-SCALING

Tab activation makes it possible to activate a specific tab in response to some event. To activate a custom tab on the Access 2013 Ribbon, use the `ActivateTab` method of the IRibbonUI object by passing to it the ID of the custom string. For example, to activate the Edu Systems tab you created in this chapter, try the following statement in the Immediate window while any of the default Access tabs is active:

```
objRibbon.ActivateTab "custTabEdu"
```

Recall that `objRibbon` is the module-level Public variable we declared earlier for accessing the IRibbonUI object. To activate a built-in tab, use the `ActivateTabMso` method. For example, the following statement activates the Create tab:

```
objRibbon.ActivateTabMso "TabCreate"
```

Finally, there is also a special `ActivateTabQ` method used to activate a tab shared between multiple add-ins. In addition to the tabID, this method requires that you specify the namespace of the add-in. The syntax is shown here

```
expression.ActivateTabQ(tabID As String, namespace as String)
```

where `expression` returns an IRibbonUI object. Keep in mind that tab activation applies only to tabs that are visible.

Group auto-scaling enables custom Ribbon groups to change their layout when the user resizes the window (see Figures 29.44 and 29.45). You can enable auto-scaling by setting the autoScale attribute of the <group> tab to `true` as in the following:

```
<group id="ImagesGroup" label="Special Features" autoScale="true">
```

Notice that the value of the autoScale attribute is entered in lowercase. Autoscaling is set on a per-group basis.

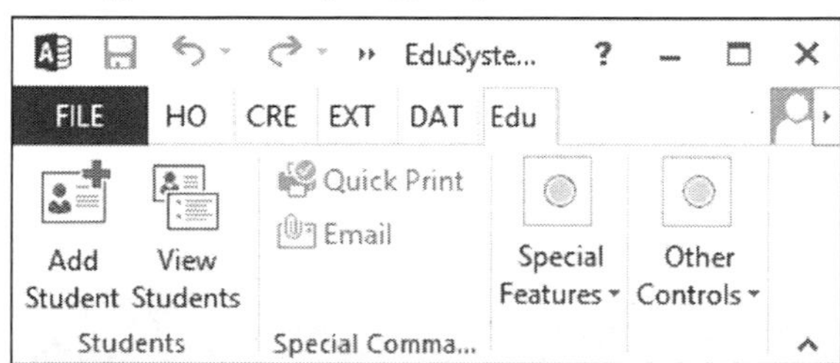

FIGURE 29.44. The commands in the Special Features and Other Controls groups of the Ribbon are automatically compressed to a single button when the Access application window is made smaller. To change the icon that appears when the group is compressed, assign an image to the group itself.

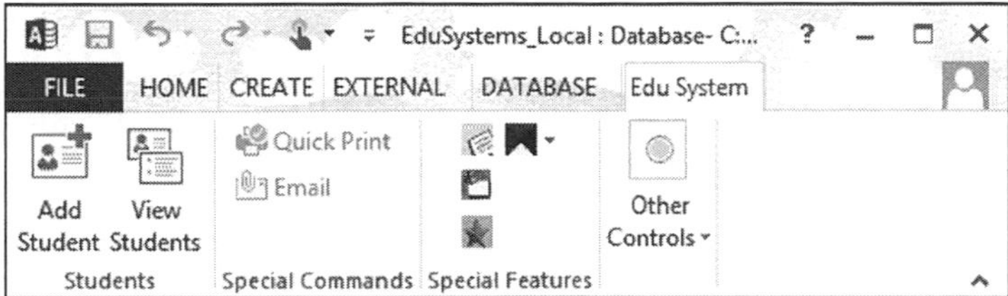

FIGURE 29.45. When you set the autoScale attribute to true, the group of controls in Special Features will change its layout to best fit the resized window.

You will find the Ribbon customizations discussed in this section in the EduSystems_16_WithAutoScale.txt file in your Access2013_HandsOn folder.

CUSTOMIZING THE BACKSTAGE VIEW

The Access File tab provides an entry point to a part of the Office UI known as Backstage View. This view is specifically designed for working with a database as a whole. It contains commands known as *Fast commands* that provide quick access to common functionality such as saving, opening, or closing a database. Here you also find the Exit command for exiting Access and the Options command for customizing numerous Access features. In addition to Fast commands, the navigation bar on the left hand side of the Backstage View includes several tabs that group related tasks. For example, clicking the Print tab in the navigation bar displays all the information related to the installed printers and allows you to easily access and change many of the print settings. The Info tab organizes tasks related to compacting and repairing a database and encrypting it with a password. As an Access developer already familiar with Ribbon UI customization, you should feel very comfortable customizing the Backstage View. Like the Ribbon, the Backstage View uses XML markup. The Backstage View is a perfect place to include custom solutions that present summaries of business processes or workflows (see the sidebar with links to Microsoft documents that will walk you through the process of customizing the Office 2013 Backstage View). In this section you'll perform a couple of simple operations in the Backstage View to get your feet wet so that you can later move on to more advanced customizations with the downloads recommended in the sidebar.

Backstage View Development

For an advanced introduction to the Backstage View, you may want to download the following Microsoft papers that apply to the current and previous versions of Microsoft Office:
Customizing the Office 2010 Backstage View for Developers:

http://msdn.microsoft.com/en-us/library/ee815851.aspx

Dynamically Changing the Visibility of Groups and Controls in the Office 2010 Backstage View:

http://msdn.microsoft.com/en-us/library/ff645396.aspx

The Backstage View XML markup should be entered between <backstage> </backstage> elements within the <customui> </customui> tags and below any Ribbon customization markup. The following XML markup adds a custom button named Synchronize and a custom tab named Endless Possibilities to the Backstage View:

```
<backstage>
<button id="btnSync" label="Synchronize" imageMso="SyncNow"
isDefinitive="true"
insertBeforeMso="FileClose" onAction="onActionCopyToArchive" />
<tab id="mySpecialTab" label="Endless Possibilities"
insertAfterMso="TabRecent">
<firstColumn>
<group id="grp01" label="Home Group"
helperText="This is group 1 help text">
<topItems>
<button id="myButton1" label="My button" />
</topItems>
</group>
<group id="gr02" label="Cheat Sheet">
<topItems>
<button id="myButton2" label="Cheat Ideas" />
</topItems>
<bottomItems>
<layoutContainer id="set1" layoutChildren="horizontal" >
<editBox id="item1" label="Cheat Item 1" />
<editBox id="item2" label="Cheat Item 2" />
</layoutContainer>
</bottomItems>
</group>
</firstColumn>
<secondColumn>
<group id="grpHyperlinks"
label="Frequently Accessed Websites" visible="true">
```

```
<primaryItem>
<button id="top1" label="Primary Button"
imageMso="HyperlinkProperties" />
</primaryItem>
<topItems>
<hyperlink id="msft" label="Microsoft"
getTarget="onActionExecHyperlink" />
<layoutContainer id="set2" layoutChildren="vertical" >
<hyperlink id="YouTube" label="YouTube"
getTarget="onActionExecHyperlink" />
<hyperlink id="amazon" label="Amazon"
getTarget="onActionExecHyperlink" />
<hyperlink id="merc" label="Mercury Learning and Information"
getTarget="onActionExecHyperlink" />
</layoutContainer>
</topItems>
</group>
</secondColumn>
</tab>
</backstage>
```

You will find the preceding Backstage View customization in the EduSystems_17_withBackstageView.txt file in your Access2013_HandsOn folder. The resulting Backstage customization is shown in Figure 29.46.

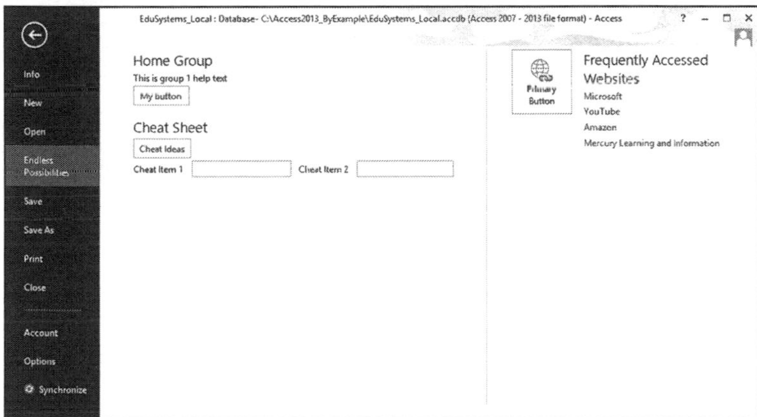

FIGURE 29.46. The Backstage View is highly customizable. The Synchronize button and the Endless Possibilities tab were created by adding custom XML markup to the USysRibbons system table.

In the preceding example XML markup, the <button> element is used to incorporate into the Backstage View navigation bar a custom command labeled Synchronize:

```
<button id="btnSync" label="Synchronize" imageMso="SyncNow"
isDefinitive="true" insertBeforeMso="FileClose" onAction="onActionCopyTo
Archive" />
```

The <button> element contains the isDefinitive attribute. When this attribute is set to `true`, clicking the button will trigger the callback procedure defined in the onAction attribute and then automatically close the Backstage View. The onAction callback for the custom Synchronize button is shown here. The callback calls the CreateDbCopy procedure that allows you to make a copy of the specified database. Be sure to enter the procedure code in the VBA code module of the EduSystems_Local database.

```
Sub onActionCopyToArchive(ctl As IRibbonControl)
  CreateDbCopy
End Sub

Sub CreateDbCopy()
  Dim fso As Object
  Dim dbName As String
  Dim dbNewName As String

  On Error GoTo ErrorHandler

  Set fso = CreateObject("Scripting.FileSystemObject")

  dbName = InputBox("Enter the name of the database " & _
    "you want to copy: " & _
    "(C:\Access2013_ByExample\Chap25.accdb)", _
    "Create a copy of")

  If dbName = "" Then Exit Sub
  If Dir(dbName) = "" Then
    MsgBox dbName & " was not found. " & Chr(13) _
    & "Check the database name or path."
    Exit Sub
  End If

  dbNewName = InputBox("Enter the name for the " & _
    "copied database:" & Chr(13) & _
    "(C:\Access2013_ByExample\Chap25Ver2.accdb)", _
    "Save As")
  If dbNewName = "" Then Exit Sub

  If Dir(dbNewName) <> "" Then
    Kill dbNewName
  End If

  fso.CopyFile dbName, dbNewName
  Set fso = Nothing
  Exit Sub
ErrorHandler:
  MsgBox Err.Number & ":" & Err.Description
End Sub
```

The Backstage View XML markup also adds to the Backstage View navigation bar a custom tab labeled Endless Possibilities. Each <tab> element can have one or more columns. Our example contains two columns. Each tab can contain multiple <group> elements. Here we have two groups in the first column and one group in the second column. The Backstage group can contain different types of controls. You can group the controls into the following three types of sections:

<primary item>	This element is used to specify the most important item in the group. The primary item control can be a button or a menu with buttons, toggle buttons, checkboxes, or another menu.
<topItems>	This element defines controls that will appear at the top of the group.
<bottomItems>	This element defines the controls that will appear at the bottom of the group.

The layout of controls in the Backstage View is defined using the <layoutContainer> element. This element's layoutChildren attribute can define the layout of controls as horizontal or vertical. The second column of our example XML markup uses the onActionExecHyperlink callback procedure for the hyperlinks shown in Figure 29.46. Enter this procedure in the VBA code module of the EduSystems_Local database:

```
Sub onActionExecHyperlink(ctl As IRibbonControl, _
  ByRef target)
  Select Case ctl.ID
  Case "YouTube"
    target = "http://www.YouTube.com"
  Case "amazon"
    target = "http://www.amazon.com"
  Case "merc"
    target = "http://www.merclearning.com"
  Case "msft"
    target = "http://www.Microsoft.com"
  Case Else
    MsgBox "You clicked control id " & ctl.ID & _
    " that has not been programmed!"
  End Select
End Sub
```

Hiding Backstage Buttons and Tabs

The following XML will hide the Options button in the Backstage View navigation bar:

```
<button idMso="ApplicationOptionsDialog" visible="false" />
```

The Backstage View uses the following button IDs: `FileSave`, `FileSaveAs`, `FileOpen`, `FileClose`, `ApplicationOptionsDialog`, and `FileExit`.

To hide the Info tab in the Backstage View, use this markup:

```
<tab idMso="TabInfo" visible="false" />
```

The Backstage View tab IDs are as follows: `TabInfo`, `TabRecent`, `TabNew`, `TabPrint`, `TabShare`, and `TabHelp`.

Things to Remember when Customizing the Backstage View.

- The maximum number of allowed tabs is 255.
- You cannot reorder built-in tabs.
- You can add your custom tab before or after the built-in tab.
- You cannot modify the column layout of any built-in tab.
- You cannot reorder built-in groups; however, you can specify the order of groups you create.

CUSTOMIZING THE QUICK ACCESS TOOLBAR (QAT)

The Quick Access toolbar that appears just above the File tab gives application users quick access to tools they use most frequently. These tools can be easily added to the toolbar by selecting More Commands from the Customize Quick Access Toolbar drop-down menu. The Quick Access toolbar can only be customized in the start from scratch mode by setting the startFromScratch attribute to `true` in the Ribbon XML customization file:

```
<ribbon startFromScratch="true">
```

The preceding XML markup will hide all built-in tabs. You must add your own custom tabs as demonstrated earlier in this chapter. Quick Access toolbar modifications are specified using the <qat> element. Within this element you should use the <documentControls> element to specify the controls that you want to appear in the Quick Access toolbar. The following XML markup creates the custom Quick Access toolbar shown in Figure 29.47. You will find this code in the

CustomUI_ QAT.txt file located in your Access2013_HandsOn folder.

```
<customUI
xmlns="http://schemas.microsoft.com/office/2009/07/customui" >
<ribbon startFromScratch="true">
  <qat>
    <documentControls>
      <button id="btnCalc2" label="Calculator"
      imageMso="SadFace" onAction="OpenCalculator" />
      <button idMso="FilePrintQuick" />
    </documentControls>
  </qat>
</ribbon>
</customUI>
```

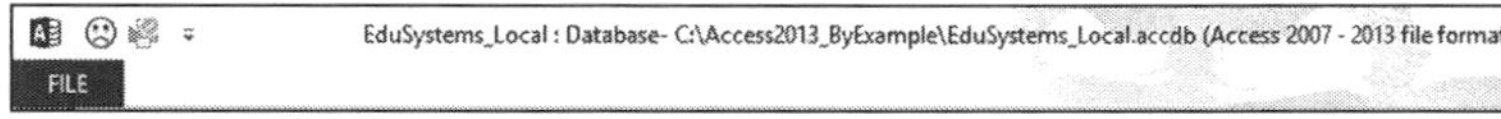

FIGURE 29.47. Customized Quick Access toolbar.

The button labeled Calculator that is represented by the SadFace image calls the OpenCalculator procedure shown here:

```
Public Sub OpenCalculator(ctl As IRibbonControl)
    Shell "Calc.exe", vbNormalFocus
End Sub
```

Enter this procedure in the VBA code module of the EduSystems_Local.accdb database.

CHAPTER SUMMARY

This chapter introduced you to using and customizing the user interface in Access 2013. After a short overview of the initial Microsoft Access screen and the Quick Access toolbar, we looked at numerous features of the Access Navigation pane. You learned how to use the Navigation pane to access and organize your database objects by using both manual techniques and VBA code. Next, we briefly covered the Ribbon interface to get you warmed up and ready for the Ribbon customization exercises. You learned how to create XML Ribbon customization markup and load it in your database by using the `LoadCustomUI` method of the Application object or via a special Access system table called USysRibbons. You also learned how Ribbon customizations can be assigned to forms or reports. You spent quite a bit of time in this chapter familiarizing yourself with various controls that can be added to the Ribbon and writing callback procedures in order to set your controls' attributes at runtime. In addition to Ribbon customizations, you learned how to modify the Quick Access toolbar.

While this chapter introduced many controls and features of the Ribbon, it did not attempt to cover all there is to know about this interface. After all, this book is about VBA programming in Access in general, not just the Ribbon. The knowledge and experience you gained in this chapter can be applied to customizing the Ribbon in all of the Microsoft Office 2013 applications.

In the next chapter, we will take a look at Access templates and macros.

Part 8 VBA and Macros

Writing VBA code is not the only way to provide rich functionality to your Access database users. Macros have long been used to enhance the user experience without writing a single line of VBA code. The Macro Designer allows you to include complex logic, business rules, and error handling in your macros.

In this part of the book, you are introduced to three types of macros that you can create in Access 2013. In addition, you learn how to convert macros to VBA and get started with built-in templates that extensively use macros.

Chapter 30 Macros and Templates

When programming Access applications, there are two other areas of Access that you need to be acquainted with: macros and templates. Macros in Access have been around longer than the Visual Basic for Applications language. When Access 2 came out in 1992, it included a macro language called Access Basic that contained a subset of Visual Basic 2.0's core syntax. Access 95 replaced Access Basic with Visual Basic for Applications, but until Access 97, macros were the most common means of automating database tasks. When Access 2000 came out, many successful macro users had already moved to the new programming platform to take advantage of the language model that offered more control over Access. In fact, in versions 2000 through 2003, Microsoft recommended VBA to automate Access applications, and macros were supported mainly for backward compatibility.

The outlook on macros changed with the release of Access 2007. After performing some extensive research, Microsoft found out that many users were intimidated by the programming environment that Access provided but were quite successful at creating macros. It seems that it is much simpler to pick a macro action and set a couple of parameters than it is to write VBA code. Because most of the Access applications created by end users are loaded with macros, Microsoft decided to improve the "macro experience" in Access 2007 by adding event handling, temporary variables (TempVars), better error handling, and a new type of macro called an embedded macro. In Access 2010, Microsoft has added a Macro sandbox, which is related to the security model introduced in the 2007 release. Access 2010 also brought a powerful enhancement known as data macros. This chapter focuses on the macro features available in Access 2013. After we've discussed macros, we will take a look at the .accdt file format used with Access desktop database templates.

MACROS OR VBA?

You can use both VBA and macros to automate your Microsoft Access applications. While macros have become very powerful in Access applications, whether you use macros or VBA will depend on what you want to do. Macros can perform just about any task you can do with the Access user interface by using the keyboard or the mouse. They provide an easy way of opening and closing various Access objects (tables, queries, forms, and reports). You can also use them to automate repetitive tasks, execute commands on the Access Ribbons, set values for form and report controls, import and export spreadsheet and text files, display informative messages, or even sound a beep. With data macros you can also enforce business rules at a table level. These are just a few examples of what macros can do.

What macros cannot do is create and manipulate database objects the way we did in VBA earlier in this book by using DAO or ADO, or step through the records in a recordset and perform an operation on each record. You need to write VBA code to perform these types of operations. You must also use VBA when you need to pass parameters to your Visual Basic procedures, call dynamic link libraries (DLLs), create custom functions, or find out whether a file exists on the system. Even if you don't want to get started with macros now that you know how to write code in VBA, you still need to understand how macros are used in Access 2013 as Microsoft makes extensive use of macros in their templates and the built-in Button Wizard creates embedded macros.

ACCESS 2013 MACRO SECURITY

In Microsoft's documentation, the term "macro security" applies to macros and VBA, as well as other executable content that could be harmful when allowed to run. In Chapter 1, "Writing Procedures in Modules," we specified that Access should trust any database file opened from the **C:\Access2013_ByExample** folder (see Hands-On 1.4). This enabled you to work with this book's examples without having to constantly deal with the Access security warning. However, if you attempt to open a file that contains macros and that file is not located in a trusted location, Access will determine whether to display a security alert by checking your macro settings (see Figure 30.1). You can change your macro settings at any time by following these steps:

1. Click the **File** tab, then click **Options**.
2. In the Access Options dialog box, click the **Trust Center** tab, then click **Trust Center Settings**.
3. In the Trust Center dialog box, select **Macro Settings**.

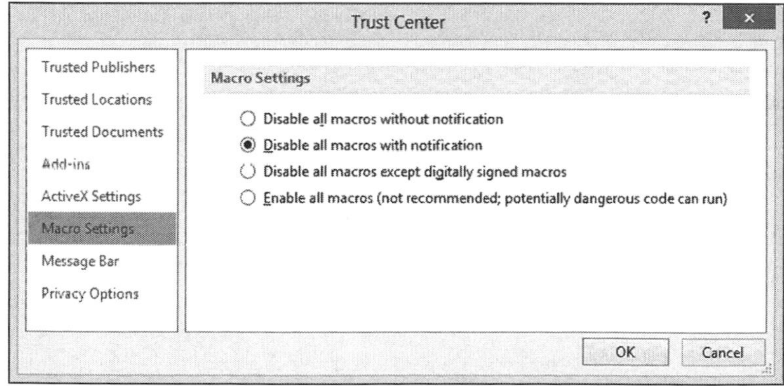

FIGURE 30.1. The Macro Settings options allow you to specify whether the macros should be disabled or allowed to run and whether you should see a notification when macros are disabled.

If the Disable all macros with notification option is selected, you may want to leave that setting as is. This option allows you to enable the disabled content only for this session by clicking the Enable Content button in the Security Warning message bar when a database file is opened.

You can access advanced security options by clicking the message text to the left of the Enable Content button in the Security Warning message bar. This will activate the Backstage view where you can click the Enable Content button to bring up a menu of additional options as shown in Figure 30.2. When you click Advanced Options, Access displays the dialog box shown in Figure 30.3.

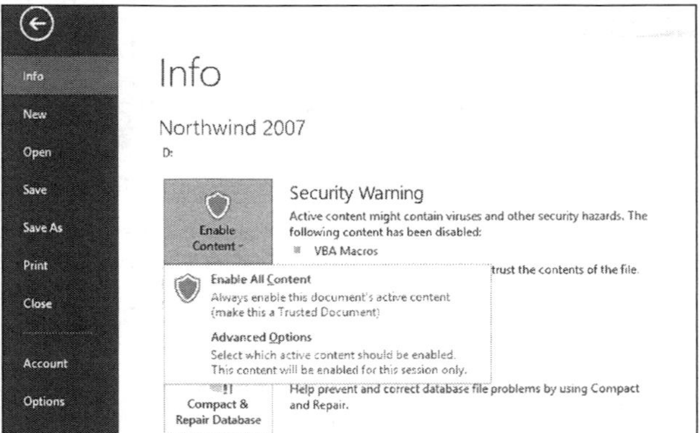

FIGURE 30.2. The Info tab in Backstage view displays information related to the Security Warning message and a brief description of the active content. By clicking on the Enable Content button, you can either enable all content in the current database or choose advanced options that allow you to specify which active content should be enabled.

FIGURE 30.3. The Microsoft Office Security Options dialog box allows you to temporarily enable disabled programming content by selecting the Enable content for this session radio button.

If you select the first radio button in Figure 30.3, Access will open the database in Sandbox mode, meaning it will turn off all executable content such as:

- VBA code and any references to it
- Unsafe expressions

  An unsafe expression contains functions that could allow a user to modify the database or gain access to resources outside the database.

- Unsafe macro actions

  These are actions that could allow a user to modify the database or gain access to resources outside the database.
- Certain types of queries such as:
  - Action Queries

    These are queries that could allow a user to make unauthorized additions, changes or deletions of database data.
  - Data Definition Language (DDL) Queries

    These are queries that are used to create or alter objects in a database, such as tables and procedures.
  - SQL Pass-Through Queries

    These queries allow a user to send commands directly to a database server that supports the Open Database Connectivity (ODBC) standard.
- ActiveX controls

These are small programs that have unrestricted access to your computer's file system that could be used to take control of your computer.

If you plan on distributing your Access database in the new . accdb file format, you can use the IsTrusted property of the CurrentProject object to test on loading whether your application has its executable content disabled. Use this property in an AutoExec macro to check whether your application can load (see the next section).

USING THE AUTOEXEC MACRO

The most important macro that every Access programmer needs to be familiar with is the AutoExec macro. This macro is not new in Access 2013; it's been with Access since the very beginning. An AutoExec macro in your Access application will automatically run when the database is opened. This is very convenient, especially when you need to check whether the rest of your application will load. Let's see how Microsoft does this in the Northwind 2007 database.

Please note files for the "Hands-On" project may be found on the companion CD-ROM.

Hands-On 30.1. Understanding and Using the AutoExec Macro

1. Copy the **C:\Access2013_ByExample\Northwind 2007.accdb** database to your desktop or any other new folder that you want to create for this hands-on exercise.
2. Double-click the copied database file to open it.
   When Access starts, notice the appearance of the Security Warning message bar with the Enable Content button.
3. Click the **Enable Content** button.
4. Dismiss the Login dialog box by clicking the "X" in the upper-right corner.
5. In the Navigation pane, select **All Access Objects** and activate the **Macros** group. Right-click the **AutoExec** macro name and choose **Design View**. Access displays the contents of the AutoExec macro, as shown in Figure 30.4.

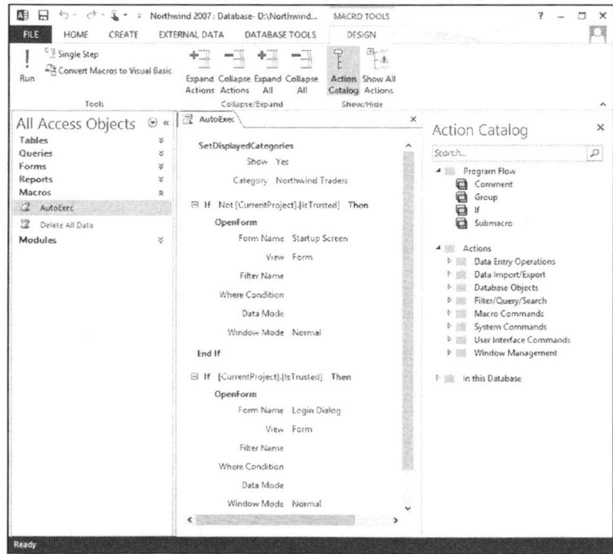

FIGURE 30.4. The contents of the AutoExec macro in the Northwind 2007 sample database as shown in Access 2013.

6. Close the AutoExec macro in the same way you close any other Access window (by clicking the "X" button in the window's upper-right corner or right-clicking the **AutoExec** tab and choosing **Close**).

Understanding Macro Actions, Arguments, and Program Flow

A macro can have more than one action, but you must specify at least one action when you create a macro. When you open the Macro Designer, the only thing you see is a drop-down list of macro actions. The macro design area will expand to show more options when you make a selection from the drop-down list. For example, if the action requires additional data, a list of arguments is displayed. Access 2013 has a long list of macro actions to pick from. If you are not sure which action to select to perform a particular task, you can browse the Action Catalog that appears to the right of the macro window (see Figure 30.4). All available macro actions are grouped by subject in the Action Catalog. You can see the description of the selected macro action at the bottom of the Action Catalog (see Figure 30.5). In addition to a hierarchical listing of macro actions, the Action Catalog contains several program flow constructs that you can apply to your macros. These are shown at the top of the Action Catalog. *Comments* should be used to document your macros. Macro *groups* make it easy to organize your macro actions in a named block that can be easily collapsed, moved, or copied. The *If* construct allows you to create macros based on a condition. Your condition could test a value in a field or evaluate the result of a function. You can use any expression that evaluates to True/False (Yes/No). To add conditional logic to your macro, double-click or drag the `If` to the macro design area. The macro actions will execute when the condition defined at the top of the `If` block is true. If the condition is not true, the action will be skipped, and the macro control will move to the next row. The actions that should not be executed when the condition is true are preceded with `Not`, as shown in Figure 30.4.

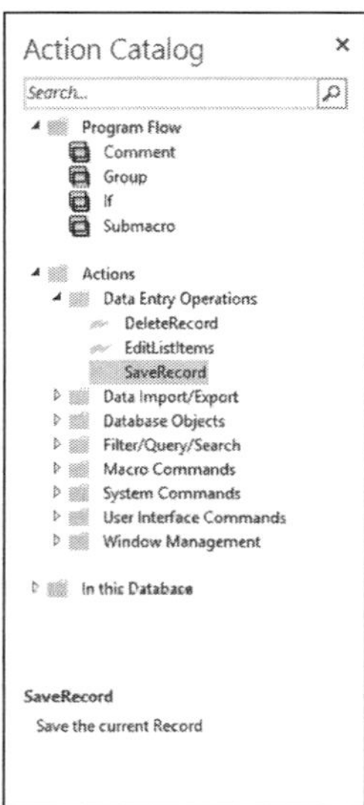

FIGURE 30.5. The Action Catalog in Access 2013. The description of the selected macro action appears at the bottom of the window.

To see all the available actions in the catalog, click the Show All Actions button on the Ribbon.

Actions that are considered unsafe are denoted by a yellow warning sign to the left of the macro action name.

The AutoExec macro included in the Northwind 2007 database uses the following macro actions:

- **SetDisplayedCategories**—This action (found in the User Interface Commands section of the Action Catalog) is used to specify which categories are displayed under Navigate to Category in the titlebar of the Navigation pane. This action has two arguments. The `Show` argument can be set to Yes to show the category name or No to hide it. The `Category` argument specifies the name of the category you want to show or hide. The Northwind 2007 database contains a custom category named Northwind Traders, so the macro starts by displaying this category in the Navigation pane.
- **OpenForm**—This action (found in the Database Objects section of the Action Catalog) is used to open any form. The form can be selected from a drop-down list when you click the Form Name box (see Figure 30.6). All forms in the current database will be shown. You can also specify the view in which the form will open. The default view is Form; you can select the view from the View drop-down box. Not all arguments need to be filled in. You can easily look up the meaning of an action's arguments by moving your mouse over the argument name. The second `If` block in the AutoExec example shown in Figure 30.6 tells Access to open the Login dialog box in Form view using the Normal window. Notice that values for some arguments (`Filter Name`, `Where Condition`, and `Data Mode`) are not provided.

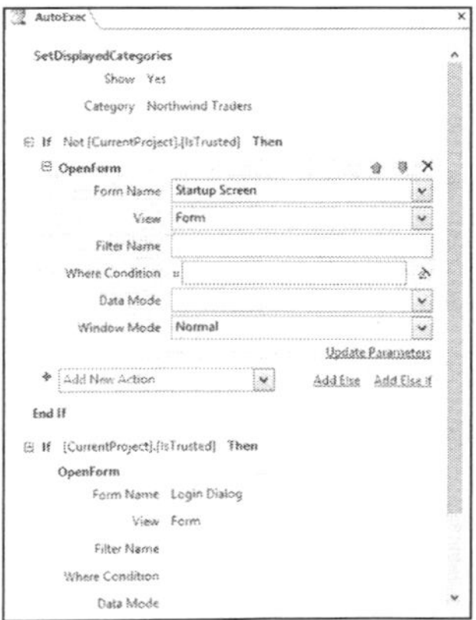

FIGURE 30.6. The AutoExec macro is shown here in Edit mode.

Whether a form opens can depend on a certain condition being met; in this example, the first `If` block tells Access to show the startup screen only if the current project (the database) is not trusted:

```
If Not [CurrentProject].[IsTrusted]
```

Notice how the IsTrusted property of the CurrentProject object is used to test whether your application has its executable content disabled. You saw this code block execute when you opened the Northwind 2007 database.

The next `If` block loads the Login dialog box if `[CurrentProject].[IsTrusted]` is true. This code block was executed when you told Access to "Enable Content" (see Step 3 in Hands-On 30.1).

To open an Access database without running the AutoExec macro, hold down the Shift key while opening the database.

CREATING AND USING MACROS IN ACCESS 2013

Access 2013 supports three types of macros: standalone macros (also used in versions of Access prior to 2007), embedded macros (introduced in Access 2007), and data macros (introduced in Access 2010). Standalone macros are visible in the Navigation pane under Macros, while embedded macros are part of the object in which they are embedded (form, report, or control) and therefore are not visible in the Navigation pane. Data macros allow developers to implement business rules in an Access application. These macros do not have a user interface; they are applied at the table level and cannot be used to open a form or a report. In the following sections, we take a closer look at each of these macro types.

Creating Standalone Macros

The AutoExec macro we looked at in the previous section is a standalone macro. Once created, this macro appears in the Navigation pane.

The general steps to create a standalone macro are as follows:

1. Click **Macro** in the Macros & Code group of the Create tab (Figure 30.7).

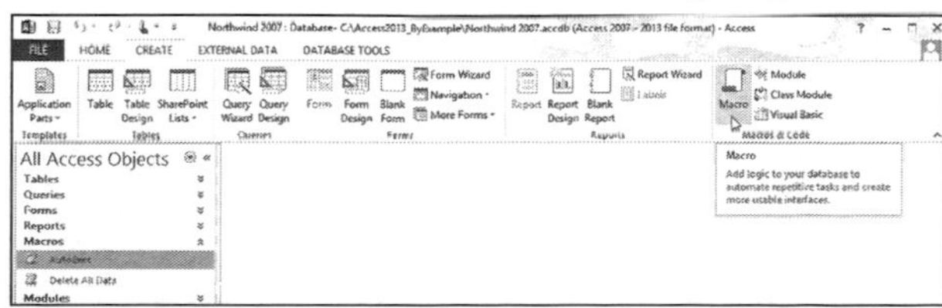

FIGURE 30.7. Creating a standalone macro.

Access displays the Macro Designer window with one drop-down box, as shown in Figure 30.8. As you can see, the Macro Designer layout has a collapsible drop-down interface.

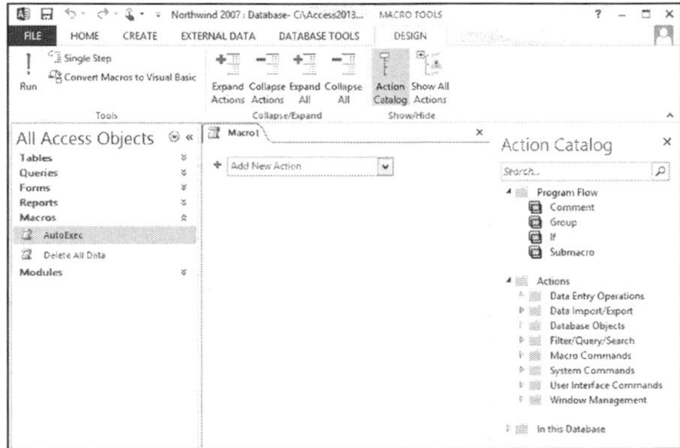

FIGURE 30.8. The initial Macro Designer window.

2. Choose the action from the drop-down list. When the Ribbon's Show All Actions button is selected, this list displays all the available macro actions. When this button is not selected, you will see a shorter list of actions that are allowed to run even if the database is not trusted.

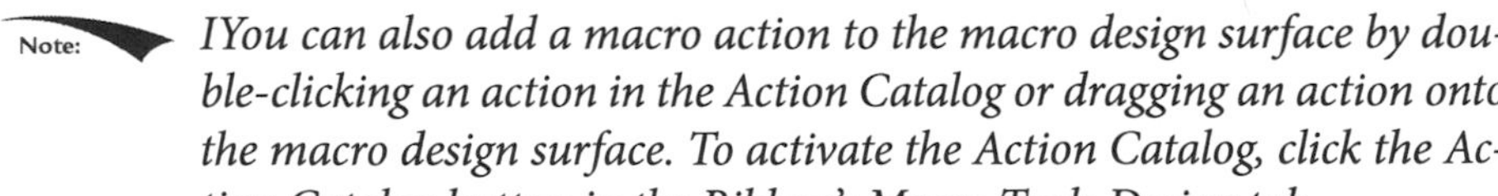

Note: *IYou can also add a macro action to the macro design surface by double-clicking an action in the Action Catalog or dragging an action onto the macro design surface. To activate the Action Catalog, click the Action Catalog button in the Ribbon's Macro Tools Design tab.*

3. If the macro action you select requires arguments, Access displays an inline dialog box where you can specify the required values (see Figure 30.9). Default argument values are prefilled for you. Notice that the code blocks are collapsible. You can expand or collapse the code areas by clicking the +/- controls to the left side of the code block or using buttons in the Collapse/Expand group of the Ribbon.

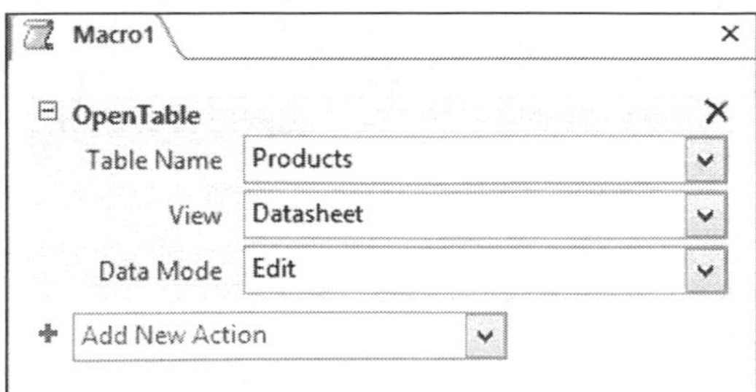

FIGURE 30.9. The OpenTable macro action opens a table. You need to specify the required arguments: the name of the table to open, the type of the view for the presentation of the data, and the Data Mode.

4. If desired, add another macro action as shown in Figure 30.10.

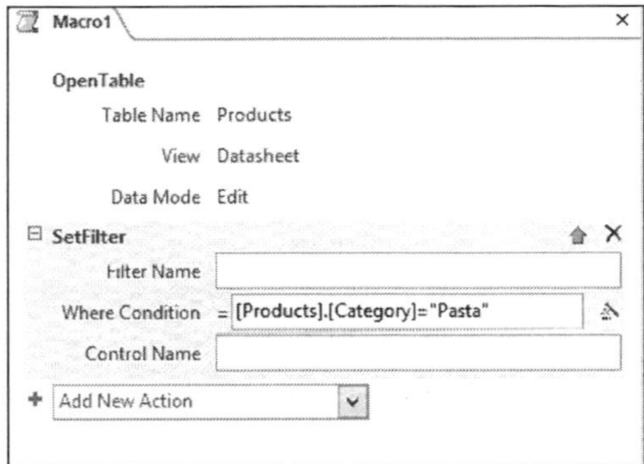

FIGURE 30.10. You can restrict the number of records in a table by using the SetFilter macro action.

5. If the macro action should be conditionally executed (see Figure 30.11), choose the `If` block from the Actions drop-down box. Type your conditional expression in the text box or click the Builder button next to the expression box to invoke the Expression Builder. Because the macro actions within the `If` block only run when the conditional expression resolves to True, the expression you enter must be of the Boolean type (True/False). If the condition evaluates to False, the action specified within the `If` block will be skipped.

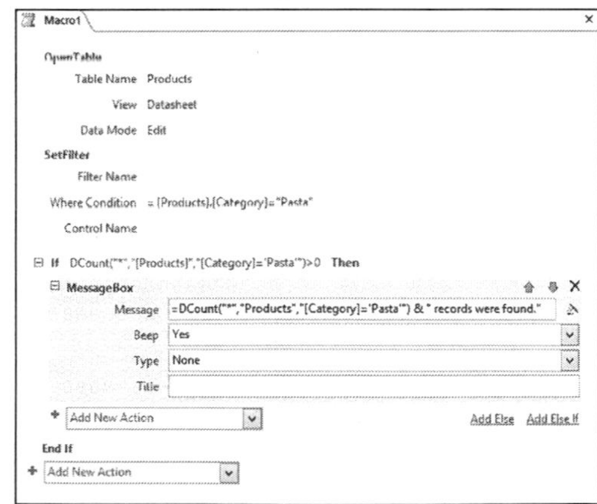

FIGURE 30.11. Adding conditions to the macro.

In Microsoft Access 2007 and earlier, you could write only simple conditional statements in the Macro Designer's Condition column. In Access 2013/2010, Macro Designer allows you to create complex conditions by using the `Else If` *and* `Else` *statements. To include these statements, click the Add Else If or Add Else hyperlinks in the lower part of the code block (see Figure 30.11).*

Expression Builder in Access 2013

Expressions are an important part of an Access application. They are used in tables, queries, forms, reports, and macros to evaluate and test data, perform calculations, manipulate character strings, and specify the logic that drives the behavior of your database application. Expressions in Access are similar to formulas and functions used in Excel. Depending on their complexity, they can contain user-defined or built-in functions, operators, identifiers, and constants. Building expressions in Access is easy thanks to the Expression Builder (see Figure 30.12). The Expression Builder offers the IntelliSense feature that provides guidance as you type an expression. If you remember syntax and available functions and properties, you can enter your expression from scratch in the provided expression box. Otherwise, you can select the expression elements, categories, and values from the appropriate panes in the lower part of the Expression Builder window. Notice in Figure 30.12 that in addition to expression elements (Functions, Constants, and Operators), the Expression Elements pane also provides quick access to the Common Expressions. These include prebuilt expressions for displaying page numbers and the current date and time.

To access the Expression Builder, click the Expression Builder icon shown in Figure 30.10.

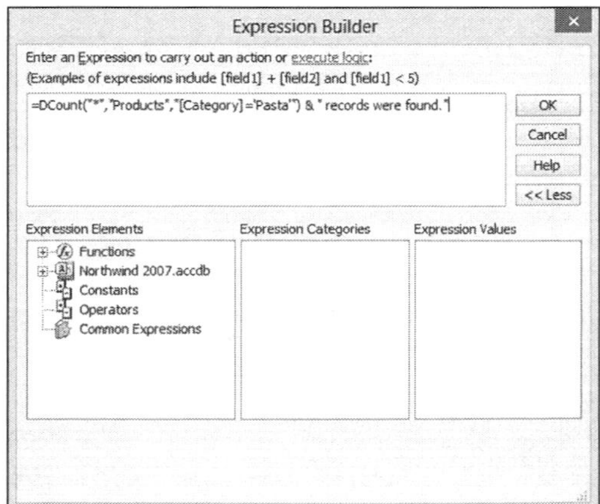

FIGURE 30.12. Building a macro expression using Expression Builder in Access 2013.

6. To make your macro actions easy to understand for yourself and others, you can add comments to the macro (see Figure 30.13). Comments are optional. To add a comment, choose **Comment** from the Actions drop-down box and

type the text in the provided box. You can also type // in an Add New Action drop-down box. Comments are easy to spot because they appear as green text. You can move the comment to the appropriate location in your macro by clicking the Move Up or Move Down arrows to the right of the comment box.

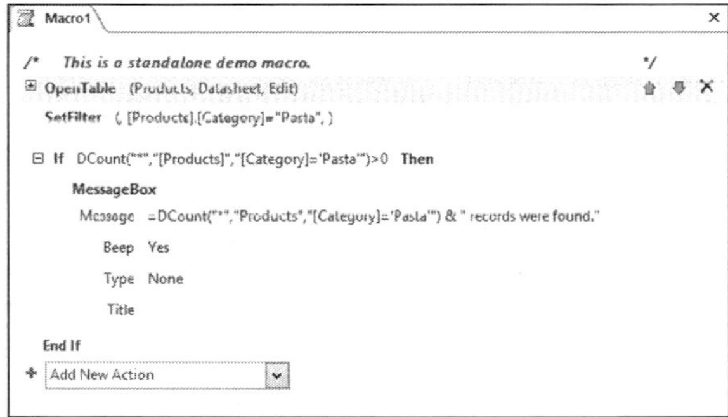

FIGURE 30.13. Comments can be added anywhere within your macro code block.

7. To add another action to your macro, select an action from the Actions drop-down. To add an action between the actions you've already entered, first select the desired action from the Actions drop-down, and then move it to the appropriate location within your macro using the Move Up or Move Down arrows.

   To delete an action, select it and click the X button. You can also right-click the action and choose **Delete** from the menu.

If you add an action that is considered "unsafe," Access displays a yellow warning sign to the left of the macro action name. An unsafe action will not execute if the database is not trusted.

8. For more complex macros you may want to use a program flow construct known as a *group*. With this construct you can put multiple actions and program flow into a group block so you can expand or collapse an entire group for better readability.
9. When you are done entering all actions for your macro, press **Ctrl+S** to save your macro, or click the **Save** button on the Quick Access toolbar. Enter the macro name in the Save As dialog box and click **OK** (see Figure 30.14).

Save As ? ×

Macro Name:

mcrOpenProductsTbl_Pasta

OK Cancel

FIGURE 30.14. Saving a macro.

10. Close the Macro Designer window. The saved macro appears in the Navigation pane.

Running Standalone Macros

You can run standalone macros from the Design view, the Navigation pane, another macro, or a VBA procedure, or in response to an event on a form, report, or control.

- **Running a macro from the Design view**—If the standalone macro is open in the Design view, you can click the Run button in the Tools group of the Design tab to run the macro.

  You can also run your macro one action at a time by selecting the Single Step button and then clicking the Run button (see "Error Handling in Macros" later in this chapter).

- **Running a macro from the Navigation pane**—A standalone macro can be run directly from the Navigation pane. Simply right-click the macro name and choose Run from the shortcut menu, but make sure you know what the macro will do before you run it. A badly designed macro could wipe out all the data in your database without asking you if you want to proceed.

  When you right-click a macro containing submacros in the Navigation pane and choose Run, only the first submacro will execute (see "Creating and Using Submacros" in the next section).

- **Running a macro from another macro**—To run a macro from another macro, you must create at least two macros. The main macro should include the RunMacro action. Set the `Macro Name` argument of this action to the name of the macro you want to run. When you run the main macro, both macros will execute.

- **Running a macro from a VBA procedure**—The `RunMacro` method of the DoCmd object carries out the RunMacro action in VBA.

  To run a standalone macro, use the following statement:

  ```
  DoCmd.RunMacro "YourMacroName"
  ```

  Optionally, you may specify how many times the macro should be run:

  ```
  DoCmd.RunMacro "YourMacroName", 2
  ```

To run a macro with submacros, use the name of the main macro followed by a period and the name of the submacro:

```
DoCmd.RunMacro "Sales.AddProducts"
```

- **Running a macro in response to an event on a form, report, or control**—A standalone macro can be bound to events for forms, reports, or controls. For example, if your form contains a button that needs to open another form and you have previously created a macro that performs this action, you can specify the macro name in the OnClick property of the button, as shown in Figure 30.15. To do this, you must open the form in Design or Layout view, click the Button control on the form, and open the property sheet. On the property sheet for the button, click the Event tab, and then click the event property for the event you want to trigger. The macro will run when you return to Form view and click the button. Notice that Access lists all the available macros when you open a drop-down list next to an event property. Macros that contain submacros are listed in two parts—the name of the standalone macro and the name of the submacro (e.g., Suppliers.Review Products).

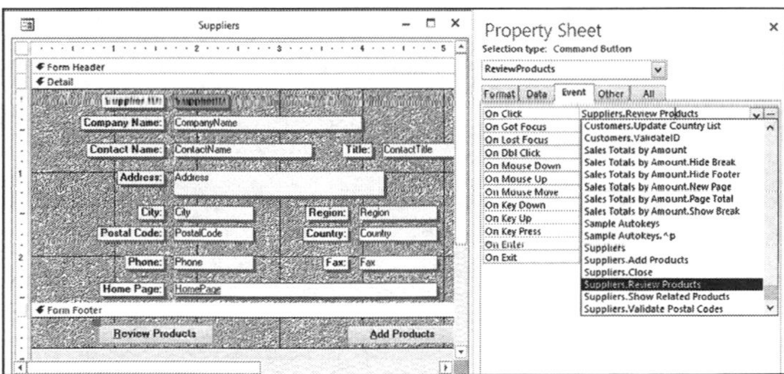

FIGURE 30.15. Binding a standalone macro to an event property. Shown here is the Suppliers form in the sample Northwind.mdb database from an earlier version of Access.

Creating and Using Submacros

Instead of having a large number of standalone macros listed in the Navigation pane, consider storing related macros together using submacros. Submacros are similar to VBA subroutines in VBE modules. Figure 30.16 shows submacros that can be attached to the Suppliers form in the Northwind.mdb database. Notice how this single macro object named Suppliers stores a number of submacros, each of which performs a different action. To create submacros within a particular macro, you must give each submacro a unique name.

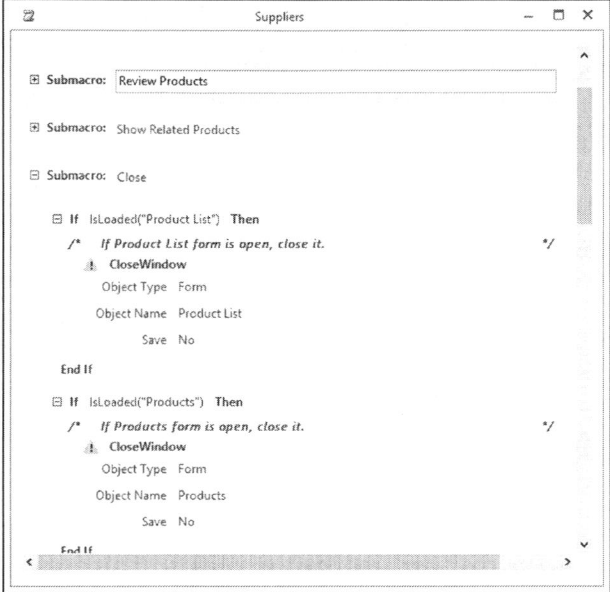

FIGURE 30.16. The Suppliers macro in the sample Northwind.mdb database contains submacros that can be used in the Suppliers form.

The general steps to create submacros are as follows:

1. Click the **Macro** button in the Macros & Code group of the Create tab (see Figure 30.7 earlier).
2. Select **Submacro** from the Add New Action drop-down list. Access enters the default name **Sub1** for your submacro (see Figure 30.17). Replace the suggested name with the desired name.

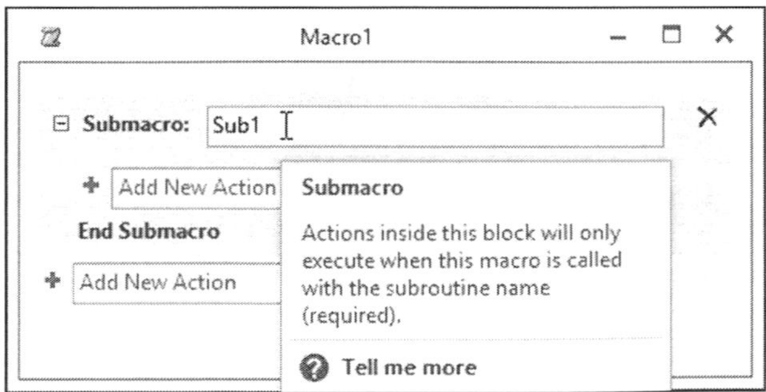

FIGURE 30.17. Creating a submacro.

3. Specify the macro actions for your submacro.
4. Add another submacro if desired and specify the actions to perform.

5. Save the macro by pressing **Ctrl+S** and typing the name for the macro. The name you specify is the name of the main macro that contains the submacros. This name will appear in the Navigation pane under Macros.
6. Close the Macro Designer window.

Recall from an earlier section that when a macro contains submacros and you right-click the macro in the Navigation pane and choose Run, only the first submacro will execute.

Submacros are frequently implemented in forms and reports. To gain a better understanding of submacros, study the Suppliers macro and the Suppliers form in the Northwind.mdb database. Another excellent example of using submacros is the Customer Labels Dialog macro attached to the Customer Labels Dialog form in the Northwind.mdb database (see Figure 30.18).

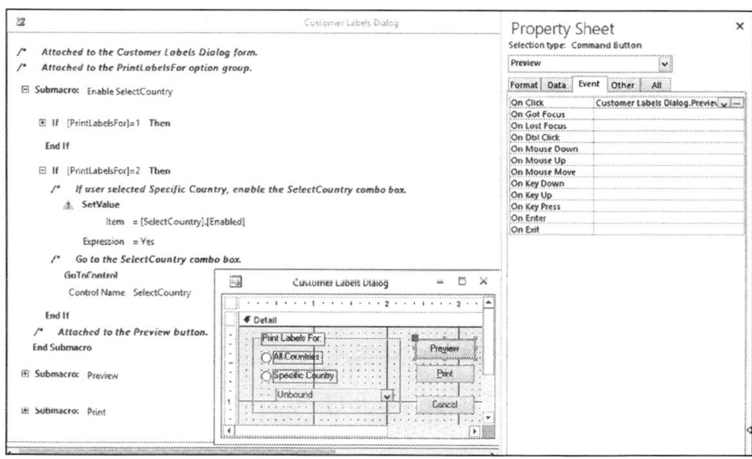

FIGURE 30.18. The implementation of submacros in the Customer Labels Dialog form in the Northwind.mdb database. The yellow warning sign in the record selector indicates that the specified action will not execute if the database is not trusted.

Creating and Using Embedded Macros

Beginning with the release of Access 2007, macros can be embedded in any of the events provided by a form, report, or control. These embedded macros are not visible in the Navigation pane.

The general steps to create an embedded macro are as follows:

1. Open a form or report in Design or Layout view.
2. Select an object to which you want to assign an embedded macro (a form, report, or control).

3. Activate the property sheet. In Design view, the Property Sheet button is located in the Tools group of the Design tab. In Layout view, you will find this button in the Tools group of the Arrange tab.
4. In the property sheet, click the **Event** tab, and then click the **Build** button (…) next to the desired property.
5. In the Choose Builder dialog box, select **Macro Builder**, then click **OK**. Access will open the same Macro Designer window that you use for creating standalone macros.
6. Choose the actions for your macro, and specify the arguments and conditions if required.
7. Press **Ctrl+S** to save your macro, and click the **Close** button in the Close group of the Design tab. Access closes the Macro Design view and enters [Embedded Macro] in the event property (see Figure 30.19).

To modify the embedded macro, click on the Build button (…) next to the property with [Embedded Macro]. Access will open the Macro Designer (Design view) where you can make the required modifications.

Keep in mind that you cannot reference an embedded macro from other macros. To reference a macro from another macro, you must create a standalone macro.

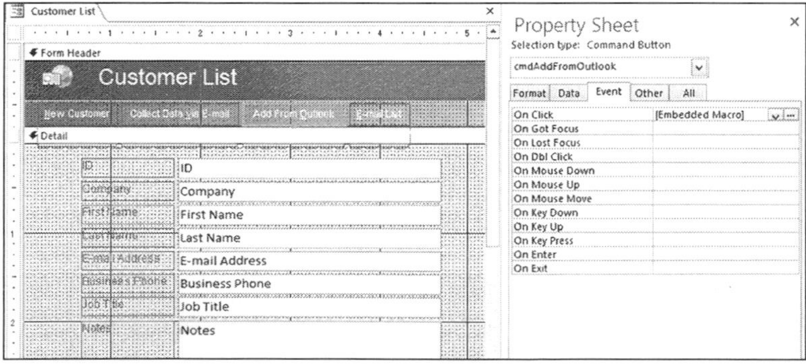

FIGURE 30.19. Assigning an embedded macro to the event property of a form's command button in the Northwind 2007.accdb database.

Copying Embedded Macros

Because embedded macros are part of the object in which they are created, the macro behind the control is also copied when you copy the form, report, or control.

You can also copy an embedded macro from one event property to another. This is possible thanks to so-called "shadow properties." What this means is that for each event property of a control, form, or report there is a "shadow" event property that contains the embedded macro for that property. For example, if your form's On Load event property is set to [Embedded Macro], then the shadow property called On Load Macro contains its embedded macro. The On Click event property has the On Click Macro property if you are using the embedded macro to trigger the On Click event. If the event property is empty, then there is no shadow property.

Hands-On 30.2 demonstrates how to use VBA to copy an embedded macro from the Shipper Details form to the Supplier List form in the Northwind 2007.accdb database.

Hands-On 30.2. Copying Embedded Macros

1. In the Northwind 2007.accdb database, open the **Supplier List** form in Design view. You may use the same version of the database that you opened in Hands-On 30.1.
2. In the Form Header section, right-click the **Home** button and choose **Copy**.
3. Right-click anywhere in the empty area of the Form Header section and choose **Paste**. The copied button appears in the upper-left corner of the Form Header section. Leave the button in this location for now until we change some of its properties. The button has the same label as the original button and a default name beginning with Command and followed by some numbers, such as Command231. You need to change the button's Name and Caption properties.
4. While the button is selected, click the **Property Sheet** button in the Tools group of the Design tab. Click the **All** tab and change the button's Name property to **cmdClose** and the Caption property to **&Close**. The ampersand in front of the letter "C" assigns a keyboard shortcut to the button.
5. Position the Close button to the left of the Home button as shown in Figure 30.20.

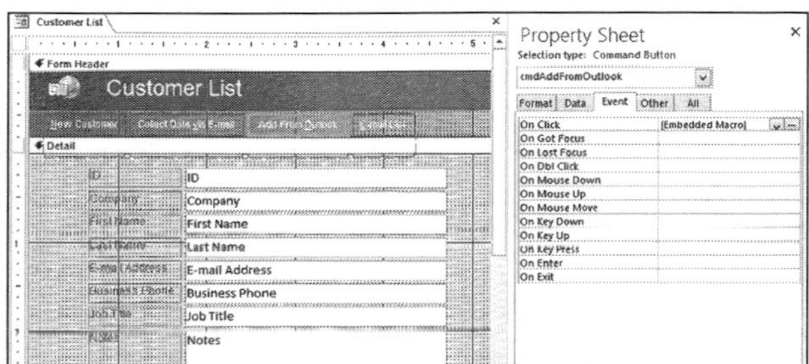

FIGURE 30.20. Use the property sheet to change the Name and Caption properties of the Close command button.

6. Press **Ctrl+S** to save the changes to the form.

7. While the Close button is selected, click the **Event** tab in the property sheet. Notice that when you copied the Home button, Access also copied the embedded macro attached to the On Click event property (see Figure 30.21). At this point, you could simply click the Build button (...) to modify this macro to have it close the Supplier List form instead of opening the Home form. However, the purpose of this exercise is to show you how to use VBA to copy an embedded macro from one property to another. We will overwrite this embedded macro with a different one by writing a VBA procedure in the next steps.

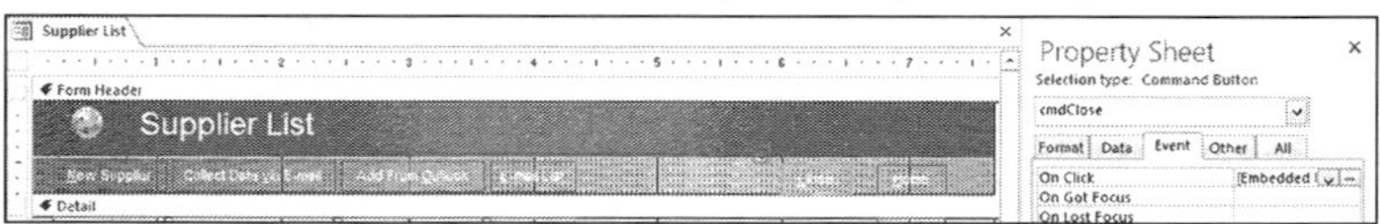

FIGURE 30.21. When you copied the Home button, the new button inherited the embedded macro assigned to the On Click event property.

8. Press **Alt+F11** to activate the Visual Basic Editor window, and choose **Insert | Module.**
9. In the module's Code window, enter the following **Copy_OnClickMacro** procedure:

```
Sub Copy_OnClickMacro()
  Dim ctl As Control

  ' open in the Design view the Supplier List form
  DoCmd.OpenForm "Supplier List", acDesign

  ' only run the code if the specified control
  ' exists on the form
  For Each ctl In Forms("Supplier List").Controls
    If TypeOf ctl Is CommandButton Then
      If StrComp(ctl.Name, "cmdClose", vbTextCompare) = 0 Then

        ' open in the Design view the Shipper Details form
        ' this form contains an embedded macro in the OnClick
        ' event of cmdClose button

        DoCmd.OpenForm "Shipper Details", acDesign

        ' copy macro from the OnClick event property of the
        ' cmdClose button on the Shipper Details form
        ' to the OnClick event property of the cmdClose button
        ' on the Supplier List form
```

```
          Forms("Supplier List").Controls("cmdClose").OnClickMacro = _
          Forms("Shipper Details").Controls("cmdClose").OnClickMacro
           DoCmd.Save acForm, "Supplier List"
           DoCmd.Close acForm, "Shipper Details"
           MsgBox "The embedded macro was successfully copied."
           Exit Sub
         End If
       End If
     Next

     MsgBox "Operation could not be performed. " & vbCrLf & _
     "Ensure that the specified control exists."
   End Sub
```

In this procedure, we begin by opening the Supplier List form in Design view and iterate through the form's controls to find out whether the form contains the control named cmdClose. We use the `TypeOf…Is` expression to specifically look for the CommandButton control. Because the Supplier List form contains several buttons, we can use the `StrComp` function to determine if we found the correct button. This function will tell us if the string specified in the second argument is found in the string specified in the first argument. The third argument of the `StrComp` function tells Access to perform the comparison of the two text strings. If the `StrComp` function returns zero (0), then we found the control we were looking for and we can proceed to open the Shipper Details form and copy the embedded macro assigned to the On Click event property of this form's cmdClose button to the On Click event property of the Supplier List's equivalent button. The following statement copies the embedded macro from the On Click event property to another On Click event property:

```
Forms("Supplier List").Controls("cmdClose").OnClickMacro = _
 Forms("Shipper Details").Controls("cmdClose").OnClickMacro
```

Once we are finished copying, we can simply exit the procedure using the early exit expression `Exit Sub`.

If the Supplier List form does not contain the button with the specified name, we display a message.

10. Run the Copy_OnClickMacro procedure.

If you followed all the steps of this hands-on exercise, you should see a message stating that the embedded macro was successfully copied. Click **OK** to close the message box. If you got a different message, check the code for any errors and ensure that the Supplier List form has the cmdClose button. Then rerun the procedure.

11. Press **Ctrl+S** to save changes in the module. Access will ask you to assign a new name to the module. Click **OK** to accept the default name.
12. Press **Alt+F11** to return to the Access window.
13. In the property sheet for the cmdClose button, click the **Build (...)** button next to the On Click event property on the Event tab. Access opens the Macro Design view, as shown in Figure 30.22. This macro will close the form when the user clicks the Close button on the Supplier List form.

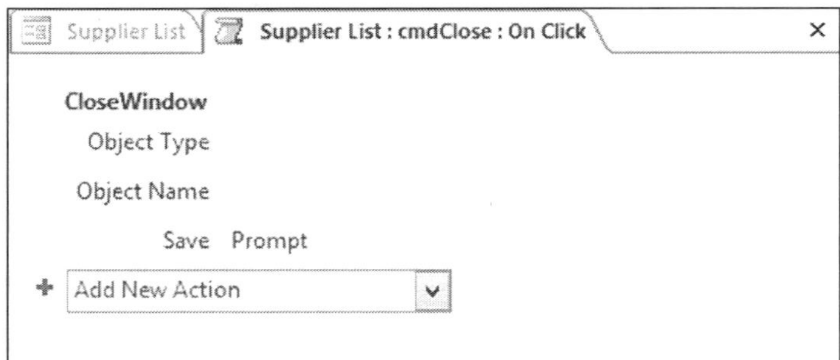

FIGURE 30.22. Examining an embedded macro after it's been copied from another event property.

14. Exit the Macro Design view.
15. Right-click the **Supplier List** tab, and choose **Form View**.
16. Click the **Close** button in the Header section of the Supplier List form to close this form.

You can see the contents of the OnClickMacro shadow property by typing the following statement in the Immediate window and pressing Enter (the form must be open for this to work):

```
?Forms("Supplier List").Controls("cmdClose").OnClickMacro
```

You should see the following output:

```
Version =196611
ColumnsShown =8
Begin
  Action ="Close"
  Argument ="-1"
  Argument =""
  Argument ="0"
End
Begin
  Comment ="_AXL:<?xml version=\"1.0\" encoding=\"UTF-16\"
  standalone=\"no\"?>\015\012<UserI"
  "nterfaceMacro For=\"cmdClose\"
  xmlns=\http://schemas.microsoft.com/office/acces
```

```
    "sservices/2009/11/application\"
    xmlns:a=\http://schemas.microsoft.com/office/ac
    "cessservices/2009/11/forms\"><S"
  End
  Begin
    Comment ="_AXL:statements><Action
    Name=\"CloseWindow\"/></Statements></UserInterfaceMacro>"
  End
```

Access has a large number of hidden properties that make it possible to get and set embedded macros. The property name begins with the name of the event property and ends with "EmMacro," such as OnClickEmMacro, AfterUpdateEmMacro, and so on. Try the following statement in the Immediate window (the form must be open for this to work), and notice that it produces the same output as the previous statement:

```
?Forms("Supplier List").Controls("cmdClose").
Properties("OnClickEmMacro").Value

Version =196611
ColumnsShown =8
Begin
  Action ="
Close"
  Argument ="-1"
  Argument =""
  Argument ="0"
End
Begin
  Comment ="_AXL:<?xml version=\"1.0\" encoding=\"UTF-16\"
  standalone=\"no\"?>\015\012<UserI"
  "nterfaceMacro For=\"cmdClose\"
  xmlns=\http://schemas.microsoft.com/office/acces
  "sservices/2009/11/application\"
  xmlns:a=\http://schemas.microsoft.com/office/ac
  "cessservices/2009/11/forms\"><S"
End
Begin
  Comment ="_AXL:statements><Action
  Name=\"CloseWindow\"/></Statements></UserInterfaceMacro>"
End
```

With this knowledge, it is easy to create a standalone macro from an embedded macro. Here's an example VBA procedure that does just that:

```
Sub SaveEmToStandalone()
  Dim strMacro As String
  Dim objFileSys As Object
  Dim objFile As Object
  Dim strFileName As String

  ' open in the Design view the form that contains
  ' the embedded macro
  DoCmd.OpenForm "Login Dialog", acDesign

  ' to write an embedded macro to a file use the
  ' Value property
  strMacro = Forms("Login Dialog"). _
  Controls("cboCurrentEmployee"). _
  Properties("AfterUpdateEmMacro").Value

  ' close the form
  DoCmd.Close acForm, "Login Dialog"

  ' Create a text file
  strFileName = "C:\Access2013_ByExample\cboAfterUpdate.txt"

  Set objFileSys = CreateObject("Scripting.FileSystemObject")
  Set objFile = objFileSys.CreateTextFile(strFileName, True)

  ' Write strMacro to the text file
  objFile.Write strMacro
  ' Close the file
  objFile.Close

  ' Use the undocumented LoadFromText method of
  ' the Application object to create a standalone macro
  ' from the text file
  Application.LoadFromText acMacro, _
   "cboEmployeeAfterUpdate", strFileName
End Sub
```

The `LoadFromText` method of the Application object makes it possible to create various Access database objects (including macros) from information that was previously saved to a text file. The `LoadFromText` method requires that you specify the object type, the object name, and the name of the text file.

After running this procedure, you should see the cboEmployeeAfterUpdate macro listed in the Navigation pane under Macros.

Working in Sandbox Mode

By default Access runs in Sandbox mode, which means that the program blocks all the expressions in field properties and controls that are considered unsafe. A *safe expression* is one that does not use functions that could be used to access drives or other resources on a user's computer to damage data or files. When Access is running in Sandbox mode, any expressions that use unsafe macro actions are marked with a yellow warning sign, as shown earlier in Figure 30.18.

Access allows you to disable Sandbox mode by setting the macro security level to low; however, for security reasons this setting is not recommended. If you trust the database and want to run unsafe expressions that the Sandbox mode blocks without having to change your current macro security, you can disable Sandbox mode by changing a Registry key. Modifying the Registry is beyond the scope of this chapter.

Generating Macros Using the Command Button Wizard

You do not have to write all your macros from scratch. Access provides a built-in tool known as the Command Button Wizard. If you are working with the ACCDB database, the wizard will generate embedded macros to open forms, run queries, find records, apply filters, or print reports. For older databases in the .mdb file formats (in order to support backward compatibility), the wizard continues to create VBA code.

Using Data Macros

Prior to Access 2010, macros could only be attached or embedded in forms and reports. Access programmers have long asked for a feature similar to SQL triggers that would enable them to automatically update data in a table or track when a record was last modified or deleted. Microsoft has answered this programming request in Access 2010 by introducing *data macros*. A data macro contains one or more actions that execute in response to a table event.

With data macros programmers can enforce complex business rules at table level. For example, by attaching a data macro to a table you can control what happens to a table's data when the user interacts with the data via an Access form. You can specify what occurs after data is inserted, updated, or deleted. For instance, you may want to verify the accuracy of table data, send an email notification to the database manager about the changes that occurred to the data, or automatically update fields in another table. By using the data macros attached to the After Insert, After Update, and After Delete events, you can check and modify records in the current table or other tables. You can use the `For Each Record` construct to iterate through a set of records in a table to update records that meet certain criteria or accumulate the totals. You can also perform specific actions before data is inserted, changed, or deleted. The Before Change data macro event will allow you to check a value in another table and, if necessary, prevent a change or insert from happening. You can use an IsInsert property to detect whether it's an insert or an update operation. You can find out whether the value of a specific field has changed by using the `Updated` function, and, if the value of a field has changed, you can use the Old property to find out the previous value of the field. Before deleting records you can use the Before Delete data macro event to determine whether the record can be deleted. You can also update an audit file to indicate that the record was deleted.

By using data macros you can guarantee that your business logic is executed even if the user modifies a record outside the forms you provide such as in a Datasheet view or by running another macro or a VBA procedure. Your data macros will run silently in the background regardless of how the data is accessed. With data macros, you no longer need to attach the same macro to a number of forms. All you need to do is add the logic to the table. Any form based on that table will inherit that logic.

In addition to *event data macros* that are triggered by table events, you can create standalone *named data macros*. Named data macros allow you to save time by incorporating the common tasks into one macro. Instead of repeating the same actions in multiple data macros, simply create a named data macro and call it from a data event. Named macros can be called using the RunDataMacro action.

Keep in mind that data macros do not have any user interface (UI); they are stored within a table itself and therefore do not show up in the Navigation pane. Do not attempt to use data macros to handle multivalue and attachment data types as they are not supported. Also, keep in mind that data macros can only be attached to events in local tables, not linked tables.

Creating a Data Macro

In the following hands-on exercise, you will work with the Purchase Order Details table in the Northwind 2007 database. You'll write a data macro to ensure that the order quantity cannot be modified if the order was already posted to inventory or the Date Received field contains a date value. The following VBA procedure has already been written by the Microsoft team to validate the Quantity field in the Form_Purchases subform for Purchase Order Details:

```
Private Sub Quantity_BeforeUpdate(Cancel As Integer)
  If Me![Posted To Inventory] Or Not IsNull(Me![Date Received]) Then
    MsgBoxOKOnly CannotModifyPurchaseQuantity
    Cancel = True
  End If
End Sub
```

While this procedure works just fine for controlling data entry operations on the form, it has no effect on data manipulations performed directly at the table level. By creating a Before Change data macro, you can ensure that this test scenario is addressed no matter how data is being accessed.

Hands-On 30.3. Creating and Testing a Data Macro

1. Start the **Northwind2007.accdb** database located in your **Access2013_ByExample** folder. Log in as **Andrew Cencini.**
2. Close the **Home** form that is automatically launched upon login.
3. Open the **Purchase Order Details** table.
4. Select the **Table** tab on the Ribbon and click the **Before Change** button (Figure 30.23). It does not matter which table field is currently selected.

Note: *Data macros can be created from the table Datasheet view or Design view (see Figure 30.23 and 30.24).*

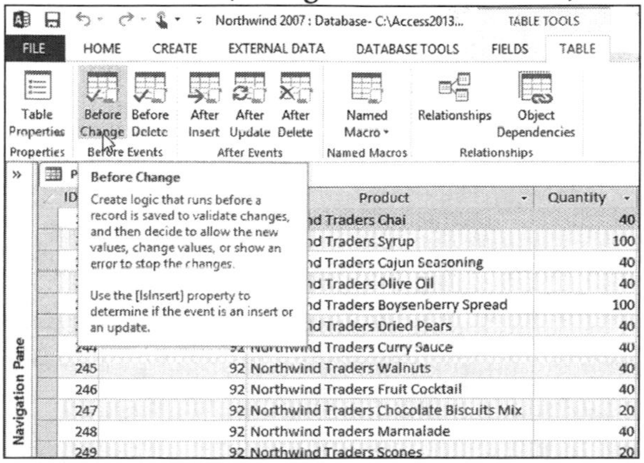

FIGURE 30.23. Creating a data macro from the Datasheet view.

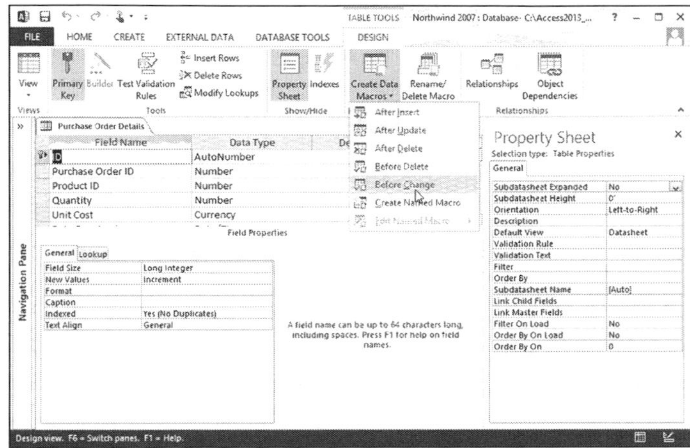

FIGURE 30.24. Creating a data macro from the Design view.

Table 30.1 lists five events that can trigger a data macro.

TABLE 30.1. Data macro events

Event Name	Event Description
Before Change	Runs before a record is about to be updated. Use it to validate changes before saving them to the table. You can include logic to allow new values or show an error to reject the changes. Use the IsInsert property to determine whether the change is an insert or an update.
Before Delete	Runs before a record is about to be deleted. You can include logic that validates the deletion and allows it, or cancels the deletion and raises an error.
After Insert	Runs after a new record has been added to the table.
After Update	Runs after a record has been edited in the table. Use the `Updated("Field Name")` function to determine if a specific field has changed. Use `Old.[Field Name]` to find out the value the field had before the record was changed.
After Delete	Runs after a record has been deleted from the table. Use `Old.[Field Name]` to find out the value the field had before the record was deleted.

5. When you click the event name, Access opens the Macro Designer (Figure 30.25).

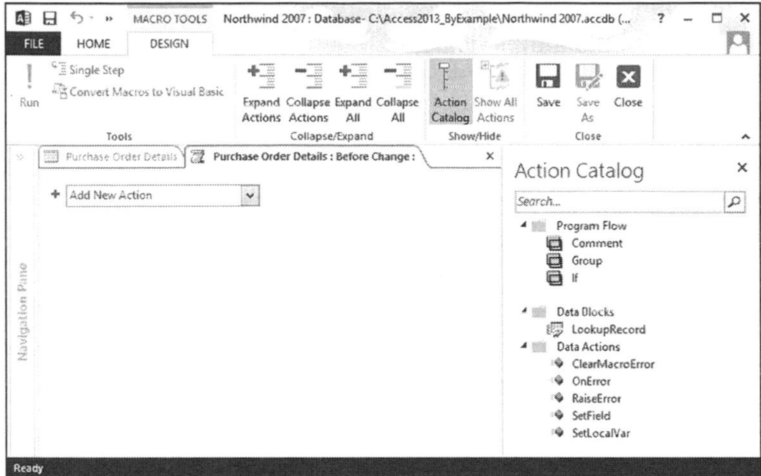

FIGURE 30.25. Macro Designer for writing data macros.

The Action Catalog shows three categories of actions that can be specified for a data macro: Program Flow, Data Blocks, and Data Actions. The actions listed in each category depend on the type of table event you have selected. When you are working with data macros, the only Program Flow constructs are comments (used for documenting your data macro), groups (used for organizing your macro), and If blocks (for applying a conditional logic). Data Blocks contain constructs that are used to perform specific operations on database records like looking up a record in a table (LookupRecord), adding a record to a table (CreateRecord), modifying an existing record in a table (EditRecord), and looping through every record in a table (ForEachRecord). Notice that only the LookupRecord data block is available for the Before Change event. When you select a construct from the Data Blocks category, you can add one or more actions and these actions will be performed as part of the data block. You can even nest data blocks. For example, you can set up a ForEachRecord data block to iterate through every record in a table and, depending on your conditional logic, create the CreateRecord data block to add a record to another table based on the found record.

The Data Actions category in the Action Catalog lists the available data actions. Some table events have more actions than others. You can find the description of an event by selecting it and then checking the bottom of the Action Catalog (see Figure 30.25).

6. Double-click the **If** construct in the Program Flow section. Access adds a conditional block as shown in Figure 30.26.

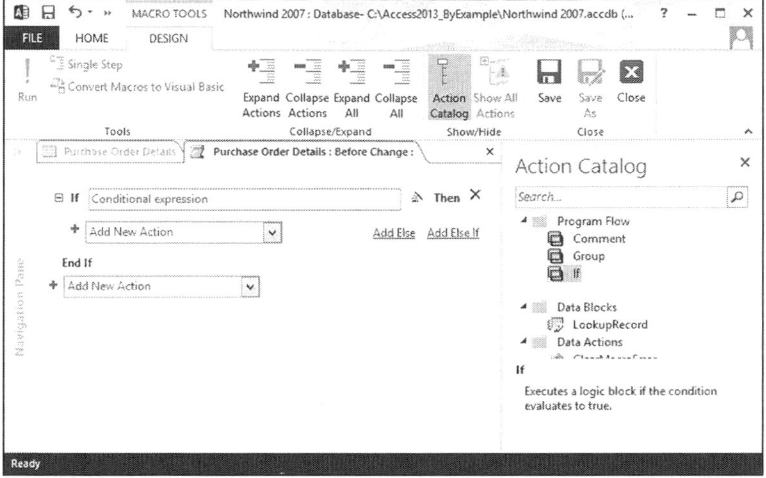

FIGURE 30.26. Adding an If block to the data macro.

7. In the If box, enter the following conditional expression on one line:
   Updated("Quantity") And ([Posted To Inventory] Or Not IsNull([Date Received]))
8. Select **SetLocalVar** from the Add New Action drop-down located within the `If...Then...End If` block. Enter **strMsg** in the Name box and **""** (an empty string) in the Expression box, as shown in Figure 30.27.

   The SetLocalVar action allows you to create a local variable. In this macro, you'll use a local variable named `strMsg` to store the error message text that you'll retrieve from the Strings table that is a part of the Northwind 2007.accdb database. Notice that the initial value of the `strMsg` variable is set to an empty string.

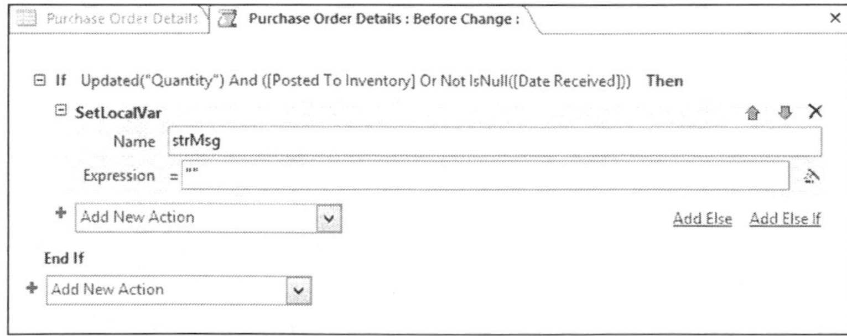

FIGURE 30.27. Adding a local variable to your data macro.

9. Select **LookupRecord** from the Add New Action drop-down located within the `If...Then...End If` block. The Macro Designer adds a LookupRecord block.

Fill in the block as depicted in Figure 30.28. Choose **Strings** from the Look Up A Record In drop-down box, and enter **[Strings].[String ID] = 31** for the Where Condition. This condition tells the macro to find the 31st record in the Strings table. Noticc that as you start typing in the Where Condition box the IntelliSense technology is at work displaying appropriate choices for you to select.

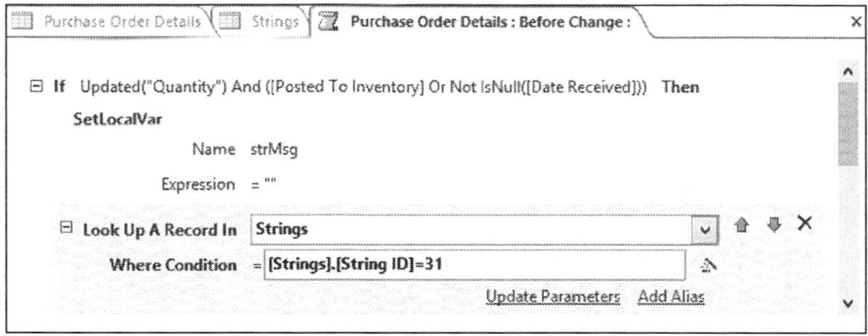

FIGURE 30.28. Adding an action to look up a record in a table.

10. Within the LookupRecord block, add a new **SetLocalVar** action. In the Name box, enter the name of the local variable **strMsg** that you declared at the beginning of the macro. In the Expression box, enter **[Strings].[String Data]**, as shown in Figure 30.29.

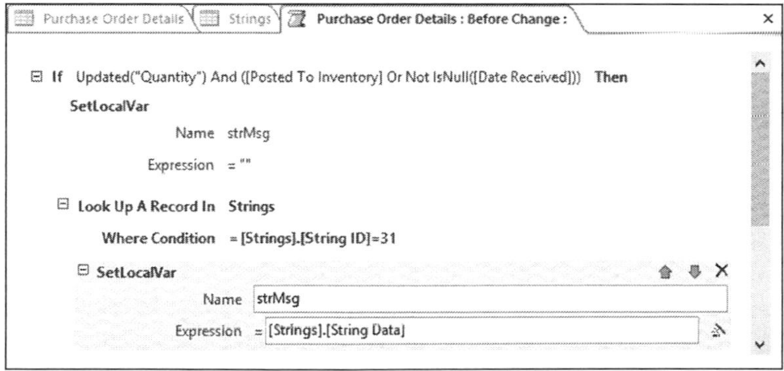

FIGURE 30.29. Storing data retrieved by the LookupRecord action in a local variable.

11. In the Add New Action drop-down box within the LookupRecord block, choose **Comment**. When a text box appears, enter the following text: **Record lookup completed.** Figure 30.30 shows the result of adding a comment. The comments appear in green italics between the /* and */ delimiters.

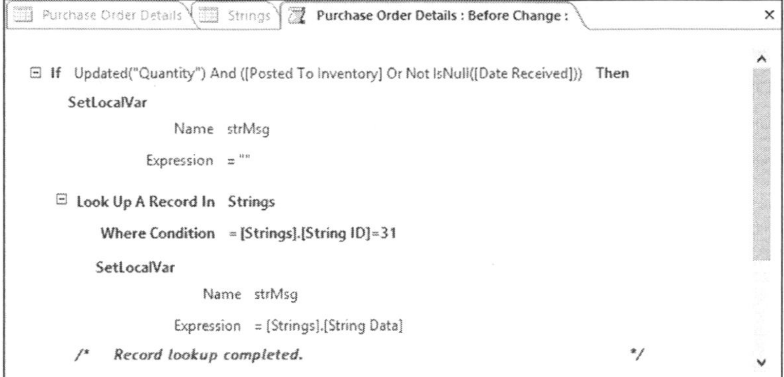

FIGURE 30.30. Adding a comment to a macro.

12. In the Add New Action drop-down box located outside the LookupRecord block, choose **RaiseError**. Enter **100** in the Error Number box and **=[strMsg]** in the Error Description box. This will tell the macro to display the text stored in the local variable `strMsg` when an error occurs. Be sure to enter the equals sign before the variable name. Figure 30.31 displays the completed macro.

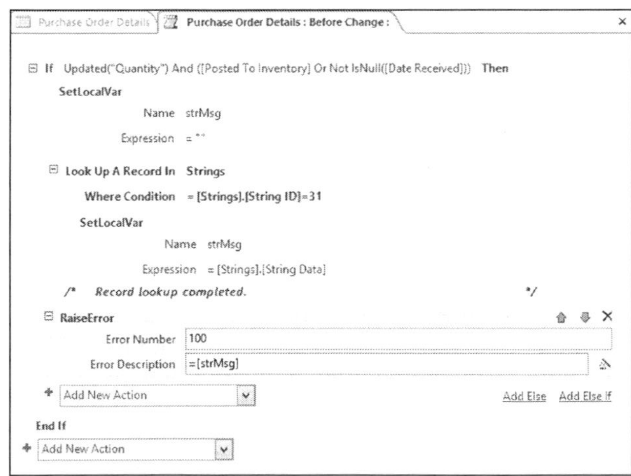

FIGURE 30.31. Adding a macro action to raise an error.

13. Click the **Save** button to save your macro, then click the **Close** button to close the Macro Desinger. Notice that when a macro is defined for a table event, the button with the event name has a yellow background (see Figure 30.32).

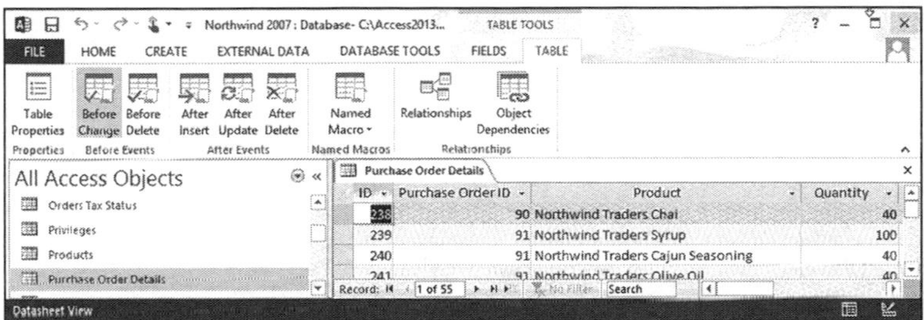

FIGURE 30.32. The highlighted Before Change button on the Ribbon indicates that there is a data macro attached to this event.

14. To test your macro, you need to perform the action that will trigger the event for which you defined the macro. In the Purchase Order Details table, enter a different value in the Quantity field for any record that has both a checkmark in the Posted To Inventory field and a value in the Date Received field. When you attempt to save the record after making a change to the Quantity field, Access displays the error message shown in Figure 30.33. The error message has been retrieved from the Strings table. Click **OK** to the message and then press the **Esc** key to exit the edit mode.

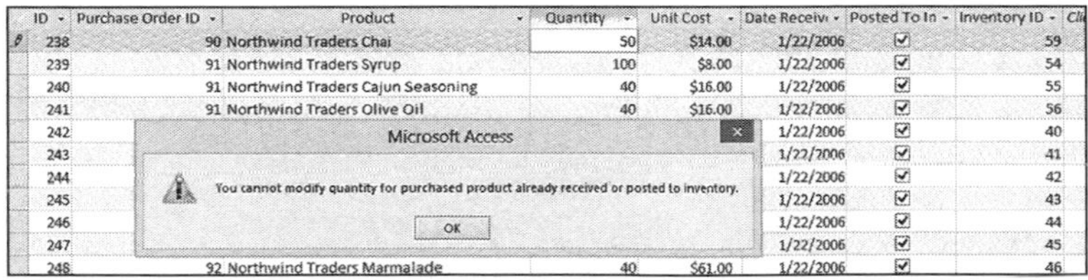

FIGURE 30.33. The error raised by the data macro assigned to the Before Change event.

A form based on a table that contains a data macro will inherit the logic defined in the table. This means that you no longer need to write separate VBA code in the form class modules to respond to events that are already handled at a table level.

Creating a Named Data Macro

As mentioned earlier, in addition to writing data macros that are triggered by a table event, you can create named data macros. You can pass arguments to these macros and call them from anywhere within your application. To create a named data macro, follow these general guidelines:

1. In the Navigation pane, double-click the desired table to open it.
2. Select the **Table** tab on the Ribbon.
3. In the Named Macros group, choose **Named Macro | Create Named Macro** (Figure 30.34). Access opens the Macro Designer as shown in Figure 30.35.

Notice that the Action Catalog lists a number of data actions that you can use in your named data macro logic.

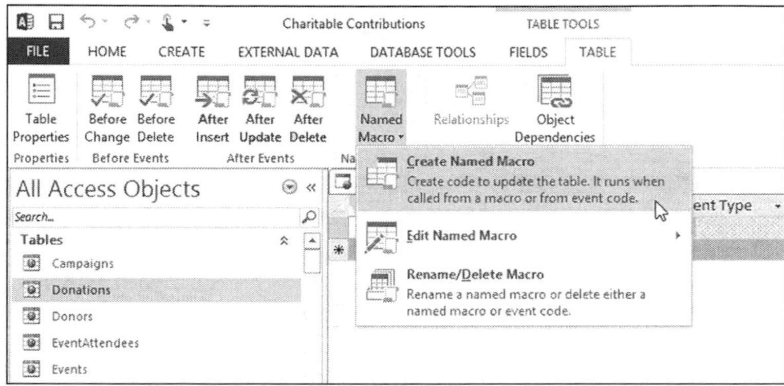

FIGURE 30.34. Creating a named data macro.

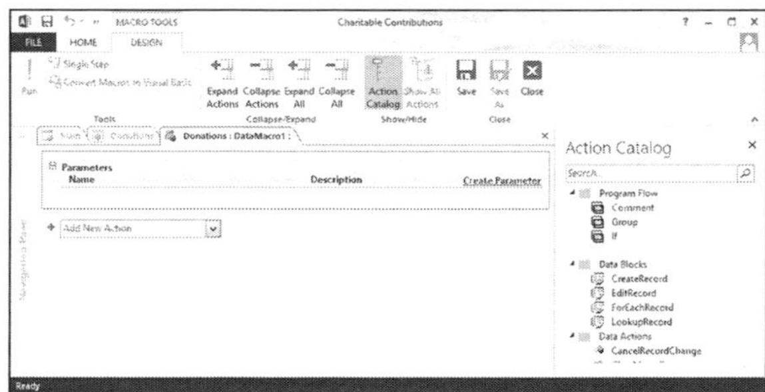

FIGURE 30.35. The Macro Designer window for creating a named data macro.

4. If you need to pass parameters to your macro, click the **Create Parameter** hyperlink at the top of the Macro Designer screen. Enter the name of the parameter in the Name box. You may also enter a description in the Description box (Figure 30.36).

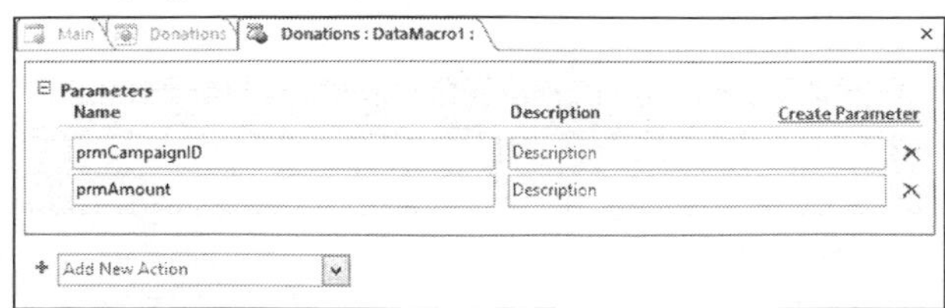

FIGURE 30.36. Specifying parameters in the named data macro.

5. Select an appropriate action from the Add New Action drop-down box to specify your macro logic. Figure 30.37 shows the completed named data macro.

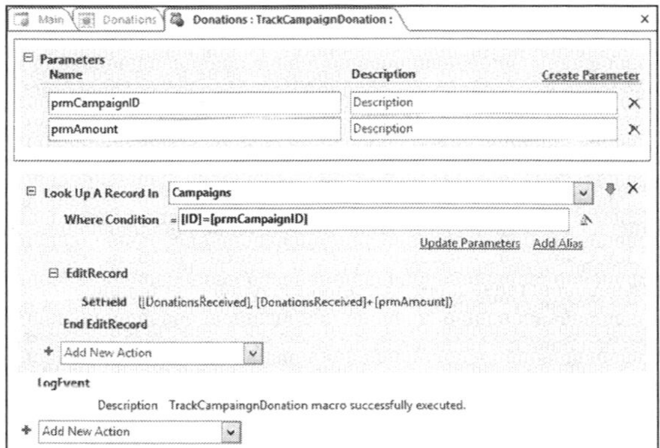

FIGURE 30.37. The completed named data macro.

The named data macro depicted in Figure 30.37 is available in the Charitable Contributions database. Follow these steps to open the database:

1. Copy the Charitable Contributions Web Database.accdb file from your **C:\Access2013_HandsOn** folder to your **C:\Access2013_ByExample** folder.
2. Double click the copied file to open it in Access.
3. In the Login window, click the **New User** hyperlink.
4. In the User Details window, enter your name in the Full Name text box, and click **Save & Close**.
5. Select your name in the Login window and click **Login**.
6. When prompted, click the **Enable Content** button in the message bar. This will activate the Login window. Select your name and click **Login**.
7. Open the Navigation pane and double-click the **Donations** table.
8. Click the **Table** tab to view the data macro and access the named macros in this table.

6. When you are done with the macro logic, save the macro by clicking the **Save As** button on the Ribbon.

Editing an Existing Named Macro

You can edit an existing named data macro by clicking the Named Macro button on the Ribbon and selecting Edit Named Macro. Access will display the list of available macros as shown in Figure 30.38.

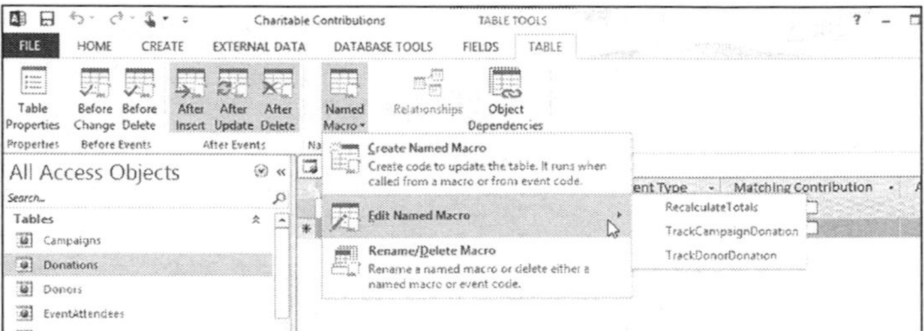

FIGURE 30.38. If the table contains named data macros, the Edit Named Macro option is highlighted.

Calling a Named Macro from Another Macro

You can run a named macro from another macro using the RunDataMacro action. Figure 30.39 shows two named data macros that are run from within the After Insert data macro in the Donations table. Notice that to run a named macro you need to:

- Specify the RunDataMacro action.
- Specify the named macro name (Donations.TrackDonorDonation, Donations.TrackCampaignDonation).
- Specify the values for the parameters that the named data macro expects.

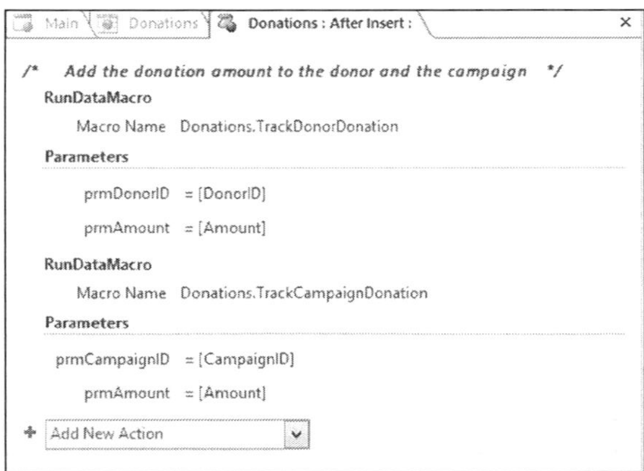

FIGURE 30.39. Running named macros from the After Insert data macro in the Donations table.

Using ReturnVars in Data Macros

A powerful feature in data macros is their ability to return values to other macros by using ReturnVars. ReturnVars can be compared to values returned by functions in VBA procedures. You can specify the ReturnVars by using the SetReturnVar action in a named data macro as depicted in Figure 30.40. After selecting the

SetReturnVar macro action from the Add New Action drop-down box, enter the name of the ReturnVar in the Name box and specify the value or expression in the Expression box. For example, to return the number of backordered inventory items, the example macro in Figure 30.40 sets up a ReturnVar named `retBackOrdered`, and sets its value in the Expression box to `[BackOrdered]`, which is the name of the field in the Inventory table. The number of the backordered items will be returned by the LookupRecord macro action for the specified ProductID. Notice that all return variables are initialized at the top of the macro.

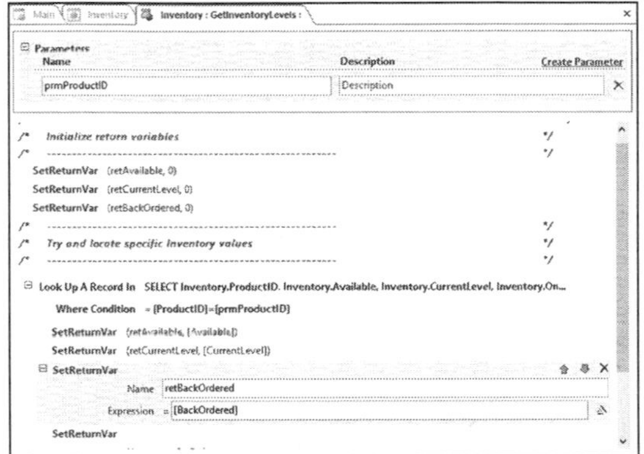

FIGURE 30.40. The named data macro GetInventoryLevels located in the Inventory table of the Northwind Web database demonstrates the use of return variables. You can open this database from the Access2013_HandsOn folder.

Access 2010 introduced a new type of a database file known as Access Web Database. You could use Access Web Databases to publish your Access data to a Microsoft SharePoint server running Access Services. Once published, your database could be used in an Internet browser. Because Access Web Database is not compatible with VBA, all programming had to be done using macros. In Access 2010, to design an Access Web Application, you had to choose File | New and click Blank Web Database. Well, this option is not available in Access 2013. While you can open, design, and publish existing Access 2010 Web databases in Access 2013 it is no longer possible to create new Access 2010 Web databases. Instead, Access 2013 offers users a new application model known as Access web app that makes it easy to create, deploy, and manage Web-based applications. You will need the knowledge of user interface (UI) macros to program actions that affect the user interface of your app, and the understanding of data macros to work with the records contained in your app. Creating and using the Access web apps is beyond the scope of this programming book as this topic warrants its own book.

To get the return value, you must first call the macro. The GetInventory-Levels macro (shown in Figure 30.40) is called from the embedded macro (see Figure 30.41) that is attached to the After Update event of chkPostedToInventory checkbox control. This control is located on the PurchaseOrderLineItems-Receiving form in the Northwind Web database.

Notice that to reference the return variable in a macro, you must use the `ReturnVars` command like this:

```
= [ReturnVars]![retBackOrdered]
```

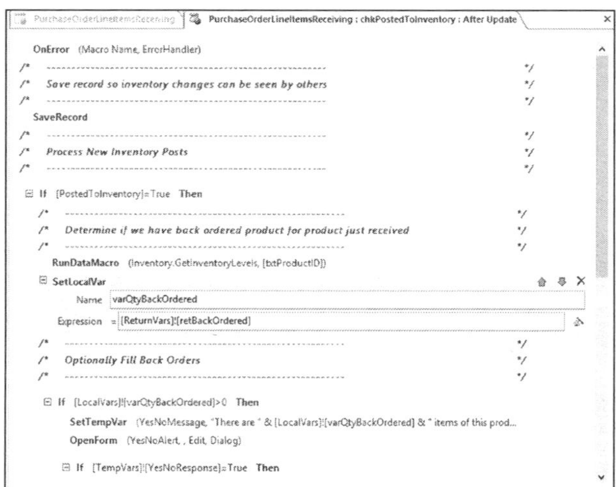

FIGURE 30.41. Referencing return variables (ReturnVars) inside a macro attached to the After Update event of a control placed on a form. Notice that the value of the return variable is being retrieved into a local variable named varQtyBackOrdered.

Tracing Data Macro Execution Errors

Access automatically writes all errors encountered during execution of your data macros in a system table called USysApplicationLog. Any failure that occurs while executing a named data macro or a data macro attached to an event will be reported in this table. By default, the USysApplicationLog table is created the first time Access encounters a data macro error. There are a couple of ways to access this table:

1. Via the Backstage view (see Figure 30.42).
   If the USysApplicationLog table is present in your database, select the **File** tab, and click the **View Application Log Table** button to open the table.

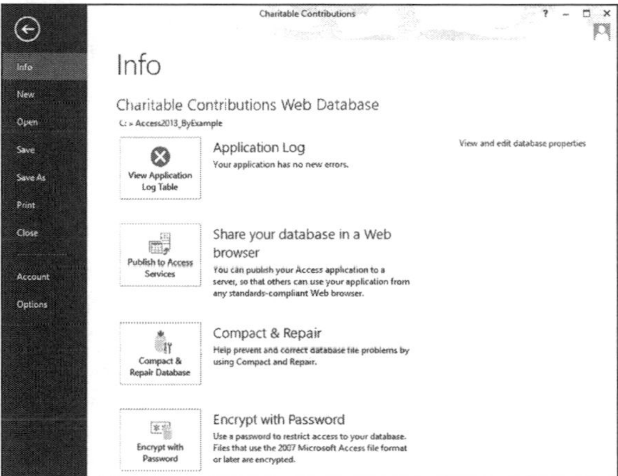

FIGURE 30.42. Accessing the USysApplicationLog table in the Backstage view.

2. Via the Navigation pane (see Figure 30.43).
   Before you can access the USysApplicationLog table from the Navigation pane you must tell Access to display system objects. To do this, select **File | Options**, and click **Current Database**. Scroll down to the Navigation section and click the **Navigation Options** button. Select the **Show System Objects** box at the bottom of the Navigation Options dialog box and click **OK**.

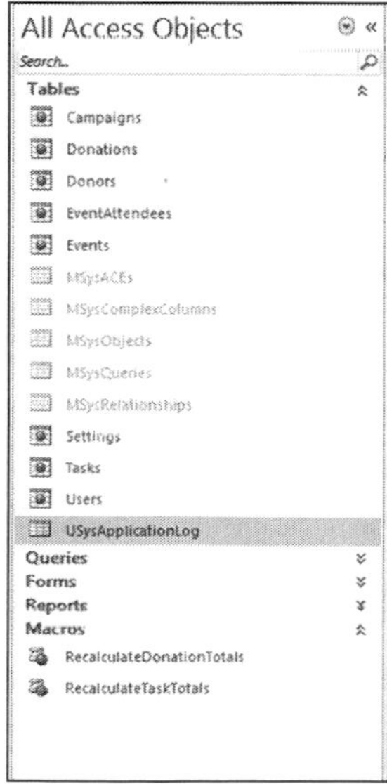

FIGURE 30.43. To open the USysApplicationLog table from the Navigation pane, you must be sure to set Access to display hidden system objects.

You can use USysApplicationLog to view the details of errors that occurred during data macro execution. Access provides a special action called LogEvent that allows you to write your own messages to the log table. You can keep track of the data macros that ran by adding the LogEvent action to the end of your named macro and setting its Description field to whatever message you want to write (see Figure 30.44).

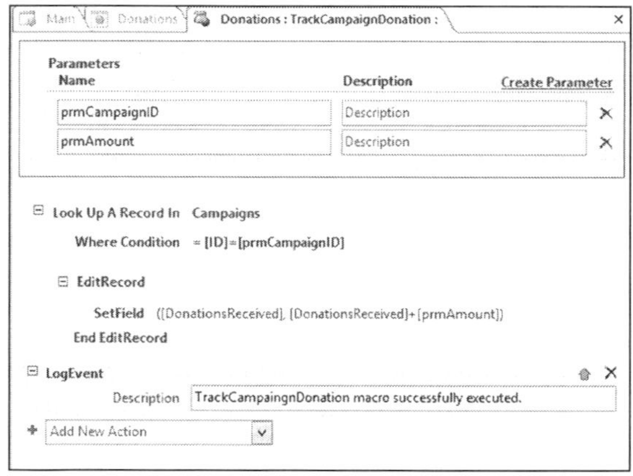

FIGURE 30.44. Adding the LogEvent action to a named data macro.

Figure 30.45 displays the contents of the USysApplicationLog table after adding data to a donations table.

USysApplicationLog

ID	SourceObjec	Data Macro I	Error Numbe	Category	Object Type	Description	Context	Created	Click to Add
1	Donations.Afte	{2FB0DDC7-78C	1	User	Macro	TrackCampaing		013 4:54:35 PM	
(New)									

FIGURE 30.45. Viewing the contents of the USysApplicationLog table.

Copying Data Macros

Access stores macros as XML. Saving your macro as XML enables you to e-mail it to someone else or create a backup copy of your macro before attempting to edit its logic.

You can copy the XML markup of your data macro to a text editor using these steps:

1. Open the macro and select the action you'd like to copy. A gray box appears around the selected action. To select all actions, press Ctrl+A.
2. Right-click the selected area and choose **Copy**.
3. Open **Notepad** and choose **Edit | Paste**.

Error Handling in Macros

Access provides special macro actions that give macros the capability to handle errors: OnError, ClearMacroError, and SingleStep. A MacroError object provides you with information about the error received and allows you to create user-friendly error messages. The OnError action is similar to the `On Error` statement in VBA. This action specifies how errors should be handled when a runtime error occurs. The OnError action has two arguments, as shown in Table 30.2.

TABLE 30.2. OnError action arguments

Arguments	Description
`Go To` (This argument is required.)	• Specifies how macros should handle errors. The `Go To` argument can be set to: `Next`, `Macro Name`, or `Fail`. `Next`—The error is recorded in the MacroError object and the execution of the macro moves to the next macro action. This is similar to the `On Error Resume Next` statement in VBA. `Macro Name`—Macro execution is passed to the macro that is named in the `Macro Name` argument. This is similar to the `On Error GoTo` statement in VBA. `Fail`—Access will stop the execution of the macro and display an error. This is similar to `On Error GoTo 0` in VBA.
`Macro Name` (This argument is optional.)	• If the `Go To` argument is set to `Macro Name`, the name of the macro in the current macro group will handle the error.

The OnError action suppresses standard error messages displayed by Access when an error occurs. When you use this action in your macro, you should use the error information saved in the MacroError object to display a user-friendly message about the error.

The MacroError object has the following properties: ActionName, Arguments, Condition, Description, MacroName, and Number. You can check the MacroError object's Number property to find out if an error occurred, as shown in Figure 30.46. If there was no error, the Number property will return zero (0). However, if `[MacroError].[Number] <> 0`, then you should handle the error right away.

By default, the MacroError object is cleared at the end of the macro execution; however, you can clear it right after the error has been handled by using the ClearMacroError action. This action will reset the error number in the MacroError object back to zero and clear other information stored in the object such as macro name, action name, condition, arguments, and description. The MacroError object contains information about only one error at a time; if more than one error occurred, only the error information about the last error can be retrieved. Therefore, when writing longer macros use the ClearMacroError action right after handling the first error so the ErrorObject will be able to capture information about the next error that might occur.

Use the StopMacro action to stop the currently running macro. In Figure 30.46, the StopMacro action is run right after the user receives the message about the macro error.

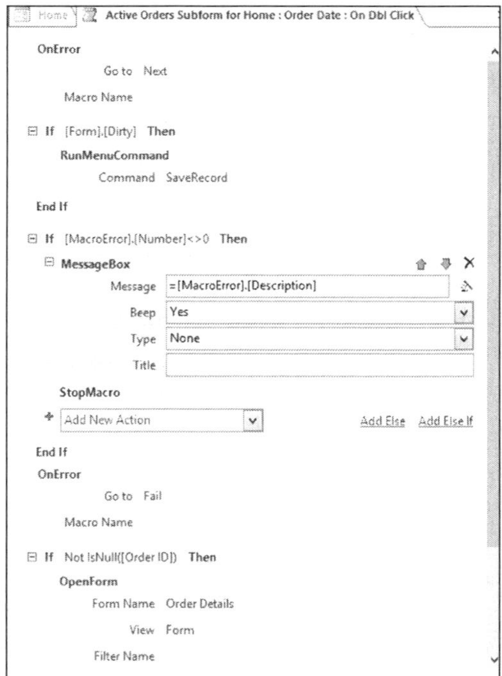

FIGURE 30.46. Error handling in an Access macro located in the Northwind 2007.accdb database.

To debug a macro that is not working properly, you can click the Single Step button in the Tools group of the Design tab and then click the Run button, or you can use the SingleStep macro action just before an action that you suspect is causing a problem. The SingleStep macro action was introduced in Access

2007. This action pauses the macro and opens the Macro Single Step dialog box (see Figure 30.47), which displays information about the current macro action (macro name, condition, action name, arguments, and error number). The Macro Single Step dialog box contains the three buttons described in Table 30.3.

TABLE 30.3. Macro Single Step dialog box buttons

Button Name	Description
Step	Move to the next macro action.
Stop All Macros	Stop the current macro and any other macros that may be running.
Continue	Use this button to exit Single Step mode and continue the normal execution of the macro.

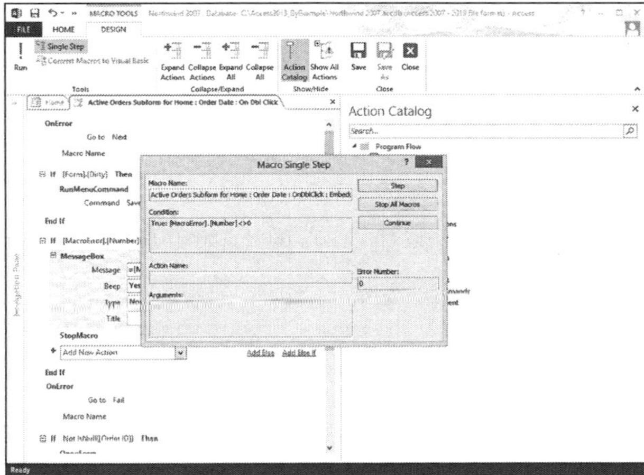

FIGURE 30.47. Debug your macros by selecting the Single Step option on the Ribbon and clicking the Run button.

If you opened the Macro Single Step dialog box using the Single Step button on the Ribbon, you must click this button again when you are done debugging your macro or the next macros that you run will also be run using Single Step mode.

Using Temporary Variables in Macros

The functionality to add temporary variables (TempVars) has been in Acccss since its 2007 release. This functionality applies to both VBA and macros. You've seen the VBA side of using the TempVar object in Chapter 3. Recall that the TempVar object of the TempVars collection allows you to get or set a value for a variable. Each TempVar object has a name and value property. In macros, there are three macro actions that relate to TempVars:

- **SetTempVar**(`name, expression`)—This macro action is used to create a new temporary variable. This variable can then be used as a condition or argument in subsequent macro actions. Temporary variables are global; therefore, you can use them in another macro, in an event procedure, or on a form or report. The first argument of the SetTempVar macro action assigns a name to the temporary variable. The second argument is the expression that Access should use to set the value for this temporary variable. You can define up to 255 temporary variables at one time.
- **RemoveTempVar**(`name`)—This macro action is used to remove the temporary variable. Use the `name` argument to provide the name of the variable to remove. It is recommended that you remove the temporary variable once you've finished working with it. If you don't remove your temporary variables, they will be removed automatically when you close the database.
- **RemoveAllTempVar**—This macro action is used to remove all temporary variables from the TempVars collection.

Figure 30.48 shows how to specify the name of a report by using a temporary variable.

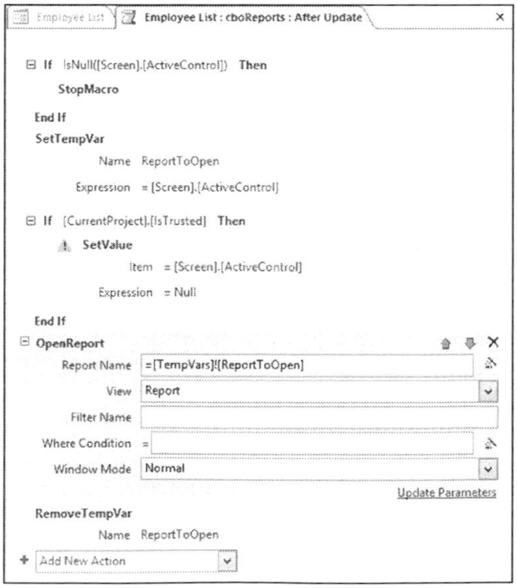

```
Employee List | Employee List : cboReports : After Update

If IsNull([Screen].[ActiveControl]) Then
    StopMacro
End If
SetTempVar
        Name  ReportToOpen
  Expression  = [Screen].[ActiveControl]
If [CurrentProject].[IsTrusted] Then
    SetValue
            Item  = [Screen].[ActiveControl]
      Expression  = Null
End If
OpenReport
     Report Name  =[TempVars]![ReportToOpen]
            View  Report
     Filter Name
 Where Condition  =
     Window Mode  Normal
                                   Update Parameters
RemoveTempVar
            Name  ReportToOpen
+ Add New Action
```

FIGURE 30.48. Using temporary variables in an Access macro.

Because both macros and VBA use the same TempVars collection, it is easy to share data between your macros and VBA procedures.

Converting Macros to VBA Code

The ability to convert standalone macros to VBA code has been available since Access 97. With the introduction of embedded macros, Access also provides a button to convert to VBA code macros stored in an event property of a form, report, or control (see Figure 30.49).

Converting a Standalone Macro to VBA

To convert a standalone macro to VBA, follow these steps:

1. In the Navigation pane under Macros, right-click the macro you want to convert, then click **Design View**.
2. In the Tools group of the Design tab, click **Convert Macros to Visual Basic** (see Figure 30.49).
3. Access will display a dialog box asking whether you want to include error handling and comments in the code (see Figure 30.50). To keep your code very simple, you can clear both checkboxes. Start the conversion process by clicking the **Convert** button.

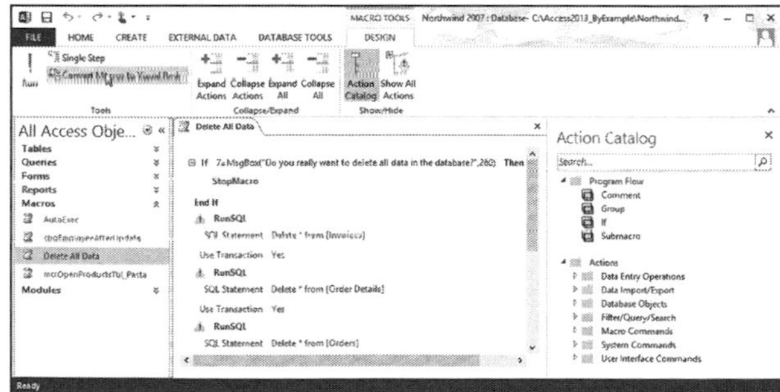

FIGURE 30.49. Converting a standalone macro to Visual Basic for Applications code.

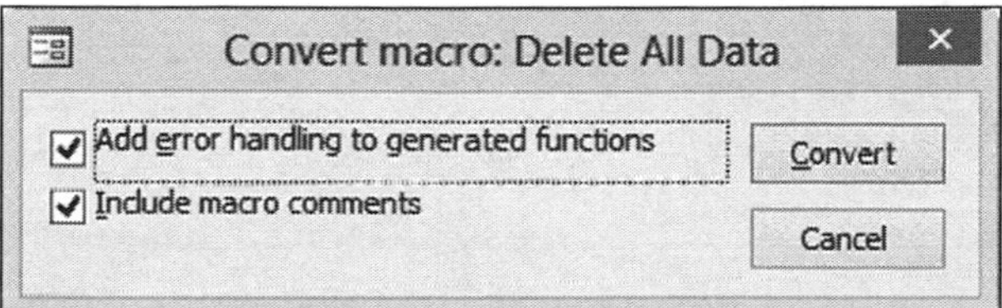

FIGURE 30.50. Access displays this dialog box when you click the Convert Macros to Visual Basic button (see Figure 30.49).

Upon completion of the macro conversion process, Access displays a message stating that the conversion is finished. Click OK to the message, and review the Modules group in the Navigation pane. You should see a separate module for the converted macro. The name of the module is Converted Macro followed by a dash and the name of the macro you converted. For example, after converting the Delete All Data macro, the name of the VBA module is Converted Macro – Delete All Data. To view the converted macro, double-click the converted module's name. This will open the Visual Basic Editor window, as shown in Figure 30.51.

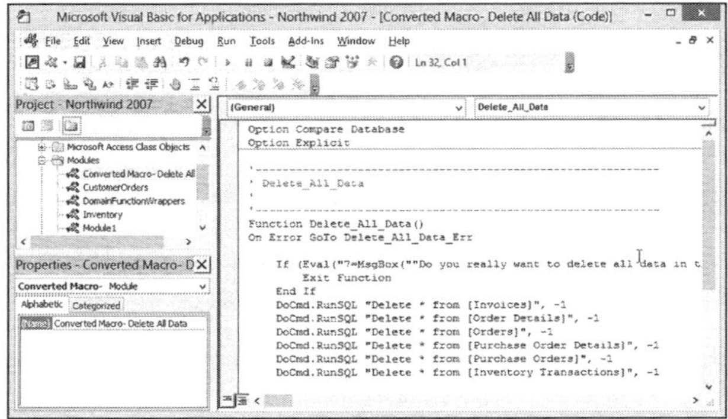

FIGURE 30.51. This VBA code was generated by Access from a standalone macro.

You can modify the code generated by the macro conversion process to suit your needs.

Converting Embedded Macros to VBA

To convert embedded macros, open the form or report in Design view You should see the button named Convert Form's Macros to Visual Basic in the Macro group, as shown in Figure 30.52.

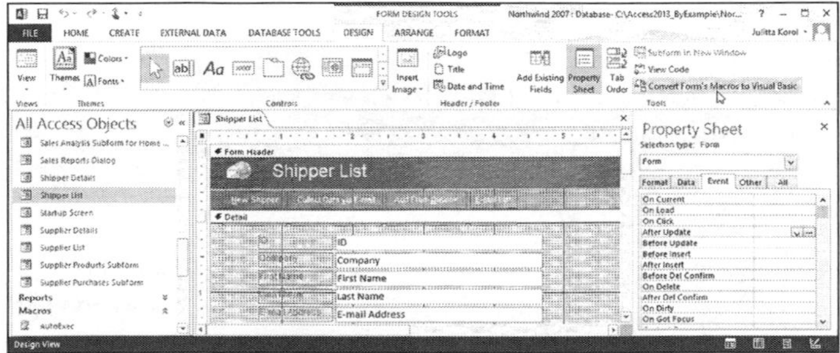

FIGURE 30.52. Converting embedded macros to VBA code using the Convert Form's Macros to Visual Basic button.

After clicking the Convert Form's Macros to Visual Basic button, Access displays the same dialog box shown earlier in the conversion process for standalone macros (see Figure 30.50). When you click the Convert button, Access begins the conversion process, and when this process completes you will see a message about the successful completion of the conversion. Click OK to the message. Next, activate the property sheet, and notice that form and control event properties that have previously been set to [Embedded Macro] now display [Event Procedure]. You can click the Build button (...) to view the VBA code. Figure 30.53 shows the VBA procedure that was generated for the embedded macro attached to the DoubleClick event of the Last Name text box control placed on a form. Notice that for each converted macro, Access writes its equivalent XML code as a comment at the top of the VBA procedure.

If after the conversion process you still want to keep the [Embedded Macro] setting in the event properties of a form, report, or control, perform these steps:

4. Save the VBA code generated by the macro conversion to a file by choosing **File | Export File** in the Visual Basic Editor window. Access will create a file with the .cls extension. To view the contents of this file, right-click its name in Windows Explorer and choose **Open With | Choose Program**. Select **Notepad** and click **OK**.
5. In Access, close the form and answer **No** when prompted for changes. Access will revert the [Event Procedure] setting in the event properties to [Embedded Macro].

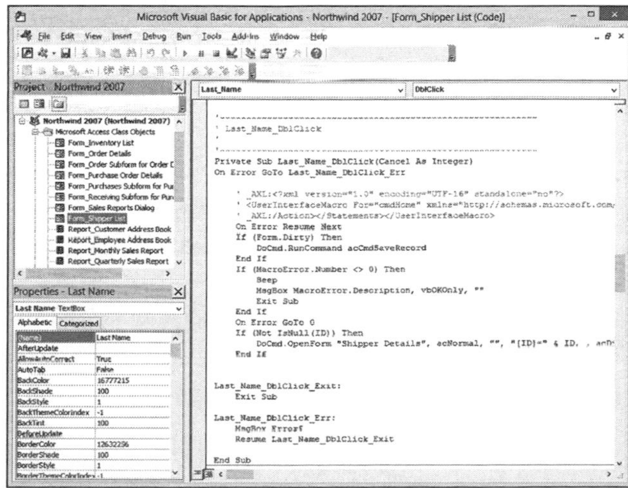

FIGURE 30.53. VBA code from a converted embedded macro.

ACCESS TEMPLATES

Access comes with several prebuilt templates that give users a head start with various types of projects. In Access 2013 the templates listed on the startup screen have built-in tables, queries, forms, and reports in various subject categories. You can use them to quickly create Access Web apps and share them on the web. Access web apps are a new type of Access file format introduced in Access 2013. This chapter focuses on the desktop template files that you can use to save time creating desktop databases. The desktop template files can be easily recognized by their .accdt file extension. By default, Access stores the template files in the C:\Users\username\AppData\Roaming\Microsoft\Templates folder. Please note that the AppData is a hidden folder and you will need to unhide in the Windows Explorer in order to access its content.

Creating a Custom Blank Database Template

When you select the Blank desktop database button in the Backstage view (File | New), and click the Create button, Access provides you with an empty database that you can customize to suit your specific needs. If you are like many users, you start your next database project by again clicking the Blank desktop database button and proceed to implement many of the same customizations that you applied to the previous database project. If you have been working like this, however, you are not taking advantage of the startup template—Blank.accdb. Instead of customizing each new blank database, simply create a new database called Blank.accdb in the template folder and customize it to include specific database properties, VBA references and custom functions, Ribbon customizations, default forms and reports, and customized controls, as well as any other special configuration settings that you normally use in your database applications. The next time you click the Blank desktop database button in the Backstage view, Access will make a copy of your Blank.accdb database so you won't need to start from scratch. Your new database will already contain the common settings that you saved in the Blank.accdb file. Moreover, if your database requirements have changed, you can create a new Blank.accdb database with settings that conform to these new requirements.

Understanding the .accdt File Format

The .accdt file format that the last three versions of Access (2013/2010/2007) use for their database templates is based on the Microsoft Office Open Packaging Convention (.opc) file format. This file format is based on the XML and ZIP archive technologies. The .opc file format is also used by the .docx, .xlsx, and .pptx file formats first introduced in Office 2007 for Word, Excel, and PowerPoint. The .opc format makes it possible to store a number of text, image, and .xml/.xsd files in a single compressed file. The .opc files can be easily opened and examined. Before you can open an Access template file (.accdt) and examine its structure, you need to add the .zip extension at the end of the filename as shown in the following steps:

1. Launch Microsoft Access 2013. On the startup screen, type **desktop asset tracking** in the search box and press Enter to begin searching the online content. You must be connected to the Internet to make it work. When Access displays the templates that matched your search criteria, click the one named Desktop contacts. Enter the name of the database as **DesktopAssetTracking.accdb** and change the folder to **C:\Access2013_ByExample**, and then click the **Create** button. Access downloads the template file and creates the specified desktop database. The resulting database is shown in Figure 30.54.

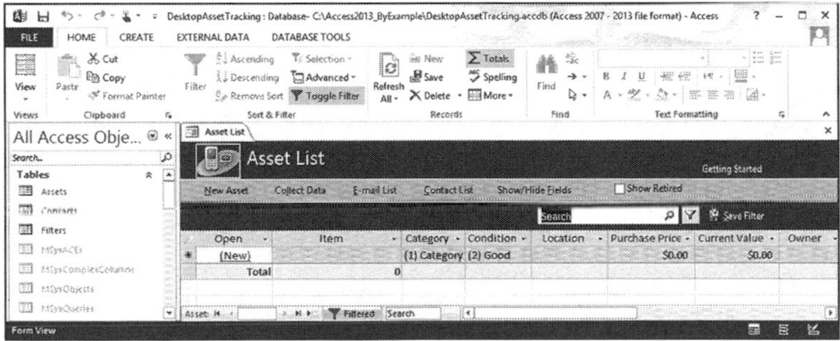

FIGURE 30.54. The Asset Tracking database is based on the desktop asset tracking template downloaded from the Microsoft Access templates archive.

2. Close the DesktopAssetTracking.accdb file and exit Microsoft Access.
3. Open your Documents folder in Windows Explorer. Notice that Access has saved the downloaded template file named **Desktop asset tracking.accdt** in this folder. If you cannot see this file, search for it on your C: drive.
4. Rename the file **Desktop asset tracking.accdt.zip** as shown in Figure 30.55.
5. When the Rename dialog box appears, click **Yes** to confirm that you want to change the filename extension. Click **Continue** if prompted to provide administrator permission to rename this file. The file format should now change to the zip archive.

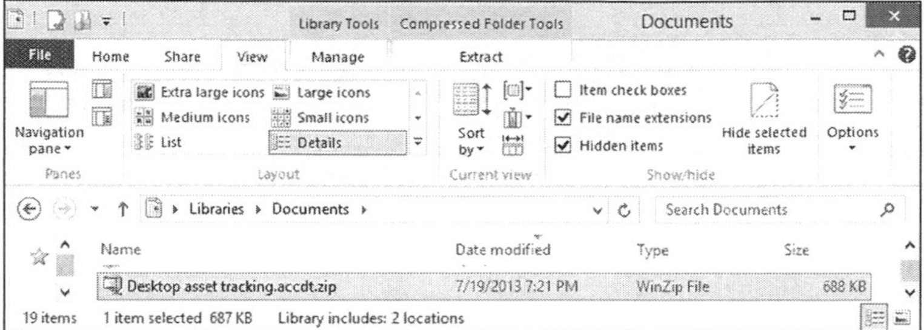

FIGURE 30.55. By adding the zip file extension to the accdt file format you can turn it into a zip archive that you can examine and modify depending on your needs.

6. To open the Desktop asset tracking .accdt.zip file, right-click the filename and choose **Open With | Compressed (zipped) Folders** or **Windows Explorer**. The folders that make up the document are shown in Windows Explorer (see Figure 30.56).

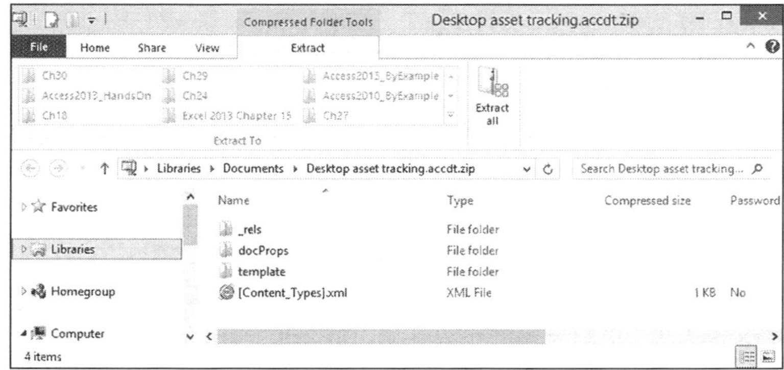

FIGURE 30.56. The directory structure of an Access template file.

Notice that the archive file contains the following three folders: _rels, docProps, and template. The _rels folder contains one .xml file with the extension .rels that defines the relationships between various files included in the file package (see Figure 30.57). Access uses this file to find out information about the template and the database.

```
<?xml version="1.0" encoding="UTF-8"?>
- <Relationships xmlns="http://schemas.openxmlformats.org/package/2006/relationships">
    <Relationship Id="rId1" Target="docProps/core.xml"
       Type="http://schemas.openxmlformats.org/package/2006/relationships/metadata/core-
       properties"/>
    <Relationship Id="Template" Target="template/template.xml"
       Type="http://schemas.microsoft.com/office/access/2005/04/template/start"/>
    <Relationship Id="rId2" Target="/docProps/Thumbnail.png"
       Type="http://schemas.openxmlformats.org/package/2006/relationships/metadata/thumbnail"/>
  </Relationships>
```

FIGURE 30.57. The contents of the .rels .xml file in the _rels folder.

You can modify the contents of the .rels file to, for example, have Access load a different preview image instead of the current Thumbnail.png image located in the docProps folder.

The docProps folder contains the Thumbnail.png image and the core.xml file that describes the core document properties such as creator name, identifier, title, description, keywords, category, version, and lastModifiedBy (see Figure 30.58).

```
<?xml version="1.0" encoding="UTF-8" standalone="true"?>
- <cp:coreProperties xmlns:xsi="http://www.w3.org/2001/XMLSchema-instance"
  xmlns:dcmitype="http://purl.org/dc/dcmitype/" xmlns:dcterms="http://purl.org/dc/terms/"
  xmlns:dc="http://purl.org/dc/elements/1.1/"
  xmlns:cp="http://schemas.openxmlformats.org/package/2006/metadata/core-properties">
     <dc:title>Assets_TP001225342</dc:title>
     <dc:description/>
     <cp:category>User Templates</cp:category>
     <dc:identifier/>
  </cp:coreProperties>
```

FIGURE 30.58. The contents of the core.xml file in the docProps folder.

When you double-click the template folder, you will see two subfolders named _rels and database, as well as a template.xml file (Figure 30.59).

« Desktop asset tracking.accdt.zip ▸ template — Search template

Favorites · Libraries · Homegroup

Name	Type	Compressed size	Password
_rels	File folder		
database	File folder		
template.xml	XML File	1 KB	No

FIGURE 30.59. The contents of the template folder.

The template.xml file contains information about the format of the template file.

You can find out a lot of information about the contents and structure of the .accdt file by opening the database folder (Figure 30.60).

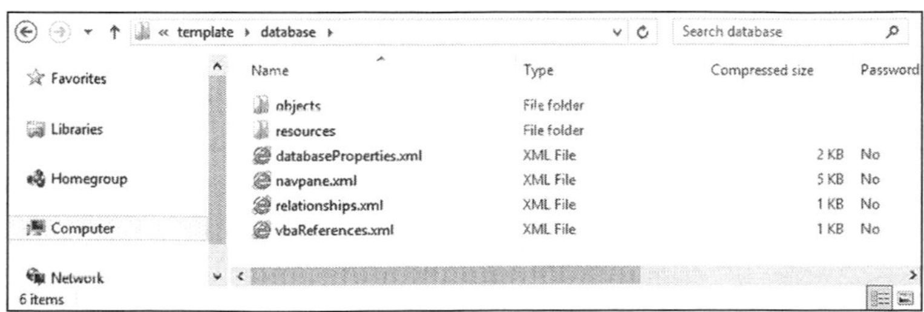

FIGURE 30.60. The contents of the database folder.

The databaseProperties.xml file in the database folder stores various database settings and properties. You can modify this file to include additional properties that need to be set by adding new nodes to the file.

The navpane.xml file contains information about the structure of the Navigation pane. It also contains the data for the Navigation pane system tables: MSysNavPaneGroupCategories, MSysNavPaneGroups, MSysNavPaneGroupToObjects, and MSysNavPaneObjectIDs.

The relationships.xml file contains the contents of the MSysRelationships system table.

The vbaReferences.xml file contains all VBA project references that Access needs to set.

In the objects folder, you will find many other files that describe different database objects (Figure 30.61).

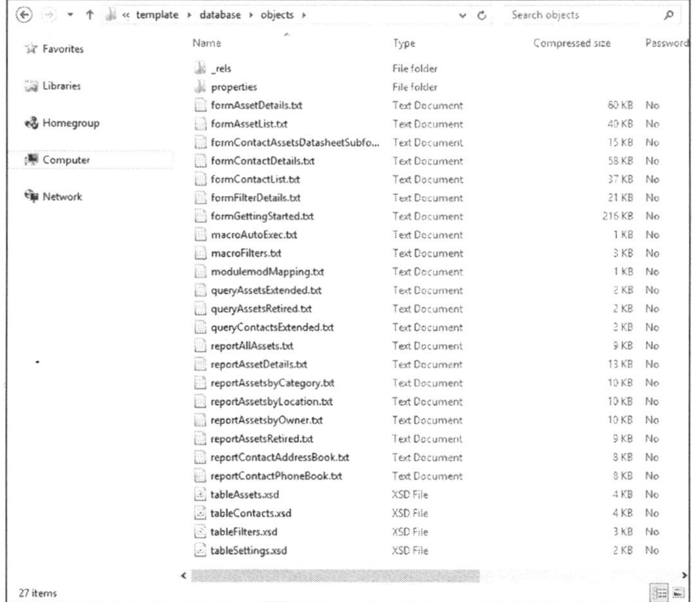

FIGURE 30.61. Files in the objects folder contain information about different database objects in the template file as well as information about sample data and properties for each object included in the template.

You can open any of the files listed in Figure 30.61 and examine the type of information being stored. Because this chapter covered macros, take a look at the macroFilters.txt file to find out how Access stores the embedded macros.

When you are done viewing the files, change the name of the Desktop asset tracking.accdt.zip file back to its original name—Desktop asset tracking.accdt.

CHAPTER SUMMARY

This chapter introduced you to working with macros in Access 2013. We took a detailed look at macro security; created standalone and embedded macros; worked with data macros; saw examples of return variables (ReturnVars), local variables, and temporary variables (TempVars), and examined the error-handling actions in macros. We also learned how standalone and embedded macros can be converted to Visual Basic code. Because Access uses embedded macros extensively in its templates, we examined the structure and contents of the .accdt file format.

This chapter concludes part VIII of the book, which focused on getting familiar with macro interface and usage of templates in Access 2013. In part IX of the book, you learn how to use your Access VBA skills to build Internet applications.

Part 9

Taking Your VBA Programming Skills to the Web

Gone are the times when working with Access required the Microsoft Access application to be installed on a user's desktop. Thanks to the development of Internet technologies, you can publish both static and dynamic Access data to the Web. In this part of the book, you learn how Active Server Pages (ASP) and Extensible Markup Language (XML) are used with Microsoft Access to develop database solutions for the World Wide Web.

Chapter 31 Access and Active Server Pages

In today's world, everyone wants to be able to access data via the company intranet or the World Wide Web. This book would not be complete without showing you how to take your skills where the demand is. So, how can you make the information stored in your Access database available for others to view or query in a Web browser? By adding some HyperText Markup Language (HTML) and Microsoft Visual Basic Scripting Edition (VBScript) to your current VBA skill set, you can start making your applications Web-ready. Microsoft Access allows you to save tables, queries, and forms to the Web as HTML or XML documents. This chapter focuses on showing you how you can create classic Active Server Pages (ASP) in order to display, query, insert, update, and delete data stored in a Microsoft Access database from a Web browser.

INTRODUCTION TO CLASSIC ASP

With Active Server Pages (ASP), a technology developed by Microsoft, and your current working knowledge of VBA, you can begin designing and programming powerful and dynamic Web applications.

The current version of ASP is 3.0, and it is available with Internet Information Services (IIS) 5.0 or higher. Active Server Pages are text files with the .asp extension. These files contain standard HTML formatting tags and embedded scripting statements. Because the default scripting language for ASP is VBScript, a subset of Visual Basic and Visual Basic for Applications, you already have many of the skills required to Web-enable your Access applications. In addition, the tools you need to make Access work with the intranet or Internet are within your reach. You don't need special tools to write your code. You can use Windows Notepad or any other text editor. So, where do you start? You can start by acquiring some knowledge of the HyperText Markup Language (HTML). There are plenty of free tutorials on the Web that can get you started. For example, for an easy, step-by-step introduction, check out the *htmlgoodies.com* Website. Because Active Server Pages are a mix of HTML and a programming language such as VBScript or JavaScript, you should learn as much as you can about each component that you will be using.

To better understand this chapter's topics, here are some terms to get acquainted with:

- **HyperText Markup Language (HTML)**—a simple, text-based language that uses special commands known as *tags* to create a document that can be viewed in a browser. HTML tags begin with a less-than sign (<) and end with a greater-than sign (>). For example, to indicate that the text should be displayed in bold letters, you simply type your text between the <b> and </b> tags like this:

  ```
  <b>This text will appear in bold letters.</b>
  ```

  Using plain HTML you can produce static Web pages with text, images, and hyperlinks to other Web pages.

- **Dynamic HTML (DHTML)**—allows the HTML tags to be changed programmatically via scripting. Use DHTML to add interactivity to your Web pages.

- **VBScript**—a scripting language based on Microsoft Visual Basic for Applications (VBA). Because this is just a subset of VBA, some of the VBA features have been removed. For example, VBScript does not support data types—every variable is a Variant. Like VBA, VBScript is an event-driven language—the VBScript code is executed in response to an event caused by a user action or the Web browser itself.

- **JavaScript**—a compact, object-oriented scripting language invented by Netscape and used for developing client and server Internet applications. This is a cross-platform language that can be embedded in other products and applications, such as Web browsers. Use JavaScript instead of VBScript in client-side scripts if you want to support browsers such as Firefox®, Chrome, or Safari®.
- **JavaScript Libraries**—are collections of JavaScript code that contain commonly used functions, animation effects, and various shortcuts that allow you to quickly accomplish common JavaScript tasks. Popular JavaScript libraries include jQuery, Dojo, Prototype, MooTools, and YUI.
- **Active Server Pages (ASP)**—also referred to as "classic" ASP (this version of ASP preceded a newer technology known as ASP.NET), Active Server Pages is a Web development technology that enables you to combine HTML, scripts, and reusable ActiveX® server components to create dynamic Web applications. ASP is not limited to a particular language. To create ASP pages, you can use scripting languages such as VBScript, JavaScript, or any language for which you have a third-party ActiveX scripting engine. While HTML pages store the actual data, Active Server Pages only store the information on how to obtain the data. How does this work? Suppose you typed the address of the ASP page in your Web browser's address bar and pressed Enter or clicked the Go button. The Web server will read the script instructions contained in the ASP page (a file with an .asp extension) and access the specified database. Once the data is obtained, the Web server will put this information into an HTML page and return that page to you in the Web browser in plain HTML code. Users never see the instructions contained in your ASP file unless they have access to the Web server and have been given the appropriate permissions to open these files. Because the Web server reads and processes the instructions in the ASP page every time your browser requests the page, the information you receive is highly dynamic. ASP allows the page to be built or customized on the fly before the page is returned to the browser. ASP is platform independent. This means that you can view ASP pages in any browser (Internet Explorer®, Firefox, Safari, Chrome, Opera™, and others).

- **ASP.NET**—(pronounced ASP DOT NET) a newer, more advanced, and feature-rich Web development technology from Microsoft that requires the Microsoft .NET Framework to be installed on users' computers. Unlike Active Server Pages (ASP), which is limited to scripting languages, .NET technology provides cross-language support (you can write and share code in many different .NET languages such as Visual Basic .NET, C#, Managed C++, JScript .NET, and J#). ASP files prepared in .NET end in .aspx, .ascx, or .asmx. ASP.NET is not an upgrade of the classic ASP; it is an entirely new infrastructure for Web development that requires learning new concepts about building Web applications and "unlearning" the concepts utilized in programming classic ASP applications. Because programming in .NET languages is quite different from writing programs in Visual Basic for Applications, it is not covered here. Instead, this chapter concentrates on ASP classic programming, which is more related to Visual Basic via its subset, VBScript.

CREATING AN ASP PAGE

In this section, you will create your first ASP page using HTML and Microsoft Visual Basic Scripting Edition (VBScript). In Hands-On 31.1, you will create an ASP page that retrieves data from the Employees table in the Northwind.mdb database.

Please note files for the "Hands-On" project may be found on the companion CD-ROM.

Hands-On 31.1. Creating an Active Server Page from a Microsoft Access Table

1. Create a new folder named **C:\Access2013_ASP_Classic** for this chapter's files.
2. Copy the **C:\Access2013_HandsOn\Northwind.mdb** database to the **C:\Access2013_ASP_Classic** folder.
3. Open Windows Notepad and type the following code:

```
<% @Language="Vbscript" %>
<%
' declare variables
Dim accessDB
```

```
Dim conn
Dim rst
Dim sql

' name of the database
accessDB = "Northwind"

' establish connection to the database
conn = "DRIVER={Microsoft Access Driver (*.mdb)};"
conn = conn & "DBQ=" & Server.MapPath(accessDB)

' Create a Recordset
Set rst = Server.CreateObject("ADODB.Recordset")

' select all records from Employees table
' for indicated fields only
sql = "SELECT FirstName, LastName, Title, City, Country "
sql = sql & "FROM Employees"

' Open Recordset (and execute SQL statement above)
' using the open connection
rst.Open sql, conn
%>

<html>
<head>
<title>Northwind Employees</title>
</head>
<body>
<table border="1">
<%
For Each fld In rst.Fields
  Response.Write ("<th>") & fld.Name & ("</th>")
Next
rst.MoveFirst
Do While Not rst.EOF
  Response.Write ("<tr>")
  For Each fld In rst.Fields
    Response.Write ("<td>") & fld.Value & ("</td>")
  Next
  Response.Write ("</tr>")
  rst.MoveNext
```

```
Loop
%>
</table>
</body>
</html>
<%
' close the Recordset
rst.Close
Set rst = Nothing
%>
```

4. Save the file as **C:\Access2013_ASP_Classic\Employees.asp** and exit Notepad. Let's spend a few minutes analyzing the ASP page you've just written. The code shown here begins by specifying a scripting language for the page with the Active Server Pages directive <% @Language="Vbscript" %>.

   The script between the <% and %> delimiters is Visual Basic script code that gets executed on the Web server. The <% says that what follows is a server-side script, not HTML. The %> indicates the end of a script segment. The script code between the <% and %> delimiters is executed on the Web server as the page is processed. Any values you want returned by the script are placed between the <% and %> delimiters.

   Similar to VBA procedures, the first step in scripting is the declaration of variables. Because all variables in VBScript are of the Variant type, you don't need to use the `As` keyword to specify the type of variable. To declare a variable, simply precede its name with the `Dim` keyword:

   ```
   Dim accessDB
   Dim conn
   Dim rst
   Dim sql
   ```

   Also, like VBA, you can declare all your variables on one line, like this:

   ```
   Dim accessDB, conn, rst, sql
   ```

   To connect with the Access database, we specify a connection string like this:

   ```
   conn = "DRIVER={Microsoft Access Driver (*.mdb)};"
   conn=conn & "DBQ=" & Server.MapPath(accessDB)
   ```

   The `DRIVER` parameter specifies the name of the driver that you are planning to use for this connection (Microsoft Access Driver (*.mdb)). The `DBQ` parameter indicates the database path. The exact path will be supplied by the `MapPath` method of the Server object:

```
Server.MapPath(accessDB)
```

You can also connect to your Access MDB database by using the OLEDB data provider as follows:

```
Set conn = Server.CreateObject("OLEDB.Connection")
conn.Open "Provider=Microsoft.Jet.OLEDB.4.0; Data Source=" _
  & Server.MapPath(accessDB)
```

To connect to an SQL server database, use the following format:

```
Set conn = Server.CreateObject("OLEDB.Connection")
conn.Open "Provider="SQLOLEDB;" & _
  "Data Source=YourServerName;" & _
  "Initial Catalog=accessDB;" & _
  "UID=yourId; Password=yourPassword;"
```

To access database records, we create the Recordset object using the `CreateObject` method of the Server object:

```
Set rst = Server.CreateObject("ADODB.Recordset")
```

After creating the recordset, we open it using the `Open` method, like this:

```
rst.Open sql, conn
```

This statement opens a set of records. The `sql` variable is set to select all the records from the Employees table for the indicated fields. The `conn` variable indicates how you will connect with the database.

The %> delimiter indicates the end of the server-side script.

Server-side and Client-side Scripting

A server-side script is the script code that runs on the Web server before the page comes down to the client (the user's machine). This script begins to run when a browser requests an ASP file from your Web server. The Web server then calls the ASP interpreter (ASP.dll), which processes the blocks of code between the <% and %> delimiter tags. After the script commands are executed, the Web page is sent to the browser. Server-side scripts cannot be readily copied because only the result of the script is returned to the browser. Users cannot view the script commands that created the page they are viewing. All they can see is the HTML source code for the page.

In addition to server-side scripts, an ASP file can contain client scripts. A client script is the script code that is processed by a browser on the user's machine while the page is viewed. Client scripts are enclosed between <script> and </script> tags. When a browser encounters a <script> tag, it sends the script that follows this tag to a scripting engine. A scripting engine is the part of the Web browser that processes the scripts. Because not all browsers can process client scripts, comment tags (<!-- and -->) are often used to make browsers that do not recognize the <script> tag ignore it.

Notice the many tags between the angle brackets: <html>, </html>, <head>, </head>, <title>, </title>, <body>, </body>, <tr>, and so on (see Table 31.1). The tags tell the browser how to display the file. For example, the <html> tag tells the browser that what follows is an HTML document. The closing </html> tag at the end of the file tells the browser that the HTML document is completed. The closing tags are denoted by placing a forward slash before the tag name (for example, </title>, </body>, </html>). Closing tags cancel the effect of the tag.

The next part of the ASP page contains HTML formatting tags that prepare a table. These tags are summarized in Table 31.1.

The table headings are read from the Fields collection of the Recordset object using the `For...Each...Next` loop. Notice that all instructions that need to be executed on the server are enclosed by the <% and %> delimiters. To enter the data returned by the server in the appropriate table cell, use the `Write` method of the ASP Response object:

```
Response.Write ("<th>") & fld.Name & ("</th>")
```

This statement will return the name of a table header. Because this instruction appears in the file between the <th> and </th> HTML formatting tags, the names of the table fields will be written as headings in bold type.

After reading the headings, the next loop reads the values of the fields in each record:

```
Response.Write ("<td>") & fld.Value & ("</td>")
```

Because this statement is located between the <td> and </td> formatting tags, each time the loop is executed, the value retrieved from the current field in a particular record will be written to the table cells.

The script ends by closing the recordset and releasing the memory used by it:

```
rst.Close
set rst = Nothing
```

Because the Web server reads and processes the instructions in the ASP page every time your browser requests the page, the information you receive is highly dynamic. ASP allows the page to be built or customized on the fly before the page is returned to the browser.

TABLE 31.1. Frequently used script delimiters and HTML tags

Delimiters and Tags	Description
<% and %>	Beginning and end of the ASP script fragment. The script code between the <% and %> delimiters will be executed on the server before the page is delivered to the user's browser.
<html> and </html>	You should place the <html> tag at the beginning of each Web page. To indicate the end of a Web page, use the closing tag: </html>. The HTML document has two main sections: <head> and <body>.
<head> and </head>	The <head> section contains information about the document such as its title, keywords, description, and stylesheet. Often Java scripts are placed in the <head> section.
<title> and </title>	Place the text you want to display in the window titlebar between these HTML tags. The <title> tag always appears within the <head> section.
<body> and </body>	The text you want to display on the Web page should be placed between these tags.
<table> and </table>	The beginning and the end of a table.
<table border= "1">	The border parameter specifies the width of the table border.
<th> and </th>	Place table headings between <th> and </th> tags. They will be automatically displayed in bold font.
<tr> and </tr>	The <tr> tag begins a new row in a table. Each table row ends with the </tr> tag.
<td> and </td>	Each table data item starts with the <td> tag and ends with the </td> tag.

Note that you are not ready yet to view the data. Before you can view the Employees data in a browser, you need to perform the following tasks:

- Install and configure Microsoft Internet Information Services (IIS). The installation instructions are presented following the next section.
- Create a virtual folder (see the section following the IIS installation instructions).

THE ASP OBJECT MODEL

ASP has its own object model consisting of the objects shown in Table 31.2.

TABLE 31.2. The ASP object model

ASP Object Name	Object Description
Request	Obtains information from a user
Response	Sends the information to the client browser
Application	Shares information for all the users of an application
Server	Creates server components and server settings
Session	Stores information pertaining to a particular visitor

The ASP objects have methods, properties, and events that can be called to manipulate various features. For example, the Response object's `Write` method allows you to write text to the client browser. The `CreateObject` method of the Server object is required to create a link between a Web page and your Access database. You will become familiar with some of these ASP objects and their properties and methods as you create the example ASP pages in this chapter.

Complete coverage of the ASP object model is beyond the scope of this book. This chapter's objective is to demonstrate how your VBA skills can be used with other Internet technologies (HTML, VBScript, and classic ASP) to programmatically access database data in an Internet browser.

INSTALLING INTERNET INFORMATION SERVICES (IIS)

Internet Information Services (IIS) is a Web server application created by Microsoft for use with the Microsoft Windows operating system. The following versions of IIS are currently in use:

- IIS 8 – Windows 8 / Windows Server 2012
- IIS 7.5 – Windows 7 / Windows Server 2008 R2
- IIS 7.0 – Windows Vista / Windows Server 2008
- IIS 6.0 – Windows Server 2003
- IIS 5.1 – Windows XP Professional

The classic version of ASP is not installed by default on IIS 7.0 and IIS 7.5. Therefore, before running the examples in this chapter, you need to enable this feature using the Control Panel. The Hands-On 31.2 exercise walks you through the process of getting your Windows 7 or 8 computer to recognize ASP classic.

Hands-On 31.2. Enabling Classic ASP on Windows 7/8

1. Open the Control Panel and click on the Programs link as shown in Figure 31.1.
2. Under Programs and Features, click **Turn Windows features on or off** as shown in Figure 31.2.

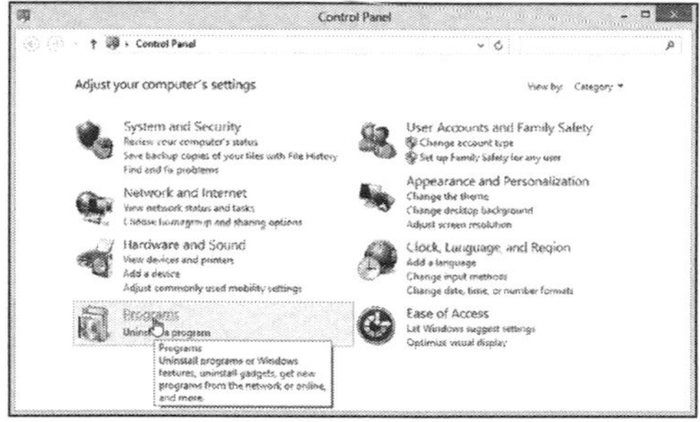

FIGURE 31.1. Enabling classic ASP in Windows 7/8 (Step 1).

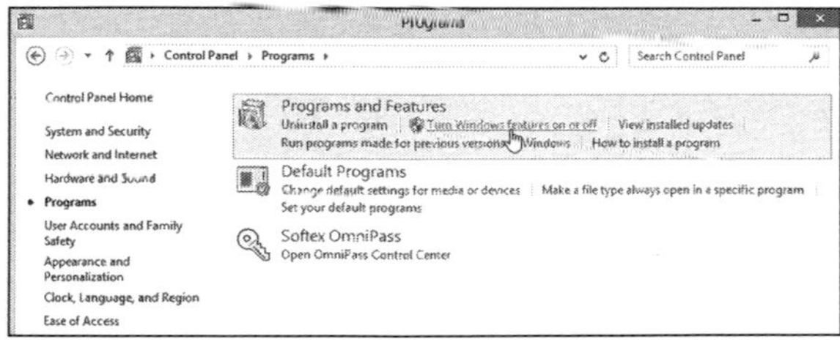

FIGURE 31.2. Enabling classic ASP in Windows 7/8 (Step 2).

3. If you are prompted with a request for permission to continue, click **Continue** in the User Account Control (UAC) window to give Windows permission to proceed.
4. Wait while Windows retrieves all the features.
5. Expand the **Internet Information Services**, then **World Wide Web Services**, and under the **Application Development Features** choose **ASP** as shown in Figure 31.3.

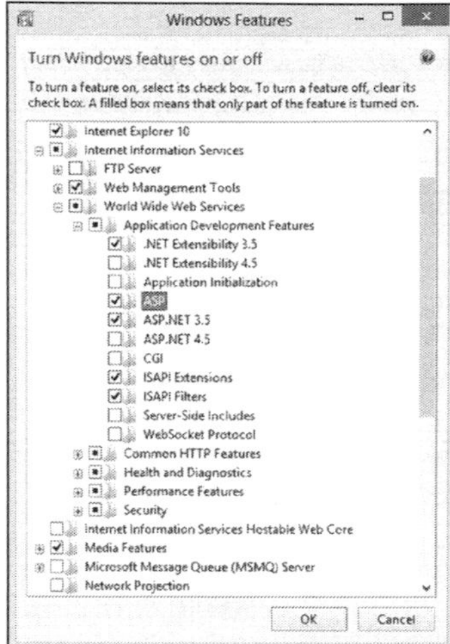

FIGURE 31.3. Enabling classic ASP in Windows 7/8 (Step 3). Make sure ASP is checked under Application Development Features.

6. After checking ASP, click **OK** and wait for Windows to apply changes. This might take several minutes. Once the features are configured, close all open Control Panel windows.
7. After completing the preceding configuration steps, you should see the folder named inetpub on your computer's system drive as shown in Figure 31.4.

After you have installed IIS, it is important that you run Windows Update to ensure that your system has the most recent security patches and bug fixes.

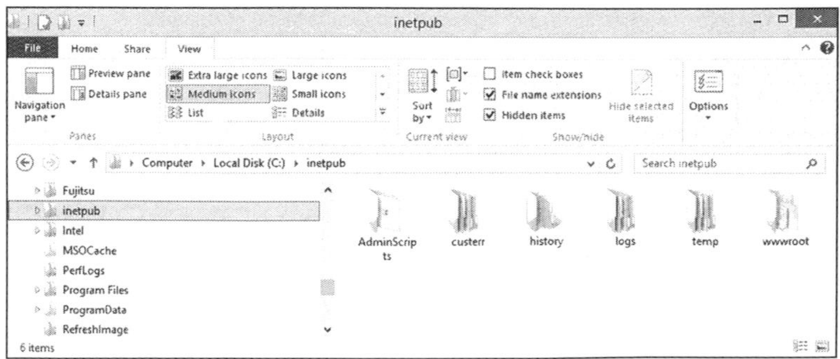

FIGURE 31.4. After you have enabled classic ASP, a new folder named inetpub appears on your computer's system drive.

CREATING A VIRTUAL DIRECTORY

The default home directory for the World Wide Web (WWW) service is \Inetpub\wwwroot. Files located in the home directory and its subdirectories are automatically available to visitors to your site. If you have Web pages in other folders on your computer and you'd like to make them available for viewing by your Website visitors, you can create virtual directories. A virtual directory appears to client browsers as if it were physically contained in the home directory.

For the purposes of this chapter, you created a directory called Access2013_ASP_Classic (see Hands-On 31.1). In Hands-On 31.3, you will designate it as a virtual directory.

Hands-On 31.3. Creating a Virtual Directory on Windows 7/8

1. Open the Control Panel, choose System and Security, and then click on Administrative Tools
2. Double-click **Internet Information Services (IIS) Manager** as shown in Figure 31.5.
3. Click **Continue** in the User Account Control (UAC) window if Windows asks you for permission to continue. Respond No to any other question.
4. Expand the tree nodes in the Connections pane on the left, right-click on **Default WebSite**, and select **Add Virtual Directory** as shown in Figure 31.6.
5. A virtual directory has an *alias*, or name that client browsers use to access that directory. An alias is often used to shorten a long directory name. In addition, an alias provides increased security. Because users do not know where your files are physically located on the server, they cannot modify them.
6. Type **NorthDB** in the Alias box as shown in Figure 31.7. Set the Physical path to point to the **C:\Access2013_ASP_Classic** folder that you created in Hands-On 31.1.
7. Click **OK** to save the changes.
8. Notice the virtual directory named NorthDB now appears under Default WebSite in the Connections pane (Figure 31.8). The middle section of the Internet Information Services (IIS) Manager displays the NorthDB Home.
9. Do not close the IIS Manager window, as you will continue with it in the next section.

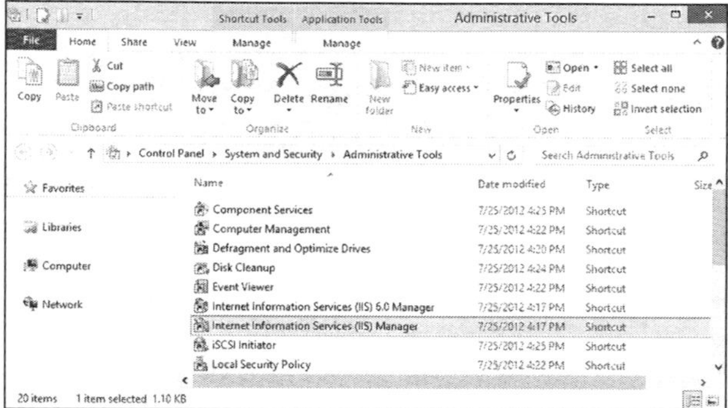

FIGURE 31.5. To set up a virtual directory on your computer, you must first activate Internet Information Services (IIS) Manager in the Administrative Tools of the Windows Control Panel.

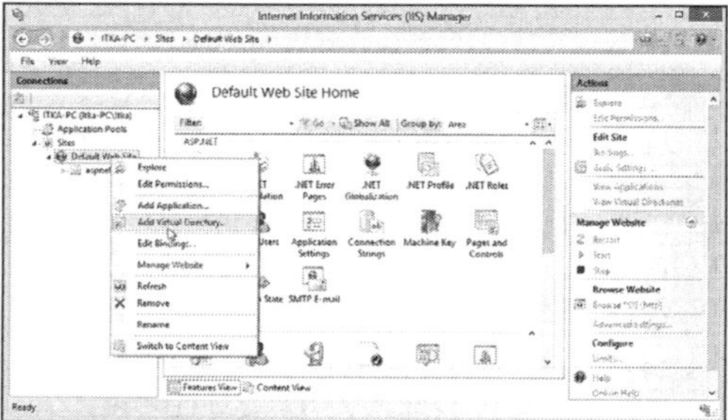

FIGURE 31.6. You can add a virtual directory by right-clicking Default Web Site in the Connections pane of the Internet Information Services (IIS) Manager window.

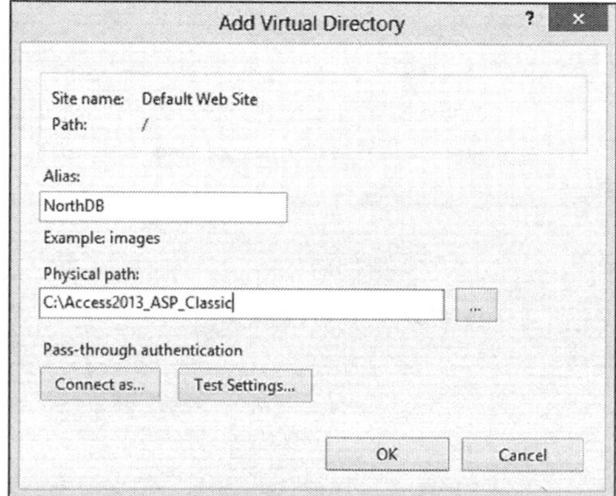

FIGURE 31.7. The Add Virtual Directory dialog box is used to specify the name and path to your Website folder (Windows 7/8). The physical folder named Access2013_ASP_Classic will be shared over the Web as NorthDB.

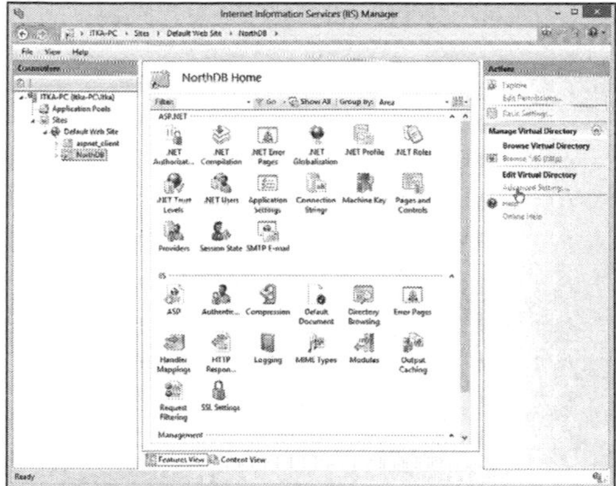

FIGURE 31.8. After creating a virtual directory, you should see it listed under Default Website in the Connections pane of the Internet Information Services (IIS) Manager window.

SETTING ASP CONFIGURATION PROPERTIES

To make it easy to debug your code and to ensure that you can use relative paths in your code, you should change a couple of default configuration properties in the IIS Manager. The following hands-on exercise walks you through the steps required to make the necessary modifications.

Hands-On 31.4. Configuring ASP Properties

1. In the Connections pane, select **Default Web Site**, and then in the middle section under IIS, double-click **ASP**.
2. Expand the Debugging Properties tree node and set the **Send Errors To Browser** property to **True**, as shown in Figure 31.9.

By default, when ASP script errors are encountered, Windows displays the following message: "An error occurred on the Server when processing the URL. Please contact the System Administrator." To prevent this error, be sure to select True next to the Send Errors To Browser property as shown in Figure 31.9.

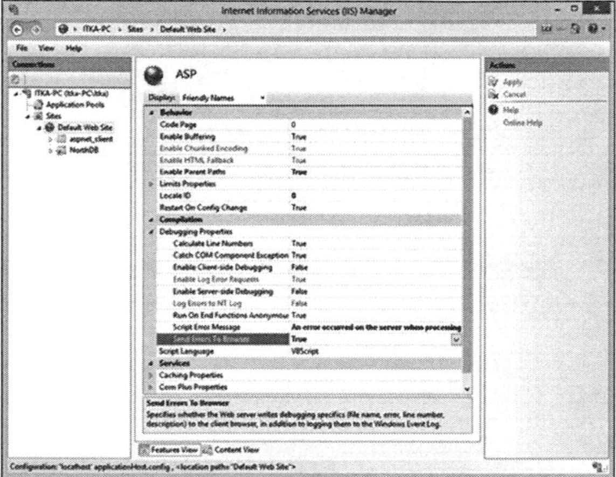

FIGURE 31.9. By setting the Send Errors To Browser property to True, you can easily troubleshoot errors when your Active Server Page encounters an error.

3. In the Behavior section, set **Enable Parent Paths** to **True** as shown in Figure 31.9.
4. Parent paths allow you to use relative addresses that contain ".." in the paths of files and folders. For example, the following line will cause an error if parent paths are disabled:

```
Response.Write Server.MapPath("../login.asp")
```

   In earlier versions of IIS, parent paths were enabled by default. In IIS 7 and above, you need to remember to enable parent paths in order to prevent errors when relative paths are used.

5. In the Actions area on the right, click **Apply** to save the changes. When changes have been successfully saved, you should see a message in the Alerts area in the right pane of the IIS Manager window that the changes have been successfully saved.
6. Close the Internet Information Services (IIS) Manager window and any Control Panel windows that are still open.

TURNING OFF FRIENDLY HTTP ERROR MESSAGES

Friendly HTTP error messages don't provide enough information for programmers to effectively troubleshoot ASP script errors. Use the following steps to uncheck the Show friendly HTTP error messages option in your browser so you will get more meaningful error messages that can help you solve your script problems.

Hands-On 31.5. Turning Off Friendly HTTP Error Messages

1. Open your Internet browser.
2. Choose **Tools | Internet Options.**
3. In the Internet Options window, click the **Advanced** tab.
4. Locate the Browsing settings and uncheck **Show friendly HTTP error messages** as shown in Figure 31.10.
5. Click **OK** to save your changes and exit the Internet Options window.

Your IIS 7/7.5 is now configured to run classic ASP scripts on your computer and provide you with meaningful error messages in case errors are encountered in your scripts at runtime.

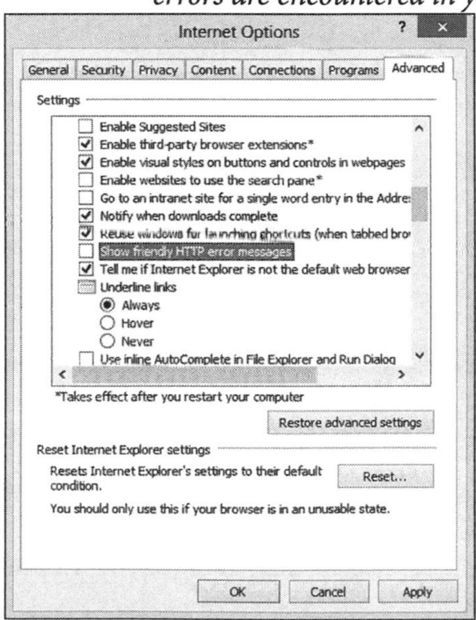

FIGURE 31.10. Turn off the Show friendly HTTP error messages option so you can see the actual Windows messages when troubleshooting your ASP scripts.

Important Note:

Your IIS is now configured to run classic ASP scripts on Windows 7/8 machine (32-bit systems). If you are working with the 64-bit system, you will need to take additional steps as follows:

a. Open the **Control Panel**, change the view to show all icons, and then click on **Administrative Tools.**

b. Double-click **Internet Information Services (IIS) Manager** as shown in Figure 31.5.
c. Expand the tree node in the Connections pane on the left, right-click the Application Pools and choose **Add Application Pool**.
d. In the name box, enter **MyClassicASP**. For the .NET Framework version choose **No Managed Code**. In the Managed Pipeline Mode drop-down, choose **Classic**. After making these selections, click **OK**.
e. The MyClassicASP entry should now appear in the Application Pool list in the middle section of the IIS Manager window. Right-click this entry and choose **Advanced Settings**.
f. In the **(General)** section of the Advanced Settings dialog, specify **True** for **Enable 32-bit Applications**.
g. Click **OK** to close the Advanced Settings dialog.
h. In the Connections pane on the left, right-click **Default Web Site**, and choose **Manage Web Site | Advanced Settings**.
i. In the Advanced Settings window, change the Application Pool to **MyClassicASP** and click **OK**.
j. Close the IIS Manager window.

For more information see the following link:

http://www.iis.net/learn/application-frameworks/running-classic-asp-applications-on-iis-7-and-iis-8

RUNNING YOUR FIRST ASP SCRIPT

Now that you've prepared the ASP file and set up the virtual directory, including the required settings, it's time to see the result of your efforts. In Hands-On 31.6, you will access the employee data from a Web browser by requesting in a browser the Active Server Page (Employees.asp) that you prepared in Hands-On 31.1.

Hands-On 31.6. Requesting an ASP Page

This hands-on exercise requires completion of Hands-On 31.1 through 31.5.

1. To ensure that all of the components you need for this chapter's examples can be quickly accessed, make sure that you have copied the sample **Northwind.mdb** database file from the **C:\Access2013_HandsOn** folder to your **Access2013_ASP_Classic** folder.

2. Launch **Internet Explorer** and type **http://localhost/NorthDB/Employees.asp** in the address box. Press **Enter** to execute the Active Server Pages (ASP) file or click the **Go** button. The contents of the Employees table should appear in your browser as shown in Figure 31.11.

FirstName	LastName	Title	City	Country
Nancy	Davolio	Sales Representative	Seattle	USA
Andrew	Fuller	Vice President, Sales	Tacoma	USA
Janet	Leverling	Sales Representative	Kirkland	USA
Margaret	Peacock	Sales Representative	Redmond	USA
Steven	Buchanan	Sales Manager	London	UK
Michael	Suyama	Sales Representative	London	UK
Robert	King	Sales Representative	London	UK
Laura	Callahan	Inside Sales Coordinator	Seattle	USA
Anne	Dodsworth	Sales Representative	London	UK

FIGURE 31.11. You can request the ASP page by typing its URL in the Web browser's address bar.

3. Right-click anywhere in the browser window and select **View Source** to view the source code (see Figure 31.12) .
4. The View Source command uses Windows Notepad to display a source file. Because the script commands contained in the ASP file are evaluated on the server before the browser receives the page, the resulting page in the browser is 100% pure HTML code. Notice that the browser does not display any of your ASP code that was surrounded by the <% and %> delimiters. The scripting code is evaluated on the server and only the resulting HTML is passed to the browser.

```


<html>
<head>
<title>Northwind Employees</title>
</head>
<body>
<table border="1">
<th>FirstName</th><th>LastName</th><th>Title</th><th>City</th><th>Country</th><tr><td>Nancy</td><td>Davolio</td><td>Sales Representative</td><td>Seattle</td><td>USA</td></tr><tr><td>Andrew</td><td>Fuller</td><td>Vice President, Sales</td><td>Tacoma</td><td>USA</td></tr><tr><td>Janet</td><td>Leverling</td><td>Sales Representative</td><td>Kirkland</td><td>USA</td></tr><tr><td>Margaret</td><td>Peacock</td><td>Sales Representative</td><td>Redmond</td><td>USA</td></tr><tr><td>Steven</td><td>Buchanan</td><td>Sales Manager</td><td>London</td><td>UK</td></tr><tr><td>Michael</td><td>Suyama</td><td>Sales Representative</td><td>London</td><td>UK</td></tr><tr><td>Robert</td><td>King</td><td>Sales Representative</td><td>London</td><td>UK</td></tr><tr><td>Laura</td><td>Callahan</td><td>Inside Sales Coordinator</td><td>Seattle</td><td>USA</td></tr><tr><td>Anne</td><td>Dodsworth</td><td>Sales Representative</td><td>London</td><td>UK</td></tr>
</table>
</body>
</html>

```

FIGURE 31.12. Viewing the source code of the ASP page.

5. Close Notepad and exit Internet Explorer.

RETRIEVING RECORDS

In the preceding sections of this chapter, we worked with the ASP page that retrieved records from the Employees table. To gain more experience in data retrieval, let's create another ASP page. The example ASP code in Hands-On 31.7 retrieves only customer names from the Customers table.

Hands-On 31.7. Creating an ASP File to Retrieve Records

This hands-on exercise requires completion of Hands-On 31.1.

1. Start **Notepad** and enter the following ASP code:
2.
```
<%@ Language="Vbscript" %>

<html>
<head>
<title>Retrieving a Recordset</title>
</head>
<body>
<%
Set conn = Server.CreateObject("ADODB.Connection")
conn.Open "DRIVER=Microsoft Access Driver (*.mdb);DBQ=" & _
 "C:\Access2013_ASP_Classic\Northwind.mdb"
Set rst = conn.Execute("SELECT CompanyName FROM Customers")
Do While Not rst.EOF
  Response.Write rst("CompanyName") & "<br/>"
  rst.MoveNext
Loop
%>
</body>
</html>
```

This code begins by specifying a scripting language for the page with the ASP directive `<%@ Language="Vbscript" %>`. Recall that the script contained within the <% and %> is Visual Basic script. This script performs the following actions:

- Creates an instance of the ADO Connection object
- Opens the connection to the Northwind.mdb database using the Microsoft Access driver (this is the DSN-less connection)

The SQL `SELECT` statement retrieves the values in the CompanyName field from the table named Customers into a Recordset object named `rst`. The SE-

LECT statement is executed with the Execute method of the Server Connection object. Notice that the instance of the Recordset object is created implicitly when the SQL statement is executed.

The Do While loop is used to output all the rows from the recordset to the browser. The Write method of the Response object outputs the value of a specific string or expression to the browser. The HTML <br/> tag is used to produce a carriage return after the value of the CompanyName field is output to the browser. Thanks to this tag, all company names are displayed on separate lines. Here, the value of the CompanyName field and a line break (
 tag) is written to the browser with the Response.Write statement like this:

```
Response.Write rst("CompanyName") & "<br/>"
```

The rst("CompanyName") part retrieves the value of the CompanyName field from the Recordset object. You can output the values from the Recordset object by using any of the following statements:

```
Response.Write rst.Fields("CompanyName")
Response.Write rst.Fields("CompanyName").Value
Response.Write rst.Fields(1)
Response.Write rst.Fields(1).Value
Response.Write rst(1)
Response.Write rst("CompanyName")
```

Because the Fields collection is the default collection of the Recordset object, you can omit the word "Fields." The MoveNext method in the next statement moves to the next record in the recordset.

3. Save the file as **C:\Access2013_ASP_Classic\GetCustomers.asp**.
4. Close Notepad.

   Now that you know what the code does, let's proceed to request this ASP page in the browser.
5. Open your Web browser and type **http://localhost/NorthDB/GetCustomers.asp** in the address bar.
6. Press **Enter** or click **Go**. When you request the GetCustomers.asp file in the browser, you get the results shown in Figure 31.13.

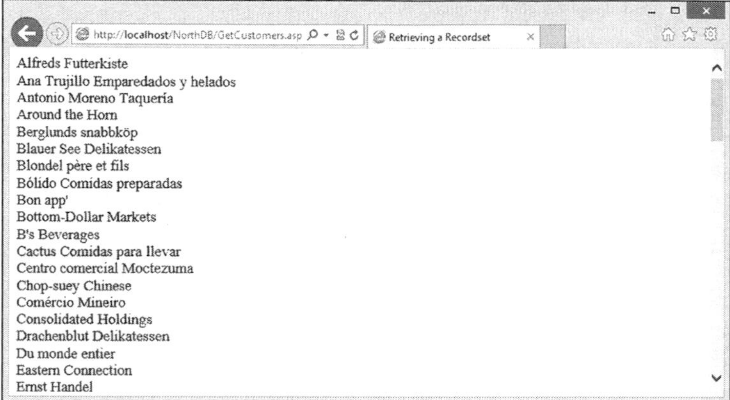

FIGURE 31.13. The ASP page created in Hands-On 31.7 displays the names of customers from the Customers table in the Northwind database.

Breaking Up a Recordset When Retrieving Records

In the preceding section, you worked with the ASP page that retrieved 91 records from the Customers table in the Northwind.mdb database. When you need to display more than a few records, it is a good idea to break up the recordset by dividing the list into multiple pages. This allows the user of your application to view a limited number of records at a time.

In Hands-On 31.8, you will create an ASP page that displays 12 customer names per page. The user will be able to move between the pages of data by clicking on the page number listed at the bottom of the page. To make the ASP page more useful, you will display the customer names as hyperlinks. Clicking on the customer name will call another ASP page that displays the customer's address as listed in the Customers table.

Hands-On 31.8. Creating a Multipage ASP File

1. Start Notepad and enter the following ASP code:

```
<%@ Language="Vbscript" %>
<html>
<head>
<title>View Few at a Time</title>
</head>
<body>
<%
Dim conn, rst, mySQL, currPage, rows, counter
Set conn = Server.CreateObject("ADODB.Connection")
```

```
conn.Open "DRIVER=Microsoft Access Driver (*.mdb);DBQ=" & _
 "c:\Access2013_ASP_Classic\Northwind.mdb"
Set rst = Server.CreateObject("ADODB.Recordset")

rst.CursorType = 3 'adOpenStatic
rst.PageSize = 12

mySQL= "SELECT * FROM Customers ORDER BY CompanyName"
rst.Open mySQL, conn
If Request.QueryString("currPage")="" Then
  currPage=1
Else
  currPage=Request.QueryString("currPage")
End If

rst.AbsolutePage=currPage
rows = 0

Response.Write ("<h2>Northwind Customers</h2>")
Response.Write ("<i>Displaying page " & currPage & " of ")
Response.Write rst.PageCount & ("</i>")
Response.Write ("<hr/>")

Do While Not rst.EOF And rows < rst.PageSize
  Response.Write ("<a href=""Address.asp?CustomerID=") & _
   rst("CustomerID") & """>"
  Response.Write rst("CompanyName") & ("</a><br/>")
  rows = rows + 1
  rst.MoveNext
Loop

Response.Write ("<hr/>")
Response.Write ("<b>Result Pages: </b>")

For counter = 1 To rst.PageCount
  Response.Write ("<a href=""PageMe.asp?currPage=") & _
   counter & """>"
  Response.Write counter & ("</a>")
  Response.Write Chr(32)
Next
rst.close
Set rst = Nothing
```

```
conn.Close
Set conn = Nothing
%>
</body>
</html>
```

2. Save the file as **C:\Access2013_ASP_Classic\PageMe.asp**.
3. Close Notepad.

Let's examine this ASP page. The scripting section begins with the declaration of variables. Because all variables are Variants in Active Server Pages, it is convenient to list them on one line:

```
Dim conn, rst, mySQL, currPage, rows, counter
```

Following the declaration of variables, the Connection object is created and the connection to the Northwind database is opened using the Microsoft Access driver.

Next, the Recordset object is created. For Recordset paging to work properly, `CursorType` must be set to `adOpenStatic`. Notice that the script uses the literal value (3) instead of the constant name `adOpenStatic`. By default, ADO enumerated constants are not defined in VBScript. However, a list of constants used with ADO is defined in the Adovbs.inc file (for VBScript) or in the Adojavas.inc file (for JScript). These files are installed in the \Program Files\Common Files\System\ado folder. To use constant names instead of their values, you can add a reference to the Adovbs.inc file at the top of your ASP page by using the `#INCLUDE FILE` directive, as shown here:

```
<%@ Language="Vbscript" %>
<!-- #INCLUDE FILE="adovbs.inc" -->
<html>
```

For the `#INCLUDE FILE` directive to work, you must copy the Adovbs.inc file to the Access2013_ASP_Classic folder. When you add this directive, you will be able to use the ADO constants instead of literal values in your VBScript. Using the enumerated constants will make your code easier to understand.

Use the PageSize property of the Recordset object to specify how many records are to be displayed on a page. Here the page is set to display 12 records:

```
rst.PageSize = 12
```

The SQL `SELECT` statement retrieves all the records in the Customers table into the recordset. We store this statement in the `mySQL` variable and proceed to open the recordset using the connection that we set up earlier:

```
rst.Open mySQL, conn
```

Next, the script retrieves the page you are currently on. If the contents of the `currPage` variable is an empty string (""), then you are on the first page.

The AbsolutePage property of the Recordset object is used to move to a particular page after opening the recordset. The AbsolutePage property identifies the page number of the current record. AbsolutePage equals 1 when the current record is the first record in the recordset.

Then the `rows` variable is initialized to zero (0). This variable limits the number of records that are displayed on a particular page.

Next, we use the `Write` method of the Response object to write a little HTML code that formats the page. For example, to format the page title we use the HTML second-level heading tag <h2> and its ending companion tag </h2> like this:

```
Response.Write ("<h2>Northwind Customers</h2>")
```

The next two `Response.Write` statements inform the user about the page number being displayed and the total number of available pages:

```
Response.Write ("<i>Displaying page " & currPage & " of ")
Response.Write rst.PageCount & ("</i>")
```

The HTML <i> tag will cause the text to appear in italics. You get the page number from the `currPage` variable and obtain the total number of pages from the PageCount property of the Recordset object.

Before we display the data, we want to draw a horizontal line on the page. This is done with the HTML <hr/> tag.

The `Do While` loop iterates through the recordset, counting the rows (records) as they are being retrieved and making sure that the number of records displayed per page is less than the specified page size. Company names are written to each page as hyperlinks using the HTML <a> anchor tag. The anchor tag uses the `href` attribute to designate a target page and forwards data to the target page when the user clicks the company name link:

```
Response.Write ("<a href=""Address.asp?CustomerID=") & _
 rst("CustomerID") & """>"
Response.Write rst("CompanyName") & ("</a><br/>")
```

The target page (Address.asp) is created in the next hands-on exercise in this chapter. A question mark (?) separates the target page from the data. The data attached to the hyperlink is a field name followed by an equals sign and the field value. When you use `Response.Write` to write the links you must pay attention to the quotes. Notice the pairs of double quotes inside the string. Each pair of double quotes ("") can be replaced with a single quote (') to make it easier to read, like this:

```
Response.Write ("<a href= 'Address.asp?CustomerID='") & _
 rst("CustomerID") & ">"
Response.Write rst("CompanyName") & ("</a><br/>")
```

The HTML <br/> tag ensures that each company name appears on a separate line.

When the value of the `rows` variable is greater than the page size, the records are output to the next page.

After all records are retrieved and placed on appropriate pages, a horizontal line is placed on the page using the HTML <hr/> tag. Following the horizontal line, a list of links to the individual pages appears with the text "Result Pages:" formatted in bold (notice the <b> and </b> HTML tags). Again, to write those page links we use the HTML <a> tag with the `href` attribute:

```
Response.Write ("<a href=""PageMe.asp?currPage=") & _
 counter & """>"
Response.Write counter & ("</a>")
```

The next statement uses the `Chr(32)` function to put a space between the page links:

```
Response.Write Chr(32)
```

Finally, the script segment ends by closing all objects and releasing the memory used. We announce the end of the file by writing two ending HTML tags:

```
</body>
</html>
```

Now that you know what the code does, let's proceed to request this page in the browser.

4. Open your browser and type **http://localhost/NorthDB/PageMe.asp** in the address bar. Press **Enter** or click **Go**. You should see the listing of Northwind customers spanning multiple pages (Figure 31.14).
5. Navigate to different pages by clicking on the page links.

Clicking on the company name does not work yet. You must create another ASP page to display the selected customer's address (see Hands-On 31.9)

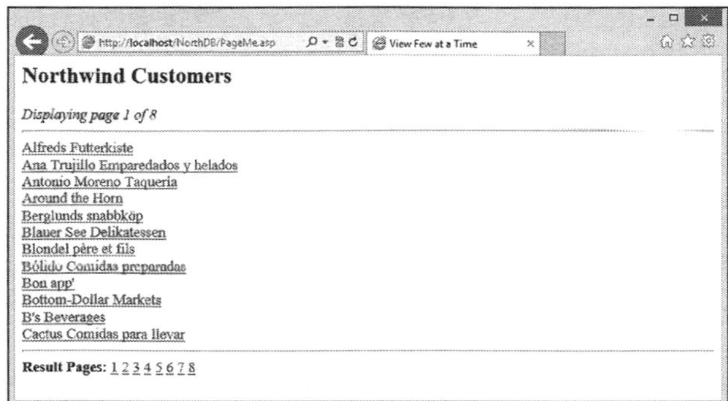

FIGURE 31.14. The result of running the ASP page titled PageMe.asp is a list of Northwind customers that is both easy to examine and to use.

Hands-On 31.9. Creating an ASP File for Loading from a Hyperlink

This hands-on exercise is required in order to use the company name hyperlinks in the PageMe.asp file created in Hands-On 31.8.

1. Start **Notepad** and enter the following ASP code:

```
<%@ Language="Vbscript" %>
<html>
<head><title>Lookup Results</title></head>
<body>
<%
Dim mySQL, myPath

CustomerID = TRIM(Request.QueryString("CustomerID"))
myPath = "C:\Access2013_ASP_Classic\Northwind.mdb"

Set conn = Server.CreateObject("ADODB.Connection")
conn.Open "Provider=Microsoft.Jet.OLEDB.4.0;" _
 & "Data Source=" & myPath

Set rst = Server.CreateObject("ADODB.Recordset")
rst.CursorType = 3 'adOpenStatic

mySQL= "SELECT * FROM Customers WHERE CustomerID='" & _
 CustomerID & "'"
rst.Open mySQL,conn

%>
```

```
<h1>Address Lookup</h1>
<i>Displaying address for
<b><%=rst("CompanyName")%></b></i>
<hr/>
<table colspan="2" align="Center">
  <tr>
    <td>Customer Id:</td>
    <td><input type="text" name="CustomerID"
        value="<%=rst("CustomerID")%>" size="5">
    </td>
  </tr>
  <tr>
    <td>Street:</td>
    <td><input type="text" name="Address"
        value="<%=rst("Address")%>" size="60">
    </td>
  </tr>
  <tr>
    <td>City:</td>
    <td><input type="text" name="City"
       value="<%=rst("City")%>" size="15">
    </td>
  </tr>
  <tr>
    <td>Region:</td>
    <td><input type="text" name="Region"
       value="<%=rst("Region")%>" size="15">
    </td>
  </tr>
  <tr>
    <td>Country:</td>
    <td><input type="text" name="Country"
      value="<%=rst("Country")%>" size="15">
    </td>
  </tr>
  <tr>
    <td>Zip:</td>
    <td><input type="text" name="PostalCode"
       value="<%=rst("PostalCode")%>" size="10">
    </td>
  </tr>
  <tr>
```

```
      <td>Phone:</td>
      <td><input type="text" name="Phone"
         value="<%=rst("Phone")%>" size="24">
      </td>
  </tr>
  <tr>
    <td>Fax:</td>
    <td><input type="text" name="Fax"
        value="<%=rst("Fax")%>" size="24">
    </td>
  </tr>
</table>
<br/><br/>
<center>
[ <a href="vbscript:history.back(1)">Go Back </a> ]
</center>

<%
rst.close
Set rst = Nothing
conn.Close
Set conn = Nothing
%>

</body>
</html>
```

The first VBScript code segment between the <% and %> delimiters connects to the sample Northwind.mdb database using the native OLEDB Provider. The SQL SELECT statement retrieves the record for the selected customer, and the information is output to the page. First, the internal title is written out and formatted using the HTML level 1 heading tag <h1>. The user is given the name of the customer whose information he is viewing. Next are the horizontal line (see the <hr/> tag) and the table structure that displays the customer information. The HTML tag <table> denotes the beginning of a table, <tr> starts a new row, and <td> indicates the table cell (where the data is displayed). Each of these tags is closed with an ending tag (</td>, </tr>, and </table>).

Once the data is written to the page, you should provide the user with a way to return to the previous page so that another customer record can be requested. The Go Back hyperlink at the bottom of the page performs the same action as clicking the Back button in the browser's toolbar:

```
<center>
[ <a href="vbscript:history.back(1)">Go Back </a> ]
</center>
```

The HTML <center> tag positions the hyperlink centered between the page margins. To make the Go Back link compatible across all browsers, you will need to replace `vbscript` with `javascript` as in the following:

```
<center>
[ <a href="javascript:history.back(1)">Go Back </a> ]
</center>
```

Now that you know what the code does, let's proceed to request this page in the browser.

2. Save the file as **C:\Access2013_ASP_Classic\Address.asp**.
3. Close Notepad.
4. In your browser's address bar, type **http://localhost/NorthDB/PageMe.asp** and press **Enter** or click **Go**. You should see the listing of Northwind customers spanning multiple pages.
5. Click a company name of your choice to view its address information. When you click a company name in the browser, the Address Lookup screen appears as illustrated in Figure 31.15.

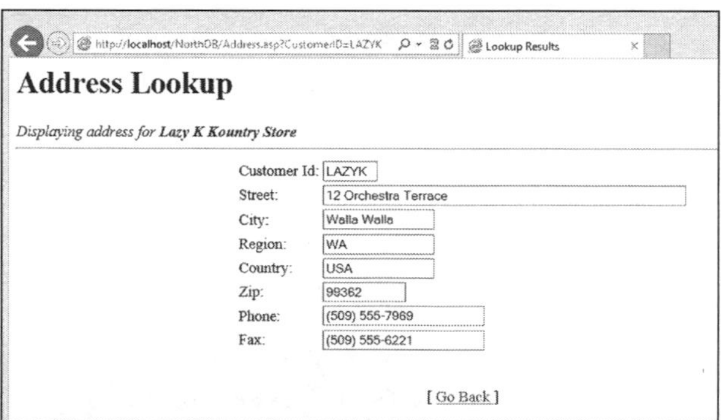

FIGURE 31.15. When you click the company name on the PageMe.asp page (see Figure 31.14), you are presented with a Web page that displays the selected company's address.

Retrieving Records with the GetRows Method

Instead of looping through a recordset to retrieve records, you can use the `GetRows` method of the Recordset object to retrieve records into a two-dimensional array. You've already seen examples of using the `GetRows` method earlier in this book. Hands-On 31.10 uses the `GetRows` method to move records from the Shippers table into an array. Once in the array, the records are written out to a table and displayed in a client browser. When you place records into an array, you can free up the Recordset and Connection objects earlier than in a loop, thus releasing valuable server resources.

Hands-On 31.10. Quick Data Retrieval

1. Start Notepad and enter the ASP code shown here:

```
<%@ Language="Vbscript" %>
<html>
<head><title>Fast Retrieve</title></head>
<body>
<%
Dim conn, rst, strSQL, myPath, fld, allRecords, RowCounter
Dim ColCounter, NumOfCols, NumOfRows, currField

strSQL = "SELECT * FROM Shippers ORDER BY ShipperId"
myPath = "C:\Access2013_ASP_Classic\Northwind.mdb"

Set conn = Server.CreateObject("ADODB.Connection")
conn.open "Provider=Microsoft.Jet.OLEDB.4.0;Data Source=" & myPath

Set rst = conn.Execute(strSQL)

Response.Write ("<table border='1'><tr>") & VbCrLf
For Each fld In rst.Fields
  Response.Write ("<td><b>") & fld.name & ("</b></td>") & VbCrLf
Next
Response.Write ("</tr>") & VbCrLf
allRecords = rst.GetRows

rst.Close
Set rst = Nothing
conn.Close
Set conn = Nothing

NumOfCols = UBound(allRecords, 1) 'columns returned
NumOfRows = UBound(allRecords, 2) 'rows returned
For RowCounter = 0 To NumOfRows
  Response.Write ("<tr>") & VbCrLf
  For ColCounter = 0 To NumOfCols
```

```
      currField = allRecords(ColCounter, RowCounter)
      If IsNull(currField) Then
        currField = currField & ("<br/>")
      ElseIf currField = "" Then
        currField="."
      End If
      Response.Write ("<td Valign='Top'>")
      Response.Write currField
      Response.Write ("</td>") & VbCrLf
    Next
    Response.Write ("</tr>") & VbCrLf
  Next
  Response.Write ("</table>")
  %>
  </body>
  </html>
```

The VBScript code here uses the OLEDB Provider to connect to the Northwind database. After executing the SQL statement, the `Write` method of the Response object is used to create a table:

```
Response.Write ("<table border='1'><tr>") & VbCrLf
```

The `VbCrLf` constant denotes a carriage return/linefeed combination. Because this constant is built into VBScript, you don't need to define it before using it. The HTML <tr> tag is used to add a table row.

Next, the `For Each...Next` loop retrieves the fields from the recordset and places the field names as table headings in the first table row. Notice how the HTML tags are embedded within the VBScript code segment. After the headings are filled in, the procedure uses the `GetRows` method of the Recordset object and places all the fetched records in the variable named `allRecords`. Because we already have all the data that we need, we close the recordset and the connection to the database. At this point the records are in a two-dimensional array. Prior to writing them into table cells, you can use the VBA `UBound` function to check how many rows and columns were retrieved. The data is placed into table cells by using the `For...Next` loop. Because some fields in a retrieved recordset may not have any data in them, you can end up with some missing HTML table cells. To avoid blank spaces in a table, the VBScript code places the HTML
 (break) tag in a table cell if the field contains a Null value:

```
currField = currField & ("<br/>")
```

You can also use a nonbreaking space ( ) for this purpose:

```
currField = currField & " "
```

This statement will make the cell border show up when the cell is empty. You can also write the following statement to ensure that there are no gaps in your table:

```
Response.Write ("<td>") & currField & " </td>"
```

In addition, if a field contains a zero-length string (""), the VBScript procedure places a dot in a table cell, so that you not only keep the structure of the table intact, but also differentiate between information that does not exist (zero-length) and information that may exist (Null). Recall that by setting the Allow Zero Length property of a table field to Yes and the Required property to No, you can enter two double quotation marks to indicate that the information does not exist. Leaving the field blank by not entering any data in it indicates that the information may exist but is not known at the time of entry.

2. Save the file as **C:\Access2013_ASP_Classic\FastRetrieve.asp**.
3. Close Notepad.
4. In your browser's address bar, type **http://localhost/NorthDB/FastRetrieve.asp** and press **Enter** or click **Go**. You should see the listing of three shipping companies placed in a table as shown in Figure 31.16.

ShipperID	CompanyName	Phone
1	Speedy Express	(503) 555-9831
2	United Package	(503) 555-3199
3	Federal Shipping	(503) 555-9931

FIGURE 31.16. The FastRetrieve ASP page retrieves records from the Shippers table using the GetRows method.

5. Open the Northwind.mdb database located in the Access2013_ASP_Classic folder and open the Shippers table in Design view. Click in the Phone field and change the **Required** property of this field to **No** and the **Allow Zero Length** property to **Yes**.
6. Save the Shippers table and open it in Datasheet view. Add **Airborne Express** as a new shipping company. Leave the Phone field for Airborne Express empty. Add **DHL** as a new shipping company. Enter 455-3333 in the Phone field for DHL.

7. Close the Shippers table and exit Microsoft Access.
8. Return to your browser and press **F5** to refresh the window or click the **Go** button to update the display. Notice the two new records you have just added.

You can change the SQL statement in the FastRetrieve.asp page to retrieve data from another Access table in the Northwind database. For example, replace

```
strSQL = "SELECT * FROM Shippers ORDER BY ShipperId"
```

With

```
strSQL = "SELECT * FROM Customers"
```

to pull data from the Customers table.

DATABASE LOOKUP USING DROP-DOWN LISTS

Access forms often use a drop-down box to look up information in a database. When you use a drop-down box, the available choices are limited, so you don't need to worry that the user will enter incorrect information. Hands-On 31.11 illustrates how you can display a drop-down listbox in a browser, load it with product names, and return product information formatted in a table.

Hands-On 31.11. Creating a Web Page with a Drop-Down Listbox

1. Start Notepad and enter the ASP code shown here:

```
<%@ Language="Vbscript" %>
<%
Dim conn, rst, strSQL
Set conn = Server.CreateObject("ADODB.Connection")
conn.ConnectionTimeout = 15
conn.CommandTimeout = 30
conn.Open "Driver={Microsoft Access Driver (*.mdb)}; DBQ=" & _
 Server.MapPath("Northwind.mdb") & ";"
Set rst = Server.CreateObject("ADODB.Recordset")
If Len(Request.QueryString("ProductID")) <> 0 Then
  strSQL="SELECT * FROM Products WHERE ProductID="
  rst.Open(strSQL & Request.QueryString("ProductID")), _
  conn, 0, 1
  If Not rst.EOF Then
    rst.MoveFirst
```

```
      Response.Write ("<html><body><table border='1'>")
      Response.Write ("<tr>")
      Response.Write ("<td><b>Product ID</b></td>")
      Response.Write ("<td><b>Product Name</b></td>")
      Response.Write ("<td><b>Quantity Per Unit</b></td>")
      Response.Write ("<td><b>Units in Stock</b></td>")
      Response.Write ("<td><b>Unit Price</b></td>")
      Response.Write ("</tr>")
      Response.Write ("<tr>")
      Response.Write ("<td align='Center'>")
      Response.Write rst.Fields("ProductID") & ("</td>")
      Response.Write ("<td align='Left'>")
      Response.Write rst.Fields("ProductName") & ("</td>")
      Response.Write ("<td align='Left'>")
      Response.Write rst.Fields("QuantityPerUnit") & ("</td>")
      Response.Write ("<td align='Center'>")
      Response.Write rst.Fields("UnitsInStock") & ("</td>")
      Response.Write ("<td align='Right'>")
      Response.Write FormatCurrency(rst.Fields("UnitPrice"),2)
      Response.Write ("</td>")
      Response.Write ("</tr>")
      Response.Write ("</table>")
    End If
    rst.Close
  End If
  rst.Open "Products", conn, 0, 1
  If Not rst.EOF Then
    rst.MoveFirst
    Response.Write ("<form action='./ProductLookup.asp' method='GET'>")
    Response.Write ("<b>Select a Product:</b><br/>")
    Response.Write ("<select name='ProductID'>")
    Response.Write ("<option></option>")
    Do While Not rst.EOF
      Response.Write ("<option value=" & rst.Fields("ProductID") & ">")
      Response.Write rst.Fields("ProductName") & ("</option>")
      rst.MoveNext
    Loop
    Response.Write ("</select>")
    Response.Write ("<input type='Submit' value='View Details'>")
    Response.Write ("</form>")
  End If
  rst.Close
```

```
Set rst = Nothing
conn.Close
Set conn = Nothing
%>
</body>
</html>
```

This VBScript code segment above begins with establishing a connection with the data source. Instead of using a fully qualified path to the Northwind database, the code shows you how to use the `MapPath` method of the ASP Server object to retrieve the path to the database. The statement `Server.MapPath("Northwind.mdb")` will return the following path: Access2013_ASP_Classic\Northwind.mdb. In fact, if you add the statement:

```
Response.Write Server.MapPath("Northwind")
```

to the preceding code, the filename with its path will appear in the browser. It is not difficult to guess that using `Server.MapPath` generates an additional request for the server to process. Therefore, when deploying your Website, you should replace `Server.MapPath` with a fully qualified path to get better performance (see the previous hands-on examples for how this is done).

Notice that before the connection to the database is opened, the following statements are used:

```
conn.ConnectionTimeout = 15
conn.CommandTimeout = 30
```

The first statement instructs the Connection object's ConnectionTimeout property to wait 15 seconds before abandoning a connection attempt and issuing an error message. In the second statement, the CommandTimeout property of the Connection object specifies how long to wait while executing a command before terminating the attempt and generating an error. The default for the ConnectionTimeout and CommandTimeout properties is 30 seconds. Using ConnectionTimeout and CommandTimeout in this example procedure is optional. Before utilizing these properties in your own database applications, make sure that the data source and the provider you are using support them.

Next, the script instantiates a Recordset object and opens it using the open connection. The recordset is opened as forward-only (0 = `adOpenForwardOnly`) and read-only (1 = `adLockReadOnly`). As mentioned earlier in this chapter, you need to add the `#INCLUDE FILE` directive at the beginning of the Active Server Pages file in order to use enumerated ADO constants.

The SQL SELECT statement contains the WHERE clause that will only pull the record for a selected product ID if the user makes a selection from the drop-down box. The data available for the selected record is then placed in a table. In this example, the table headings are hard-coded. If you don't want to hard-code the headings, you could loop through the recordset to read the field names (see the FastRetrieve.asp file created earlier for an example). After writing out the table headings, the procedure fills the table cells with data. The table will contain only one row of data because the recordset is limited to one product selected from the drop-down list. After the data is presented in a table, the Recordset object is closed.

Next, another recordset is opened. This time the code opens the entire Products table. We loop through the recordset to build a drop-down listbox. For each record, an <option> tag is created, its value is set to the ProductID field, and the text is set to the ProductName. The first entry in the drop-down list is a blank line. This effect is achieved by omitting the value and text attributes inside the HTML <option> tag:

```
<option></option>
```

The drop-down listbox is part of a form. The <form> tag is used to generate an HTML form.

Forms allow user input into the browser and act as a container for ActiveX controls. Forms can be processed via two methods: GET and POST. This example uses the GET method to send information. (See Hands-On 31.13 for an example of processing form input with the POST method.) Within a <form> and </form> block, you can insert tags representing various HTML controls. In this example, the form contains the listbox produced by the <select> tag and a command button produced by the <input> tag. When the user clicks a submit form button labeled "Get Product Details," the data gathered from the drop-down listbox is passed to the Active Server Pages file specified within the <form> tag by the action attribute.

2. Save the file as **C:\Access2013_ASP_Classic\ProductLookup.asp**.
3. Close Notepad.
4. In your browser's address bar, type **http://localhost/NorthDB/ProductLookup.asp** and press **Enter** or click **Go**. The Web page displays a drop-down box and a button as shown in Figure 31.17.
5. Open the drop-down list. When you do so, the list of products appears. Notice that the first entry in the list is a blank line.
6. Select a product from the drop-down list and click the **View Details** button. The product details appear in a table, as shown in Figure 31.18.

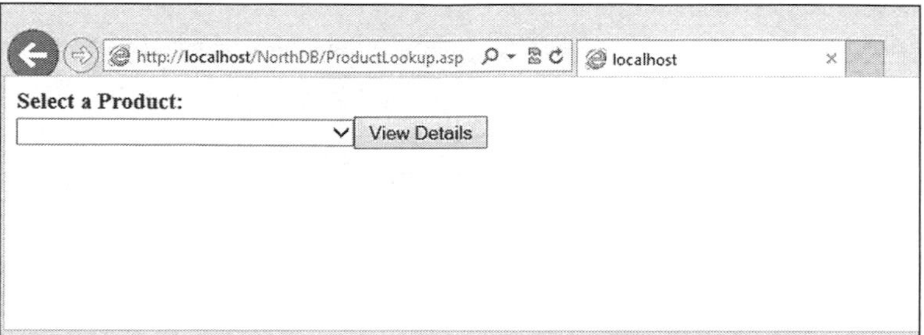

FIGURE 31.17. By using a drop-down box in a Web page, you can provide a user-friendly interface for selecting records.

http://localhost/NorthDB/ProductLookup.asp?I

localhost

Product ID	Product Name	Quantity Per Unit	Units in Stock	Unit Price
14	Tofu	40 - 100 g pkgs.	35	$23.25

Select a Product:

View Details

FIGURE 31.18. When you select a product from the drop-down list and click the View Details button, the selected product information is presented at the top of the Web page.

When you use a form with the `GET` method to send the information, the data is appended to the request for the processing page. The data being passed is visible in the address bar in your browser (see Figure 31.18). Because the data is visible, you can easily troubleshoot any problems by looking at the address bar. The drawback of using the `GET` method for sending information is that the data is not secure and it is limited in size to the maximum length of the request string.

DATABASE LOOKUP USING A MULTIPLE-SELECTION LISTBOX

In the previous section, you saw an example of looking up product information by selecting a product name from a drop-down list. At times, however, a user may want to view several products at once. To meet this requirement, you will need to create a multiple-selection listbox and process the user's selections.

Hands-On 31.12 illustrates how you can display a multiple-selection listbox in a browser, load it with the product names, and return product information formatted in a table.

Hands-On 31.12. Creating a Web Page with a Multiple-Selection Listbox

1. Start Notepad and enter the following ASP code:

```
<%@ Language=VBScript %>
<html>
<head><title>Select Multiple Products</title></head>
<body>
<%
Dim conn, rst, strSelect, strWhere, strSQL, totalItems, fld

Set conn = Server.CreateObject("ADODB.Connection")
conn.ConnectionTimeout = 15
conn.CommandTimeout = 30

conn.Open "Driver={Microsoft Access Driver (*.mdb)}; " _
 & "DBQ=" & Server.MapPath("Northwind.mdb") & ";"
Set rst = Server.CreateObject("ADODB.Recordset")

If Len(Request.QueryString("ProductID")) <> 0 Then
  strSelect="SELECT ProductID AS [ID], ProductName AS "
  strSelect=strSelect & "[Product Name], QuantityPerUnit "
  strSelect=strSelect & "AS [Qty/Unit], UnitsInStock "
  strSelect=strSelect & "AS Stock, UnitPrice AS "
  strSelect=strSelect & "[Unit Price] FROM Products "

  strWhere = "WHERE ProductID="
  strSQL = strSelect & strWhere

  totalItems = Request.QueryString("ProductID").Count
  myValues = Request.QueryString("ProductID").Item

  Response.Write ("<p/><h3><i>")
  Response.Write "The following SQL statement was used:"
  Response.Write ("</i></h3>")

  If totalItems = 1 Then
    rst.Open(strSQL & Request.QueryString("ProductID")), _
     conn, 0, 1
  %>
```

```
    <%=strSQL & Request.QueryString("ProductID") %>

    <%
    Else
      strWhere = "WHERE ProductID IN ("
      strSQL = strSelect & strWhere
      rst.Open(strSQL & myValues & ")"), conn, 0, 1
    %>
      <textarea cols="80" rows="3">
      <%=strSQL & myValues & ")" %></textarea>

    <%
    End if
  ' get table headings
    Response.Write "<p/><table Border=""1"">"
    Response.Write ("<tr>")
    For Each fld in rst.Fields
      Response.Write "<th>" & fld.Name & ("</th>")
    Next
    Response.Write "</tr>"

  ' get the data
    Do While not rst.EOF
      Response.Write ("<tr>")
      For Each fld in rst.Fields
        Response.Write ("<td>")
        If fld.Name = "UnitPrice" Then
          Response.Write FormatCurrency(fld.value,2)
        Else
          Response.Write fld.value
        End If
        Response.Write ("</td>")
      Next
      Response.Write ("</tr>")
      rst.MoveNext
    Loop
    Response.Write ("</table>")
    rst.Close
  End If

  rst.Open "Products", conn, 0, 1
```

```
If Not rst.EOF Then
   rst.MoveFirst
%>
   <form action="MultiProductLookup.asp" method="GET">
   <b><i><font size="2" face="Tahoma">
   Hold down CTRL or SHIFT <br/>
   to select multiple products:</font></i></b><br/>
   <select name="ProductID" multiple size="8">
<% Do While Not rst.EOF%>
    <option value="<%=rst.Fields("ProductID")%>">
    <%=rst.Fields("ProductName")%></option>
<%  rst.MoveNext
   Loop
%>
   </select>
   <input type="Submit" Value="Get Product(s) Details">
   </form>
<%
End If
rst.Close
Set rst = Nothing
conn.Close
Set conn = Nothing
%>
</body>
</html>
```

The preceding VBScript code segment establishes a DSN-less connection to the Northwind database by using the Microsoft Access driver and instantiates a Recordset object. Refer to the previous hands-on exercise for an explanation of the Connection object's ConnectionTimeout and CommandTimeout properties and the Server object's `MapPath` method.

The code proceeds to check whether the user has selected any items in the listbox. If at least one product was picked from the list, the procedure defines the SQL `SELECT` statement and uses the `QueryString` method of the Request object to retrieve the total number of selected products. This number is then stored in the `totalItems` variable. The next `Request.QueryString` statement retrieves the IDs of the selected items and places them in the `myValues` variable. The next statement announces that the line that follows is the SQL statement the user has selected. This statement is formatted with the HTML <h3> and <i> tags. This will make the enclosed text an italicized heading of size 3 (the largest heading is 1 and the smallest is 6). The <p/> tag designates the text as a plain paragraph.

If one product was selected in the listbox, a recordset is opened using the following statement:

```
rst.Open(strSQL & Request.QueryString("ProductID")), conn, 0, 1
```

Recall that 0 and 1 at the end of this statement indicate a forward-only and read-only recordset.

The statement:

```
<textarea cols="80" rows="3">
  <%=strSQL & myValues & ")" %></textarea>
```

will write the complete SQL statement to the <textarea> control whenever the user selects multiple products. When the user selects a single product from the listbox, the following statement:

```
<pre><%=strSQL & Request.QueryString("ProductID") %></pre>
```

will write the complete SQL statement to the body of the HTML page. When you use the HTML <pre> and </pre> tags, the text between these tags is formatted exactly as it is typed. Spaces and carriage returns are preserved. If more than one product was selected in the listbox, we need to change the contents of the `strWhere` variable to include the `IN` keyword in the `WHERE` clause. The `IN` keyword restricts the rows being selected to those rows where the column values are in the list presented in the SQL statement. Assuming that the user selected products with IDs of 1, 3, and 6 in the listbox, the following SQL statement will be generated:

```
SELECT ProductID AS [ID], ProductName AS [Product Name],
QuantityPerUnit AS [Qty/Unit], UnitsInStock AS Stock, UnitPrice
AS [Unit Price] FROM Products WHERE ProductID IN (1, 3, 6)
```

The remaining code segment creates a table in a browser. We use the `For Each... Next` loop to write out the column names to the browser:

```
For Each fld in rst.Fields
  Response.Write ("<th>") & fld.Name & ("</th>")
Next
```

The <th> tag makes a cell a table heading. This automatically makes the text bold. After populating the table with the headings, we use the `Do While` loop to write out the table rows until the end of the recordset is encountered. We must obtain field values for each column in a row. This is done with the `For Each... Next` loop like this:

```
For Each fld in rst.Fields
  Response.Write ("<td>")
  If fld.Name = "UnitPrice" Then
    Response.Write FormatCurrency(fld.value,2)
  Else
    Response.Write fld.value
  End If
  Response.Write ("</td>")
Next
```

Notice the conditional statement within the code segment. We use it to perform an additional operation on the UnitPrice field. We format this field as currency using the `FormatCurrency` function. When all the table rows are written to the browser, the table is closed with the HTML table close tag </table>, and the recordset itself is closed.

Next, the VBScript code continues by opening the recordset based on the Products table and cycling through this recordset to retrieve the product IDs and product names for inclusion in the listbox. The HTML form section contains the `multiple` keyword in the <select> tag to indicate that the listbox should be created. The size of the listbox is set to display eight products like this:

```
<select name="ProductID" multiple size="8">
```

Once we have defined the listbox we can populate it with product names using a `Do While` loop. We use the <option> tag with the Value attribute <option value=" "> to specify items in the list:

```
Do While Not rst.EOF
%>
  <option value="<%=rst.Fields("ProductID")%>">
  <%=rst.Fields("ProductName")%></option>
<%
  rst.MoveNext
Loop
```

Notice again that we set the list values outside the VBScript. For better understanding, and to practice various methods of coding, you can rewrite this code as follows:

```
Do While Not rst.EOF
  Response.Write "<option value="
```

```
    Response.Write rst.Fields("ProductID") & (">")
    Response.Write rst.Fields("ProductName") & ("</option>")
    rst.MoveNext
  Loop
```

To allow the user to submit selections to the server, the form contains the submit button labeled "Get Product(s) Details." When the user presses this button, the form data will be submitted using the `GET` method.

The procedure ends by closing both the Recordset and Connection objects and freeing up memory.

Let's test our work in the browser.

2. Save the file as **C:\Access2013_ASP_Classic\MultiProductLookup.asp**.
3. Close Notepad.
4. In your browser's address bar, type **http://localhost/NorthDB/MultiProductLookup.asp** and press **Enter** or click **Go**.
5. The browser will display a listbox. Select the items as shown in Figure 31.19 and press the **Get Product(s) Details** button. The product details will be displayed as shown in Figure 31.20.

FIGURE 31.19. You can allow users to filter the data by using a multiple-selection listbox.

http://localhost/NorthDB/MultiProductLookup.asp?Pr   Select Multiple Products

The following SQL statement was used:

```
    SELECT ProductID AS [ID], ProductName AS [Product Name], QuantityPerUnit AS
[Qty/Unit], UnitsInStock AS Stock, UnitPrice AS [Unit Price] FROM Products
WHERE ProductID IN (1, 3, 6)
```

ID	Product Name	Qty/Unit	Stock	Unit Price
1	Chai	10 boxes x 20 bags	39	18
3	Aniseed Syrup	12 - 550 ml bottles	13	10
6	Grandma's Boysenberry Spread	12 - 8 oz jars	120	25

Hold down CTRL or SHIFT to select multiple products:

Chai
Chang
Aniseed Syrup
Chef Anton's Cajun Seasoning
Chef Anton's Gumbo Mix
Grandma's Boysenberry Spread
Uncle Bob's Organic Dried Pears
Northwoods Cranberry Sauce

Get Product(s) Details

FIGURE 31.20. After selecting the products in the listbox (see Figure 31.19) and clicking on the Get Product(s) Details button, your browser displays data as shown here.

ADDING DATA TO A TABLE

You may want to use a Web page to collect data from a user and save it in Access. The following hands-on exercise creates a simple data entry form that contains two fields. The purpose of this form is to allow users to enter new shippers into the Northwind database Shippers table.

Hands-On 31.13. Creating a Data Input Page

1. Start Notepad and enter the ASP code shown here:

```
<%@ Language="Vbscript" %>
<%
Dim conn, strConn, strSQL, name, phone, goAhead
name=Request("txtCompanyName")
phone=Request("txtPhone")

For Each key In Request.Form
  If Request.Form(key) = "" Then
    If key = "txtCompanyName" Then
      Response.Write ("<font color = 'Blue'>")
      Response.Write ("Please enter the Shipper name.</font>")
    Else
      Response.Write ("<font color = 'Red'>")
      Response.Write ("Please enter the Phone number.</font>")
    End If
    goAhead = False
    Exit For
  End If
  goAhead=True
Next
If goAhead = True Then
  name=Replace(Request("txtCompanyName"),"'","''")
  If Len(name) <> 0 Or _
    Len(phone) <> 0 Then
    Set conn = Server.CreateObject("ADODB.Connection")
    strConn="Provider=Microsoft.Jet.OLEDB.4.0;Data Source="
    strConn=strConn & server.MapPath("Northwind.mdb") & ";"
    strConn=strConn & "User ID=; Password=;"

    strSQL = "INSERT INTO Shippers (CompanyName, Phone)"
    strSQL = strSQL & "Values ('" & name & "'"
```

```
    strSQL = strSQL & ",'" & phone & "')"

    With conn
      .Mode = 3
      .Open strConn
      .Execute(strSQL)
    End With
    Response.Write "<i><font color = 'Green'>" & _
     "Successfully added the following data:</i></font><hr/>"
  ' get the ShipperID (the autonumber field in the Shippers table)
    strSQL = "SELECT MAX(ShipperID) AS lastID FROM Shippers;"
    Set rst = conn.Execute(strSQL)
    Response.Write ("Shipper ID: <b>") & rst("lastID")
    Response.Write ("</b><p/>")
    rst.close
    Set rst = Nothing
    conn.Close
    Set conn = Nothing
    Response.Write ("Company Name: <b>")
    Response.Write Request("txtCompanyName") & ("</b><p/>")
    Response.Write ("Phone Number: <b>") & phone & ("</b>")

  ' clear the Shipper Name and Phone input boxes
    name = ""
    phone = ""
  End If
End If
%>
<html>
<head>
<title>Data Entry Screen</title>
</head>
<body>
<form name="form1" action="NWDataEntry.asp" method="POST">
<p/>
Shipper Name: <input type="text" name="txtCompanyName"
 value="<%=name%>" size="30" > 
Phone: <input type="text" name="txtPhone" value="<%=phone%>">
<p/><p/>
<input type="Submit" name="cmdSubmit" Value="Add Data"></p>
</form>
</body>
</html>
```

The preceding VBScript segment assigns values to the `name` and `phone` variables. These values are collected from the text fields located on the HTML form. To collect information from a form, use the `Request.Form("name")` command, where `name` is the name of the form field (text box, checkbox, etc.). The VB-Script here uses the abbreviated form of the `Request.Form` command:

```
Request("txtCompanyName")
```

To remove leading and trailing spaces that users often enter in text fields, use the `TRIM` function as follows:

```
name =TRIM(Request("txtCompanyName")
```

The `For Each...Next` loop validates user input prior to sending information to the server. It's a good idea to write validation scripts to check for such things as whether the user entered a valid number or whether a text box was left empty. This example only checks whether any of the text fields are empty. Data validation should be performed on the client side to reduce server loads and improve response time. Notice how we check the values of the form elements with the `For Each...Next` loop:

```
For Each key In Request.Form
  If Request.Form(key) = "" Then
    If key = "txtCompanyName" Then
      Response.Write ("<font color = 'Blue'>")
      Response.Write ("Please enter the Shipper name.</font>")
    Else
      Response.Write ("<font color = 'Red'>")
      Response.Write ("Please enter the Phone number.</font>")
    End If
    goAhead = False
    Exit For
  End If
  goAhead=True
Next
```

This code iterates through the Forms collection to check whether the user has entered any data in the CompanyName and Phone fields and displays a message in a different color if any of the text fields were left blank. If both text fields were filled in, the `goAhead` variable is set to True, and the procedure continues. Because the company name that the user entered may contain an apostrophe, an error could occur when the value is inserted into the SQL statement. To avoid the error, the procedure uses the `Replace` function to replace one apostrophe with two apostrophes in the user-supplied text:

```
name=Replace(Request("txtCompanyName"),"'","''")
```

Provided that the length of the strings contained in the `name` or `phone` variables is not equal to zero (0), the connection is established to the Northwind database, and the SQL `INSERT INTO` statement is executed. This statement inserts a new record into the Shippers table and places the contents of the `name` and `phone` variables into the CompanyName and Phone fields. Next, the procedure uses the green font color to inform the user about the successful addition of the data. Another SQL statement is executed to retrieve the ID of the newly added record, and the `Response.Write` statement displays the ShipperID for the user to see in the browser.

After retrieving the value of the ShipperID field, the Recordset and Connection objects are closed. Next, we write out the user-supplied shipper name and phone number to the browser, and clear the `name` and `phone` variables so that the form's input boxes display no data.

The remaining section of the ASP page contains HTML tags that generate a form where the user can enter the shipping company and phone number, and includes a button for submitting information to a Web server. Notice that the form's Action argument refers to the file named NWDataEntry.asp. When the user submits data that he or she entered in the form's text fields by clicking the Add Data button, the browser will use the `POST` method to send the information to the ASP file on the server, in this case NWDataEntry.asp.

An ASP file can create a form that posts information to itself (as shown in this example) or to another ASP file. By using the `POST` method, you can send an almost unlimited number of characters to the Web server. The `POST` method is also more secure than the `GET` method because the information passed to the server does not appear in the browser's address bar. (Refer to the previous hands-on exercise for an example of processing form input with the `GET` method.) Notice how the values of the `name` and `phone` variables are retrieved:

```
Value = "<%=name%">
Value = "<%=phone%">
```

2. Save the file as **C:\Access2013_ASP_Classic\NWDataEntry.asp**.
3. Close Notepad.
4. In your browser's address bar, type **http://localhost/NorthDB/NWDataEntry.asp** and press **Enter** or click **Go**.
5. Enter the data shown in Figure 31.21 and press the **Add Data** button.

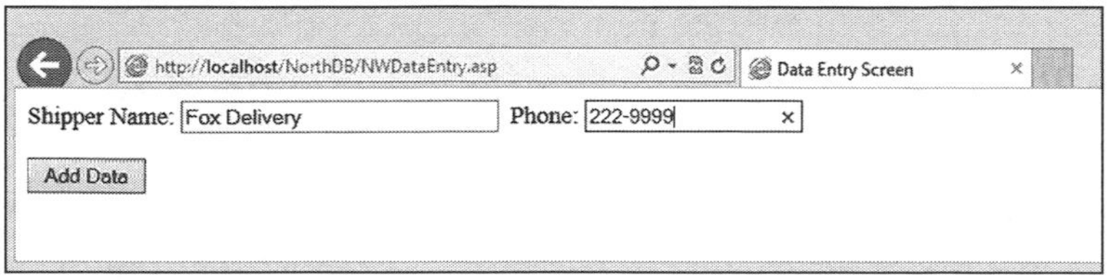

FIGURE 31.21. When you request the ASP page prepared in Hands-On 31.13, you are presented with the data entry screen for the Northwind database Shippers table.

When you enter data in the Shipper Name text box and click the Add Data button, the browser displays the data that was inserted into the Shippers table and allows you to make more additions by clearing out previous values from the input boxes (Figure 31.22).

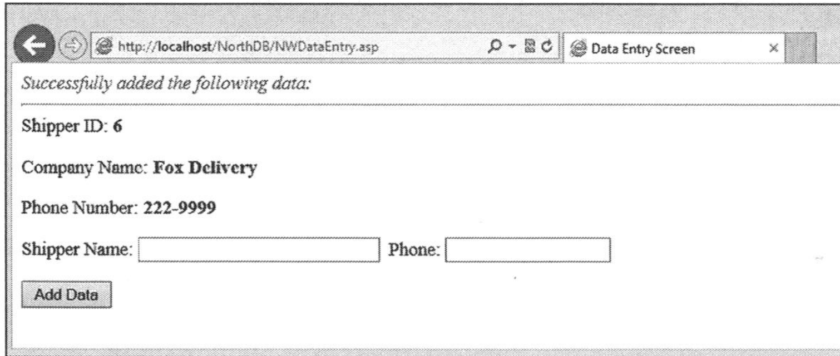

FIGURE 31.22. Notice that the browser displays the Shipper ID of the newly added record as well as the data entered in the text boxes prior to the Add Data button being pressed (see Figure 31.21). You can continue adding new data by typing new values into the text boxes and clicking the Add Data button.

MODIFYING A RECORD

You can display a record in a browser and allow the user to edit the data. Changes made to the data can then be submitted to the server for processing. The easiest and quickest way to modify a record is by executing the SQL `UPDATE` statement.

The following hands-on exercise creates an ASP page where the user can select a product to update from a drop-down list. After clicking on the Retrieve Data button, the selected product's current price and units in stock are retrieved from the Products table. The retrieved data is placed in text boxes inside a table structure. The user can edit the data in the retrieved fields and insert the changes to the database table by clicking the Update Data button.

Hands-On 31.14. Creating a Page for Data Modification

1. Start Notepad and enter the following ASP code:

```
<%@ Language="Vbscript" %>
<html>
<head><title>Update Product Information</title></head>
<%
Set conn = Server.CreateObject("ADODB.Connection")
conn.Open "DRIVER=Microsoft Access Driver (*.mdb);DBQ=" & _
 "C:\Access2013_ASP_Classic\Northwind.mdb"

set rst = Server.CreateObject("ADODB.Recordset")

If Len(Request.QueryString("ProductID")) <> 0 Then
  strSQL="SELECT * FROM Products WHERE ProductID="
  rst.Open(strSQL & Request.QueryString("ProductID")), _
  conn, 0, 1
  If Not rst.EOF Then
    rst.MoveFirst
%>
  <body>
  <form id="form2" name="form2" action="UpdateProduct.asp"
   method="POST">
  <input type="hidden" name="txtProductID"
   value="<% =rst("ProductID") %>">
  <input type="hidden" name="txtProductName"
   Value="<% =rst("ProductName") %>">

  <center><h4><% =rst("ProductName") %>
   (Product ID=<%=rst("ProductID")%>)</h4></center><p/>
  <table border="0" cellspacing="4" cellpadding="4">
    <tr>
      <td width="200" colspan="2"><font color="Blue">
       Unit Price ($):</font></td>
      <td bgcolor="#00FF00">
      <input type="text" name="UnitPrice"
       value="<% =rst("UnitPrice") %>"</td>
    </tr>
    <tr>
    <td colspan="2"><font color="Blue">
     Units In Stock:</font></td>
    <td bgcolor=#00FF00">
```

```
      <input type="text" name="UnitsInStock"
       value="<%=rst("UnitsInStock") %>"</td>
    </tr>
    <tr>
      <td colspan="2"></td>
      <td><input type="submit" value="Update Data"
        id="submit2" name="submit2"></td>
    </tr>
</table><hr/>
</form>
<%
End If
rst.Close
End If
If Not IsEmpty(Request.Form("submit2")) Then
  If Request.Form("UnitPrice")= "" or _
    Request.Form("UnitsInStock") = "" Then
    Response.Write ("<b><font color='Red'>")
    Response.Write ("You cannot leave any fields blank. Try again.")
    Response.Write ("</b></font>")
  Else
    strSQL = "UPDATE Products SET " _
    & "UnitPrice = '" & Request.Form("UnitPrice")& "', " _
    & "UnitsInStock = '" & Request.Form("UnitsInStock")& "' " _
    & "WHERE ProductID = " & Request.Form("txtProductID")
    conn.Execute strSQL
    Response.Write "The following Update statement was "
    Response.Write "executed for <b>" & _
      Request.Form("txtProductName")
    Response.Write ("</b><br/>")
    Response.Write ("<pre>" & strSQL & "</pre><br/>")
  End If
End If
strSql = "SELECT * FROM [Products] ORDER BY [ProductName]"
Set rst = conn.Execute (strSql)
If Not rst.EOF Then
  rst.MoveFirst
%>
  <form action="UpdateProduct.asp" method="GET">
    <table>
      <tr>
        <td><b>Select a Product to Update</b></td>
```

```
        <td><SELECT Name="ProductID">
          <option></option>
<%
  Do While Not rst.EOF
    Response.Write "<option Value='" & rst("ProductID") & "'> " & _
    rst("ProductName") & "</option>"
    rst.MoveNext
  Loop
End If
rst.Close
set rst = Nothing
conn.Close
set conn = Nothing
%>
<td><input type="submit" value="Retrieve Data"
id="submit1" name="submit1"></td>
</select></td>
</tr>
</table><hr/>
</form></body>
</html>
```

Notice that the preceding ASP page contains two HTML forms: Form1 and Form2.

Form1 (whose code appears at the bottom of the ASP page) displays a drop-down list of products for the user to select. This form uses the GET method to send data to the server. This means that you will see the query string in the browser's address bar once you click the Retrieve Data button (see Figures 32.23 and 32.24).

Form2 (whose code appears higher in the ASP page) displays two text boxes with Unit Price and Units in Stock values for the product that was selected from the drop-down list on Form1. The user can modify the data in these text boxes. This form uses the POST method to send the information to the server. The submitted information will not be visible in the browser's address bar. This form will be submitted to itself after the user clicks the Update Data button. Two hidden text fields are placed on Form2 to store information about the retrieved Product ID and Product Name:

```
<input type="hidden" name="txtProductID"
   value="<% =rst("ProductID") %>">
<input type="hidden" name="txtProductName"
   Value="<% =rst("ProductName") %>">
```

In this example, the information stored in hidden fields is used by the VBScript code later in the ASP file to create an SQL `UPDATE` statement and write an information message in the browser. Hidden form fields are often used with the `POST` method to hide information from the user.

The first VBScript code segment establishes a connection to the Northwind database and creates an instance of the Recordset object. Next, we check whether a selection was made from the drop-down list. If the user made a product selection and clicked the Retrieve Data button, we open the recordset:

```
rst.Open(strSQL & Request.QueryString("ProductID")), conn, 0, 1
```

The `Open` method of the Recordset object is used to issue an SQL `SELECT` statement with the `WHERE` clause that specifies which record should be retrieved. We placed the `SELECT` statement in the `strSQL` variable. The `Open` method also specifies the connection to the database (`conn`), the cursor type (`adOpenForwardOnly` = 0), and the lock type (`adLockReadOnly` = 1). The recordset is opened to retrieve only the data that the user is allowed to modify. The data is placed in a table (see the HTML code segment). Once the data is retrieved, the recordset is closed.

The next VBScript code segment runs after the user clicks the Update Data button on Form2. When the form is posted, all controls, including the command buttons, are posted with it. You can use the `IsEmpty` function to find out if the user clicked the command button:

```
If Not IsEmpty(Request.Form("submit2")) Then
```

Prior to submitting the data to the server for insertion into the Products table, the code checks whether the Unit Price and Units in Stock text boxes contain any data. If either of these fields is empty (the user may have erased the data completely), a validation message is sent to the browser and the user must request the product again from the drop-down list if he wants to continue. On the other hand, if there is data in both text fields (even if the user has not made any changes to the original data), clicking the Update button on the form will send the SQL `UPDATE` statement to the server. As a result, the user will see the name of the product he updated together with the SQL `UPDATE` statement that was executed.

The last code segment creates a recordset to populate a drop-down list with product names. You should already be familiar with this code as it was demonstrated in the previous hands-on exercise.

2. Save the file as **C:\Access2013_ASP_Classic\UpdateProduct.asp**.
3. Close Notepad.

4. In your browser's address bar, type **http://localhost/NorthDB/UpdateProduct.asp** and press **Enter** or click **Go.**
5. Select a product as shown in Figure 31.23 and click the **Retrieve Data** button. The resulting page displays the product information, as shown in Figure 31.24.

FIGURE 31.23. When you request the UpdateProduct.asp file in your browser, a screen appears with a drop-down list where you can select a product you want to update.

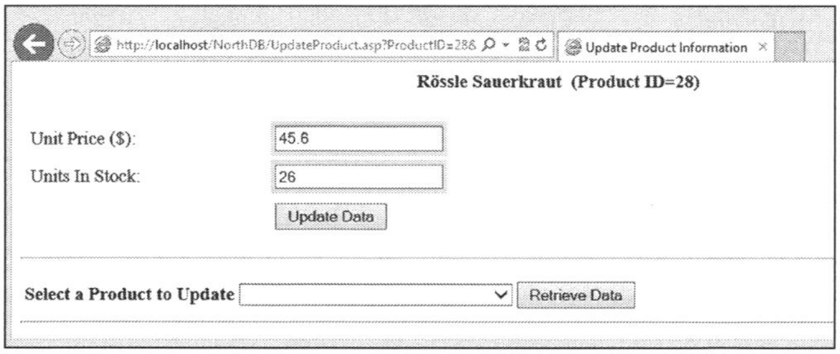

FIGURE 31.24. When you select a product from the drop-down list and click the Retrieve Data button, the selected product's unit price and units in stock data are retrieved from the Products table and placed at the top of the page.

You can modify the original data in the text boxes and click the Update Data button. If you click the Update Data button when information is missing in the Unit Price or Units In Stock text boxes, you are prompted to enter the data and try again. If you click the Update Data button while the Unit Price and Units In Stock text boxes are not empty, the `UPDATE` statement is executed on the server and the submitted changes are inserted in the Products table. See the page in Figure 31.25 confirming the update request.

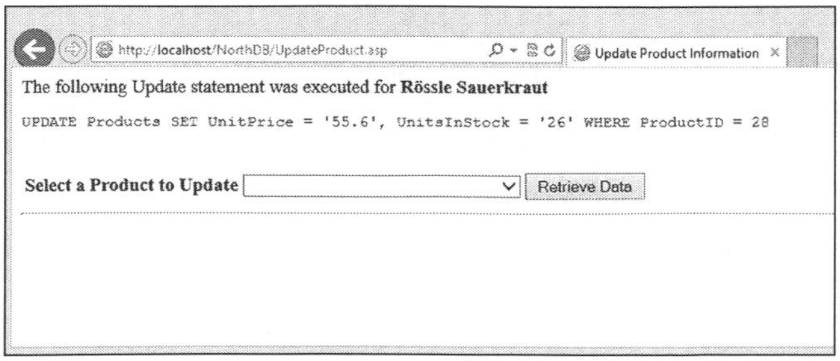

FIGURE 31.25. After submitting the product modification, you are presented with the confirmation page with the `UPDATE` statement that was executed and you are given an opportunity to continue by retrieving other products for modification.

DELETING A RECORD

When you need to delete a record, you can use the SQL DELETE statement. When writing a VBScript to handle the delete request, it's always a good idea to check for the following conditions:

- Did the user specify a record to delete? The user may have pressed the submit button without typing the record ID in the provided text box.
- Does the provided record ID exist in the table? This question is particularly important when the user is expected to type the record ID in a text box instead of selecting it from the drop-down list.
- What happens when the record the user wants to delete has related records in other tables? As you know, Access will not allow you to delete records when the referential integrity rules are enforced.

The next hands-on exercise demonstrates how to delete a shipper from the Shippers table.

Hands-On 31.15. Creating Pages that Allow Record Deletions

This hands-on exercise uses two files for performing the delete operation. The first file is the HTML form that will submit the data to the second file, which is the ASP page.

1. Start Notepad and enter the ASP code shown here:

```
<html>
<head>
<title>DELETE DEMO</title>
</head>
<body>
<form name="DeleteShipperForm" method="GET"
action="DeleteShipper.asp"
<input type="Hidden" name="Action" value="Delete">
Please enter the Shipper ID you want to delete
<input type="Text" Size="6" name="ShipperID">
<input type="Submit" name="Delete" value="Submit">
</form>
</body>
</html>
```

   In this HTML page, the form's Action argument will call the ASP page named DeleteShipper.asp upon clicking the Submit button (see Figure 31.26).

2. Save the file as **C:\Access2013_ASP_Classic\DeleteShipper.html**.

3. Close Notepad.
4. Start Notepad and enter the ASP code shown here:

```
<%@ Language="Vbscript" %>
<%
Dim conn
Dim mydbFile
Dim myShipper
Dim strSQL

Set conn = Server.CreateObject("ADODB.Connection")
mydbFile=Server.MapPath("Northwind.mdb")

conn.Open "Driver={Microsoft Access Driver (*.mdb)}; " & _
 "DBQ=" & mydbFile & ";"
myShipper = Cstr(Request.QueryString("ShipperID"))
strSQL = "DELETE * FROM Shippers WHERE "
strSQL = strSQL & "ShipperID = " & myShipper

If myShipper <>"" Then
  Set rst = Server.CreateObject("ADODB.Recordset")
  rst.Open "Shippers", conn, 3
  rst.Find "ShipperID = " & myShipper
  If rst.EOF Then
    Response.Write ("The Shipper ID ") & myShipper
    Response.Write (" does not exist.")
  Else
    On Error Resume Next
    conn.Execute strSQL
    If conn.Errors.Count > 0 Then
      Response.Write "Error Number: " & err.Number & ("<p/>")
      Response.Write "Error Description: " & _
         err.Description & ("<p>")
    Else
      Response.Write ("<h2>The Shipper ID ") & myShipper & _
       " was deleted.</h2>"
    End If
    rst.close
    Set rst = Nothing
  End If
Else
  Response.Write "The Shipper ID was not supplied."
End If
```

```
%>
<html>
<head><title>DELETE SHIPPER</title></head>
<body>
<hr/>
<a href="DeleteShipper.html">
Please click here to return.</a>
</body>
</html>
```

The VBScript code segment shown here establishes a connection to the data source and stores the ShipperID value in the `myShipper` variable. If the variable is not empty, then the code proceeds to create an instance of the Recordset object and opens the Shippers table. The recordset is opened using the static cursor (`adOpenStatic`) represented by the value 3 in the following statement:

```
rst.Open "Shippers", conn, 3
```

Recall that the static cursor retrieves all the data as it was at a point in time and is particularly desirable when you need to find data. The next statement uses the `Find` method to check whether the supplied ShipperID exists in the Shippers table:

```
rst.Find "ShipperID = " & myShipper
```

Next, the `If...Then...Else` statement decides what information should be returned to the browser. When the EOF property of the Recordset object is True, the recordset contains no records. In this situation you want to tell the user that there is no such record in the table. However, if the record is found in the Shippers table, the SQL `DELETE` statement is executed:

```
strSQL = "DELETE * FROM Shippers WHERE "
strSQL = strSQL & "ShipperID = " & myShipper
conn.Execute strSQL
```

As noted at the beginning of this section, a user may enter a ShipperID that has related records in other tables. Because this situation will certainly result in an error, the VBScript is instructed to ignore the error and continue with the next line of code:

```
On Error Resume Next
```

The next line of code is another `If...Then...Else` block statement that sends a text message to the browser depending on whether or not the error was generated. The code displays the error number and description if the user picked a

ShipperID that cannot be deleted. You may want to replace this code section with a more user-friendly message. If there is no error, then the browser will display a message that the record was deleted. The text of this message is formatted in large letters using the HTML level 2 heading tag <h2>.

Next, the Recordset object is closed. And now we are back at the first `If...Then...Else` statement block where the `Else` part is executed if the user happened to click the Submit button without first typing in the ShipperID to delete.

The final part of the ASP page shown here creates a hyperlink to allow the user to navigate back to the HTML form (DeleteShipper.html). To create a hyperlink, use the following format:

```
<a href="address">displaytext</a>
```

where *address* is the name of the HTML file you want to activate and *displaytext* is the text that the user should click on.

5. Save the file as **C:\Access2013_ASP_Classic\DeleteShipper.asp**.
6. Close Notepad.
7. In your browser's address bar, type **http://localhost/NorthDB/DeleteShipper.html** and press **Enter** or click **Go**. Your screen should resemble Figure 31.26.

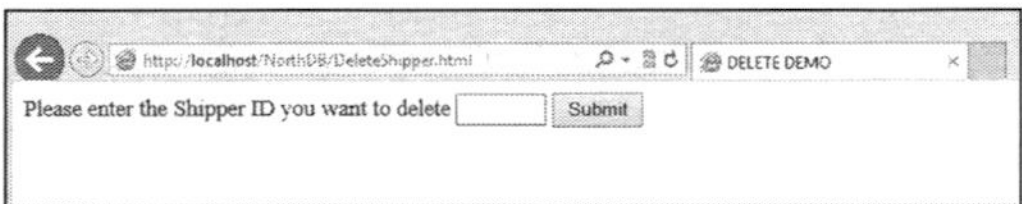

FIGURE 31.26. This HTML page is used for submitting information to an ASP page.

8. Click the **Submit** button without typing anything in the provided text box. You should see a message informing you that the ShipperID was not supplied. Also, there is a link to allow you to return to the previous page.
9. Click the hyperlink to return to the previous page and enter **999** in the text box, then click the **Submit** button. Because this ShipperID does not exist in the Shippers table, you are again informed about the problem and provided a way to return to the previous page.
10. Click the hyperlink to return to the previous page. Enter the ShipperID that you inserted into the Shippers table in Hands-On 31.13 and click the **Submit** button. If you don't have a shipper record to delete, add a new record to the Shippers table and delete it using this process. When you type in a ShipperID that exists in the Shippers table but is not referenced in other tables, you get the screen that confirms the deletion (Figure 31.27).

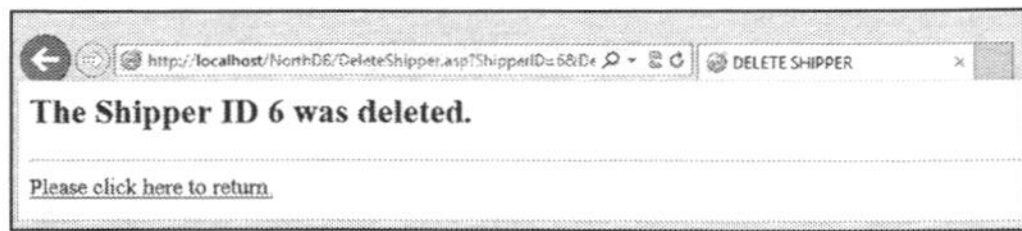

FIGURE 31.27. This screen confirms a deletion of the shipper record having the ID of 4.

When you enter a ShipperID that is referenced in other tables, Access will not allow you to delete that Shipper's record:

Error Number: -2147467259
Error Description: [Microsoft][ODBC Microsoft Access Driver]
The record cannot be deleted or changed because table 'Orders' includes related records.

To see this error in action, try to delete the shipper with an ID of 1.

You can trap the error −2147467259 in your VBScript code to display a user-friendly message.

CREATING A WEB USER INTERFACE FOR DATABASE ACCESS

Now that we've developed some sample ASP pages, let's see how you can put them together so that they can be easily accessed from a Web browser. Instead of typing the ASP filename in the browser's address bar every time you want to perform a particular operation, you can use hyperlinks and frames to organize the data. For example, take a look at Figure 31.28, which displays the Internet Explorer window divided into three areas. You can navigate between individual hands-on examples created in this chapter by clicking on the text in the left pane. When you click a hyperlink, the main portion of the window on the right-hand side will fill with the data retrieved from a database or display an interface to obtain the data (see Figure 31.29).

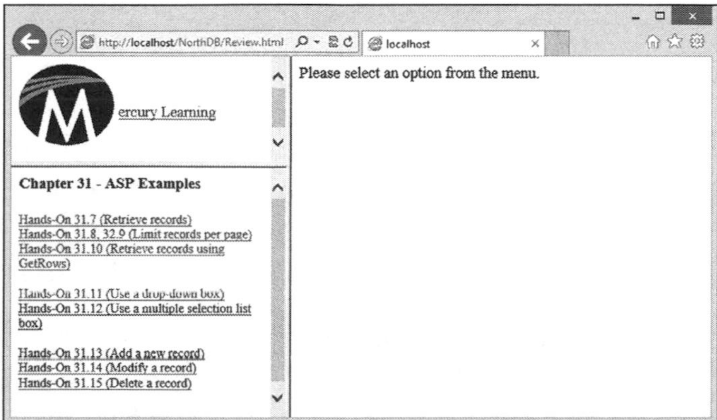

FIGURE 31.28. This page allows easy access to the hands-on examples on ASP programming introduced in this chapter.

Hands-On 31.16 walks you through the steps required to create the user interface shown in Figure 31.28.

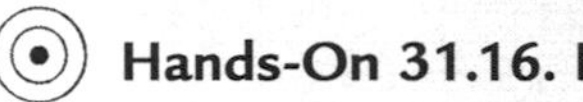

Hands-On 31.16. Providing Easy Access to Data with Frames

This hands-on exercise creates and uses the following four HTML files: Review.html, Logo.html, Examples.html, and Results.html (see Table 31.3).

TABLE 31.3. HTML files used in Hands-On 31.16

Filename	Purpose
Review.html	Creates a page containing two frames and breaks one frame into two rows (resulting in three areas visible on the screen). Specifies what should be displayed within each frame and tells browsers that do not support frames to ignore them.
Logo.html	Places a company logo and a hyperlink to navigate to the company Website.
Examples.html	Creates jumps to this chapter's hands-on examples.
Results.html	Used for dumping information from the database or displaying the interface to obtain the data. Displays an information message to the user when the page is first opened.

1. Start Notepad and enter the following HTML code:

```
<html>
<frameset cols="290,*">
  <frameset rows="110,*">
    <frame src="Logo.html">
    <frame src="Examples.html">
  </frameset>
  <frame src="Results.html" name="myDisplay">
</frameset>
<noframes>
  This page requires frames to be viewed.
</noframes>
</html>
```

This HTML code demonstrates how to divide a page into several areas by using frames. Each frame is controlled by its own HTML file.

You create a frame by using the HTML <frameset> and <frame> tags with various parameters. A frame can contain other frames. The page shown in Figure 31.28 has two frames. The left frame is broken into two rows.

The statement:

```
<frameset cols="290",*">
```

divides the page into two columns. The left column is 290 pixels wide, and the right column occupies the remainder of the screen (denoted by the asterisk). A frameset can contain both frames and other framesets.
Next, the left column is divided into two rows, like this:

```
<frameset rows="110",*">
```

The top row is 110 pixels high, and the bottom row occupies the remaining portion of the frame. The top frame points to the Logo.html file. The source file for the lower frame is Examples.html:

```
<frame src="Logo.html">
<frame src="Examples.html">
```

The two <frame> tags provide the information about each frame. The `src` attribute defines the source data for the frame.
Next, we define the source data for the frame on the right side:

```
<frame src="Results.html" name="myDisplay">
```

The Results.html file is created later in this hands-on exercise. We use the `name` attribute to define the frame's name so that we can refer to this frame later.

Each frameset ends with the </frameset> tag. Browsers that do not support frames will display an information message located between the <noframes> and </noframes> tags:

```
<noframes>
 This page requires frames to be viewed.
</noframes>
```

2. Save the file as **C:\Access2013_ASP_Classic\Review.html**.
3. Choose **File | New** to create a new document in Notepad.
4. Enter the following HTML code:

```
<html>
<body>
<img align="Middle" width="100" height="100"
alt="Visit us today!" src="mercury-logo-m.jpg">
<font color="Blue">
<a href="http://www.merclearning.com">ercury Learning
</a></font>
</body>
</html>
```

The preceding HTML code places an image (mercury-logo-m.jpg, available on the CD-ROM) in the top row of the frame on the left side using the <img> tag. The `align` parameter tells the browser to position the text that follows in the middle. The `width` and `height` parameters determine the size of the image in pixels. The text placed to the right of the image is a hyperlink. Clicking on it will take the user to the Mercury Learning Website.

5. Save the file as **C:\Access2013_ASP_Classic\Logo.html**.
6. Copy the image file **mercury-logo-m.jpg** from the Access2013_HandsOn folder to your Access2013_ASP_Classic folder.
7. Choose **File | New** to create a new document in Notepad.
8. Enter the following HTML code:

```
<html>
<head>
<base target ="myDisplay">
</head>
<body>
<h4>Chapter 31 - ASP Examples</h4>
<p/>
<font size="-1">
<a href="GetCustomers.asp" target="myDisplay">
Hands-On 31.7 (Retrieve records)
</a><br/>
<a href="PageMe.asp">
Hands-On 31.8, 32.9 (Limit records per page)
</a><br/>
<a href="FastRetrieve.asp">
Hands-On 31.10 (Retrieve records using GetRows)
</a><br/><p/>
<a href="ProductLookup.asp">
Hands-On 31.11 (Use a drop-down box)
</a><br/>
<a href="MultiProductLookup.asp">
Hands-On 31.12 (Use a multiple selection list box)
</a><br/><p/>
<a href="NWDataEntry.asp">
Hands-On 31.13 (Add a new record)
</a><br/>
<a href="UpdateProduct.asp">
Hands-On 31.14 (Modify a record)
</a><br/>
```

```
<a href="DeleteShipper.html">
Hands-On 31.15 (Delete a record)
</a>
</font><p/>
</body>
</html>
```

The code in the Examples.txt file is straightforward. It contains a set of jumps to different hands-on examples. Each of these examples has a corresponding ASP file in the Access2013_ASP_Classic folder. To control where the requested information should be displayed, use the <base> tag:

```
<base target="myDisplay">
```

The `target` attribute is set to the name of the frame where the information should appear. When you place the <base target> tag at the beginning of the file, all of the links will display in the same frame.

Here's how we create the first hyperlink:

```
<a href="GetCustomers.asp" target="myDisplay">
  Hands-On 31.7 (Retrieve records)
</a><br/>
```

The <a> tag defines a hypertext link. The `href` attribute specifies the associated URL. In other words, when the user clicks on the "Hands-On 31.7 (Retrieve records)" hyperlink, the GetCustomers.asp file will be requested and its output will be placed in the specified target (right-hand frame).

9. Save the file as **C:\Access2013_ASP_Classic\Examples.html**.
10. Choose **File | New** to create a new document in Notepad.
11. Enter the following HTML code:

```
<html>
  Please select an option from the menu.
</html>
```

12. Save the file as **C:\Access2013_ASP_Classic\Results.html**. The Results.html file will display the information text when the Review.html file is first opened in the browser (refer to Figure 31.28 at the beginning of this section).
13. Close Notepad.

You now should have four HTML files as outlined at the beginning of this hands-on exercise. Let's test our user interface.

14. In your browser's address bar, type **http://localhost/NorthDB/Review.html** and press **Enter** or click **Go**. Your screen should resemble Figure 31.28, shown earlier in this chapter.

15. Verify the results of each hands-on exercise by clicking on its link (see Figure 31.29).

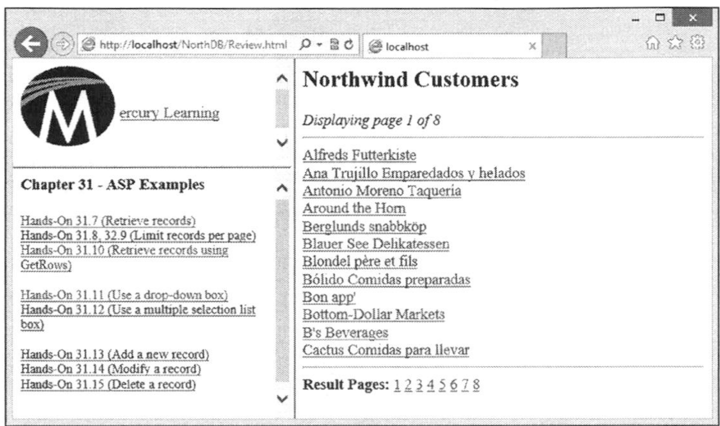

FIGURE 31.29. When you click a hyperlink in the frame on the left, the right-hand frame will display the requested information as illustrated here.

CHAPTER SUMMARY

This chapter has introduced you to the world of Web development by using a server-side scripting technology from Microsoft known as Active Server Pages (ASP). You learned how to use VBScript, a subset of VBA, to quickly extract data from a database and present it to a user in a standard HTML page. You also learned how to submit Action queries to insert, update, and delete a database record. You've seen two coding styles: one that mixes HTML and script commands, and one that returns HTML text to the browser by using the ASP built-in Response object and its `Write` method. By working through several hands-on examples, you've seen that making your application Web-ready is not rocket science. ASP scripts are quite easy to write if you understand VBA statements and have already worked with ActiveX Data Objects (ADO). Although it was easy to get started, this chapter did not attempt to teach you all there is to know about HTML, ASP, or VBScript. If you decide to move into the lucrative and exciting world of Web development, you should learn more about these subjects. Numerous books and articles have been written on the topics we've only touched on in this chapter, as well as many other topics that are necessary to create truly functional Internet applications.

In the next chapter, you explore another Internet technology known as Extensible Markup Language (XML) and learn how it is integrated with Access 2013.

Chapter 32 XML Features in Access 2013

If you need to deliver information over the Web or you want to store, share, and exchange data between different applications regardless of the operating system or programming language used, you need to become familiar with Extensible Markup Language (XML). Imagine these two scenarios where your combined knowledge of Access and XML will come in handy:

- You have just received a file in XML format and you need to merge its data with an existing Access table, or perhaps create a new table.
- You have been asked to provide a data dump from your Access database in XML format.

XML is a complex language that cannot be covered within the pages of one chapter; however, this chapter will get you started using XML with Access 2013.

WHAT IS XML?

In the previous chapter, you learned how HTML (HyperText Markup Language) uses tags to format data on a Web page. Like HTML, Extensible Markup Language uses markup tags; however, its tags serve a different purpose—they are used to describe data content. HTML uses fixed, noncustomizable tags to provide formatting instructions that should be applied to the data. XML is *extensible*, which means that it is not restricted to a set of predefined tags. XML allows you to invent your own tags in order to define and describe data stored in a wide range of documents. The XML parser does not care what tags you use; it only needs to be able to find the tags and confirm that the XML document is well formed. A document that follows the formatting rules for XML is considered a well-formed document (see the section titled "What Is a Well-Formed XML Document?").

What Is a Parser?

If you want to read, update, create, or manipulate any XML document, you will need an XML parser. A *parser* is a software engine, usually a dynamic-link library (DLL), that can read and extract data from XML. Microsoft Internet Explorer 5 and above have a built-in XML parser (MSXML.DLL, MSXML2.DLL, MSXML3.DLL, MSXML4.DLL, MSXML5.DLL, and MSXML6.DLL) that is capable of reading well-formed documents and detecting those that are not. MSXML has its own object model, known as DOM (Document Object Model), that you can use from VBA to quickly and easily extract information from an XML document. The Microsoft XML parser has been renamed Microsoft XML Core Services (MSXML).

To ensure that you are working with the most recent XML parser, check out the following link:

http://www.microsoft.com/downloads/details.aspx?FamilyID=993C0BCF-3BCF-4009-BE21-27E85E1857B1&displaylang=en

An XML document must also be valid. When a document is *valid*, it follows the predefined rules for valid data. These rules are defined in a Document Type Definition (DTD) or a schema, which is written in XML. DTD is an old method of data validation. Later in this chapter you will see how Access uses a schema to determine the types of elements and attributes an XML document should contain, how these elements and attributes should be named, whether they're optional or required, their data types and default values, and the relationship between the elements.

Because of its extensibility, XML makes it easy to describe any data structure and send it anywhere across the Web using common protocols such as HTTP (HyperText Transfer Protocol) or FTP (File Transfer Protocol). Although XML was designed specifically for delivering information over the World Wide Web, it is being utilized in other areas, such as storing, sharing, and exchanging data. Because XML is stored in plain text files, it can be read by many types of applications, independent of the operating system or hardware.

What Is a Well-Formed XML Document?

An XML document must have one root element. While in HTML the root element is always <html>, in an XML document you can name your root element anything you want. Element names must begin with a letter or underscore character. The root element must enclose all other elements, and elements must be properly nested. The XML data must be hierarchical; the beginning and ending tags cannot overlap.

```
<Employee>
  <Employee ID>090909</Employee ID>
</Employee>
```

All element tags must be closed (a beginning tag must be followed by an ending tag):

```
<Sessions>5</Sessions>
```

You can use shortcuts, such as a single slash (/), to end the tag so you don't have to type the full tag name. For example, if the current <Sessions> element is empty (does not have a value), you could use the following tag:

```
<Sessions />
```

Tag names are case-sensitive: The tags <Title> and </Title> aren't equivalent to <TITLE> and </TITLE>.

For example, the following line:

```
<Title>Beginning VBA Programming</Title>
```

is not the same as:

```
<TITLE>Beginning VBA Programming</TITLE>
```

All attributes must be inside quotation marks:

```
<Course ID="VBAEX1"/>
```

You cannot have more than one attribute with the same name within the same element. If the <Course> element has two ID attributes, they must be written separately, as shown here:

```
<Course ID="VBAEX1"/>
<Course ID="VBAEX2"/>
```

The main goals of XML are the separation of content from presentation and data portability. It is important to understand that XML was designed to address the limitations of HTML and not to replace it. One of these limitations is the inability of HTML to identify data. By using XML tags you can give meaning to the data in the document and provide a consistent way of identifying each item of data. By separating content from presentation and structuring data based on its meaning, we are finally able to create documents that are easy to reuse, manipulate, and search.

XML SUPPORT IN ACCESS 2013

Microsoft Access has supported XML since its 2002 release. In Access 2013, you can import and export XML data by using buttons available in the Import and Export areas on the External Data tab or you can do this programmatically with VBA. Additionally, Access 2013 has the capability to export related tables to a single XML file. When importing XML data, you can create multiple tables from a single XML document and schema. Unfortunately, the parent–child relationships between the tables are not maintained; you must create them yourself. You can also specify a custom schema during the export or import of XML data.

EXPORTING XML DATA

You can export tables, queries, forms, and/or reports to XML files from an Access database (MDB or ACCDB) file. There is no XML support for macros or modules. When you export a form or report, you actually export the data from the form or report's underlying table or query.

Access uses a special XML vocabulary known as ReportML for representing its objects as XML data. *ReportML* is an XML file that contains tags describing properties, methods, events, and attributes of the Access object being exported.

This file is generated automatically by Access when you begin the export process and is used by Access to generate the final output files.

To allow XML data to be viewed in browsers in a user-friendly format, ReportML relies on a rather complicated stylesheet that contains formatting instructions. We examine stylesheets later in this chapter.

After the formatting instructions contained in the stylesheet have been applied to the XML file, the ReportML file is automatically deleted.

No matter what Access object you need to export to XML, you always follow the same procedure:

- To export all the data, select the appropriate object (table, query, report, or form) in the database window and choose External Data | XML File, or right-click the object name in the Navigation pane and select Export | XML File from the shortcut menu.
- To export a single record or a filtered or sorted set of records, open the appropriate object and follow these steps:

You want to...	Step 1	Step 2
Export a single record	Select that record	Choose External Data | XML File, specify the name of the export file you want to create, and click OK. Click the More Options button and in the Records to Export area, select Current Record.
Export filtered records	Apply a filter to the records	Choose External Data | XML File and select the appropriate options.
Export records in a predefined order	Arrange records in the order you want	Choose External Data | XML File and select the appropriate options.

The following hands-on exercise demonstrates how to use the Export command to save the Shippers table in XML format.

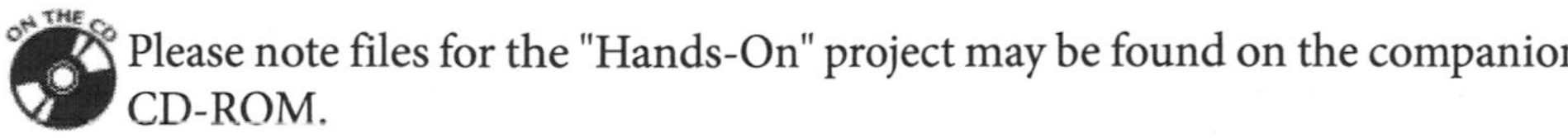

Please note files for the "Hands-On" project may be found on the companion CD-ROM.

Hands-On 32.1. Exporting an Access Table to an XML File

1. Use Windows Explorer to create a new folder named **C:\Access2013_XML** for this chapter's practice files.
2. You will now set the Access2013_XML folder as a virtual directory.

- Refer to Chapter 31 on how to set up the Internet Information Services on your computer. When you have configured your computer, choose **Control Panel | Administrative Tools | Internet Information Services (IIS) Manager**. In the Connections pane, double-click on your computer, and then double-click the **Sites** folder. Right-click **Default Web Site** and choose **Add Virtual Directory**. In the Add Virtual Directory dialog box, type **xml** in the Alias box, and enter **C:\Access2013_XML** in the Physical path box. Click **OK** to exit the dialog box. You should see a new folder named xml under the Default Web Site in the Connections pane. Close the Internet Information Services (IIS) Manager.

3. Open the sample **C:\Access2013_HandsOn\Northwind 2007.accdb** database. Login as Andrew Cencini.
4. In the Navigation pane, select the **Shippers** table, and choose **External Data**. In the Export Group, choose **XML File.**
5. In the File name box, enter C:**\Access2013_XML\Shippers.xml** and click **OK**.

In the Export XML dialog box that appears, there are three checkboxes (see Figure 32.1). The first one, which is selected by default, will cause Access to generate an XML file containing the data from the Shippers table. The second checkbox specifies that Access should create an XSD file with the data definition. The third checkbox tells Access to generate the stylesheet (XSL) file that will contain formatting specifications.

FIGURE 32.1. The Export XML dialog box displays three checkboxes; the first one is selected by default. The More Options button allows for more customization.

6. Select all the checkboxes and click **OK** to proceed with the export.
7. When the export operation completes, Access displays the Export - XML File window where you are given a chance to save the export steps so that you can repeat them in the future without using the wizard.
8. Click **Close** to exit the Export - XML File window without saving the export steps.

Understanding the XML Data File

In Hands-On 32.1 you prepared the Shippers.xml file. Let's switch to Windows Explorer and examine the contents of your Access2013_XML folder.

Hands-On 32.2. Examining the Contents of an XML Data File

1. Open Windows Explorer and switch to the **C:\Access2013_XML** folder. Figure 32.2 displays the contents of the Access2013_XML folder after exporting the Shippers table to XML in Hands-On 32.1.

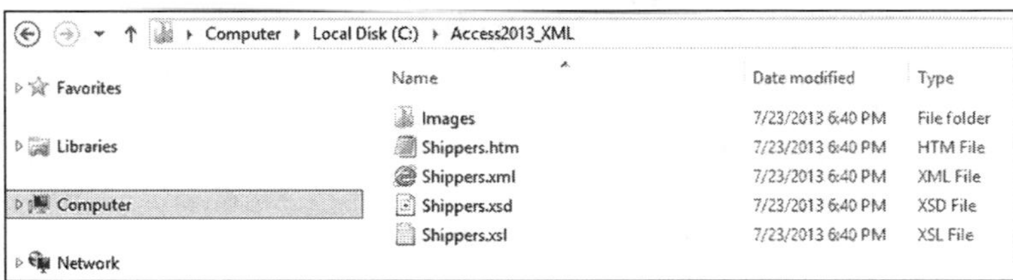

FIGURE 32.2. After exporting the Shippers table to XML with all three checkboxes selected in the Export XML dialog box, Access creates four files and Images folder.

2. Highlight the Shippers XML document (**Shippers.xml**) and choose **Open With** from the File menu. Select **Internet Explorer**. Access displays the Shippers data in XML format as shown in Figure 32.3.

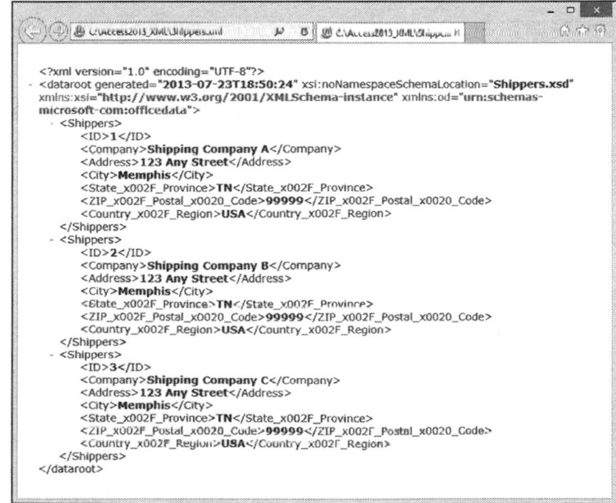

FIGURE 32.3. The tree-like structure of the XML document.

When you open an XML file in Internet Explorer, you can see the hierarchical layout of an XML document very clearly. The plus and minus (+/–) signs make it possible to display the document as a collapsible tree. The first line in the XML file is a processing instruction. Processing instructions begin with <? and end with ?>. The XML document begins with a processing instruction that contains an XML declaration:

```
<?xml version="1.0" encoding="UTF-8" ?>
```

The version attribute (`version="1.0"`) tells the XML processor that the document conforms to version 1.0 of the XML specification. The encoding attribute (`encoding="UTF-8"`) indicates character sets to Web browsers. By default, XML documents use the UTF-8 encoding of Unicode.

Character Encodings in XML

The encoding declaration in the XML document identifies which encoding is used to represent the characters in the document. UTF-8 encoding allows the use of non-ASCII characters, regardless of the language of the user's operating system and browser or the language version of Office. When you use UTF-8 or UTF-16 character encoding, an encoding declaration is optional. XML parsers can determine automatically if a document uses UTF-8 or UTF-16 Unicode encoding.

The second line in the XML document is a dataroot element:

```
<dataroot xmlns:od="urn:schemas-microsoft-com:officedata"
xmlns:xsi="http://www.w3.org/2001/XMLSchema-instance" xsi:noNamespaceS
chemaLocation="Shippers.xsd" generated="2013-07-23T18:50:24">
```

The dataroot element tag defines two namespaces:

```
xmlns:od="urn:schemas-microsoft-com:officedata"
xmlns:xsi="http://www.w3.org/2001/XMLSchema-instance"
```

A *namespace* is a collection of names in which each name is unique. The XML namespaces are used in XML documents to ensure that element names do not conflict with one another and are unique within a particular set of names (a namespace).

For example, the <TITLE> tag will certainly have a different meaning and content in an XML document generated from the Books table than the <TITLE> element used to describe the courtesy titles of your customers. If the two XML documents containing the <TITLE> tag were to be merged, there would be an element name conflict. Therefore, in order to distinguish between tags that have the same names but need to be processed differently, namespaces are used.

The attribute *xmlns* is an XML keyword for a namespace declaration. The namespace is identified by a Uniform Resource Identifier (URI)—either a Uniform Resource Locator (URL) or a Uniform Resource Name (URN). The URI used as an XML namespace name is simply an identifier; it is not guaranteed to point to anything. Most namespaces use URIs for the namespace names because URIs are guaranteed to be unique. The use of a namespace is identified via a name prefix, which is mapped to a URI to select a namespace.

For example, in the context of the Shippers.xml document, the "od" prefix is associated with the urn:schemas-microsoft-com:officedata namespace and the "xsi" prefix identifies the http://www.w3.org/2001/XMLSchema-instance namespace. These prefixes may be associated with other namespaces outside of this particular XML document. Notice that the prefix is separated from the xmlns attribute with a colon and the URI is used as the value of the attribute.

In addition to namespaces, the dataroot element specifies where to find the schema. This is done by using two attributes: the location of a schema file that defines the rules of an XML document and the date the file was generated.

An XML document's data is contained in elements. An element consists of the following three parts:

- Start tag—Contains the element's name (e.g., <ID>)
- Element data—Represents the actual data (e.g., 1)
- End tag—Contains the element's name preceded by a slash (e.g., </ID>)

If you click on the minus sign in front of the dataroot element, you will notice that the dataroot element encloses all the elements in the Access XML file. Each element in a tree structure is called a *node*.

The dataroot node contains child nodes for each row of the Shippers table. Notice that the table name is used for each element representing a row. You can expand or collapse any row element by clicking on the plus or minus sign (+/–) in front of the element tag name.

Within row elements, there is a separate element for each column (ID, Company, and so on). Notice that each XML element contains a start tag, the element data, and the end tag:

```
<Shippers>
  <ID>1</ID>
  <Company>Shipping Company A</Company>
```

```
    <Address>123 Any Street</Address>
    <City>Memphis</City>
    <State_x002F_Province>TN</State_x002F_Province>
    <ZIP_x002F_Postal_x0020_Code>99999</ZIP_x002F_Postal_x0020_Code>
    <Country_x002F_Region>USA</Country_x002F_Region>
</Shippers>
```

The ID, Company, Address, City, State_x002F_Province, Zip_x002F_Postal_x0020_Code, and Country_x002F_Region elements are children of the Shippers element. In turn, each Shippers element is a child of the dataroot element. XML documents can be nested to any depth as long as each inner node is entirely contained within the outer node.

3. Close the browser containing the Shippers.xml file.

Understanding the XML Schema File

Now that you have familiarized yourself with the structure of an XML document, let's look at another type of XML file that was created by Access during the export to XML process—the XML schema file (XSD).

Schema files describe XML data using the XML Schema Definition (XSD) language and allow the XML parser to validate the XML document. An XML document that conforms to the structure of the schema is said to be *valid*.

Here are some examples of the types of information that can be found in an XML schema file:

- Elements that are allowed in a given XML document
- Data types of allowed elements
- Number of allowed occurrences of a given element
- Attributes that can be associated with a given element
- Default values for attributes
- Child elements of other elements
- The sequence and number of child elements

Hands-On 32.3. Examining the Contents of an XML Schema File

1. Open Windows Explorer and switch to the **C:\Access2013_XML** folder containing the files generated in Hands-On 32.1 (see Figure 32.2).
2. Use Notepad to open the **Shippers.xsd** file located in the Access2013_XML folder. Access displays the contents of the Shippers.xsd file as shown in Figure 32.4.

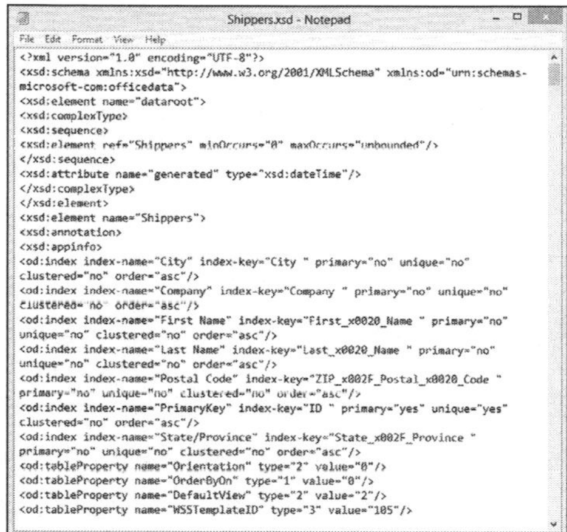

FIGURE 32.4. The schema file shown here defines the data in the Shippers.xml document.

If you take a look at the Shippers.xsd file currently open in Notepad, you will notice a number of XSD declarations and commands that begin with the <xsd> tag followed by a colon and the name of the command. You will also notice the names of the elements and attributes that are allowed in the Shippers.xml file as well as the data types for each element.

The names of the data types begin with the "od" prefix followed by a colon. For example:

`od:jetType="text"`	Defines the Jet data type for an element
`od:sqlSType="nvarchar"`	Defines the Microsoft SQL Server data type for an element
`od:autounique="yes"`	Defines a Boolean data type for an auto-incremented identity column
`od:nonNullable="yes"`	Indicates whether or not a column can contain a Null value

The schema file also specifies the number of times an element can be used in a document based on the schema. This is done via the minOccurs and maxOccurs attributes.

3. Close Notepad and the Shippers.xsd file.

To find out more about XML schemas, check out the following links:
http://www.w3.org/TR/xmlschema-0/
http://www.w3.org/TR/xmlschema-1/
http://www.w3.org/TR/xmlschema-2/

Understanding the XSL Transformation Files

When you examined the contents of the Shippers.xml document earlier in this chapter you may have noticed that the file did not contain any formatting instructions. Although it is easy to display the XML file in the browser, end users expect to see documents that are nicely formatted. To meet their expectations, the raw XML data is formatted with the Extensible Stylesheet Language (XSL).

When you exported the Shippers table to XML and selected the Presentation of your data (XSL) checkbox in the Export XML dialog box (see Hands-On 32.1), Access generated an XSL file. Extensible Stylesheet Language is a transformation style language that uses XSL Transformations (XSLT) to create templates that are applied to the source document data to create the target document. The target document can be another XML document, an HTML page, or even a text-based file.

XSL files include all the XSLT transforms that are needed to define how the data is to be presented. Transformations allow you to change the order of elements and selectively process elements. Later in this chapter you will learn how to create XSL files with XSLT transforms to display only selected fields from the Access-generated XML documents. There is no limit to the number of stylesheets that can be used with a particular XML document. By creating more than one XSL file, you can present different formats of the same XML document to various users.

Hands-On 32.4. Examining the Contents of an XSL File

1. Right-click the **Shippers.xsl** file located in the **C:\Access2013_XML** folder and choose **Open with | Internet Explorer**. Access displays the contents of the Shippers.xsl file as shown in Figure 32.5.

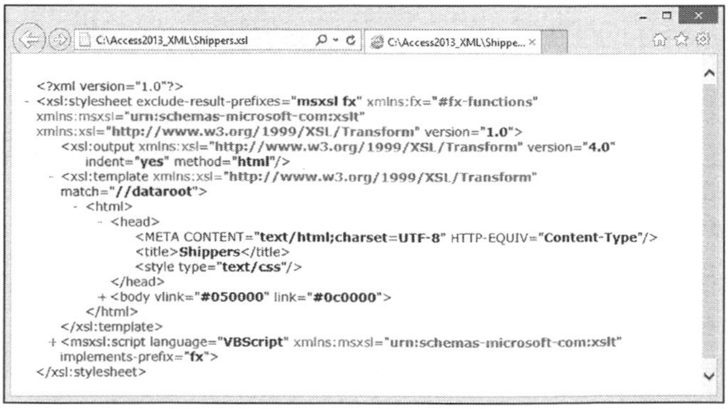

```
C:\Access2013_XML\Shippers.xsl    C:\Access2013_XML\Shippe...

<?xml version="1.0"?>
- <xsl:stylesheet exclude-result-prefixes="msxsl fx" xmlns:fx="#fx-functions"
  xmlns:msxsl="urn:schemas-microsoft-com:xslt"
  xmlns:xsl="http://www.w3.org/1999/XSL/Transform" version="1.0">
    <xsl:output xmlns:xsl="http://www.w3.org/1999/XSL/Transform" version="4.0"
      indent="yes" method="html"/>
  - <xsl:template xmlns:xsl="http://www.w3.org/1999/XSL/Transform"
    match="//dataroot">
    - <html>
      - <head>
          <META CONTENT="text/html;charset=UTF-8" HTTP-EQUIV="Content-Type"/>
          <title>Shippers</title>
          <style type="text/css"/>
        </head>
      + <body vlink="#050000" link="#0c0000">
      </html>
    </xsl:template>
  + <msxsl:script language="VBScript" xmlns:msxsl="urn:schemas-microsoft-com:xslt"
    implements-prefix="fx">
</xsl:stylesheet>
```

FIGURE 32.5. The XSL stylesheet document is just another XML document that contains HTML formatting instructions and XSLT formatting elements for transforming raw XML data into HTML.

When you scroll through the contents of the Shippers.xsl file you will notice a number of XSLT formatting elements such as <xsl:template>, <xsl:for-each>, and <xsl:value-of>. You will also find many HTML formatting instructions such as <head>, <title>, <body>, <table>, <colgroup>, <col>, <tbody>, <tr>, <td>, <div>, and <strong>.

The first line of the stylesheet code declares that this is an XML document that follows the XML 1.0 standard (version). An XSL document is a type of XML document. While XML documents store data, XSL documents specify how the data should be displayed.

The second line declares the namespace that will be used to identify the tags in the XSL document. (See the "Understanding the XML Data File" section earlier in this chapter for more information about namespaces.) The third line specifies that HTML should be used to display the data.

The next line is the beginning of the formatting section. Before we look at the XSLT tags, you need to know that XSL documents use templates to perform transformations of XML documents. The XSL stylesheet can contain one or more XSLT templates. You can think of templates as special blocks of code that apply to one or more XML tags. Templates contain rules for displaying a particular set of elements in the XML document. The use of templates is made possible via special formatting tags.

Notice that the Shippers.xsl file contains the <xsl:template> tag to define a template for the entire document. The <xsl:template> element has a match attribute. The value of the match attribute indicates the nodes (elements) for which this template is appropriate.

For example, the special pattern "//" in the match attribute tells the XSL processor that this is the template for the document root:

```
<xsl:template match="//dataroot"
xmlns:xsl="http://www.w3.org/1999/XSL/Transform">
```

The template ends with the </xsl:template> closing tag. Following the definition of the template, standard HTML tags are used to format the document. Next, the XSLT formatting instruction <xsl:for-each> (Figure 32.6) tells the XSL processor to do something every time it finds a pattern. The pattern follows the select attribute.

For example:

```
<xsl:for-each select="Shippers">
```

tells the XML processor to loop through the <Shippers> elements. The loop is closed with a closing loop tag:

```
</xsl:for-each>
```

The XSLT formatting instruction <xsl:value-of> tells the XSL processor to retrieve the value of the tag specified in the select attribute. For example:

```
<xsl:value-of select="ID">
```

tells the XML processor to select the ID column. Because this formatting instruction is located below the <xsl:for-each> tag, the XSL processor will retrieve the value of the ID column for each Shippers element. The select attribute uses the XML Path language (XPath) expression to locate the child elements to be processed.

If you scroll down the Shippers.xsl file, you will also notice that Access has generated a number of VBScript functions to evaluate expressions. To prevent the XSL processor from parsing these functions, the function section is placed within the CDATA directive.

2. Close the browser containing the Shippers.xsl file.

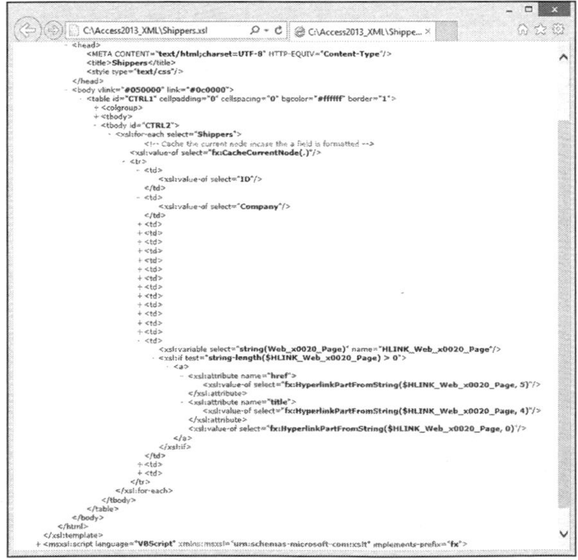

FIGURE 32.6. The XSLT formatting instructions in the Shippers.xsl file.

What Exactly Is XPath?

XPath is a query language used to create expressions for finding data in the XML data file. These expressions can manipulate strings, numbers, and Boolean values. They can also be used to navigate an XML tree structure and process its elements with XSLT instructions. XPath is designed to be used by XSL Transformations (XSLT). With XPath expressions, you can easily identify and extract from the XML document specific elements (nodes) based on their type, name, values, or the relationship of a node to other nodes. When preparing stylesheets for transforming your XML documents into HTML, you will often use various XPath expressions in the select attribute.

For more information about Extensible Stylesheet Language (XSL), visit the following link: http://www.w3.org/TR/xsl/.

Viewing XML Documents Formatted with Stylesheets

When you exported the Shippers table to XML format, Access applied XSLT transforms to turn the XML data into an HTML file so that you can view formatted data in the browser (see Figure 32.8). To display the Access-generated Shippers.htm file in Internet Explorer, type ***http://localhost/xml/Shippers.htm*** in the address box and press Enter.

If you open the Shippers.htm file in Internet Explorer and the page is blank, you will need to check if there were any parsing errors on loading either the XML or style sheet. To troubleshoot this issue while in Windows Explorer, press F12 to open the developer tools. Choose Script, and press the Start debugging button (see Figure 32.7). The button name will change to Stop debugging. If there are any errors you should see them listed in the window's right-hand pane.

To correct the reported warnings and errors, do the following:

1. Open Shippers.htm file in Notepad.
2. Enter <!DOCTYPE html> on the first line.
3. Move the </BODY> ending tag on the line just before the </HTML> ending tag.
4. In the ApplyTransform function, replace the lines:

```
Document.Open "text/html"
Document.Write objData.TransformNode(objStyle)
```

with the following (ensuring that the Document is spelled with a lowercase “d”):

```
document.Open "text/html"
document.Write objData.TransformNode(objStyle)
```

5. Press Ctrl+S to save the changes to the Shippers.htm.
6. Close Notepad.
7. Reload the ***http://localhost/xml/Shippers.htm*** to view the Shippers table.

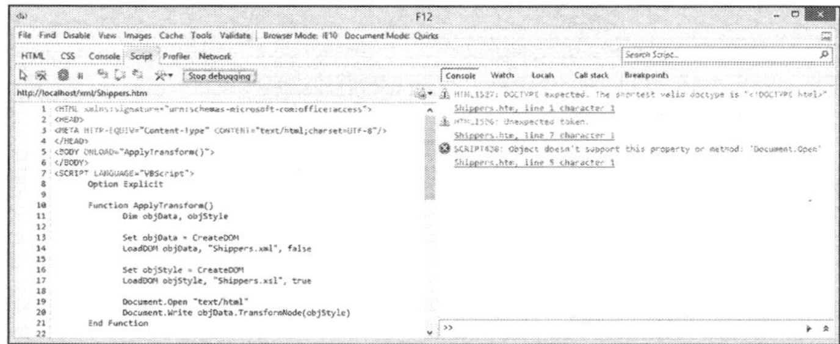

FIGURE 32.7. At the time of writing, the HTM file generated by Access 2013 via the Export to XML File does not load in Internet Explorer due to parsing errors.

ID	Company	Last Name	First Name	E-mail Address	Job Title	Business Phone	Home Phone	Mobile Phone	Fax Number	Address	City	State/Province	ZIP/Postal Code	Country/Region	Web Page	Notes	Attachments
1	Shipping Company A									123 Any Street	Memphis	TN	99999	USA			
2	Shipping Company B									123 Any Street	Memphis	TN	99999	USA			
3	Shipping Company C									123 Any Street	Memphis	TN	99999	USA			

FIGURE 32.8. This HTML file was created from XML data by using XSLT.

To display only the fields containing data, export data to the XML file based on a query. For example, the following SQL statement can be used to create the qryShippers query:

```
SELECT Shippers.[ID], Shippers.[Company], Shippers.[Address],
  Shippers.[State/Province], Shippers.[ZIP/Postal Code],
  Shippers.[Country/Region]
FROM Shippers;
```

You can then right-click the qryShippers query in the Navigation pane and choose Export | XML File. The resulting HTML file based on the qryShippers query is shown in Figure 32.9.

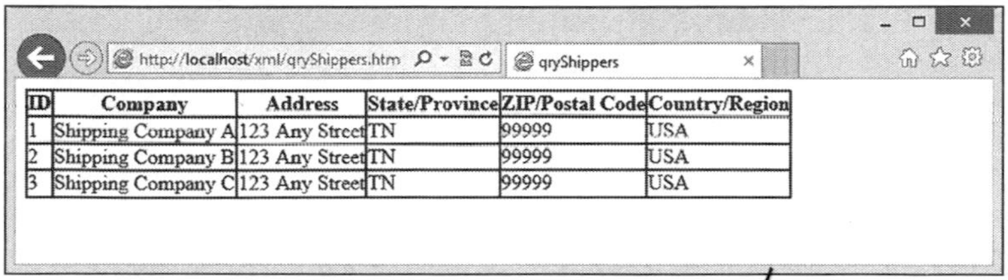

http://localhost/xml/qryShippers.htm qryShippers

ID	Company	Address	State/Province	ZIP/Postal Code	Country/Region
1	Shipping Company A	123 Any Street	TN	99999	USA
2	Shipping Company B	123 Any Street	TN	99999	USA
3	Shipping Company C	123 Any Street	TN	99999	USA

FIGURE 32.9. This HTML file was created from XML data by using XSLT.

Advanced XML Export Options

When you exported the Shippers table to XML format, you may have noticed the More Options button in the Export XML dialog box (see Figure 32.1 at the beginning of this chapter). Pressing this button opens a window with three tabs as shown in Figure 32.10. Each tab groups options for the types of XML objects that you can export. The Data tab contains options for the XML document, the Schema tab lists options for the XSD document, and the Presentation tab provides options for generating the XSL document.

Data Export Options

The options shown on the Data tab (see Figure 32.9) control the data that is exported to the XML documents. These options are grouped into three main areas.

The Data to Export section displays data that you may want to export. In this particular scenario, the Customers table has been chosen for export. Because this table is directly related to the Orders table in the Northwind 2007 database, the Orders table is displayed as a child node of Customers. The Orders table is related to the Order Details table and so on. Clicking on the plus sign in front of the [Lookup Data] node will display the names of tables that provide lookup information for the main tables. By clicking on the checkbox you may export just the table that you originally requested or you can export the customers' data along with all the orders, and perhaps include lookup information.

Below the Data to Export section is the Export Location area that shows the filename for the XML document that will be created when you click the OK button. You can change the location of this document by using the Browse button. Simply navigate to the folder where you want to save the XML file. You can also change the name of the document by replacing the name shown in the text box with another name.

The area to the right of the Data to Export section allows you to specify which records you want to export. This area contains three option buttons that allow you to export all records, filtered records, or the current record. Notice that only one option is enabled in Figure 32.10.

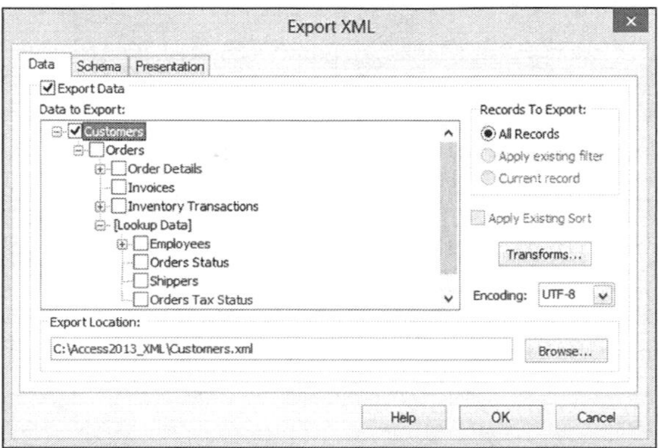

FIGURE 32.10. Use the Data tab in the Export XML window to set advanced data options.

When you highlight the table to export in the database window and then choose the Export command from the File menu, only the All Records option button will be enabled in the Records To Export section. Opening the table prior to choosing the Export command tells Access to enable the All Records and Current record option buttons. And if you open the table and apply a filter to the data, then select the Export command, Access will enable the Apply existing filter option button in addition to the other two buttons.

The other options on the Data tab are Apply Existing Sort, Transforms, and Encoding. The Apply Existing Sort checkbox is enabled if the exported object is open and a sort is applied. Access will export the data in the specified order. Clicking the Transforms button allows you to select a custom XSL transform file to apply to the data during export. You can choose from the transforms you have written or received with the XML data. Use the Encoding drop-down list to select UTF-8 or UTF-16 encoding for the exported XML. The default is UTF-8.

When you export an object from an Access database file, Access exports static data. This means that the exported object is not automatically updated when the data changes. If the data in the Access database has changed since you exported an Access object to an XML data file, you will need to re-export the object so the new data is available to the client application.

Exporting live data is supported by Access data projects (adp file format) in Access 2010. Support for ADP was removed in Access 2013, therefore additional options related to Access data projects are not discussed here.

Schema Export Options

The options shown on the Schema tab (see Figure 32.11) control the way the schema file for the object is exported. Advanced schema options are presented in two sections: Export Schema and Export Location.

The Export Schema section has two checkboxes. By selecting the Export Schema checkbox you indicate that you want to export the object's schema as an XSD file. This selection is the same as choosing the Schema of the data (XSD) option in the first Export XML dialog box (see Figure 32.1). The checkboxes under Export Schema allow you to specify whether you want to include primary key and index information in the XSD schema file, and whether to export all table and field properties.

The Export Location section has two option buttons that allow you to specify whether you want the schema information to be embedded in the exported XML data document or stored in a separate schema file. You can enter the filename in the provided text box and specify the location of the schema file by clicking the Browse button.

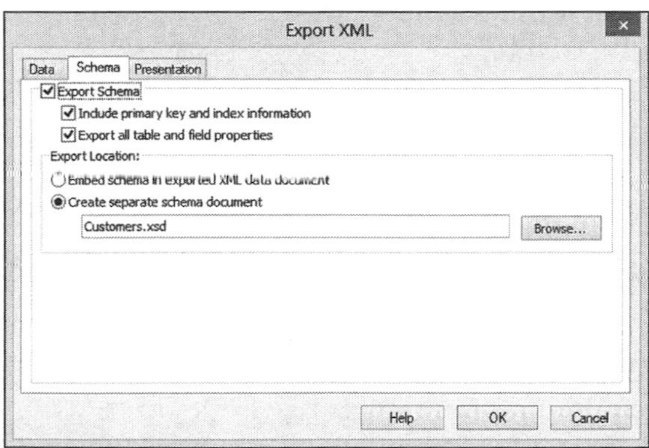

FIGURE 32.11. Use the Schema tab in the Export XML dialog box to set advanced schema options.

Presentation Export Options

The selections on the Presentation tab (see Figure 32.12) specify available options for the XSL files. The Export Presentation (HTML 4.0 Sample XSL) checkbox allows you to indicate whether you want to export the object's presentation. Choose the Client (HTML) option in the Run from section if you want the presentation to run on the client. Access will create an HTML file with the script necessary to perform the transform. The script will be executed on the client machine. While this selection reduces the load on the server, a client application will need to download a few files (HTML document, XML data file, and XSD schema file) to present the data in the browser. If the XSL file is going to be placed on the Web server and called from an ASP page, choose the Server (ASP) option. By choosing this option, only the final HTML is downloaded to the client.

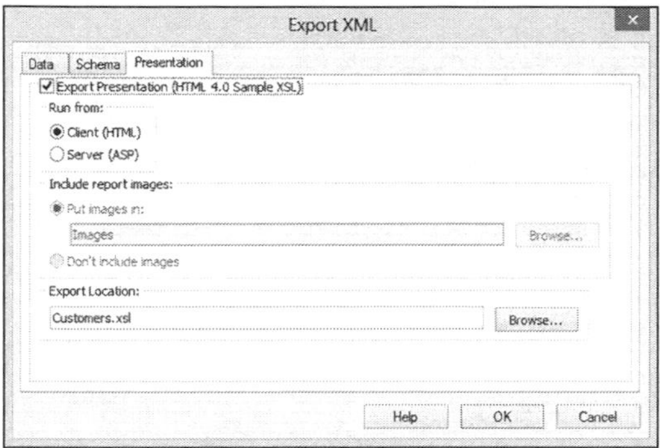

FIGURE 32.12. Use the Presentation tab in the Export XML dialog box to set advanced presentation options.

If the exported presentation includes pictures, you can indicate whether to include them in the output by clicking the appropriate option button in the Include report images section. If you choose to include the images, Access will create separate image files and link them with the HTML file. By default, the image files are stored in the Images folder of the main export folder. To place them in another location, click the Browse button to specify the folder name.

The Export Location section allows you to specify the name and location of the export files. When you export a presentation file, Access creates two files: an XSL file that includes all the XSLT transforms needed to define how the data is presented, and a simple HTML file that contains properly formatted data from the exported object and not the raw data with XML tags. The HTML file contains a snapshot of the data as it existed during the export process.

APPLYING XSLT TRANSFORMS TO EXPORTED DATA

When exporting Access data to XML format, you can use custom transformation files (XSL) to modify the data after you export it. Hands-On 32.5 demonstrates how to create a custom stylesheet for use after export. This stylesheet assumes that for each customer in the Customers table we want to display only selected columns from the Orders table. You learn how to apply this custom stylesheet in Hands-On 32.6. Take a quick look at Figure 32.15 later in this chapter to see the final outcome.

Hands-On 32.5. Creating a Custom Transformation File

1. Open Notepad and enter the following statements:

```
<?xml version="1.0" encoding="UTF-8"?>
<xsl:stylesheet  version="1.0"  xmlns:xsl="http://www.w3.org/1999/XSL/
Transform">
<xsl:output method="html" version="4.0" indent="yes"/>
<xsl:template match="dataroot">
  <html>
  <body>
  <h2><font name="Verdana">Orders by Customer</font></h2>
  <p></p>
  <xsl:apply-templates select="Customers"/>
  </body>
  </html>
</xsl:template>
<xsl:template match="Customers">
<table>
  <tr>
  <td bgColor="#FFCC33">
  <font color="#000000">
  <xsl:value-of select="CustomerID"/>
  </font>
  </td>
  <td><b>
  <xsl:value-of select="CompanyName"/>
  </b></td>
  </tr>
</table>
<table cellpadding="5" cellspacing="5">
  <tr bgColor="black">
  <td bgcolor="black" width="10px"></td>
  <td><font color="white" size="1">Order ID</font></td>
  <td><font color="white" size="1">Order Date</font></td>
  <td><font color="white" size="1">Shipped Date</font></td>
  <td><font color="white" size="1">Required Date</font></td>
  <td><font color="white" size="1">Freight</font></td>
  </tr>
  <xsl:apply-templates select="Orders"/>
</table>
</xsl:template>
```

```
<xsl:template match="Orders">
  <tr>
  <td bgcolor="black" width="10px"></td>
  <td><xsl:value-of select="OrderID"/></td>
  <td><xsl:value-of select="substring(OrderDate, 1, 10)"/></td>
  <td><xsl:value-of select="substring(ShippedDate, 1, 10)"/></td>
  <td><xsl:value-of select="substring(RequiredDate, 1, 10)"/></td>
  <td>$<xsl:value-of select="format-number(Freight,'####0.00')"/></td>
  </tr>
</xsl:template>
</xsl:stylesheet>
```

2. Save the file as **C:\Access2013_XML\ListCustOrders.xsl**. You must include the file extension to ensure that the file is not saved as text.
3. Close Notepad.

Let's now proceed to analyze the contents of the ListCustOrders.xsl file that will be used to transform XML to HTML in our next hands-on exercise. Notice that because the XSLT stylesheet is an XML document, we started out with a standard XML declaration:

```
<?xml version="1.0" encoding="UTF-8"?>
```

Next, we defined the namespace for the stylesheet and declared its prefix like this:

```
<xsl:stylesheet version="1.0"
xmlns:xsl="http://www.w3.org/1999/XSL/Transform">
```

On the third line we indicated that XSLT should transform the XML into HTML by using the <xsl:output> tag as follows:

```
<xsl:output method="html" version="4.0" indent="yes"/>
```

In the preceding XML, the <xsl:output> tag has three attributes: method, version, and indent. The method attribute specifies the format of the output. This can be XML, HTML, or text. The version attribute sets the version number for the output format. The indent attribute, which is set to "yes" in this example, indicates that the XML should be indented. This will make the final XML document more readable when viewed in the browser.

The remaining part of the XSL file contains transformation instructions for the XML document element nodes. We begin by creating the root template. The <xsl:template> tag initiates a template within a stylesheet. Because a template must indicate which nodes you want to use, we supplied the node information by using the tag's match attribute, like this:

```
<xsl:template match="dataroot">
```

This tells the XSLT processor to extract the XML document's root node. The root node provides a base node upon which we will build our Web page. Notice that in the root template we included the <html> and <body> tags to create the structure of the final document and used HTML tags such as <h2>, <font>, and <p> to add the required formatting to our Web page. In the root template, we are also telling the XSLT processor that it should apply the template rules found in the Customers template (defined further down in the file):

```
<xsl:apply-templates select="Customers"/>
```

When the XSLT processor encounters the <xsl:apply-templates> instruction, it will proceed to the following line:

```
<xsl:template match="Customers">
```

This line marks the beginning of the Customers template rule. Within it there are HTML tags as well as other XSLT processing instructions. For example, to output the CustomerID we use the <xsl:value-of> tag with the select attribute like this:

```
<xsl:value-of select="CustomerID"/>
```

Because the <xsl:value-of> tag does not have any content, you must end it with the forward slash (/). Notice that we placed the value of the CustomerID field in a table cell. Using the same approach we output the CompanyName:

```
<xsl:value-of select="CompanyName"/>
```

Next, we defined the column headings for the Orders table. For a special effect, we added to the output a 10-pixel-wide dummy column with a black background:

```
<td bgcolor="black" width="10px"></td>
```

We also told the XSLT processor to apply the Orders template:

```
<xsl:apply-templates select="Orders"/>
```

The Orders template rules indicate how to extract values for each of the defined column headings. This is done by using the <xsl:value-of> tag with the select attribute, like this:

```
<td><xsl:value-of select="OrderID"/></td>
<td><xsl:value-of select="substring(OrderDate, 1, 10)"/></td>
<td><xsl:value-of select="substring(ShippedDate, 1, 10)"/></td>
<td><xsl:value-of select="substring(RequiredDate, 1, 10)"/></td>
<td>$<xsl:value-of select="format-number(Freight,'####0.00')"/></td>
```

To obtain only the date portion from the OrderDate, ShippedDate, and RequiredDate columns, we use the XPath `substring` function in the select attribute. This function has the same syntax as the VBA `Mid` function, allowing you to extract a specified number of characters from a string starting at a specific position. The format of the `substring` function is as follows:

```
substring(string, startpos, length)
```

startpos is the position of the first character to extract, and *length* represents the number of characters to be returned from *string*. Therefore, the expression

```
<xsl:value-of select="substring(OrderDate, 1, 10)"/>
```

tells the XSLT processor to retrieve only the first 10 characters from the value found in the OrderDate column.

Notice also that to correctly format the Freight column we used the format-number XPath expression like this:

```
<xsl:value-of select="format-number(Freight,'####0.00')"/>
```

This tells the XSLT processor to format the value found in the Freight column as a number using two decimal places. Notice that the dollar sign cannot be a part of the XPath expression. It is appended to the final output as shown here:

```
<td>$<xsl:value-of select="format-number(Freight,'####0.00')"/>
</td>
```

Notice that each of the defined template rules ends with the </xsl:template> ending tag and the stylesheet itself ends with the </xsl:stylesheet> tag.

This concludes our hands-on example of how you can make your own custom stylesheets. While this is a basic stylesheet to get you started, in real life you will probably want to create stylesheets that allow:

- Batch-processing nodes (<xsl:for-each> tag with the select attribute)
- Conditional processing of nodes (<xsl:if> tag with the test attribute)
- Decisions based on conditions (<xsl:choose> tag and <xsl:when> tag with the test attribute)
- Sorting nodes before processing (<xsl:sort> tag with the select attribute)

For more information about XSL Transformations (XSLT), visit the following link: http://www.w3.org/TR/xslt#section-Applying-Template-Rules.

Now that you have a custom stylesheet, what do you do with it? Hands-On 32.6 demonstrates how to export data from an Access table directly to an HTML file and apply a custom transform so that only certain columns are displayed in the browser.

Hands-On 32.6. Exporting Data and Applying a Custom XSL File

1. Open the **C:\Access2013_HandsOn\Northwind.mdb** database.
2. In the Navigation pane, right-click the **Customers** table and choose **Export | XML File.**
3. In the File name box, enter **C:\Access2013_XML\ListCustOrders.xml**, and click **OK.**
4. In the Export XML dialog box, make sure that the first two checkboxes are selected. Click the **More Options** button.
5. In the Data to Export area, the Customers table is automatically selected. Click the checkbox next to the **Orders** table to include it in the export.
6. Click the **Transforms** button.
7. In the Export Transforms window that appears, click the **Add** button.
8. Access displays the Add New Transform window. Switch to the **C:\Access2013_XML** folder and select the **ListCustOrders.xsl** file that you created in the previous hands-on exercise. Click the **Add** button to add this file to the list of transforms. The transformation file appears in the list as shown in Figure 32.13.

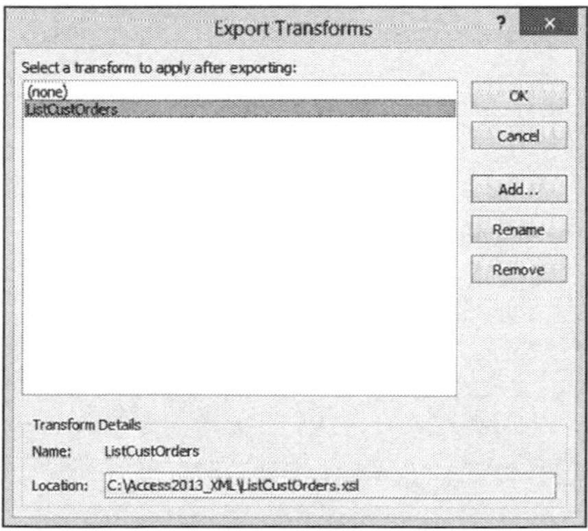

FIGURE 32.13. Use this window to indicate a transformation file (stylesheet) to be used after export.

9. In the Export Transforms window, click **OK.**
10. Back in the Export XML dialog box, change the file extension from xml to **html** as shown in Figure 32.14.

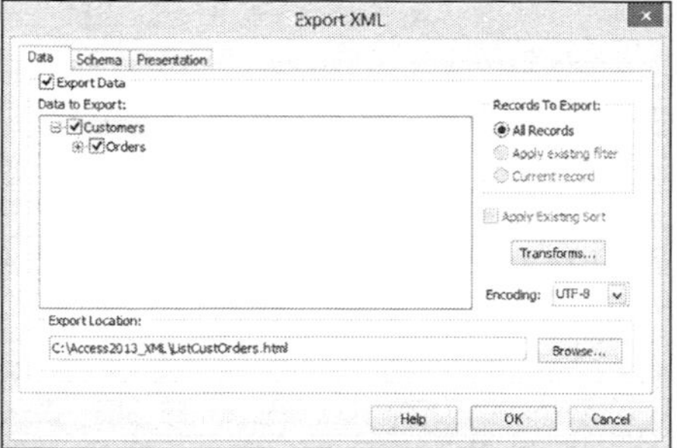

FIGURE 32.14. To export XML data directly to the HTML file, you must choose the transformation file using the Transforms button and change the file extension from xml to html.

11. Click the **OK** button to begin the export.
12. Upon successful export operation, click **Close**.

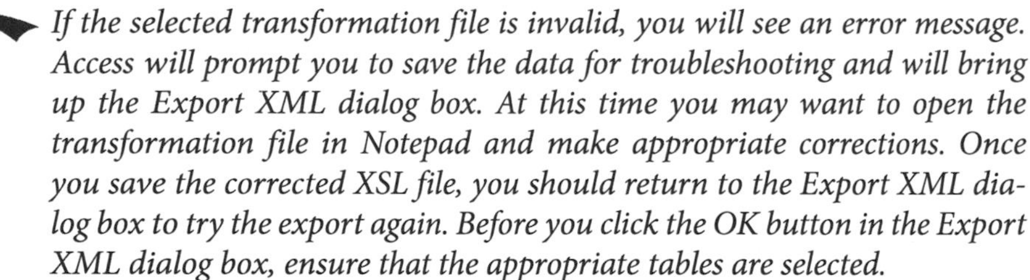

Note: *If the selected transformation file is invalid, you will see an error message. Access will prompt you to save the data for troubleshooting and will bring up the Export XML dialog box. At this time you may want to open the transformation file in Notepad and make appropriate corrections. Once you save the corrected XSL file, you should return to the Export XML dialog box to try the export again. Before you click the OK button in the Export XML dialog box, ensure that the appropriate tables are selected.*

13. Close the Northwind database and exit Microsoft Access.
14. In your browser's address bar, type **http://localhost/xml/ListCustOrders.html** and press **Enter**. The final result of applying the custom transformation file is shown in Figure 32.15.
15. Close the browser window.

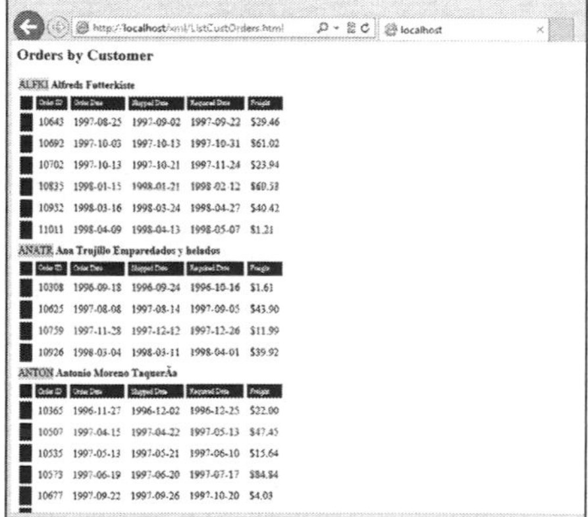

FIGURE 32.15. XML data can be formatted any way you like by applying a custom transformation (see Figures 32.13 and 32.14).

IMPORTING XML DATA

You can use the Access built-in Import command to import an XML data or XML schema document to a database. When you import structure or data from an XML file, Access assigns the Text data type to all the fields in a table. However, when you import structure from an XSD schema file, each field is assigned a data type that closely matches the data type specified in the schema. You can change the data types after importing data or a table structure as long as the fields' data allows such a change.

When you import a schema, Access creates a new empty table with the structure of the imported schema. Earlier in this chapter, when you exported the Shippers table to XML format, Access also created the schema of that table. Hands-On 32.7 shows how to import this schema document to a new Access database.

Hands-On 32.7. Importing a Schema File (XSD) to an Access Database

1. Create a new Access database named **C:\Access2013_ByExample\Chap32.accdb**.
2. In the Access window, choose **External Data**, and then select **XML File** from the **Import & Link** group.
3. In the Get External Data - XML File window, type **C:\Access2013_XML\Shippers.xsd** in the File name box and click **OK**. Access displays the Import XML dialog box as shown in Figure 32.16.

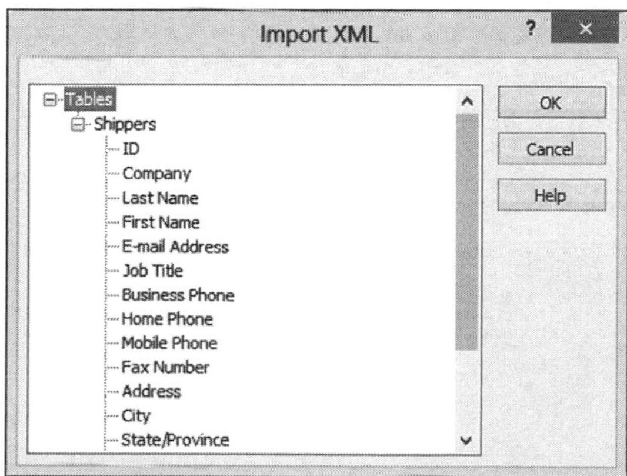

FIGURE 32.16. When importing a schema file to an Access database, the Import XML dialog box displays the table name and its columns as defined in the schema.

Notice that you cannot indicate which columns you would like to import. Access always imports the entire XSD file.

4. Click **OK** to perform the import. When the import operation is completed, click **Close**. The Shippers table appears in the Navigation pane of the Access window. Figure 32.17 shows this table opened in Design view.
5. Close the Chap32.accdb database.

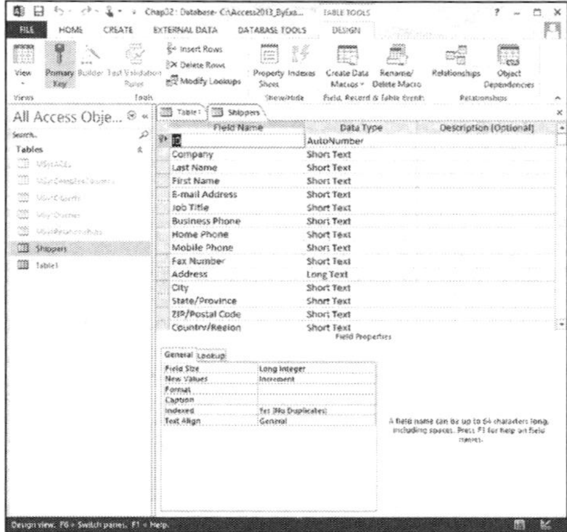

FIGURE 32.17. The Shippers table was created by importing the Shippers.xsd schema file.

When importing an XML data file to an Access database, you can use the Import Options section to specify whether you want to import structure only, import structure and data, or append data to an existing table (see Figure 32.18). When you append data to an existing table, Access compares the structure of the imported table with the table structures that are already in the database. If Access cannot find a table structure matching the imported table, the data is placed in a new table; otherwise, it is appended to the existing table. You can also click the Transform button in the Import XML dialog box to specify a transformation file that you want to apply when the XML data is imported.

It is important to point out that when XML data is imported to an Access database, it is not linked with the original XML file. This means that to refresh the data in the table, you need to repeat the import process.

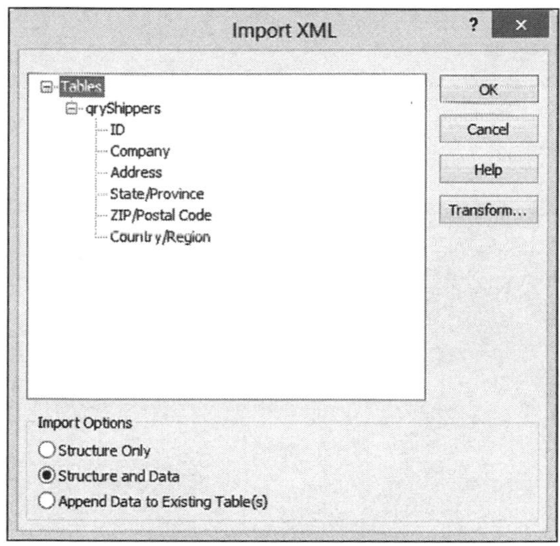

FIGURE 32.18. Importing XML data (qryShippers.xml) to the Chap32.accdb database.

The following project demonstrates how to import XML data to an Access database and modify the data before import using a transformation file. We will perform the tasks outlined here:

- Create a custom transformation file to be used after the XML data import
- Export the Customers table and the related Orders table to an XML file
- Import to an Access database only two columns from the Customers table and five columns from the Orders table

Custom Project 32.1. Importing XML Data to an Access Database and Applying a Transform

Part 1: Creating a Custom Transformation File to Be Used after the XML Data Import

1. Open Notepad and enter the following statements:

```
<?xml version="1.0" encoding="UTF-8"?>
<xsl:stylesheet  version="1.0"  xmlns:xsl="http://www.w3.org/1999/XSL/
Transform">
<xsl:output method="html" version="4.0" indent="yes"/>
<xsl:template match="dataroot">
  <html>
  <body>
```

```
    <table>
    <xsl:apply-templates select="Customers"/>
    </table>
    <table>
    <xsl:apply-templates select="Customers/Orders"/>
    </table>
    </body>
    </html>
  </xsl:template>
  <xsl:template match="Customers">
    <Customer>
    <CustomerID>
    <xsl:value-of select="CustomerID"/>
    </CustomerID>
    <CompanyName>
    <xsl:value-of select="CompanyName"/>
    </CompanyName>
    </Customer>
  </xsl:template>
  <xsl:template match="Customers/Orders">
    <Order>
    <OrderID>
    <xsl:value-of select="OrderID"/>
    </OrderID>
    <OrderDate>
    <xsl:value-of select="substring(OrderDate, 1, 10)"/>
    </OrderDate>
    <ShippedDate>
    <xsl:value-of select="substring(ShippedDate, 1, 10)"/>
    </ShippedDate>
    <RequiredDate>
    <xsl:value-of select="substring(RequiredDate, 1, 10)"/>
    </RequiredDate>
    <Freight>
    <xsl:value-of select="format-number(Freight,'####0.00')"/>
    </Freight>
    </Order>
  </xsl:template>
  </xsl:stylesheet>
```

2. Save the file as **C:\Access2013_XML\CustomerOrders.xsl**. You must include the file extension to ensure that the file is not saved as text.

3. Close Notepad.

   Since you've already created a similar stylesheet in Hands-On 32.5, the contents of the CustomerOrders.xsl file should be recognizable. All that's different here are the <Customer> and <Order> tags that specify the names of Access tables where we want to place our XML data. When importing data, tables are named according to the name of the XML element being imported. If the Access database already has a table with the specified name, a number is appended to the name.

Part 2: Exporting the Customers and Related Orders Tables to an XML File

1. Open the **C:\Access2013_HandsOn\Northwind.mdb** database. In the Navigation pane, right-click the **Customers** table and choose **Export | XML File.**
2. In the Export - XML File window, type **C:\Access2013_XML\CustomerOrders.xml** in the File name box and click **OK.**
3. Access displays the Export XML dialog box with three checkboxes; the first two checkboxes should be selected. Click the **More Options** button.
4. In the Data to Export area of the Export XML dialog box, select the checkbox next to the **Orders** table. The Customers and Orders tables should both be selected.
5. Click **OK** to perform the export of all the records in the selected tables. When the export operation is completed, click **Close.**
6. Close the Northwind.mdb database file.

Part 3: Importing to an Access Database Only Two Columns from the Customers Table and Five Columns from the Orders Table

1. Open the **C:\Access2013_XML\Chap32.accdb** database file that you created in Hands-On 32.7.
2. In the Access window, choose **External Data**, and then select **XML File** from the **Import & Link** group.
3. In the Get External Data - XML File window, type **C:\Access2013_XML\CustomerOrders.xml** in the File name box and click **OK.**

   Access displays the Import XML window with the file's Customers and Orders tables listed. By expanding nodes in the tree structure you can see the columns in each table, but you cannot indicate which columns to import, as Access always imports the entire file by default. You can, however, tell Access to perform a custom XSLT transform to import only the columns needed.
4. In the Import XML window, click the **Transform** button.

5. In the Import Transforms window that appears, click the **Add** button to apply a transform before importing.
   Access displays the Add New Transform window. Switch to the **Access2013_XML** folder and select the **CustomerOrders.xsl** file that you created in Part 1 of this project. Click the **Add** button to add this file to the list of transforms.
6. In the Import Transforms window, click **OK**.
7. Back in the Import XML window, make sure that the **Structure and Data** option button is selected under Import Options and click **OK**. When Access finishes importing the **C:\Access2013_XML\CustomerOrders.xml** document, click **Close**.
8. In the Navigation pane of the Access window, notice the appearance of two new tables: Customer and Order. Open both tables and check their contents. As you can see, Access has applied the custom stylesheet before importing the data and only the columns specified in the stylesheet were imported (see Figure 32.19).
9. Open the **Order** table in Design view. Notice that all the fields in this table have been assigned the Text data type. After importing data or table structure you can change the fields' data types.
10. Change the data type of the OrderDate, ShippedDate, and RequiredDate columns to **Date/Time** and the Freight column's data type to **Currency** to match the original Orders table.
11. Save the modified Order table and close the Chap32.accdb database file.

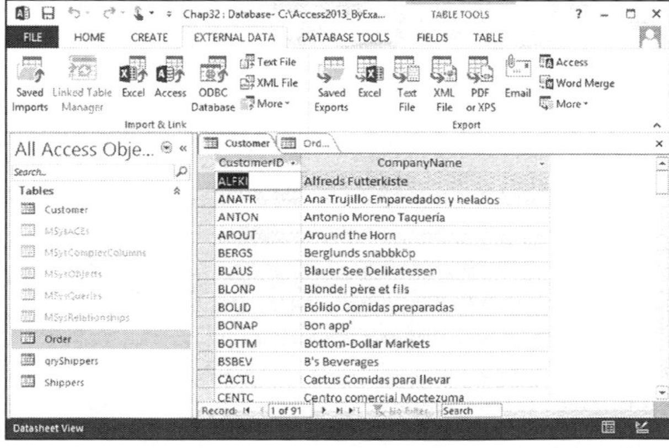

FIGURE 32.19. Applying a custom transformation file before XML data import to limit the number of columns of data imported to Access database tables.

PROGRAMMATICALLY EXPORTING TO AND IMPORTING FROM XML

Now that you've mastered the use of Microsoft Access 2013 built-in commands for exporting and importing XML data, let's look at what tools are available for programmers who want to perform these XML operations via code. In the following sections of this chapter, you will learn how to work with XML using:

- The `ExportXML` and `ImportXML` methods from the Microsoft Access 15.0 Object Library
- The `TransformXML` method

Exporting to XML Using the ExportXML Method

Use the Microsoft Access 15.0 Object Library `ExportXML` method of the Application object to export XML data, schemas (XSD), and presentation information (XSL) from a Microsoft Access database, Microsoft SQL Server 2000 Desktop Engine (MSDE 2000), or Microsoft SQL Server 6.5 or later.

The `ExportXML` method takes a number of arguments, which are shown in Table 32.1.

TABLE 32.1. Arguments of the ExportXML method (in order of appearance)

Argument Type	Data Type / Description
`ObjectType` (required)	`AcExportXMLObjectType` Use one of the following constants: Constant — Value `acExportForm` — 2 `acExportFunction` — 10 `acExportQuery` — 1 `acExportReport` — 3 `acExportServerView` — 7 `acExportStoredProcedure` — 9 `acExportTable` — 0 Specifies the type of Access object to export. The constant values 10, 7, and 9 are used only with Microsoft Access projects.
`DataSource` (required)	String Indicates the name of the Access object specified in the `ObjectType` argument.

(Contd.)

Argument Type	Data Type / Description
`DataTarget` (optional)	String Specifies the path and filename for the exported data. Omit this argument only if you don't want the data to be exported.
`SchemaTarget` (optional)	String Specifies the path and filename for the exported schema information. Omit this argument only if you don't want the schema to be exported to a separate file.
`PresentationTarget` (optional)	String Specifies the path and filename for the exported presentation information. Omit this argument only if you don't want the presentation information to be exported.
`ImageTarget` (optional)	String Specifies the path for the exported images. Omit this argument if you don't want to export images.
`Encoding` (optional)	`AcExportXMLEncoding` Use one of the following constants: Constant Value `acUTF16` 1 `acUTF8` 0 The default is `acUTF8`. Specifies the text encoding for the exported data.
`OtherFlags` (optional)	`AcExportXMLOtherFlags` Use one or more of the following constants: Constant Value `acEmbedSchema` 1 `acExcludePrimaryKeyAndIndexes` 2 `acExportAllTableAndFieldProperties` 32 `acLiveReportSource` 8 `acPersistReportML` 16 `acRunFromServer` 4 Specifies behaviors associated with exporting to XML. Values can be added to specify a combination of behaviors. Here are the meanings of the constants: (1) Write schema information into a separate document specified by the `DataTarget` argument. This value takes precedence over the `SchemaTarget` argument. (2) Does not export primary key and index schema properties. (32) The exported schema contains properties of the table and its fields.

(Contd.)

Argument Type	Data Type / Description
	(8) Used only when exporting reports bound to SQL Server 2000. Will create a live link to a Microsoft SQL Server database. (16) Persists the exported object's ReportML file. (4) Used only when exporting reports. Creates an Active Server Pages (ASP) or HTML wrapper. The default is HTML.
`WhereCondition` (optional)	String Specifies a subset of records to export.
`AdditionalData` (optional)	`AdditionalData` `AdditionalData` is an Access object that represents the collection of tables and queries that will be included with the parent table that is exported by the `ExportXML` method (see Hands-On 32.8). Specifies additional tables to export. This argument is ignored if the `OtherFlags` argument is set to `acLiveReportSource` (8).

In its simplest form, the `ExportXML` method looks like this:

```
Application.ExportXML ObjectType:=acExportTable, _
DataSource:="Customers", _
DataTarget:= "C:\Access2013_XML\North_Customers.xml"
```

The preceding statement, when typed on a single line (without the underscore characters) in the Visual Basic Editor's Immediate window or inside a VBA procedure stub in a Visual Basic module, will render the Customers table in the XML format in the North_Customers.xml file.

Using the arguments described in Table 32.1, you can easily write the command to export the XML Products table with its schema and presentation information placed in separate files:

```
Application.ExportXML ObjectType:=acExportTable, _
 DataSource:="Products", _
 DataTarget:= "C:\Access2013_XML\North_Products.xml", _
 SchemaTarget:= "C:\Access2013_XML\North_ProdSchema.xsd", _
 PresentationTarget:= "C:\Access2013 XML\North_ProdReport.xsl"
```

To export a specific customer's data to an XML data file, use the following statement:

```
Application.ExportXML ObjectType:=acExportTable, _
 DataSource:="Customers", _
 DataTarget:="C:\Access2013_XML\OneCustomer.xml", _
 WhereCondition:="CustomerID = 'GROSR'"
```

To try out the preceding statements, open the Northwind.mdb database, switch to the Visual Basic Editor window, insert a new standard module and type each statement inside a Visual Basic procedure named Test_ExportToXML. After executing the procedure, locate and check out the newly created XML files in your C:\Access2013_XML folder.

Hands-On 32.8 demonstrates how to export to XML three tables: Customers, Orders, and Order Details.

Hands-On 32.8. Exporting Multiple Tables to an XML Data File

1. In the **C:\Access2013_ByExample\Chap32.accdb** database, switch to the Visual Basic Editor window.
2. Choose **Insert | Module** to add a standard module to the current VBA project.
3. In the module's Code window, enter the following **Export_CustomerOrderDetails** procedure:

```
Sub Export_CustomerOrderDetails()
  Dim objAppl As New Access.Application
  Dim objOtherTbls As AdditionalData
  Dim strDBName As String
  strDBName = "C:\Access2013_HandsOn\Northwind.mdb"
  On Error GoTo ErrorHandler
  objAppl.OpenCurrentDatabase (strDBName)
  objAppl.Visible = False
  Set objOtherTbls = objAppl.CreateAdditionalData
  ' include the Orders and OrderDetails tables in export
  objOtherTbls.Add "Orders"
  objOtherTbls.Add "Order Details"
  ' export Customers, Orders, and Order Details table into
  ' one XML data file
  objAppl.ExportXML ObjectType:=acExportTable, _
  DataSource:="Customers", _
  DataTarget:="C:\Access2013_XML\CustomerOrdersDetails.xml", _
  AdditionalData:=objOtherTbls
  MsgBox "Export operation completed successfully."
Exit_Here:
  On Error Resume Next
  objAppl.CloseCurrentDatabase
  Set objAppl = Nothing
  Exit Sub
```

```
ErrorHandler:
  MsgBox Err.Number & ": " & Err.Description
  Resume Exit_Here
End Sub
```

The Application object refers to the active Microsoft Access application, which in this case is the Chap32.accdb database where you wrote the procedure code shown here. Because this database does not contain the tables we want to export, we used the `New` keyword to create a new instance of the Microsoft Access Application object and then opened another Access database (Northwind.mdb) using the `OpenCurrentDatabase` method. You can use the `OpenCurrentDatabase` method to open an existing Microsoft Access database as the current database.

Using the AdditionalData object, you can export any set of Access tables to an XML data file. To use this object, perform the following:

- Declare an object variable as AdditionalData:

  ```
  Dim objOtherTbls As AdditionalData
  ```

- Create the AdditionalData object using the `CreateAdditionalData` method of the Application object and set the object variable to the newly created object:

  ```
  Set objOtherTbls = objAppl.CreateAdditionalData
  ```

- Use the AdditionalData object's `Add` method to add table names to the object:

  ```
  objOtherTbls.Add "Orders"
  objOtherTbls.Add "Order Details"
  ```

- Pass the AdditionalData object to the `ExportXML` method:

  ```
  objAppl.ExportXML ObjectType:=acExportTable, _
  DataSource:="Customers", _
  DataTarget:="C:\Access2013_XML\CustomerOrdersDetails.xml", _
  AdditionalData:=objOtherTbls
  ```

4. Place the insertion point anywhere within the Export_CustomerOrderDetails procedure code and choose **Run | Run Sub/UserForm**. Access executes the procedure code and displays a message.
5. Click **OK** to clear the informational message.
6. Switch to Windows Explorer and locate and open the **C:\Access2013_XML\CustomerOrdersDetails.xml** file. Notice that all the requested data was placed into one file.

7. Exit Windows Explorer.

Now that you know how to use VBA to export Access tables to XML, let's see how Access handles other objects. Custom Project 32.2 demonstrates how to export the Invoice report from the Northwind.mdb database to an XML file together with the presentation information and images.

Custom Project 32.2. Exporting an Access Report to an XML Data File with ASP

This project requires prior completion of the Hands-On 32.1 exercise.

Part 1: Creating a VBA Procedure to Export Invoice Data

1. In the Visual Basic Editor window, choose **Insert | Module** to add a standard module to the current VBA project.
2. In the module's Code window, enter the following **Export_InvoiceReport** procedure:

```
Sub Export_InvoiceReport()
  Dim objAppl As New Access.Application
  Dim strDBName As String
  strDBName = "C:\Access2013_HandsOn\Northwind.mdb"
  On Error GoTo ErrorHandler
  objAppl.OpenCurrentDatabase (strDBName)
  objAppl.Visible = False
  objAppl.ExportXML ObjectType:=acExportReport, _
   DataSource:="Invoice", _
   DataTarget:="C:\Access2013_XML\Invoice.xml", _
   PresentationTarget:="C:\Access2013_XML\Invoice.xsl", _
   ImageTarget:="C:\Access2013_XML", _
   WhereCondition:="OrderID=11075"
  MsgBox "Export operation completed successfully."
Exit_Here:
  On Error Resume Next
  objAppl.CloseCurrentDatabase
  Set objAppl = Nothing
  Exit Sub
ErrorHandler:
  MsgBox Err.Number & ": " & Err.Description
  Resume Exit_Here
End Sub
```

Take a look at the last two arguments of the `ExportXML` method used in this procedure. `ImageTarget` specifies that images displayed on the Invoice report are to be placed in the Access2013_XML folder. The `WhereCondition` argument specifies that we want only the data for Order 11075.

Part 2: Executing the VBA Code to Export Data

1. Place the insertion point anywhere within the Export_InvoiceReport procedure code and choose **Run | Run Sub/UserForm**. Access executes the procedure code and displays a message.
2. Click **OK** to clear the informational message.
3. Switch to Windows Explorer and open the **C:\Access2013_XML** folder.
4. Notice that Access has created a number of files: Invoice.xsl (stylesheet), Invoice.xml (XML document), Invoice.htm (HTML document), and two image files (PictureLogo.bmp and NameLogo.bmp).

Part 3: Viewing the Invoice Page in the Browser

1. In the browser's address bar, type **http://localhost/xml/Invoice.htm** and press **Enter.**
2. Important Note: If you see a blank page, see the Note preceding Figure 32.7 earlier in this chapter.
3. The file opens up as shown in Figure 32.20. Notice that the invoice displayed in the browser is an exact image of the report displayed in the Access user interface. Obviously there are some spacing issues that did not exist in Access 2010 when the same report was rendered in the browser.
4. Close the browser.

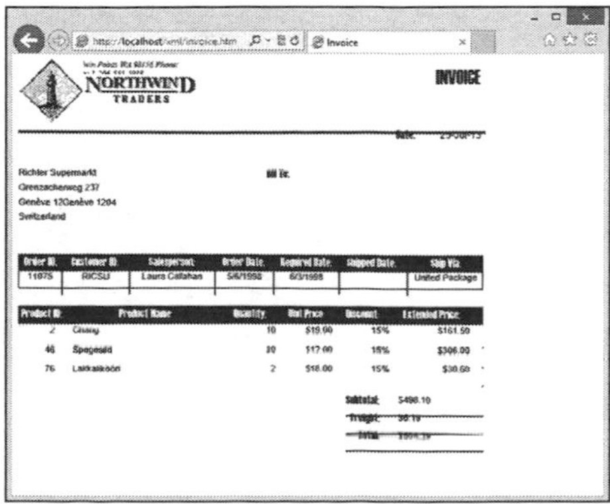

FIGURE 32.20. Viewing the exported invoice report in Internet Explorer.

Part 4: Examining the Content of the Invoice.htm File

5. In Windows Explorer, right-click the **Invoice.htm** file and choose **Open with | Notepad**. The content of the Invoice file is shown in Figure 32.21.

```
<!DOCTYPE html>
<HTML xmlns:signature="urn:schemas-microsoft-com:office:access">
<HEAD>
<META HTTP-EQUIV="Content-Type" CONTENT="text/html;charset=UTF-8"/>
</HEAD>
<BODY ONLOAD="ApplyTransform()">
<SCRIPT LANGUAGE="VBScript">
        Option Explicit

        Function ApplyTransform()
                Dim objData, objStyle

                Set objData = CreateDOM
                LoadDOM objData, "Invoice.xml", false

                Set objStyle = CreateDOM
                LoadDOM objStyle, "Invoice.xsl", true

                document.Open "text/html"
                document.Write objData.TransformNode(objStyle)
        End Function

        Function CreateDOM()
                On Error Resume Next
                Dim tmpDOM

                Set tmpDOM = Nothing
                Set tmpDOM = CreateObject("MSXML2.DOMDocument.6.0")
                If tmpDOM Is Nothing Then
                        Set tmpDOM = CreateObject("MSXML2.DOMDocument.3.0")
                End If
                If tmpDOM Is Nothing Then
                        Set tmpDOM = CreateObject("MSXML2.DOMDocument")
                End If

                Set CreateDOM = tmpDOM
        End Function

        Function LoadDOM(objDOM, strXMLFile, isXSL)
                On Error Resume Next

                objDOM.Async = False
                objDOM.Load strXMLFile
                If (objDOM.ParseError.ErrorCode <> 0) Then
                        MsgBox objDOM.ParseError.Reason
                End If

                If (isXSL) Then
                        objDOM.setProperty "AllowXsltScript", true
                End If
        End Function
</BODY>
</SCRIPT>
</HTML>
```

FIGURE 32.21. The contents of the Invoice.html.

Notice that when the HTML page loads, it executes the VBScript `ApplyTransform` function:

```
<BODY ONLOAD="ApplyTransform()">
```

The VBScript code uses a software component called the XML Document Object Model (DOM). The DOM offers methods and properties for working with XML programmatically, allowing you to output and transform the XML data. The DOMDocument object is the top level of the XML DOM hierarchy and represents a tree structure composed of nodes. You can navigate through this tree structure and manipulate the data contained in the nodes by using various methods and properties. Because every XML object is created and accessed from DOMDocument, you must first create the DOMDocument object in order to work with an XML document.

The `ApplyTransform` function begins by setting an object variable (`objData`) to an instance of DOMDocument that's returned by a custom `CreateDOM` function:

```
Set objData = CreateDOM
```

Notice in the `CreateDOM` function that appears at the bottom of the VBScript code that a reference to the DOMDocument is set via the `CreateObject` method of the Server object. Because different versions of the MSXML parser may be installed on a client machine (DOMDocument5, DOMDocument4, DOMDocument, etc.), the function attempts to instantiate the DOMDocument object using the most recent version. If such a version is not found, it looks for older versions of the MSXML parser that may exist. It is extremely important that only one version of the DOMDocument is used, since mixing DOMDocument objects from different versions of the MSXML parser can cause ugly errors.

Once the DOMDocument object has been instantiated, the `LoadDOM` function listed at the bottom of the page is called. This function expects two parameters: `objectDOM`, which is the `objData` variable referencing the DOMDocument, and `strXMLFile`, which is the name of the file to load into the DOMDocument object. To ensure that Internet Explorer waits until all the data is loaded before rendering the rest of the page, the Async property of the DOMDocument is set to False:

```
objDOM.Async = False
objDOM.Load strXMLFile
```

The `Load` method is used to load the supplied file into the `objData` object variable. This method returns True if it successfully loaded the data and False otherwise. If there is a problem with loading, a description of the error is returned in a message box.

The Document object of XML DOM exposes a parseError object that allows you to check whether there was an error when loading the XML file or stylesheet. The ParseError object has the properties listed in Table 32.2.

TABLE 32.2. The ParseError object properties

Property	Description
errorCode	Error number of the error that occurred.
filepos	Character position within the file where the error occurred.
line	Line number where the error occurred.
linepos	Character position within the line where the error occurred.
reason	Text description of the error.
srcText	The source (text) of the line where the error occurred.
url	URL or path of the file that was loaded.

After loading the Invoice.xml data file into the DOM software component, the `ApplyTransform` function repeats the same process for the Invoice.xsl file. After both files are successfully loaded, the transform is applied to the data using the `TransformNode` method:

```
document.Write objData.TransformNode(objStyle)
```

The `TransformNode` method performs the transformation by applying the XSL stylesheet to the XML data file. The result is the HTML document displayed in the browser as shown in Figure 32.20 earlier in this chapter.

6. Close the Invoice.htm file and exit Notepad.

Transforming XML Data with the TransformXML Method

So far you've learned how to use stylesheets to transform XML data files to HTML formatting in order to create a Web page. While rendering XML files into HTML for display in a Web browser is the most popular use of stylesheets, XML data files can also be transformed into other XML files using the XSLT transforms.

In this section, we will learn how to use the `TransformXML` method to apply an XSL stylesheet to an XML data file and write the resulting XML to another XML data file.

The `TransformXML` method takes a number of arguments, which are presented in Table 32.3. In its simplest form, the `TransformXML` method looks like this:

```
Application.TransformXML DataSource:="C:\Access2013_XML\InternalCon-
tacts.xml",
  TransformSource:="C:\Access2013_XML\Extensions.xsl", _
  OutputTarget:="C:\Access2013_XML\EmpExtensions.xml"
```

The preceding statement can be used inside a VBA procedure stub to programmatically apply the specified stylesheet.

TABLE 32.3. Arguments of the `TransformXML` method (in order of appearance)

Argument Type	Data Type	Description
`DataSource` (required)	String	Specifies the full path of the XML data file that will be transformed.
`TransformSource` (required)	String	Specifies the full path of the XSL stylesheet to apply to the XML data file specified in the `DataSource` argument.
`OutputTarget` (required)	String	Specifies the full path of the resulting XML data file after applying the XSL stylesheet.

(Contd.)

Argument Type	Data Type	Description
`WellFormedXMLOutput` (optional)	Boolean	Set this argument to True to create a well-formed XML document. Set this argument to False to encode the resulting XML file in UTF-16 format. The default is False.
`ScriptOption` (optional)	`AcTransformX-MLScriptOption` Use one of the following constants: Constant Value `acDisableScript` 2 `acEnableScript` 0 `acPromptScript` 1	Use this argument to specify the action that should be taken if the XSL file contains scripting code. `acPromptScript` is the default.

Custom Project 32.3 demonstrates how to transform an XML data file into another XML file. We will start by creating a custom stylesheet named Extensions.xsl that will transform the InternalContacts.xml file (generated from the Northwind.mdb database Employees table) into an XML file named EmpExtensions.xml. Next, we will write a VBA procedure that exports the XML source file and performs the transformation. Finally, we will import the resulting XML data file into Access.

Custom Project 32.3. Applying a Stylesheet to an XML Data File with the TransformXML Method

Part 1: Creating a Custom Stylesheet for Transforming an XML Source File into Another XML Data File

1. Open Notepad and enter the following statements:

```
<?xml version="1.0"?>
<xsl:stylesheet  version="1.0"  xmlns:xsl="http://www.w3.org/1999/XSL/
Transform">
<xsl:output method="xml" indent="yes"/>
<xsl:template match="/">
<dataroot>
<xsl:for-each select="//Employees">
<Extensions>
  <LastName>
  <xsl:value-of select="LastName" />
```

```
    </LastName>
    <FirstName>
    <xsl:value-of select="FirstName" />
    </FirstName>
    <Extension>
    <xsl:value-of select="Extension" />
    </Extension>
</Extensions>
</xsl:for-each>
</dataroot>
</xsl:template>
</xsl:stylesheet>
```

Take a look at the preceding stylesheet and notice that we have asked the XSL processor to produce the output in XML format:

```
<xsl:output method="xml" indent="yes"/>
```

Next, we used the following instruction:

```
<xsl:template match="/">
```

This instruction defines a template for the entire document. The special pattern "/" in the match attribute tells the XSL processor that this is a template for the document root. Because each XML document must have a root node, we proceeded to define <dataroot> as the document root. You can use any name you want for this purpose. Next, we told the XSL processor to get all the Employees nodes from the source XML data file:

```
<xsl:for-each select="//Employees">
```

The first forward slash in the preceding instruction represents the XML document root. This is the same as:

```
<xsl:for-each select="dataroot/Employees">
```

Next, we proceed to extract data from the required nodes. We are only interested in three columns from the source XML data file: FirstName, LastName, and Extension. We create the necessary elements using the <xsl:value-of> tag with the select attribute specifying the element name:

```
<LastName>
  <xsl:value-of select="LastName" />
</LastName>
<FirstName>
```

```
    <xsl:value-of select="FirstName" />
  </FirstName>
  <Extension>
    <xsl:value-of select="Extension" />
  </Extension>
```

We tell the XSL processor to place the defined elements under the <Extensions> node. When importing the resulting XML file to Access, Access will create an Extensions table with three columns: LastName, FirstName, and Extension. You can use any name you want when specifying the container node for your elements.

To finish the stylesheet, we must write the necessary closing tags:

```
  </xsl:for-each>

  </dataroot>
  </xsl:template>
  </xsl:stylesheet>
```

2. Save the file as **C:\Access2013_XML\Extensions.xsl**. You must include the .xsl file extension to ensure that the file is not saved as text.
3. Close Notepad.

Now that we've got the stylesheet for our transformation, we can write a VBA procedure to actually export the source data and perform the transformation.

Part 2: Writing a VBA Procedure to Export and Transform Data

1. In the Chap32.accdb database, choose **External Data** and click **Access** in the Import & Link group. In the File name box, enter **C:\Access2013_HandsOn\Northwind.mdb** and click **OK**. In the Import Object dialog box, select **Employees** and click **OK**. Click **Close** to exit the External Data dialog box. You should see the Employees table in the Navigation pane.
2. In the Visual Basic Editor window, choose **Insert | Module** to add a standard module to the current VBA project.
3. In the module's Code window, enter the following **Transform_Employees** procedure:

```
Sub Transform_Employees()
  ' use the ExportXML method to create a source XML data file
  Application.ExportXML ObjectType:=acExportTable, _
   DataSource:="Employees", _
   DataTarget:="C:\Access2013_XML\InternalContacts.xml"
  MsgBox "The export operation completed successfully."
  ' use the TransformXML method to apply the stylesheet
```

```
    ' that transforms the source XML data file into
    ' another XML data file
    Application.TransformXML _
     DataSource:="C:\Access2013_XML\InternalContacts.xml", _
     TransformSource:="C:\Access2013_XML\Extensions.xsl", _
     OutputTarget:="C:\Access2013_XML\EmpExtensions.xml", _
     WellFormedXMLOutput:=False
    MsgBox "The transform operation completed successfully."
End Sub
```

The first part of this procedure exports the Employees table from the Northwind database to an XML file named InternalContacts.xml. The second part of this procedure applies the Extensions.xsl stylesheet prepared in Part 1 of this custom project to the InternalContacts.xml data file. The resulting XML document after the transformation is named EmpExtensions.xml. A portion of this file is shown in Figure 32.22.

4. Run the **Transform_Employees** procedure.

FIGURE 32.22. Partial contents of the EmpExtensions.xml file.

After transforming our source XML data file into another XML document, you can bring it into Access with the External Data | Import command (see Part 3).

Part 3: Importing the Transformed XML Data File to Access

1. In the database window, choose **External Data | Import XML File.**
2. In the File name box, type **C:\Access2013_XML\EmpExtensions.xml** and click **OK**. Access displays the Import XML dialog box. In the Import XML dialog box, click **OK** to perform the import. Click **Close** to exit the Import XML window.

3. In the Navigation pane of the Access window, notice the appearance of the Extensions table. Open the **Extensions** table to examine its contents.
4. Close the Extensions table.

A nice thing about XSLT transformations is that you can apply different stylesheets to the same XML data file to create and view the resulting document in different formats.

For example, let's assume that in the Extensions table you'd like to combine the LastName and FirstName columns into one column and sort the data by last name. You could create the following Extensions_SortByEmp.xsl stylesheet and apply it to the InternalContacts.xml file to get the desired XML output:

```
<?xml version="1.0"?>
<xsl:stylesheet version="1.0"
xmlns:xsl="http://www.w3.org/1999/XSL/Transform">
<xsl:output method="xml" indent="yes"/>
  <xsl:template match="/">
  <dataroot>
  <xsl:apply-templates select="dataroot/Employees">
  <xsl:sort select="LastName" order="ascending" />
  </xsl:apply-templates>
  </dataroot>
  </xsl:template>
  <xsl:template match="//Employees">
  <Extensions>
  <FullName>
  <xsl:value-of select="LastName" />
  <xsl:text>, </xsl:text>
  <xsl:value-of select="FirstName" />
  </FullName>
  <Extension>
  <xsl:value-of select="Extension" />
  </Extension>
  </Extensions>
  </xsl:template>
</xsl:stylesheet>
```

The preceding stylesheet uses the <xsl:apply-templates> tag to tell the XSL processor to select the child elements of the dataroot/Employees node. For each child element, it will find in the stylesheet the matching template rule and process it:

```
<xsl:apply-templates select="dataroot/Employees">
```

```
    <xsl:sort select="LastName" order="ascending" />
  </xsl:apply-templates>
```

The <xsl:sort> tag specifies how the resulting XML document should be sorted. The select attribute of this tag is set to LastName, indicating that the file should be sorted by the LastName element. The order attribute defines the sort order as ascending.

Next, in this stylesheet you can see the template rule that begins with the <xsl:template> tag. Its match attribute specifies which nodes in the document tree the template rule should process:

```
<xsl:template match="//Employees">
```

The //Employees expression in the match attribute is equivalent to dataroot/Employees.

Next, you need to define the document node in the output file as Extensions, and proceed to define its child elements as FullName and Extension:

```
<Extensions>
  <FullName>
  <xsl:value-of select="LastName" />
  <xsl:text>, </xsl:text>
  <xsl:value-of select="FirstName" />
  </FullName>
  <Extension>
  <xsl:value-of select="Extension" />
  </Extension>
</Extensions>
```

The FullName element should contain the last name of the employee followed by a space and the first name. You can obtain the values of these fields with the <xsl:value-of> tag and use the <xsl:text> </xsl:text> tag pair to output a comma followed by a space between the last name and first name. Since there is nothing special about the Extension element, you can simply use the <xsl:value-of> tag to obtain this element's value.

Finally, complete the template and the stylesheet with the required closing tags:

```
</xsl:template>
</xsl:stylesheet>
```

To apply the preceding stylesheet to the source XML file, you could write the following VBA procedure:

```
Sub Transform_ContactsSort()
  Dim objAppl As New Access.Application
```

```
        Dim strDBName As String
        strDBName = "C:\Access2013_HandsOn\Northwind.mdb"
        On Error GoTo ErrorHandler
        objAppl.OpenCurrentDatabase (strDBName)
        ' use the ExportXML method to create a source XML data file
        objAppl.ExportXML ObjectType:=acExportTable, _
         DataSource:="Employees", _
         DataTarget:="C:\Access2013_XML\InternalContacts.xml"
        ' use the TransformXML method to apply the stylesheet that
        ' transforms the source XML data file into another XML
        ' data file
    objAppl.TransformXML _
      DataSource:="C:\Access2013_XML\InternalContacts.xml", _
      TransformSource:="C:\Access2013_XML\" & _
      "Extensions_SortByEmp.xsl", _
      OutputTarget:="C:\Access2013_XML\EmpExtensions.xml", _
         WellFormedXMLOutput:=False
    Exit_Here:
        On Error Resume Next
        objAppl.CloseCurrentDatabase
        Set objAppl = Nothing
        Exit Sub
    ErrorHandler:
        MsgBox Err.Number & ": " & Err.Description
        Resume Exit_Here
    End Sub
```

After importing the EmpExtensions.xml file to Access, you should see the Extensions1 table in the database window. When opened, this table displays a sorted list of employees with their extensions (see Figure 32.23).

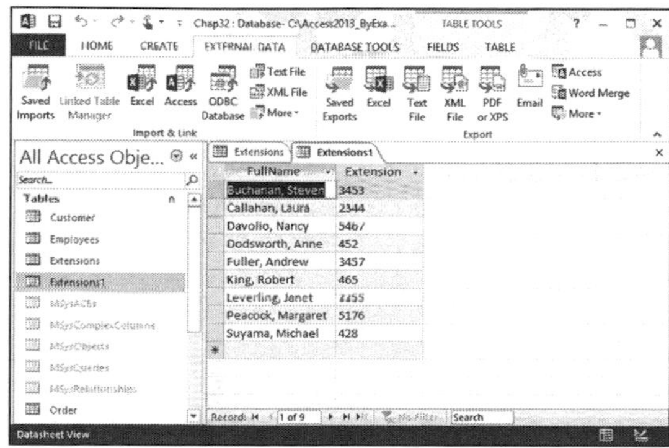

FIGURE 32.23. Extensions table after it was reformatted with another stylesheet.

Importing to XML Using the ImportXML Method

Use the `ImportXML` method to programmatically import an XML data file and/or schema file. The `ImportXML` method takes two arguments, as shown in Table 32.4.

TABLE 32.4. Arguments of the ImportXML method (in order of appearance)

Argument Type	Data Type	Description
`DataSource` (required)	String	Specifies the full path of the XML file to import.
`ImportOptions` (optional)	`acImportXMLOption` Use one of the following constants: Constant — Value `acAppendData` — 2 `acStructureAndData` — 1 `acStructureOnly` — 0	Specifies whether to import structure only (0), import structure and data (1) (default), or append data (2).

The following procedure will import the structure of the Extensions table from the EmpExtensions.xml file:

```
Sub Import_XMLFile()
  Application.ImportXML _
   DataSource:="c:\Access2013_XML\EmpExtensions.xml", _
   ImportOptions:=acStructureOnly
End Sub
```

MANIPULATING XML DOCUMENTS PROGRAMMATICALLY

You can create, access, and manipulate XML documents programmatically using the XML Document Object Model (DOM). The DOM has objects, properties, and methods for interacting with XML documents.

To use the XML DOM from your VBA procedures, take a few minutes now to set up a reference to the MSXML Object Library using the following steps:

1. Switch to the Visual Basic Editor window in Chap32.accdb and choose **Tools | References.**
2. In the References window, select **Microsoft XML, v6.0** (see Figure 32.24) and click **OK.**

   If you don't have version 6.0 installed, select the lower version of this object type library or upgrade your browser to the higher version so that the most recent library is available.

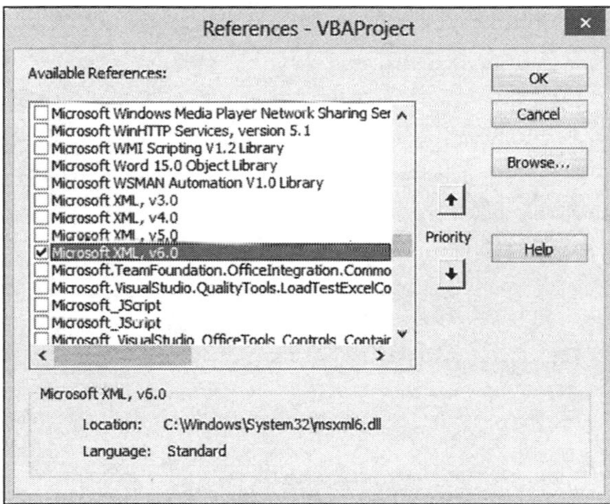

FIGURE 32.24. To work with XML documents programmatically, you need to establish a reference to the Microsoft XML object type library.

3. Now that you have the reference set, open the Object Browser (press F2) and examine XML DOM's objects, methods, and properties (see Figure 32.25).

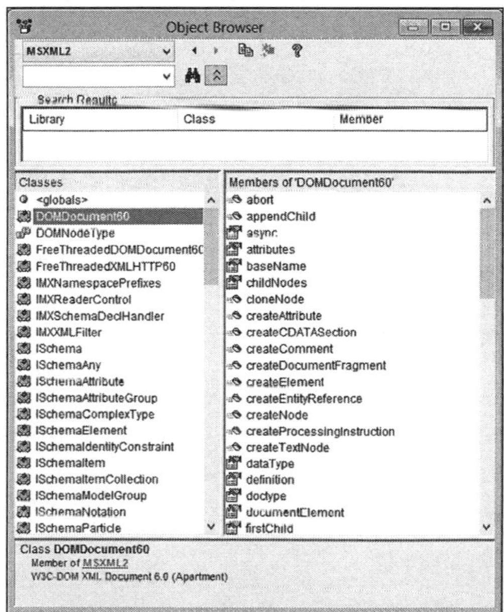

FIGURE 32.25. To view objects, properties, and methods exposed by the XML DOM, open the Object Browser after setting up a reference to the Microsoft XML object type library (see Figure 32.24).

4. Close the Object Browser window.

As mentioned earlier in this chapter (see Part 4 of Custom Project 32.2), the DOMDocument object is the top level of the XML DOM object hierarchy. This object represents a tree structure composed of nodes. You can navigate through this tree structure and manipulate the data contained in the nodes by using various methods and properties. The hands-on exercises in the following sections demonstrate how to read and manipulate XML documents by using VBA procedures.

Loading and Retrieving the Contents of an XML File

Hands-On 32.9 shows how to open an XML data file and retrieve both the raw data and the actual text stored in nodes.

Hands-On 32.9. Loading and Retrieving the Contents of an XML File

1. In the Visual Basic Editor window of the Chap32.accdb database, choose **Insert | Module** to add a new standard module to the current VBA project.
2. In the module's Code window, enter the following **ReadXMLDoc** procedure:

For this procedure to work correctly, you must set up the reference to the Microsoft XML object type library as instructed at the beginning of this section.

```
Sub ReadXMLDoc()
  Dim xmldoc As MSXML2.DOMDocument60
  Set xmldoc = New MSXML2.DOMDocument60
  xmldoc.Async = False
  If xmldoc.Load("C:\Access2013_XML\Shippers.xml") Then
    Debug.Print xmldoc.XML
    ' Debug.Print xmldoc.Text
  End If
End Sub
```

To work with an XML document, we begin by creating an instance of the DOMDocument object as follows:

```
Dim xmldoc As MSXML2.DOMDocument60
Set xmldoc = New MSXML2.DOMDocument60
```

MSXML uses an asynchronous loading mechanism by default for working with documents. Asynchronous loading allows you to perform other tasks during long database operations, such as providing feedback to the user as MSXML parses the XML file or giving the user the chance to cancel the operation.

Before calling the `Load` method, however, it's a good idea to set the Async property of the DOMDocument object to False to ensure that the XML file is fully loaded before other statements are executed. The `Load` method returns True if it successfully loaded the data and False otherwise. Having loaded the XML data into a DOMDocument object, you can use the XML property to retrieve the raw data or use the Text property to obtain the text stored in document nodes.

3. Position the insertion point anywhere within the code of the ReadXMLDoc procedure and choose **Run | Run Sub/UserForm**. The procedure executes and writes the contents of the XML file into the Immediate window as shown in Figure 32.26.
4. In the code of the ReadXMLDoc procedure, comment the first `Debug.Print` statement and uncomment the second statement that reads **Debug.Print xmldoc.Text.**
5. Run the ReadXMLDoc procedure again. This time the Immediate window shows the entry as one long line of text:

```
1 Shipping Company A 123 Any Street Memphis TN 99999 USA 2 Shipping Company B 123 Any Street Memphis TN 99999 USA 3 Shipping Company C 123 Any Street Memphis TN 99999 USA
```

Immediate

```
<?xml version="1.0"?>
<dataroot xmlns:od="urn:schemas-microsoft-com:officedata" xmlns:xsi="http://www.w3.org/
    <Shippers>
        <ID>1</ID>
        <Company>Shipping Company A</Company>
        <Address>123 Any Street</Address>
        <City>Memphis</City>
        <State_x002F_Province>TN</State_x002F_Province>
        <ZIP_x002F_Postal_x0020_Code>99999</ZIP_x002F_Postal_x0020_Code>
        <Country_x002F_Region>USA</Country_x002F_Region>
    </Shippers>
    <Shippers>
        <ID>2</ID>
        <Company>Shipping Company B</Company>
        <Address>123 Any Street</Address>
        <City>Memphis</City>
        <State_x002F_Province>TN</State_x002F_Province>
        <ZIP_x002F_Postal_x0020_Code>99999</ZIP_x002F_Postal_x0020_Code>
        <Country_x002F_Region>USA</Country_x002F_Region>
    </Shippers>
    <Shippers>
        <ID>3</ID>
        <Company>Shipping Company C</Company>
        <Address>123 Any Street</Address>
        <City>Memphis</City>
        <State_x002F_Province>TN</State_x002F_Province>
        <ZIP_x002F_Postal_x0020_Code>99999</ZIP_x002F_Postal_x0020_Code>
        <Country_x002F_Region>USA</Country_x002F_Region>
    </Shippers>
</dataroot>
```

FIGURE 32.26. By using the XML property of the DOMDocument object you can retrieve the raw data from an XML file.

Working with XML Document Nodes

As you already know, the XML DOM represents a tree-based hierarchy of nodes. An XML document can contain nodes of different types. For example, an XML document can include a document node that provides access to the entire XML document or one or more element nodes representing individual elements. Some nodes represent comments and processing instructions in the XML document, and others hold the text content of a tag. To determine the type of node, use the nodeType property of the IXMLDOMNode object. Node types are identified by either a text string or a constant.

For example, the node representing an element can be referred to as NODE_ELEMENT or 1, while the node representing the comment is named NODE_COMMENT or 8. See the MSXML2 Library in the Object Browser for the names of other node types.

In addition to node types, nodes can have parent, child, and sibling nodes. The hasChildNodes method lets you determine if a DOMDocument object has child nodes. There's also a childNodes property, which simplifies retrieving a collection of child nodes. Before you start looping through the collection of child nodes, it's a good idea to use the length property of the IXMLDOMNode object to determine how many elements the collection contains.

The following hands-on exercise uses the Shippers.xml file to demonstrate how to work with XML document nodes.

Hands-On 32.10. Working with XML Document Nodes

1. In the same module of the Visual Basic Editor window where you entered the ReadXMLDoc procedure in the previous hands-on exercise, enter the following **LearnAboutNodes** procedure:

```
Sub LearnAboutNodes()
  Dim xmldoc As MSXML2.DOMDocument60
  Dim xmlNode As MSXML2.IXMLDOMNode
  Set xmldoc = New MSXML2.DOMDocument60
  xmldoc.Async = False
  xmldoc.Load ("C:\Access2013_XML\Shippers.xml")
  If xmldoc.hasChildNodes Then
    Debug.Print "Number of child Nodes: " & _
      xmldoc.childNodes.length
    For Each xmlNode In xmldoc.childNodes
      Debug.Print "Node name:" & xmlNode.nodeName
      Debug.Print vbTab & "Type:" & xmlNode.nodeTypeString _
       & "(" & xmlNode.nodeType & ")"
      Debug.Print vbTab & "Text: " & xmlNode.Text
    Next xmlNode
  End If
  Set xmldoc = Nothing
End Sub
```

Notice that this procedure uses the hasChildNodes property of the DOMDocument object to check whether there are any child nodes in the loaded XML file. If child nodes are found, the length property of the childNodes collection returns the total number of child nodes found. Next, the procedure loops through the childNodes collection and retrieves the node name using the nodeName property of the IXMLDOMNode object.
The nodeTypeString property returns the string version of the node type (for example, processing instruction, element, text, etc.) and the nodeType property is used to return the enumeration value. Finally, the Text property of the IXMLDOMNode object retrieves the node text.

2. Position the insertion point anywhere within the code of the LearnAboutNodes procedure and choose **Run | Run Sub/UserForm**. Running the LearnAboutNodes procedure produces the following output:

```
Number of child Nodes: 2
Node name:xml
  Type:processinginstruction(7)
  Text: version="1.0" encoding="UTF-8"
Node name:dataroot
  Type:element(1)
  Text: 1 Shipping Company A 123 Any Street Memphis TN 99999 USA 2
Shipping Company B 123 Any Street Memphis TN 99999 USA 3 Shipping Company
C 123 Any Street Memphis TN 99999 USA
```

Retrieving Information from Element Nodes

Let's assume that you want to read the information from only the text element nodes. Use the `getElementsByTagName` method of the DOMDocument object to retrieve an IXMLDOMNodeList object containing all the element nodes. This method takes one argument specifying the tag name to search for. To search for all the element nodes, use "*" as the tag to search for.

The following hands-on exercise demonstrates how to obtain data from XML document element nodes.

Hands-On 32.11. Retrieving Information from Element Nodes

1. In the Visual Basic Editor Code window, enter the following **IterateThruElements** procedure below the last procedure code you entered in Hands-On 32.10:

```
Sub IterateThruElements()
  Dim xmldoc As MSXML2.DOMDocument60
```

```
  Dim xmlNode As MSXML2.IXMLDOMNode
  Dim xmlNodeList As MSXML2.IXMLDOMNodeList
  Dim myNode As MSXML2.IXMLDOMNode
  Set xmldoc = New MSXML2.DOMDocument60
  xmldoc.Async = False
  xmldoc.Load ("C:\Access2013_XML\Shippers.xml")
  Set xmlNodeList = xmldoc.getElementsByTagName("*")
  For Each xmlNode In xmlNodeList
    For Each myNode In xmlNode.childNodes
      If myNode.nodeType = NODE_TEXT Then
        Debug.Print xmlNode.nodeName & "=" & xmlNode.Text
      End If
    Next myNode
  Next xmlNode
  Set xmldoc = Nothing
End Sub
```

The IterateThruElements procedure retrieves the XML document name and the corresponding text for all the text elements in the Shippers.xml file. Notice that this procedure uses two `For Each...Next` loops. The first one (the outer loop) iterates through the entire collection of element nodes. The second one (the inner loop) uses the nodeType property to find only those element nodes that contain a single text node.

2. Position the insertion point anywhere within the code of the IterateThruElements procedure and choose **Run | Run Sub/UserForm**. Running the IterateThruElements procedure produces the following results:

```
ID=1
Company=Shipping Company A
Address=123 Any Street
City=Memphis
State_x002F_Province=TN
ZIP_x002F_Postal_x0020_Code=99999
Country_x002F_Region=USA
ID=2
Company=Shipping Company B
Address=123 Any Street
City=Memphis
State_x002F_Province=TN
ZIP_x002F_Postal_x0020_Code=99999
Country_x002F_Region=USA
ID=3
```

```
Company=Shipping Company C
Address=123 Any Street
City=Memphis
State_x002F_Province=TN
ZIP_x002F_Postal_x0020_Code=99999
Country_x002F_Region=USA
```

Retrieving Specific Information from Element Nodes

You can list all the nodes that match a specified criterion by using the `selectNodes` method. The following hands-on exercise prints to the Immediate window the text for all Company nodes that exist in the Shippers.xml file. The //Company criterion of the `selectNodes` method looks for the element named Company at any level within the tree structure of the nodes.

Hands-On 32.12. Retrieving Specific Information from Element Nodes

1. In the Visual Basic Editor Code window, in the same module where you entered previous procedures, enter the following **SelectNodesByCriteria** procedure:

```
Sub SelectNodesByCriteria()
  Dim xmldoc As MSXML2.DOMDocument60
  Dim xmlNodeList As MSXML2.IXMLDOMNodeList
  Dim myNode As MSXML2.IXMLDOMNode
  Set xmldoc = New MSXML2.DOMDocument60
  xmldoc.Async = False
  xmldoc.Load ("C:\Access2013_XML\Shippers.xml")
  Set xmlNodeList = xmldoc.selectNodes("//Company")
  If Not (xmlNodeList Is Nothing) Then
    For Each myNode In xmlNodeList
      Debug.Print myNode.Text
      If myNode.Text = "Shipping Company A" Then
        myNode.Text = "Airborne Express"
        xmldoc.Save "C:\Access2013_XML\Shippers.xml"
      End If
    Next myNode
  End If
  Set xmldoc = Nothing
End Sub
```

The SelectNodesByCriteria procedure creates the IXMLDOMNodeList object that represents a collection of child nodes. The `selectNodes` method applies the specified pattern to this node's context and returns the list of matching nodes as IXMLDOMNodeList. The expression used by the `selectNodes` method specifies that all the Company element nodes should be included in the node list.

You can use the `Is Nothing` conditional expression to find out whether a matching element was found in the loaded XML file. If the matching elements were found in the IXMLDOMNodeList, the procedure iterates through the node list and prints each element node text to the Immediate window. In addition, if the node element's text value is Shipping Company A, the procedure replaces this value with Airborne Express. The `Save` method of the DOMDocument is used to save the changes in the Shippers.xml file.

---

2. Position the insertion point anywhere within the code of the SelectNodesByCriteria procedure and choose **Run | Run Sub/UserForm**. Running the SelectNodesByCriteria procedure produces the following results:

```
Shipping Company A
Shipping Company B
Shipping Company C
```

When you run this procedure again, you should see the following output:

```
Airborne Express
Shipping Company B
Shipping Company C
```

Retrieving the First Matching Node

If all you want to do is retrieve the first node that meets the specified criterion, use the `SelectSingleNode` method of the DOMDocument object. For this method's argument specify the string representing the node you'd like to find. For example, the following procedure finds the first node that matches the criterion //Company in the Shippers.xml file:

```
Sub SelectSingleNode()
  Dim xmldoc As MSXML2.DOMDocument60
  Dim xmlSingleNode As MSXML2.IXMLDOMNode
  Set xmldoc = New MSXML2.DOMDocument60
  xmldoc.Async = False
  xmldoc.Load ("C:\Access2013_XML\Shippers.xml")
  Set xmlSingleNode = xmldoc.SelectSingleNode("//Company")
  If xmlSingleNode Is Nothing Then
    Debug.Print "No nodes selected."
```

```
    Else
      Debug.Print xmlSingleNode.Text
    End If
    Set xmlDoc = Nothing
End Sub
```

The XML DOM provides a number of other methods that make it possible to programmatically add or delete elements in the XML document tree structure. Covering all of the details of the XML DOM Object Model is beyond the scope of this chapter . When you are ready for more information on this subject, visit the following Website: *http://www.w3.org/DOM/*.

USING ACTIVEX DATA OBJECTS WITH XML

In Chapter 17, you learned how to save ADO Recordsets to disk using the Advanced Data TableGram (`adPersistADTG`) format. This section expands on what you already know about ADO Recordsets by showing you how to use ADO with XML. Since the release of ADO version 2.5 (in 2000), it is possible to save all types of recordsets to disk as XML using the Extensible Markup Language (`adPersistXML`) format.

Saving an ADO Recordset as XML to Disk

To save an ADO Recordset to a disk file as XML, use the `Save` method of the Recordset object with the `adPersistXML` constant. Hands-On 32.13 demonstrates how to create an XML file from ADO.

Hands-On 32.13. Creating an XML Document from ADO

1. In the Visual Basic Editor window of the Chap32.accdb database, choose **Insert | Module** to add a new standard module to the current VBA project.
2. Choose **Tools | References** to open the References dialog box. Check the box next to **Microsoft ActiveX Object Library 6.1** (or a lower version) and click **OK**.
3. In the module's Code window, enter the following **SaveRst_ToXMLwithADO** procedure:

```
Sub SaveRst_ToXMLwithADO()
    Dim rst As ADODB.Recordset
    Dim conn As New ADODB.Connection
    Const strConn = "Provider=Microsoft.Jet.OLEDB.4.0;" _
```

```
   & "Data Source=C:\Access2013_HandsOn\Northwind.mdb"
  ' open a connection to the database
  conn.Open strConn
  ' execute an SQL SELECT statement against the database
  Set rst = conn.Execute("SELECT * FROM Products")
  ' delete the file if it exists
  On Error Resume Next
  Kill "C:\Access2013_XML\Products_AttribCentric.xml"
  ' save the recordset as an XML file
  rst.Save "C:\Access2013_XML\Products_AttribCentric.xml", _
   adPersistXML
  ' cleanup
  Set rst = Nothing
  Set conn = Nothing
End Sub
```

This procedure begins by establishing a connection to the sample Northwind database using the ADO Connection object. Next, it executes an SQL `SELECT` statement against the database to retrieve all of the records from the Products table. Once the records are placed in a recordset, the `Save` method is called to store the recordset to a disk file using the `adPersistXML` format. If the disk file already exists, the procedure deletes the existing file using the VBA `Kill` statement. The `On Error Resume Next` statement bypasses the `Kill` statement if the file you are going to create does not yet exist.

4. Position the insertion point anywhere within the code of the procedure and choose **Run | Run Sub/UserForm**.
5. Open the **C:\Access2013_XML\Products_AttribCentric.xml** file created by the **SaveRst_ToXMLwithADO** procedure and examine its content.

The Web browser displays the raw XML as shown in Figure 32.27. Notice that the content of this file looks different from other XML files you generated in this chapter. The reason for this is that XML that is persisted from ADO Recordsets is created in attribute-centric XML. Microsoft Access supports only element-centric XML. Therefore, in order to import to Access an XML file created from ADO, you must first create and apply an XSLT transformation to the source document. The stylesheet you create should convert the attribute-centric XML to element-centric XML that Access can handle (see Hands-On 32.14).

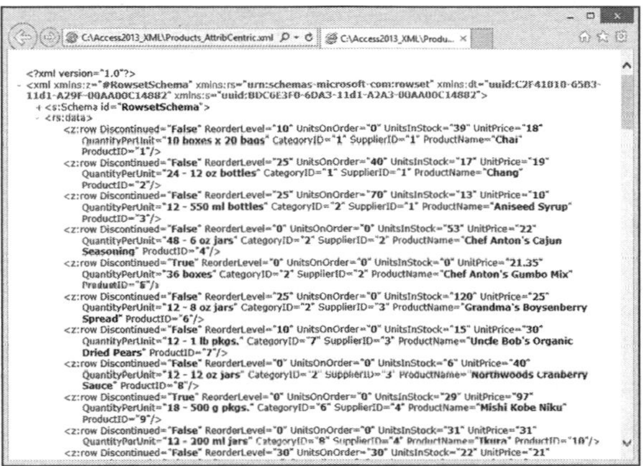

FIGURE 32.27. Saving a recordset to an XML file with ADO produces an attribute-centric XML file.

Attribute-Centric and Element-Centric XML

In the XML file generated in Hands-On 32.13 (see Figure 32.27) notice below the XML document's root tag two child nodes: <s:Schema> and <rs:data>.

The schema node describes the structure of the recordset, while the data node holds the actual data. Inside the <s:Schema id="RowsetSchema"> and </s:Schema> tags, ADO places information about each column: field name, position, data type and length, nullability, and whether the column is writable. Each field is represented by the <s:AttributeType> element. Notice that the value of the name attribute is the field name. The <s:AttributeType> element also has a child element, <s:datatype>, which holds information about its data type (integer, number, string, etc.) and the maximum field length.

Below the schema definition is the actual data. The ADO schema represents each record using the <z:row> tag. The fields in a record are expressed as attributes of the <z:row> element. Every XML attribute is assigned a value that is enclosed in a pair of single or double quotation marks; however, if the value of a field in a record is Null, the attribute on the <z:row> is not created. Notice that each record is written in the following format:

```
<z:row ProductID='1' ProductName='Chai' SupplierID='1'
CategoryID='1' QuantityPerUnit='10 boxes x 20 bags'
UnitPrice='18' UnitsInStock='39' UnitsOnOrder='0'
ReorderLevel='10' Discontinued='False'/>
```

The preceding code fragment is attribute-centric XML that Access cannot import. To make the XML file compatible with Access, you should have each record written out as follows:

```
<Product>
  <ProductID>1</ProductID>
  <ProductName>Chai</ProductName>
  <SupplierID>1</SupplierID>
  <CategoryID>1</CategoryID>
  <QuantityPerUnit>10 boxes x 20 bags</QuantityPerUnit>
  <UnitPrice>18</UnitPrice>
  <UnitsInStock>39</UnitsInStock>
  <UnitsOnOrder>0</UnitsOnOrder>
  <ReorderLevel>10</ReorderLevel>
  <Discontinued>False</Discontinued>
</Product>
```

This code fragment represents element-centric XML. Each record is wrapped in a <Product> tag, and each field is an element under the <Product> tag.

Changing the Type of an XML File

Because it is much easier to work with element-centric XML files (and Microsoft Access does not support attribute-centric XML), you must write an XSL stylesheet to transform an attribute-centric XML file to an element-centric XML file before you can import an XML file created from an ADO Recordset to Access.

The following hands-on exercise demonstrates how to write a stylesheet to convert an XML document from attribute-centric to element-centric.

Hands-On 32.14. Creating a Stylesheet to Convert Attribute-Centric XML to Element-Centric XML

1. Open Notepad and type the following stylesheet code:

```
<xsl:stylesheet version="1.0"
  xmlns:xsl="http://www.w3.org/1999/XSL/Transform"
  xmlns:rs="urn:schemas-microsoft-com:rowset">
<xsl:output method="xml" encoding="UTF-8" />
  <xsl:template match="/">
  <!-- root element for the XML output -->
  <Products xmlns:z="#RowsetSchema">
  <xsl:for-each select="/xml/rs:data/z:row">
  <Product>
```

```
      <xsl:for-each select="@*">
      <xsl:element name="{name()}">
      <xsl:value-of select="."/>
      </xsl:element>
      </xsl:for-each>
      </Product>
      </xsl:for-each>
      </Products>
      </xsl:template>
    </xsl:stylesheet>
```

2. Save this stylesheet as **C:\Access2013_XML\AttribToElem.xsl**. Be sure to include the .xsl extension so the file is not saved as text. We will use this stylesheet for the transformation in the next hands-on exercise.

Notice in the preceding stylesheet that the "@*" wildcard matches all attribute nodes. Each time the <z:row> tag is encountered, an element named <Product> will be created. And for each attribute, the attribute name will be converted to the element name using the built-in XPath `name( )` function. Expressions in curly braces are evaluated and converted to strings. The `select="."` returns the current value of the attribute being read.

See the next section on how to apply this stylesheet to the XML document.

Applying an XSL Stylesheet

Now that you've created the stylesheet to transform an attribute-centric XML file into an element-centric file, you can use the `transformNodeToObject` method of the DOMDocument object to apply the stylesheet to the Products_AttribCentric.xml file created in Hands-On 32.13. The hands-on exercise that follows demonstrates how to do this. In addition, the procedure in this exercise will import the converted ADO XML file to Access.

Hands-On 32.15. Applying a Stylesheet to an ADO XML Document and Importing It to Access

1. Enter the following procedure below the procedure code you created in Hands-On 32.13:

```
Sub ApplyStyleSheetAndImport()
  Dim myXMLDoc As New MSXML2.DOMDocument60
  Dim myXSLDoc As New MSXML2.DOMDocument60
  Dim newXMLDoc As New MSXML2.DOMDocument60
  Dim strXMLFile As String
```

```
  strXMLFile = "C:\Access2013_XML\Products_AttribCentric.xml"
  myXMLDoc.Async = False
  If myXMLDoc.Load(strXMLFile) Then
    myXSLDoc.Load "C:\Access2013_XML\AttribToElem.xsl"
    ' apply the transformation
    If Not myXSLDoc Is Nothing Then
      myXMLDoc.transformNodeToObject myXSLDoc, newXMLDoc

      ' save the output in a new file
      newXMLDoc.Save "C:\Access2013_XML\Products_Converted.xml"

      ' import to Access
      Application.ImportXML _
      "C:\Access2013_XML\Products_Converted.xml"
    End If
  End If
  ' cleanup
  Set myXMLDoc = Nothing
  Set myXSLDoc = Nothing
  Set newXMLDoc = Nothing
End Sub
```

This procedure begins by loading both the Products_AttribCentric.xml file (created in Hands-On 32.13) and the AttribToElem.xsl stylesheet (created in Hands-On 32.14) into the DOMDocument object. Next, the stylesheet is applied to the source file by using the `transformNodeToObject` method. This method is applied to a node in the source XML document's tree and takes two arguments. The first argument is a stylesheet in the form of a DOMDocument node. The second argument is another DOMDocument node that will hold the result of the transformation. Next, the result of the transformation is saved to a file (Products_Converted.xml) and the file is imported to Access using the `ImportXML` method, which was introduced earlier in this chapter.

2. Run the ApplyStyleSheetAndImport procedure.
3. Open the **C:\Access2013_XML\Products_Converted.xml** file. Notice that the Products_Converted.xml file content is now element-centric XML (see Figure 32.28).
4. In the Access window, locate and open the table named **Product**.

The Product table was created by the `ImportXML` method in the Apply-StyleSheetAndImport procedure.

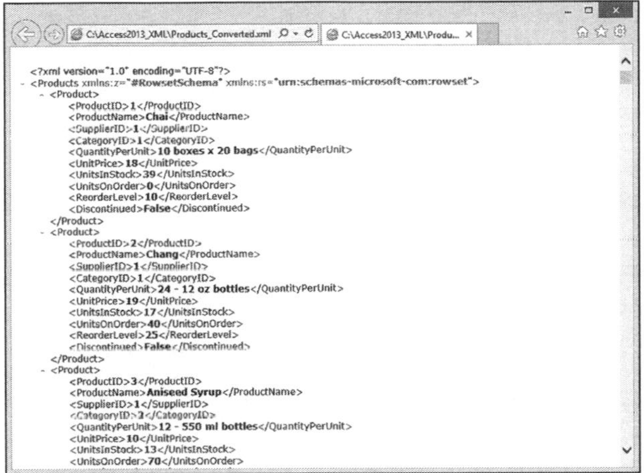

FIGURE 32.28 This element-centric XML file is a result of applying a stylesheet to the attribute-centric ADO Recordset that was saved to an XML file.

Transforming Attribute-Centric XML Data into an HTML Table

As you've seen in earlier examples, creating an XML file from an ADO Recordset results in generated output that contains attribute-centric XML. To import this type of output to Access you had to create a special stylesheet and apply the transformation to convert the attribute-centric XML to the element-centric XML that Access supports. But what if you simply want to display the XML file created from an ADO Recordset in a Web browser? You can create a generic XSL stylesheet that draws a simple HTML table for the users when they open the XML attribute-centric file in their browser.

Hands-On 32.16 demonstrates how to create a stylesheet to transform the attribute-centric XML file that we created in Hands-On 32.13 into HTML. Hands-On 32.17 performs the transformation by inserting a reference to the XSL stylesheet into the XML document.

Hands-On 32.16. Creating a Generic Stylesheet to Transform an Attribute-Centric XML File into HTML

1. Open Notepad and type the following stylesheet code:

```
<?xml version="1.0"?>
<xsl:stylesheet version="1.0"
xmlns:xsl="http://www.w3.org/1999/XSL/Transform"
xmlns:s='uuid:BDC6E3F0-6DA3-11d1-A2A3-00AA00C14882'
xmlns:dt='uuid:C2F41010-65B3-11d1-A29F-00AA00C14882'
```

```
xmlns:rs='urn:schemas-microsoft-com:rowset'
xmlns:z='#RowsetSchema'
xmlns:html="http://www.w3.org/TR/REC-html40">
<xsl:template match="/">
<html>
<head>
<title>Using Stylesheet to convert attribute based
 XML to HTML</title>
<style type="text/css">
.myHSet { font-Family:verdana; font-Size:9px; color:blue; }
.myBSet { font-Family:Garamond; font-Size:8px; }
</style>
</head>
<body>
<table width="100%" border="1">
<xsl:for-each
select="xml/s:Schema/s:ElementType/s:AttributeType">
<th class="myHSet">
<xsl:value-of select="@name" />
</th>
</xsl:for-each>
<xsl:for-each select="xml/rs:data/z:row">
<tr>
<xsl:for-each select="@*">
<td class="myBSet" valign="top">
<xsl:value-of select="."/>
</td>
</xsl:for-each>
</tr>
</xsl:for-each>
</table>
</body>
</html>
</xsl:template>
</xsl:stylesheet>
```

The preceding stylesheet uses the feature known as Cascading Stylesheets (CSS) to format the HTML table. A style comprises different properties—bold, italic, font size and font weight, color, etc.—that you want to apply to particular text (titles, headers, body, etc.) and assigns a common name to these properties. Thus, in this stylesheet, two styles are defined. A style named myHSet is applied to the table headings, and a style named myBSet is used for formatting the text in the body of the table. Using styles is very convenient. If you don't like the formatting, you can simply change the style definition and get a new look instantly. Notice that to define a style you must type a period and a class name. Using letters and numbers, you can define any name for your style class. After the class name, you need to type the definition for the class between curly braces { }.

```
<style type="text/css">
  .myHSet { font-Family:Verdana; font-Size:9px; color:blue; }
  .myBSet { font-Family:Garamond; font-Size:8px; }
</style>
```

Notice that the definition of the class includes the name of the property followed by a colon and the property value. Properties are separated by a semicolon. A semicolon is also placed before the ending curly brace (}). A style class can be applied to any HTML tag.

The example stylesheet uses template-based processing. The following instruction defines a template for the entire document:

```
<xsl:template match="/">
```

The code between the opening and closing tags will be processed for all tags whose names match the value of the match attribute. In other words, you want the pattern matching to be applied to the entire document (/).

Next, a loop is used to write out the column headings. To do this, you must move through all the AttributeType elements of the root element, outputting the name attribute's value like this:

```
<xsl:for-each
select="xml/s:Schema/s:ElementType/s:AttributeType">
  <th class="myHSet">
  <xsl:value-of select="@name" />
  </th>
</xsl:for-each>
```

An attribute's name is always preceded by @.

Next, another loop runs through all the <z:row> elements representing actual records:

```
<xsl:for-each select="xml/rs:data/z:row">
```

All the attributes of any <z:row> element are enumerated:

```
<xsl:for-each select="@*">
  <td class="myBSet" valign="top">
  <xsl:value-of select="." />
  </td>
</xsl:for-each>
```

The string "@*" denotes any attribute. For each attribute found under the <z:row> element, you need to match the attribute name with its corresponding value. Notice the period in the <xsl:value-of> tag. The period represents the node that XSLT is currently working with. In summary, the preceding code fragment tells the XSLT processor to display the value of the current node during the iteration of the <z:row> attributes.

2. Save the stylesheet as **AttribToHTML.xsl**. Be sure to include the .xsl extension so the file is not saved as text.
3. Close Notepad.
4. Open the **AttribToHTML.xsl** file in the browser to test whether it is well formed. If you made any errors while typing the stylesheet code, you must correct the problems before going on to the next section.
5. Close the browser.

Now that you are finished with the stylesheet, you need to link the XML and XSL files. You can do this by adding a reference to a stylesheet in your XML document as shown in Hands-On 32.17.

Hands-On 32.17. Linking the Attribute-Centric XML File with the Generic Stylesheet and Displaying the Transformed File in a Web Browser

1. Save the Products_AttribCentric.xml file as **Products_AttribCentric_2.xml**.
2. Open the **Products_AttribCentric_2.xml** file with Notepad.
3. Type the following definition in the first line of this file:

   ```
   <?xml-stylesheet type="text/xsl" href="AttribToHTML.xsl"?>
   ```

   This instruction establishes a reference to the XSL file.
4. Save the changes made to the Products_AttribCentric_2.xml file and close Notepad.
5. Open the **Products_AttribCentric_2.xml** file in your browser. You should see the data formatted in a table (see Figure 32.29).

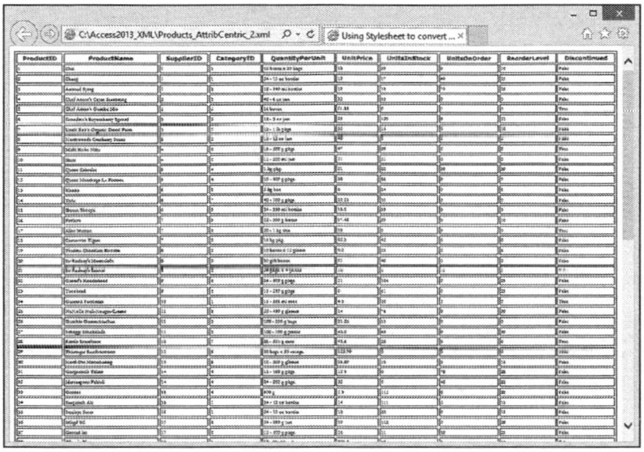

FIGURE 32.29. You can apply a generic stylesheet to an XML document generated by the ADO to display the data in a simple HTML table.

Loading an XML Document in Excel

After saving an ADO Recordset to an XML file on disk (see Hands-On 32.13 earlier in this chapter), you can load it into a desired application and read it as if it were a database. To gain access to the records saved in the XML file, use the `Open` method of the Recordset object and specify the filename, including its path and the persisted recordset service provider as `Provider=MSPersist`. The following hands-on exercise demonstrates how to open a persisted recordset and write its data to an Excel workbook.

Hands-On 32.18. From Access to Excel: Loading an XML File into an Excel Workbook

1. In the Visual Basic Editor window, choose **Insert | Module** to add a new standard module to the current VBA project.
2. Choose **Tools | References** and click the checkbox next to the **Microsoft Excel 15.0 Object Library (or its earlier version)**. Click **OK** to exit the References dialog box.
3. In the module's Code window, enter the following **OpenAdoFile** procedure:

```
Sub OpenAdoFile()
  Dim rst As ADODB.Recordset
  Dim objExcel As Excel.Application
  Dim wkb As Excel.Workbook
  Dim wks As Excel.Worksheet
```

```
    Dim StartRange As Excel.Range
    Dim h As Integer
    Set rst = New ADODB.Recordset
    ' open your XML file and load it
    rst.Open "C:\Access2013_XML\Products_AttribCentric.xml", _
     "Provider=MSPersist"
    ' display the number of records
    MsgBox "There are " & rst.RecordCount & " records " & _
     "in this file."
    Set objExcel = New Excel.Application
    ' create a new Excel workbook
    Set wkb = objExcel.Workbooks.Add
    ' set a reference to the ActiveSheet
    Set wks = wkb.ActiveSheet
    ' make Excel application window visible
    objExcel.Visible = True
    ' copy field names as headings to the 1st row
    ' of the worksheet
    For h = 1 To rst.Fields.Count
    wks.Cells(1, h).Value = rst.Fields(h - 1).Name
    Next
    ' specify the cell range to receive the data (A2)
    Set StartRange = wks.Cells(2, 1)
    ' copy the records from the recordset beginning in cell A2
    StartRange.CopyFromRecordset rst
    ' autofit the columns to make the data fit
    wks.Range("A1").CurrentRegion.Select
    wks.Columns.AutoFit
    ' save the workbook
    wkb.SaveAs "C:\Access2013_XML\ExcelReport.xls"
    Set objExcel = Nothing
    Set rst = Nothing
  End Sub
```

This procedure is well commented, so we will skip its analysis and proceed to the next step.

4. Run the OpenAdoFile procedure.
   When the procedure is complete, the Excel application window should be visible with the ExcelReport.xls workbook file displaying products retrieved from the XML file (see Figure 32.30).
5. Close the Excel workbook and exit Excel.

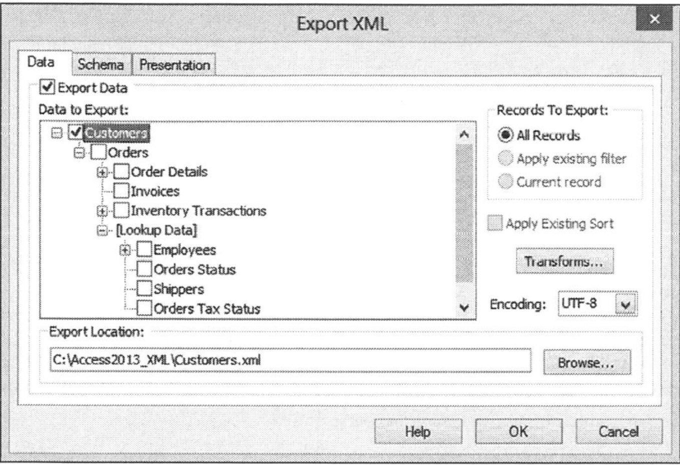

FIGURE 32.30. An ADO Recordset persisted to an XML file is now opened in Excel.

CHAPTER SUMMARY

This chapter has shown you that Microsoft Access 2013 makes it easy to work with XML files. Using Access built-in commands and/or VBA programming code, you can both export Access data to an XML file and import an XML file and display the file as an Access table. You learned what XML is and how it is structured. After working through the examples in this chapter, it's easy to see that XML supplies you with numerous ways to accomplish a specific task. Because XML is stored in plain text files, it can be read by many types of applications, independent of the operating system or hardware. You learned how to transform data from XML to HTML and from one XML format to another. You explored the ADO Recordset methods suitable for working with XML programmatically and were introduced to XSL stylesheets and XSLT transformations.

All of the methods and techniques you've studied here will take time to sink in. XML is not like VBA. It is not very independent and needs many supporting technologies to assist it in its work. So don't despair if you don't understand something right away. Learning XML requires learning many other new concepts (like XSLT, XPath, schemas, etc.) at the same time. Take XML step by step by experimenting with it. The time that you invest in studying this technology will not be wasted. XML has been around for quite a while and is here to stay. The three main reasons why you should really consider using XML are as follows:

- XML separates content from presentation.

  If you are planning to design Web pages, you no longer need to make changes to your HTML files when the data changes. Because the data is kept in separate files, it's easy to make modifications.

- XML is perfect for sharing and exchanging data.

  You no longer have to worry about whether your data needs to be processed by a system that's not compatible with yours. Because all systems can work with text files (and XML documents are simply text files), you can share and exchange your data without a headache.

- XML can be used as a database.

  You no longer need a database system to have a database. What a great value!

This chapter concludes the final part of this book, which focused on working with an Access database over the Internet by writing Active Server Pages and XML files.

INDEX

E

F

H

I

K

L

M

N

O

P

Q